Acoustical Imaging

Volume 21

Edited by

Joie Pierce Jones

University of California at Irvine
Irvine, California

SPRINGER SCIENCE+BUSINESS MEDIA, LLC

The Library of Congress cataloged the first volume of this series as follows:

International Symposium on Acoustical Holography.

    Acoustical holography; proceedings. v. 1–
New York, Plenum Press, 1967–

v. illus. (part col.), ports. 24 cm.

    Editors: 1967– . A. F. Metherell and L. Larmore (1967 with H. M. A. el-Sum)
    Symposium for 1967– held at the Douglas Advanced Research Laboratories, Huntington
Beach, Calif.

    1. Acoustic holography—Congresses—Collected works. I. Metherell. Alexander A., ed.
II. Larmore, Lewis, ed. III. el-Sum, Hussein Mohammed Amin, ed. IV. Douglas Advanced
Resarch Laboratories, v. Title.
QC244.5.I.5   69-12533

Proceedings of the 21st International Symposium on Acoustical Imaging,
held March 28–30, 1994, in Laguna Beach, California

ISBN 978-1-4613-5797-1   ISBN 978-1-4615-1943-0 (eBook)
DOI 10.1007/978-1-4615-1943-0

© 1995 Springer Science+Business Media New York
Originally published by Plenum Press in 1995

10 9 8 7 6 5 4 3 2 1

INTERNATIONAL ADVISORY COMMITTEE
FOR THE
21ST INTERNATIONAL SYMPOSIUM ON ACOUSTICAL IMAGING

Professor Joie Pierce Jones (USA)
(Chairman of the Symposium and Editor of the Proceedings)
* * * *

Professor Pierre Alais (France)
Professor Yoshinao Aoki (Japan)
Dr. Walter Arnold (Germany)
Professor Abdullah Atalar (Turkey)
Dr. Jeff Bamber (UK)
Dr. Valentin Burov (Russia)
Professor G. Busse (Germany)
Professor Noriyoshi Chubachi (Japan)
Professor Adam Dziewonski (USA)
Professor Helmut Ermert (Germany)
Mr. Ken Erikson (USA)
Dr. Yin Feng (China)
Dr. Katherine Ferrara (USA)
Professor Leonard Ferrari (USA)
Professor Mathias Fink (France)
Professor Ian Galton (USA)
Dr. Woon Siong Gan (Singapore)
Dr. Jim Greenleaf (USA)
Professor Benli Gu (China)
Professor C.R. Hill (UK)
Dr. Hugh Jones (Canada)
Dr. Larry Kessler (USA)
Dr. Fred Lizzi (USA)
Professor Hua Lee (USA)
Dr. Sidney Leeman (UK)
Professor Leonardo Masotti (Italy)
Professor Vernon Newhouse (USA)
Professor Bill O'Brien (USA)
Professor Song Bai Park (Korea)
Dr. John Powers (USA)
Professor Takuso Sato (Japan)
Dr. John Thijssen (The Netherlands)
Professor Piero Tortoli (Italy)
Professor Chen Tsai (USA)
Professor Olaf von Ramm (USA)
Professor Bob Waag (USA)
Professor Glen Wade (USA)
Professor P.N.T. Wells (UK)

Acoustical Imaging
Volume 21

Acoustical Imaging

Recent Volumes in This Series:

A Continuation Order Plan is available for this series. A continuation order will bring delivery of each
new volume immediately upon publication. Volumes are billed only upon actual shipment. For further
information please contact the publisher.

PREFACE

This volume represents the proceedings of the 21st International Symposium on Acoustical Imaging, which was held at the Surf and Sand Hotel in Laguna Beach, California, March 28-30, 1994. These unique and highly interdisciplinary series of symposiums have met at intervals of roughly 18 months over the past 30 some years. In general these meetings are devoted to all aspects and all fields of imaging that use acoustics. The meetings are usually small, with 100 to 200 participants, and stimulate useful interchanges across disciplines. These are the only regular meetings where the major researchers in all areas of acoustical imaging can come together to interchange ideas and new concepts. The Acoustical Imaging Symposiums have long been regarded as the premier meeting of this type in the general field of acoustics. The highly regarded and carefully edited proceedings have been published regularly by Plenum Press. I am proud and honored to serve as editor of the 21st volume in this series.

The 21st Symposium was attended by well over 100 participants from some 18 countries. During the three day symposium, 94 scientific presentations were given, 66 as formal lectures and 28 in a poster format. Sufficient time was available during the conference, both following the presentations and informally during meals and breaks, for active discussions among all participants. Over 80 of the presentations have been selected for inclusion in these proceedings. Among these are three invited tutorials designed to provide an overview in three very different areas of development. The invited presentations include "Time Reversal Mirrors" by Mathias Fink from the University of Paris, "Diffraction Tomography Revisited" by Sidney Leeman and colleagues from King's College London, and "Piezocomposite Materials for Acoustical Imaging Transducers" by Wally Smith from the Office of Naval Research in the United States.

As editor and meeting chairman, I owe a substantial debt of gratitude to many individuals and organizations for helping to make the symposium and this volume a success. First, of course, thanks must go to the authors and their research groups who contributed such excellent material to this meeting. This series of symposiums has always been characterized by presentations of the highest quality. Such quality is particularly evident this year. Even a casual reading of this volume will demonstrate the great diversity and high standards that characterize the ongoing activities in acoustical imaging. Those critics who thought that acoustics began and ended with Lord Rayleigh have been proven sorely wrong.

Much thanks must also go to the International Advisory Committee. This group, whose membership is detailed following this Preface, reviewed the abstracts, solicited participants as well as funding, offered good advice, and assisted in all aspects of the meeting organization. Particular thanks go to several chairmen of previous symposiums for their wise council and practical suggestions. These include Hua Lee, Glen Wade, Helmut Ermert, Larry Kessler, and John Powers.

These symposiums would be impossible to organize without outside financial support. Various units within the University of California have been most generous in spite of a very difficult economic period. At the Irvine Campus, the Office of Research and Graduate Studies, the Department of Radiological Sciences, and the Department of Electrical and Computer Engineering were all very supportive. At the Santa Barbara Campus, the Center for High Speed Image Processing was equally generous. Special thanks go to the National Science Foundation for their continuing support of these symposiums and to Dr. George Lea, in particular, at NSF for his long standing encouragement and enthusiasm.

Finally, if the reader will permit a personal note, the editor would like to thank his wife Becky Jones, for assisting in almost every aspect of this symposium. The efficient organization and relaxed ambiance of the meeting was due in large part to her skills and tireless effort. As the editor's companion and best friend for many years, she made my task always bearable and most times even fun.

The 22nd International Symposium on Acoustical Imaging will be held in Florence, Italy, September 4-6, 1995. Professor Piero Tortoli of the University of Florence will serve as Chairman.

Joie Pierce Jones

University of California at Irvine

CONTENTS

WAVE PROPAGATION

INVERSE SCATTERING

TRANSDUCERS AND ARRAYS

NOVEL IMAGING METHODS AND SYSTEMS

TISSUE CHARACTERIZATION

DOPPLER, COLOR-FLOW, AND VELOCITY IMAGING

ACOUSTICAL MICROSCOPY

MATERIALS CHARACTERIZATION

GEOPHYSICAL AND INDUSTRIAL APPLICATIONS

UNDERWATER ACOUSTICS

TIME REVERSAL MIRRORS

Mathias Fink

Laboratoire Ondes et Acoustique
ESPCI - University Paris VII
10 rue Vauquelin, Paris
75005, France

INTRODUCTION

Time reversal invariance is a very rich and powerful concept in classical and quantum mechanics. However, simple experimental evidence of this concept is difficult to obtained in the field of classical mechanics. In the field of acoustic waves, where time reversal invariance also occurs, it may be observed much more simply with large ultrasonic piezo-electric transducer arrays. If time reversal mirrors are of interest as a tool in fundamental physics, their applications in problems using large ultrasonic transducer arrays are numerous. It concern medical applications (imaging, lithotripsy and hyperthermia) as well as non destructive testing and underwater acoustics. In these applications, that require a very narrow beam, high focusing performances are needed. However, large arrays may suffer from strong geometrical distortions which should be corrected. Besides, the focusing of ultrasonic waves becomes a difficult operation as soon as the medium of propagation contains heterogeneities (spatial variations of the compressibility and/or the density in biological tissues and in sea water, liquid-solid interfaces in non destructive testing). In such conditions, the acoustic beam can be distorted and redirected, so that optimal focusing cannot be achieved. Compensation for these two kinds of distortions must be self-adaptative because the designer has no a priori knowledge on them. The use of a Time Reversal Mirror represents an original solution to this problem[1,2].

In the time reversal process, we take advantage of the properties of piezoelectric transducers, i.e., their transmit and receive capabilities, their linearity, and the capability of instantaneous measurement of the temporal pressure waveforms. The pressure field $p(r_j, t)$ detected with a set of transducer elements located at positions r_j is digitized and stored during a time interval T. The pressure field is then resynthetised by the same transducers in a reversed temporal chronology (last in, first out). This is equivalent to the transmission of $p(r_j, T-t)$. A time reversal mirror consists of a one-or-two dimensional transducer array.

Each transducer is connected to its own electronic circuitry that consists of a receiving amplifier, an A/D converter, a storage memory and most importantly a programmable transmitter able to synthetise a time-reversed version of the stored signal.

Such a time reversal mirror allows one to convert a divergent wave issued from an acoustic source into a convergent wave focusing on the source. Unlike an ordinary mirror that produces the virtual image of an acoustic object, the TRM produces a real acoustic image of the initial source. This process can be used to focus on a reflective target that behaves as an acoustic source after being insonified. The method works even if there is an inhomogeneous medium between the target and the mirror.

The TRM is a generalization of an optical phase conjugated mirror[3] (PCM) in the sense that it applies to pulsed broadband signals rather than to monochromatic ones. A comparaison of these concepts shows that in contrast to PCM that works in continuous mode, TRM processing permits choice of any temporal window to be time reversed. This allows operation in an iterative mode. In multitarget media, this process converges on the most reflective target. In the case of an extended target, automatic resonances can be achieved[4].

This resarch was first aimed at overcoming lithotripsy (biological stones destruction) limitations. In state-of-the-art lithotripsy, the determination of stone position is achieved with either an ultrasonic scanner or an X-ray imaging unit. Although the stone position may be accurately known with X-ray systems, the focusing of the destructive ultrasonic wave, through inhomogeneous tissue remains difficult. The sound speed inhomogeneities can distort and redirect the ultrasonic beam. An even more crucial problem is related to stone motion due to breathing. The movement amplitude can be as large as 2 cm and stone tracking is needed for efficient therapy. The time reversal process can solve these problems. The goal is to locate a given reflecting target among others, as for example, a stone in its surroundings (other stones and organs walls). The region of interest is insonified by the transducer array. The reflected field is sensed on the whole array, time reversed and retransmitted. As this process is iterated, the ultrasonic beam will select the target with the highest reflectivity. If the target is spatially extended the process will converge on one spot whose dimension depends only on the TRM geometry and on the wavelength. A high amplification of the last iteration can be used in stone destruction.

Another application of TRM techniques has been developed in our laboratory in the field of non destructive testing of metallurgical samples. Current ultrasonic inspection techniques require that the sample is immersed in a water tank. Very small metallurgical defects have to be detected through liquid-solid interfaces of any geometrical shape and the incident beam can be strongly distorted by the geometry of these liquid-solid interfaces. Focused transducers with special lens design have been built to overcome these limitations. Such transducers have a focusing lens designed to make all the propagation times between the transducer surface and the focal point equal, thus providing an optimum focusing at this point. However these transducers are designed to focus ideally on only one point of the solid. Unfortunately, industrial inspections on thick samples require many different transducers and are quite expensive and time consuming. The use of Time Reversal Mirror (TRM) represents an original solution to this problem. It realizes in real time a focusing process matched to the defect shape and to the propagation medium.

This paper introduces the time reversal approach through a discussion of the different techniques used for focusing in inhomogeneous media. In contrast to time delay focusing techniques that can only correct distortions produced by a thin aberrator located close to the transducer array, TRM focusing is shown to compensate for distortion where ever the aberrator position.

It is shown that the TRM focusing is optimal in the sense that it realizes the spatial-temporal matched filter to the propagation transfer function through inhomogeneous media. It is a self-adaptive technique which compensate for any geometrical distortions of the array structure as well as for distortions due to the propagation through inhomogeneous media.

Experimental results and various applications are described for 64 and 128 channels time reversal mirrors and demonstrate the possibility of robust focusing through an inhomogeneous medium, as well as the efficiency of the iterative mode for focusing on the most reflective target.

ADAPTATIVE TIME DELAY FOCUSING TECHNIQUES

Time delay focusing techniques are based on the use of transducer arrays. The pressure signals coming from an acoustic source are sensed by each transducer element and then digitized. A cross-correlation algorithm is next used to estimate the time delay between signals from neighbouring array elements. These delays determine the optimal time delay characteristic required to focus on the source[5,6]. Another technique[7] consists in the maximization of the energy of the summed signals with respect to the time delay law. These two techniques are similar in the sense that the energy maximization can be viewed as the maximization of the cross-correlation between all pairs of signals at the same time. These two techniques are currently being investigated to determine the proper time delay for focusing in medical ultrasonic imaging where the inhomogeneous medium is the tissue path.

However, these techniques rarely give rise to optimal focusing. The first problem comes from the nature of the reflective target on which we want to focus. The ideal target is point-like and behaves as the source of a spherical wave that is distorted in its propagation through the aberrating medium. In practice, medical ultrasound scanners work in the near field and the sources may not be point-like. For example, a kidney stone or a bone has a finite size. Then elementary signals coming from such targets can be very different from one transducer to the other and cross-correlation results will thus be unaccurate.

In more frequent situations, the region of interest does not contain any highly reflective target. This is, for example, the case in imaging of the abdomen where sound speed aberrations in the body can degrade the focusing characteristics of the ultrasonic beam at frequencies in the low megahertz range. Here, the single reflective target is replaced by the many scatterers that comprise the region being imaged. The adequate focusing time delay law can be estimated for such scattering media with a cross-correlation technique[5,8]. This possibility results from a remarkable property of the scattered pressure field that results from the summation of the individual echoes reflected on each scatterer. The field propagates with a spatial correlation function width that increases proportionaly to the propagation distance. This property, which is described by the Van Cittert Zernike theorem[9] means that, neighbouring transducers, as long as they are sufficiently far from the scattering region, will sense echographic signals that are highly correlated. An aberrating layer will shift these signals and the cross-correlation technique will indeed find the proper delay.

Although the problem of the target nature seems to be under control, the nature of the inhomogeneous medium seems to be more critical. An underlying assumption of the cross-correlation technique is that all the effects of the inhomogeneity on the spherical wave front sum up as a simple distortion of the wave front shape, i.e., the aberrator only modifies the propagation delay between the source and the elementary transducers. Then, knowledge of the proper time delay characteristic is sufficient for accurate focusing in the receive or transmit modes. However, this hypothesis is very restrictive ; it is valid only when *the aberrator is thin and located very close to the array*). In optical astronomy, this hypothesis is valid as long as the athmospheric turbulence is confined close to the telescope and far from the stars. In most medical applications this hypothesis is false : we always work in the near field of the transducers and the inhomogeneities are distributed over the whole volume. A wave propagating in such an inhomogeneous medium is not only delayed, but its spatial and temporal shape is also distorted through refraction, diffraction and multi-scattering. The adaptative delay-line focusing technique no longer works when the array probe-aberrator distance increases[10]. In these situations a more general approach is needed. It consists of a time reversal process that takes into account all of the information recorded from the medium (such as time-delay and waveform modifications of the individual signals recorded on each transducer element).

TIME REVERSAL OF ULTRASONIC FIELDS : BASIC PRINCIPLES

The time reversal processes represent an original solution to these different problems. Time reversal processing take advantage of the invariance of the wave equation under a time-reversal operation. This means that if we want to focus in the transmit mode through any inhomogeneous medium or through a liquid-solid interface, it is enough to record the distorted wavefield coming from a source (active or passive) located at the desired focal point and to time-reverse this field. The time-reversed wavefield backpropagates through the inhomogeneities and optimally refocuses on the source.

Time reversal invariance of wave equation is valid only, in the frequency range, where the medium can be assimilated to a lossless propagating medium. In a fluid medium with compressibility $\kappa(\mathbf{r})$ and density $\rho(\mathbf{r})$ that vary with space, the acoustic pressure field wave equation reads

$$\kappa(\mathbf{r})\frac{\partial^2 p}{\partial t^2} = \nabla(\frac{\nabla p}{\rho(\mathbf{r})}) \qquad (1)$$

Looking at this propagation equation, we note that it has a special behavior with respect to the temporal variable t ; indeed, it contains only a second-order time-derivative operator. This property is the starting point of the time reversal principle, valid only for a lossless propagation medium. As an immediate consequence, if $p(\mathbf{r},t)$ is a pressure field solution of the propagation equation, then $p(\mathbf{r}, - t)$ is another solution of the problem. This property is specific to the invariance under a time reversal operation. In the case of an isotropic solid sample immersed in the fluid, the acoustical field is described in the solid by the acoustic displacement field $\mathbf{u}(\mathbf{r},t)$ according to another time reversal invariant wave equation

$$\rho_s\frac{\partial^2 \mathbf{u}}{\partial t^2} =(\lambda+2\mu)\nabla(\nabla\mathbf{u}.)-\mu\nabla \wedge \nabla \mathbf{u} \wedge \mathbf{u} \qquad (2)$$

where λ and μ are the Lame coefficients and r_s is the solid density. Strain and stress continuities allows, from the knowledge of $\mathbf{u}(\mathbf{r},t)$ in the solid to determine a unique solution $p(\mathbf{r},t)$ for the pressure field in the liquid. As an immediate consequence of the time reversal invariance of eq 2, if $\mathbf{u}(\mathbf{r},t)$ is the displacement field in the solid, then $\mathbf{u}(\mathbf{r},-t)$ is another solution of the problem and is is related through the strain and stress continuities to the time-reversed pressure solution $p(\mathbf{r},-t)$ in the fluid.

The Time Reversal Cavity

In any propagation experiment, initial conditions (the acoustic sources and the boundary conditions) determine a unique solution p(**r**,t) in the fluid. Our goal, in time-reversal experiments, is to modify the initial conditions in order to generate the dual solution p(**r**,-t). However, due to causality requirements, p(**r**,-t) is not an experimentaly valid solution. Therefore, we will limit ourselves to the generation of p(**r**,T-t).

One solution to the difficult problem of generating the time reversed solution consists in measuring during a time interval T the pressure p(**r**,t) in the whole 3-D volume (T is long enough so that the pressure field vanishes for t > T), and then retransmit in all the volume p(r,T-t). This solution is unrealistic since it requires sampling the whole volume with transmit-receive probes. A more realistic solution take advantage of the Huygens principle : the wavefield in any point of a volume can be predicted by the knowledge of the field and its normal derivative on a closed surface surrounding the volume. Therefore, the time reversal operation is reduced from a 3-D volume to a 2-D surface. Indeed, the knowledge of the pressure field and its normal derivative at any point of a closed surface is sufficient to predict the pressure field inside of this surface[11].

Starting from this point of view, the focusing on a target in an inhomogeneous fluid medium can be treated in the following way :

A point-like source located in an inhomogeneous medium creates a spherical wavefront that is distorted after propagation in the medium. We consider a closed surface surrounding the object source and the inhomogeneities, and we assume that we are able to measure the pressure field and its normal derivative at any point on the closed surface. In a second step, we assume that we are able to create secondary sources (monopole and dipole sources) on the surface that correspond to the time-reversal of the signals measured during the first step. As a result of the secondary sources created on the surface a time-reversed pressure field back-propagates inside the surface and is distorted by the interaction with the inhomogeneities.

Under such conditions, it can be shown[12,13] that the time-reversed pressure field is focused on the initial source position. It is important to note that the time-reversal technique provides a better focusing than the correlation technique ; in particular, it is not necessary to assume that the inhomogeneities are only located in the neighborhood of the transducer array, and to assume that the effects of inhomogeneities reduce to a simple time-delay varying from one transducer element to another.

The efficiency of time-reversal techniques obtained from closed cavities has been analysed for weakly and strongly inhomogeneous media[13], and it is shown that the time-reversed pressure field is focused on the initial source position. However, the finite spectral bandwidth of ultrasonic signals restricts resolution because the spatial scales of inhomogeneities that are smaller than the minimum wavelength, are blurred. Therefore, the generation of p(**r**,T-t) in the whole volume is not perfect.

The Time Reversal Mirror

The time reversal cavity is, of course, an ideal concept for focusing through inhomogeneous media. Unfortunately, such a cavity is difficult to realize in practice. The strongest limitation is linked to the difficulty of surrounding the focal region by a set of transducers. In medical applications, as well as, in non-destructive testing, we are usually to work in pulse-echo mode and the probe is located only on one side of the region of interest. This mode of operation is more practical and allows focusing from an array of transducers. In this case, the time reversal cavity may be replaced by a time reversal mirror. Such a mirror can be plane or prefocused, one-dimensional or two-dimensional. However, its ability to focus through aberrating media may be compared with that of closed cavity.

Other limitations of TRM, even in homogeneous media, are the same as those observed in classical focusing with delay-line techniques :

- (i) - Diffraction effects act as low-pass filter on the spatial frequency spectrum of any wavefield. The resulting image of a point is a spot with dimension that depend on the wavelength.

- (ii) - The limited dimension of the TRM induces a point spread function whose width is related, as in classical focusing techniques, to the angular aperture of the mirror observed from the focal point.

- (iii) - The spatial sampling of the TRM by a set of transducers introduces grating lobes. These lobes can be avoided by using an array pitch of the order of $\lambda/2$, where λ is the central wavelength of the pressure field. However, such a fine sampling is not necessary if the TRM is prefocused on the region of interest (cylindrical or spherical mirror).

- (iv) - The temporal sampling of the data recorded and transmitted by the TRM has to be comparable to that of the time delay law in classical time delay focusing. A maximum rate of $T/8$ (T central period) is needed to avoid secondary lobes.

As illustrated in Figure 1, time reversal focusing on a target through inhomogeneous media requires three steps. The first step consists of transmitting a wavefront, through the inhomogeneous medium, from the array to the target. The target generates a scattered pressure field that propagates through the inhomogeneous medium and is distorted. The second step is a recording step; the backscattered pressure field is recorded by the transducer array. In the last step, the transducer array synthesizes the time reversed field. This pressure field propagates through the aberrating medium and focuses on the target.

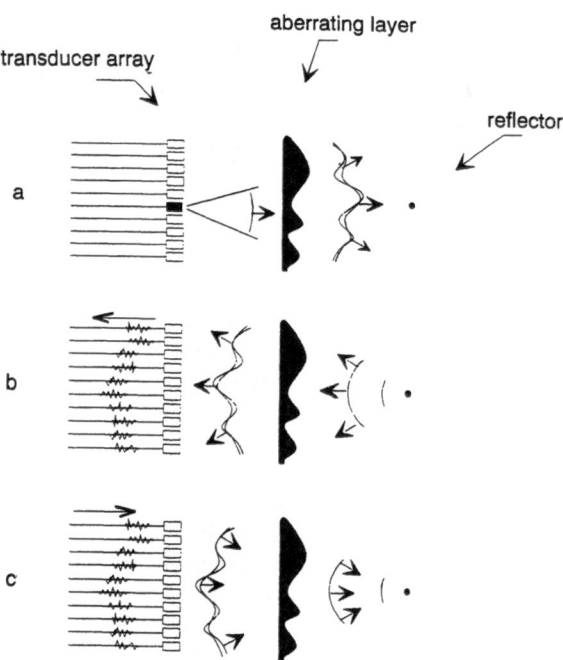

Figure 1. Time Reversal focusing process

TIME REVERSAL OF ULTRASONIC FIELDS : EXPERIMENTAL RESULTS

TRM experimental arrangement

Experimental data on time reversal mirrors have been obtained by F.Wu and J.L Thomas[14] with different real time electronic prototypes made of 64 or 128 channels working in transmit-receive mode. The transmission module is made of 128 programmable transmitters. Each programmable transmitter is driven by a 32 Kbytes buffer memory through 12 bits D/A converter operating at 40 MHz sampling rate. Each converter is followed by a linear power-amplifier. The transmitter delivers 30 V peak-to-peak voltage to a 50 Ω transducer impedance. In the transmit mode the 128 transmitters work simultaneously and are connected to 128 elements of 1D or 2D transducer arrays.

Two steps are required in the receive mode : amplification and A/D conversion. We use a set of 16 A/D converters through a multiplexer, so that recording a complete set of A lines requires 8 consecutive emissions. A set of 16 logarithmic amplifiers allows the recording of a 90 dB instantaneous dynamic range through the 16 A/D converters. Recorded data are digitized at a sampling rate of 40 MHz with an 10-bit dynamic range. Exponentation of the data is made in each buffer memory to correct the logarithmic operation before the time reversal operation. A Compaq 486 computer controls the time-reversal process and the matching of the dynamic range between the receive and the transmit mode. The prototype allows a complete 128 channel time reversal operation in less than 10 ms and different 1D and 2D arrays were designed for these experiments.

Focusing Through a Random Aberrating Layer

In this set of experiments, we used a 1D plane Time Reversal Mirror made of a plane linear array. The array pitch is 0.75 mm and the working frequency is 3.5 Mhz (TRM1). The target is observed through a strongly aberrating medium. The experiment is carried out for different TRM-aberrator distances. The aberrating medium is made of a rubber layer which thickness is randomly modulated. The ultrasonic velocity in the rubber is about 1200 m/s. The layer is shaped with a random profile in the lateral direction and a uniform thickness in elevation. The random profile induces a random shift on the layer transit time, whose standard deviation is about 0.15 µs compared to the acoustic period 0.3 µs ofthe array. The set of experiments was done in two steps. The first step investigates the distortions introduced by the layer on cylindrical beamforming experiments. Cylindrical beamforming delays are computed in order to focus through homogeneous water at 90 mm from the surface of the array and along the array center line. 64 elements transmit an identical pulse through the calculated bank of delay lines and a needle point hydrophone scans the focal plane (z = 90 mm) for different positions of the aberrating layer. Figure 2 shows the directivity pattern measured in the focal plane (maximum of the pressure field at z = 90 mm) when the aberrator is located at different depths from the linear array. These figures correspond to an array-aberrator distance d of respectively 0 mm , 25 mm and 50 mm. The appearance of these figures clearly shows that the maximum defocusing effect is observed for d = 0 mm, which corresponds to the aberrator against the array. The beam is widely spread. The -6 dB lateral resolution is about 6 mm and the side lobe level is very high. The -6 dB lateral resolution corresponds to the focusing of an apparent transmitting aperture smaller than the 48 mm long array aperture. Figure 2 shows that the beam spreading is reduced when the probe-aberrator distance increases, since the focused beam intercepts a smaller area of the aberrator. The second step of the experiment illustrates the time reversal focusing process. The hydrophone is used as a point-like reflector located at a

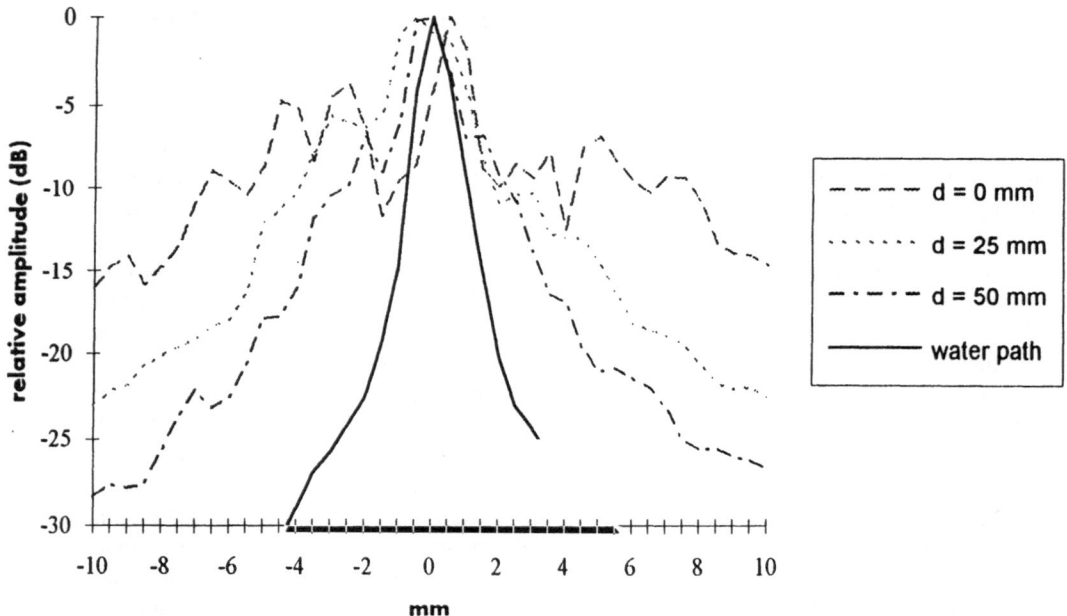

Figure 2. Directivity Patterns in the focal plane obtained by cylindrical focusing trough water and with an aberrating layer

depth z = 90 mm. For each TRM-aberrator distance, the point reflector is illuminated through the aberrating layer by the incident beams corresponding to the pressure fields of Figure 2 . The echoes from the hydrophone are distorted by the rubber layer and are recorded on the array (see for example figure 3a for d = 0 mm and figure 3b for d= 25 mm). The corresponding signals are time-reversed and reemitted. The time-reversed waves propagate through the aberrator and the time-reversed pressure field is measured by scanning the focal plane with the same hydrophone now working in the receive mode. Figure 4 shows the resulting focal beams for d = 0, 25 and 50 mm. The directivity patterns are similar and correspond to a - 6 dB lateral resolution about 1.5 mm. These results demonstrate the efficiency of time reversal focusing to compensate the distortions induced by aberrators whatever the probe-aberrator distance. Figure 5 shows the results obtained with the adaptative time delay focusing technique. The time delay focusing law is obtained by the cross correlation technique from the data of figure 3. We can observe that for large distances adaptative time delay focusing does not help.

In the last part of this section, we wish to discuss a very fundamental point about focusing processes through inhomogeneous media. Focusing with TRM seems to be a complicated technique because it requires to transmit a different waveform from each transducer element. While, in the time-delay technique, it reduces to transmit the same pulse within a time-delay from each element. Time-delay focusing is a valid technique only to correct for the effect of an aberrating layer located close to the array. Indeed, thin random layer acts only as a random propagating delay. A spherical wave originating from a point source is then delayed by the propagation through the aberrating layer. The individual pressure signals recorded along the output plane of the aberrator are only delayed. Their shape are identical, only their arrival time are randomly modified. This is clearly shown on

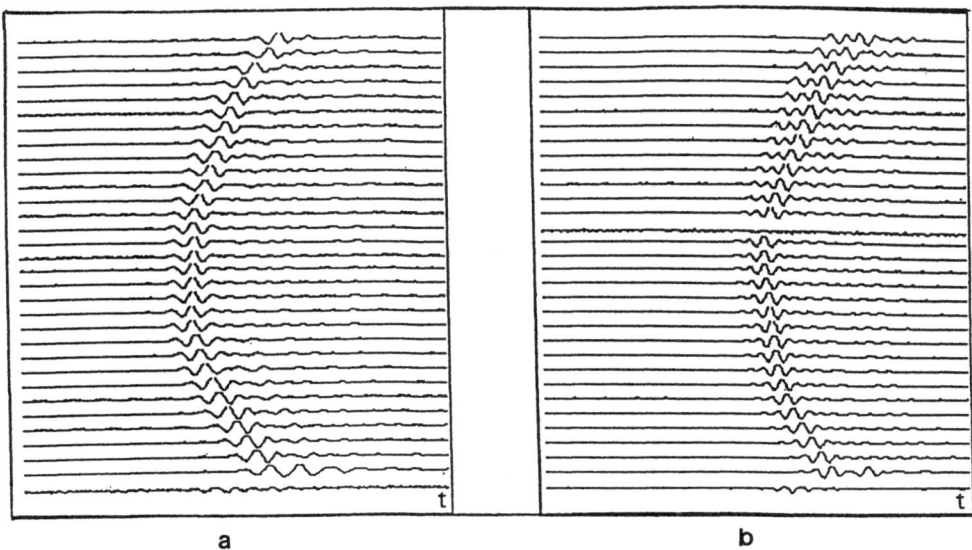

a   b

Figure 3. Echographic signals (a) d = 0 mm. (b) d = 25 mm

figure 3a which corresponds to the hydrophone echoes propagating through the aberrator and recorded when the array-aberrator distance d = 0 mm. In this case, transmitting the time-reversed of figure 3a or transmitting an identical pulse through matched delay-lines will give the same focusing results.

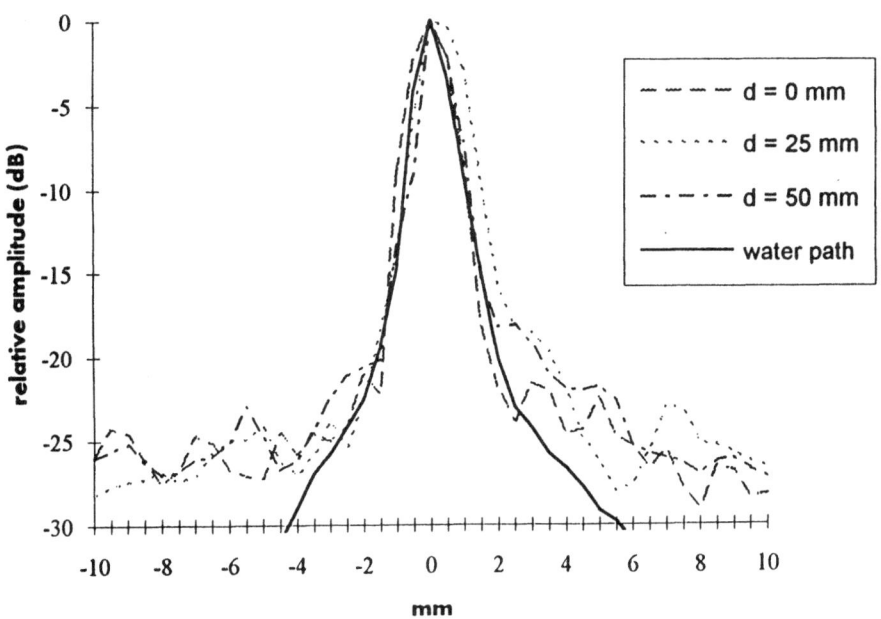

Figure 4. Directivity patterns obtained by time reversal focusing

However, when the array-aberrator distance increases, the recorded pressure signals originating from a point source are not only delayed, but their shapes are no longer similar. Figure 3b shows the signals recorded by the array from the hydrophone when d = 25 mm. The strong distortions of the recorded signals and the complicated shape of this pattern results from the propagation of the pressure field from the output plane of the layer towards the array recording plane (d = 25 mm). Although the pressure signals are identical within a time delay on the output layer plane, the interference between all the Huygens's wavelets originating from this plane results in a complicated pattern which has lost the shape invariance. In this case, time reversal focusing is the only valid technique to refocus. Delay-line focusing techniques loose their efficiency when the probe-aberrrator distance increases.

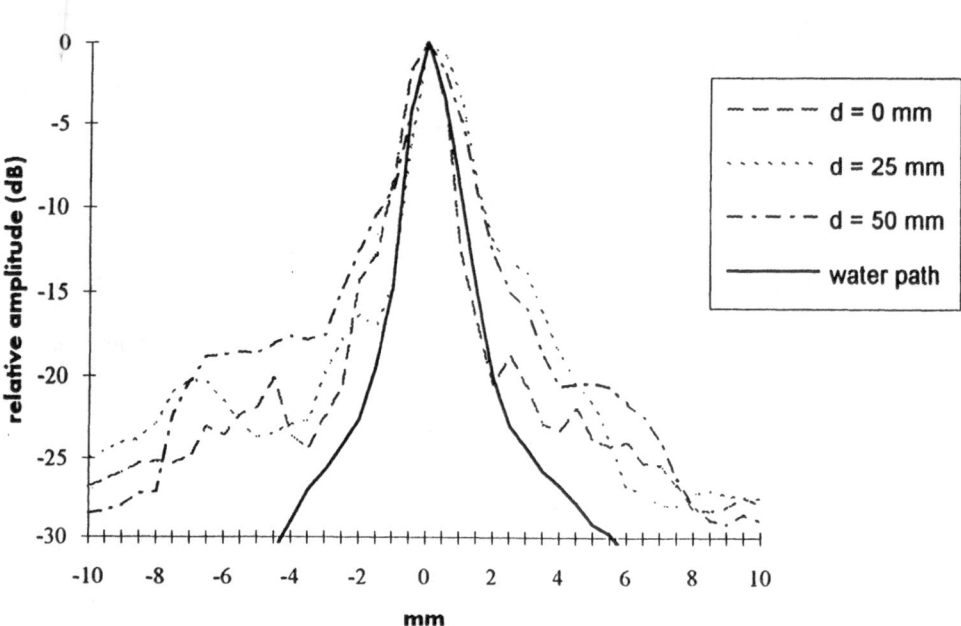

Figure 5. Directivity patterns obtained by adaptative time delay focusing

Autoadaptative Focusing in Solid Media. Applications to NDT

These experiments were performed with a 2D prefocused array working at 3 MHz (TRM2) by N. Chakroun[15]. TRM2 is a prefocused spherical mirror with a 200 mm radius of curvature made of 121 elements. The central element is a plane disk and the others are annular sector elements. The transducers are distributed according to a six annuli structure of respectively 1, 8, 16, 24, 32 and 40 elements. The total aperture is equal to 60 mm immersed in a water tank. The samples of interest were made of titanium. Titanium is a noisy medium in the sense that it gives rise to a strong ultrasonic speckle noise related to an heterogeneous microstructure. The block of titanium is located in front of the mirror. The working frequency of TRM2 is 3 MHz (λ =0.5 mm in the water and λ =2 mm for longitudinal wave in the titanium sample). The experiments were performed to detect a particular type of defect which can be observed in titanium : the hard-α. The hard-α is a small defect of irregular and unknown shape and of small acoustic impedance contrast. It can be characterised by its amplitude with classical control technique. This defect has a backscattered amplitude field equivalent to a 0.4 mm diameter flat bottom hole. It is located inside a block of titanium set at 20 mm depth and 5 mm off axis . In the first step, TRM2 illuminated a sector containing the hard-α defect. The large illuminating beam was transmitted by a single transducer element located at the center of the array. After the first illumination, the echoes from the block were recorded. Figure 6a (reception 1) presents the recorded data in grey level for each of the 121 transducers elements. The data are represented in B mode, where the horizontal axis represents the time arrival of the echo (depth) and the vertical axis the transducer numbers. On these data we see the two echoes coming from the interfaces between titanium and water. Between these echoes we can notice the titanium speckle noise. In the second step of the experiment we select for each of the recorded data a 2 μs time-window after the front face echo. These windows are time reversed and retransmitted. The time-reversed waves propagate and refocused on the hard-α and we recorded again the echoes coming from the block. Figure 6b (reception 2) presents the new recorded data. We notice always the echoes from the two interfaces, but between these echoes an oscillating line appears. This line corresponds to the echoes from the hard α received by the TRM elements. The amplitude of the oscillation corresponds to

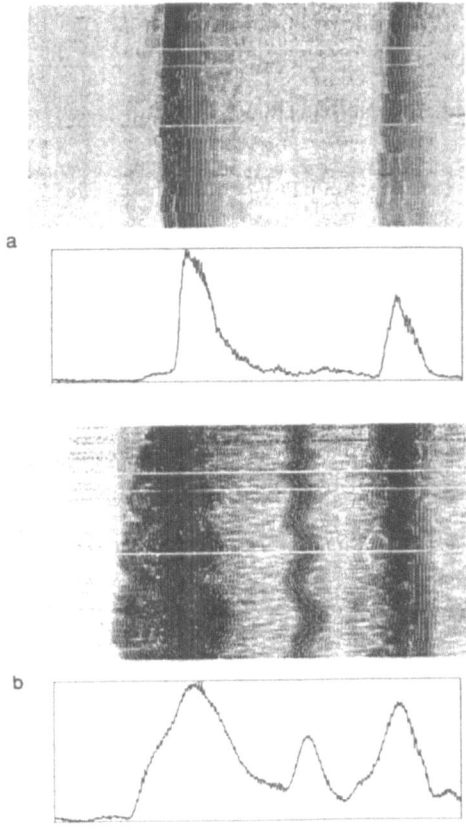

Figure 6. (a) B mode presentation and summation after the first illumination. (b) B mode presentation and summation after one time reversal transmission

an off axis defect, whose wave front intercepts obliquely the 2D array. We can add all this data for all the elements and the corresponding sum is presented on the lower part of the figures. A 40 dB improvement of the signal to speckle noise ratio is observed in this figure compared to the target level before the time reversal process. The important point to be noticed here is the different behavior of time reversal versus the echographic speckle noise compared to hard-α echo. In titanium the speckle noise is originated from the very thin microstructure whose dimension is of the order of $\lambda/100$. A time reversal process cannot time reversed this kind of waveform due to the loss of information during the propagation with a 3.5 Mhz central frequency. Besides the small coherent echographic signal originated from the defect can be time reversed with efficiency, and after one TRM process the level of this echo compared to the one of speckle noise has increased of nearly 40 dB. This kind of experiment can be conducted for different positions of the hard-α in the illuminating field. The hard-α is automatically detected in a section of more then 4 cm^2 around the mirror axis. These results demonstrate the ability of TRM to compensate for distorsion induced by liquid-solid interfaces observed at oblical incidence, and to select coherent echo from incoherent background noise.

THE ITERATIVE TIME REVERSAL MODE.

One major advantage of the TRM is the ability to choose the origin and the duration of the signals to be time reversed. This is done through the definition of the temporal window that selects the data to be time reversed. When the medium of interest contains several reflectors, the time reversal technique cannot directly be focused on one point. Indeed, if the medium contains two targets of different reflectivity illuminated by a short pulse, the time reversal of the echoes reflected from these targets will generate two wavefronts refocused on each target. The mirror produces the real acoustic images of the two reflectors on themselves. The highest amplitude wavefront illuminates the most reflective target, while the weakest wavefront illuminates the second target. However, what we have described is, in fact, only valid if we neglect the multiple scattering processes which can occur between the two targets. In order to avoid these multiple scattered waves, we can select echoes within a particular time reversal window. In this case, the time reversal process can be iterated[1,2,16]. After the first time-reversed illumination, the weakest target will be illuminated more weakly and will reflect a wavefront much fainter than the one coming from the strongest target. After some iterations, the process will converge and produce a wavefront focused on the most reflective target. This process converges if the target separation is sufficient to avoid the illumination of one target by the real acoustic image of the other target [16,17]. In the case of multi-target medium the convergence problem is more complicated and has been studied with great care by C. Prada[18]

Although this simple presentation implies the iterative mode is very attractive, one can argue a contradiction exists between the concept of iteration and the physical principle of time reversal invariance. Indeed, the complete time reversal of an acoustic "scene" results in the time-reversed scene. Therefore, the iteration of the time reversal operation gives stationnary results, a contradiction with wave field modification after each iteration. In fact, the contradiction is only apparent. A complete time reversal operation requires a closed time reversal cavity surrounding the acoustic scene and a recording time T long enough to take into account all the multiple scattered waves. In our technique, we utilize only a finite spatial aperture and short temporal windows. Therefore, some information is lost. This information loss gives the iterative mode its target selection capabilities.

Experiments have been performed with TRM1 to demonstrate the ability of TRM

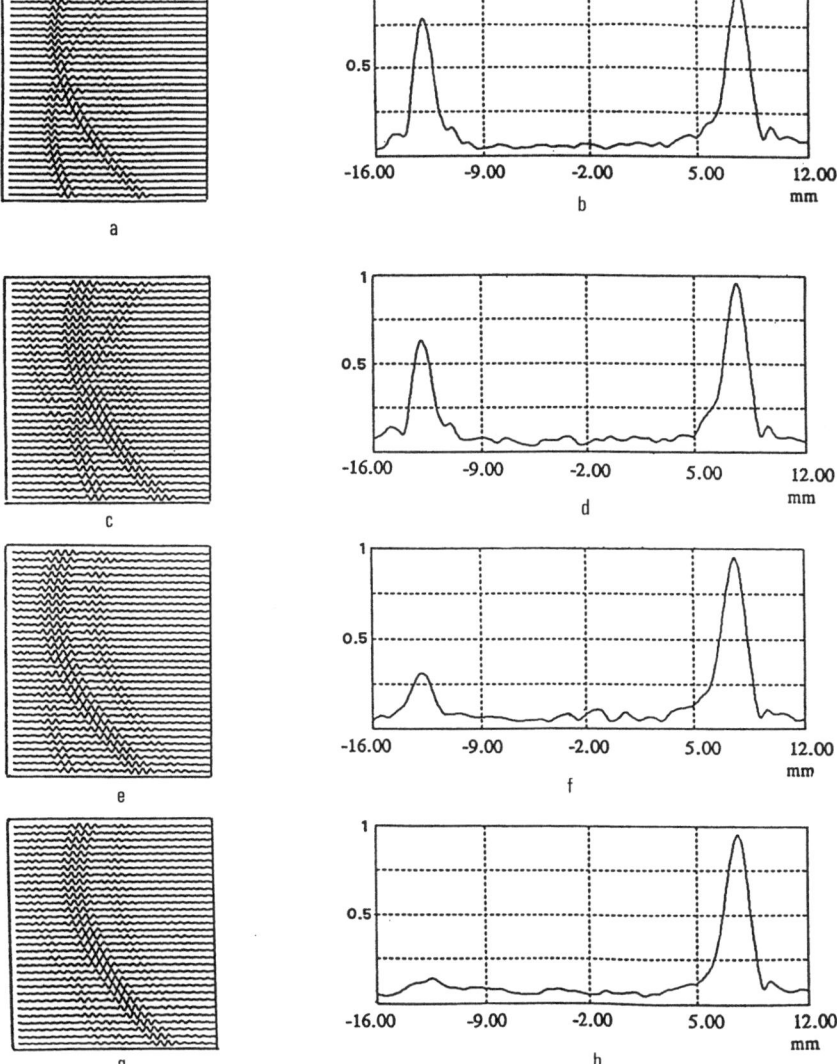

Figure 7. Selective focusing by time reversal iteration in the case of two wires

iterative mode to select the most reflective target in a multi target medium. The medium is made of two different wires situated at a depth $z = 110$ mm from the array. The two wires are parallel to the long axis of each array element and they are respectively located on both sides of the array axis. A 1.5 mm diameter copper wire is located at 7 mm away from the array axis and a 0.7 mm brass wire is located at -13 mm from the axis.

In the first step TRM1 illuminated an angular sector containing the two wires. The illuminated beam was transmitted by a single transducer element located at the array center. The directivity pattern of a single element is wide. After the first illumination, the echoes from the two targets were recorded. Figure 7a shows the recorded data corresponding to two individual wave fronts pointing at the two wires. The recorded signals were then time-reversed and retransmitted. The time-reversed waves propagated and the new directivity pattern was measured by scanning the plane z = 110 mm with the hydrophone. Figure 7b represents the directivity pattern which clearly shows two maxima corresponding to the two target locations. The pressure field reachs a higher value at the location of the copper wire whose scatttering cross section is larger than the one of the brass wire. The process was iterated : the new echoes from the wires were recorded, time-reversed and retransmitted, etc. As the process was iterated the brass wire which reflects less energy received a weaker time-reversed wave. This is clearly shown on Figures 7b, 7d and 7h which correspond to the directivity pattern of the first, second, and third iterations. The brass wire is no more illuminated after the last illumination. Figures 7c, 7e and 7g represent the reflected wavefronts recorded by the array after the first, second, and third iterations. At the end of the iteration process (fig 7g) only the echoes from the most reflective target remain.

Two particular points may be emphasized:

(i) Figures 7a, 7c, 7e and 7g show the echo duration becomes longer after each iteration. This is due to the multiple convolutions of the pressure signals by the acoustoelectric impulse response of the transducer elements.

(ii) The echo pattern shown on Figure 7a is more complex than the simple description made previously. In fact, the two wave fronts are followed by weaker replica. These replica correspond to surface waves propagating around the wires and generated by mode conversion of the incident wave into surface waves. The wires cannot be assumed to point-like targets. They behave as extended sources .

When the process is iterated, target resonances can intensify by the time reversal procedure and Figure 7g shows two replicas appearing symmetrically before and after the principal wave front.

These experiments demonstrate the ability of TRM iterative mode to select the most reflective target in a multi target medium. This may viewed as a learnig process that selects among several wave fronts the one coming from the most important reflector.

CONCLUSION

Time reversal of ultrasonic fields is a new tool in applied physics. Time reversal mirrors are made of large transducer arrays connected to large storage memories. They allow a very efficient way to focus an ultrasonic wave through inhomogeneous media.. Their applications are numerous. It concerns medical applications (imaging, lithotripsy and hyperthermia) as well as non destructive testing and underwater acoustics.

Experimental results obtained with 64 and 128 channel time reversal mirrors demonstrate the possibilities of this technique and several companies are now working on the developement of ultrasonic time reversal mirrors. However time reversal mirrors are not only devoted to applied physics. They can also be used in fundamental physics to solve some problems in the field of wave propagation. Inverse problems can be solved through the use of a TRM. TRMs can also be used to study multiple scattering processes in heterogeneous media and to give some new approachs in the problem of wave localisation.

REFERENCES

1.   M. Fink, C. Prada, F. Wu, D. Cassereau, "Self focusing with time reversal mirror in inhomogeneous media," Proceedings of IEEE Ultrasonics Symposium 1989, Montreal, Vol 2, pp 681-686 (1989)

2.   M. Fink, "Time Reversal of Ultrasonic Fields : Basic Principles" IEEE Trans. Ultrason. Ferroelec. Freq. Contr, Vol 39 (5) pp 555-566 (1992)

3.   D.M .Pepper, "Non linear optical phase conjugation," in *Laser Handbook*, Vol 4, North-Holland Physics Publishing, Amsterdam , pp 333-485 (1988)

4.   J.L Thomas, P. Roux and M. Fink, "Inverse problem in wave scattering with an acoustic time reversal mirror", Phys Rev Letters, Vol 72, 5, (janv 1994)

5.   S.W.Flax, M. O'Donnel, "Phase aberration correction using signals from point reflectors and diffuse scatterers : Basic principles," IEEE Trans. on Ultras. Ferroelec. and Freq. Control. Vol 35 pp 758-767 (1988)

6.   M. O'Donnel and S.W. Flax, "Phase aberration correction using signals from point reflectors and diffuse scatterers: Measurements," IEEE Trans. on Ultras. Ferroelec. and Freq. Control. Vol 35 pp 768-774 (1988).

7.   L.Nock, G.E. Trahey, SW.Smith, "Phase aberration correction in medical ultrasound using speckle brightness as a quality factor," J. Acoust. Soc. Am. Vol 85 pp 1819-1833 (1989)

8.   R.Mallart, M.Fink, "Sound speed fluctuations in medical ultrasound imaging. Comparaison between different correction algorithms," Proceedings of the 19th International Symposium on Acoustical Imaging" Plenum Press (April 1991)

9.   R.Mallart, M.Fink, "The Van Cittert- Zernike Theorem in pulse-echo measurements,". J. Acoust. Soc. Am . 90 (5) pp 2718-2727 (1991)

10.   F. Wu, M. Fink, R. Mallart, J.L. Thomas, N. Chakroun. D. Cassereau, C. Prada, "Optimal Focusing through aberrating media : a comparaison between time reversal mirror and time delay correction techniques," Proceedings of IEEE Ultrasonics Symposium 1991, Orlando

11.   R.P. Porter and A.J. Devaney, "Generalized Holography and the inverse source problems," J. Opt. Soc. Am 72 pp327-330 (1982)

12.   D. Cassereau, F. Wu, M. Fink. "Limits of Self-focusing using closed time-reversal cavities and mirrors- Theory and experiment," Proceedings IEEE Ultrasonics Symposium 1990 Hawaï, pp 1613-1618, Dec 1990

13.   D.Cassereau and M.Fink, "Time-Reversal of ultrasonic fields : theory of the closed time-reversal cavity," IEEE Trans. Ultrason. Ferroelec. Freq. Contr.,Vol 39 (5) pp 579-592 (1992)

14.   F. Wu ,J.L Thomas and M. Fink, "Time Reversal of Ultrasonic Fields : Experimental results," IEEE Trans. Ultrason. Ferroelec. Freq. Contr,Vol 39 (5) pp 567- 578 (1992)

15.   N. Chakroun, M. Fink, F. Wu, " Ultrasonic non destructive testing with time reversal mirrors," Proceedings IEEE Ultrasonics Symposium 1992, Tucson

16.   C. Prada, F. Wu, M. Fink, "The Iterative Time Reversal Mirror : A solution to self focusing in pulse-echo mode," J.Acoust.Soc.Am, 90 (2), pp 1119-1129 (1991)

17.   C. Prada, "Retournement temporel des ondes ultrasonores," Thèse de Doctorat de l'Université Paris VII, Juin 1991

18.   C.Prada, J.L. Thomas and M.Fink, "The iterative time reversal process : analysis of the convergence" submitted to JASA

THEORETICAL STUDY OF FOCUSING TECHNIQUES THROUGH PLANE INTERFACES: COMPARISON BETWEEN TIME-REVERSAL METHODS AND FERMAT'S SURFACE TECHNIQUES

D. Cassereau and M. Fink

Laboratorire Ondes et Acoustique, Université Paris VII, ESPCI
10 rue Vauquelin, 75231 Paris Cedex 05, France

NOTATIONS

- c_1, c_2, ρ_1 and ρ_2 are the velocities and densities of the two fluids,
- c_{ij} and ρ_{ij} are dimensionless ratios defined by $c_{ij} = c_i/c_j$ and $\rho_{ij} = \rho_i/\rho_j$, $i, j \in [1, 2]$,
- f, f_x and f_y are the temporal and spatial frequencies linked to t, x and y via Fourier transforms,
- f_r is defined by $f_r^2 = f_x^2 + f_y^2$,
- ν_i is a complex defined by $\nu_i^2 = 4\pi^2(f^2/c_i^2 - f_r^2)$, $i \in [1, 2]$.

INTRODUCTION

In this paper, we are interested in focusing a pressure field through a plane interface separating two fluids with different acoustical properties. The expected focal point is located in fluid 1 at the origin of the spatial coordinate system, the interface is in the plane $z = h > 0$, and the emitting (and receiving) surface is in fluid 2 in the plane $z = Z > h$.

The first method is based on the time-reserval principle: it can be described as a two-step process [1-4]. We first consider an active source located at the expected focal point and compute the pressure field transmitted through the interface in the plane $z = Z$. We suppose that we are able to measure the pressure and its normal derivative at every point on the receiving surface. During the second step, the initial source is removed or remains passive, and we create secondary sources (monopole and dipole sources) of the emitting surface that correspond to the time-reversal of the components measured during the first step [1-4]. In this case, the receiving/emitting surface will be called Time-Reversal Mirror (TRM) using the Time-Reversal Principle (TRP).

The second method is based on the Fermat's surface technique: as above, we create secondary sources of the surface of the emitter, but these secondary sources reduce to a time-delay law (the amplitude is constant) that compensates the dispersion of time of propagation between the different points on the emitter and the expected focal point, computed along a geometrical path according to the Snell's law [5,6]. In this case, the emitting surface will be called Fermat's Surface Mirror (FSM) using the Fermat's Surface Principle (FSP).

The TRP takes into account the whole information coming from the source through the interface, while the FSP reduces to a time-delay law. Furthermore, the TRP is adaptative, while the FSP explicitly depends on the focal point and the interface. In this paper, we compare the efficiency of these two different approaches in focusing the acoustical beam through the plane interface. We also analyse the influence of the radiation condition on the emitting surface.

The geometrical configuration is illustrated in Fig. 1.

Acoustical Imaging, Vol. 21, Edited by
J.P. Jones, Plenum Press, New York, 1995

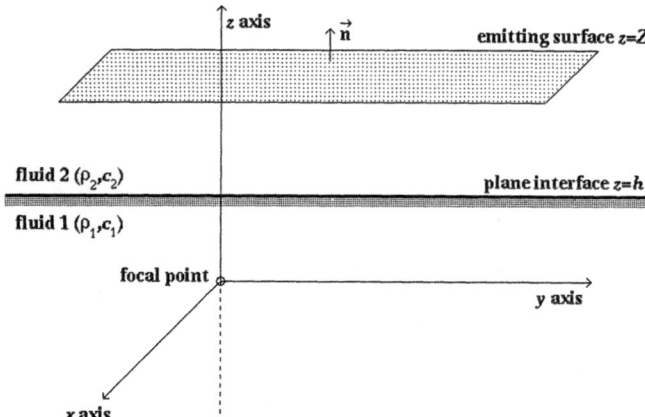

Figure 1: Geometrical configuration of interest. The focal point is at the origin of the spatial coordinate system, the interface is in the plane $z = h > 0$ and the emitting surface in the plane $z = Z > h$, $\vec{n}$ is normal to the emitting surface.

THE TIME-REVERSAL APPROACH

We first consider the active source in fluid 1: it generates an incident pressure field $p_i(x, y, z, t)$ that satisfies the wave equation

$$(\nabla^2 - c_1^{-2}\partial_{tt})p_i(x, y, z, t) = -\phi(t)\delta(x)\delta(y)\delta(z), \tag{1}$$

where $\phi(t)$ describes the temporal variations of the source excitation. Since $p_i(x, y, z, t)$ only corresponds to the incident field (and not to the total field in fluid 1), we do not take into account the interface and $p_i(x, y, z, t)$ reduces to a divergent spherical wave [4,7]; it can be written as

$$p_i(x, y, z, t) = \phi(t - R/c_1)/4\pi R, \tag{2}$$

with $R = (x^2 + y^2 + z^2)^{1/2}$. Taking the 3D Fourier transform of (2) over x, y and t, we obtain a decomposition of the incident field into plane monochromatic components. This decomposition can be written near the interface (in the range $0 \leq z \leq h$) for positive values of the frequency f as [4,7]

$$\tilde{P}_i(f_x, f_y, z, f) = A_1 \exp(j\nu_1|z|)\tilde{\phi}(f), \tag{3}$$

with $A_1 = j/2\nu_1$. The interaction of this incident pressure field with the interface yields the generation of reflected and transmitted components. Writting the continuity of the pressure and normal displacement at the interface, the transmitted field can be classically obtained as follows [5-7]:

$$\tilde{P}_t(f_x, f_y, z, f) = A_1 T_{12} \exp(j\nu_2 z)\tilde{\phi}(f), \tag{4}$$

where T_{12} is the transmission coefficient from fluid 1 to fluid 2 defined by [5-7]

$$T_{12} = \frac{2\nu_1}{\nu_1 + \rho_{12}\nu_2} \exp\left[j(\nu_1 - \nu_2)h\right]. \tag{5}$$

In order to simplify the forward mathematics, we suppose that the TRM is infinite, thus the transform performed during the second step reduces to a complex conjugation in the frequency domain. The initial source is removed or remains passive. As in previous works, the secondary sources created on the surface of the TRM (monopole and dipole sources) result from a complex conjugation of the transmitted pressure and its normal derivative measured during the first step. Since the normal direction $\vec{n}$ to the TRM is oriented in direction of increasing values of z, we obtain [3,4]

$$\begin{cases} \tilde{\Sigma}_1(f_x, f_y, f) = A_1^* T_{12}^* \exp(j\nu_2 Z)^* \tilde{\phi}(f)^*, \\ \tilde{\Sigma}_0(f_x, f_y, f) = -j\nu_2^* A_1^* T_{12}^* \exp(j\nu_2 Z)^* \tilde{\phi}(f)^*. \end{cases} \tag{6}$$

These secondary sources on the surface of the TRM yield the generation of a diffracted pressure field $\tilde{P}_d(f_x, f_y, z, f)$ in fluid 2 that is transmitted through the interface, therefore resulting in a time-reversed pressure field $\tilde{P}_{tr}(f_x, f_y, z, f)$ in fluid 1. We are now interested in the computation of $\tilde{P}_{tr}(f_x, f_y, z, f)$ depending on the radiation condition on the surface of the TRM.

Case I: the TRM is mounted in an infinite rigid planar baffle (it measures, time-reverses and re-emits the normal derivative of the field). In this case, we classically observe that the diffracted pressure field only depends on $\tilde{\Sigma}_0(f_x, f_y, f)$ and can be written as follows [4,8]

$$\tilde{P}_d(f_x, f_y, z, f) = 2\tilde{\Sigma}_0(f_x, f_y, f) \times A_2 \exp\left[j\nu_2(Z - z)\right], \qquad (7a)$$

with $A_2 = j/2\nu_2$. In this equation, the second term corresponds to the 3D Fourier transform over x, y and t of the impulse free-space Green's function of fluid 2, evaluated for a source at Z and an observation point at z. The field given in (7a) is now incident at the interface, therefore resulting in a transmitted time-reversed pressure field that can be written, after some computation steps that we do not present in detail, as follows

$$\tilde{P}_{tr}(f_x, f_y, z, f) = \underbrace{\frac{-j}{2\nu_1^*} \exp(-j\nu_1 z)\tilde{\phi}(f)^*}_{\text{Term 1}} \times \underbrace{\frac{\nu_2^*}{\nu_2} T_{21} T_{12}^* |\exp(j\nu_2 Z)|^2}_{\text{Term 2}}. \qquad (7b)$$

In this equation, T_{21} is the transmission coefficient from fluid 2 to fluid 1. By elementary symmetry, it can be obtained from the expression of T_{12}, interchanging c_1 and c_2, ρ_1 and ρ_2, and reversing the direction of propagation in the exponential function. Looking at (7b), we note that the time-reversed pressure field can be decomposed into the product of two terms. The first term is identical, up to a complex conjugation, to the incident field created by the source during the first step, if ν_1 is real. The second term can be understood as a correction term that includes the radiation conditions on the surface of the TRM and the transmission effects through the interface. Similar results have been obtained in the particular case of a single homogeneous fluid. Indeed, the TRP yields the reconstruction of the propagative components of the initial field only, while the evanescent components are lost as soon as the propagation distances are great compared to the wavelength [4]. This lack of evanescent components yields a more narrow angular spectrum, thus a broader focal pattern. This effect is directly related to purely imaginary values of ν_1 or ν_2 in the exponential functions. Since these complex values are both present in exponential functions in (7b), the truncation due to the lack of evanescent components now arises as soon as $f_r > f/\max(c_1, c_2)$. This is due to the fact that the wave fields must propagate in both fluids during steps 1 and 2, and the size of the focal pattern is directly related to the theoretical limitations resulting from the fastest fluid.

Case II: the TRM is mounted in an infinite soft planar baffle (it measures, time-reverses and re-emits the field). In this case, we classically observe that the diffracted pressure field only depends on $\tilde{\Sigma}_1(f_x, f_y, f)$ and can be written as follows [9]

$$\tilde{P}_d(f_x, f_y, z, f) = -2\tilde{\Sigma}_1(f_x, f_y, f) \times \frac{\partial}{\partial z_s} A_2 \exp\left[j\nu_2(z_s - z)\right]\Big|_{z_s = Z}. \qquad (8a)$$

In this equation, the second term corresponds to the 3D Fourier transform over x, y and t of the normal derivative (with respect to $\vec{n}$) of the impulse free-space Green's function of fluid 2, evaluated for a source at Z and an observation point at z. Similarly to the previous case, the resulting time-reversed pressure field near the intial source position can be written as follows

$$\tilde{P}_{tr}(f_x, f_y, z, f) = \underbrace{\frac{-j}{2\nu_1^*} \exp(-j\nu_1 z)\tilde{\phi}(f)^*}_{\text{Term 1}} \times \underbrace{T_{21} T_{12}^* |\exp(j\nu_2 Z)|^2}_{\text{Term 2}}. \qquad (8b)$$

As above, the time-reversed pressure field can be decomposed into the product of two terms. The first one is unchanged, while the second one is modified due to the new radiation conditions on the surface of the TRM. In fact, the only difference between cases I and II consists in the ratio ν_2^*/ν_2 that either equates $+1$ or -1, corresponding to propagating or evanescent components in fluid 2. Thus, the solutions obtained in these two cases are identical, unless the spatial frequencies f_x and f_y correspond to evanescent modes in fluid 2. This result is in complete agreement with a similar conclusion in the case of a single unbounded homogeneous fluid [4].

Case III: the TRM behaves as in free-space (it measures, time-reverses and re-emits both the field and its normal derivative). For a single homogeneous fluid, we proved that the time-reversed pressure field reduces to half the sum of the solutions obtained in cases I and II. Since this only results from the diffraction formalism, it remains valid here [4].

NUMERICAL RESULTS

Starting from the equations presented in this paper, we have developped a software to compute the pressure field generated near the expected focal point using the TRP and FSP. Although the equations are given in the monochromatic regime, the computation is performed in the time domain with a sampling frequency of 20 MHz. The excitation function $\phi(t)$ is a sinusoisal burst with a Gaussian enveloppe, the central frequency is 1 MHz and the relative -6 dB bandwidth is 100 % [4]. The pressure field is computed in the plane $z = 0$, parallel to the interface, that contains the focal point. The figures only represent the maximum amplitude of the temporal signals obtained at the different observation points. The acoustical parameters of the two fluids c_1, c_2, ρ_1 and ρ_2 are 6000 m/s, 1500 m/s, 4500 kg/m^3 and 1000 kg/m^3, respectively, corresponding to a focal point in the fastest medium and an emitter in water. The interface is at h=15 mm and the emitter at Z=20 mm. In all the following, the curves obtained using the TRP and FSP are shown simultaneously, the periodically marked curves corresponding to the FSP.

Since the propagation distances are great compared to the wavelengths, we could not observe any significant differences between the different radiation conditions on the surface of the emitter. Thus, the computations have been performed using (7a) and (7b).

Figure 2 represents the results obtained with an emitter of infinite size. We clearly see that the field is focused on the expected position, and that the amplitude of the signal decreases rapidly with the distance between the observation point and the focal point. The size of the focal pattern, measured at -6 dB, is about 7 mm, while the theoretical limitation at 1 MHz, due to diffraction, is 6 mm. Comparing the two focusing methods, we observe that the mainlobe sizes are identical, and that the sidelobes level is much higher using the FPS (we obtain a difference of about 10 dB at ± 30 mm). This is due to the fact that the FSP considers a constant amplitude in emission, while the TRP also takes into account the amplitude variations due to diffraction and refraction of the field generated by the active source in step 1.

In Fig. 3, we now consider a square emitter of dimension 40 mm and a focal point on its symmetry axis. Once again, the pressure field is focused on the expected position, but the focal pattern is broader (about 8.5 mm at -6 dB) and the decrease of the amplitude is not as rapid as in the previous case. The differences between the TRP and FSP, observed in Fig. 2, remain clearly visible in Fig. 3.

In Fig. 4, we consider the same configuration as in Fig. 3, but the focal point is now 10 mm apart from the symmetry axis of the emitter in the x and y directions. We observe that the pressure field is effectively focused on the expected position, but the mainlobe size increases (about 9.5 mm at -6 dB): this is due to the fact that we now try to focus on a target that is not located on the symmetry axis of the emitter, and this is not an optimal configuration. Nevertheless, this result proves that the TRP is efficient enough to be able to focus, even in inadequate situations. The field obtained using the FSP shows a similar mainlobe and a higher sidelobes level. We also observe that the focal pattern is not symmetric, and that the asymmetry effect is enhanced in the case of the FSP.

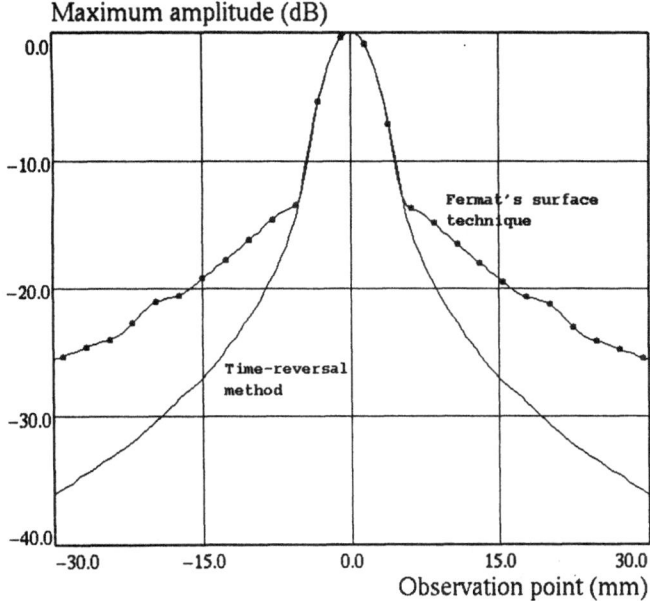

Figure 2. Maximum amplitude of the temporal signals observed in the plane $z = 0$ with the TRP (continuous curve) and the FSP (marked curve) for an emitting surface of infinite size. The FSP yields a higher sidelobes level compared to the TRP.

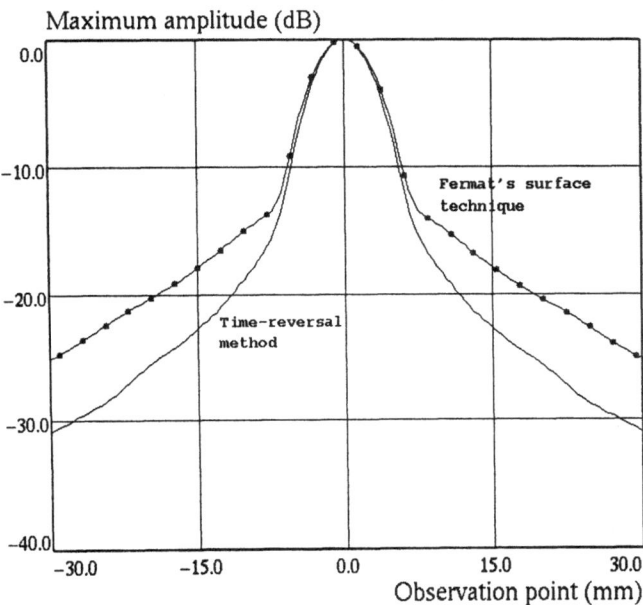

Figure 3. Maximum amplitude of the temporal signals observed in the plane $z = 0$ with the TRP (continuous curve) and the FSP (marked curve) for a square emitting surface of dimension 40 mm. The focal point is on the symmetry axis of the emitter. The FSP yields a higher sidelobes level compared to the TRP.

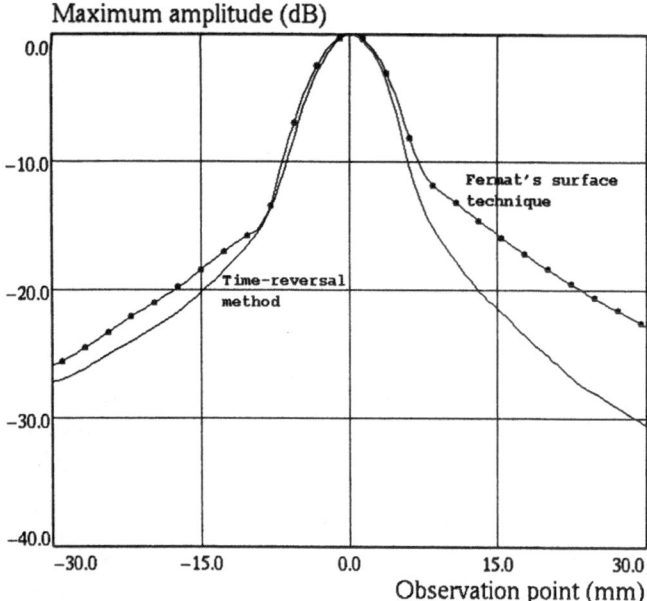

Figure 4. Maximum amplitude of the temporal signals observed in the plane $z = 0$ with the TRP (continuous curve) and the FSP (marked curve) for a square emitting surface of dimension 40 mm. The focal point is 10 mm apart from the symmetry axis of the emitter in the x and y directions. The focal patterns are asymmetric, particularly in the case of the FSP.

CONCLUSION

In this paper, we have shown that the TRP and FSP are both efficient to optimize focusing through a plane interface separating two fluids. One great advantage of time-reversal is that the principle is dynamically adaptative to different focal points in front of the emitting surface, while the Fermat's surface technique explicitly depends on the interface and spatial position of the focal point. If we compare the mainlobe size of the focal pattern resulting from these two techniques, we do not observe any significant difference. Nevertheless, the FSP results in much higher sidelobes that the TRP, and in focal patterns that are more asymmetric if the focal point is not located on the symmetry axis of the emitting surface. The high sidelobes level can be very embarrassing, particularly for imaging applications, since they yield the generation of unexpected echoes that perturb the images. This is a classical property of Fermat's surface transducers, and the results presented here show a great improvement in focusing techniques using time-reversal. We are now working on an extension of this analysis to the case of an interface separating a fluid and a solid, in order to study the influence of mode conversions due to the longitudinal and transverse modes that can exist simultaneously in a solid medium.

REFERENCES

1. M. Fink, "Time-reversal of ultrasonic fields − Part I: Basic Principles", IEEE Trans. Ultrason. Ferroelec. Freq. Contr. **39**, 555-566 (1992).
2. F. Wu, J.L. Thomas and M. Fink, "Time-reversal of ultrasonic fields − Part II: Experimental results", IEEE Trans. Ultrason. Ferroelec. Freq. Contr. **39**, 567-578 (1992).
3. D. Cassereau and M. Fink, "Time-reversal of ultrasonic fields − Part III: Theory of the closed time-reversal cavity", IEEE Trans. Ultrason. Ferroelec. Freq. Contr. **39**, 579-592 (1992).
4. D. Cassereau and M. Fink, "Focusing with plane time-reversal mirrors: an efficient alternative to closed cavities", J. Acoust. Soc. Am. **94**, 2373-2386 (1993).
5. P. M. Morse and K. U. Ingard, *Theoretical acoustics*, New York, McGraw Hill (1968).
6. L. M. Brekhovskikh, *Waves in layered media*, New York, Academic Press (1960).
7. D. Cassereau and D. Guyomar, "Reflection of an impulse spherical wave at a plane interface separating two fluids", J. Acoust. Soc. Am. **92**, 1706-1720 (1992).
8. G. R. Harris, "Transient field of a baffled planar piston having an arbitrary vibration amplitude distribution", J. Acoust. Soc. Am. **70**, 186-204 (1981).
9. M. Born and E. Wolf, *Principles of optics*, New York, Pergamon Press (1975).

LAMB WAVE PROPAGATION IN MULTILAYERED ANISOTROPIC SOLIDS AND ITS APPLICATION TOWARDS IMAGING MATERIAL DEFECTS

W. Yang[1] and T. Kundu[2]

[1] Graduate Research Assistant
[2] Associate Professor
Department of Civil Engineering and Engineering Mechanics
University of Arizona, Tucson, AZ 85721

ABSTRACT

In this paper the elastic wave propagation analysis in a multilayered plate is presented. Individual layers in the multilayered specimen is modeled as elastic, homogeneous and anisotropic. From this analysis the incident angle and frequency of the ultrasonic signal for Lamb wave generation in a multilayered composite plate are obtained. Displacement and stress fields inside the plate are computed for different Lamb wave modes. It shows that for detecting and imaging defects in a specific layer certain Lamb wave modes are more effective than others.

INTRODUCTION

Ultrasonic technique has become one of the most popular nondestructive testing techniques because of its versatility and ease of operation. Most of the times it can easily detect internal crack and inclusion type defects in homogeneous and layered materials. However, it has its own shortcomings. It is not very effective in detecting cracks which are vertical to the plate surface. This is because the ultrasonic signal is not reflected by the crack when the signal propagation direction is parallel to the crack surface. However, by acoustic microscopy one can detect vertical cracks, located close to the specimen surface, because these cracks obstruct the path of leaky Rayleigh waves generated by the acoustic microscope lens[1]. However, when these vertical cracks are located at a greater depth acoustic microscope cannot detect them because Rayleigh waves do not penetrate deep into the material. Lamb waves on the other hand propagate through the entire plate and can be an effective tool to detect these deep vertical cracks. Kundu and Blodgett[2] studied Lamb wave propagation in isotropic layers; in this paper that study is extended to include anisotropic layers.

Elastic wave propagation in an unidirectional fiber reinforced composite plate has been studied by some investigators[3,4]. Those studies are extended to include multilayered anisotropic solids in this paper.

PROBLEM FORMULATION

Solution to the elastic wave propagation problem in a transversely isotropic solid plate immersed in a liquid is available in the literature[3,4] and is not repeated here. Extending that formulation to the multilayered plate problem involves some new manipulations which are presented here. A plate is made of several layers. In each layer the fiber directions can be different. Let us introduce $x_1 x_2 x_3$ coordinate system such that x_1 coincides with the fiber direction of the top layer and x_3 is measured along the thickness direction of the plate, see figure 1. The second layer has a different fiber direction. Let the angle between the fiber direction x_1' of the second layer and that of the first layer (x_1) be equal to θ. The direction of propagation of the plane incident wave has a vertical component in x_3 direction and a horizontal component (in x_1'' direction) which may or may not coincide with any fiber direction. The angle between x_1 and x_1'' is denoted by ϕ.

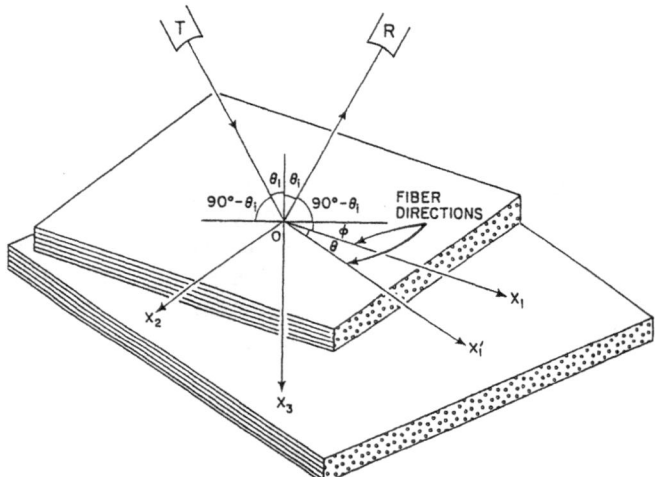

Figure 1. Problem geometry - fiber directions in two different layers are different.

The stress-strain relation in an unidirectional fiber reinforced layer is given by[5],

$$
\begin{Bmatrix} \sigma_{11} \\ \sigma_{22} \\ \sigma_{33} \\ \sigma_{23} \\ \sigma_{31} \\ \sigma_{12} \end{Bmatrix} = \begin{pmatrix} c_{11} & c_{12} & c_{12} & 0 & 0 & 0 \\ c_{12} & c_{22} & c_{23} & 0 & 0 & 0 \\ c_{12} & c_{23} & c_{22} & 0 & 0 & 0 \\ 0 & 0 & 0 & c_{44} & 0 & 0 \\ 0 & 0 & 0 & 0 & c_{55} & 0 \\ 0 & 0 & 0 & 0 & 0 & c_{55} \end{pmatrix} \begin{Bmatrix} u_{1,1} \\ u_{2,2} \\ u_{3,3} \\ u_{2,3} + u_{3,2} \\ u_{3,1} + u_{1,3} \\ u_{1,2} + u_{2,1} \end{Bmatrix} \qquad (1)
$$

when x_1 direction coincides with the fiber direction. In equation (1), $c_{44} = (c_{22^-}$

c_{23})/2. For a general layer whose fiber direction x_1' is different from x_1 the stress-strain relationship given in equation (1) must be expressed in the $x_1'x_2'x_3'$ coordinate system so that the stiffness matrix for this layer has a form similar to the one given in equation (1). If one defines $x_1'x_2'x_3'$ as the local coordinate associated with an individual layer and $x_1x_2x_3$ as the global coordinate for the entire problem geometry then by coordinate transformation one can relate the stress displacement vectors {S'} and {S} in these two coordinate systems in the following manner.

$$\{S'\} = [B]\{S\} \tag{2}$$

where

$$[B] = \begin{pmatrix} \cos\theta & \sin\theta & 0 & 0 & 0 & 0 \\ -\sin\theta & \cos\theta & 0 & 0 & 0 & 0 \\ 0 & 0 & 1 & 0 & 0 & 0 \\ 0 & 0 & 0 & \cos\theta & \sin\theta & 0 \\ 0 & 0 & 0 & -\sin\theta & \cos\theta & 0 \\ 0 & 0 & 0 & 0 & 0 & 1 \end{pmatrix} \tag{3}$$

and

$$\{S'\} = [\,u_1' \quad u_2' \quad u_3' \quad \sigma_{31}' \quad \sigma_{32}' \quad \sigma_{33}'\,]^T$$

$$\{S\} = [\,u_1 \quad u_2 \quad u_3 \quad \sigma_{31} \quad \sigma_{32} \quad \sigma_{33}\,]^T \tag{4}$$

The stress-displacement vector $\{S^i\}$ of the i-th layer can be expressed in terms of the vector of the coefficients of the wave potentials $\{C^i\}$ in the i-th layer in the form,

$$\{S^i\} = [B^i]^T[Q^i][E^i]\{C^i\} \tag{5}$$

where superscript i in vectors and matrices of equation (5) indicates that they are associated with the i-th layer. $[B^i]$ is defined in equation (3), $[Q^i]$ and $[E^i]$ are 6x6 matrices whose elements are functions of incident wave frequency, layer dimensions and layer properties. Expressions of these matrices are available in references 3 and 4, and $\{C^i\}$ is defined by

$$\{C\} = [\,A_1^+ \quad A_2^+ \quad A_3^+ \quad A_1^- \quad A_2^- \quad A_3^-\,]^T \tag{6}$$

where superscripts "+" and "-" of $A_{1,2,3}$ indicate the coefficients of wave potentials associated with downgoing and upgoing waves respectively.

If one uses superscripts "+" and "-" to indicate the position at the "bottom" and "top" surface of a layer respectively then the continuity condition at the interface can be expressed as

$$\{S^{i+}\} = \{S^{(i+1)-}\} \tag{7}$$

Then following Thomson-Haskell[6] or "propagator matrix" technique one can relate stress-displacement vectors associated with the top and bottom surfaces of the plate in the following form

$$\{S(0)\} = [G]\{S(h)\} \tag{8}$$

where [G] is a 6x6 matrix product obtained by multiplying 6x6 layer matrices associated with each layer. Then unknown quantities are determined by satisfying the top and bottom surface boundary conditions.

NUMERICAL RESULTS

Results are presented for two specimens. Both specimens have two layers, each layer is 0.5 mm thick. In specimen 1 the fiber directions in the two layers are 45° apart [$\theta = 45^O$], in specimen 2 both layers have same fiber directions [$\theta = 0^O$]. Hence, specimen 2 is effectively a single layer plate, 1 mm thick. The elastic constants associated with each layer are, c_{11} = 160.73 GPa, c_{12} = 6.44 GPa, c_{22} = 13.92 GPa, c_{23} = 6.92 GPa, c_{44} = 3.5 GPa, c_{55} = 7.07 GPa, and density = 1.578 gm/cc. These values are identical to those given in reference 3.

A plane acoustic wave of frequency 1 MHz is incident on the top surface of the plate with an incident angle θ_i. The horizontal component of the incident wave direction makes an angle ϕ with the x_1 axis. For our calculations we take $\phi = 0^O$, in other words the wave direction coincides with the fiber direction of the top layer as shown in figure 1. Solid and dotted lines in figures 2 and 3 correspond to the plane wave reflection and transmission coefficients (R and T) of specimens 1 and 2 respectively. It should be noted that, as expected, dips in R coincide with the peaks in T; these dips and peaks correspond to the plate wave modes or Lamb wave modes. In other words, incident angles corresponding to these dips and peaks produce plate waves. It should be noted here that the two-layer-plate (specimen 1, figure 2) shows more plate wave modes than the single layer plate (specimen 2, figure 3). Displacement and stress fields corresponding to these dips and peaks of R and T have been computed and some of theses results are shown in this section.

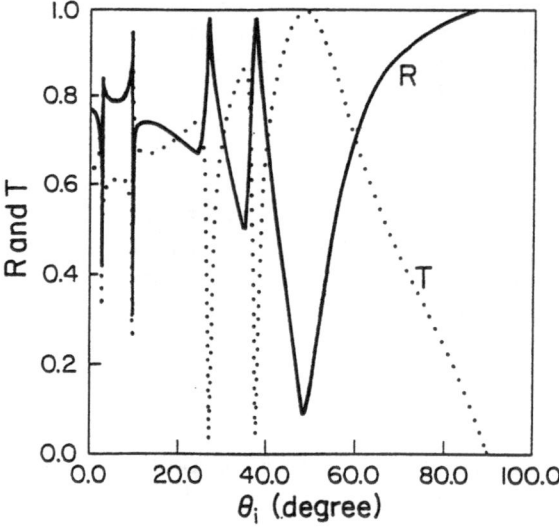

Figure 2. Plane wave reflection (R, solid line) and transmission (T, dotted line) coefficients as a function of the incident angle in specimen 1 - two layers have fibers in two different directions.

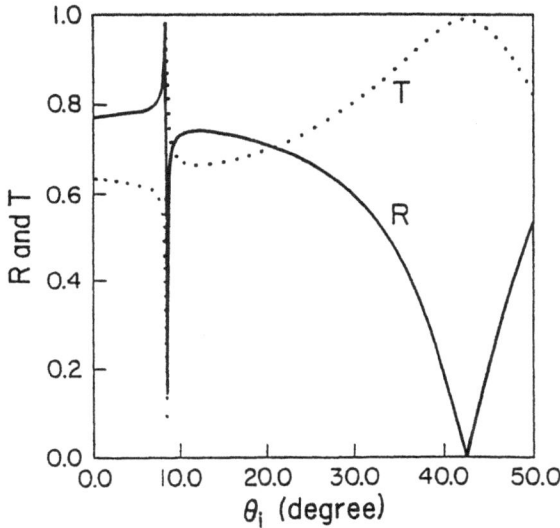

Figure 3. Plane wave reflection (R, solid line) and transmission (T, dotted line) coefficients as a function of the incident angle in specimen 2 - unidirectional fiber reinforced composite plate.

In figure 2, two very sharp peaks and dips in R and T are observed at 9.5997° and 9.7398°. Displacement and stress components along the depth of the plate for these two angles of incidence are plotted in figures 4 and 5. Solid, dotted and dashed lines correspond to u_1, u_2, and u_3 in displacement plots and σ_{13}, σ_{23}, and σ_{33} in stress plots respectively. This small variation (about 0.14°) of the incident angle does not significantly alter the shape of the variation of horizontal displacement and shear stress components, but the vertical displacement and normal stress variations are altered significantly. It should be noted that u_3 and σ_{33} at the bottom surface of the plate, $x_3 = 1$ mm, are almost equal to zero in figure 4, that corresponds to a dip in T, and they are significantly higher in figure 5, that corresponds to a peak in T. These results are in consistence with one's expectation, that lower transmission coefficient should produce less excitation at the bottom surface.

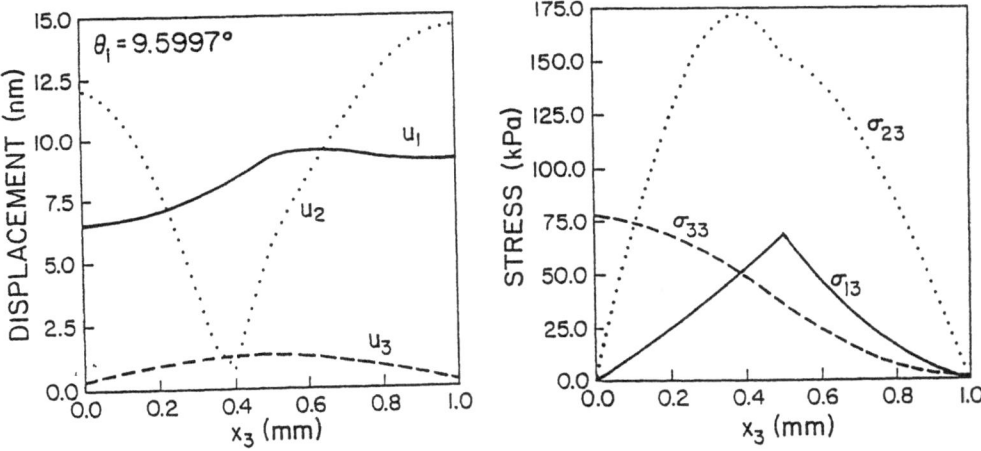

Figure 4. Displacement (left figure) and stress (right figure) components as a function of depth (x_3) in specimen 1, angle of incidence = 9.5997°.

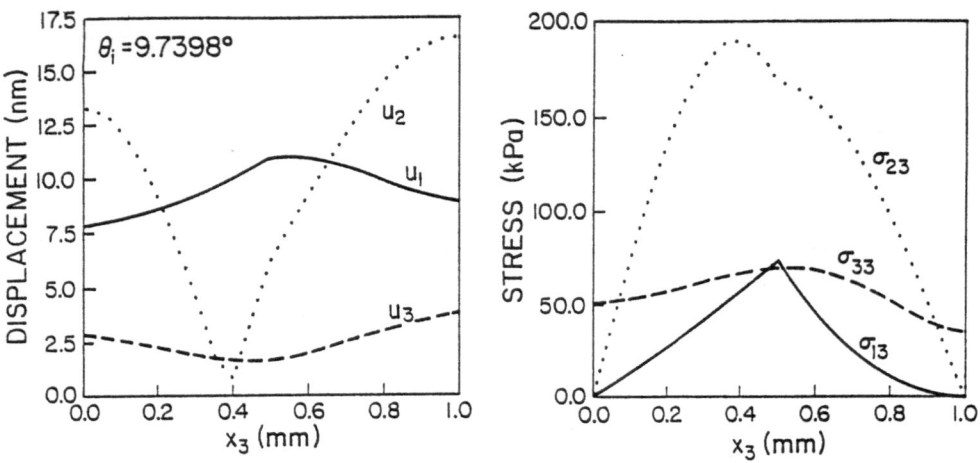

Figure 5. Displacement (left figure) and stress (right figure) components as a function of depth (x_3) in specimen 1, angle of incidence = 9.7398°.

Displacement and stress components for 27.1873° incident angle are plotted in figure 6. It is interesting to note that u_1 and σ_{13} is significantly stronger in the bottom layer , z > 0.5 mm. Hence, if one can decouple the effects of u_1 and σ_{13} from other displacement and stress components and sense those separately then defects at the bottom layer can be easily detected and imaged by exciting the plate at this mode.

Figure 7 shows displacement and stress components for 37.6977° incidence. In this case also u_1 is much stronger in the bottom layer. σ_{13} on the other hand is weak in both top and bottom layers and very strong at the interface. Hence, for detecting interface defects, such as delamination, debonding or interface corrosion this angle of incidence is very effective.

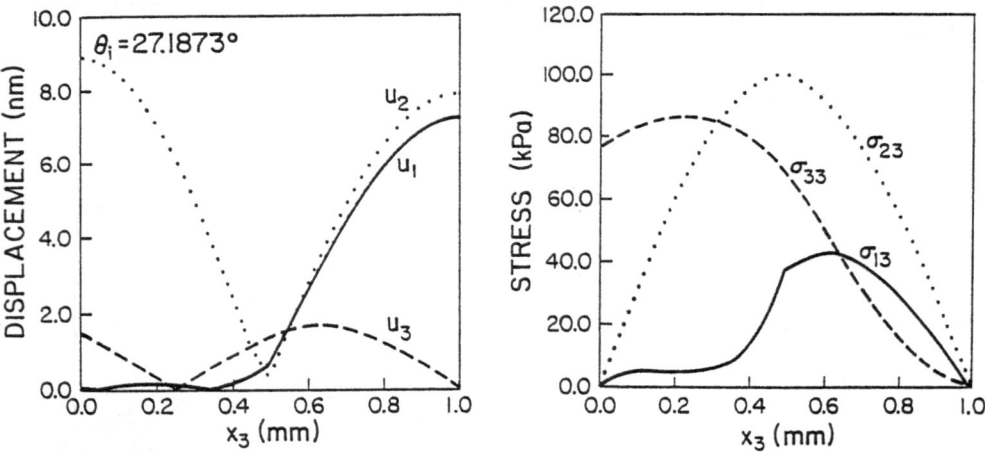

Figure 6. Displacement (left figure) and stress (right figure) components as a function of depth (x_3) in specimen 1, angle of incidence = 27.1873°.

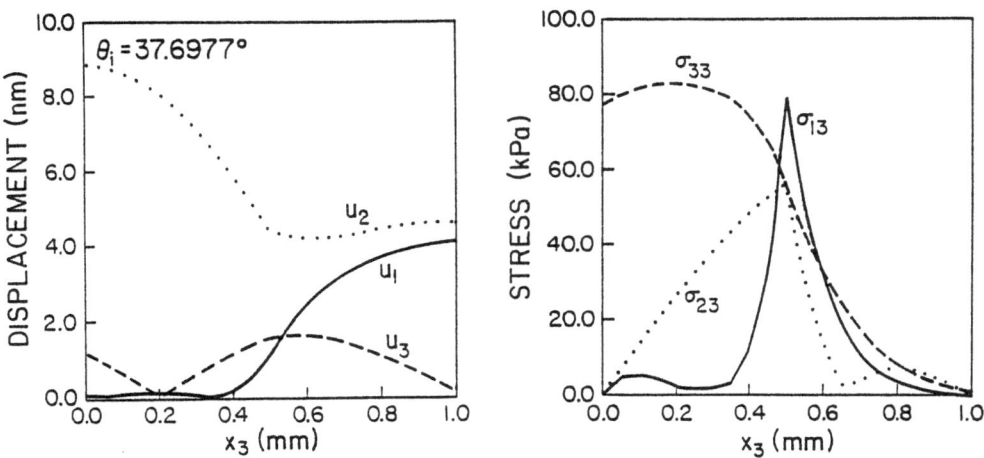

Figure 7. Displacement (left figure) and stress (right figure) components as a function of depth (x_3) in specimen 1, angle of incidence = 37.6977°.

In figure 4 peaks and dips in R are observed at 8.5025° and 8.6767°. At these two angles displacements and stresses are plotted in figures 8 and 9. u_2 and σ_{23} are zero in figures 8 and 9 because of the material symmetry and excitation symmetry about the x_1 axis. One can observe that u_3 and σ_{33} at the bottom surface are comparatively stronger in figure 9 than in figure 8. This is expected since figures 8 and 9 correspond to a dip and a peak in T respectively.

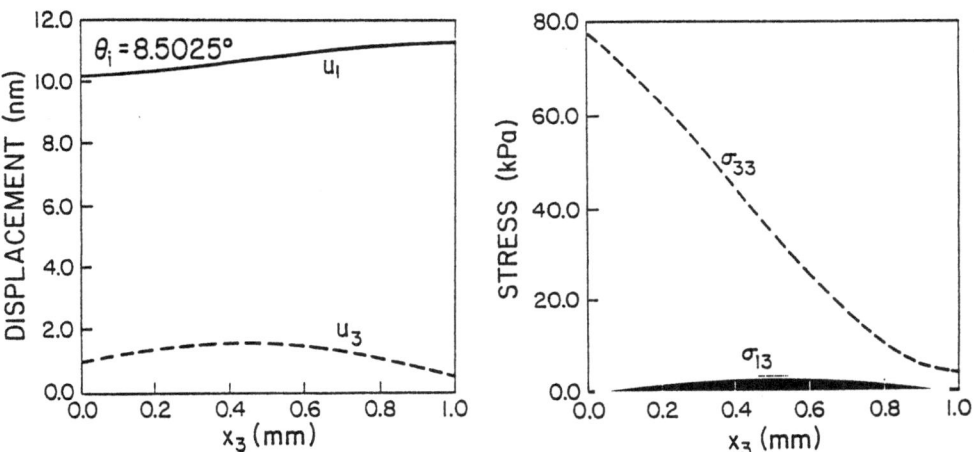

Figure 8. Displacement (left figure) and stress (right figure) components as a function of depth (x_3) in specimen 2, angle of incidence = 8.5025°.

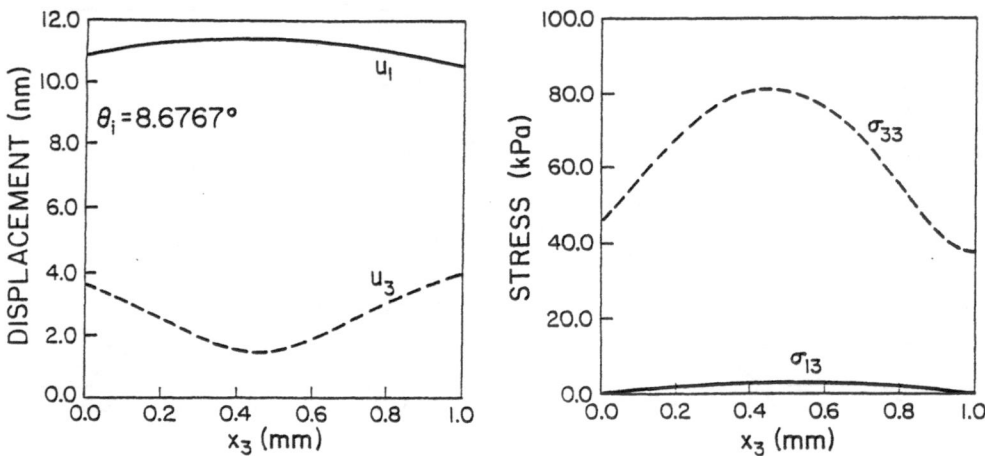

Figure 9. Displacement (left figure) and stress (right figure) components as a function of depth (x_3) in specimen 2, angle of incidence = 8.6767°.

CONCLUDING REMARKS

Preliminary calculations presented in this paper show that in layered composite plates it is possible to generate different Lamb wave modes which produce displacement and stress components of different strengths in different layers. One can effectively use these plate wave modes to detect and image material defects at various depths inside different layers.

ACKNOWLEDGMENT

This research was carried out under NSF research grant MSS-9310528.

REFERENCES

1. J.M.R. Weaver, M.B. Somekh, G.A.D. Briggs, S.D. Peck and C. Illet, Application of the scanning acoustic microscope to the study of materials science, IEEE Trans. Sonics and Ultrason., SU32:302(1985).
2. T. Kundu and M. Blodgett, Detection of material defects in layered solids using Lamb waves, in: Review of Progress in Quantitative NDE, Plenum Press, Proc. QNDE'93 Conf., Brunswick, Maine, Aug.1-6 (1993).
3. A.K. Mal, C.C. Yin and Y. Bar-Cohen, Ultrasonic nondestructive evaluation of cracked composite laminate, Comp. Engr., 1:85(1991).
4. T. Kundu and B. Maxfield, A new technique to detect surface wave and measure their velocities, in: Acousto-Optics and Acoustic Microscopy, S.M. Gracewski and T. Kundu, eds., ASME, New York, AMD140:81(1993).
5. R.M. Christesen. "Mechanics of Composites Materials", Ch.4, John Wiley, New York (1981).
6. T. Kundu and A.K. Mal, Elastic waves in a multilayered solid due to a dislocation source, Wave Motion, 7:459(1985).

A COMPARISON OF THE TRANSIENT PROPAGATION PROPERTIES OF GAUSSIAN AND BESSEL WAVES

John P. Powers, William Reid, John G. Upton,
and Ray Van de Veire

Naval Postgraduate School
Department of Electrical and Computer Engineering, Code EC/Po
833 Dyer Road, Room 437
Monterey CA 93943-5121

INTRODUCTION

Low–diffraction waves [1–5] have become of interest in ultrasound systems because of their longer depth of field for use in imaging and pulse–echo applications. Continuous-wave (CW) Bessel waves of infinite extent suffer no diffraction spreading [6–9]. Continuous spatially truncated Bessel waves also have less spreading than CW spatially truncated Gaussian waves. Here we use a computer-based simulation to investigate the propagation properties of *pulsed* Gaussian and Bessel waves with circularly finite extent. If the width of the Gaussian or Bessel wave is a and the diameter of the circle that truncates the wave is d, the ratio of d/a determines the propagation properties of the pulsed wave. Our propagation simulation uses fast Fourier spatial transforms to rapidly calculate the spatial impulse response wave, $h(x, y, z, t)$, at a location z in front of the source. The complete temporal response can be found by convolving the impulse response $h(x, y, z, t)$ with the time excitation waveform $T(t)$. The predicted propagation patterns are presented to compare the behavior of the Gaussian and Bessel waves. For small ratios of d/a, the Gaussian and Bessel excitations can be made quite similar and the resulting diffraction patterns are also nearly the same. For large values of d/a, it is more difficult for the source functions to mimic each other and the wave patterns are quite different. In particular, in this regime of operation, the Bessel waves shows significant sidelobes, as the Bessel wave begins to have both positive and negative excitations. In the Gaussian wave, these sidelobes are absent, due to the smooth continuous nature of the Gaussian spatial excitation.

Figure 1 shows the geometry of the problem. The source is assumed to be located in a planar rigidly baffled region shown at the bottom of the figure. The normal velocity of the source is assumed to be known; we want to find the excited wave at the observation point (x, y, z) (or in the entire parallel plane located a distance z away from the source plane) as a function of time. The medium is assumed to be linear and homogeneous and has a velocity of 1500 m/s (i.e., that of water).

Acoustical Imaging, Vol. 21, Edited by
J.P. Jones, Plenum Press, New York, 1995

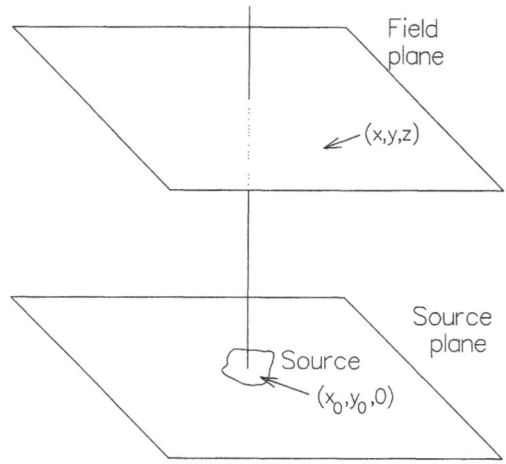

Figure 1. Geometry of source and receiving plane.

REVIEW OF PROPAGATION SIMULATION TECHNIQUE

The propagation simulation technique [10, 11] is based on linear systems theory. We assume that the medium is lossless, linear, and homogeneous and that the source is surrounded by a rigid baffle. The first set of assumptions assures that propagation is a linear operation. The velocity of the source is assumed to be separable in space and time and is given by $v(x_0, y_0, 0, t) = s(x, y)T(t)$.

Figure 2 shows the block diagram approach to modeling propagation. In part (a) of the figure, we apply a point source spatial excitation with a temporal impulse, $\delta(x, y)\delta(t)$. By definition the resulting wave at the observation point is the *impulse response* of the propagation operation. Mathematically, this response is also the Green's function of the problem, $g(x, y, z, t)$. The Green's function for propagation into the half-space from a rigid baffle is known to be $g(x, y, z, t) = \delta(ct - R)/2\pi R$ where $R = \sqrt{x^2 + y^2 + z^2}$.

When the propagation operation is excited by a temporal impulse with an arbitrary spatial excitation, $s(x, y)\delta(t)$, as shown in part (b) of Figure 2, linear systems theory predicts that the result, $h(x, y, z, t)$, is the spatial convolution of the impulse response, $g(x, y, z, t)$, with the spatial portion of the excitation, $s(x, y)$, or

$$h(x, y, z, t) = s(x, y) \underset{x}{*} \underset{y}{*} \delta(t). \tag{1}$$

Following the literature, we will call this result the *spatial impulse response* (i.e., the response to an arbitrary spatial excitation with a temporal impulse).

Finally, when the propagation operation is initiated by an arbitrary, separable function of space and time, $s(x, y)T(t)$, linear systems theory predicts that the result, $\phi(x, y, z, t)$, will be the temporal convolution of $T(t)$ with the spatial impulse response, $h(x, y, z, t)$, or

$$\phi(x, y, z, t) = T(t) \underset{t}{*} h(x, y, z, t) = s(x, y)T(t) \underset{x}{*} \underset{y}{*} \underset{t}{*} g(x, y, z, t). \tag{2}$$

The prospect of computing the triple convolution of Eq. 2 is daunting. Instead, we prefer to enter into the spatial transform domain in order to represent the double spatial

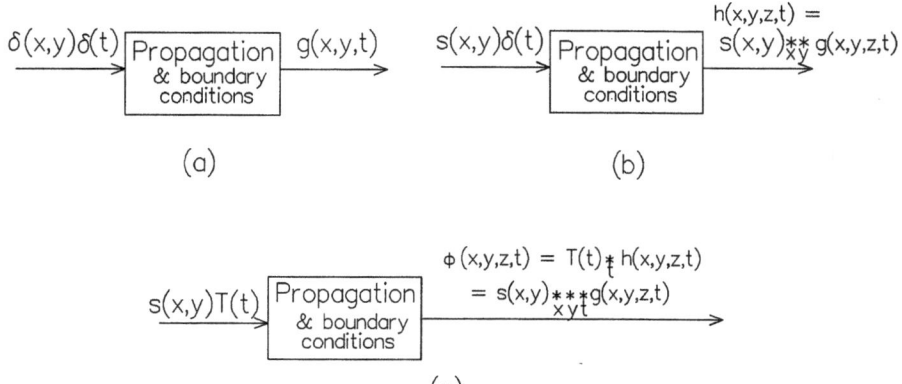

Figure 2. Block diagram explanation of propagation model. (a) Impulse response, (b) spatial impulse response, and (c) general to modeling the propagation.

convolutions as a multiplication. The two-dimensional spatial transform $\tilde{a}(f_x, f_y)$ of a function $a(x, y)$ is defined by

$$\tilde{a}(f_x, f_y) \quad = \quad \mathcal{F}\{a(x,y)\} = \int_{-\infty}^{\infty} \int_{-\infty}^{\infty} a(x,y) \, e^{+j2\pi(f_x x + f_y y)} \, dx \, dy \; . \tag{3}$$

The inverse two-dimensional spatial transform is

$$a(x, y) \quad = \quad \mathcal{F}^{-1}\{\tilde{a}(f_x, f_y)\} = \int_{-\infty}^{\infty} \int_{-\infty}^{\infty} \tilde{a}(f_x, f_y) \, e^{-j2\pi(f_x x + f_y y)} \, df_x \, df_y \; . \tag{4}$$

Taking the two-dimensional spatial transform of Eq. 2 gives

$$\tilde{\phi}(f_x, f_y, z, t) \quad = \quad T(t) \overset{*}{_t} \, [\tilde{s}(f_x, f_y) \tilde{g}(f_x, f_y, z, t)] \; , \tag{5}$$

and taking the inverse transform provides

$$\phi(x, y, z, t) \quad = \quad T(t) \overset{*}{_t} \, \mathcal{F}^{-1}\{\tilde{s}(f_x, f_y) \tilde{g}(f_x, f_y, z, t)\} \; . \tag{6}$$

The transform of the propagation impulse response, $\tilde{g}$, is known as the *transfer function* of propagation. For the rigid baffle and propagation into the half-space of lossless media, this transfer function is known [10] to be

$$\tilde{g}(f_x, f_y, z, t) \quad = \quad J_0\left(\rho\sqrt{c^2 t^2 - z^2}\right) H(ct - z) \tag{7}$$

where $\rho = \sqrt{f_x^2 + f_y^2}$ and $H(\cdot)$ is the Heaviside step function. From this function we can picture propagation as a time-varying spatial filter that begins as an all-pass spatial filter and then increasingly becomes a low-pass spatial filter.

The method for simulating propagation, then, is

1. Find the two-dimensional Fourier transform of $s(x, y)$.

2. For each desired value of z and t, multiply the result with $\tilde{g}(f_x, f_y, z, t)$ as expressed in Eq. 7.

3. Take the inverse two-dimensional inverse transform of the product to find the spatial impulse response, $h(x, y, z, t)$.

4. If desired, find the output for various $T(t)$ by convolving $T(t)$ with $h(x, y, z, t)$.

5. If desired, find the wave pressure $p(x, y, x, t)$ from

$$p(x, y, z, t) = \phi_0 \frac{\partial \phi}{\partial t}, \tag{8}$$

where ρ_0 is the density of the medium.

This simulation technique has been implemented in Fortran [12] and in MATLAB [13–15]. The following studies were produced with the MATLAB models.

NUMERICAL SIMULATIONS

We now turn our attention to the simulation of the Gaussian and Bessel excitations. The equation for a symmetric Gaussian excitation that is truncated by a circle is

$$s_G(r; \sigma, d) = \begin{cases} e^{r^2/\sigma^2} & \text{if } r < d/2 \\ 0 & \text{if } r \geq d/2 \end{cases} \tag{9}$$

where r is the radial distance from the z axis, σ is the "radius" of the curve at the $1/e-$ amplitude points, and d is the diameter of the truncation circle. (In our simulations, $d = 51$ samples; the sample points are located 2.5 mm apart [14,15].)

Similarly, the equation for the Bessel excitation is

$$s_B(r; a, d) = \begin{cases} J_0(ar) & \text{if } r < d/2 \\ 0 & \text{if } r \geq d/2 \end{cases} \tag{10}$$

where a is a scaling factor that controls the width of the Bessel function.

In our simulation, we chose to make the values of the spatial excitation functions equal at $r = 0$ and at the half–maximum points (i.e., at the radius where each function is equal to one–half of its maximum value). For the chosen functions this relates a to σ by $a = 1.291943/\sigma$; we will refer only to the value of σ from this point on. Figure 3 shows a cross-section of the Gaussian and Bessel excitations for $\sigma = 2$ (i.e., narrow waves within the truncation circle). The Gaussian smoothly diminishes while the Bessel wave undergoes its oscillatory behavior before being truncated.

Figures 4 and 5 show perspective views of these spatial excitation functions for $\sigma = 2$ and $\sigma = 16$, respectively. For $\sigma = 16$, the waves are appreciably wider within the truncation circle and closely resemble each other, as seen in the cross-section representation of the right side of Figure 3.

The calculated outputs for these pairs of input functions are shown in Figures 6 and 7. The resultant wave, $h(x, y, z, t)$, in our simulation is a 128x128x64 data array. Only one value of y is chosen for the plot, that corresponding to the cross-section located at $y = 0$ (i.e., we plot $h(x, 0, z, t)$). The other 127 cross-sections are not plotted. For our results, we have chosen the location at $z = 10$ cm and selected a time span of approximately 0.4 milliseconds. The time interval begins just before $t = z/c$, i.e., just before the first arrival of the wave at the observation plane. It is noted that we also calculated the fields for values of σ between the values of 2 and 16, but do not show the results in order to conserve space.

We observe from Figure 7 that the waves are very similar. Upon looking at Figure 5 and the right figure of Figure 3, we observe that the input spatial excitation functions

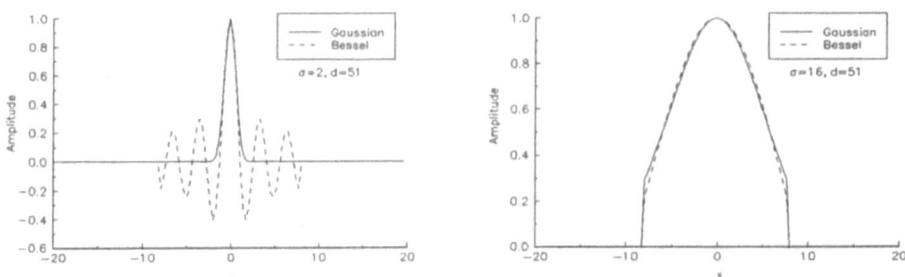

Figure 3. Cross-section of the input Gaussian and Bessel functions for (left) $\sigma = 2$ and (right) $\sigma = 16$. The curves are matched at $x = 0$ and at the half-maximum widths.

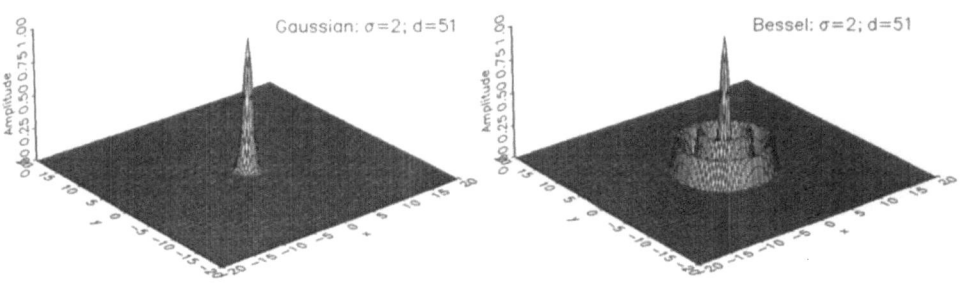

Figure 4. Perspective views of (left) the Gaussian spatial excitation function and (right) the Bessel excitation function for $\sigma = 2$.

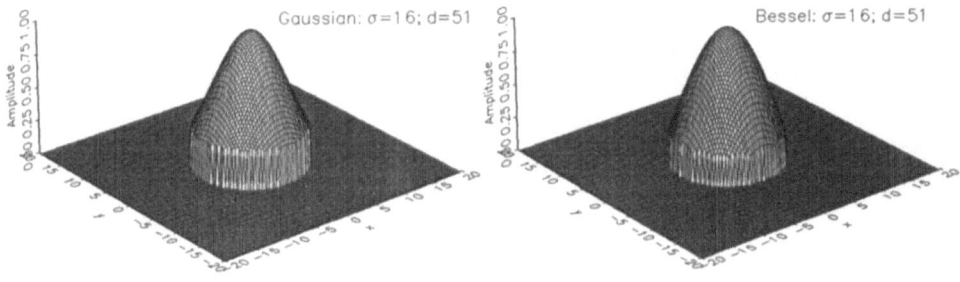

Figure 5. Perspective views of (left) the Gaussian spatial excitation function and (right) the Bessel excitation function for $\sigma = 16$.

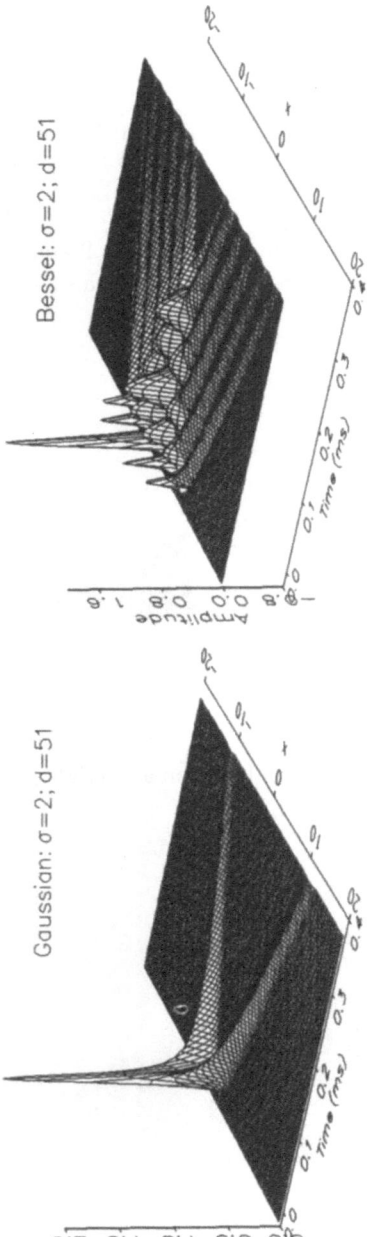

Figure 6. Perspective views of a cross-section, $h(x, 0, 10 \text{ cm}, t)$, for (a) the Gaussian function and (b) the Bessel function for $\sigma = 2$.

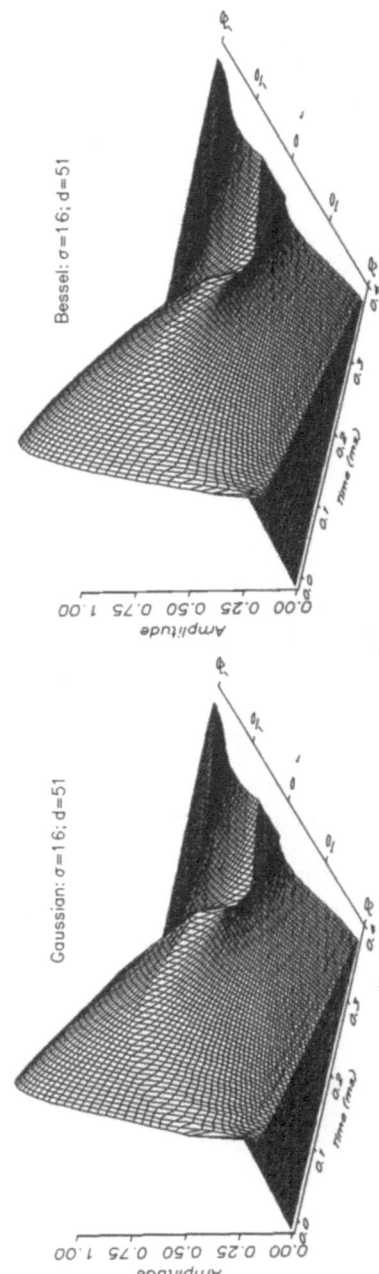

Figure 7. Perspective views of a cross-section, $h(x, 0, 10 \text{ cm}, t)$, for (a) the Gaussian function and (b) the Bessel function for $\sigma = 16$.

36

are almost the same for this (comparatively) large value of σ/d, hence it is not surprising that the spatial impulse responses are almost the same.

On the other hand, Figure 6 shows that the waves are quite dissimilar for small values of σ/d. This is intuitively expected after inspection of Figure 5 and the left side of Figure 3 since the excitations are quite different. In particular, we note from Figure 6 that the Bessel excitation produces fairly large sidelobes. (The negative lobes cannot be seen in this perspective view.) The depth of field along the t–axis centerline ($x = 0$) is longer for the Bessel excitation, but also suffers from a nonuniform amplitude. The Gaussian excitation has the benefit of producing no sidelobes and falls abruptly along the time-axis centerline.

SUMMARY

We have calculated the spatial impulse response of two spatially excited sources, a Gaussian spatial excitation and a Bessel spatial excitation. Both sources were spatially truncated by a circular aperture of diameter d. The excitation sources were equal-valued at their peaks and their half–maximum widths were equal. (This assumption related the Bessel scaling factor a to the Gaussian width parameter σ.) For large values of σ/d, the excitations and the spatial impulses were almost the same. For small values of σ/d, the excitations and the spatial impulse response differed widely. The Gaussian wave was characterized by the absence of any sidelobes and an abrupt decrease along the time–axis centerline. The Bessel–wave spatial impulse response exhibited both positive and negative sidelobes and a varying–amplitude wave along the time–axis centerline.

ACKNOWLEDGEMENTS

This work was sponsored by the Direct–Funded Research Program of the Naval Postgraduate School with the cooperation of the Office of Naval Research.

REFERENCES

[1] J. Brittingham, "Focus wave modes in homogeneous Maxwell's equations: transverse electric mode," *J. Applied Physics*, vol. 54, pp. 1179–1189, 1983.

[2] J. Lu and J. F. Greenleaf, "Ultrasonic nondiffracting transducer for medical imaging," *IEEE Trans. on Ultrasonics, Ferroelectrics, and Frequency Control*, vol. 37, no. 5, pp. 438–447, 1990.

[3] J. Lu and J. F. Greenleaf, "Nondiffracting X waves — exact solutions to free–space scalar wave equation and their finite aperture realizations," *IEEE Trans. on Ultrasonics, Ferroelectrics, and Frequency Control*, vol. 39, no. 1, pp. 19–31, 1992.

[4] J. Lu and J. F. Greenleaf, "Experimental verification of nondiffracting X waves," *IEEE Trans. on Ultrasonics, Ferroelectrics, and Frequency Control*, vol. 39, no. 3, pp. 441–446, 1992.

[5] J. Lu and J. F. Greenleaf, "Sidelobe reduction for limited diffraction pulse–echo systems," *IEEE Trans. on Ultrasonics, Ferroelectrics, and Frequency Control*, vol. 40, no. 6, pp. 735–746, 1993.

[6] J. Durnin, "Exact solutions for nondiffracting beams I. The scalar theory," *J. Optical Society of America A*, vol. 4, no. 4, pp. 651–654, 1987.

[7] J. Durnin, J. Miceli, and J. Eberly, "Diffraction–free beams," *Physical Review Letters*, vol. 58, pp. 1499–1501, 1987.

[8] J. Durnin, J. Miceli, and J. Eberly, "Comparison of Bessel and Gaussian beams," *Optics Letters*, vol. 13, pp. 79–80, 1988.

[9] P. Kiełczyński and W. Pajewski, "Acoustic field of Gaussian and Bessel transducers," *J. Acoustical Society of America*, vol. 94, no. 3, pp. 1719–1721, 1993.

[10] D. Guyomar and J. Powers, "A Fourier approach to diffraction of pulsed ultrasonic waves in lossless media," *J. Acoustical Society of America*, vol. 82, no. 1, pp. 354–359, 1987.

[11] D. Guyomar and J. P. Powers, "Boundary effects on transient radiation fields from vibrating surfaces," *J. Acoustical Society of America*, vol. 77, no. 3, pp. 907–915, 1985.

[12] T. Merrill, *A transfer function approach to scalar wave propagation in lossy and lossless media*, Master's thesis, Naval Postgraduate School, Monterey, California, March 1987.

[13] J. Upton, *Microcomputer simulation of a Fourier approach to optical wave propagation*, Master's thesis, Naval Postgraduate School, Monterey, California, March 1992.

[14] W. R. Reid, *Microcomputer simulation of a Fourier approach to ultrasonic wave propagation*, Master's thesis, Naval Postgraduate School, Monterey, California, December 1992.

[15] J. P. Powers, "Acoustic propagation modeling using MATLAB," Tech. Rep. NPS EC–93–104, Naval Postgraduate School, Monterey, California, September 1993.

ULTRASONIC PULSE REFLECTION FROM
A GRADED BOUNDARY PLANE SURFACE

Sidney Lees, Michael Weber

Bioengineering Department
Forsyth Dental Center
140 Fenway
Boston MA 02115

Most often when the reflection ratio of a sonic wave at a boundary is calculated it is assumed that the boundary between the two media is infinitely sharp. It is also assumed that the media are homogeneous and isotropic. In this paper the situation is considered where the properties of the surface of the second medium increase with penetration to a limiting value, which defines the graded boundary. In particular consider the structure described in Fig 1, where the density and velocity increase linearly with depth. The specific acoustic impedance increases as the square of the depth.

$$\rho = \rho_0 + k_0 x$$
$$c = c_0 + k_1 x \qquad (1)$$
$$Z = \rho c$$

L = thickness of boundary

$0 \leq x \leq L$

Fig 1

The graded boundary can be closely simulated by a sequence of very thin homogeneous layers in which the sonic properties increases slightly in each successive layer.

Consider the boundary between two homogeneous isotropic media

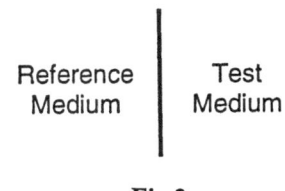

Reference Medium | Test Medium

Fig 2

The wave equation is $\dfrac{\partial^2 d}{\partial t^2} = c^2 \dfrac{\partial^2 d}{\partial x^2}$ $\qquad$ (2)

The Laplace transform of this equation with respect to time, assuming zero initial conditions is

$$\frac{p^2}{c^2} D = \frac{\partial^2 D}{\partial x^2} \qquad (3)$$

where p is the Laplace variable and D is the transform of d. The equation is satisfied by the expression

$$D = F(p) exp(-px/c) + G(p) exp(px/c) \qquad (4)$$

which expressed in time is

$$d = f(t - x/c) + g(t - x/c) \qquad (5)$$

by interpreting exp(-px/c) as a time delay. The functions f and g represent forward and reverse propagating waves.

At the boundary the normal displacement and the normal stress must be the same on both sides, which can be shown to result in the expression

$$\begin{bmatrix} \beta_0 & 1 \\ -\beta_0 Z_0 & Z_0 \end{bmatrix} \begin{bmatrix} F_0 \\ G_0 \end{bmatrix} = \begin{bmatrix} \beta_0 & 1 \\ -\beta_0 Z_1 & Z_1 \end{bmatrix} \begin{bmatrix} F_1 \\ G_1 \end{bmatrix} \qquad (6)$$

$$\beta_0 = exp(-2px/c), \quad Z_0 = \rho_0 c_0, \quad Z_1 = \rho_1 c_1$$

Now consider a series of boundaries

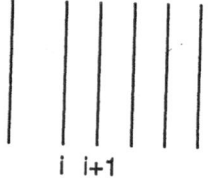

i i+1

Fig 3

At each boundary the equation is

$$\begin{bmatrix} \beta_i & 1 \\ -\beta_i Z_i & Z_i \end{bmatrix} \begin{bmatrix} F_i \\ G_i \end{bmatrix} = \begin{bmatrix} \beta_i & 1 \\ -\beta_i Z_{i+1} & Z_{i+1} \end{bmatrix} \begin{bmatrix} F_{i+1} \\ G_{i+1} \end{bmatrix} \qquad (7)$$

so that

$$\begin{bmatrix} F_i \\ G_i \end{bmatrix} = \begin{bmatrix} \beta_i & 1 \\ -\beta_i Z_i & Z_i \end{bmatrix}^{-1} \begin{bmatrix} \beta_i & 1 \\ -\beta_i Z_{i+1} & Z_{i+1} \end{bmatrix} \begin{bmatrix} F_{i+1} \\ G_{i+1} \end{bmatrix} \qquad (8)$$

Define

$$A_i = \begin{bmatrix} \beta_i & 1 \\ -\beta_i Z_i & Z_i \end{bmatrix}, \quad B_i = \begin{bmatrix} \beta_i & 1 \\ -\beta_i Z_{i+1} & Z_{i+1} \end{bmatrix}, \qquad (9)$$

$$C_i = A_i^{-1} B_i$$

$$\begin{bmatrix} F_0 \\ G_0 \end{bmatrix} = C_0 \begin{bmatrix} F_1 \\ G_1 \end{bmatrix} = C_0 C_1 \begin{bmatrix} F_2 \\ G_2 \end{bmatrix} = C_0 \ldots C_{n-1} \begin{bmatrix} F_n \\ G_n \end{bmatrix} \tag{10}$$

$$\begin{bmatrix} F_0 \\ G_0 \end{bmatrix} = D \begin{bmatrix} F_n \\ G_n \end{bmatrix}, \qquad D = \prod_0^{n-1} C_i = \begin{bmatrix} a_{11} & a_{12} \\ a_{21} & a_{22} \end{bmatrix} \tag{11}$$

The ensemble of matrices characterizing the individual layers condenses into a single four element matrix in which the elements are complex quantities. Consider the last layer to be semi-infinite so that there is no rearward propagating wave in the *nth* layer, or $G_n = 0$.

$$\begin{bmatrix} F_0 \\ G_0 \end{bmatrix} = \begin{bmatrix} a_{11} & a_{12} \\ a_{21} & a_{22} \end{bmatrix} \begin{bmatrix} F_n \\ 0 \end{bmatrix} \tag{12}$$

$$F_0 = a_{11} F_n, \quad G_0 = a_{21} F_n, \qquad G_0 = \frac{a_{21}}{a_{11}} \cdot F_0 \tag{13}$$

In the expression, $\beta_i = \exp(-2 p_i x_i / c_i)$,, take $p_i = j\omega_i$;

then G_0 is a frequency function which can be evaluated once the values of r_i and c_i are known. This requires D to be calculated from which a_{21} and a_{11} are presented in a

table of complex values of ω_i.

Eqs 12 and 13 relate the ultrasonic reflection from the graded boundary for normal incidence to the input wave form. It is the first step in computing the echo as outlined in the following procedure.

1. Develop an expression for wave propagation through the graded boundary as a function of frequency.

2. Express the input pulse as a frequency function, for example as a sum of harmonic waves, not necessarily the Fourier series for the pulse.

3. Calculate the amplitude and phase for each harmonic component after reflection.

4. Sum the harmonic sequence emerging from the graded boundary as a time series

5. Represent the output graphically.

6. Calculate the deviation from the ideal reflection and represent the deviation graphically.

The work leading to Eq(13) represents step 1 of the procedure. In the second step the input function is expressed as a frequency function. The input function used in the calculations in this paper to illustrate the technique is the typical ultrasonic W wave form defined in Fig 4.

$$\omega_0 = \frac{\pi}{(T/3)}$$

$$W = 0, \qquad t \le -\frac{T}{2}$$

$$= (1/2)\cos\omega_0 t, \qquad -\frac{T}{2} \le t \le -\frac{T}{6}$$

$$= \cos\omega_0 t, \qquad -\frac{T}{6} \le t \le \frac{T}{6}$$

$$= (1/2)\cos\omega_0 t, \qquad \frac{T}{6} \le t \le \frac{T}{2}$$

$$= 0 \qquad \frac{T}{2} \le t$$

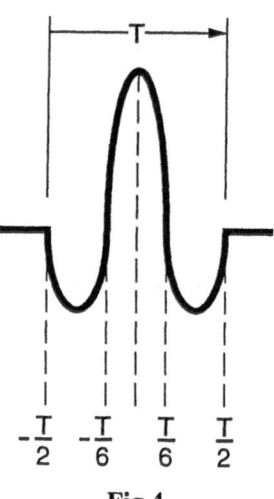

Fig 4

The Fourier transform of the symmetrical wave centered on zero is

$$W = \frac{1}{2\pi}\left[\frac{\omega_0}{\omega_0^2 - \omega^2}\right]\left[\cos\omega T/6 - \cos\omega T/2\right] \tag{14}$$

When the function is convoluted with an impulse at $t = T/2$, the W-function is shifted to the right to start at $t = 0$. The shifted expression is

$$W = \left[\frac{1}{2\pi}\right]\left[\frac{\omega_0}{\omega_0^2 - \omega^2}\right]\left[\cos\omega T/6 - \cos\omega T/2\right]\exp(-j\omega T/2) \tag{15}$$

The input pulse can be represented by a sum of harmonic functions, not the Fourier series, using an arbitrary number of terms , as seen in Fig 5. Forty terms were used here.

$$F_0 = \sum_{-N}^{N} F_i \cos(\omega_i t + \phi_i), \qquad N = 40 \tag{16}$$

$$F_i = |W(\omega_i)|, \qquad \phi_i = arg(W(\omega_i)) \tag{17}$$

$$for \quad -\omega_{max} \le \omega_i \le \omega_{max}$$

The expression for the echo takes the form

$$G_0 = \frac{a_{12}}{a_{11}} F_0(\omega_i) = \sum_{-N}^{N} \frac{a_{12}}{a_{11}} F_i \cos(\omega_i t + \phi_i) \tag{18}$$

Specific values for the number of terms in Eq(19) and the increment in frequency were adopted. Take

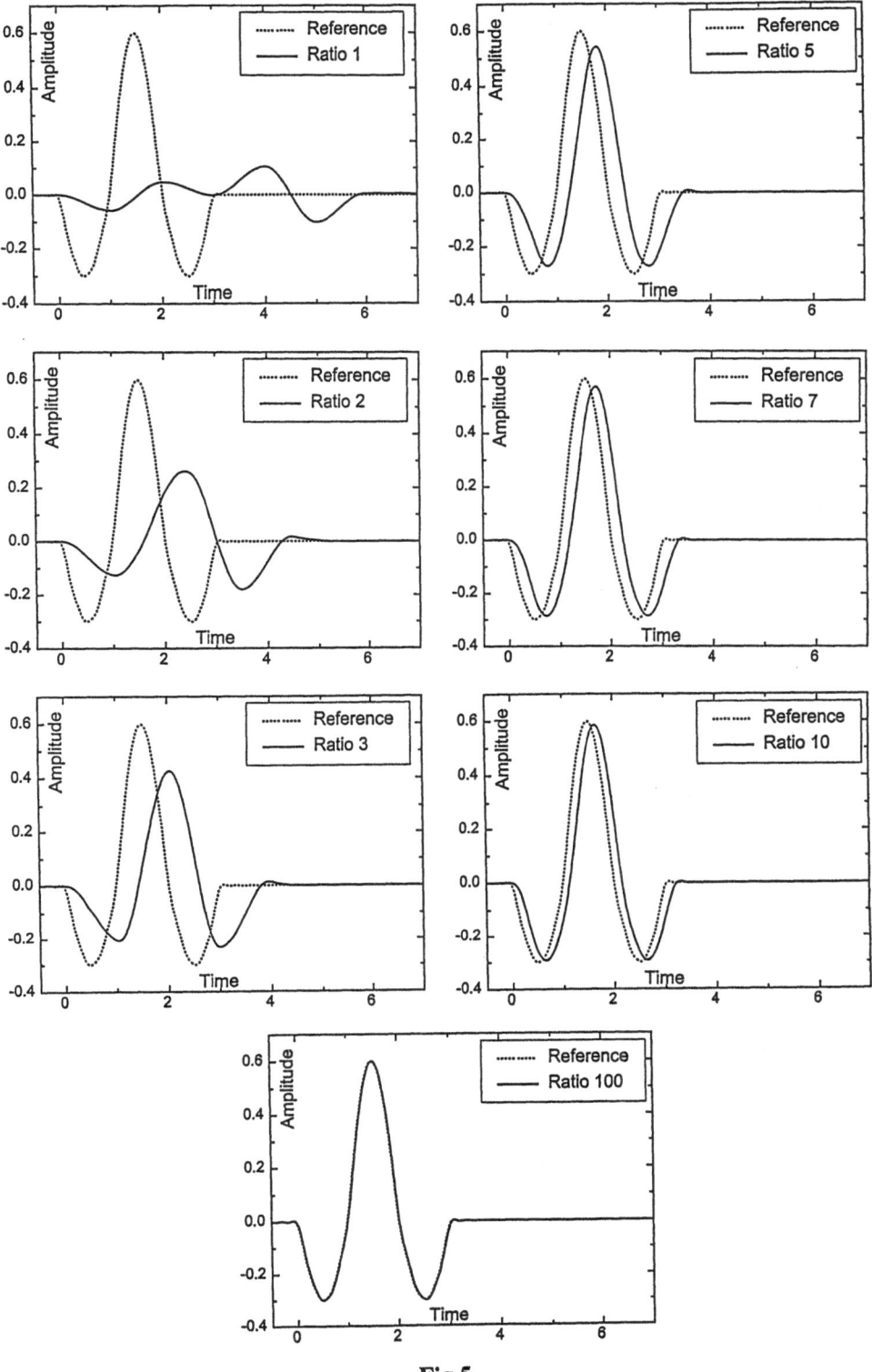

Fig 5

$$\omega_{max} = \frac{8\pi}{(T/3)}, \qquad \Delta\omega = 0.2\frac{\pi}{(T/3)}, \qquad \omega_{max} = 40\Delta\omega$$

$$\omega_i = (1-40)\Delta\omega, \qquad i = 0,\cdots,80$$

(19)

Again in order to illustrate the method and to show how the graded boundary can affect the echo, the reference medium was assumed to have a density, ρ, of 11.0 gm/cc and a sonic velocity, c, of 1.5 km/s. The corresponding values for the second medium were taken to be 2.0 and 3.0 respectively, like that for bone.

The effect of the graded boundary is characterized by its thickness compared to the length of the pulse in the reference medium. It is expressed as the ratio, R, the pulse length to the thickness of the graded boundary.

$$R = \frac{c_0 T}{L}, \quad c_0 = \text{sonic velocity in reference medium}$$

T = pulse duration, L = gaded boundary thickness = $M\Delta x$

Δx = thickness of each layer, M = number of layers

(20)

$$Tc_0 = RM\Delta x$$

Choose Δx in terms of the smallest wavelength of the harmonic sequence in Eq (18).

$$\text{Take} \quad \Delta x = \psi\lambda_{min} = \psi\frac{2\pi c_0}{\omega_{max}} \quad T = RM\psi\frac{2\pi}{\omega_{max}}, \qquad M = \frac{T}{\psi R}\frac{\omega_{max}}{2\pi} = \frac{12}{\psi R}$$

(21)

For R = 1, ψ = 0.02, M = 600

To maintain the same number of layers for R> 1, adjust ψ so that

$$\psi R = 0.02$$

A number of examples were calculated using the parameters cited above and shown in Fig 5 for R equal to 1,2,3,5,7,10 and 100. The time unit is T/3. The time intervals used in the calculation was 0.02T/3, i.e. 50 points per time unit. The calculation spans from - 0.5 to 7.0 time units.

The reflection ratio for the infinitely sharp boundary is 0.60. When the pulse length is the same as the thickness of the graded boundary (R=1) the echo is so badly warped it is difficult to identify it as originating from the input pulse. However the original pulse shape can be readily identified when R is 2, although there is significant distortion. The echo peak greatly lags the ideal echo and the wave form has been decreased slowly as R increases, but even for R equal to 10 there is significant lag. The ratio, R, must be at least 50 before the lag becomes almost zero.

The next four figures provide quantitative measures for the deviations of the echoes. Fig. 6 shows how the lag of the peak is dependent on R. The pulse length must be less than 0.1 of the boundary thickness for the lag to be reasonably small. In Fig. 7 the quantity shown

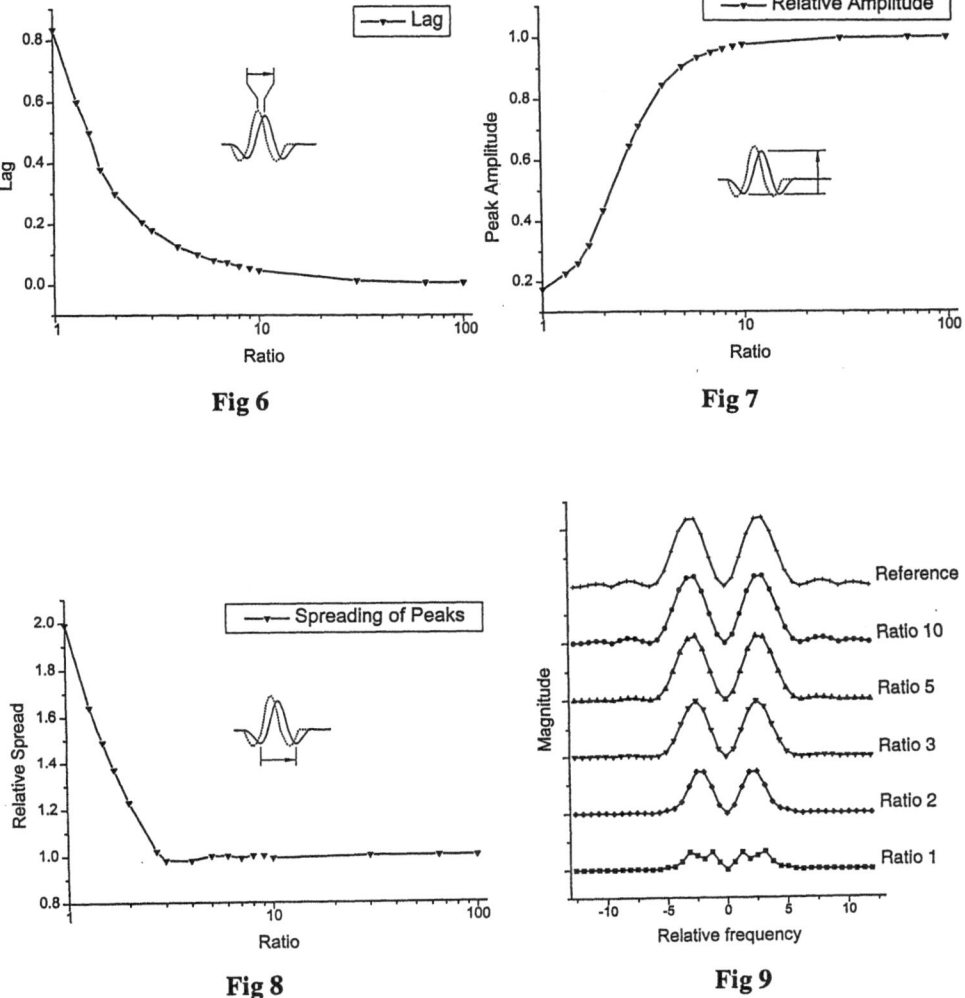

Fig 6

Fig 7

Fig 8

Fig 9

is the ratio of the peak amplitude of the echo to that of the ideal case. The peak is measured from the minumum to the maximum. Again it is necessary that the pulse length be no more than 0.1 as much as the thickness of the graded boundary. In Fig. 8 the peak spread is represented. The ordinate is the ratio of the duration between the two minima of the echo to that of the reference. This parameter can be ignored when R is 3.

The frequency spectrum represented by the magnitude of each harmonic component is presented in Fig. 9. The central frequency is the same for each R value and the peaks are at the same frequency. It is the magnitude distribution alone which is deteriorated where the pulse length is comparable to the boundary thickness.

In conclusion it is observed that the pulse echo can be severly distorted by the graded boundary in such a way that the results resemble the echo seen for a dissipative medium or which is nonlinear. This may be a problem in biological systems where the media are far from homogeneous.

Achnowledgement

The work was supported in part by The National Institute on Aging, Grant AGO 2325.

SIMULATION CALCULATIONS FOR MONOFREQUENT SOUND FIELDS IN LAYERED MEDIA

Elfgard Kühnicke

Dresden University of Technology
Institut of Technical Acoustics
D-01062 Dresden, Germany

PROBLEM

As to optimize the ultrasonic transducer parameters and to interpret the measured results, knowledge about the sound field is required. That is relevant both in the fields of non-destructive testing (NDT) and of ultrasonic diagnostics in medicine. In many cases, the sound wave transmitted by the transducer passes layers of different impedances until it reaches the object to be tested and examined, resp. As the active element of the transducer has a finite aperture, which mainly is located not parallel to the surface of the specimen, and as the occuring interfaces are partly curved, the calculation of the sound field for such three-dimensional elastodynamic problems is not possible in the form of a closed solution.

This paper presents a semianalytical, monofrequent method to calculate the sound field of ultrasonic transducers applied to layered media.

Previous sound field calculations for angle beam transducers and dual element transducers with Plexiglass wedges in non-destructive testing assumed that wave propagation in the Plexiglass wedge may be simulated by a plane wave. This model was used to study problems in transducer optimization for duals coupled to a plane specimen surface[1], or to a curved one[2]. The sound field of foil transducers used with the immersion technology was also studied[3]. Programs based on this approximation require only short computation times so that they are well suited for transducer design calculations when delay, also the path between active element and test piece, is small which often is the case with many problems in non-destructive testing.

The novel programs are used for judging the validity of the plane wave approximation and for studying the occuring effects. These programs are the basis for calculations where more than one interface is passed by the sound wave.

Examples from the ultrasonic diagnostics of the eye and from non-destructive testing are presented. The passage of the sound field through the different layers is calculated, and focussing and defocussing actions due to different interfaces are studied.

MODEL FOR MONOFREQUENT SOUND FIELDS

The problem is solved step by step by decomposing the solid into layers, i.e. the sound field is calculated in each layer, and the stress distribution calculated for an interface is used as exitation for the sound field in the adjoining layer. It is assumed that at least one of these two layers is free from shear stress or that between two media which are not free from shear stress a thin layer of water exists. So no shear force but only forces acting vertically to the interface are transfered.

Procedure of these Calculations:

The piezoelement is uniformly covered with point sources where the distance between these forces (force-to-force distance) is about 1/7 the wavelength of the adjoining layer.
The displacement u in the interior of the layer results from the summation via the field of all point sources of the entry surface

$$u(x,y,z) = \sum \frac{1}{r_i} F_{N,i} * P(\alpha') * e^{(j*k_L*r_i)}$$

where k_L - the wave number in the layer, r_i - the distance between the point under consideration and source point, α' - the angle between r and the normal in the source point, and $P(\alpha')$ - directional field of a point source. The directional field of a point source $P(\alpha')$ is a Fraunhofer approximation for a normal force acting on the surface of a half space.
The normal stress σ_{zz} on the interface of the layer results in

$$\sigma_{zz}(x,y,z) = \sum \frac{1}{r_i} F_{N,i} * P(\alpha') * T(\beta) * e^{(j*k_L*r_i)}$$

The normal stress on the interface differs from the displacement in its interior by the so-called transmission factor $T(\beta)$ where $T(\beta)$ is dependent on the angle between the local vector from source point to point under consideration and the normal on the interface at the point under consideration. On the basis of the calculated stress distribution, now the sound field can be determined analogously in the adjoining layer.
Zeros in the characteristic functions of a directional field and in the transmission factor functions result from occuring surface waves.

CALCULATION EXAMPLES FOR NON-DESTRUCTIVE TESTING

The non-destructive testing of components made e.g. of steel uses angle beam transducers with Plexiglass wedges and therefore sound fields in and behind a Plexiglass layer are considered. Such transducers are coupled with the aid of a thin liquid layer.
Figure 1 shows the sound field of the longitudinal wave transmitted in Plexiglass (1a), in steel behind a Plexiglass path of 5mm, with a plane interface (1b) and with a curved interface having a radius of curvature r=40mm (1c) in the cross sectional surface of a shaft below the middle of the element. This element is a rectangular one with the dimensions a=12mm in axial direction of the shaft and b=6mm and a frequency of f=4MHz. As is seen in Figure 1, the element is not symmetrical to the z-axis. It is laterally shifted by an amount of one radius. The z-coordinate represents the distance to the element.
As known, the sound field of a transducer is divided into two zones, the near field and the far field. The location of the last maximum is known as the near field distance N. For a piston with a diameter d the near field distance is calculated with the equation $N=d^2/4\lambda$ where λ is the wavelength.
As in Plexiglass the longitudinal wave has a shorter wavelength than in steel, the near field is more extended (please compare Figures 1a and 1b). Consequently, the sensitivity zone in

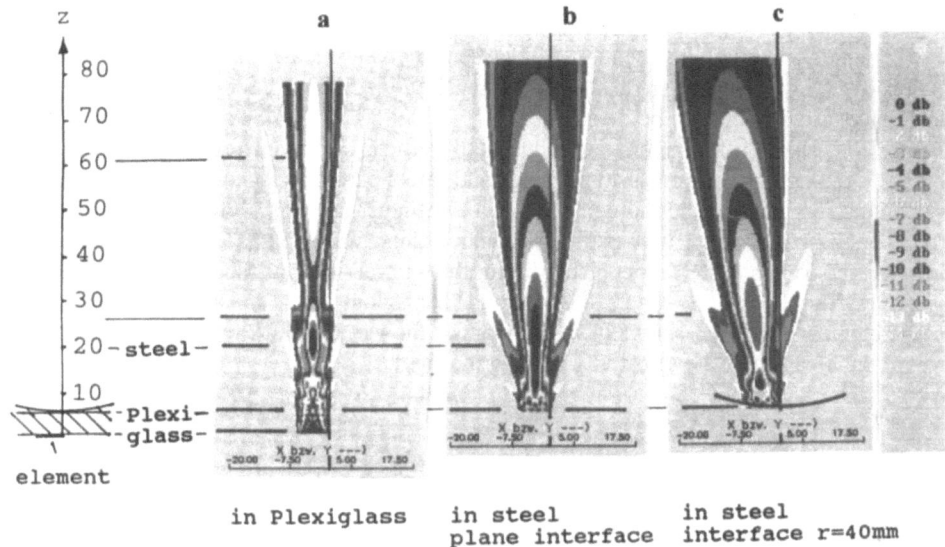

Figure 1. NDT - Position of the sensitivity zone for a single element
(transducer: a=12mm, b=6mm, f=4MHz)

Plexiglass is more distant from the element than in steel. Figure 1c shows that a convex interface between Plexiglass and steel causes the sound field to become more divergent. Also a curved interface may produce refraction, so that the beam is deflected.

A dual element transducer contains two elements in a single housing. Its sound field is obtained from the superposed sound fields of the transmitter and the receiver elements. The two elements are arranged to form a "roof", that means they are angled to one another, to create a crossed-beam sound path in the test material. The cross point is determined both by the roof angle and by the curvature of the entry surface.

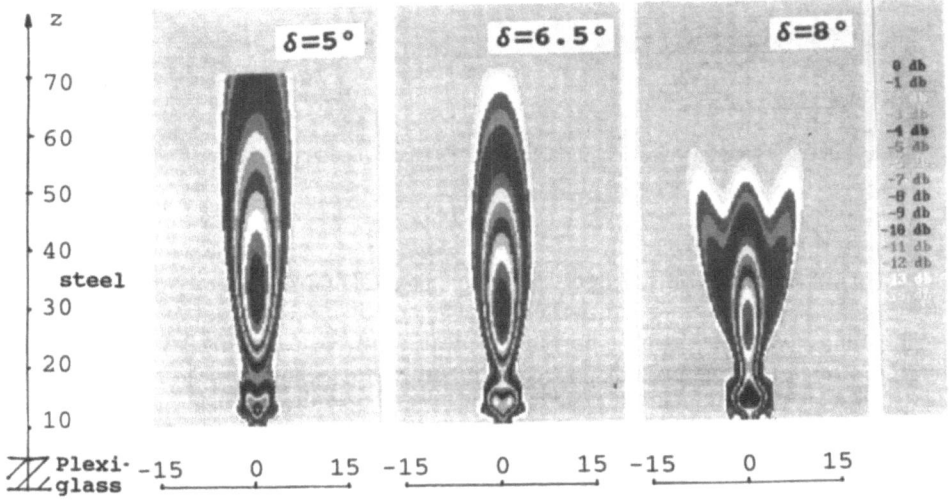

Figure 2. NDT - Sound field in shafts r=40mm for duals behind a Plexiglass delay
(transducer: a=12mm, b=6mm, f=4MHz)

Figure 2 shows the calculation for a dual element transducer with the above elements at various roof angles δ for a steel shaft with the radius of curvature r=40mm. Variation of the roof angle changes the location of the sensitivity zone, i.e. the radius of curvature requires to choose the roof angle so that the sensitivity zone will be located in that depth in which flaws are expected to be detected.

For testing shafts, the immersion technique often uses a curved foil-element transducer with the radius of curvature r_F=40mm. The sound field enters the specimen through a water layer. Figure 3 shows the sound field for a foil transducer in water, vertical sound incidence into a round section steel with radius r=40mm behind water paths of 10mm and 15mm, resp. As the stress acts perpendicularly to the curved foil surface, no defocussing is observed with the water-to-steel transmission. This is contrary to the sound incidence from a plane element. Figure 3 shows in which manner the variation of the water path allows to arrange the location of the sensitivity zone in the shaft.

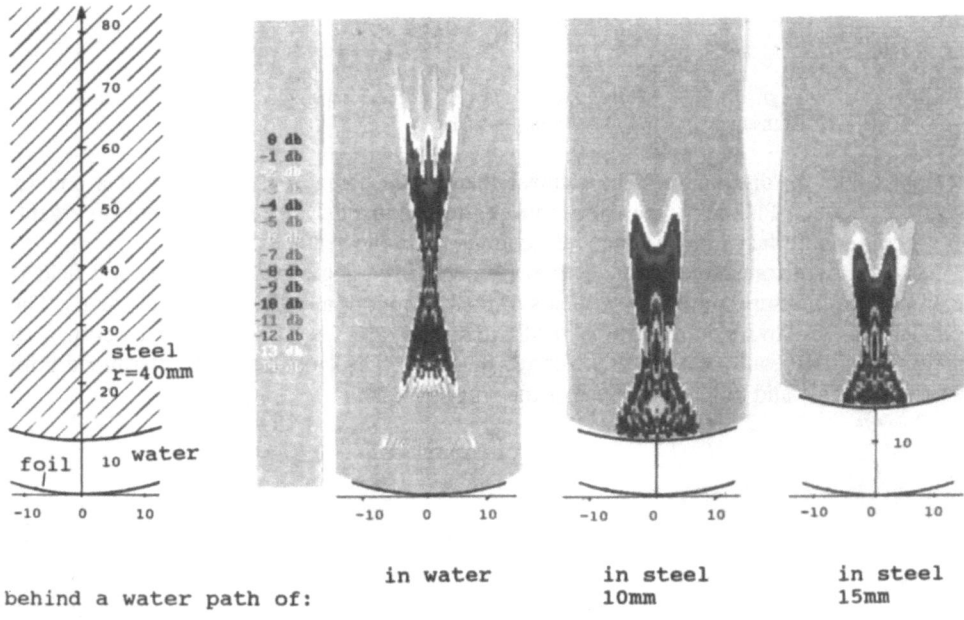

behind a water path of: in water   in steel   in steel
                                     10mm   15mm

Figure 3. NDT - Sound field calculation for a foil transducer (a=6mm, b=22mm, r_F=40mm, f=4MHz)

CALCULATION EXAMPLES FOR ULTRASONIC DIAGNOSTICS IN MEDICINE

Ultrasonic diagnostics applied to the eye deals with the following layers which show different radii of curvature and impedances; viz. water delay, cornea, aqua oculi, lens, vitreous body. The problem to be solved is to determine the sound field on the retina level for eyes with different dimensions and for different coupling conditions of the transducer.

The following calculations are based on a square element with the dimension of 5*5mm² and a frequency of f=8MHz as well as on cylindrical interfaces and a water path of 10mm. The z-component now represents the distance to the cornea. The densities ρ, the wave propagation velocities c and the thickness of the layers and also the radii of interfaces curvature are given in Figure 4. These parameters are taken from Buschmann and Trier[4].

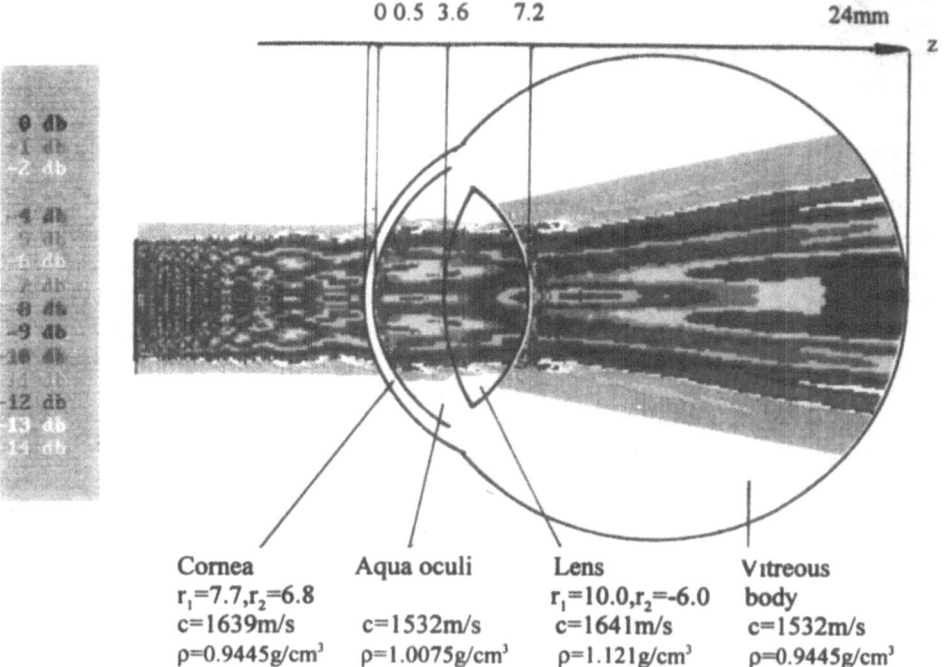

Figure 4. Ultrasonic diagnostics in medicine - sound field in a human eye behind a water path of 10mm (transducer: d=5mm, f=8MHz)

Figure 4 represents the sound field through a human schematic eye. Figure 5 shows the normal stress distribution on the retina at the distances between cornea and retina of 24mm (5b) and 21mm (5c). In comparison with the sound field on the retina in Figure 5b the Figure 5a shows the sound field in water at the same location. In Figure 6 the longitudinal sections of the same sound field are shown after transmission through the partial interfaces, where the interface to the next layer was neglected at first to allow to observe exactly the location of the sensitivity zone and the divergence of the beam in the layer concerned.

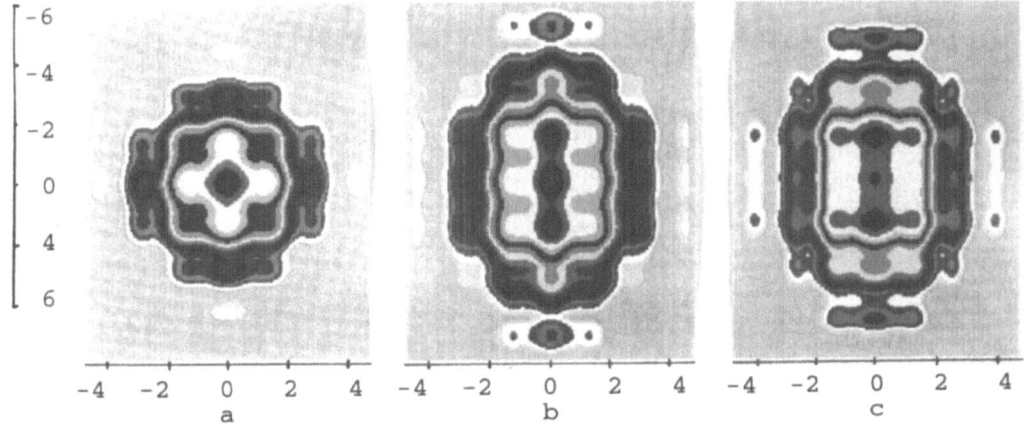

Figure 5. Ultrasonic diagnostics in medicine - Cross-section of the sound field at the distance z
a)in water z=24mm, b) on the retina z=24mm, c) on the retina z=21mm

Figure 6d shows the longitudinal section of the sound field behind the lens. The distance between retina and lens depends on the size of the eye. From the longitudinal section 6d the appearence of the expected sound field on the retina (cross section) can be estimated.

The sound field in water (Figures 6a) shows a very extended near field being in conformity with the considerations on the piston-type transducer with a diameter of d=5mm and a frequency of f=8MHz for which the near field length results in N=33mm. When the sound field enters the interface between water and cornea, it is defocussed (6b) due to the convex interface curvature which again is compensated by the focussing effect of the concave interface curvature between cornea and the aqua oculi (6c). Both the interfaces of the lens cause defocussing. The longitudinal section of the sound field behind the lens (figure 6d) shows that defocussing due to the passage of sound field through the different layers of the eye causes splitting of the sensitivity zone, and therefore secondary structures (also marked) occur in the sound field behind the lens. I.e., additionally to the sensitivity zone on the symmetry axis S_1 there are off-axis sensitivity zones S_2, which are less distant to the transducer. At the retina distance, according to the schematic eye of 24mm plotted, the retina intersects each of these three sensitivity zones. The appropriate sound field on the retina is represented in Figure 5b.

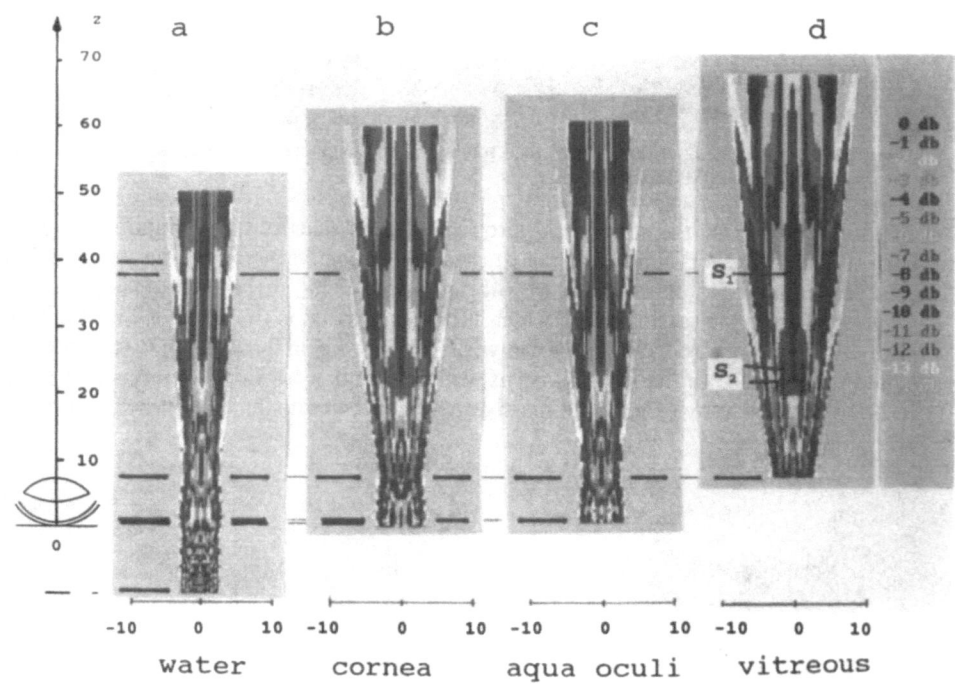

Figure 6. Ultrasonic diagnostics in medicine - Sound field affected by different eye elements (transducer: d=5mm, f=8MHz; water delay of 10mm)

With a hyperopic eye, i.e. the eye being shorter and the retina still less distant to the lens, the off-axis sensitivity zones have higher intensities than the on-axis ones (Figure 5c). When the received signal is assumed to come from the eye axis an erroneous conclusion will result. To avoid mistakes due to the secondary structures the sensitivity zone has to be arranged less distant to the lens, i.e. the main structure has to be located on the retina level. This can be achieved either by increasing the water path, by decreasing the element size connected with a decreased near field length, or by using a focussing transducer.

With different dimensions of the eyes, the sound field on the retina level differs in size and shape of the sensitivity zone. Investigations on the influence of different coupling conditions were also made. Variation of the water delay between element and eye and also an unsymmetrically attached element causes an alteration of the sound field. The variation in size and shape by shifting the element is more important than by the variation in the dimension of the human eye. This shows that exact coupling conditions are necessary.

SUMMARY

In medical use as well as in non-destructive testing, similar problems will arise regarding the design optimization of the transducer and evaluation problems. The presented simulation model allows to determine the sound field emitted by the transducer element, which may be curved, and thus the location and size of the sensitivity zone, after the field has passed several layers of the test object.

On the one hand, simulation calculations will be helpful for the transducer design optimization and for selecting the most applicable transducer type as well as for choosing the most practicable testing procedure. It has been pointed out that the location of the sensitivity zone inside the test object can be varied by varying the delay distance or the roof angle in the case of a dual element transducer. Curved foil transducers will fit best the curved test object surfaces.

On the other hand, these simulation calculations give an approach to evaluate measured signals. The appearance of the sound field depends on material properties (wave velocity, density) as well as on geometric parameters (layer thickness, radius of interface curvature) of the layers passed by the sound field. Secondary structures and also variation of coupling conditions can cause erroneous evaluations.

ACKNOWLEDGEMENT

The study "Calculation examples for ultrasonic diagnostics in medicine" is a part of the research project "Optisch-akustische Biometrie", supported by Deutsche Forschungs-gemeinschaft under contract No. Tr 195/2-1, Project leader H.-G.Trier, University of Bonn.

REFERENCES

1. E. Kühnicke, Modell für SE-Prüfköpfe, Seminar "Modelle und Theorien für die Ultraschallprüfung", 5./6.November 1989 in Berlin, Proceedings of the DGZfP Vol.23: p.89 (1989)
2. E. Kühnicke: Gekrümmte Oberflächen brauchen optimierte Prüfköpfe, Materialprüfung 35: 128 (1993)
3. E. Kühnicke: Einsatz von Ultraschallprüfköpfen zur Prüfung von Wellen und Bauteilen mit gekrümmter Oberfläche - Schallfeldmodellierungen, Annual conference on NDT 17.-18.Mai 1993 in Garmisch-Partenkirchen, Proceedings of the DGZfP Vol.37: p.183 (1993)
4. W.Buschmann, H.G.Trier: "Ophthalmologische Ultraschalldiagnostik", Springer-Verlag, Berlin Heidelberg, (1989)

EFFECT OF NONLINEARITY ON PROPAGATION OF SHEAR ACOUSTIC WAVES ALONG THE INTERFACE BETWEEN TWO CRYSTALS

V.I.Gorentsveig, Yu.S.Kivshar, and E.S.Syrkin

Institute for Low Temperature Physics & Engineering
47 Lenin Avenue, Kharkov, 310164, Ukraine

Introduction

Surface and interface acoustic waves find wide application in fundamental research and in technology, such as seismology, signal processing, and nondestrustive testing.

For a case of a large amplitude excitation the nonlinearity of elastic properties of the medium is substancial, especially for surface waves because of energy concentration at the surface.

Generally, surface acoustic waves in crystals are three component waves. However, for a certain crystal types and for a special choice of the surface plane there can propagate in a definite direction two-component wave of a vertical (Rayleigh) polarization and one-component wave of the shear-horizontal (SH) polarization, independently.[1]

Rayleigh wave is strongly localized at the surface. The amplitude of the wave decreases exponentially into the bulk of the crystall with a characteristic length of order of a wavelength. Account of small nonlinear terms in the wave equation does not lead to drastical changes in the character of the wave, and yields such effects as higher harmonics generation.[2,3]

For the SH polarization there is the only solution of the wave equation within the framework of the linear theory of elasticity to satisfy the boundary condition at the free surface. That is the wave of a constant amplitude. But this solution does not satisfy the requirement for surface waves to decay into the bulk of the crystal. However, this bulk wave is unstable in the sense that it may become a surface wave upon an insignificant change in properties of the medium or of the surface, such as piezoelectricity,[4,5] spatial dispersion,[6,7] or surface distortions.[8] For real crystals the length of localization of the SH waves at the surface due to those effects is of order of hundreds wavelengths.

Apparently, in the linear theory SH elastic waves can not propagate along the ideal interface between to medium.[9] Surface distortions can yield existence of the SH wave weakly localized at the interface.[10]

It appears that the nonlinearity of elastic properties of the medium can lead to a much stronger localization of the SH wave upon the increase of its amplitude, both for the cases of surface[11-13] and interface[14,15] waves. The characteristic length of the nonlinear localization can be of order of ten wavelengths.

In the present paper we consider the nonlinear SH waves propagating along the ideal interface between two elastic media. These waves arise in a threshold manner when the amplitude maximum of the wave exceeds a critical value. Localization of the waves near the interface is entirely due to the nonlinearity of the medium. Such waves can exist under definite relationships between the second, third and fourth orders elastic moduli of the media. For some cases there is top restriction for the maximum of the wave amplitude.

1. Nonlinear wave equation

We consider the case of propagation of SH wave along the interface between two crystals without excitation of Rayleigh

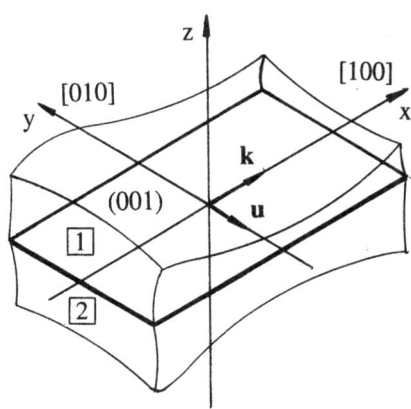

Fig. 1. Geometry of the system for propagation of shear elastic waves along the interface between two crystals.

waves. The simplest case is propagation of the wave in the [100] direction (x axis) in cubic symmetry crystals which coincide with their (100) planes (z = 0) (see Fig.1). The displacement vector **u** for the wave is paralel to the [010] direction (y axis) and depends on the two coordinates, x and z.

The wave equation for the y projection u of the displacement with the nonlinear terms taken into account has the following form for each medium:[16]

$$c^{-2}u_{tt} = u_{xx} + u_{zz} + (\alpha u_x^2 + \beta u_z^2)u_{xx} + (\beta u_x^2 + \alpha u_z^2)u_{zz}$$

$$+ 4\beta u_x u_z u_{xz} \quad , \tag{1.1}$$

where the subscripts t, x and z indicate the derivatives with respect to these variables; c is the transverse velocity of sound in the medium, $c = (C_{44}/\rho)^{1/2}$, where ρ is the density; α and β are the dimensionless coefficients of nonlinearity (their values can be of order 10^3):[17]

$$\alpha = (C_{4444}/2 + 3C_{344} + 3C_{11}/2)/C_{44} \quad , \tag{1.2a}$$

$$\beta = (C_{4466}/2 + C_{144} + 2C_{456} + C_{44} + C_{12}/2)/C_{44} \quad ; \tag{1.2b}$$

C_{ik}, C_{ikl}, C_{iklm} are the elastic moduli of the second, third and fourth order for the crystal of cubic symmetry, correspondingly.

Equation (1.1) is obtained by the expansion of the elastic energy in a power of the distortions[18] under the conditions,

$$\alpha u_x^2 \, , \; \beta u_x^2 \, , \; \alpha u_z^2 \, , \; \beta u_z^2 \, \ll \, 1 \; , \tag{1.3}$$

which provide smallness of the nonlinear terms in Eq. (1.1) in comparison with the linear ones (u_{xx} and u_{zz}).

Surface wave should satisfy the condition of decrease into the bulk of the crystal: $u \rightarrow 0$ as $|z| \rightarrow \infty$.

Assuming the dependence of u on z to be much slower than the wave dependence of u on x,

$$u_z^2 \, \ll \, u_x^2 \; , \tag{1.4}$$

and to be $|\beta| \le |\alpha|$ (e.g., for an isotropic medium $\beta = \alpha/3$),[11] we reduce Eq. (1.1) to the following form:

$$c^{-2}u_{tt} - u_{xx} - u_{zz} = \alpha u_x^2 u_{xx} \; ; \tag{1.5}$$

and all requirements (1.3) reduce to the first of them.

2. Transversely localized waves

Considering the stationary surface wave we seek for the solution of Eq. (1.5) in the form of the series in terms of harmonics:

$$u = \sum_{n=1}^{\infty} U_n(z) \cos[n(kx - \omega t)] \; , \tag{2.1}$$

where $U_n(z)$ is real amplitude of the n-th harmonic; frequency ω and wave number k are fixed parameters of the wave.

Neglecting the influence of higher harmonics on the fundamental one (wich makes the main contribution to the wave)[19] we obtain the stationary nonlinear Schrodinger equation for the basic harmonic amplitude, $U \equiv U_1$:

$$U_{zz} + (\omega^2/c^2 - k^2)U - \alpha k^4 U^3/4 = 0 \; . \tag{2.2}$$

The conditions (1.4) and (1.3) of applicability of Eq. (1.5), thus of Eq. (2.2), yield the following inequalities for the amplitude:

$$U_z^2 \ll (kU)^2 \ll |\alpha|^{-1} \quad . \tag{2.3}$$

The solution of Eq. (2.2), decreasing when $|z| \rightarrow \infty$, is of a various form depending on the sign of the parameter of nonlinearity for the medium:

$$\text{when} \quad \alpha < 0 \quad \text{then} \quad U = V\text{sech}[(z - Z)/l] \quad , \tag{2.4}$$

$$\text{when} \quad \alpha < 0 \quad \text{then} \quad U = V\text{cosech}[|z - Z|/l] \quad , \tag{2.5}$$

where a characteristic length l of the amplitude decrease into the bulk of the crystal is introduced:

$$l^{-2} = k^2 - \omega^2/c^2 \quad . \tag{2.6}$$

An amplitude parameter V is related to the length l:

$$|\alpha|(kV)^2/8 \equiv (kl)^{-2} \quad ; \tag{2.7}$$

due to the relation (2.7) the conditions (2.3) of applicability of equation (2.2) reduce to a single inequality,

$$(kl)^2 \gg 1 \quad . \tag{2.8}$$

The condition (2.8) can be easily satisfied, supposing the deformations to be within the strength limits ($kV < 10^{-3}$), and to be $|\alpha| < 10^3$.

The solution (2.4) of the case $\alpha < 0$ (so-called focusing nonlinearity) is localized and continuous on the entire z axis; the point of maximum of the solution is Z, the value of the maximum is V.

In a case of a single focusing medium the boundary condition at a free surface, $u_z|_{z = 0} = 0$, is satisfied for the nonlinear solution (2.4) by situating of a maximum of an arbitrary value on the surface. The obtained SH surface wave is localized at the surface entirely due to nonlinearity of the medium.[11-13]

The nonlinear localization depth (see Eq. (2.7)) is inversely proportional to the amplitude maximum. As for existing crystals the nonlinear parameter α can be of order of 10^3, then for really reachable shear wave deformations ($kU < 10^{-3}$) the length of the nonlinear localization can be of order of ten wavelength.

The solution (2.5) of the case $\alpha > 0$ (defocusing nonlinearity) possess singularity at the point Z.

3 Boundary conditions

The boundary conditions at the interface are specified by the requirements of the continuity of the shear displacement,

$$u\big|_{z=+0} = u\big|_{z=-0} \quad , \tag{3.1}$$

and of the shear stress,

$$\sigma_{yz}\big|_{z=+0} = \sigma_{yz}\big|_{z=-0} \quad . \tag{3.2}$$

The expression for the shear stress in a medium at the surface transversal to z axis, accounting the nonlinear terms of the same order as the wave equation (1.1), is:[16]

$$\sigma_{yz} = \mu u_z (1 \pm \beta u_x^2 + \alpha u_z^2 / 3) \quad , \tag{3.3}$$

where $\mu \equiv C_{44}$, is a shear elastic modulus of the medium.

Let us numerate the media (using subscripts for parameters of the media) in such a way that

$$c_1 > c_2 \quad , \tag{3.4}$$

and z axis to be directed into the first medium ($z > 0$).

Owing to conditions (1.3), we can use the linear limit of the expression (3.3), getting the linear boundary conditions at the interface:

$$\mu_1 u_z \big|_{z=+0} = \mu_2 u_z \big|_{z=-0} \quad . \tag{3.5}$$

The conditions of the same form (3.1) and (3.5) for the fundamental harmonic in Eq. (2.1) constitute the set of two equations for obtaining the coordinates (Z_1 and Z_2) of the characteristic points for the two media for the amplitude function U(z) for the interface wave.

Possible cases of existence of the interface waves correspond to the cases of solution existence for the set of equations.

4. Additional relations for the interface waves

We consider the united shear wave propagating along the interface in the both media adjoining to one another.

The amplitude of the basic harmonic for the wave is expressed by one of the functions (2.4) or (2.5) for each medium, in accordance to the sign of the nonlinear paramater α of the medium, and is characterized by a specific amplitude parameter (V_1 or V_2) and the corresponding localization length (l_1 or l_2) (see Eq. (2.7)).

The length of localization (l_1 or l_2) for the wave in each medium is related by the formula (2.6), containing the corresponding sound velocity (c_1 or c_2), to the same for the both media frequency ω and wave number k. Because of that, the lengths of localization of the interface wave in the two media should be related to one another:

$$k^2 - l_2^{-2} = S(k^2 - l_1^{-2}) \quad , \qquad (4.1)$$

where

$$S = c_1^2 / c_2^2 \geq 1 \quad , \qquad (4.2)$$

in accordance to our choice (3.4) numerating of the media.

Consequently, the following inequality takes place:

$$l_1^{-2} \geq l_2^{-2} \quad . \qquad (4.3)$$

Due to relation (4.1) only one (V_1 or V_2) of the two amplitude parameters retains free in the expressions (2.4) and (2.5) for the amplitude of the united interface wave in the two media.

Because of the requirements of localization and continuity of the function U(z) giving solution for the two half-spaces, at least one of the media should be defocusing (see Eq. (2.4)), the point of the maximum (z_{max} = Z) to be situated in corresponding half-space; the singularity point Z for a defocusing medium, if another medium is of such a type, should be situated out of occupied half-space.

For obtaining restrictions for the existence of the nonlinear interface wave correctly, we should take the absolute maximum U_{max} of the wave amplitude as a free parameter, wich coincides with the amplitude parameter (V_1 or V_2) of the solution for that focusing medium where the maximum is situated.

The dispersion law of the interface wave can be obtained using the relations (2.6) and (2.7) for that medium, and it includes the free parameter U_{max} :

$$\omega^2 = c_m^2 k^2 (1 - |\alpha| k^2 U_{max}^2 /8) \quad , \tag{4.4}$$

where m being 1 or 2 is the number of the medium.

A value of the amplitude of the wave on the interface (z = 0) can be found using the solution (2.4) in that medium:

$$U_0 = U_{max} \operatorname{sech}(Z_{max}/l_m) \quad . \tag{4.5}$$

The case when the nonlinearity of another medium could be neglected (the so-called case of a linear medium) can be considered as the limit as $\alpha \rightarrow 0$ for that medium in all expressions. The form for amplitude U(z) in the medium is

$$U = U_0 \exp[z/l] \quad . \tag{4.6}$$

Because of restrictive relation (4.1), a special testing is required to check the statement,

$$l_1^{-2}, l_2^{-2} > 0 \quad , \tag{4.7}$$

which is necessary for existence of a wave solution localized at the interface between two media (see Eq. (2.2)); due to relation (4.3) it is sufficient check only the second of the inequalities (4.7).

5. Waves at the interface between two focusing media

Let us consider the case when both of the media are focusing:

$$\alpha_{1,2} < 0 \quad . \tag{5.1}$$

Then the amplitude $U(z)$ is of the form (2.4) in the both media (see Fig. 2), and we obtain the following equations for the coordinates Z_1 and Z_2 ($Z_1 Z_2 \geq 0$):

$$\text{ch}^2(Z_1/l_1) = (M - A)/(M - L) \quad , \tag{5.2}$$

$$\text{ch}^2(Z_2/l_2) = (L/A)(M - A)/(M - L) \quad , \tag{5.3}$$

where the designations are introduced for the parameters,

$$M = \mu_1^2/\mu_2^2 \quad , \tag{5.4}$$

$$A = |\alpha_2|/|\alpha_1| \quad , \tag{5.5}$$

and for a ratio of localization depths (see relation (4.3)),

$$L = l_1^2/l_2^2 \leq 1 \quad . \tag{5.6}$$

The ratio of the amplitude parameters for two media can be expressed in terms of L and A, accordingly to Eq. (2.7):

$$V_2^2/V_1^2 = L/A \quad . \tag{5.7}$$

Generally, the only one of the two possible configurations for the amplitude $U(z)$, composed from the congruent pairs of the parts of the curves $U(z)$ for the separate media (see Fig.2), should be stable, having an absolute maximum situated in the media with the smaller amplitude parameter:

$$U_{max} = \min(V_1, V_2) \quad . \tag{5.8}$$

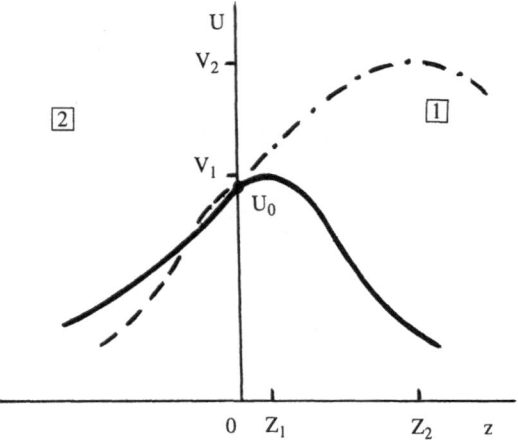

Fig.2. Profile of the shear elastic wave localized at the interface between two focusing nonlinear media.

The variable L is a function of a single free parameter U_{max}.

For the existence of solutions for the Eqs. (5.2) and (5.3) their right-hand sides should exceed the value of 1. There are four cases of the media parameters relations satisfying the requirement.

Case 1 : $A < 1 \leq M$.   (5.9)

Then we obtain the restriction for the variable L depending on the free parameter:

$L \geq A$.   (5.10)

Because of the inequality (5.10) the relation between the amplitude parameters (see Eq.(5.7)) is $V_2 \geq V_1$. Thus, the point of the maximum should be situated in the first media ($Z_{max} = Z_1 \geq 0$), and $U_{max} = V_1$, in accordance to the rule (5.8). So, the amplitude parameter V_1 should be considered as a free parameter for the case under consideration. As the length l_1 is related to V_1 we can express the variables of interest terms of l_1.

Using relation (4.1), we obtain the following expressions for l_2 and L:

$$(kl_2)^{-2} = S(kl_1)^{-2} - (S - 1) , \tag{5.11}$$

$$L = S - (S - 1)(kl_1)^2 . \tag{5.12}$$

Thus, the requirement (5.10) can be presented as the restriction for l_1 (accounting relation $S > A$):

$$(kl_1)^{-2} \geq (S - 1)/(S - A) , \tag{5.13}$$

Using the relation (5.11) we can express the compulsory requirement (4.7) as the unequality for l_1 ,

$$(kl_1)^{-2} > (S - 1)/S , \tag{5.14}$$

which appears to be provided by a stronger inequality (5.13).

The condition (5.13) means that for existence of the interface wave the amplitude maximum U_1 (see Eq. (2.7)) should exceed the definite threshold value U_{thr}, where

$$U^2_{thr} = 8(S - 1)/(S - A)|\alpha_1|k^2 . \tag{5.15}$$

Coordinates Z_1 and Z_2 can be expressed in terms of the free parameter $V_1 = U_{max}$ via the lengh l_1 (using the relations (5.11), (5.12), and then (2.7)).

The amplitude U_0 on the interface can be expressed as a function of U_{max} , using Eq. (4.5) with $m = 1$ and Eq. (5.2):

$$U_0 = U_{max}[(M - L)/(M - A)]^{1/2} . \tag{5.16}$$

When $U_{max} = U_{thr}$, the case corresponding to the equality in the relation (5.10), then (see Eqs. (5.2) and (5.3)) the maximum is situated on the interface, $z_{max} = 0$.

Let us note that to provide the condition (2.8) of applicability of the approximation yielding the analitical solution under the restriction (5.13), the parameters of the media should satisfy the following relation:

$$S - A \gg S - 1 . \tag{5.17}$$

The opposite case could mean not the absence of the interface wave, but just more complicated solution.

Case 2 :   A < M < 1 .   (5.18)

Then we obtain the same requirement (5.10), and the additional restriction for the variable L:

L < M .   (5.19)

Because of this, all conclusions and relations from (5.11) to (5.17) of the previous case are valid, and the following restriction is added because of the inequality (5.19) (accouning the relation S > M):

$$(kl_1)^{-2} < (S - 1)/(S - M)$$ (5.20)

The condition (5.20) means that for existence of the interface wave the amplitude maximum V_1 (see Eq. (2.7)) should not exceed the definite top value U_{top}, where

$$U^2_{top} = 8(S - 1)/(S - M)|\alpha_1|k^2 .$$ (5.21)

When U_{max} tends to U_{top}, that corresponds to the tendency of inequality (5.19) to the equality, then the region of the localization of the wave moves infinitely away from the interface: $z_{max} \longrightarrow \infty$ (see Eqs. (5.2) and (5.3)).

Case 3 :   M < A < 1 .   (5.22)

Then, for the variable L we obtain the restrictions inversed to the inequalities (5.19) and (5.10).

Consequently, the relation between the amplitude parameters is $V_2 \geq V_1$, and the point of maximum should be situated in the second media ($Z_{max} = Z_2 \leq 0$), and $U_{max} = V_2$. So, the amplitude parameter V_2 should be considered as a free parameter of the problem under consideration. As the length l_2 is related to V_2 we can express the variables of interest in terms of l_2.

Using the relation (4.1) we obtain the following expressions for l_1 and L:

$$(kl_1)^{-2} = [(kl_2)^{-2} - (S - 1)]/S \ , \tag{5.23}$$

$$L = S/[1 - (S - 1)(kl_2)^2] \ . \tag{5.24}$$

Thus, the limitations for L mentioned above yield the following restrictions for l_2:

$$M(S - 1)/(S - M) < (kl_2)^{-2} \le A(S - 1)/(S - A) \tag{5.25}$$

Let us note that the compulsory requirement (4.7) is provided by the left-hand one of inequalities (5.25) due to positivness of its left-hand side (see Eq. (5.22)).

The conditions (5.25) mean that for existence of the interface wave the amplitude maximum V_2 (see Eq. (2.7)) should exceed the threshold value U'_{thr}, but should not exceed the top value U'_{top}, where (compare to Eqs. (5.15) and (5.21)).

$$U'^2_{thr} = 8M(S - 1)/(S - M)|\alpha_2|k^2 = (M/A)U^2_{top} \ . \tag{5.26}$$

$$U'^2_{top} = 8A(S - 1)/(S - A)|\alpha_2|k^2 = U^2_{thr} \ . \tag{5.27}$$

Coordinates Z_1 and Z_2, and an amplitude U_0 on the interface can be expressed as a function of a free parameter V_2 analagously to the Case 1.

One can also investigate that when U_{max} tends to U'_{thr} then the region of the localization of the wave moves infinitely away from the interface: $Z_{max} \rightarrow \infty$, and when $U_{max} = U'_{top}$ then the maximum is situated on the interface: $z_{max} = 0$.

Let us note that to provide the condition (2.8) of applicability of the approximation yielding the analitical solution under the restrictions (5.25) the parameters of the media should satisfy the following relation:

$$S - M \gg S - 1 \ . \tag{5.28}$$

Case 4 : $M < 1 < A$. \hfill (5.29)

Then we obtain the only restriction, L > M, which coincides with one of the limitations of the Case 3, while the other requirement to L of that case is satisfied now for any possible values of the variable L (see relation (5.6)). Thus, we can repeat all conclusions made for the Case 3, excluding the requirement given by the right-hand one of the relations (5.25). So, only the threshold condition for the amplitude maximum retains: $V_2 > U'_{thr}$.

6. Waves at the interface between focusing and defocusing media

For this case there are two options depending on which one of the two media (first or second) is focusing.

Option I : $\quad \alpha_1 < 0 \quad, \quad \alpha_2 > 0 \quad,$ $\hfill (6.1)$

i.e. the medium with a lower sound velocity is defocusing.

Then the solution U(z) is of the form (2.4) in the first medium, containing a point Z_1 of an amplitude maximum ($U_{max} = V_1$), and of the form (2.5) in the second medium with the point Z_2 of singularity situated out of the medium: $Z_1, Z_2 > 0$ (see Fig.3). So, the amplitude parameter V_1 should be considered as a free parameter.

The equations for the coordinates Z_1 and Z_2 are following:

$$ch^2(Z_1/l_1) = (M + A)/(M - L) \quad, \hfill (6.2)$$

$$ch^2(Z_2/l_2) = (M/A)(L + A)/(M - L) \quad, \hfill (6.3)$$

where the relation (2.7) is used.

The expression for the amplitude U_0 on the interface is the following (see Eq. (4.5) with m = 1, and Eq. (6.2)):

$$U_0 = U_{max}[(M - L)/(M + A)]^{1/2} \quad, \hfill (6.4)$$

where L is a function of U_{max} via l_1 (see Eq. (5.12)).

There are two cases for existence of solutions for the Eqs. (6.2) and (6.3).

Case 1 : $\quad M > 1 \quad.$ $\hfill (6.5)$

Then there is the only restriction (5.14) for the length l_1

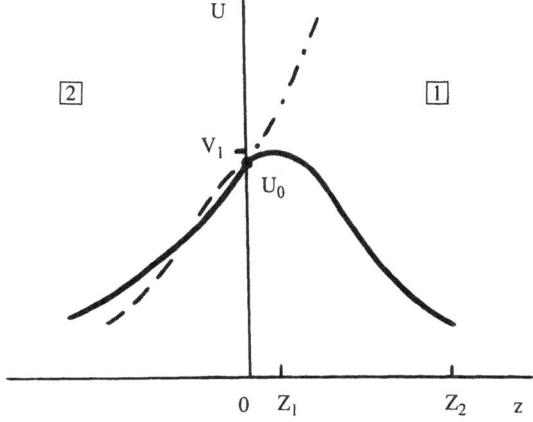

Fig.3. Profile of the shear elastic wave localized at the interface between two nonlinear media, one fucusing (medium 1) and another defocusing (medium 2).

following the requirement (4.7), which yields the threshold condition, $U_{max} \geq U''_{thr}$, where

$$U''^2_{thr} = 8(S - 1)/S|\alpha_1|k^2 \quad . \tag{6.6}$$

<u>Case 2</u> : $M < 1$. \hfill (6.8)

Then there is the restriction (5.19) for the variable L, which yields the top condition, $U_{max} < U_{top}$ (see Eq. (5.21)). Besides, the threshold condition, $U_{max} \geq U''_{thr}$, retains.

<u>Option II</u> : $\alpha_1 > 0$, $\alpha_2 > 0$, \hfill (6.9)

i.e. the medium with a higher sound velocity is defocusing.

Then the solution U(z) is of the form (2.4) in the second medium, containing a point Z_2 of an amplitude maximum ($U_{max} = V_2$), and of the form (2.5) in the first medium with the point Z_1 of singularity situated out of the medium: $Z_1, Z_2 < 0$. So, the amplitude parameter V_2 should be considered as a free parameter.

The equations for the coordinates Z_1 and Z_2 are following:

$$ch^2(Z_1/l_1) = (L + A)/(L - M) \quad , \tag{6.10}$$

$$ch^2(Z_2/l_2) = (L/A)(M + A)/(L - M) \quad . \tag{6.11}$$

There are the only restriction for parameters, $M < 1$, and the requirement to the value of L the same as one of the two requirements for the Case 3, which yields the threshold condition, $U_{max} > U'_{thr}$ (see Eq. (5.26)).

The expression for the amplitude U_0 on the interface is the following (see Eq. (4.5) with $m = 2$, and Eq. (6.11)):

$$U_0 = U_{max}[(A/L)(L - M)/(M + A)]^{1/2} \quad , \tag{6.12}$$

where L is a function of U_{max} via l_1 (see Eqs. (5.24), (2.7)).

Conclusions

We have shown that in the case of the large amplitude excitation there can exist one-component shear-horizontal acoustic waves propagating along the interface of two elastic media. These waves are localized at the interface due to the nonlinearity of the medium. The characteristic length of the localization is inversely poportional to the maximum of the wave amplitude and can be of the order of ten wavelengths.

Such waves can exist under definite relationships between the second, the third, and the fourth orders elastic moduli of the media.

The nonlinear shear-horizontal waves can propagate even along the ideal interface, the case when the interface waves of such polarization can not exist within the framework of the linear theory of elasicity. The SH interface waves arise in a threshold manner when the amplitude maximum exceeds certain value depending on the parameters of the media. For some cases of the relations between the parameters there is a top restriction for the maximum of the amplitude; the further increase of the energy of excitation can yield disappiarance of the interface wave.

References

1. I.A.Viktorov, "Surface Acoustic Waves in Solids" [in Russian], Nauka, Moscow (1981).

2. V.I.Pavlov and I.Yu.Solodov, Nonlinear effects for surface elastic waves in solids, *Sov. Phys. Solid State* **19**:1727 (1977).

3. D.F.Parker and F.M.Talbot, Analysis and computation for nonlinear elastic surface waves of permanent form, *J. of Elasticity* **15**:389 (1985).

4. J.J.Blustein, A new surface wave in piezoelectric materials, *Appl. Phys. Lett.* **13**:412 (1968).

5. Yu.V.Gulyaev, Surface electroacoustic waves in solids, *Sov. Phys JETP Lett.* **9**:37 (1969).

6. G.P.Alldredge, Shear-horizontal surface waves on the (001) face of cubic crystals, *Phys. Lett.* **A41**:281 (1972).

7. I.M.Gel'fgat, Effect of spatial dispersion for non-Rayleigh surface waves in crystals, *Sov. Phys. Solid State* **22**:1640 (1980).

8. V.R.Velasco and P.Garcia-Moliner, Surface effects in elastic surface waves, *Phys. Scr.* **20**:111 (1973).

9. I.M.Gel'fgat and E.S.Syrkin, A problem of existance of transverse acoustic waves localized at the plane interface of two solid media, *Sov. Phys. Acoust.* **28**:256 (1982).

10. Yu.A.Kosevich and E.S.Syrkin, Capillary effects and elastic vibrations at the plane defect in crystal, *Sov. Phys. Crystallogr.* **33**:797 (1988).

11. V.G.Mozhaev, A new type of surface acoustic waves in solids due to nonlinear elasticity, *Phys. Lett.* **A139**:333 (1989).

12. V.I.Gorentsveig, Yu.S.Kivshar, A.M.Kosevich,and E.S.Syrkin Nonlinear surface elastic modes in crystals, *Phys. Lett.* **A144**:479 (1990).

13. V.I.Gorentsveig, Yu.S.Kivshar, A.M.Kosevich,and E.S.Syrkin Self-modulated nonlinear shear surface acoustic waves in crystals, *Sov. J. Low Temp. Phys.* **16**:833 (1990).

14. V.I.Gorentsveig, Yu.S.Kivshar, and E.S.Syrkin, Nonlinear shear waves localized at the interface of two elastic media, *Sov. Tech. Phys. Lett.* **16**:824 (1990).

15. Yu.A.Kosevich, Nonlinear shear surface waves at interfaces and planar defects of crystalls, *Phys. Lett.* **A146**:529 (1990).

16. A.A.Maradudin, Surface acoustic waves on real surface, in: "Physics of Phonons", T.Paszkiewicz, ed., Springer-Verlag, Berlin (1987).

17. I.N.Frantsevich, F.F.Voronov, and S.A.Bakuta, "Elastic Constants and Elastic Moduli of Metals and Nonmetals" [in Russian], Naukova Dumka, Kiev (1982).

18. R.N.Terston, in: "Physical Acoustics", W.P.Mason and R.N.Terston, ed., Academic Press, New York-London (1970).

19. V.I.Gorentsveig, Generation of higher harmonics of nonlinear shear surface acoustic waves, *Phys. Rev.* **B**, in press (1994).

DIFFRACTION TOMOGRAPHY REVISITED

Sidney Leeman[1], Andrew J. Healey[1], Mark Betts[1], and Joie P. Jones[2]

[1]Department of Medical Engineering and Physics
King's College School of Medicine and Dentistry
London SE22 8PT, U.K.
[2]Department of Radiological Sciences
University of California irvine
Irvine, CA 92717

INTRODUCTION

Difffraction tomography has been investigated for many years, and certainly, at one time, the expectations for the technique were optimistic [Mueller, Kaveh and Wade, 1979]. It is perhaps appropriate to examine now, in a non-mathematical way, whether the method still warrants our attention, and whether some of the criticisms that have been levelled at it in the past are, indeed, valid ones. In this discussion, the emphasis will be placed on medical applications: this, at least, simplifies the problems, inasmuch as shear wave propagation may be neglected when attempting to image soft tissues.

What exactly is being done when ultrasound imaging, in general, and diffraction tomography, in particular, are implemented? Essentially, there is an unknown "object" (= patient, in medical applications!), and it is desired to uncover some information about its interior. A (potentially) known ultrasound wave is contrived to penetrate into the object, where it interacts, and the waves which result from this interaction are measured outside the object. From the known input, and the measured output, it is desired to construct, if only approximately, a map of one, or more, of the parameters which describe(s) that interaction. That, in essence, is the image, and different configurations for the 'in' and 'out' schemes have, traditionally, been called different (ultrasound) imaging techniques.

The 'in'/'out' arrangement may be called the 'data acquisition configuration', and describes the 'experimental' (= measurement) setup. Knowledge of the interaction is really based on theoretical notions, that is to say, on the physical model that is constructed for the ultrasound propagation in the object. Finally, the measurements have to be converted into an image, and this presentation will focus more specifically on imaging methods where some computation - or inversion algorithm - is required to transform the data into a meaningful representation of the object.

The main aim is to have a one to-one mapping between the distribution of an interaction parameter in the object, and the displayed image 'matrix'. In general, two features are much discussed: 'resolution' - which is a measure of the fine detail displayed in the image; and 'distortion' - which indicates the geometric accuracy of the final display. A third feature, not often mentioned, is 'fidelity', which is the quantitative accuracy to which an individual scattering parameter is imaged. It has already been pointed out [Leeman and Jones, 1984] that these three features are, to some extent, independent -- it is, in fact, possible to have a high resolution, but geometrically distorted, image, which shows only low fidelity to the interaction parameter being imaged (consider a high resolution conventional B-mode image).

DIFFRACTION TOMOGRAPHY

What then, *is* diffraction tomography? When the data acquisition is such that only backscattered waves are measured, then some familiar imaging methods are defined (B-mode imaging, Doppler imaging), as well as more esoteric techniques such as reflectivity tomography and impediography. If only the forward, transmitted wave is measured, then computerised ultrasound tomography ('CUT') is defined. From this point of view, diffraction tomography refers to the case where the general angle-scattered field is measured. Indeed, this definition suggests that *all* ultrasound imaging may be regarded as being diffraction tomography, with CUT and conventional backscattering techniques regarded as special cases.

Consider now a hypothetical, but commonly assumed, diffraction tomography setup: a number of assumptions are made in order to develop an inversion algorithm, and it is convenient to list these under the following headings.

Physical model

A Helmholtz type wave equation (linear, longitudinal waves, with velocity fluctuations only) is often assumed to describe the ultrasound propagation in the (static) medium.

Data acquisition configuration

The incident wave is assumed plane, with a fixed direction of input. The scattered wave is measured over all angles, in the far-field.

Computational model

The first Born or Rytov approximation is commonly assumed (neglect multiple scattering). Conventionally, reflections at the boundary of the object are disregarded.

We will focus on a few of the assumptions listed here, and examine them in rather more detail. It should be emphasised, perhaps, that space constrains the view taken to be descriptive, rather than technically detailed.

THE PHYSICAL MODEL

A number of different physical models have been investigated as suitable descriptions for the ultrasound / soft tissue interaction. However, some of these are mathematically complex, and it is by no means clear that simpler, phenomenological models will not be equally suitable for use in diffraction tomography studies.

The Helmholtz equation

The most favoured model is that based on the Helmholtz-type wave equation:

$$\nabla^2 U(\mathbf{r}) + k^2 U(\mathbf{r}) = -n(\mathbf{r})k^2 U(\mathbf{r}) \tag{1}$$

where $U(\mathbf{r})$ denotes the ultrasound field at location $\mathbf{r}$, k is the (constant) wave number, and $n(\mathbf{r})$ is the scattering density. Occasionally, a similar model is chosen, but without the k^2 in the scattering term (RHS of equation). This linear model is really appropriate only for the case of a medium exhibiting velocity fluctuations, and is of limited validity in medical applications. Since velocity fluctuations, in practice, are caused by either density or compressibility fluctuations, or both, and because the density and compressibility fluctuations exhibit different angle-scattering behaviour (the former is a dipole, the second a monopole scatterer), this model really contains two different interaction parameters which enter the diffraction tomography inversion algorithm. Thus, two independent data sets would be needed in order to obtain the two quantitative images (one of density, and the other of compressibility) -- thereby maintaining , in principle, image fidelity.

It is certainly possible to devise data acquisition configurations that will unscramble the two scattering functions, and the experimental setups are not unduly cumbersome. The first experiment measures the forward scattering, which contains contributions from both the density and compressibility fluctuations; and the second experiment measures only the $\pi/2$ scattering, which originates from compressibility fluctuations only. In order to obtain complete data sets, measurements have to be effected for all input field directions -- or, equivalently, rotating the object. The above arguments hold, strictly, only if the first Born approximation ('1BA') holds.

DATA ACQUISITION CONFIGURATION

The angular spectrum and directivity signature

Diffraction tomography theories are usually framed in terms of an incident monochromatic plane wave. While this is considered an 'inessential' simplification, it is by no means straightforward to incorporate a more realistic input wave into the inversion procedure, which will require a specification of the full, three dimensional, plane-wave decomposition of the incident beam/pulse. One approach to this is via the angular spectrum description, which requires measurement by a point hydrophone in a plane, and then involves Fourier transformation of the 2-D point measurements to obtain its various components [Schafer and Lewin, 1989]. In general, it is fair to state that the angular spectrum is most suited for the description of continuous wave fields, and that the description becomes rather more cumbersome for pulses. Moreover, the theory necessitates the introduction of non-physical evanescent waves, and it is by no means clear how these are to formally handled within diffraction tomography treatments.

The directivity signature may be regarded as a generalisation of the angular spectrum concept, and is a much more appropriate description for pulsed fields [Leeman and Costa, 1993]. It indicates the decomposition of a general field, $p(\mathbf{r},t)$, into its travelling, continuous plane wave elements, and is defined in terms of the generalised Fourier transform of the field:

$$p(\mathbf{r},t) = \int d^3\mathbf{k}\, F(\mathbf{k})\, exp[i(\mathbf{k} \bullet \mathbf{r} - \omega t)] \tag{2}$$

with $k = 2\pi/\lambda$, $\omega = ck$, $k = |\mathbf{k}|$, λ the wavelength, and c the wave velocity. The directivity signature, $F(\mathbf{k})$, has a number of advantages over the angular spectrum. Its plane wave components may be measured directly with a 'large aperture hydrophone', and are consequently all physically allowed waves (no evanescent waves are admitted into the formalism). Rapid extrapolation of the measured field may be made, over relatively large distances, with minimal computational overheads, in the temporal domain (Fig. 1). Moreover, the description may be naturally extended to lossy media, and even for nonlinear propagation. The fundamental reason why the directivity signature is more appropriate than the angular spectrum for the measurement of the input field is that it is defined with respect to different initial/boundary conditions. The angular spectrum requires the field values to be specified, for all time, on a plane, while the directivity signature proceeds from a knowledge of the field throughout a spatial volume at some initial instant.

Far field measurement

The inversion theory is really phrased in terms of the scattering amplitude, which tells how the outgoing scattered wave at infinity is modulated in amplitude and phase. Thus, far-field measurements are often attempted, despite the practical difficulties. If such measurements are not possible, then the experimental data will have to be either extrapolated outwards (to 'infinity') to regain the scattering amplitude, or extrapolated inwards to attempt a reconstruction by the 'filtered backpropagation' technique. In this context it is interesting to note that large aperture hydrophone techniques are again appropriate: even when used in the near field, the output of such a hydrophone may be shown to be proportional to the value of the scattering amplitude for the scattering direction indicated by the normal to the plane of the hydrophone.

THE COMPUTATIONAL MODEL

The Helmholtz equation may be expressed in its integral formulation, as

$$U(\mathbf{r}) = U_0(\mathbf{r}) + k^2 \int d^3\mathbf{r}'\, G(\mathbf{r},\mathbf{r}')n(\mathbf{r}')U(\mathbf{r}') \tag{3}$$

where U_0 is the 'incident field' (the wave that propagates when the scattering density is zero), and G denotes the (two-point) 'free space' Green's function, $\{exp(ik|\mathbf{r}-\mathbf{r}'|)\}/4\pi|\mathbf{r}-\mathbf{r}'|$. Equ. 3 may be symbolically written as

$$[1-K]U = U_0 \tag{4}$$

where K is an appropriate operator. Clearly, the solution to Equ. (4) may be formally written as

$$U = [1-K]^{-1}U_0 = U_0 + KU_0 + K^2U_0 + \ldots\ldots \tag{5}$$

Figure 1 : Pulse extrapolation by means of the directivity signature. The directivity signature of the transient field from a commercial, focused (1 - 4 cm), 5.0 MHz was measured at 5cm from the transducer face, in .6 deg increments, with a large aperture hydrophone. The horizontal bar shows the transducer face diameter (6mm). The envelopes of the extrapolated pulses, in a plane, at three different distances are shown: top at 0cm, middle at 3.5cm, and, bottom, at 9cm. The vertical and horizontal axes are displayed to the same scale, but the pulses are not shown at their true ranges. Note that the expected features of the field appear, including the appropriate wave front curvatures at both the transducer face and the far field , as well as the enhanced amplitude and reduced pulse dimensions near the nominal focal point. Calculations were effected via a rapid time-domain algorithm on an IBM-compatible PC, but axial symmetry of the field was assumed in order to further reduce the computational overhead.

where the inverse operator has been written as a Born-Neumann expansion. The desired scattering solution for Equ. (5) is obtained by using the asymptotic ($r \to \infty$) form for G, and this procedure establishes a relationship between the measured scattering amplitude, $\Phi(k,\theta,\varphi)$, and the unknown scattering density. If the Born-Neumann expansion does not converge, other solutions for the operator $[1-K]^{-1}$ will have to be found (e.g., the Fredholm method).

Boundary reflections

The above theory conveniently neglects to explicitly take into account the reflections of the incident wave at the boundary of the object. While, formally, it presents no great difficulty to take this into account, the actual measurement of the reflectivity of the surface may pose a number of practical problems. It is interesting to point out that large aperture hydrophones, constructed from thin PVDF film, may also have a role to play in this regard.

The Born and Rytov approximations

The 1BA is obtained by retaining only the first order term in the Born-Neumann expansion. The resultant linear relationship between Φ and n is then easily solved for n by Fourier techniques, provided that the measured data provide an adequate region of support for Φ. Indeed, the various ways in which the Fourier space support of the scattering amplitude may be (experimentally) filled, have been, in the past, occasionally been hailed as new imaging methods, rather than different ways of implementing diffraction tomography. It is clear that the 1BA cannot be valid unless the Born-Neumann expansion converges, and it may be shown that a necessary condition for this is given as $\Delta kR/2 < 1$, with Δ^2 the maximum magnitude of the scattering density, assumed to be enclosed within a spherical volume of (least upper bound) radius R.

In certain cases, the 1BA may be perfectly adequate. Even when higher order Born terms need to be taken into account, a 1BA inversion of the scattering data has the potential to recover an object which correctly indicates the presence and shape of voids. But, in general, the suspicion that higher order terms in the Born-Neumann expansion cannot be dropped, has tended to dampen enthusiasm for a computational model based on the 1BA.

The Rytov approximation ('RA') is a much-used alternative to the 1BA, and may be seen to be equivalent to writing Equ. 5 as:

$$U = U_0 . \sum_{q=0}^{\infty} (KU_0)^q/(q!U_0^q) \tag{6}$$

This expansion may be formally recovered from Equ. 5 via the replacement

$$K^q U_0 \to (q!U_0)^{-1}(KU_0)^q \tag{7}$$

Thus, the RA may be written in terms of the 1BA alone.

Even though it was developed primarily to describe the forward propagating field, the consensus of opinion is that the RA is a more appropriate approximation to use, in general, than the 1BA. But geometrical considerations, rather than scattering strength, may be important when comparing the validity of the two approximations: while the 1BA is expected to be poor when the scattering object is very extended, the RA may be expected to be compromised when the scattering object has sharp corners [Leeman, Chandler, Ferrari and Seggie, 1987]. This problem should be borne in mind when resorting to the RA in some NDE applications.

Since the RA essentially consists of writing the field as an exponentiated version of the 1BA, the recovery of the scattering distribution from the scattering measurements may be established by the the the application of Fourier techniques, and presents no more computational diffficulty than when using the 1BA.

Multiple scattering

The higher order terms in the Born-Neumann expansion represent the contribution of multiple scattering processes to the outgoing field. While the 1BA, by definition, neglects such terms, it may be seen from Equs. 6 & 7 that the RA does incorporate these, in a somewhat corrupted form.

It has been suggested that diffraction tomography is inherently flawed because any inversion procedure, which computes the scattering density from the measured outgoing field, will give ambiguous results when there is significant contribution from multiple scattering terms. However, the generality of such a statement may be disproved by the demonstration of the following counter-example [Leeman, Chandler and Ferrari, 1987].

Consider a (rotationally symmetric) scattering density which may be expressed as:

$$n(r) = \int_{\mu}^{\infty} ds\ N(s)\ \{exp(-sr)\}/r\ , \qquad \text{with} \int_{\mu}^{\infty} ds\ |N(s)|\ \text{finite} \qquad (8)$$

Note that such a representation implies that the object is infinitely extended, with no restrictions on the scattering strength. Thus the 1BA is not applicable, in general.

Each term in the Born-Neumann expansion for the field scattered from an object in the above class may be explicitly calculated, and its analyticity properties in a complex ξ-space investigated, where $\xi \equiv 2k^2(1 - \cos\theta)$, with θ the scattering angle. It may be shown that the mth Born term exhibits a branch cut starting at $(m\mu)^2$. The discontinuity across each successive branch cut, as calculated via analytic continuation of data from the measured 'physical region' ($0 \leq \xi \leq 4k^2$), allows a progressively better approximation to the scattering density to be made. The importance of this example is not to suggest that it necessarily presents a practical inversion procedure, but to demonstrate that, in principle, exact inversion may be implemented (in certain cases, at least) even in the presence of strong multiple scattering.

CONCLUSIONS

This somewhat selective overview of diffraction tomography, with longitudinal waves, has suggested that not all criticisms of diffraction tomography are necessarily valid. The discussion has been structured in terms of three key aspects of the technique: the physical model underlying the method, the data acquisition configuration which determines the experimental design, and the computational model which establishes the inversion algorithm.

There have been relatively few attempts to develop the technique for more realistic physical models than the Helmholtz equation. Some 'inessential approximations' made in developing the theory (such as a plane wave input field) may be quite difficult to relax when 'real' experiments are performed. It may be argued that the emphasis placed on theory has tended to hold back the development of field measurement techniques which potentially simplify the method. In this context, the use of large aperture hydrophones may be of some value: in particular they considerably simplify the measurement of the directivity signature, which overcomes some of the limitations of the angular spectrum description of pulsed fields. There has been little exploitation of computational models that extend beyond the first Born or Rytov approximations, but the validity conditions for the latter, in particular, may need to be investigated further. It does not appear to be true that diffraction tomography is intrinsically flawed when multiple scattering effects are evident.

Medical applications are limited to the imaging of only a few appropriate sites, and it is much more likely that the first breakthrough will come in some NDE application. But it may be some years before truly exciting imaging systems become available.

ACKNOWLEDGEMENTS

The Royal Society and The Wellcome Trust are thanked for equipment grants which enabled much of the development of large aperture hydrophones to be carried out.

REFERENCES

Leeman, S. and Costa, E.T., 1993, Large Aperture Hydrophones for Field Measurement and Calibration, *in*: "Acoustic Sensing and Imaging", 294, IEE, London.

Leeman, S., Chandler, P.E., Ferrari, L.A., 1987, Diffraction tomography with multiple scattering, *in*: "Acoustical Imaging 15", H.W. Jones, ed., 29, Plenum Press, New York.

Leeman,S., Chandler, P.E., Ferrari, L.A. and Seggie, D.A., 1987, Validity of the Born and Rytov approximations, *in*: "Progress in Underwater Acoustics", H. Merklinger, ed., 35, Plenum Press, New York.

Leeman, S. and Jones, J.P., 1984, Tissue information from ultrasound scattering, *in*: "Acoustical Imaging 17", M. Kaveh, ed., 13, Plenum Press, New York.

Mueller, R.K., Kaveh, M. and Wade, G., 1979, Reconstruction tomography and applications to ultrasonics, *IEEE Proceedings*, 67:567.

Schafer, M.E. and Lewin, P.A., 1989, Transducer characterization using the angular spectrum method, *J. Acoust. Soc. Am.*, 85(5):2202.

APPLICATIONS OF SOLID-ANGLE FUNCTIONS IN ACOUSTICAL SCATTERING

John M. Richardson, Glen Wade, and George Goebel

University of California
Santa Barbara, California 93106, U.S.A.

INTRODUCTION

It is known that, in the far-field (FF) regime, the pulse-echo (PE) scattering of acoustical waves from a weak, internally uniform inhomogeneity can be simply represented in terms of the planar area function (PArF) of the scatterer. The PArF can be obtained by moving an imaginary cutting-plane, oriented parallel to the incident wavefront, through the scatterer and by recording the areas of the cross-sections as a function of distance. After the PArF is re-expressed as a function of the round-trip time, the impulse response function (IRF) is obtained by taking the second time derivative and multiplying by a known constant. In the case of a strong scatterer a similar procedure is employed, except the shadow zone is regarded as an extension of the object. In an earlier paper (Richardson and Wade, 1992), the PArF was discussed at some length, but a generalization, called the solid-angle function (SAnF) approach, was discussed only briefly. The purpose of the present paper is to give a full treatment of the latter approach.

The SAnF methodology and its applications to a variety of acoustical scattering problems is based upon the following assumptions:

(1) The scattering experiments involve only PE measurements with a single transducer operating in both transmit and receive modes.
(2) The transducer is omnidirectional over an appropriate solid angle.
(3) The scatterer is either very weak* or very strong*.
(4) In the former case the scatterer is internally uniform, isotropic and lossless.
(5) In either case the host medium is uniform, isotropic and lossless.
(6) The incident acoustic wave has a longitudinal polarization.

These assumptions, of course, limit the scope of the present treatment. However, many aspects of the scope limitation do not represent inherent limitations of general SAnF methodology.

REVIEW OF THE PLANAR AREA FUNCTION APPROACH

The older PArF methodology is discussed here with the objective of providing a background for the subsequent discussion of the SAnF approach. A secondary objective is to provide a basis for the comparison between the two approaches. In the subsequent

---

*The term "weak" means that the acoustical properties of the scatterer differ only slightly from those of the host medium and the term "strong" means the opposite, to such an extent that the scatterer is almost totally reflective.

sections we will discuss the SAnF and its application to a weakly scattering sphere, a weakly scattering cube, and strongly scattering sphere.

Theory of the PArF Approach to Scattering

The PArF approach has antecedents reaching back many decades. The earliest version of this methodology (of which we are aware) involves a relatively primitive application of a restricted PArF to the FF, PE scattering of electromagnetic waves from conducting bodies (i.e., strong scatterers) using the Kirchhoff approximation (Ruck etal., 1970). The above use of the term "restricted" refers to the fact that only the illuminated part of the body surface was used in the construction of the area function. Later progress was made (Richardson and Cohen-Tenoudji, 1991) in extending the area function to include the entire body with the shadow zone regarded as an extension of the body.

The case of weak scatterers, for which the Born approximation is valid, has been investigated (Rose and Richardson, 1981) as a domain for the application of the PArF methodology. Most of the discussion in this section deals with this case.

In Fig. 1, we give a schematic representation of an experiment in which a PE scattering measurement is made in the FF regime for a finite bandwidth (BW). In this figure a transducer is placed at a

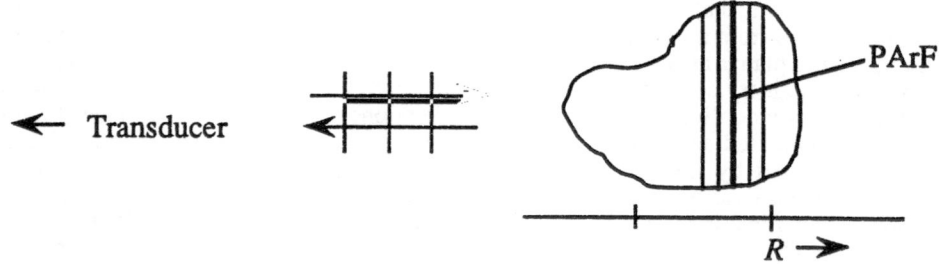

Figure 1. Schematic representation of the PArF pertaining to a weak scatterer in a PE scattering experiment in the FF regime with finite bandwidth.

great distance to the left (in fact, off the page). The PArF is partially indicated by a series of vertical lines representing successive positions of the cutting plane parallel to the incident wavefront. We will represent the PArF by the symbol $A_{pl}(R)$, where R is the distance from the transducer to a particular cutting plane. It is desirable to re-express $A_{pl}(R)$ as a function of t, the round-trip time, i.e.,

$$A_{pl}(R) = A_{pl}\left(\frac{1}{2}ct\right),\tag{1}$$

where c is the speed of acoustical waves in the host medium. It can be shown (Rose and Richardson, 1981) that the impulse response function (IRF), represented by the symbol I(t) is related to the PArF by the simple expression

$$I(t) = \alpha\frac{d^2}{dt^2}A_{pl}\left(\frac{1}{2}ct\right)\tag{2}$$

where α is a known constant dependent upon the acoustical properties of the scatterer and the host medium. The reader is reminded that the more commonly used scattering amplitude is the Fourier transform of I(t). In the case of a strong scatterer, a slightly modified version of Eq. (2) is used in which the α is replaced by a different constant α' (dependent only upon the acoustical properties of the host medium) and the shadow zone is regarded as an extension of the object.

Examples

In the ensuing subsections we present results for three cases: (1) the weakly scattering sphere, (2) the weakly scattering cube with various orientations, and (3) the strongly scattering sphere.

Weakly Scattering Sphere

In this subsection we present, for this case, graphs of the PArF and its first and second time derivatives in Fig. 2. The reader is reminded that the latter is proportional to the IRF.

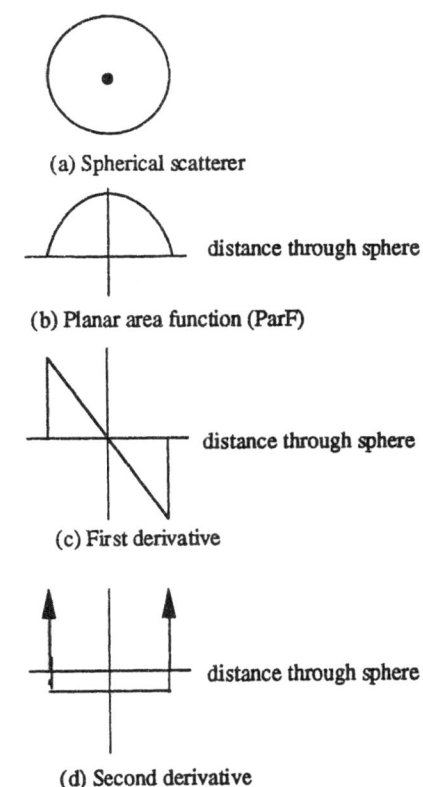

(a) Spherical scatterer

(b) Planar area function (ParF)

(c) First derivative

(d) Second derivative

Figure 2. Weakly-scattering sphere in the far field regime.

Weakly Scattering Cube

In this subsection we present. in Fig. 3, graphs of the IRF for the cube with three incident directions [100], [110], and [111].

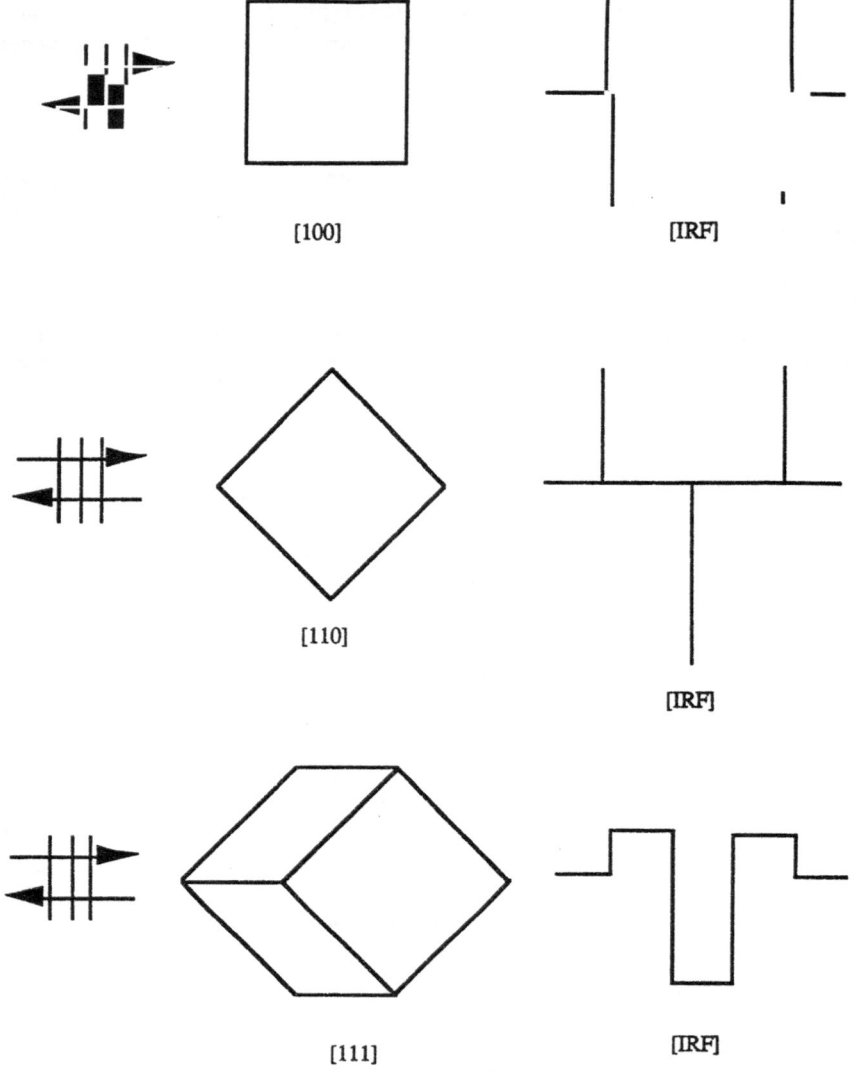

Figure 3. Far-field PE scattering from a weak, internally uniform inhomogeneity with a cubical boundary.

Strongly Scattering Sphere

In this subsection we present in Fig. 4 graphical representations of the PArF for this case and its first and second time derivatives. In relation to the weakly scattering sphere, this methodology is practically the same except for the fact that the shadow zone is regarded as an extension of the sphere and the constant α in Eq. (2) is replaced by α'.

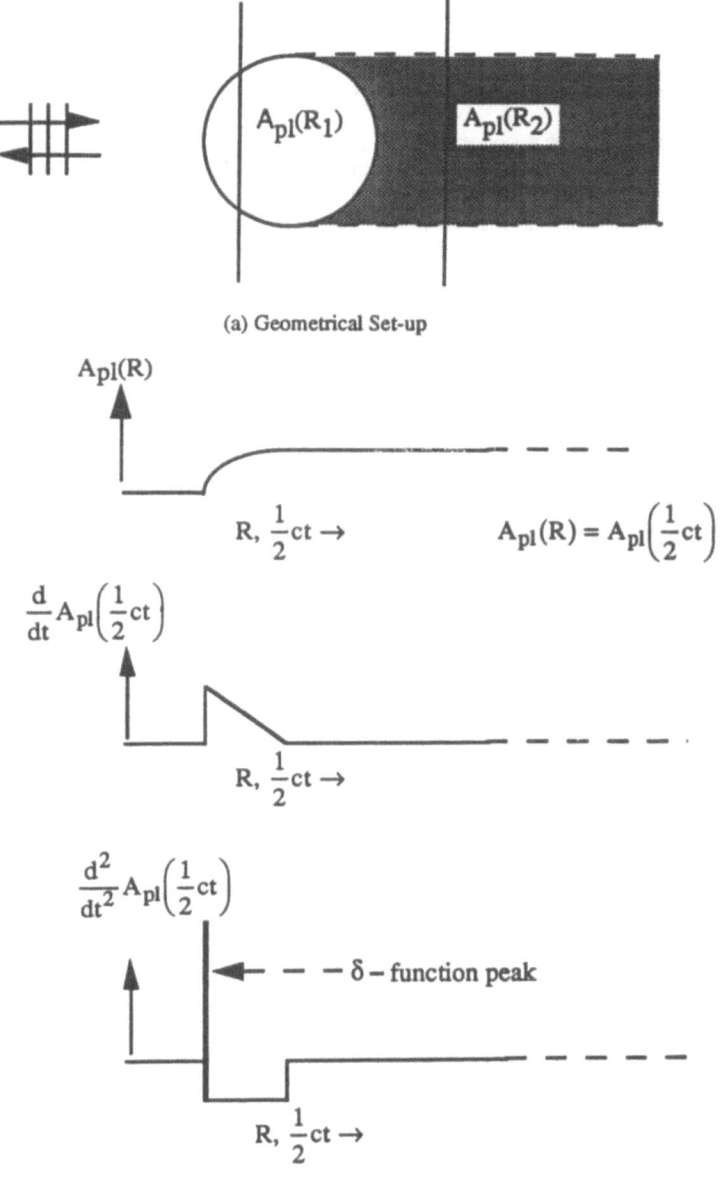

(a) Geometrical Set-up

$$A_{pl}(R) = A_{pl}\left(\frac{1}{2}ct\right)$$

(b) The PArF and its Derivatives

Figure 4. Use of the planar area function (PArF) in the Kirchhoff approximation applied to a strongly scattering sphere.

Comments

One striking characteristic of this methodology is that the determination of PArF requires far less effort than that involved in the conventional methods of scattering theory. For instance, the examples considered above were solved by inspection, except for the cube with the [111] incident direction, which required some elementary algebra. Furthermore, the approach has an intuitive appeal in most cases.

SPHERICAL AREA FUNCTIONS AND SOLID-ANGLE FUNCTIONS

We now consider a generalized form of the area function approach that is valid for all transducer positions (outside of the scatterer domain, of course) and for all BW's. This is the solid-angle function (SAnF) approach, which is the main concern of this paper. It is clear that the PArF approach is a special limiting case of the above SAnF approach. We will start with the spherical area function (SArF) as shown in Fig. 5 along with the SAnF, which is closely related. In the figure we also show a spherical segment of radius R centered at the point T (the position of the transmit-receive transducer) whose surface is cutting through the scatterer. The area of the cross-section is clearly the SArF evaluated at the radius R. Relative to the point T the SArF subtends a solid angle corresponding to the SAnF. Letting SArF be represented by the symbol $A_{sph}(R)$ and SAnF, by $\Omega(R)$, we obtain the obvious result

$$\Omega(R) = R^{-2} A_{sph}(R). \tag{3}$$

It can be shown that the IRF is given by the simple relation

$$I(t) = \beta \frac{d^2}{dt^2} \Omega\left(\frac{1}{2} ct\right) \tag{4}$$

where β is a known constant dependent upon the acoustical properties of scatterer and the host medium. Eq. (4) is clearly

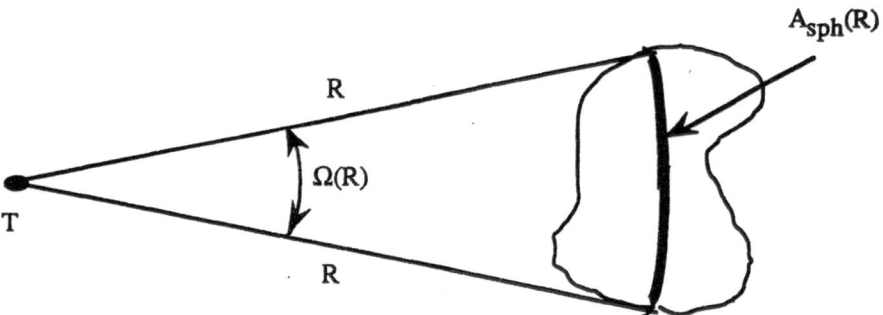

Figure 5. Geometrical set-up for the definitions of $A_{sph}(R)$ and $\Omega(R)$

analogous to Eq. (2), however on the basis of this analogy one might be tempted use the SArF on the r. h. side of (4) instead of SAnF, which would be wrong.

The above relation has been derived rigorously from first principles with the use of the Born approximation. This work will be discussed in a separate paper (Richardson. to be published). Hence the above result is rigorously equivalent to the Born approximation for any weak, internally uniform and isotropic, scatterer with any transducer position (outside of the scatterer) and any frequency (or, equivalently, any BW).

In the case of strongly scattering objects for which the Kirchhoff approximation is valid, Eq. (4) still holds but with a different constant β' which depends only upon the acoustical properties of the host medium. As in the FF regime the shadow zone is regarded as an extension of the object. However here the shadow is defined as the set of points in space not having lines of sight to the transducer. These modified results are rigorously equivalent to the Kirchhoff approximation for any strong scatterer with any transducer position (outside of the object) and any BW. While the Born approximation is asymptotically exact in the limit of small property deviations, the Kirchhoff approximation is not exact under any circumstances with a long time domain.

FILTERED IMPULSE RESPONSE FUNCTIONS

Due to the commutability of convolution operations we can shift the time differentiations to the filter function to be defined later. In the time-domain we obtain from Eq. (2) or (4) the result

$$\bar{I} = F(t)*I(t) = \int dt' F(t-t')I(t')$$

$$= \int dt' F(t-t') \frac{d^2}{dt^2} \left\{ \begin{array}{c} \alpha A_{pl}\left(\frac{1}{2}ct\right) \\ \beta \Omega\left(\frac{1}{2}ct\right) \end{array} \right\}$$

$$= \int dt' \left[\frac{d^2}{dt^2} F(t-t') \right] \left\{ \begin{array}{c} \alpha A_{pl}\left(\frac{1}{2}ct\right) \\ \beta \Omega\left(\frac{1}{2}ct\right) \end{array} \right\} \tag{5}$$

where F(t) is the filter function. It is to be noted that, on the last line of Eq. (5), the time differentiations are operating on the time-shifted filter function F(t - t'). Assuming that the latter is sufficiently nonsingular, the final integral is free of singularities.

In the frequency domain, we obtain Eq. (2) or (4) in the form

$$a(\omega) = F\,I(t) = -\int dt \exp(-i\omega t)\,I(t)$$

$$= -\omega^2 \int dt \exp(-i\omega t) \left\{ \begin{array}{c} \alpha A_{pl}\left(\frac{1}{2}ct\right) \\ \beta \Omega\left(\frac{1}{2}ct\right) \end{array} \right\} \tag{6}$$

where $a(\omega)$ is the scattering amplitude. The filtered version is given by

$$\bar{a}(\omega) = F(\omega)a(\omega) \tag{7}$$

where $F(\omega) = F\,F(t)$.

COMPUTATION OF SOLID-ANGLE FUNCTIONS

We turn now to the problem of determining SAnF's from object models. There are two approaches for a specifies object model with a given transducer location: an analytic calculation or a numerical computation. Overall, the determination by either approach is more difficult for the general (not-FF) regime than it is for the FF regime.

Some Analytical Results for the SArF and SAnF

We consider two simple weakly scattering objects, i.e. a semi-infinite half-space and a sphere, shown in Fig. 6 with their geometrical set-ups.

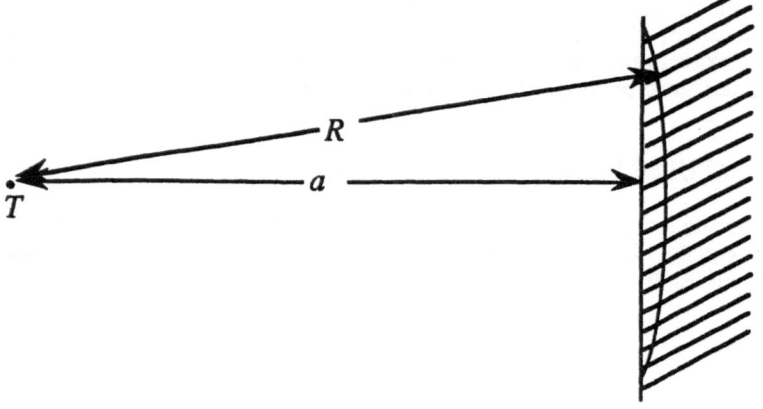

(a) Semi-infinite scatterer

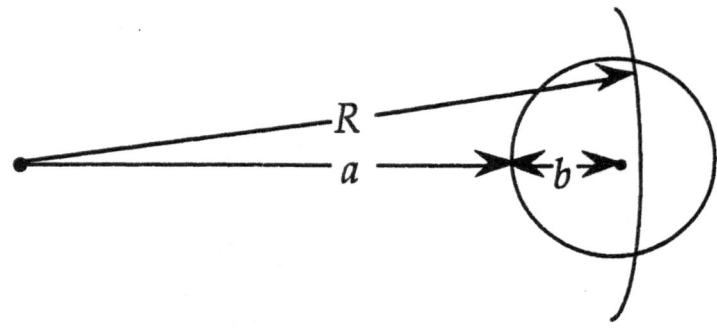

(b) Spherical scatterer

Figure 6. Analytical results for A_{sph}

The first object involves transducer placed at a distance a from its flat surface. The spherical area function (SArF) represented by the symbol $A_{sph}(R)$ is given by

$$A_{sph}(R) = 2\pi R^2 (1 - \frac{a}{R}) \qquad (8)$$

if R>a and equals 0 otherwise. This result is remarkable for the fact that there is no bounded solution in the FF regime. The second object, a sphere with a radius b with a distance a from the transducer to the nearest part of the sphere, is shown in the lower part of the figure. Its SArF is given by

$$A_{sph}(R) = 2\pi R^2 \left(1 - \frac{\alpha}{R}\right)$$

$$\alpha = \frac{1}{2}\left(a + b + \frac{R^2 - b^2}{a + b}\right) \qquad (9)$$

if a<R<a + 2b and equals 0 otherwise. It is easily seen that this result approaches Eq. (8) the sphere becomes infinitely large with a held constant.

The SAnF's and IRF's can easily be obtained by the use of Eqs. (3) and(4).

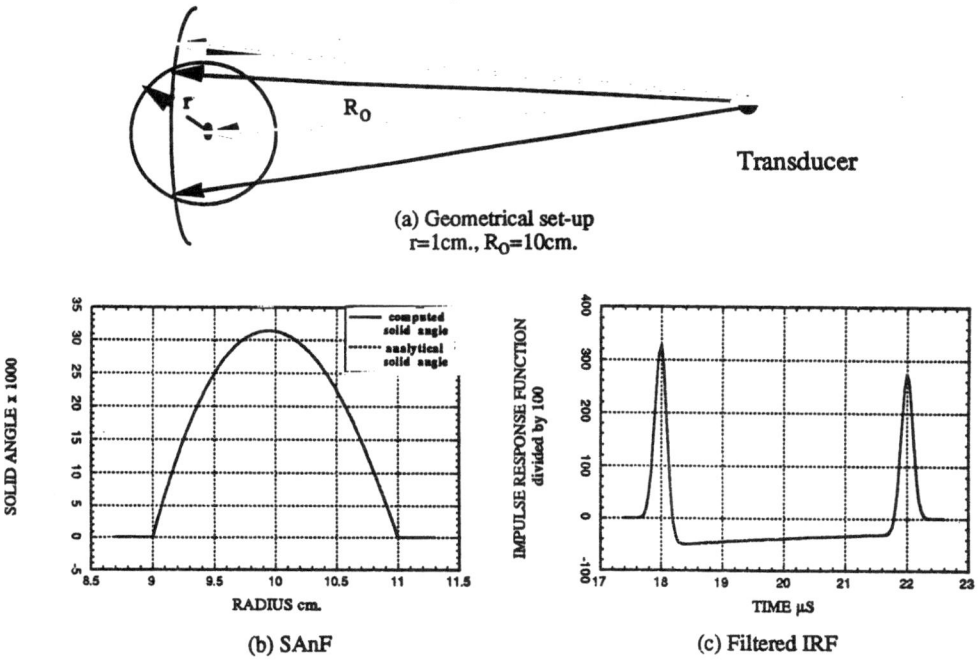

(a) Geometrical set-up
r=1cm., R_O=10cm.

(b) SAnF

(c) Filtered IRF

Figure 7. Weakly Scattering Sphere

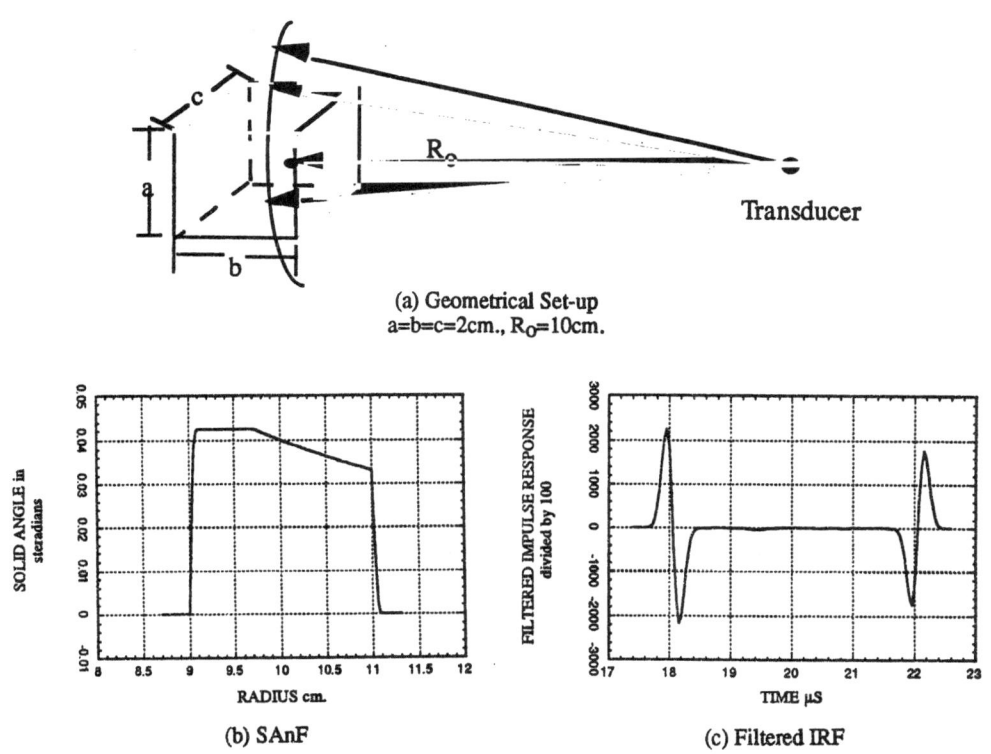

(a) Geometrical Set-up
a=b=c=2cm., R_O=10cm.

(b) SAnF

(c) Filtered IRF

Figure 8. Scattering in [100] direction

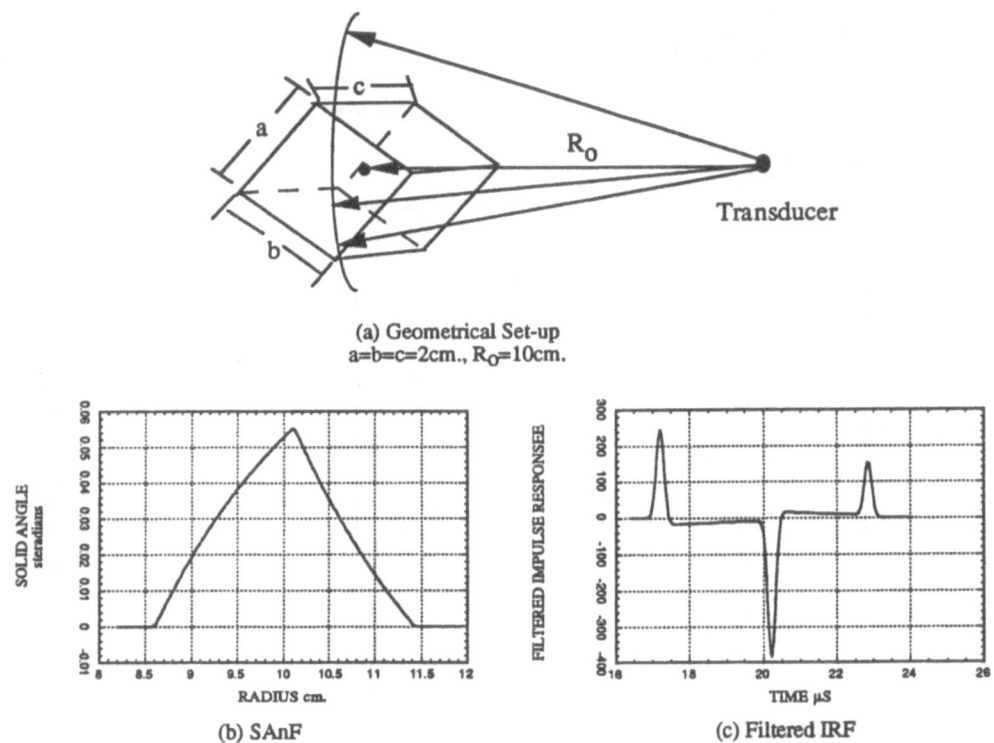

(a) Geometrical Set-up
a=b=c=2cm., R_O=10cm.

(b) SAnF

(c) Filtered IRF

Figure 9. Scattering in the [110] direction.

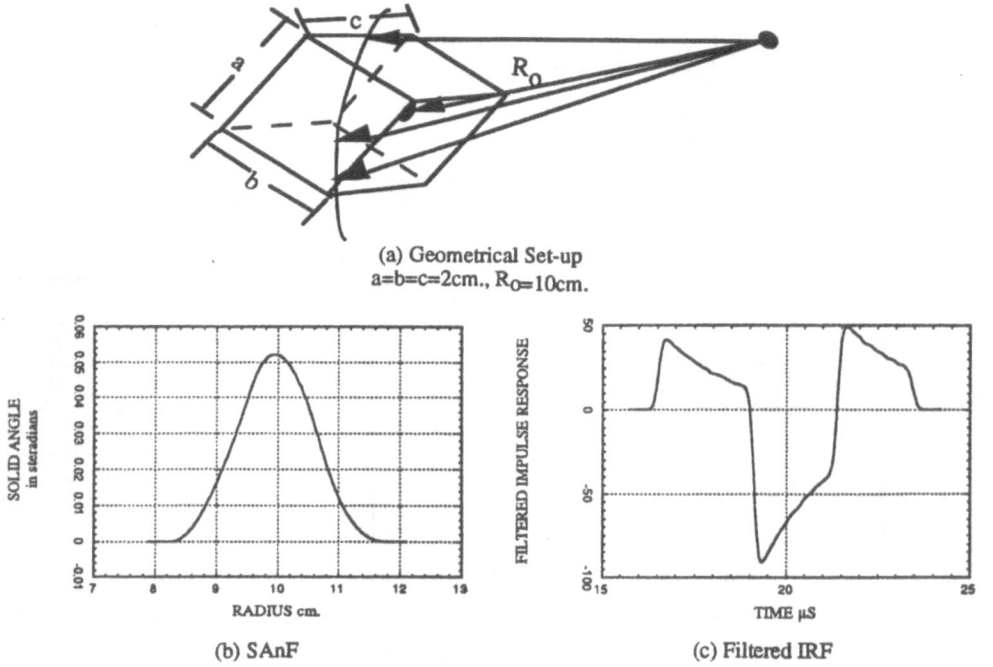

(a) Geometrical Set-up
a=b=c=2cm., R_O=10cm.

(b) SAnF

(c) Filtered IRF

Figure 10. Scattering in the [111] direction.

The Numerical Computation of the SAnF and IRF

The analytical approach is practical only for simple objects (except in the FF regime). Even the cube presents very tedious analytical problems. In the majority of cases numerical methods are advisable. We have developed a general algorithm to be applied to any object for which an indicator (characteristic) function can be formulated. In the interest of brevity we will give a detailed discussion in a separate paper (Richardson, Goebel, and Wade, to be published). However we will note here a few essential features of the algorithm: (1) a formulated indicator function (as we have already mentioned), (2) a specified position of the transducer, and (3) a planar grid of square pixels through the centers of which radii of length R (centered at the transducer) pass and thus define spherical area elements on the sphere of radius R. From these features one can sum the elementary areas situated inside the object.

EXAMPLES OF NUMERICAL RESULTS

In the ensuing discussion we present results of the numerical computation of the solid-angle function (SAnF) and the filtered form of the IRF. The purpose of filtering is to avoid the amplified computer round off errors introduced by the direct double time-differentiation of SAnF's. We consider three cases: a weakly scattering sphere, a weakly scattering cube with several orientations, and a strongly scattering sphere. These results can be compared directly with a similar sequence of results in the FF regime presented in an earlier part of this paper. The value R_0 represents the distance of the transducer to the center of the corresponding object. In the filtering operation we have used a filter function F(t) with a specified "standard deviation" of 4.24 μS.

Weakly Scattering Sphere

In Fig. 7 we show the numerical results for this case. In part (b) we show a comparison of the analytical and numerical approaches to the SAnF (only for this case) in which the agreement is so good that the dashed and solid lines are indistinguishable verifying that the inevitable numerical errors an be maintained at a low level. Here the filtered form of the IRF has a similarity to the earlier results in the FF regime but with distortions and smoothing of an obvious nature.

Weakly Scattering Cube with Various Incident Directions

In Figs. 8, 9, and 10 we show the numerical results for this case with different orientations corresponding to the wave directions: [100], [110], and [111]. The IRF's here should compared with those shown in Fig. 3 for the FF regime.

Strongly Scattering Sphere

In Fig. 11 we show the numerical results for the strongly scattering sphere. These results are similar in many ways to those in the FF regime except for one important feature, namely the existence of the conical shadow zone in the present treatment. Because of the use of the SAnF instead of the SArF the two end results are very close.

SUMMARY

The solid-angle function (SAnF) is obtained from the spherical area function (SArF) by dividing the latter by R^2. The SArF is obtained by taking a sphere of radius R centered at the transducer and recording the spherical area of the spherical segment lying within the scatterer. Re-expressing the SAnF in terms of the round-trip time we obtain the result that the impulse response function (IRF) is proportional to the second time-derivative of the SAnF.

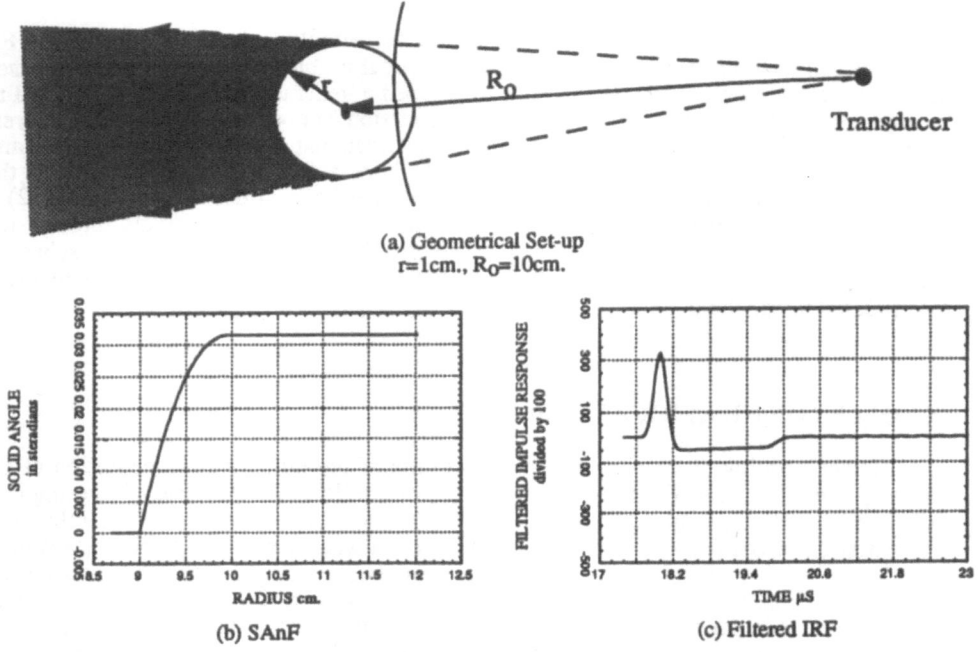

(a) Geometrical Set-up
r=1cm., R_0=10cm.

(b) SAnF

(c) Filtered IRF

Figure 11. Strongly scattering sphere

These results are valid for weakly scattering objects that are internally uniform and isotropic and for strongly scattering objects for which the shadow zones are regarded as extensions of the objects. Here a shadow zone is the set of points that do not have lines-of-sight to the transducer. In all cases the host medium is assumed to be uniform and isotropic. The SAnF approach to weakly scattering objects is exactly equivalent to the Born approximation and the approach to strongly scattering objects is exactly equivalent to the Kirchhoff approximation for all meaningful transducer positions and all frequencies. We have developed analytic and numerical methods for determining SAnF's from object models.

REFERENCES

Richardson, J. M., 1994, "Theory of the solid-angle functions and their application to scattering," to be submitted for publication.

Richardson, J. M., and Cohen-Tenoudji, F., 1994, "Area-function formulation of the scattering of acoustic and electromagnetic waves," to be published.

Richardson, J. M., Goebel, G., and Wade, G., 1994, "Numerical computation of solid-angle functions," to be submitted for publication.

Richardson, J. M., and Wade, G., 1992, "Applications of area solid-angle functions in acoustics," Acoustical Imaging Vol. 19, edited by H. Ermert and H. Harjes, pp. 167-173, Plenum, New York.

Rose, J., and Richardson, J. M., 1981, "Time-domain Born approximation," Proc. DARPA/AFWL Rev. of Progress in NDE, 383-388.

Ruck, G. T., Barrick, D. E., Stuart, W. D., and Krichbaum, C. K Radar Cross-Section Handbook, 50-66, Plenum, New York.

COMPUTERIZED SIMULATIONS FOR REFLECTIVE DIFFRACTED TOMOGRAPHY BY ITERATION METHOD

Xiu-ping Tao, Jian-chun Cheng, Song-ru Fang and Shu-yi Zhang

Institute of Acoustics and Laboratory of Modern Acoustics
Nanjing University, Nanjing, 210093, China

INTRODUCTION

In the past decade, the diffracted computerized tomography using acoustic wave was studied extensively. Various transmitting-receiving configurations have been presented, such as transmission[1] and reflection[2] tomography. In those configurations, the transmitter-receiver systems usually must be rotated through 360°. Unfortunately, in some practical conditions, such as in geophysical exploration, the system can not be rotated and the receiver can only detect the reflected wave at the earth surface where the transmitter is located.

In this paper, we present a reflective diffracted tomography technique for cross-sectional reconstruction of acoustic parameters of an object. Fig.1 illustrates the data collection configuration, in which the transmitter and receiver arrays are located at the same line, and for all the positions of the transmitters we measure the scattered field for all the positions of the receivers. The requirement that the object be viewed over 360° is replaced by the scan of the transmitter and receiver at the same side of the object. This data collection strategy offers an advantage over those of transmission and reflection tomography techniques in Refs.1 and 2. We have developed reconstuction formulas for this configration, and analyzed the frequency coverage using Fourier diffraction theorem. It is found that the coverage is very uncomplete, especially in the low frequency area. Therefore, we have empolyed an iteration method under certain priori-knowledge to get the unknown low frequencies and improve the image quality. The computer simulations show the validity of the algorithm.

BASIC FORMULA

Consider an inhomogeneous scattering object located in the homogeneous half-space region in some depth under the surface, as depicted in Fig.1, in which the transmitting and receiving arrays are located collinearly on the surface above the object. The acoustic field $\phi(\hat{r}, \hat{r}_T; t)$ at point $\hat{r} = (x, y)$ generated by the transmitter at $\hat{r}_T = (x_T, 0)$ satisfies the equation

$$\kappa(\hat{r})\frac{\partial^2\phi(\hat{r},\hat{r}_T;t)}{\partial t^2}-\hat{\nabla}\cdot[\sigma(\hat{r})\hat{\nabla}\phi(\hat{r},\hat{r}_T;t)]=f(t)\delta(\hat{r}-\hat{r}_T) \tag{1}$$

where κ is the compressibility of the medium, and $\sigma=1/\rho$, ρ is the density of the medium, f(t) is the source function. Here we have assumed that the medium is nonviscoelastic and κ is independent of the frequency.

Taking the Fourier transform of Eq.(1) for time variable t, yields

$$\omega^2\kappa(\hat{r})\phi(\hat{r},\hat{r}_T;\omega)+\hat{\nabla}\cdot[\sigma(\hat{r})\hat{\nabla}\phi(\hat{r},\hat{r}_T;\omega)]=-f(\omega)\delta(\hat{r}-\hat{r}_T) \tag{2}$$

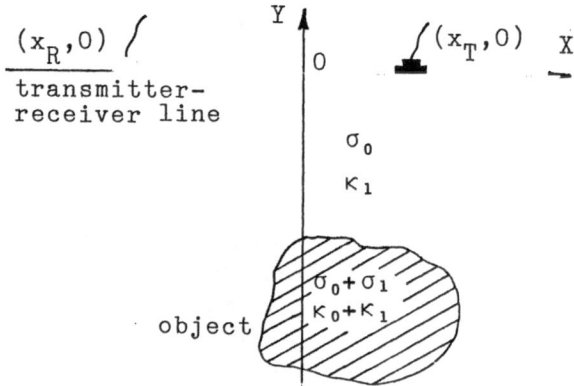

Figure 1. Configuration of the transmitter and receiver arays.

The total acoustic field can be expressed by the superposition of the incident field $\phi_i(\hat{r},\hat{r}_T;\omega)$ and the scattered field $\phi_s(\hat{r},\hat{r}_T;\omega)$,

$$\phi(\hat{r},\hat{r}_T;\omega)=\phi_i(\hat{r},\hat{r}_T;\omega)+\phi_s(\hat{r},\hat{r}_T;\omega) \tag{3}$$

The $\phi_i(\hat{r},\hat{r}_T;\omega)$ and $\phi_s(\hat{r},\hat{r}_T;\omega)$ satisfy, respectively,

$$\nabla^2\phi_i(\hat{r},\hat{r}_T;\omega)+k_0^2\phi_i(\hat{r},\hat{r}_T;\omega)=-\frac{f(\omega)}{\sigma_0}\delta(\hat{r}-\hat{r}_T) \tag{4}$$

$$\nabla^2\phi_s(\hat{r},\hat{r}_T;\omega)+k_0^2\phi_s(\hat{r},\hat{r}_T;\omega)=-[k_0^2\frac{\kappa_1(\hat{r})}{\kappa_0}\phi(\hat{r},\hat{r}_T;\omega)$$

$$+\hat{\nabla}\cdot[\frac{\sigma_1(\hat{r})}{\sigma_0}\hat{\nabla}\phi(\hat{r},\hat{r}_T;\omega)]] \tag{5}$$

where

$$\sigma(\hat{r}) = \sigma_0 + \sigma_1(\hat{r}) \tag{6}$$

$$\kappa(\hat{r}) = \kappa_0 + \kappa_1(\hat{r})$$

and $c^2_o = \sigma_o/\kappa_o$, $k_0 = \omega/c_0$, c_0 is the acoustic velocity.

By employing the Green function method, one can obtain the scattered field received at $\hat{r}_R = (x_R, 0)$,

$$\phi_s(\hat{r}_R,\hat{r}_T;\omega) = -\int_s [k_0^2 \frac{\kappa_1(\hat{r}')}{\kappa_0} \phi(\hat{r}',\hat{r}_T;\omega)$$

$$+\hat{\nabla}\cdot(\frac{\sigma_1(\hat{r}')}{\sigma_0}\hat{\nabla}\phi(\hat{r}',\hat{r}_T;\omega))]G(\hat{r}',\hat{r}_R;\omega)d\hat{r}' \tag{7}$$

Here the Green function $G(\hat{r}',\hat{r};\omega)$ is defined

$$\nabla^2 G(\hat{r}',\hat{r};\omega) + k_0^2 G(\hat{r}',\hat{r};\omega) = \delta(\hat{r}-\hat{r}') \tag{8}$$

To accomplish the linearization, the Born approximation is used,

$$\phi_s(\hat{r}_R,\hat{r}_T;\omega) = -\int_s [k_0^2 \frac{\kappa_1(\hat{r}')}{\kappa_0} \phi_i(\hat{r}',\hat{r}_T;\omega)$$

$$+\hat{\nabla}\cdot(\frac{\sigma_1(\hat{r}')}{\sigma_0}\hat{\nabla}\phi_i(\hat{r}',\hat{r}_T;\omega))]G(\hat{r}',\hat{r}_R;\omega)d\hat{r}' \tag{9}$$

Taking the Fourier Transform of the Eq.(9) for the space variables x_T and x_R, yields

$$\phi_s(k_R,k_T;\omega) = -\int_s [k_0^2 \frac{\kappa(\hat{r}')}{\kappa_0} \phi_i(\hat{r}',k_T;\omega)G(\hat{r}',k_R;\omega)$$

$$-\frac{\sigma_1}{\sigma_0}\hat{\nabla}\phi_i(\hat{r}',k_T;\omega)\cdot\hat{\nabla}G(\hat{r}',k_R;\omega)]d\hat{r}' \tag{10}$$

On the other hand, by solving the Eqs.(4) and (8), one can get $\phi_i(x',y',k_T;\omega)$ and $G(x',y',k_R;\omega)$

$$G(x,y,k_R;\omega) = \frac{1}{i\alpha_R}\exp(i\alpha_R y)\exp(-ik_R x) \tag{11}$$

$$\phi_i(x,y,k_T;\omega) = -\frac{f(\omega)}{\sigma_0}\frac{1}{i\alpha_T}\exp(i\alpha_T y)\exp(-ixk_T) \tag{12}$$

where we have assumed that evanescent waves can be ignored and k_R, k_T vary from $-k_0$ to k_0, and

$$\alpha_R = \sqrt{k_0^2 - k_R^2} \qquad (13)$$

$$\alpha_T = \sqrt{k_0^2 - k_T^2} \qquad (14)$$

Inserting $G(x',y',k_R;\omega)$ and $\phi_i(x',y',k_T;\omega)$ into Eq.(10), one can obtain,

$$\frac{\sigma_0}{f(\omega)}\phi_s(k_R,k_T;\omega) = -k_0^2 \frac{1}{\alpha_R \alpha_T}\gamma_\kappa(k_x,k_y) - (\frac{k_R k_T}{\alpha_R \alpha_T} + 1)\gamma_\sigma(k_x,k_y) \qquad (15)$$

where

$$\gamma_\kappa(k_X,k_Y) = \int_s \gamma_\kappa(x',y')\exp[-i(k_R+k_T)x' + i(\alpha_R+\alpha_T)y']dx'dy' \qquad (16)$$

$$\gamma_\sigma(k_X,k_Y) = \int_s \gamma_\sigma(x',y')\exp[-i(k_R+k_T)x' + i(\alpha_R+\alpha_T)y']dx'dy' \qquad (17)$$

and $\gamma_\kappa(x,y) = \kappa_1(x,y)/\kappa_o$, $\gamma_\sigma(x,y) = \sigma_1(x,y)/\sigma_o$.

Now, we have obtained the relation between the Fourier transform of the scattered field from the object and the Fourier transform of the variations of the object parameters. For simplicity, we assume constant density and $f(\omega)/\sigma_0 = 1$. And thus we have

$$\phi_s(k_R,k_T;\omega) = -k_0^2 \frac{1}{\alpha_R \alpha_T}\gamma_\kappa(k_X,k_Y) \qquad (18)$$

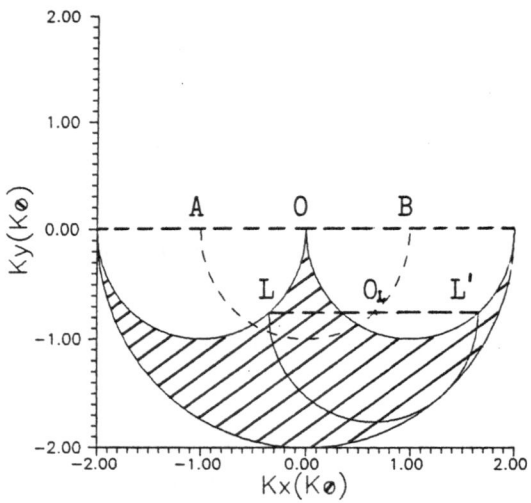

Figure 2. Frequency domain coverage.

In fact, we can reconstruct the two parameters with two sets of data generated at two distinct frequencies[4].

FREQUENCY COVERAGE AND ITERATION PROCESS

The frequency domain coverage obtained from the data collected by the method described in last section is illustrated in Fig.2. The coverage is obtained as follows

$$k_X = k_R + k_T \tag{19}$$

$$k_Y = -(\alpha_R + \alpha_T) \tag{20}$$

For a fixed k_R(and thus α_R), as k_T changes from $-k_0$ to k_0, the locus defined by Eq.(19) and Eq.(20) is a downward semicircle LL' which has a radius equal to k_0 and is centered at $O_L(k_R, -\alpha_R)$. As k_R varies, the center O_L moves along another downward semicircle AB with radius k_0 and center at $(0,0)$. As k_R varies from $-k_0$ to k_0, the set of semicircles covers a shaded region shown in Fig.2.

With such uncomplete coverage, it is obvious we can not get good image. We can refer to the technique proposed by Sato[1]. By means of combining data from the projections with a priori object information using iterative revisions in both image and transform spaces, an accurate image of an object can be achieved[3].

If we assume zeros for the missing points, the quality of the image will suffer. However, the image can be improved by using a priori information about the object. Sometimes we know the outer boundary of the object prior to reconstruction. We may also know the range of variation of the acoustic parameters of interest. This knowledge can be used to eliminate errors caused by the lack of data. Iterative revision between object space and transform space can be carried out using both data and a priori knowledge. A improved image can be obtained by performing the following steps:

1) Choosing a discrete set of transmitter positions and measuring the reflected scattered field $\phi_s(\hat{r}_R, \hat{r}_T; t)$ from the object at each of these positions.

2) Calculating the Fourier transform $\phi_s(k_R, k_T; \omega)$ to obtain $\gamma_\kappa(k_X, k_Y)$.

3) Employing inverse Fourier transform to calculate $\gamma_\kappa(x,y)$.

$$\gamma_\kappa(x,y) = -\int_s k_0^{-2} \alpha_R \alpha_T \phi_s(k_R, k_T; \omega) \exp[i(xk_X + yk_Y)] dk_X dk_Y \tag{21}$$

$$= -\int_s k_0^{-2} \alpha_R \alpha_T \phi_s(k_R, k_T; \omega) \exp[ix(k_R + k_T) - iy(\alpha_R + \alpha_T)] H_1 dk_R dk_T$$

where H_1 is Jacobian,

$$H_1 = abs(\frac{k_T}{\alpha_T} - \frac{k_R}{\alpha_R}) \tag{22}$$

4) Using a priori knowledge about the object to correct $\gamma_\kappa(x,y)$ and getting the first modified image $\gamma_\kappa^{(1)}(x,y)$.

5) Taking the Fourier transform of the first image to obtain values in the region not covered by $\gamma_\kappa(k_X, k_Y)$ in step 2).

6) Repeating steps 3) through 5) in a first iteration loop. Then performing further iteration loops as desired.

Note that in step 2), we get values of $\gamma_\kappa(k_X,k_Y)$ only on the arcs semilar to LL' in Fig.2, from the values of the Fourier transform of each projection. The region which is covered in the Fourier domain is limited. By using iteration loops we can obtain approximate values for the points in the region not covered.

COMPUTER SIMULATION

We present the computer simulations of the reconstructing algorithm. In these simulations we have computed the amplitude of the reflected acoustic field applying Fourier diffraction theorem, in which Born approximation is used. The data contain random noise. The signal to noise ratio is about 30dB. Instead of using ellipse-shaped objects[5], we reconstruct objects with irregular shape.

The Eq.(21) can be discretized using standard digital signal processing methods. We assume the insonifing frequency is k_0. The sampling theorem requires that the sampling intervals in x and y be no greater than

$$\delta = \frac{\pi}{k_0} = \frac{\lambda}{2} \tag{23}$$

where λ is the wavelength of the insonifing wave. In its discrete version, let the object domain in the x-y plane be represented by an N^2 matrix with x and y axes sampled with the same interval δ. We also assume that the transmitting-receiving line is sampled with the same interval. N transmitting locations are used, and for each transmitting site there are N receiving locations on the line. Digital signal processing considerations dictate that if spatial sampling interval is given by Eq.(23), then the sampling interval to be used in the frequency domain should be

$$\Delta = \frac{2k_0}{N} \tag{24}$$

This interval will be used for the sampled representation of k_R and k_T. With such discretization Eq.(21) is rewritten as

$$\gamma_\kappa(m,n) = -\frac{4k_0^2}{N^2} \sum_i \sum_j k_0^{-2} \alpha_i \alpha_j \phi_s(k_i,k_j;\omega) \tag{25}$$

$$\exp[ix_m(k_i+k_j) - iy_n(\alpha_i+\alpha_j)]H_1$$

In our simulations, we get $\phi_s(k_R,k_T;\omega)$ through Fourier diffraction theorem. For an object to be tested, we calculate its discrete Fourier transform $\gamma_\kappa(k_X,k_Y)$. Then we have $\phi_s(k_R,k_T;\omega)$ through Eq.(18). Performing the summation in Eq.(25), we have the reconstructed object image. So we can image irregular objects instead of regular ones.

After we get $\gamma_\kappa^{(1)}(X,Y)$ through step-by-step summation, we revise it with priori-knowledge, and then calculate its Fourier transform through fast Fourier transform(FFT). To avoid truncating error, we use a Hanning window after revision. Now in the frequency domain, the original coverage S_1 and the FFT result S_2 is superposed. Set the FFT values in the originally covered region zero and keep the FFT values out of the region. For the new coverage, we calculate its inverse Fourier transform

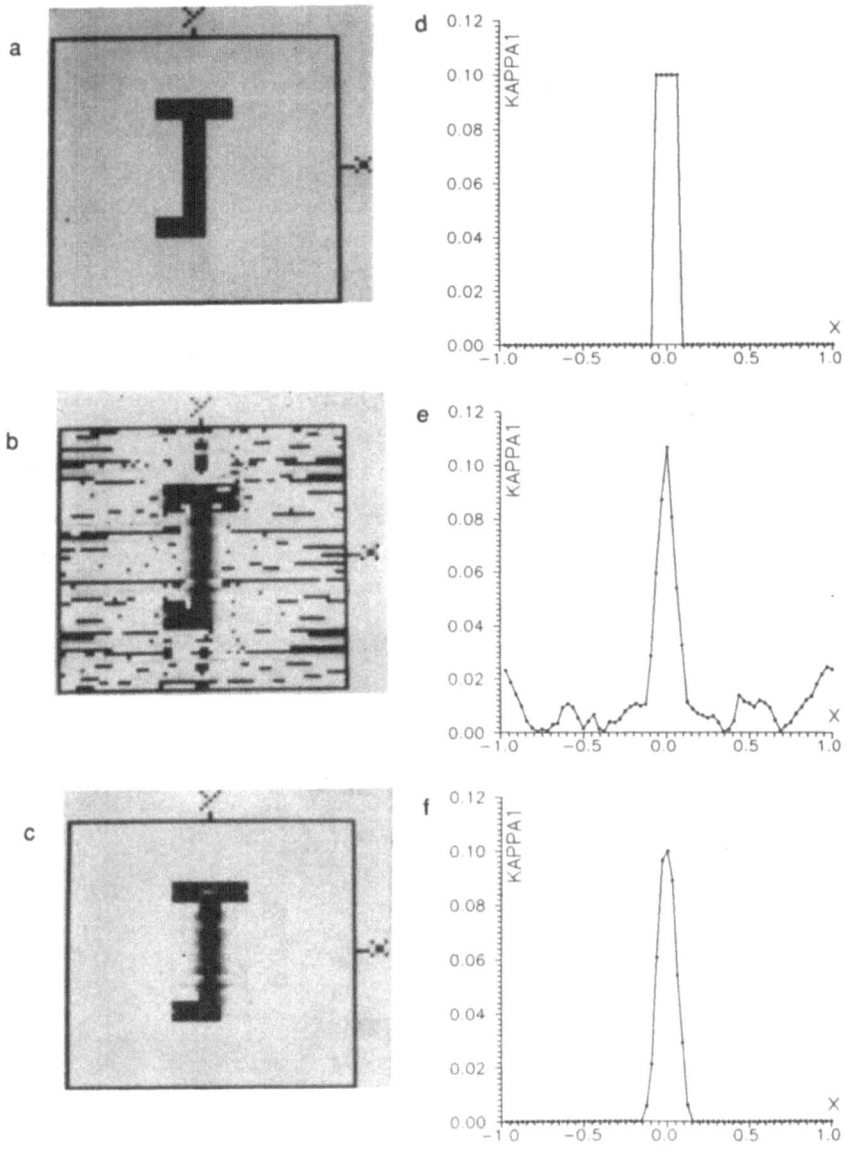

Figure 3. The images for computerized simulations of the algorithm: the object (a), no iteration (b), with five loops of iteration (c). The numerical comparison on the line y=0 through images, (d) is from (a), (e) from (b), and (f) is from (c).

$$\gamma_\kappa(x,y) = \int_{s1+s2} \gamma_\kappa(k_x,k_y) \exp[i(xk_x+yk_y)]dk_x dk_y \qquad (26)$$

$$= \gamma_\kappa^{(1)}(x,y) + \int_{s2} \gamma_\kappa(k_x,k_y) \exp[i(xk_x+yk_y)]dk_x dk_y$$

The second part can be calculated by inverse FFT. And a Hanning window is used to avoid Gibbs phenomena. Then we get $\gamma_\kappa^{(2)}(x,y)$ after one complete loop. Repeating the procedure described above for several times, we can get an accurate image. Fig.3 shows the numerical simulation results of the algorithm presented. We set $\kappa_1 = 0.1\kappa_0$.

CONCLUSION

We have presented a new technique for reflective diffracted tomography that requires only the scan of the transducers on the same side of the object, which is usually the seismic exploration situation. The Fourier diffraction theorem related to this case is derived by using Born approximation. The frequency coverage is analyzed and it is found to be quite limited. We demonstrate that good reconstructed images can be obtained by applying a priori knowledge in an iterative approach.

REFERENCES

1. T.Sato, S.J.Norton, M.Linzer, O.Ikeda and M.Hirama, Tomographic image reconstruction from limited projections using iterative revisions in image and transform spaces, *Appl.Opt.*, 20:395(1981).
2. B.A.Roterts and A.C.Kak, Reflection mode diffraction tomography, *Ultrasonic Imaging*, 7:300(1985).
3. C.Q.Lan, K.K.Xu, and G.Wade, Limited angle diffraction tomography and its application to planar scanning systems, *IEEE Trans.on Sonics Ultrason.*, 32(1):9(1985).
4. A.J.Devaney, Variable density acoustic tomography, *J.Acoust.Soc.Am.*78(1):120(1985).
5. D.Nahamoo, S.X.Pan, and A.C.Kak, Synthetic aperture diffraction tomography and its interpolation-free computer implementation, *IEEE Trans.Sonics Ultrason.*,31(4): 218(1984).

EXACT INVERSE SCATTERING SOLUTIONS IN MULTI-DIMENSIONS

(PERSPECTIVES OF USING IN ACOUSTICAL IMAGING)

V.A.Burov and O.D.Rumiantseva

Department of Physics, Moscow State University
Moscow, 119899, Russia

INTRODUCTION

The aim of the present report is not to propose new algorithms, but to consider perspective and possibilities of practical realization of some algorithms for the solution of the inverse scattering problem being developed lately. These algorithms have been intended for the solution of both the inverse potential scattering problems and, especially recently, for the inverse wave scattering problems (the acoustical problems, in the first place). Some modern functional methods of reconstructing the space scatterer structure are analyzed in the report. The mathematical support for the acoustical ultrasonic tomograph can be an example of one of the most actual application spheres of these methods. The interest for reconstructing the sufficiently strong scatterers is not casual. In medical tomography problems the respective variation of phase velocity is

$\Delta c/c_0 < 5 \times 10^{-2}$ and reconstructed objects sizes are L $\simeq$ 20cm, so that the parameter $(\Delta c/c_0)k_0 L$ characterizing the scatterer power rarely occurs less than 1.

THE MONOCHROMATIC INVERSE SCATTERING PROBLEM

It is the fact that wave fields ψ probing the scatterer $v(r)$ subject to reconstruction are monochromatic that allows to consider both potential and acoustical inverse scattering problems within the general scheme. Then $v(r)$ is present as a functional coefficient in the Helmholtz equation for an inhomogeneous medium:

$$\Delta \psi + k_0^2 \, \psi = v(r) \, \psi,$$

where k_0 is the wave number. In the acoustical problems $v(r)$ carries the information about inhomogeneities of the phase sound velocity, medium density and wave absorption coefficient. For monochromatic methods the background medium, in which the finite scatterer $v(r)$ is located, is assumed to be homogeneous and non absorbing. The initial data to reconstruct $v(\mathbf{r})$ are the scattering amplitude $f(\mathbf{k}, \mathbf{1})$ for all directions of scattered wave in the far zone $\mathbf{1}$ at each fixed wave vector $\mathbf{k}$ of the incident plane wave. Here $\mathbf{k}, \mathbf{1} \in \mathbb{R}^n$, n is the problem dimension (n = 2, 3), $\mathbf{k}^2 = \mathbf{1}^2 = k_o{}^2$. Relations allowing to transit from the inverse problem solution in the physical domain of real vectors $\mathbf{k}, \mathbf{1} \in \mathbb{R}^n$ to "non-physical" domain of complex vectors $\mathbf{k} = \mathbf{k}_R + i\,\mathbf{k}_I$, $\mathbf{1} = \mathbf{1}_R + i\,\mathbf{1}_I$, are the base of these methods. Foundation for such transition is the possibility to generalize the scattering amplitude $f(\mathbf{k}, \mathbf{1})$ of physical wave fields $\psi(\mathbf{r}, \mathbf{k})$ ($\mathbf{k}, \mathbf{1} \in \mathbb{R}^n$) that meets requirements of the Lippman-Shwinger's – type equation

$$\psi(\mathbf{r}, \mathbf{k}) = \exp(i\,\mathbf{k}\,\mathbf{r}) + \int G_o^+(\mathbf{r} - \mathbf{r'}, \mathbf{k})\,\psi(\mathbf{r'}, \mathbf{k})\,v(\mathbf{r'})\,d\mathbf{r'}$$
$$\mathbf{r'} \in \mathbb{R}^n$$

with the classical Green function $G_o^+(\mathbf{r}, \mathbf{k})$ in the complex wave vectors case. The generalized scattering amplitude $h(\mathbf{k}, \mathbf{1})$ ($\mathbf{k}, \mathbf{1} \in \mathbb{C}^n$) corresponds to fictitious wave fields meeting requirements of the analogous equation of the same type, but in the complex $\mathbf{k}$-space with the Green function $G(\mathbf{r}, \mathbf{k})$ introduced by L.D.Faddeev.[1] The limit values of the functions $h(\mathbf{k}, \mathbf{1})$ and $G(\mathbf{r}, \mathbf{k})$ are obtained at $|\mathbf{k}_I| \to 0$ and critically depend on the value of the angle between the vectors $\mathbf{k}_R$ and $\mathbf{k}_I$ if their reciprocal orientation is fixed. In particular, the classical Green functions $G_o^\pm(\mathbf{r}, \mathbf{k} = \mathbf{k}_R)$ corresponding to the coming and delayed potentials are obtained at $|\mathbf{k}_I| \to 0$, if $\mathbf{k}_I$ and $\mathbf{k}_R$ are either collinear or anti-collinear:

$$G_o^\pm(\mathbf{r}, \mathbf{k} = \mathbf{k}_R) = \exp(i\,\mathbf{k}_R\,\mathbf{r})\,\lim_{\varepsilon \to +0} G\left(\mathbf{r}, \mathbf{k} = \mathbf{k}_R \pm i\,\varepsilon\,\frac{\mathbf{k}_R}{|\mathbf{k}_R|}\right);$$

$$f(\mathbf{k} = \mathbf{k}_R, \mathbf{1} = \mathbf{1}_R) = \lim_{\varepsilon \to +0} h\left(\mathbf{k} = \mathbf{k}_R + i\,\varepsilon\,\frac{\mathbf{k}_R}{|\mathbf{k}_R|}, \mathbf{1}_R\right).$$

One-to-one correspondence between the physical and generalized scattering amplitudes $f(\mathbf{k}, \mathbf{1})$ and $h(\mathbf{k}, \mathbf{1})$ takes place.[1,2] On the base of the relations binding these two functions some reconstruction algorithms from the monochromatic data have been developed. So, only such complex vectors $\mathbf{k}$ and $\mathbf{1}$ figure in the algorithms, for which the relations

$$\mathbf{k}^2 = \mathbf{1}^2 = k_o{}^2; \qquad \mathbf{k}, \mathbf{1} \in \mathbb{C}^n \qquad (1)$$

continue to be valid. Because of k_o is real the requirements (1) mean, first, orthogonality of $(\mathbf{k}_R, \mathbf{1}_R)$ and $(\mathbf{k}_I, \mathbf{1}_I)$ respectively and, secondly, a unique relationship between their absolute values:

$$\mathbf{k}_R \perp \mathbf{k}_I, \quad \mathbf{k}_R^2 - \mathbf{k}_I^2 = k_o{}^2; \qquad \mathbf{1}_R \perp \mathbf{1}_I, \quad \mathbf{1}_R^2 - \mathbf{1}_I^2 = k_o{}^2.$$

Asymptotic Methods in the Three-Dimension

To reconstruct the space Fourier transformation $\tilde{v}$ of the scatterer $v(\mathbf{r})$ by means of the relation

$$\tilde{v}(-\mathbf{p}) = \lim_{|\mathbf{k}| \to \infty, \ \mathbf{k}^2 = k_0^2, \ \mathbf{p}^2 = 2\mathbf{k}\mathbf{p}} h(\mathbf{k}, \mathbf{k}-\mathbf{p}) \tag{2}$$

is proposed.[2,3,4] At fixed $\mathbf{k}$, $\mathbf{1} \in \mathbb{C}^n$ the space spectrum $\tilde{v}$ is reconstructed on the space frequency $-\mathbf{p} \in \mathbb{R}^3$, $\mathbf{p} = \mathbf{k} - \mathbf{1} = \mathbf{k}_R - \mathbf{1}_R$. The conditions $\mathbf{k}^2 = k_0^2$ and $\mathbf{p}^2 = 2\mathbf{k}\mathbf{p}$ are equivalent to ones (1) at $\mathbf{k}_I = \mathbf{1}_I$. The equation (2) is the consequence of the Green-Faddeev function property:

$$G(\mathbf{r}, \mathbf{k}) \to 0 \quad \text{at} \quad |\mathbf{k}| = (\mathbf{k}_R^2 + \mathbf{k}_I^2)^{1/2} \to \infty, \quad \mathbf{k}^2 = \mathbf{k}_R^2 - \mathbf{k}_I^2 = k_0^2 , \tag{3}$$

from which $\psi(\mathbf{r}, \mathbf{k}) \to \exp(i \ \mathbf{k} \ \mathbf{r})$. That is why at large $|\mathbf{k}|$ the problem begins to be quasi-Born, just as it takes place for the potential scattering problem when particles energy tends to the infinity.[5] It should be noted that though the property (3) is valid for any n-dimensional space one can not use (2) to reconstruct $\tilde{v}$ at $n = 2$. It results from the fact that the conditions (1) and condition $\mathbf{k}_I = \mathbf{1}_I$ are permissible at fixed $\mathbf{k} \in \mathbb{C}^2$ only for $\mathbf{1} = \pm\mathbf{k}$, i.e. for $\mathbf{p} = 0$, $2\mathbf{k}_R$.

Let's analyze utilization bounds of the algorithm (2) at $n=3$ that are conditioned by the requirement $|\mathbf{k}| \to \infty$. For weak, Born, scatterers the relation (2) is valid even at $\mathbf{k}_I = 0$. As the scatterer power increases the rescattering effects must be necessarily taken into consideration. It is "weakening" of the rescattering effects for which the imaginary addition $\mathbf{k}_I \neq 0$ is needed; the stronger scatterer, the larger $|\mathbf{k}_I|$ being required to reach the asymptotic equality $\tilde{v}(\mathbf{1}-\mathbf{k}) = h(\mathbf{k}, \mathbf{1})$. We have estimated a minimum value $|\mathbf{k}_I| = |\mathbf{k}_I^{min}|$ at which the condition $|\mathbf{k}| \to \infty$ may be considered as a satisfied one. Any scatterer $v(\mathbf{r})$ belonging to the certain class of scatterers with given physical properties can be presented as sum of "large scale" component v_{ls} ($v_{ls} \approx$ const$_r$ within the scattering area R) and "fluctuating" one $v_f(\mathbf{r})$ varying with characteristic linear scale Δx around the value v_{ls}: $v(\mathbf{r}) = v_{ls} + v_f(\mathbf{r})$. Then $|\mathbf{k}_I^{min}|$ must simultaneously meet two requirements:

$$\frac{|v_{lc}|}{k_0 |\mathbf{k}^{min}|} \frac{L}{\lambda_0} \ll 1, \qquad \frac{|v_f|}{k_0 |\mathbf{k}^{min}|} \frac{\Delta x}{\lambda_0} \left(\frac{L}{\Delta x}\right)^{1/2} \ll 1 , \tag{4}$$

where L is the scatterer size, λ_0 is the wave length, $|v_f| = \max_{\mathbf{r} \in R} |v_f(\mathbf{r})|$, $|\mathbf{k}^{min}| = \{ k_0^2 + 2 (\mathbf{k}_I^{min})^2 \}^{1/2}$.

On the other side, the question of the noise-resistance stability of the scheme (2) has been researched. The noise-resistance is defined by necessity to recalculate the function $h(\mathbf{k}, \mathbf{l})$ ($\mathbf{k}, \mathbf{l} \in \mathbb{C}^3$) proceeding from measured experimentally $f(\mathbf{k}, \mathbf{l})$ ($\mathbf{k}, \mathbf{l} \in \mathbb{R}^3$) by means of equations presented in the article,[2] for example. The implemented analysis [6] has shown that the stable reconstruction of $h(\mathbf{k}, \mathbf{l})$ (and $\tilde{v}$ from (2)) was possible only under the condition

$$\left|\mathbf{k}_I\right| L \lesssim 1 . \tag{5}$$

And in the case of $\left|\mathbf{k}_I\right| L \gg 1$ the errors of the function $h(\mathbf{k}, \mathbf{l})$ and scatterer estimation increase according to the exponential law.

Thus, the algorithm (2) is effectively applicable only for comparatively weak scatterers with sufficiently small wave sizes L/λ_0 and not great deviation of wave parameters, while the conditions (4) and (5) for $\mathbf{k}_I = \mathbf{k}_I^{min}$ are still compatible:

$$\left|v_{ls}\right| \ll k_0/L ; \qquad \left|v_f\right| \ll k_0 / (L \, \Delta x)^{1/2} .$$

The more scatterer size, the less its power must be.

The data redundancy of dimensional type in the three-dimensional inverse scattering problem is due, in the present case, to the fact that the whole multitude of the vectors pair $\mathbf{k}$ and $\mathbf{l}$ correspond to the same vector $\mathbf{p} = \mathbf{k} - \mathbf{l}$. This fact allows to increase the stability threshold of the solution $\tilde{v}$ in $(\left|\mathbf{k}_I^{min}\right| L)^{1/2}$ times. However, the exponential character of the reconstruction errors growth remains dominating as before and the noise influence is very active. Nevertheless, the described character of the solution instability is, in the present case, a shortcoming of the concrete algorithm (2) and does not reflect the specific character of the three-dimensional monochromatic inverse scattering problem.

The Two-Dimensional Inverse Monochromatic Inverse Scattering Problem and Its Stability

One-to-one reflection of two-dimensional complex vectors $\mathbf{k}$ and $\mathbf{l}$ on the complex plane allows to use Cauchy's formula and the Riemann non-local problem formalism for creating the approximate and exact schemes of the mean power scatterer reconstruction.

The approximate algorithm. This algorithm proposes the reconstruction of two-dimensional scatterers which create such a small back scattering of inhomogeneous waves for any probing direction that it may be ignored,[7] i.e.

$$h(\mathbf{k}, \mathbf{l} = -\mathbf{k}^*) \equiv 0 \qquad \forall \, \mathbf{k} \in \mathbb{C}^2 . \tag{6}$$

The approximate scheme connects in a linear manner the scatterer $v(\mathbf{r})$ at any fixed point $\mathbf{r} = \{x, y\}$ with a function $\mathcal{K}(\mathbf{r}, \varphi)$ characterizing the generalized inner wave field:

$$v(\mathbf{r}) = -\frac{k_0}{2\pi} \left(i \, \frac{\partial}{\partial x} + \frac{\partial}{\partial y} \right) \int_0^{2\pi} \mathcal{K}(\mathbf{r}, \varphi) \, \exp(i \, \varphi) \, d\varphi . \tag{7}$$

According to its physical sense the function $\mathcal{K}(\mathbf{r}, \varphi) = \mathcal{K}(\mathbf{r}, \mathbf{k}_R)$ is a difference between two limit values of the fictitious field $\psi(\mathbf{r}, \mathbf{k})$ at $\mathbf{k}_I \to 0$, $\mathbf{k}_I \perp \mathbf{k}_R$; these limit values correspond to two orientations $\mathbf{k}_I$ with respect to $\mathbf{k}_R$.[8] Therefore an estimation $\mathcal{K}(\mathbf{r}, \varphi)$ doesn't demand the calculation of all values $h(\mathbf{k}, \mathbf{l})$ ($\mathbf{k}, \mathbf{l} \in \mathbb{C}^2$), but only of the two limit values of this function. The function $\mathcal{K}(\mathbf{r}, \varphi)$ is ultimately estimated on the base of the integral equations, in which the physical scattering amplitude $f(\mathbf{k}, \mathbf{l})$ ($\mathbf{k}, \mathbf{l} \in \mathbb{R}^2$) comes into both the operator and the right side. It is very important that the researched reconstructed scheme taking into account multiple scattering is, at the same time, linear with regard to $v(\mathbf{r})$. We have found out that a mechanism of such nonlinear members compensation is provided by two factors: 1) the restriction of the space scatterer spectrum width, the more rigid, the stronger scatterer:

$$\tilde{v}(\mathbf{q}) = 0, \qquad |\mathbf{q}| \geq \frac{2 \, k_0}{m} \qquad (8)$$

(m is the number of scattering acts taken into consideration), what is equivalent to the requirement (6); 2) special properties of the generalized Green-Faddeev function $G(\mathbf{r}, \mathbf{k})$ at $\mathbf{k} = \mathbf{k}_R + i \, \mathbf{k}_I$, $\mathbf{k}_R \perp \mathbf{k}_I$, $\mathbf{k}_I \to 0$.[8]

The analyzed approximate algorithm proves to be greatly perspective for utilization in practice because of sufficiently high noise-resistance of the solution and the comparatively non great member of computing operations: scatterer reconstruction in N^2 space points demands the order of N^4 operations.[6,8] In this connection the algorithm has been comprehensively researched.[9] During reconstruction of the scatterer with the finite size as its power grows and the space spectrum expends the certain errors prove to occur because the compensation mechanism loses capacity for work. This fact results from the appearance in the secondary sources' space spectrum components lying outside the $2k_0$-radius circle. In view of this, the simple procedure is proposed to stabilize the solution and reduce natural noises influence to the minimum. Also, the value diapason of monochromatic wave frequency has been founded, in which reconstruction proves to be the most effective. For biological tissues it lies in the limits from dozens kHz to a few MHz, what coincides with the frequency diapason being really applied in medical diagnostics. Waves absorption being negligible small in background, the algorithm is valid for dissipative scatterers as well.

The exact reconstruction algorithm. The exact algorithm for two-dimensional mean power scatterers with an arbitrary space spectrum $\tilde{v}$ has also been developed.[10] For these scatterers the requirement (8) is violated in principle. Then the procedure of the reconstruction $v(\mathbf{r})$ demands calculations from $f(\mathbf{k}, \mathbf{l})$ ($\mathbf{k}, \mathbf{l} \in \mathbb{R}^2$) the function $h(\mathbf{k}, \mathbf{l} = -\mathbf{k}^*)$ ($\mathbf{k} \in \mathbb{C}^2$), responsible for back scattering the inhomogeneous waves. This circumstance leads to sharp decreasing the noise-resistance of the solution according to the requirement (5), as well as for the asymptotic method (1) of the three-dimensional scatterer reconstruction. For example, even in the Born approximation the reconstruction stability is provided for scatterers, whose space spectrum is mainly concentrated within the $2 (k_o^2 + L^{-2})^{1/2}$-radius circle. The correction $L^{-2} \ll k_o^2$ because of $L \gg \lambda_o$ in tomographic experiments. There is an evident analogy with the analytical continuation of spectrum of a function having a finite support: the function spectrum being known in the certain frequency domain, it can be defined for all other frequencies. However, the analytical continuation operation has a very strong instability. For this reason the scatterer spectrum extension is in practice possible only for one or two sampling points.

As the power of the scatterer having a wide space spectrum increases, the instability of the scatterer reconstruction also increases, because a fine scale structure of the secondary sources and their wave fields within the scattering domain becomes more complicated. The appearance of such structure results from the fact that at every rescattering act the secondary sources spectrum is defined by convolution of the scatterer spectral density with the same inner field spectral density.[8]

The higher the order of the rescattering, the more complicated is the space structure of the inner wave field. Consequently, the instability of the inner wave field reconstruction also becomes stronger, because the complicated structure of this field is feebly shown in the observed external wave field structure. While increasing the scatterer power, the restriction (8) on its space spectrum becomes more rigid. A similar situation occurs for the iterative methods of solution too.[8]

Thus, for the inverse two-dimensional problems at monochromatic observation regime two classes of scatterers yield to reconstruction in the best manner. The first class is presented by weak scatterers which may possess sufficiently wide space spectrum. The second one - by mean power scatterers with non complicated smooth space structure providing the localization of the main space spectrum part in the narrow diapason. At the same time, the reconstruction with satisfactory accuracy of two-dimensional contrasting and complicated scatterers is not really realized in the monochromatic regime due to the redundancy absence of the problem. For these scatterers reconstruction the redundancy scattering data given by the polychromatic observation regime are required.

The Three-Dimensional Monochromatic Inverse Scattering Problem

For practice the case of the three-dimensional scatterer is the most interesting. It is important that instability effects will appear weaker (even when the inhomogeneous waves back scattering is available) thanks to redundant scattering data provided by the problem dimensions. This fact could permit to essentially widen in comparison with the two-dimensional monochromatic problem the class of scatterers, for which a satisfactory reconstruction is practically realizable. In this connection the generalization of algorithms being considered on the three-dimensional space would be important. The generalization is not trivial as the direct one-to-one reflection of the three-dimensional wave vectors $\mathbf{k}$ and $\mathbf{l}$ on a complex plane does not exist. Structure of a functional algorithm to reconstruct the three-dimensional scatterer in monochromatic regime can be depicted so:[3,4,11]

$$\tilde{v}(-\mathbf{p} = \mathbf{l}-\mathbf{k}) = h(\mathbf{k},\mathbf{l}) + Z_1\{ h(\mathbf{k},\mathbf{l}) \big|_{\mathbf{k}_I \neq 0}\} + Z_2\{ h(\mathbf{k},\mathbf{l}) \big|_{\mathbf{k}_I \to 0}\}. \quad (9)$$

Here $h(\mathbf{k}, \mathbf{l})$ is the "Born" member giving the estimation $\tilde{v}(-\mathbf{p})$ in the weak scatterer case; Z_1 and Z_2 are correcting members taking into consideration the waves rescattering. The member Z_1 proposes to integrate the nonlinear combination of the functions h(k, 1) over the multitude ("volume") of complex vectors k + 1 ($1_I = k_I$, $k_I \perp k_R$, $1_I \perp 1_R$) at $k_I \neq 0$; the number Z_2 is formed by the linear combination of the limit values h(k, 1) at $k_I \to 0$. In this respect the algorithm (9) has something in common with the exact two-dimensional algorithm discussed above. To successfully solve the problem by this method (9) and to appreciate its possibilities one must find a connection between the main scatterer parameters (its size, power, space spectrum width) and the volume of the complex vectors domain, in which the generalized scattering amplitude h(k, 1) must be preliminary evaluated, as well as it has already taken place in the exact two-dimensional algorithm. The question, for which scatterers the "volume" member Z_1 may be ignored is turning, thereby, into some analogue of the approximate (simplified) algorithm (7) and also remains unanswered for a while.

THE INVERSE SCATTERING PROBLEM IN PULSE REGIME

During the last years an aspiration to developed methods suitable for solution of different physical nature problems has been observed. So, a formal generalizations of the Newton-Marchenko equation[5] has been got to solve the multi-dimensional acoustical problems of any dimension in a near-field.[12,4] Impulse observation regime is proposed, what is very perspective from practical point of view. In the three-dimensional case, for example, this equation looks so:

$$G^+(t,\mathbf{r},\mathbf{y}) = \int_S d^2\hat{z}\,|z|^2 \int dt'\left(G^+(t'-t,\,\mathbf{z},\mathbf{r})\ \frac{\partial}{\partial n_z}G^+(t',\mathbf{z},\mathbf{y}) - \right.$$

$$\left. - \frac{\partial}{\partial n_z}G^+(t'-t,\,\mathbf{z},\mathbf{r})\ G^+(t',\mathbf{z},\mathbf{y}) \right), \qquad t \geq \tau; \tag{10}$$

$$G^+(t,\mathbf{r},\mathbf{y}) = 0, \qquad t < \tau. \tag{10a}$$

Here $d^2\hat{z}$ is an element of a solid angle ($|\hat{z}|=1$); S is a surface surrounding the finite scattering domain R; $G^+(t, \mathbf{r}, \mathbf{y})$ is the delayed Green function for medium containing a scatterer $\varepsilon(\mathbf{r})$:

$$\left(\frac{\partial^2}{\partial r^2} - \left(\frac{1}{c_0^2} + \varepsilon(\mathbf{r}) \right) \frac{\partial^2}{\partial t^2} \right) G^+(t,\mathbf{r},\mathbf{y}) = -\,\delta(t)\,\delta(\mathbf{r}-\mathbf{y})\ . \tag{11}$$

While obtaining the equation (10) the causality principle (10a) and analogous one $G^-(t, \mathbf{r}, \mathbf{y}) = 0$, $t > \tau$ for the coming Green function have been critically used. Delay time $\tau \leq |\mathbf{r}-\mathbf{y}|/\,c_m$, where $c_m = \max\limits_{\mathbf{r} \in \mathbb{R}^3} c(\mathbf{r})$.

Thus, the equation (10) allows to connect the total field $G^+(t, \mathbf{r}, \mathbf{y})$

inside the domain $r \in R$ with the scattering data $G^+(t, r, y)$ got by means of point sources $y \in S$ and point receivers $z \in S$. Reconstruction $G^+(t, r, y)$ is equivalent to that of the scatterer $\varepsilon(r)$ itself.

Another form of recording (10) is to bring (10) to the Fredholm equation of the second kind by subtracting from (10) the similar equation for the Green function of the background G_0^+:

$$G_{sc}^+(t,r,y) - \int_S d^2\hat{z} |z|^2 \int dt' \left(G_{sc}^+(t'-t, z,r) \frac{\partial}{\partial n_z} G^+(t',z,y) - \right.$$

$$\left. - \frac{\partial}{\partial n_z} G_{sc}^+(t'-t, z, r) \; G^+(t', z, y) \right) =$$

$$= \int_S d^2\hat{z} |z|^2 \int dt' \left(G_0^+(t'-t, z, r) \frac{\partial}{\partial n_z} G_{sc}^+(t', z, y) - \right.$$

$$\left. - \frac{\partial}{\partial n_z} G_0^+(t'-t, z, r) \; G_{sc}^+(t', z, y) \right), \qquad t \geq \tau; \qquad (12)$$

$$G_{sc}^+(t, r, y) = 0, \qquad t < \tau.$$

The function $G^+(t, z, y)$ being measured and all functions at the right side of (12) being known, the scattered field $G^+(t,r,y) - G_0^+(t,r,y) \equiv G_{sc}^+(t,r,y)$ is unknown in the record (12). Little is known about the solution of the equations (10) or (12) in acoustical case, growth of scattering power as frequency increases making this question far more complicated. Because of this, an important practical problem is the possibility to use the Newton-Marchenko equation in the case of probing fields differing from δ-pulses. The analysis made has shown that the equations (10) or (12) may be modified to the case of *probing pulses with a finite duration* τ_i *and an arbitrary spectrum* $A(\omega)$. Multiplication of ω-space analogue of equation (10) or (12) by $A^2(\omega)$ and the following t-presentation return lead to equation of the type (10) or (12), in which the first item $G^+(t,r,y)$ is replaced by the function $G_{A^2}^+(t,r,y)$ with the spectrum $A^2(\omega)$, and all other integrand functions at the left and right sides - by the corresponding functions G_A^+ with the spectrum $A(\omega)$. If $\tau_i < 2|r-y|/c_m$, i.e. the surface S is moved away from the scattering domain R to a certain distance, then the obtained integral equation remains exact and the causality principle of the type (10a) takes place for $t < \tau \leq \{|r-y|/c_m - \tau_i/2\}$. However, if there are frequencies ω, the respective contribution from which to the spectrum $A(\omega)$ is little, then the reconstruction procedure of the inner field becomes ill-posed because of the necessity to reconstruct the functions G_A^+ by means of the function $G_{A^2}^+$ during solution.

In the simplest case of a rectangle spectrum $A(\omega)$ removing the ill-posed solution problem the destruction of the exact limitation of the duration introduces an approximation to the equation of the type (10) or (12). It is a consequence of the fact that the probing signal duration becomes theoretically infinite and superposition of the time supports of the functions $G^+_{A_2}(t,r,y)$ and $G^-_{A_2}(t,r,y)$ appears. Then for a chosen delay τ_i the error in the equation of the type (10) or (12) is conditioned by neglect of this superposition.

To solve the equation of the type (10) for the finite duration pulse is possible by iterative method; here as an initial guess of unknown functions at the right side of (10) their values for background space are accepted. To solve the equation of the type (12) is possible both iteratively and directly. In any case, the numerical solution demands to make these equations algebraic and to estimate *the number of computing operations*. With such a situation the right choice of sampling steps defined by widths of the temporal and space spectra of the inner field $G^+(t,r,y)$ is important.[9] Let for the two-dimensional problem the number of sampling points for the variable t is N_t, for the scattering domain $r \in R$ - N^2_x, for the sources domain $y \in S$ - $N_y \simeq N_x$, for the receivers domain $z \in S$ - $N_z \simeq N_x$. Note, that when rescatterings are being taken into consideration the maximum space sampling step of the inner field must be finer than that of the scatterer itself. It results from widening the space spectrum of the secondary sources in comparison to that of the scatterer. Then the solution of the discrete analogue of the equation (12) demands $\simeq N^2_t N^4_x$ operations (in the three-dimensional problem $\simeq N^2_t N^7_x$ of those). However, it is expediently to solve (12) in the frequency ω-presentation,[12] which demands $\simeq (\log N_t) N_t N^4_x$ operations in the two-dimensional problem and $\simeq (\log N_t) N_t N^7_x$ operations in the three--dimensional one. So, sufficiently powerful computers being available, the solution of the three-dimensional inverse scattering problem for not very complicated scatterer structure ($N_t \simeq N_x \simeq 10 \div 30$) by the method (10) or (12) is real.

Looking for the inner field $G^+(t, r, y)$ simultaneously requires solving the algebraic equation (10) or (12) for all sampling time points, for which the scattered signal is present. Nevertheless, the scatterer $\varepsilon(r)$ itself may be evaluated from the solution $G^+(t,r,y)$ at any one fixed parameter y and one parameter t. There are many formulae of evaluation,[5,12,13] for example, directly from the wave equation (11):

$$\varepsilon(r) = -\frac{1}{c_0^2} + \frac{\dfrac{\partial^2}{\partial r^2} G^+(t, r, y)}{\dfrac{\partial^2}{\partial t^2} G^+(t, r, y)} \quad , \quad r \in R. \tag{13}$$

Yet all these formulae are proposed without any connection with statistic aspects of the problem, which should be necessarily taken into account in practice. By this reason the detailed research problem of the statistic dependence of the inner field estimation errors and the scatterer estimation errors connected with them arises. This problem is so much nontrivial that the relationship between these two kinds of errors is nonlinear (see (13)). Absence of noises correlation in the initial scattering data results in the errors of the reconstructed inner field $G^+(t,r,y)$ for different sources points $y \in S$ to be also slightly correlative between themselves. This fact allows to make the secondary statistic treatment of the scatterer estimations obtained for different values y and t. Therefore, a combined use of all values $G^+(t, r, y)$ increases exactness of statistical scatterer estimation, i.e. it leads to a stability growth in the final scatterer estimation.

In spite of the estimations having been obtained above the detailed analysis of relationship and reciprocal influence of the inverse scattering problem parameters (the scatterer size, its power, space spectrum width and form) and parameters of the scattering data obtaining scheme (the middle frequency and duration of probing pulse, the number of points of sources and receivers, character of space and time sampling of intermediate solutions for the inner field) remains actual. For the present one can only say that as the scatterer power increases and the inner field space spectrum widens, the number of independent scattering data (i.e. the number of independent sampling points over y, z, r, as well as over t due to the increase of the scattered field duration) grows. It results in catastrophic growth of the computing operations number, but at the same time, the redundancy number of the intermediate data (the inner field) increases, that improves the final scatterer estimation. Besides, the decrease of the probing pulse duration diminishes a domain volume within the scatterer making a simultaneous contribution in the scattering processes; this fact limits intensity and width of the inner scattered field space spectrum. Finally, in the problems of medical diagnostics, geoacoustics and so on the total set of the scattering data (i.e. for all points of sources and receivers) is not succeeded to obtain always. Because of this the restoration of the missing scattering data by the characterization equations becomes actual.

REFERENCES

1. L.D.Faddeev, Inverse problem of quantum scattering theory. II ",
   in: "Itogi Nauki i Techniki, Sov. Prob. Mat.",
   R.V.Gamkrelidze, ed., VINITI, Moscow (1974) [in Russian].
   (Transl. *J. of Soviet Math.* 5:334 (1976)).
2. R.G.Novikov, Multi-dimensional inverse spectral problem for
   the equation $-\Delta\psi + (v(x) - E u(x))\psi = 0$. *Func. Analysis
   and Its Applic.* 22:263 (1988).
3. A.I.Nachman, Reconstructions from boundary measurements, *Annals
   of Math.*, 128:531 (1988).
4.   M.Cheney , A review of multidimensional inverse potential
   scattering, in: "Inverse Problems in Partial Differential
   Equations", D.Colton, ed., SIAM, Philadelphia (1990).
5. R.G.Newton, The Marchenko and Gel'fand methods in the inverse
   scattering problem in one and three dimensions, in: "Conf.
   on Inverse Scattering: Theory and Appl.", J.B.Bednar, ed.,
   SIAM, Philadelphia (1983).
6. V.A.Burov, and O.D.Rumiantseva, The solution stability and
   restrictions on the space scatterer spectrum in the
   two-dimensional monochromatic inverse scattering problem, in:
   "Ill-Posed Problems in Natural Sciences", A.N.Tikhonov, ed.,
   TVP, Moscow (1992).
7. R.G.Novikov, Reconstruction of two-dimensional Schrödinger
   operator from the given scattering amplitude at fixed energy,
   Theoret. and Mathem. Physics, 66:154 (1986).

8. V.A.Burov, and O.D.Rumiantseva, The functional-analytical methods
   for the scalar inverse scattering problems, *in*: "Anal.
   Methods for Opt. Tomogr.", G.G.Levin, ed., Proceed. SPIE (1992).
9. V.A.Burov, and O.D.Rumiantseva, Solution of two-dimensional
   acoustical inverse scattering problem based on functional-
   -analytical methods II. Effective application domain, *Acoust.
   Phys.*, 39:419 (1993).
10. R.G.Novikov, The two-dimensional inverse scattering problem
    at fixed positive energy and the Riemann generalized non-local
    problem, *C.R.Acad.Sci.Paris*, 312:675 (1991).
11. R.G.Novikov, and G.M.Henkin, $\bar{\partial}$-equation in multi-dimensional
    inverse scattering problem, *Russian Mathem. Survey*, 42:109 (1987).
12. D.Budreck, and J.H.Rose, Three-dimensional inverse scattering in
    anisotropic elastic media, *Inverse Problems*, 6:331 (1990).
13. M.Cheney, G.Beylkin, E.Somersalo, and R.Burridge, Three-dimensional
    inverse scattering for the wave equation with variable speed:
    near-far formulae using point sources, *Inverse Problems*,
    5:1 (1989).

RECONSTRUCTING THE SHAPE OF AN OBJECT IN AN ISOTROPIC SOLID

V.A.Burov   and I.P.Prudnikova

Department of Physics, Moscow State University
Moscow, 119899, Russia

INTRODUCTION

The problem of shape determination of inclusion in solid is actual for nondestructive ultrasound control of constructions and materials, and belongs to the class of inverse problems. At present the most success is achieved in solving of inverse scattering problems for scalar waves, that is in shape determination of boundary scatterer in fluid from the scattered field measured outside the object. Optimization methods worked out permit to take into account rescattering of high orders by computational means[1,2,3,4].

The method of scatterer shape reconstructing in solid introduced in the report is a generalization of optimization method for determination the shape of boundary inclusion in fluid[4,5]. The key point of this method is an approximation of scattered field by field of sources located on some given surface inside the object. Such approximation permits to get relatively simple integral equations connecting scattered field and object shape function. At all stages of the solution of the problem all kinds of wave motion and their transformations are automatically taken into account. This circumstance has a particular importance for the problem of cavity shape determination in solid so far as rise of surface waves on the object boundary and their rescattering call forth the possibility of reconstruction of shadow side of object. In case of scattering in fluid the reconstruction is possible only when multi directional scattering data is used.

STATEMENT OF THE INVERSE SCATTERING PROBLEM FOR CAVITY IN SOLID

This section contains derivation of the integral relation between observed field scattered by cavity in isotropic infinite elastic media and characteristic function $\gamma(\mathbf{r})$ describing shape of the object. Then incorrect inverse problem is solved by means of optimization method based on this relation.

Let cavity with smooth boundary (in sense of continuity of the first derivatives) occupies domain D with boundary ∂D in homogeneous solid. The

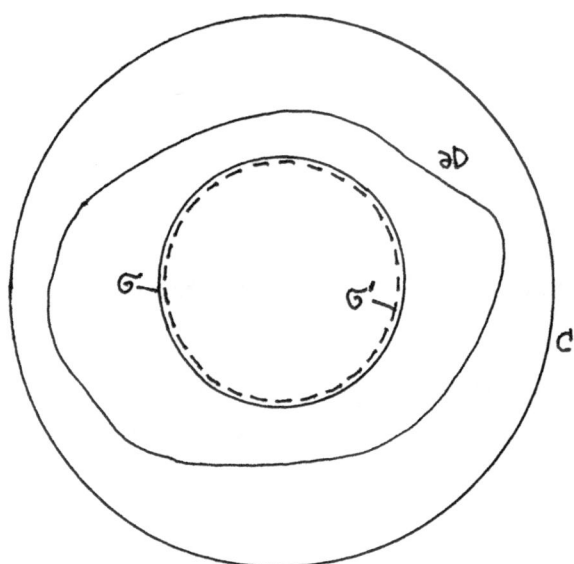

Figure 1.Configuration of the area of object boundary
searching.

obstacle is sounded by plane monochromatic longitudinal ultrasound wave
with displacement vector $\mathbf{u}(t,\mathbf{r}) = \mathbf{A}\ e^{\ i(k_o\mathbf{r}-wt)}$. The field of
displacements satisfies wave equation:

$$(\lambda +\mu)\partial_j\partial_i u_i(\mathbf{r})+ \mu\partial_i\partial_i u_j(\mathbf{r}) + \rho\omega^2 u_j(\mathbf{r}) = -\rho f_j(\mathbf{r}); \tag{1}$$

where ρ denotes the density of elastic media, $\mathbf{f}(\mathbf{r})$ - vector of incident
field sources body force, λ and μ - Lame parameters of media.

For the problem in such statement an integral equation can be
obtained analogously to integral Green's formula for internal domain V
with boundary S for scalar waves[6] :

$$\int_S \left\{ \mathbf{t}(\mathbf{r}')\cdot G(\mathbf{r}/\mathbf{r}')\ -\mathbf{u}(\mathbf{r}')\cdot[\ \hat{\mathbf{n}}'\cdot\Sigma(\mathbf{r}/\mathbf{r}')]\right\} dS' +$$

$$\int_V \rho\ \mathbf{f}(\mathbf{r}')G(\mathbf{r}\ /\mathbf{r}')\ d\mathbf{r}' = \left\{ \begin{array}{ll} \mathbf{u}(\mathbf{r}) & ,\mathbf{r}\in V; \\ 0 & ,\mathbf{r}\notin V; \end{array} \right. \tag{2}$$

where notations are used: $\mathbf{t}(\mathbf{r})$ - tension vector, $G(\mathbf{r}/\mathbf{r}')$ - Green's tensor
of the second rank, $\Sigma(\mathbf{r}/\mathbf{r}')$ - Green's tensor of the third rank, defined
by equation:

$$\Sigma_{lmn} = \lambda\ \delta_{lm}\partial_k G_{kn} + \mu\ (\ \partial_l G_{mn} + \partial_m G_{lm}); \tag{3}$$

In order to apply (2) to problem discussed, we turn now to the
consideration of the theorem on approximation of scattered field by

antenna potential[7] . The theorem says that density of secondary sources, placed on same arbitrary smooth curve or surface inside the object can be chosen so that the field induced outside the object is closed to the field scattered by real boundary with any given accuracy. At this point necessary geometrical constructions must be carried out (for the sake of simplicity, two dimensional problem will be further discussed). Let secondary sources, approximately describing scattered field outside D, to be located on the known circumference σ' and characterized by density $F(r)$. Let also the position of another two circumference to be a priori known: the first one, C contains boundary of the object inside it, and the other, σ , on the contrary, is internal respectively to boundary ∂D looked for. The curve of secondary sources σ' lies inside σ. All this circumferences have common center (see fig. 1). Further domain R between C and σ is named area of object boundary searching. Scattering from the localized boundary inhomogeneity represented by (2) relative to domain D/V_σ (where V_σ is an area bounded by circumference σ) is given by equation:

$$\int_{D/V_\sigma} \nabla \cdot (\ u(r') \cdot \Sigma(r/r')\ )\ dr' = \int_\sigma t(r') \cdot G(r/r')\ dS';$$
(4)

For treating the problem of scattering from boundary inclusion, it is convenient to describe shape of the object by characteristic function $\gamma(r)$. In accordance with configuration shown at fig.1, $\gamma(r)$ is equal to unit inside D/V_σ and zero outside it. Then, if $\gamma(r)$ is involved in the left integral in (4), one can consider the integration limits given beforehand, that is, instead of unknown domain D/V_σ, area of boundary searching R is used:

$$\int_R \gamma(r')\ \nabla \cdot (\ u(r') \cdot \Sigma(r/r')\ )\ dr' = -\int_\sigma t(r') \cdot G(r/r')\,dS';$$
(5)

Equation (5) connects the total field $u(r')$ and $t(r')$ inside R with the shape function of the object $\gamma(r)$ looked for. The possibility to obtain field $u(r')$ and $t(r')$ inside R from observed one is included in method of representation of boundary scattering by secondary sources, introduced above. Namely, the field inside is valued by means of scattered field approximation by antenna potential $F(r)$, placed on beforehand known circumference σ'. Thus, equation (5) can be considered as relation between magnitudes γ to be find and observed field.

Removal of secondary sources of scattered field from boundary into the obstacle allows to operate mathematically with physically nonexisting scattered field inside.

If center of polar framework is placed in common center of σ, σ' and C, density of scattered field sources can be written in form:

$$F(r) = F(\varphi)\ \delta(r-r_0);$$
(6)

where r_0 - radius of circumference of secondary sources localization σ'. So scattered field in any point characterized by radius-vector $|r| > r_0$, is represented by:

$$u(\mathbf{r}) = \int_0^{2\pi} G(\mathbf{r},\mathbf{r}_0,\varphi)\ F(\varphi)\ r_0\ d\varphi; \tag{7}$$

In order to accomplish transformation (7) one should separate variables r and φ in Green's tensor, and then realize integration on φ. It can be done by expanding Green's tensor in series based on the system of longitudinal and transverse vector wave functions:

$$\begin{cases} \psi_i^n\ (\mathbf{r}) = \dfrac{\varepsilon_n}{2}\ \nabla_i H_n^{(1)}\ (k_p r)\begin{Bmatrix} \cos n\vartheta \\ \sin n\vartheta \end{Bmatrix}\ , & \mathbf{r} = (r,\vartheta) \\[3mm] \phi_i^n\ (\mathbf{r}) = \dfrac{\varepsilon_n}{2}\ (\nabla \times \hat{\mathbf{z}})_i H_n^{(1)}(k_s r)\begin{Bmatrix} \cos n\vartheta \\ \sin n\vartheta \end{Bmatrix}, & \varepsilon_n = \begin{cases} 1,\ n=0 \\ 2,\ n\neq0 \end{cases} \end{cases} \tag{8}$$

Here operator $(\nabla\times\hat{\mathbf{z}})_i$ denotes i-th component of vector product of operator ∇ and unit vector on axis OZ perpendicular to the plane (r,φ); $H_n^{(1)}(x)$ – Hankel function of the first kind, order n. Then expansion of tensor G can be written in well-known form:

$$G_{ij}(r,\vartheta,r_0,\varphi) = \sum_{n=0}^{\infty} \psi_i^n\ (\mathbf{r}_>)\ \mathrm{Re}\ \psi_j^n\ (\mathbf{r}_<)\ +\ \sum_{n=0}^{\infty} \phi_i^n(\mathbf{r}_>)\ \mathrm{Re}\ \phi_j^n(\mathbf{r}_<); \tag{9}$$

To emphasize separate dependence of tensor G from r and φ, one can rewrite (9) in suitable manner:

$$G_{ij}(r,\vartheta,r_0,\varphi) = \sum_{n=0}^{\infty} \frac{\varepsilon_n}{2}\ A_{ij}^n(r,r_0)\ \Phi_{ij}^n(\varphi,\vartheta); \tag{10}$$

Indices $i,j = 1,2$ - corresponds to polar coordinates r and φ. There is no summing up on indices i and j on the right of the expression. Angular component of tensor G in (10) is given by:

$$\Phi_{ij}^n = \begin{pmatrix} \cos n(\vartheta-\varphi) & \sin n(\varphi-\vartheta) \\[2mm] \sin n(\vartheta-\varphi) & \cos n(\vartheta-\varphi) \end{pmatrix}; \tag{11}$$

The explicit form of functions Φ_{ij}^n allows to obtain system of linear equations separately for each number of harmonic n, connecting Fourier coefficients of secondary sources vector density

$$^nF_j^{c,s} = \frac{1}{\pi} \int_0^{2\pi} F_j(\varphi) \begin{pmatrix} \cos n\varphi \\ \sin n\varphi \end{pmatrix} d\varphi; \tag{12}$$

and Fourier coefficients of scattered field on fixed circumference with radius r:

116

$$
{}_{n}u_i^{c,s} = \frac{1}{\pi} \int_0^{2\pi} u_i^{scat}(r,\vartheta) \begin{pmatrix} \cos n\vartheta \\ \sin n\vartheta \end{pmatrix} d\vartheta \; ; \qquad (13)
$$

Indices s and c on the left denote reference to sin- or cos-coefficient of expansion, and index "scat" on the right - to scattered field. It is convenient to introduce four-component vectors F^n and u^n instead of set of Fourier coefficients of harmonic n (the set consists of four elements $\{F_i^{c,s}\}^n$, $\{u_i^{c,s}\}^n$, $i = r, \varphi$). Now the set of systems of equations can be written in matrix form:

$$
u^n(r) = {}^n\hat{A}(r) \, F^n \; ; \qquad n = 0,\ldots,\infty \; ; \qquad (14)
$$

Matrix ${}^n\hat{A}(r)$ includes radial components of tensor G from (10) and depends on radius of circumference where scattered field is determined and radius of secondary sources circumference. Each system of equations (14) permits to obtain components of the n-th scattered field harmonic from harmonic of the secondary sources density with the same number, or vice versa. Let scattered field to be measured on the circumference with radius R, and Fourier coefficients of its expansion in series to be calculated. Then Fourier coefficients of scattered field on any other radius r ($|r| > r_0$) can be determined separately for each harmonic number n:

$$
u^n(r) = {}^n\hat{A}(r) \; {}^n\hat{A}^{-1}(R) \; u^n(R) \; ; \qquad n = 0,\ldots,\infty \; ; \qquad (15)
$$

The first derivatives of scattered field inside area of object boundary searching R can be obtained by the same way from the derivatives of the scattered field observed.

Above it was assumed that the center of observed field circumference coincides with the center of polar framework, and is placed inside the object. This unnecessary requirement was introduced only in order to make clear the essence of the method and can be readily removed.

Thus, relations (5) and (15) allow to start solving the problem of reconstruction of boundary obstacle in solid.

ALGORITHM OF RECONSTRUCTING SHAPE OF CAVITY IN SOLID

Optimization method for solving the problem, proposed below, consists in determination of object shape function by minimization of functional including functional of declination and some stabilizing penalty member. Penalty member is constructed using additional information about inhomogeneity investigated. So far as characteristic function $\gamma(r)$ is used for description of the object, condition of belonging $\gamma(r)$ to the class of binary functions can be involved into solution as such additional information. Mathematically the condition put onto $\gamma(r)$ can be written as: $\gamma(r) - \gamma^2(r) = 0$ for $r \in R$, so penalty functional tooks the form:

$$
\Phi_2[\gamma] = \zeta \int_R \left(\gamma(r) - \gamma^2(r) \right)^2 dr \; ; \qquad (16)
$$

Weight ζ is brought in to operate speed of convergence of iterative solution process . It should be noted that dimension of characteristic function exceeds dimension of unknown boundary of object by unit. Seems it is required a correspondent abundance of initial scattering data. Important consequence of penalty member introduction into optimization method is a significant reducing of the abundance.

Integral equation (5) is the base of declination functional constructing:

$$\Phi_1[\gamma] = \sum_{t \in T} \int_P \left| \int_R \gamma(r') \nabla \cdot (u^t(r') \cdot \Sigma(r/r')) \ dr' + \int_\sigma t^t(r') \cdot G(r/r') dS' \right|^2 dr_p; \qquad (17)$$

So far as (5) is satisfied for any external point r_p then operation of average on some domain P should be involved, so that scattered field has been described entirely. Domain P can be named area of conditional receiving and should not necessary coincide with real domain of receivers location. Besides summing over the domain P, a set of directions of incident field is introduced into functional in order to increase amount of initial data , and so to provide better stability of the algorithm. For the sake of convenience the set is denoted as domain T.

In accordance with presented configuration of the area of object boundary searching it is convenient to carry out sampling in polar system of coordinates. Thus we have the correspondence: $r \to r_n$, $n = 1,..,N$, $\varphi \to \varphi_m$, $m = 1,..,M$, $\gamma(r) \to \gamma_{mn}$. Then equation (5) can be written as:

$$\sum_{n=1}^{N} \sum_{m=1}^{M} \gamma_{mn} \int_{V_{mn}} \nabla \cdot (u(r') \cdot \Sigma(r/r') \ dr' = - \int_\sigma t(r') \cdot G(r/r') \ dr'; \qquad (18)$$

$$\sum_{mn} \gamma_{mn} X_{mn}^{ptl} = - P^{ptl}$$

$$p = 1,\ldots,P, \ t = 1,\ldots,T, \ l = 1,2.$$

where index l refers to components r and φ of vector equation, p denotes external point chosen, t – number of direction of incident field. V_{mn} – elementary cell in polar coordinates. Integrals X_{mn}^{ptl} and P^{tl} should be calculated with sufficient accuracy, method of the calculation is not discussed here. Functional (16-17) is transformed into the function of multi variables after sampling:

$$\Phi[\gamma] = \sum_{l=1}^{2} \sum_{p=1}^{P} \sum_{t=1}^{T} \left| \sum_{ij} \gamma_{ij} X_{ij}^{ptl} + P^{ptl} \right|^2 + \zeta \sum_{ij} (\gamma_{ij} - \gamma_{ij}^2)^2 = min; \qquad (19)$$

Minimum condition of (19) can be obtained in form of system of nonlinear equations:

$$\sum_{ij} \gamma_{ij} \left(\sum_{ptl} Re \ [X_{ij}^{ptl} X_{km}^{ptl}] + \zeta \delta_{ik} \delta_{jm} \right) = -\sum_{ptl} Re \ [P^{ptl} X_{km}^{ptl}] + \zeta \gamma_{km}^2 (3 - 2\gamma_{km}); \qquad (20)$$

$$k = 1,\ldots,N, \ m = 1,\ldots,M.$$

Solution of this system of equations is realized by organizing external and internal cycles of iterations. The external one consists in linearization of the system (nonlinear member $\gamma_{km}^2(3-2\gamma_{km})$ is calculated using values γ_{km} from preceding iteration). The internal cycle – solution of linear system of equation by extrapolated Richardson's method.

Applying the method described, numerical modelling of the solution has been carried out for parameters corresponding to the following real problem of the shape reconstruction of a circular cavity with radius equal to 3 cm (area of object boundary searching has limits: r = [2cm ,4.5cm], φ = [0,2π]). Wave numbers were assumed to be 5.0 cm^{-1} and 7.91 cm^{-1} respectively for longitudinal and transverse waves. Modelling was realized for two schemes of radiating: when object is sounded from one direction (shadow part of boundary has an angular size π) and when object is consequently sounded from three directions (φ = 3π/4, π, 5π/4, nonsonified part has an angular size π/2). In both cases values γ_{ij}= 0.5 were taken as initial for all (i,j). The result of object shape reconstruction by method presented is shown on fig.2. Thus both

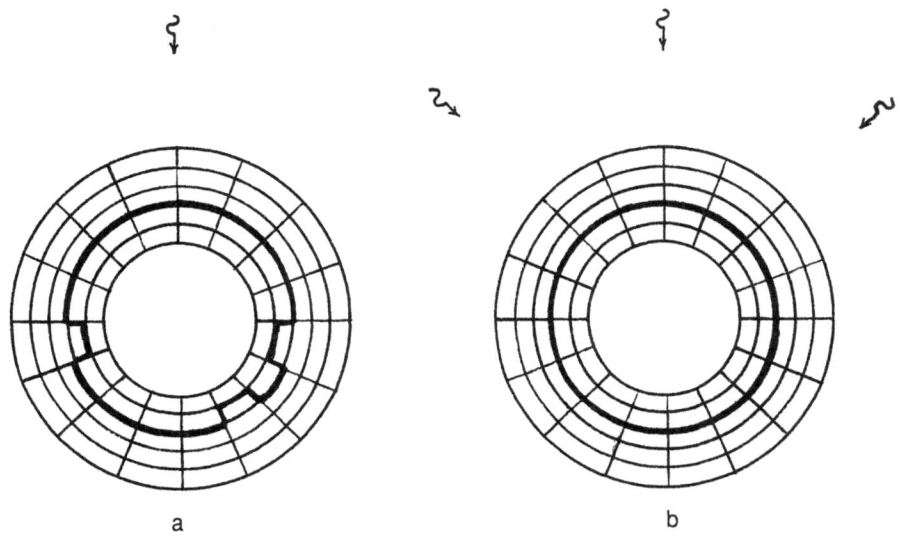

a   b

Figure 2. The results of reconstruction of circular cavity , a – one direction of sonification, b – three directions of sonification.

illuminated and shadow side of object can be satisfactorily reconstructed by the method suggested.

Theoretically the method hasn't taken into account any limit relations concerning frequency. But optimal work of algorithm, from computational point of view,is achieved on intermediate frequencies ka = 1..10 (k – wave number, a – size of object). When frequency is increased to values ka>>10 , computational expenditure grows up significantly (because of reconstructing more detailed structure of boundary). For low frequencies (ka << 1) the method of continuation of scattered field inside the object looses stability, so it is unexpedient to use them.

The method of continuation of scattered field inside area of object boundary searching can be modified so that initial scattered data for field calculation should be measured only on some part of the circumference of receiving. So the possibility of one-side reconstruction can be achieved when Fourier-coefficients of scattered field are determined by the operation of analitical continuation (see, for example, for scalar waves[2]). However, "mechanical" involving of the internal field obtained by such way in reconstructing algorithm presented above causes some dificulties. The first is significant unstability of reconstruction process due to accumulation of experimental and computational errors. So it is expedient to correct scattered field on each step , that is ,to organize simultaneous searching of both object shape function and field scattered by the object. This means that initial functional should included one more set of variables:

$$
\Phi_1[\gamma,u]=\sum_{t\in T}\int_P\int_R \left| \int \gamma(r')\nabla\cdot(u^t(r')\cdot\Sigma(r/r'))\ dr'+\int_\sigma t^t(r')\cdot G(r/r')dS' \right|^2 dr_p + \tag{21}
$$

$$
+ \int_{P'} \left| h(r) - \sum_{n=0}^{\infty} u_n^{c,s} \begin{pmatrix} \cos n\varphi \\ \sin n\varphi \end{pmatrix} \right|^2 dr ;
$$

Here h(r) - measured scattered field, P' - domain of real receiving (in the contrary to conventional one), it is a part of the circumference of receiving. Coefficients of expansion of scattered field on the circumference of receiving are considered as values to be determined (as well as $\gamma(r)$). This algorithm is more complicated in comparison with presented above and is now under investigation.

REFERENCES

1. D.Colton, Inverse scattering problem for time harmonic acoustic waves, *SIAM Review.*26:323 (1984).
2. T.S.Angell, R.E.Kleinman, and G.F.Roach, An inverse transmission problem for the Helmholtz equation, *Inv.Prob.*3:149 (1987).
3. K.Onishi, Numerical method for inverse scattering problems in two-dimensional scalar field, *in* "Ill posed problems in natural sciences", A.N.Tikhonov,ed., TVP, Moscow (1992).
4. V.A.Burov, A.V.Glaskov, A.A.Gorunov, I.P.Prudnikova, O.D.Rumiantseva, and E.Y.Tagunov, Digital and physical modelling of two-dimensional boundary inverse problems of a scalar waves scattering, *Acoustical Physics.* 36:466 (1990).
5. V.A.Burov, I.P.Prudnikova, and N.S.Sirotkina, Inverse problem of ultrasonic scattering by boundary inhomogeneity in isotropic solids,*Acoustical Physics.* 38:555 (1992).
6. Y.H.Pao, and V.Varatharajulu, Huygen's principle, radiation conditions and integral formulas for the scattering of elastic waves, *J.Acoust.Soc.Am.*59:1361 (1976).
7. V.V.Kravtsov, The approximation of functions of a number of variables by antenna potential, *Sov.Phys.Doc.* 22:23 (1977).

PIEZOCOMPOSITE MATERIALS

FOR ACOUSTICAL IMAGING TRANSDUCERS

Wallace Arden Smith

Materials Division
Office of Naval Research
800 North Quincy Street
Arlington, Virginia 22217-5660 U.S.A.

ABSTRACT

This overview describes the application of composite piezoelectric materials in acoustical imaging transducers. Attention is focused on one composite structure which has found particularly fruitful applications in acoustical imaging—the 1-3 piezocomposite structure consisting of long thin piezoceramic rods held parallel to each other by a polymer matrix. These piezocomposites may be viewed as 'new' materials with a set of 'effective' homogeneous materials properties whenever the spatial scale of the constituents in the composite structure is smaller than the wavelength of sound. From knowledge of the sub-wavelength 'microstructure,' we obtain an intuitive understanding of the origin of the piezocomposite's properties, and see how the material properties can be tuned to specific transducer application needs. Commercial applications in medical ultrasonic diagnostics, non-destructive testing, and undersea imaging illustrate the practical exploitation of 1-3 piezocomposites.

INTRODUCTION

Essential to the formation of acoustical images is the transducer that generates the probe pulses and detects the returning echoes. At the heart of most acoustical imaging transducers lies a piezoelectric material which performs the electromechanical energy conversion. The past decade has witnessed the emergence of a new class of piezoelectric materials—composite piezoelectrics—that has found many fruitful applications in acoustical imaging transducers. These composite piezoelectrics are made by combining a piezoelectrically active constituent (typically, a piezoelectric ceramic) with a piezoelectrically passive material (typically, a polymer) to form a new material (the composite piezoelectric) whose properties differ markedly from those of it constituents.

Even in the early days of piezoelectric ceramics, trials were made to form a flexible piezoelectric material by mixing ceramic powders of barium titanate and lead zirconate titanate with rubbers.[1] The first report of elastomers containing piezoceramic powders that showed technologically useful properties[2] spurred a Navy sponsored research program to explore the potential of composite piezoelectrics with a goal of improved hydrophone sensors in mind. This research program was carried out at the Materials Research Laboratory of The Pennsylvania State University by Professors Robert E. Newnham and L. Eric Cross, teamed with students and coworkers—we owe much of our understanding of piezocomposites to this pioneering effort.[3-5] Stimulated by this research on hydrophone materials, a modest industrially funded effort at Penn State studied the utility of piezocomposites for pulse-echo ultrasonic transducers with applications in medical diagnostic imaging in mind; the Penn State researchers soon identified[6] the 1-3 piezocomposite structure, illustrated in Figure 1, as the most promising candidate for pulse-echo ultrasonic applications. The first publications detailing the benefits to medical ultrasonic imaging were presented at the 1984 IEEE Ultrasonics Symposium simultaneous by a Penn State/Stanford/Philips collaboration[7-11] and by Hitachi Central Research Laboratory.[12] There must have been other parallel paths of evolution—unknown to me—as piezocomposites soon appeared in many ultrasonic transducer products: single element transducers, annular arrays, and linear arrays. Indeed, at this Acoustical Imaging Symposium in 1983, an annular array was described[13] in which many deep grooves had been cut, and subsequently filled with a polymer, to suppress unwanted lateral modes in the array elements; while not called a piezocomposite material, this procedure produced a structure quite similar to that shown in Figure 1.

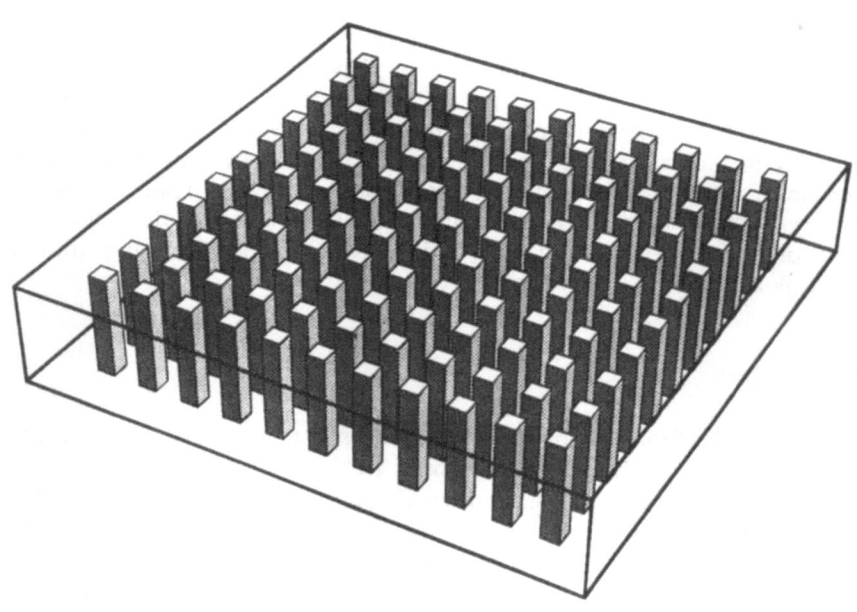

Figure 1. Schematic representation of the 1-3 piezocomposite structure. Long thin rods of piezoceramic are held parallel to each other by a passive polymer matrix.

While there are many possible ways to combine an active piezoceramic with a passive polymer,[14] this review will focus on the 1-3* composite structure shown in Figure 1. The teleological cause for this focus is that precisely this 1-3 piezocomposite structure has found the most fruitful applications in pulse-echo acoustical imaging. The next section describes the materials properties of these piezocomposites; the aim is to provide a simple intuitive picture for the macroscopic properties in terms of the properties of the constituents, their relative proportion, and the spatial scale of the composite microstructure. The third section is devoted to the transducer properties—those inherent in the macroscopic material properties plus additional technological aspects. The fourth section illustrates the breadth of practical applications with examples from medical diagnostic, non-destructive testing, and undersea acoustical imaging.

MATERIAL PROPERTIES

Effective Material Parameters

This section presents a simple physical picture for the "effective" material properties of a 1-3 piezocomposite, specifically, its dielectric, elastic and piezoelectric parameters.[16] We will discuss the composite's properties in terms of the material properties of the constituents—active piezoceramic, passive polymer—and their relative proportions. The properties for the composite will describe the composite as a homogeneous "effective medium" through this set of parameters.

When is such an "effective medium" description valid? In a composite medium, so long as all acoustic waves excited have wavelengths longer than the composite microstructure, the sound waves will, in effect, average the behavior of the medium over dimensions the scale of a wavelength—the sound field will not see the subwavelength microstructure of the composite. We, of course, can use our knowledge of the composite's substructure to determine what the appropriate "average" material properties are, but the acoustic waves will see the composite as a homogeneous medium possessing these "effective" material parameters. In addition, we only need a model for the 1-3 composites whose domain of validity encompasses their use in acoustical imaging transducers. For our purposes, the piezocomposites will be large thin sheets, much as shown in Figure 1, with metal electrodes applied to the top and bottom surfaces; this piezoelectric layer will be used as a thickness mode resonator in a frequency band around its thickness mode resonance frequency.

Composite Microstructure Scale

How fine must the composite microstructure be? This is not an easy question: the issue is important and difficult to answer succinctly. A fine microstructure is more challenging—and more costly—to make, so one would like to use as coarse a microstructure as possible. The performance of a piezocomposite will degrade gracefully in some respects, but rather catastrophically in others, if the microstructure is too coarse. The answer depends on what you are willing to pay and what transducer characteristics are important in your application.

There are two useful viewpoints on piezocomposite spatial scale. First, from a frequency domain perspective, the phenomena of lateral stopband resonances provides

---

   * The designation "1-3" refers to the fact that the active piezoceramic phase is connected in 1 dimension, that is, through the thickness of the plate, while the passive polymer phase is connected in all 3 dimensions.[3,15]

one answer.[7,8,17] When thickness mode vibrations in the composite plate are excited, either by applying a voltage across the electrodes or by an incident sound wave, acoustic waves running in the plane of the plate—Lamb waves—are also excited. These lateral running waves are reflected in the lateral periodicity of the composite microstructure and standing wave resonances are set up. If such a resonance occurs in the operating frequency band of the transducer, the desired thickness mode resonance is corrupted by unwanted coupling to resonant lateral running waves; long ringdown times result. The solution is to make the lateral structure "sufficiently" fine that the wavelengths of lateral stopband resonances are so short that their frequency is pushed well above the transducer's operating frequency—about twice the thickness mode resonance frequency is a useful criterion.[18]

A second perspective on piezocomposite spatial scale is provided when we consider the motion of the piezocomposite layer in the time domain. We want the piezocomposite plate to move—in response to an exciting voltage or an incident sound wave—in a uniform thickness mode oscillation. Both the polymer and ceramic must move up and down together. In transmission, the motion of the piezoceramic rods must be transferred to the side—to the surrounding polymer. In reception, the force on the polymer must be transmitted laterally to the piezoceramic rods as only they produce the electric signal. The dimensions over which this lateral force transfer is effective is proportional to the shear stiffness of the polymer. To ensure effective lateral transfer of force between the ceramic and polymer phases, a useful criterion is to make the spacing between ceramic rods about one quarter of a shear wavelength in the polymer.[19]

These two perspectives are not really different. They are, in fact, just the frequency domain and time domain sides of the same coin. For the rest of this section we shall leave the issue of composite spatial scale behind, assuming that the scale is "sufficiently" fine that discussion of homogeneous "effective" material properties for the composite makes sense.

Density

The simplest property of the composite to obtain is its density, ρ, which is just the average of the ceramic's density, ρ^c, and the polymer's density, ρ^p, weighted by the volume fraction of ceramic, v, and polymer, $(1-v)$, respectively, namely,

$$\rho = v \, \rho^c + (1-v) \, \rho^p . \tag{1}$$

This is an exact expression. This simple linear interpolation between the polymer's density and the ceramic's density is shown in the top curve in Figure 2 where the composite's material properties are plotted versus the ceramic volume fraction.

Dielectric Permittivity

Electrically the composite plate, shown in Figure 1, appears like a capacitor in which the ceramic and polymer phases make parallel contributions to the capacitance. Thus, the dielectric permittivity of the composite plate, ε,* also nearly interpolates linearly between the permittivity of the polymer, ε^p, and that of the ceramic, ε^c, as the ceramic volume fraction, v, increases from 0% to 100%. The permittivity of a typical piezoceramic ranges from 100's to 1000's, while the permittivity of a typical polymer is

---

* Strictly speaking ε_{33}^S in a coordinate system in which the plate lies in the x-y plane and the electrodes are on the faces of the plate.

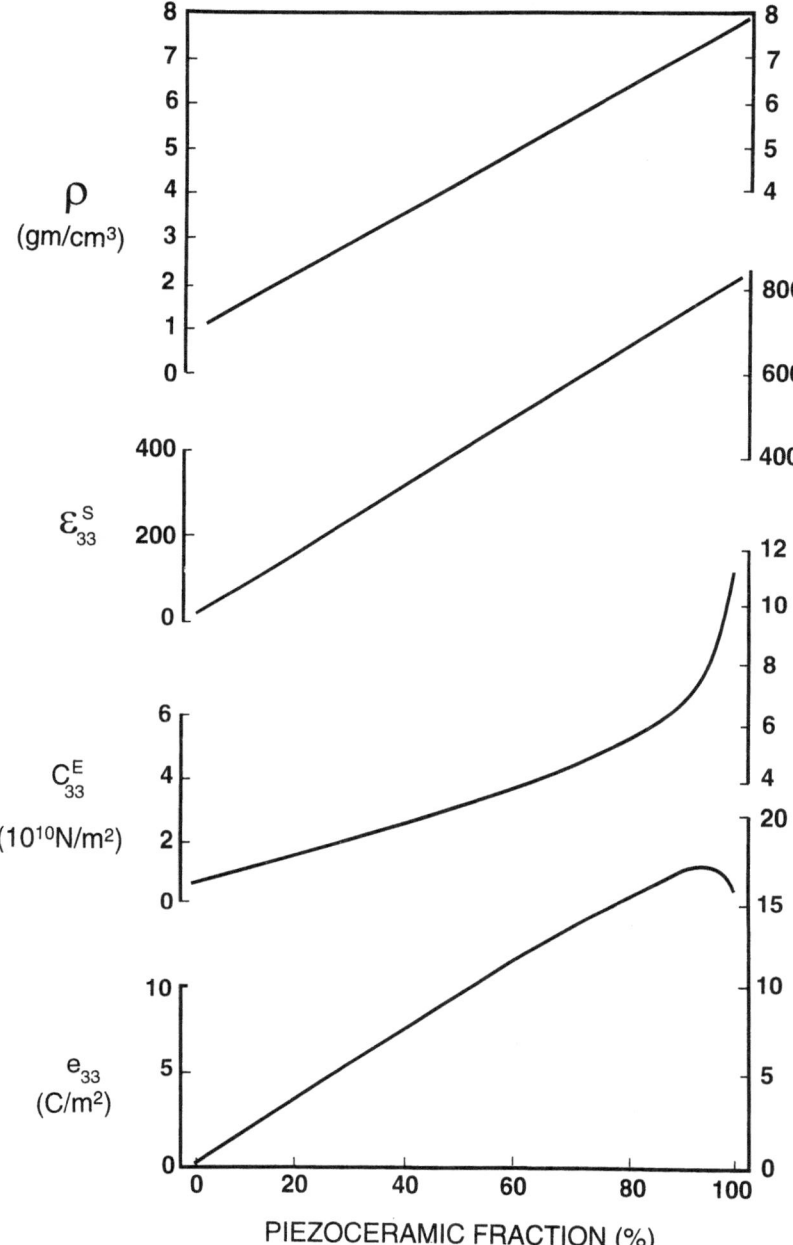

Figure 2. The variation with ceramic fraction of a piezocomposite's material parameters: density, ρ; dielectric permittivity, ε; elastic stiffness, c; and piezoelectric constant, e.

only about 10. Because of this large difference in permittivities, the fringe fields are quite weak, so this linear dependence is also nearly exact. Indeed, most of the electric flux passes through the ceramic phase, and the polymer's contribution to the composite's permittivity can be neglected except at exceedingly small ceramic fraction, that is,

$$\varepsilon \approx v \ \varepsilon^c + (1-v) \ \varepsilon^p \approx v \ \varepsilon^c. \tag{2}$$

This approximation is so good that minor deviations are not even visible on the scale on which material properties are plotted in Figure 2.

Elastic Stiffness

The ceramic fraction dependence of the elastic stiffness of the ceramic plate, c,† can be understood in terms of a parallel springs picture, much as the dielectric constant is understood from a parallel capacitors picture. This leads to a composite stiffness which interpolates nearly linearly between the polymer's stiffness and the ceramic's stiffness as the ceramic fraction increases, namely,

$$c \approx v \ c^c + (1-v) \ c^p. \tag{3}$$

The deviation from strict linearity is, however, evident in Figure 2. This deviation stems from an important physical effect. Namely, at low ceramic fraction, the ceramic rods are free to expand laterally against the much softer polymer becoming softer themselves, but as the composite's ceramic fraction approaches 100% there simply isn't enough polymer left to absorb the ceramic rods' lateral motion. At high volume fraction then, the rods become laterally clamped, consequently making a stiffer contribution to the overall stiffness of the composite plate. The upward swoop in the composite's stiffness is due to partial lateral clamping of the ceramic rods at high volume fraction. Even with an exceedingly soft polymer, this lateral stiffening effect will not disappear, its onset will be just delayed to higher ceramic content.

Piezoelectric Coefficient

It is instructive to consider the composite's piezoelectric coefficient, e,† which measures the charge produced on the composite plate's faces for a given change in the thickness of the plate, when the plate does not expand or contract laterally; this coefficient measures the strength of the piezoelectric interaction in circumstances most relevant to thickness mode oscillations. A good first approximation to the composite's piezoelectric response is again, a simple volume fraction weighted average of the contributions of each constituent, except that the polymer makes no contribution as it is, by hypothesis, piezoelectrically inactive, namely,

$$e \approx v \ e^c \tag{4}$$

Deviations from strict proportionality at high ceramic fraction are caused by the same physical effect as the deviations in the elastic stiffness. At low ceramic content the piezoelectric rods make a somewhat greater contribution than one might first estimate,

---

† Strictly speaking c_{33}^E and e_{33}, respectively.

because they are free to bulge or contract laterally with the polymer taking up the slack; it is only the composite as a whole that is laterally clamped, not the individual constituents. *This enhancement of the piezoelectric response by unclamping the sideways motion of the piezoceramic rods provides the dominant benefit in the 1-3 piezocomposite structure.* However, as the ceramic content nears 100%, the lateral clamping of the rods takes over, and the charge coefficient returns to the value for a clamped ceramic plate.

Electrical and Mechanical Losses

In this simple discussion, the composite and its constituents have been described as lossless—both electrically and mechanically. This is both a desirable state and one that can be readily achieved. In a well designed transducer, one does not want to throw away energy into heat. Piezoceramics with electrical loss tangents of a percent or so and mechanical Q's near one hundred or better are readily selected. Polymers have even lower electrical losses, but *can* have appreciable mechanical losses; here, care must be used in choosing a polymer that has low losses. It can be done. Piezocomposites with mechanical losses of two percent have been made.

TRANSDUCER PROPERTIES

From the piezocomposite material parameters, one can readily calculate the material properties relevant to its performance as a thickness mode transducer, namely, the electric impedance, the acoustic impedance, the longitudinal velocity, and the electromechanical coupling.

Electric Impedance

The electrical impedance of the piezoelectric layer in an acoustic transducer is dominated by the capacitance—there is negligible inductance. That capacitance is just given by the area of the layer divide by its thickness—pure geometrical factors—times the piezolayer's dielectric permittivity. We have already described how the permittivity depends on the constituent properties and proportion in Equation (2); it is just the ceramic fraction times the permittivity of the piezoceramic. Thus, the composite's dielectric constant will always be strictly less than that of the piezoceramic used to make it. This can be a drawback, especially, in small, high-frequency arrays where the geometrical factors make the element capacitance low even with a pure piezoceramic. To get the composite's permittivity as high as possible one should use a "soft" piezoceramic—dielectric constants near 5000 are achievable—and a reasonably large ceramic fraction.

Acoustic Impedance

The specific acoustic impedance, Z, of any material can be calculated directly from its density, ρ, and elastic stiffness, c, namely,

$$Z = \sqrt{c\,\rho} \qquad\qquad (5)$$

The variation with ceramic fraction of the composite's specific acoustic impedance, Z—calculated from c and ρ given above—is plotted in Figure 3. As the ceramic content increases, the composite's acoustic impedances increases monotonically from that of the polymer, typically 3 to 4 Mrayl, to that of the ceramic, typically over 30 Mrayl. The

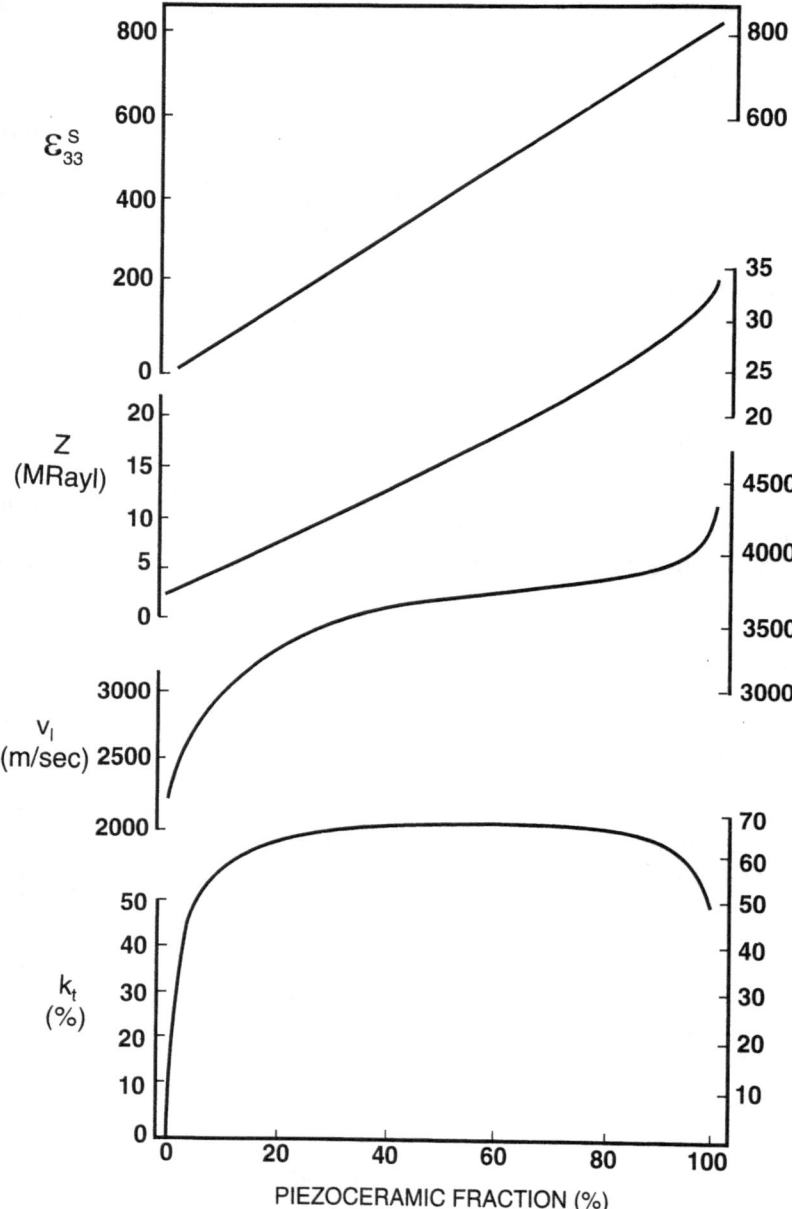

Figure 3. The variation with ceramic fraction of a piezocomposite's transducer parameters: dielectric permittivity, ε; specific acoustic impedance, Z; longitudinal velocity, v_l; and thickness mode electromechanical coupling constant, k_t.

piezocomposites provide here a major benefit: the ability to adjust the acoustic impedance over nearly an order of magnitude. Even when the imaging medium cannot be exactly matched—as is the case with tissue and water which both stand at about 1.5 Mrayl—reducing the mismatch between transducer and medium is an advance. One needs fewer acoustic matching layers and the tolerance is broadened both on their thickness and impedance; good acoustic coupling to the medium is simpler and cheaper to achieve. Optimizing the impedance match to water or tissue pulls the balance towards an elastically softer polymer and a smaller ceramic content; this latter desire balances the desire for higher ceramic content provided by the typical electrical impedance matching need noted above. A trade-off must be made.

Longitudinal Velocity

The longitudinal velocity, v_l, of a material is also determined by its density, ρ, and elastic stiffness, c, namely,

$$v_l = \sqrt{c/\rho} \tag{6}$$

This quantity, plotted in Figure 3, also monotonically increases from it value in the polymer, typically about 2000 m/sec, to its ceramic value, typically over 4500 m/sec. This is a relatively narrow range, but velocity tuning helps reduce refraction at the face of the transducer and the imaging medium. For water or tissue, with a longitudinal velocity of 1500 m/sec, a soft polymer and low ceramic content are again preferred.

Electromechanical Coupling

A material's electromechanical coupling factor—arguably the single most important characteristic of a piezoelectric transducer material—can also be deduced directly from its piezoelectric, elastic, and dielectric parameters. The square of the coupling factor is a direct measure of the material's ability to convert electrical energy into acoustic and vice versa. While many of the shortfalls of a piezoelectric can be remedied in other ways—acoustic matching layers fix acoustic impedance problems, or expensive electronics can compensate for a short fall in electric impedance—the piezoelectric layer is the only place where electromechanical energy conversion occurs. There are no technological fixes for low electromechanical coupling.

The electromechanical coupling also determines the major performance characteristic of an acoustical imaging transducer: its bandwidth. Low electrical and mechanical losses coupled with good electrical and acoustic impedance matching can minimize a transducer's insertion loss, but only at the matching frequency. The bandwidth for effective electromechanical energy transduction varies directly with the piezoelectric's coupling factor. In imaging applications range resolution is frequently determined by pulse length: a broad bandwidth yields a short pulse which means good range resolution. Without a reasonable coupling factor, bandwidth can be gained by introducing losses which eliminate the transducer's peak sensitivity. This bandwidth is purchased at the expense of sensitivity—hardly a good bargain. Thus, the piezoelectric's coupling factor impacts a combination of critical issues for acoustical imaging transducers: resolution and sensitivity.

Figure 3 shows the variation of the composite's thickness mode electromechanical coupling factor, k_t, with the composite's ceramic fraction. This plot shows the single most important transducer performance benefit of piezocomposites: *a piezocomposite's thickness mode coupling constant, k_t, can exceed even that of its constituent ceramic, k_t—almost equaling that ceramic's rod extensional mode coupling factor, k_{33}.* Basically, as noted above, this occurs because the ceramic in the composite is a long thin rod with little lateral restraint. While the composite plate is laterally clamped, the rods are nearly free to move laterally. These weakly constrained rods are simply more effective at electromechanical energy conversion than the laterally clamped ceramic in a large piezoceramic disk.

Density

Piezocomposites are always less dense than their constituent ceramic. For large underwater transducers this is a big plus, as neutral buoyancy is desirable. Even for small ultrasonic transducers, reduced weight provides systems benefits in mechanical sector scanners: a lighter transducer places less demands on the motor that must move it, particularly in real-time sector scanners that reverse motion many times per second.

Formability

Depending on the polymer used, piezocomposites can range from flexible to quite rigid. With flexible polymers, the transducer's shape can be defined by the shape of the backing; with the right polymer, even a rigid piezocomposite plate can be formed during processing, typically at elevated temperatures. This formability permits quite complex shapes for focusing or steering the acoustic beam. Moreover, the forming process is quite economical compared with procedures for shaping solid ceramic plates.

Lateral Modes

The thickness mode oscillations of a piezoelectric plate are true normal modes, strictly speaking, only in an infinite plate. In a finite plate driven to oscillate in a thickness-like fashion, the edges of the plate radiate running waves towards the inside of the plate. When these running waves reach the other side of the plate they are reflected off that edge back towards their origin. For the appropriate frequency the reflected wave will be in phase with the original wave, and a standing wave resonance is set up in these running waves. Thus, in a finite piezoelectric plate, the thickness mode resonance is coupled to lateral mode resonances. These undesired lateral modes rob energy from the desired thickness oscillations and lead to long tails in a transducer's impulse response. In a 1-3 piezocomposite plate, the coupling of lateral modes to the thickness mode can be strongly suppressed because the lateral running waves can be made heavily damped motions. Two physical mechanisms contribute to this damping of lateral running waves.[7,8,17,20] First, most polymers provide a strong attenuation of shear motion. The lateral running waves correspond to motion out of the plane of the plate with a propagation vector in the plane; within the polymer phase, this represents a shearing motion that leads to strong losses. Second, the spacing of the periodic array of rods in the plate can be tuned to resonantly reflect running waves whose frequency lies near the thickness mode resonance of the plate. This Bragg reflection in the array of rods attenuates the running wave without dissipating energy into heat. By proper design of the composite material—choice of polymer and rod spacing—the running waves can be strongly damped, thus effectively decoupling spurious lateral mode resonances from the desired thickness mode resonance.

Array Crosstalk

The strong attenuation in lateral running waves also contributes to the reduction of interelement crosstalk in arrays. Indeed, the sideways propagation of acoustic energy in a piezocomposite plate can be made so low that the elements of an array can be defined by patterning the electrodes alone—the piezoelectric layer does not need to be cut to effectively isolate array elements. Figure 4 illustrates a linear array defined by the electrode pattern alone. In addition to the strong attenuation of lateral running waves described above, there is a third physical effect that contributes to low crosstalk in uncut piezocomposite arrays: the elastic anisotropy of the piezocomposite. For waves propagating in a 1-3 piezocomposite medium along the direction of the rods, the continuous ceramic phase stiffens the medium so that the acoustic velocity—for both shear and longitudinal waves—is only somewhat lower than that of the piezoceramic. However, for waves propagating perpendicular to the rods, the much softer polymer phase makes the dominant contribution, resulting in much lower velocities in both shear and longitudinal waves. In such a strongly anisotropic medium, the lateral running waves—consisting of coupled shear and longitudinal waves bouncing back and forth between the top and bottom of the plate—have their energy flow bent strongly towards the direction of the rods. Thus as these waves carry energy to the sides, the energy makes many bounces off the front and back of the transducer providing many opportunities to radiate out into the medium.

Performance Trade-offs

This section summarizes many advantages of 1-3 piezocomposites—lower acoustic impedance, lower acoustic velocity, higher electromechanical coupling, lower density, formability, weak coupling of lateral modes, and low crosstalk in undiced arrays. However, all of these enhancements cannot be achieved simultaneously. Trade-offs must be made. For example, a good acoustic impedance and velocity match to water or tissue is achieved at very low volume fraction of piezoceramic, but here the electromechanical coupling drops off and the low dielectric constant makes electrical impedance matching difficult. The need to make such trade-offs provides stimulating intellectual challenges to the materials/transducer engineer—and ensures stable long-term employment.

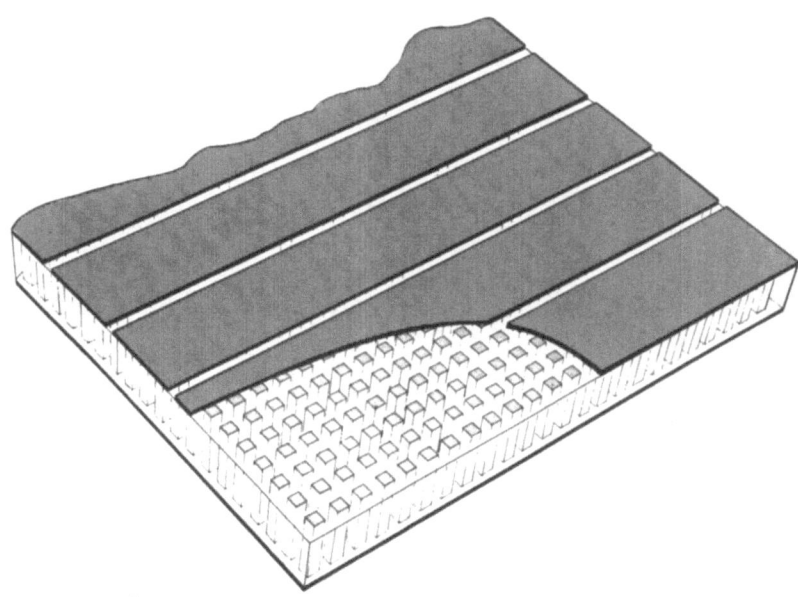

Figure 4. Piezocomposite linear array transducer with elements defined by the electrode pattern alone.

APPLICATIONS

While there is a substantial research literature describing the properties of ultrasonic imaging transducers made using piezocomposites, citations to commercial product brochures gives a more concrete depiction of the reality of these applications.

Figure 5 illustrates the performance of transducers produced by Echo Ultrasound[21] We see here the very compact impulse response afforded by the piezocomposites, as well as the absence of lateral resonances.

The absence of lateral modes in annular arrays is a particular strong point with piezocomposites. Good acoustic design in an annular array dictates that the width of the outer rings should be comparable to the piezolayer thickness, while good isolation between ring elements is also essential. Cutting a piezoceramic disk to reduce interelement crosstalk produces ceramic rings so narrow that the desired thickness mode is strongly coupled to lateral resonances resulting in long ringdown times. With the

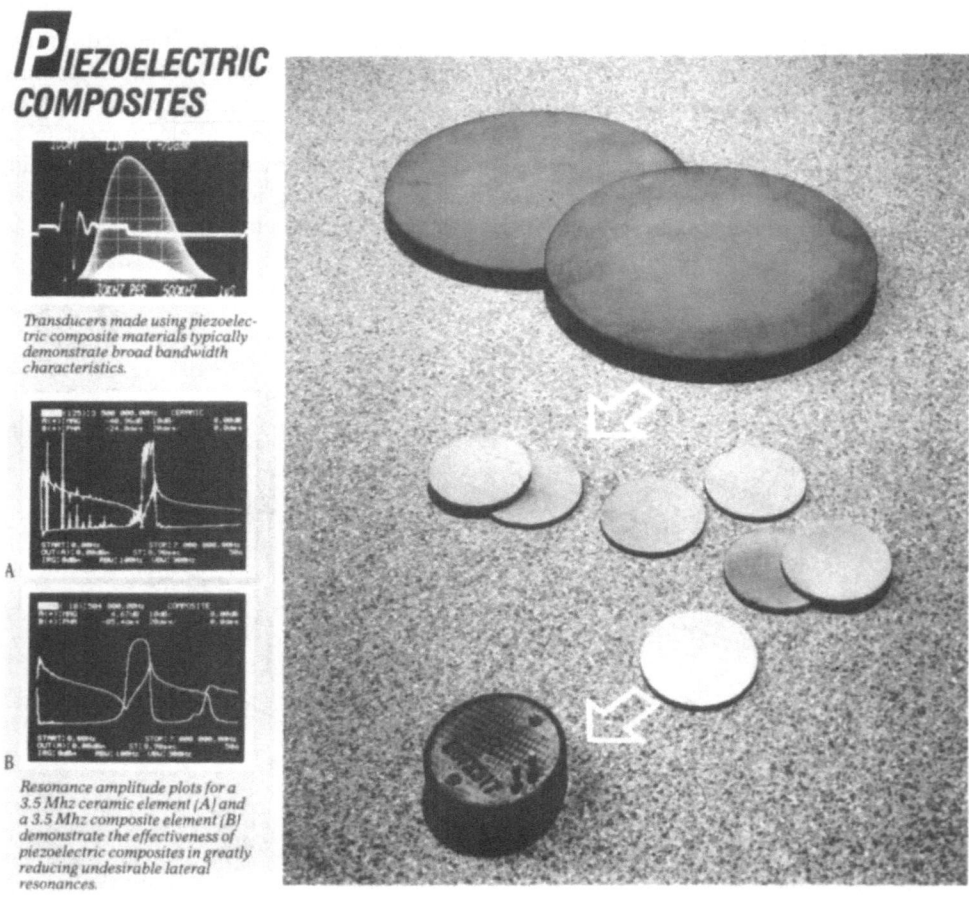

Figure 5. Echo Ultrasound product literature describing piezocomposites for medical imaging transducers.

piezocomposite disk the annular rings can be defined by the electrode pattern alone; the interelement isolation is excellent and the thickness mode resonance is uncorrupted by coupling to any lateral modes. Product literature from Precision Acoustic Devices,[22] shown in Figure 6, graphically illustrates this point: it describes an eight element annular array with 6 dB insertion loss, greater that 75% bandwidth, a 7 λ 40dB pulse length, and less than 35 dB of interelement crosstalk. These annular array transducers are used in mechanical sector scanners for medical diagnostic imaging.

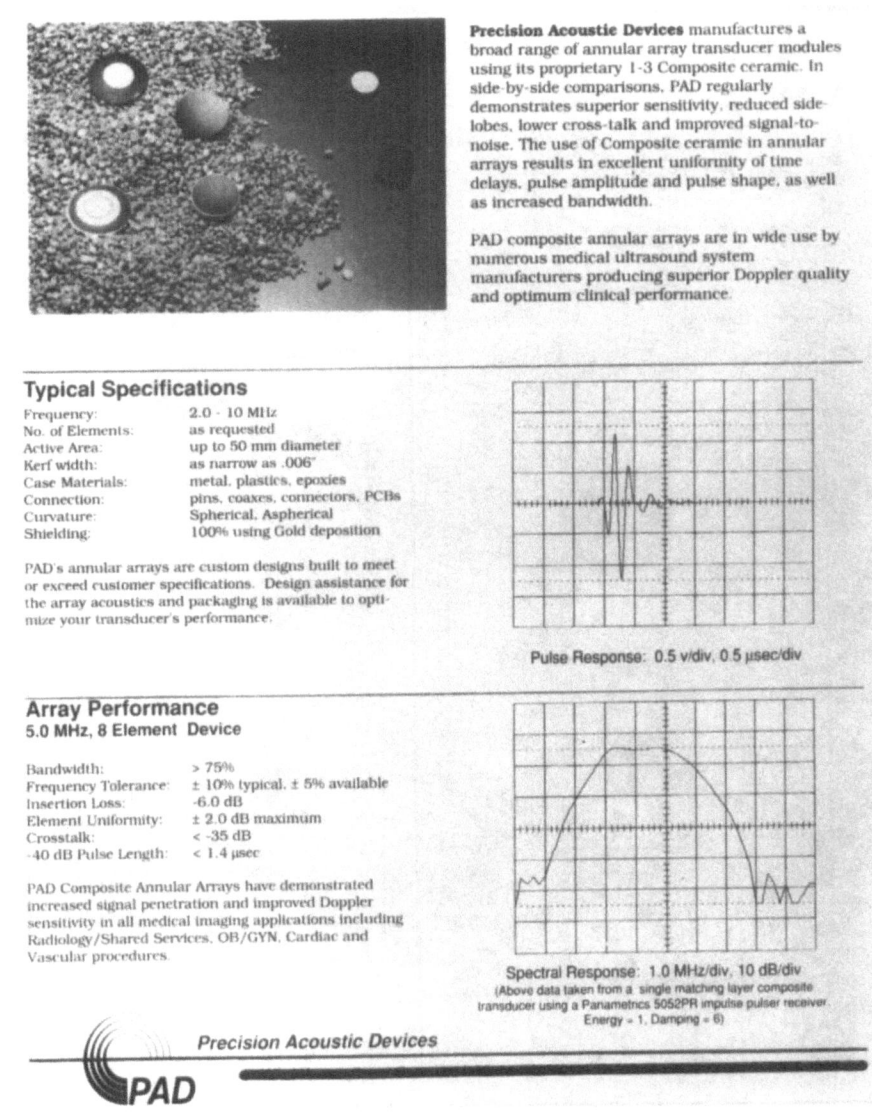

Precision Acoustic Devices manufactures a broad range of annular array transducer modules using its proprietary 1-3 Composite ceramic. In side-by-side comparisons, PAD regularly demonstrates superior sensitivity, reduced side-lobes, lower cross-talk and improved signal-to-noise. The use of Composite ceramic in annular arrays results in excellent uniformity of time delays, pulse amplitude and pulse shape, as well as increased bandwidth.

PAD composite annular arrays are in wide use by numerous medical ultrasound system manufacturers producing superior Doppler quality and optimum clinical performance.

Typical Specifications

Frequency:	2.0 - 10 MHz
No. of Elements:	as requested
Active Area:	up to 50 mm diameter
Kerf width:	as narrow as .006"
Case Materials:	metal, plastics, epoxies
Connection:	pins, coaxes, connectors, PCBs
Curvature:	Spherical, Aspherical
Shielding:	100% using Gold deposition

PAD's annular arrays are custom designs built to meet or exceed customer specifications. Design assistance for the array acoustics and packaging is available to optimize your transducer's performance.

Pulse Response: 0.5 v/div, 0.5 µsec/div

Array Performance
5.0 MHz, 8 Element Device

Bandwidth:	> 75%
Frequency Tolerance:	± 10% typical, ± 5% available
Insertion Loss:	-6.0 dB
Element Uniformity:	± 2.0 dB maximum
Crosstalk:	< -35 dB
-40 dB Pulse Length:	< 1.4 µsec

PAD Composite Annular Arrays have demonstrated increased signal penetration and improved Doppler sensitivity in all medical imaging applications including Radiology/Shared Services, OB/GYN, Cardiac and Vascular procedures.

Spectral Response: 1.0 MHz/div, 10 dB/div
(Above data taken from a single matching layer composite transducer using a Panametrics 5052PR impulse pulser receiver.
Energy = 1, Damping = 6)

Precision Acoustic Devices

PAD

Figure 6. Precision Acoustic Devices product literature showing piezocomposite annular array transducers.

The next illustration is also drawn from medical ultrasonic imaging, but focuses on linear arrays with convex curvature. Figure 7, extracted from the literature of Diasonics,[23] describes how the excellent acoustic impedance matching to tissue afforded by the piezocomposites facilitates the reduction of reverberations between the strong scatterers in the skin and the transducer face, thus producing a clear near-field image.

Acoustic non-destructive testing presents transducer problems similar to medical ultrasonics. We see in Figure 8, extracted from the product literature of Imasonic,[24] that piezocomposites play a useful role here too. This depicts single element piezocomposites transducers, ranging in frequency from 500 kHz to 15 MHz, and notes that they can readily be formed with the desired aspherical curvature to provide a sharp focus.

Matched Impedance Transducers – beyond wide bandwidth.

Matched Impedance (MI) transducers represent a major advancement in transducer technology that complements the power and precision of Confocal Imaging.

Diasonics' new series of transducers incorporate custom fabricated composite materials with low acoustic impedance that is more closely matched to that of the body. This improves the effectiveness of Confocal Imaging and dramatically enhances the clarity of the near field.

Finally, the lower acoustic impedance provides an inherently wide bandwidth, which provides excellent resolution in the near field and uncompromised penetration in the far field.

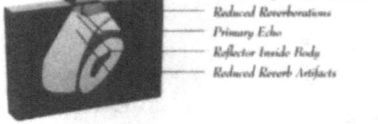

Conventional Wide-band Technology

- Conventional Wide-band Transducer
- Reverberations
- Primary Echo
- Reflector Inside Body
- Reverb Artifacts

Matched Impedance Technology

- Matched Impedance Transducer
- Reduced Reverberations
- Primary Echo
- Reflector Inside Body
- Reduced Reverb Artifacts

Matched Impedance Technology includes "soft" composite materials that reduce reverberations between tissue interfaces and the transducer face, virtually eliminating near field haze.

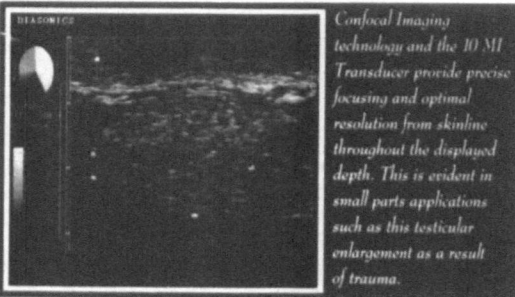

Confocal Imaging technology and the 10 MI Transducer provide precise focusing and optimal resolution from skinline throughout the displayed depth. This is evident in small parts applications such as this testicular enlargement as a result of trauma.

Although they represent a major advancement in technology, our new MI transducers are just a part of a wide and growing selection of application-specific transducers.

Figure 7. Diasonics product literature describing piezocomposites for medical imaging transducers.

BROCHURE # IMA451

ULTRASONIC TRANSDUCERS WITH PIEZOCOMPOSITE TECHNOLOGY

STANDARD IMMERSION TRANSDUCERS – IM SERIES

Benefiting from over 10 years of experience in the development and manufacturing of new piezocomposite technology, IMASONIC proposes the IM series ultrasonic transducers for non destructive testing by immersion technics.

The utilization of this new technology provides the best sensitivity/bandwidth compromise available on the market. The sensitivity gain is between 10 and 20 dB compared to the transducers realized with a traditional technology, and then the bandwidth is between 60 and 100 %.

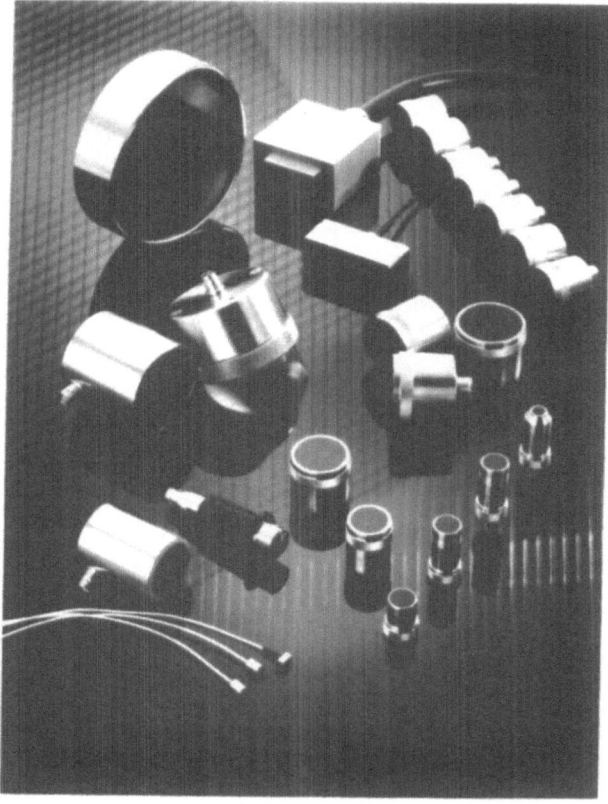

Aspherical transducer (left of background)
Immersion and contact transducers (middle left)
Linear arrays (middle of background)
Immersion transducers IM series (right side)
High temperature, delay line and miniaturiazed transducers (left of the foreground)

Figure 8. Imasonic product literature showing piezocomposite transducers for non-destructive testing.

The final illustration, Figure 9, from Fugro UDI,[25] shows an ultrasonic imaging system used to guide a tethered undersea vehicle used by the North Sea oil industry. Here a mechanically scanned transducer made from piezocomposites makes a clear image at 200 kHz, 500 kHz, or 1 MHz at ranges up to 300m, 100m or 40m respectively. Two piezocomposite transducers are contained in the scanhead: a large transmitter that emits a narrow pulse, and a smaller receiver with a wider beam to make allowance for the rotation of the scanhead during the reception interval. Using two transducer elements allows one to optimize the piezocomposite material separately for transmission and reception—both the aperture size and material properties. While the requirements—far-field imaging—are quite different from medical and non-destructive testing, both the basic material properties and the ability to tailor them over broad ranges commend piezocomposites in this application also.

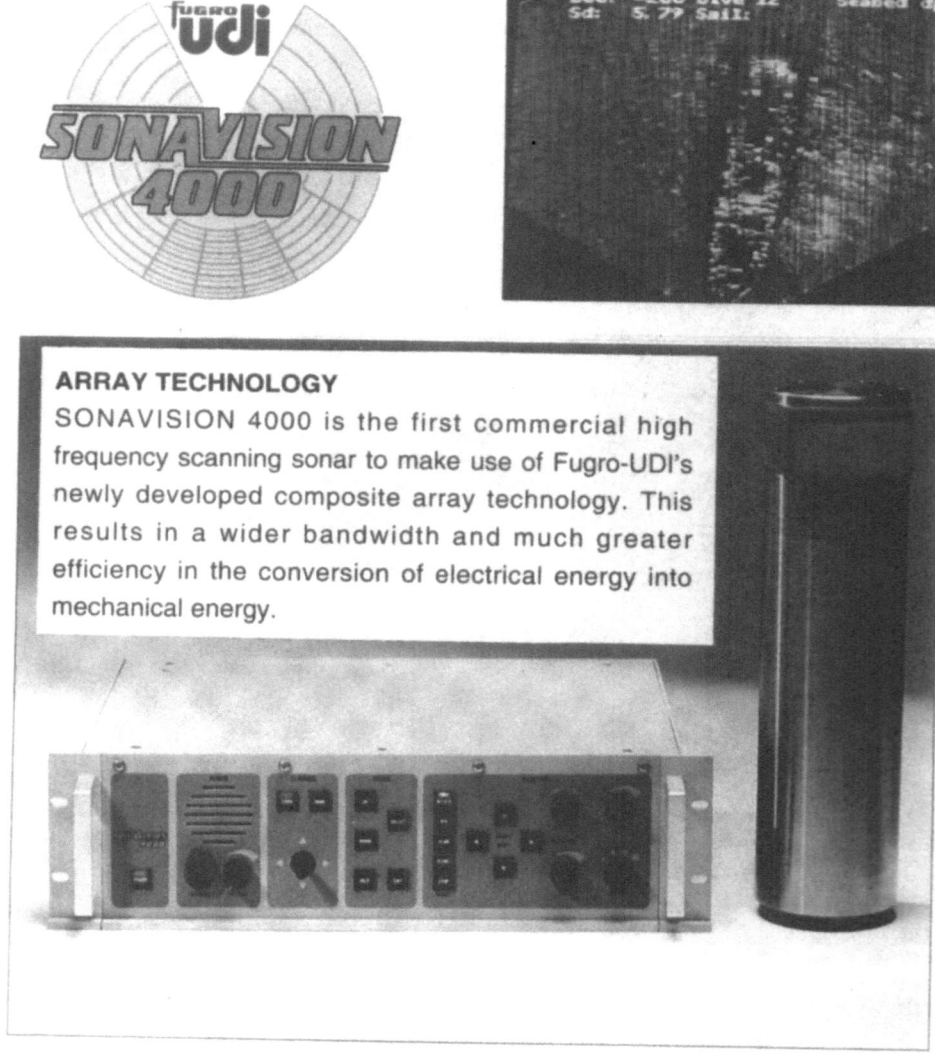

Figure 9. Fugro UDI product literature showing an undersea imager using piezocomposite transducers.

DISCUSSION

This broad overview of the properties and applications of 1-3 composite piezoelectrics has highlighted some of the advantages this class of materials offers the transducer designer. These highlights show how the design flexibility inherent in piezocomposites commends their use in a broad range of acoustical imaging transducers. The compact format of this paper allows me only to whet the desires of scientists and engineers, not provide the total story. The references should help the interested reader towards a more complete story.[26] A letter or phone call to the author will inundate the inquirer with more information about piezocomposites than a sensible person wants.

ACKNOWLEDGEMENTS

For the information illustrating the practical applications of piezocomposites, I am indebted to Dr. Hal Kunkel of Echo Ultrasound, Mr. Edward Kopp of Blatek Incorporated, Drs. S. Omar Ishrak and Mehmet Salahi of Diasonics, Dr. Gerard Fleury of Imasonic, and Dr. Colin MacLean of Fugro UDI.

REFERENCES

1. Paul L. Smith, Private Communication.

2. W. B. Harrison, "Flexible Piezoelectric Organic Composites," pages 257-268 in *Proceedings of the Workshop on Sonar Transducer Materials,* edited by P. L. Smith and R. C. Pohanka, Naval Research Laboratory, Washington, D.C., February 1976. Presented at the Workshop on Sonar Transducer Materials, Naval Research Laboratory, Washington, D.C., 13-14 November 1975.

3. R. E. Newnham, D. P. Skinner, and L. E. Cross, "Connectivity and Piezoelectric-Pyroelectric Composites," *Materials Research Bulletin,* vol. 13, pp. 525-536, 1978.

4. D. P. Skinner, R. E. Newnham, and L. E. Cross, "Flexible Composite Transducers," *Materials Research Bulletin,* vol. 13, pp. 599-607, 1978.

5. R. E. Newnham, L. J. Bowen, K. A. Klicker, and L. E. Cross, "Composite Piezoelectric Transducers," *Materials in Engineering,* vol. 2, pp. 93-106, 1980.

6. T. R. Gururaja, W. A. Schulze, T. R. Shrout, A. Safari, L. Webster and L. E. Cross, "High Frequency Applications of PZT/Polymer Composite Materials," *Ferroelectrics* 39, 1245-1248 (1981).

7. T. R. Gururaja, W. A. Schulze, L. E. Cross, R. E. Newnham, B. A. Auld and J. Wang, "Resonant Modes in Piezoelectric PZT Rod-Polymer Composite Materials," Proceedings of the 1984 IEEE Ultrasonics Symposium 523-527 (1984). Presented at the 1984 IEEE Ultrasonics Symposium, Dallas, Texas, 14-16 November 1984.

8. B. A. Auld and Y. Wang, "Acoustic Wave Vibrations in Periodic Composite Plates," Proceedings of the 1984 IEEE Ultrasonics Symposium 528-532 (1984). Presented at the 1984 IEEE Ultrasonics Symposium, Dallas, Texas, 14-16 November 1984.

9. T. R. Gururaja, W. A. Schulze, L. E. Cross and R. E. Newnham, "Ultrasonic Properties of Piezoelectric PZT Rod-Polymer Composites," Proceedings of the 1984 IEEE Ultrasonics Symposium 533-538 (1984). Presented at the 1984 IEEE Ultrasonics Symposium, Dallas, Texas, 14-16 November 1984.

10. W. A. Smith, A. A. Shaulov and B. M. Singer, "Properties of Composite Piezoelectric Materials for Ultrasonic Transducers," Proceedings of the 1984 IEEE Ultrasonics Symposium 539-544 (1984). Presented at the 1984 IEEE Ultrasonics Symposium, Dallas, Texas, 14-16 November 1984.

11. A. A. Shaulov, W. A. Smith and B. M. Singer, "Performance of Ultrasonic Transducers Made from Composite Piezoelectric Materials," Proceedings of the 1984 IEEE Ultrasonics Symposium 545-548 (1984). Presented at the 1984 IEEE Ultrasonics Symposium, Dallas, Texas, 14-16 November 1984.

12. H. Takeuchi, C. Nakaya and K. Katakura, "Medical Ultrasonic Probe Using PZT/Polymer Composite," Proceedings of the 1984 IEEE Ultrasonics Symposium 507-510 (1984). Presented at the 1984 IEEE Ultrasonics Symposium, Dallas, Texas, 14-16 November 1984.

13. P. Alias, P. Challande, C. Kammoun, B. Nouailhas and F. Pons, "A New Technique for Realizing Annular Arrays or Complex Shaped Transducers," pages 357-368, in *Acoustic Imaging* **13**, edited by M. Kaveh, R. K. Mueller and J. F. Greenleaf, Plenum Press, New York, New York (1984). Presented at the Thirteenth Acoustic Imaging Symposium, Minneapolis, Minnesota, 26-28 October 1983.

14. A recent comprehensive review appears in T. R. Gururaja, A. Safari, R. E. Newnham, and L. E. Cross, "Piezoelectric Ceramic-Polymer Composites for Transducer Applications," in *Electronic Ceramics,* edited by L. M. Levinson, pp. 92-128, Marcel Dekker, New York, New York, 1987.

15. S. M. Pilgrim, R. E. Newnham, and L. L. Rohlfing, "An Extension of the Composite Nomenclature Scheme," *Materials Research Bulletin*, vol. 22, pp. 677-684 (1987).

16. W. A. Smith and B. A. Auld, "Modeling 1-3 Composite Piezoelectrics: Thickness-Mode Oscillations," *IEEE Transactions on Ultrasonics, Ferroelectrics, and Frequency Control,* vol. 38, pp. 40-47, 1991.

17. B. A. Auld, H. A. Kunkel, Y. A. Shui, and Y. Wang, "Dynamic Behavior of Periodic Piezoelectric Composites," Proceedings of the 1983 IEEE Ultrasonics Symposium pp. 554-558, 1983.

18. W. A. Smith, A. A. Shaulov, and B. A. Auld, "Design of Piezocomposites for Ultrasonic Transducers," *Ferroelectrics* vol. 91, pp. 155-162 (1989). Presented at the European Conference on Applications of Polar Dielectrics/International Symposium on Applications of Ferroelectrics, Zurich, Switzerland, 28 August - 1 September 1988.

19. C. G. Oakley, "Analysis and Development of Piezoelectric Composites for Medical Ultrasound Transducer Applications," Ph. D. Thesis, The Pennsylvania State University, May 1991.

20. Yuzhong Wang, "Waves and Vibrations in Elastic Superlattice Composites," Ph. D. Thesis, Stanford University, December 1986.

21. Echo Ultrasound, Rural Delivery 2, Box 118, Reedsville, Pennsylvania 17084-9772, (717) 667-3266, (717) 667-6843 fax.

22. Precision Acoustic Devices, Blatek Incorporated, 2820 East College Avenue, State College, Pennsylvania 16801, (814) 231 2085, (814) 231 2087 fax.

23. Diasonics, 1565 Barber Lane, Milpitas, California 95035, (408) 432-9000, (408) 922-1174 fax.

24. Imasonic, 15, rue Alain Savary, 25000 Besancon, France, (081) 80 51 71, (081) 80 17 21 fax.

25. Fugro UDI, Bridge of Don Industrial Estate, Denmore Road, Aberdeen AB2 8JW United Kingdom, (0224) 703 551, (0224) 821 339 fax.

26. W. A. Smith, "The Role of Piezocomposites in Ultrasonic Transducers," *Proceedings of the IEEE Ultrasonics Symposium* 755-766 (1989) contains an extensive set of references; a more up to date listing is available from the author.

A NEW THREE-DIMENSIONAL MODEL FOR CIRCULAR PIEZOELECTRIC TRANSDUCERS

A. Iula, N. Lamberti, G. Caliano and M. Pappalardo

Università di Salerno
Dip. di Ingegneria dell'Informazione ed Ingegneria Elettrica
Via Ponte Don Melillo, I-84084 Fisciano (SA), Italy

INTRODUCTION

Piezoelectric ceramic disks are usually employed as active components in ultrasonic transducers for acoustical imaging and not destructive testing. Thickness mode is mainly need for all the applications in which high frequencies are required. For inspection of highly scattering and absorbing materials, however, lower frequencies must be used. To this end two solutions can be adopted: the Langevin composite transducer or a single disk working on its radial mode. In fact, with this vibration mode a significant strain also occurs in the transverse direction due to the elastic coupling. In this work, we develop a new three-dimensional model which takes both radial and thickness displacements into account; it can be used to optimize the geometry of the disk in order to maximize the thickness displacement of the radial mode. The vibration of a piezoceramic disk is described, in terms of cylindrical coordinates, by a system of two coupled differential wave equations with coupled boundary conditions. The solution of this problem is very difficult and approximation methods are usually required. An approximate solution was obtained by Brissaud [1], who proposed an approximated 3-D model based on the assumption that the displacements along the radial and thickness directions are dependent only on the related coordinate. In this model the stress free boundary conditions are satisfied only in an approximate way. Our approach also consists in considering two orthogonal wave functions as solutions of the differential equations, each depending only on the corresponding axis but, for the thickness mode, we choose a more general solution which permits to consider different transmission media for each external surface. As far as the boundaries are concerned, we apply integral conditions, obtaining an approximate model of the external behavior of the disk. Following this approximation we compute modified elastic and elastoelectric constants for the material. With this approach we model the plate in the frequency domain as a four ports system with one electric and three mechanical ports, one for each external surface.

THE MATHEMATICAL MODEL

Consider the circular piezoceramic plate shown in fig. 1, polarized along thickness. In this analysis we will use cylindrical coordinates whose origin is in the center of the plate. An a.c. voltage is applied across the electrodes which are obtained metallizing the two major surfaces of the disk. The E_r and E_θ components of the electric field vanish on these two surfaces and we assume that they are also zero throughout the plate. Further we assume, due to axially symmetry, that $\partial\ /\partial\theta = 0$, that the radial and thickness displacements u_r and u_z

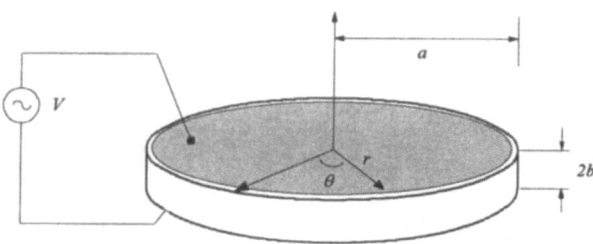

Fig. 1. Geometry of the piezoceramic disk.

are uncoupled i.e. $u_r = u_r(r)$ and $u_z = u_z(z)$ and finally that $u_\theta = 0$. These hypotheses yield $S_{rz} = S_{r\theta} = S_{z\theta} = 0$ and the constitutive equations reduce to:

$$T_{rr} = c_{11}^D S_{rr} + c_{12}^D S_{\theta\theta} + c_{13}^D S_{zz} - h_{31} D_z$$
$$T_{\theta\theta} = c_{12}^D S_{rr} + c_{11}^D S_{\theta\theta} + c_{13}^D S_{zz} - h_{31} D_z$$
$$T_{zz} = c_{13}^D S_{rr} + c_{13}^D S_{\theta\theta} + c_{33}^D S_{zz} - h_{33} D_z \tag{1}$$
$$E_z = -h_{31} S_{rr} - h_{31} S_{\theta\theta} - h_{33} S_{zz} + 1/\varepsilon_{33}^S D_z$$

The strain tensor components are given by:

$$S_{rr} = \frac{\partial u_r}{\partial r}; \ \ S_{\theta\theta} = \frac{u_r}{r}; \ S_{zz} = \frac{\partial u_z}{\partial z} \tag{2}$$

The wave equations for the radial and thickness mode are given respectively by:

$$\frac{\partial T_{rr}}{\partial r} + \frac{T_{rr} - T_{\theta\theta}}{r} = \rho \frac{\partial^2 u_r}{\partial t^2}$$

$$\frac{\partial T_{zz}}{\partial z} = \rho \frac{\partial^2 u_z}{\partial t^2} \tag{3}$$

With the above hypotheses the displacements computed by (1), (2) and (3) are c_{13}^D independent; this causes a significant error in the propagation velocity evaluation and therefore in the resonant frequencies of the radial mode. In the one dimensional model of the thin plate proposed by Meitzler [2], c_{11}^D and c_{12}^D are modified taking into account that $T_{zz} = 0$ throughout the plate. This assumption can be adopted also for thick plates and its validity will be shown comparing experimental and numerical results. With this final hypothesis the wave equation in radial direction becomes:

$$c_{11}^{D'} \left[\frac{\partial^2 u_r}{\partial r^2} + \frac{1}{r} \frac{\partial u_r}{\partial r} - \frac{u_r}{r} \right] = \rho \frac{\partial^2 u_r}{\partial t^2} \tag{4}$$

where $c_{11}^{D'} = c_{11}^D - \frac{c_{13}^{D\,2}}{c_{33}^D}$

The wave equation in thickness direction is:

$$c_{33}^D \frac{\partial^2 u_z}{\partial z^2} = \rho \frac{\partial^2 u_z}{\partial t^2} \tag{5}$$

For harmonic waves the more general solutions of eqs. (4) and (5) are

$$u_r = AJ_1(k_1 r)e^{j\omega t}$$

$$u_z = \left[B\sin(k_3 z) + C\cos(k_3 z) \right]e^{j\omega t} \tag{6}$$

where $v_1 = \sqrt{\dfrac{c_{11}^{D'}}{\rho}}$; $k_1 = \dfrac{\omega}{v_1}$; $v_3 = \sqrt{\dfrac{c_{33}^{D}}{\rho}}$; $k_3 = \dfrac{\omega}{v_3}$.

The A, B and C constants can be computed imposing the continuity of the velocities between the external surfaces of the plate and the surrounding media i.e.:

$$\left.\frac{\partial u_r}{\partial t}\right|_{r=a} = u_1; \qquad \left.\frac{\partial u_z}{\partial t}\right|_{z=b} = u_2; \qquad \left.\frac{\partial u_z}{\partial t}\right|_{z=-b} = u_3 \tag{7}$$

With these boundary conditions the wave functions become:

$$u_r = \frac{u_1}{j\omega J_1(k_1 a)} J_1(k_1 r)$$

$$u_z = \left[\frac{(u_2 - u_3)}{2\sin(k_3 b)}\sin(k_3 z) + \frac{(u_2 + u_3)}{2\sin(k_3 b)}\cos(k_3 z) \right] \tag{8}$$

The external behavior of the plate is computed imposing the continuity between the stresses and the forces on its external surfaces:

$$2\pi a \int_{-b}^{b} T_{rr}(a)dz = -F_1$$

$$2\pi \int_{0}^{a} T_{zz}(b)rdr = -F_2 \tag{9}$$

$$2\pi \int_{0}^{a} T_{zz}(-b)rdr = -F_3$$

Taking into account the classical relation between the current I and the electric displacement D_z and the relation between V and E_z we obtain the following equations:

$$F_1 = \frac{Z_1}{j}\left[\left(\frac{J_0(k_1 a) - J_1(k_1 a)/k_1 a}{J_1(k_1 a)} \right) + \frac{c_{12}^D}{c_{11}^D} \right]u_1 + \frac{2\pi a c_{13}^D}{j\omega}\left[u_2 + u_3\right] + \frac{h_{31}}{j\omega}I$$

$$F_2 = \frac{2\pi a c_{13}^D}{j\omega} u_1 + \frac{Z_3}{j}\left[\frac{u_2}{tg(k_3 2b)} + \frac{u_3}{sin(k_3 2b)}\right] + \frac{h_{33}}{j\omega} I$$

(10)

$$F_3 = \frac{2\pi a c_{13}^D}{j\omega} u_1 + \frac{Z_3}{j}\left[\frac{u_2}{sin(k_3 2b)} + \frac{u_3}{tg(k_3 2b)}\right] + \frac{h_{33}}{j\omega} I$$

$$V = \frac{h_{31}}{j\omega} u_1 + \frac{h_{33}}{j\omega}[u_2 + u_3] + \frac{1}{j\omega C_0} I$$

where $Z_1 = \rho v_1 4\pi ab$ and $Z_3 = \rho v_3 \pi a^2$ can be seen as the piezoceramic acoustic impedance along the r and z directions respectively and where $C_0 = \varepsilon_{33}^S \pi a^2 / 2b$ is the so called piezoceramic "clamped capacity".

Eqs. (10) can be written in a matrix form:

$$\mathbf{F} = \mathbf{Au} \tag{11}$$

where $\mathbf{F}$ is the vector of the forces F_i and of the voltage V, $\mathbf{u}$ is the vector of the velocities u_i and of the current I and $\mathbf{A}$ is the 4×4 matrix representing the element model in the frequency domain. Loading the mechanical ports with the acoustic impedance of the surrounding media and applying an a.c. voltage to the electric port we can characterize as a transducer the external behavior of the piezoceramic plates computing the electrical impedance and the transmission and reception transfer functions. As a matter of fact this last function is defined as:

$$TTF = \frac{F_i}{V} \tag{12}$$

Our model makes it possible to compute separate transfer functions for each external surface with arbitrary acoustic loads.

MODEL VALIDATION

In order to verify the validity of the proposed model, the electrical impedance of different PZT-5 piezoceramic disks are measured and compared with the computed results. Measurements are carried out with an impedance bridge HP4194A. Figs. 2a and 2b show two of these comparisons for diameter to thickness ratio (2a/2b) equals to 5.7 and 3.5 respectively. As it is possible to see the model is able to predict resonance and anti resonance frequencies with good accuracy both for the first radial mode and for the first thickness mode which are the most used in practical applications. The model also shows other radial modes between these two principal ones which do not agree with the experimental results; this is probably due to the presence , in the disk, of other mode, not taken into account by the model, such as the "edge modes" first reported by Shaw [3].

NUMERICAL RESULTS

Fig. 3 shows the computed resonance frequency spectra of a PZT-5 ceramic disk with a/b ranging from 1 to 25 and thickness of 10 mm. As usual, the frequency axis is normalized multiplying by the plate thickness. Coupling between thickness and radial modes is clearly evident from the spectra. Fig. 4a show the TTF of a simulated transducer made of a PZT5 disk with radius 17.5 mm and thickness 10 mm. Radiation is in water and different backing conditions are considered. The first radial mode shows a bandwidth smaller than thickness mode but with good response. If the disk is laterally backed, the bandwidth of the first radial mode increases significantly at the expense of the response, as it is shown in fig. 4b. Finally, fig 5 shows the maximum value of the TTF of the first radial and of the first thickness mode for different diameter to thickness ratios. As it is possible to see the radial mode has and optimum value for a/b=8.5.

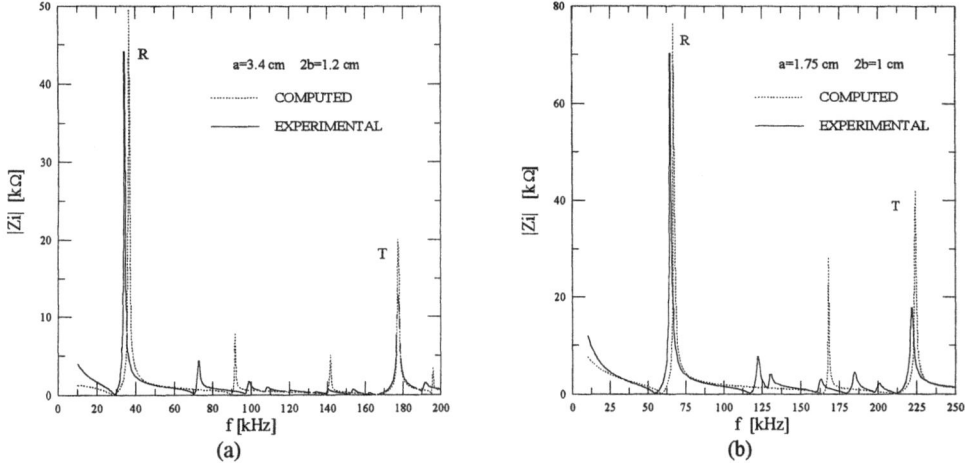

Fig. 2. Comparison between computed and experimental impedance for a disk with (a) a/b=5.7 and (b) a/b=3.5

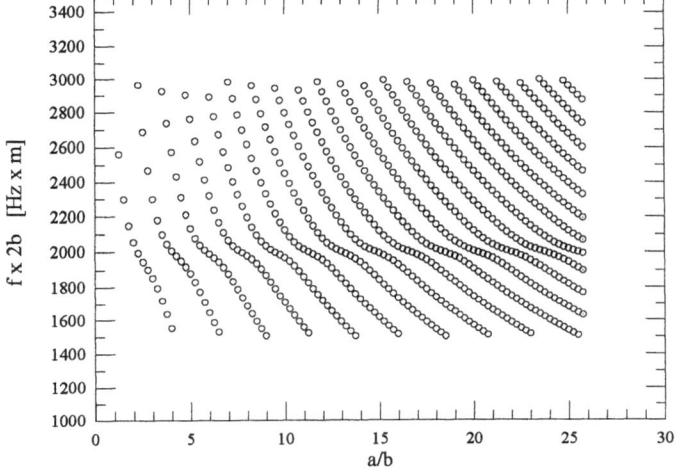

Fig. 3. Computed resonant frequency spectra of a PZT-5 disk.

CONCLUSION

A mathematical model of a piezoceramic disk with not negligible thickness was derived. The model takes into account both radial and thickness modes and it is able to handle interaction with external media. For this reason it can be used for transducer design. The model predicts a particular value of the diameter to thickness ratio able to maximize the thickness displacements of the radial mode. This optimum value corresponds to a radial frequency about seven times lower than that of the thickness mode. Following this indication low frequency transducers with good response can be designed. If appropriate lateral backing is employed good bandwidth can also be obtained. A significant advantage of this kind of transducer compared to the typical Langevin low frequency type is that it offers a much larger radiating area.

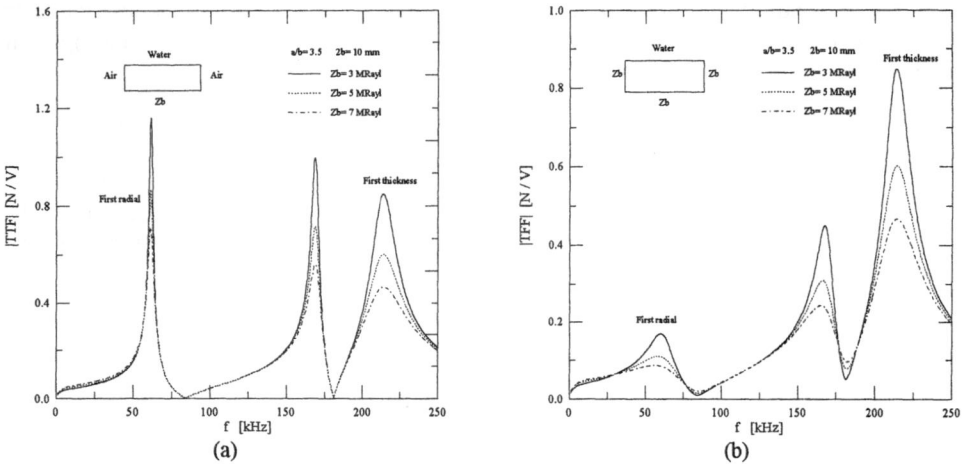

Fig. 4. Computed TTF of a PZT-5 disk radiating in water with different backing conditions: (a) laterally free, (b) laterally backed.

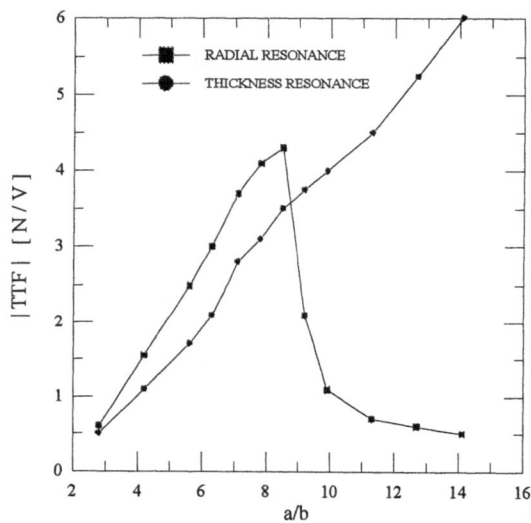

Fig. 5. TTF maximum value of radial and thickness mode for different a/b ratios.

REFERENCES

[1] M. Brissaud: "Characterization of Piezoceramics", *IEEE Transaction on Ultrasonics, Ferroelectrics and Frequency Control*, vol. 38, no. 6, pp. 603-617, November 1991.

[2] A.H. Meitzler, H.M. O'Brian, JR., and H.F. Tiersten: "Definition and Measurement of Radial Mode Coupling Factors in Piezoelectric Ceramic Materials with Large Variation in Poisson's Ratio", *IEEE Transaction on Sonics and Ultrasonics*, vol. SU-20, no. 3, pp. 233-239, July 1973.

[3] E. A. G. Shaw: "On the resonant vibration of thick barium titanate disks", *Journ. Acoust. Soc. Am.*, vol 28-1, pp. 38-50, January 1956.

COMPARISON OF SIDELOBES OF LIMITED
DIFFRACTION BEAMS AND LOCALIZED WAVES

Jian-yu Lu and James F. Greenleaf

Biodynamics Research Unit, Department of Physiology and Biophysics
Mayo Clinic/Foundation
Rochester, MN 55905

INTRODUCTION

Limited Diffraction Beams

Limited diffraction beams are a class of non-spreading solutions to the isotropic/homogeneous scalar wave equation. The first limited diffraction beam, called Bessel beam, was discovered by Durnin in 1987.[1] Later, Lu and Greenleaf discovered families of limited diffraction beams[2,3] that include all the limited diffraction beams known previously, in addition to an infinity of new beams. One family of limited diffraction beams has an X-like shape along the beam axis and was termed X wave. X waves are different from the Bessel beam because they have multiple frequencies.[2]

With an infinite aperture and energy, limited diffraction beams would propagate to infinite distance without spreading. Even if produced with a finite aperture and energy, they have a very large depth of field. Because of this advantage, limited diffraction beams could have applications in medical imaging,[4,5] tissue characterization,[6] nondestructive evaluation of materials[7] and other wave related areas such as electromagnetics[8] and optics.[9] A recent review of limited diffraction beams is given in Ref. 10.

Localized Waves

Although limited diffraction beams have a large depth of field, their sidelobes are larger than conventional focused beams at their focuses. Sidelobes may lower the contrast in medical imaging and increase the sampling volume in tissue characterization.

Localized waves were first discovered by Brittingham in 1983.[11] They were further developed by Ziolkowski et al.[12,13] and Donnelly et al.[14,15,16] Localized waves are also non-spreading and can propagate with only local deformations in their waveforms. Under ideal conditions, localized waves have lower sidelobes than limited diffraction beams. In this paper, we will study the conditions under which localized waves, specifically, subsonic (or subluminal) localized waves developed by Donnelly et al.,[15] have lower sidelobes.

X WAVE AND SUBSONIC (SUBLUMINAL) LOCALIZED WAVE

Many limited diffraction beams and localized waves have been discovered.[10] Sidelobes of different limited diffraction beams are similar, they decay in the order of $1/\sqrt{r}$, where $r = \sqrt{x^2 + y^2}$ is a radial distance perpendicular to the beam axis. Sidelobes of localized waves decay in the order of $1/r$. Because sidelobes of localized waves are also similar to each other, for simplicity, we study only the subsonic (or subluminal) localized wave[15] in the following. To study the difference between localized waves and limited diffraction beams, we will use a limited diffraction beam (the zeroth-order X wave[2]) for comparison.

The zeroth-order X wave is given by[2]

$$\Phi_{XBB_0} = \frac{a_0}{\sqrt{(r \sin \zeta)^2 + [a_0 - i(z\cos\zeta - ct)]^2}} \ ,$$ (1)

where Φ_{XBB_0} represents acoustic pressure or velocity potential (or scalar electric or magnetic field strength in electromagnetics), a_0 is a constant that determines the decay speed of the high frequency components of the X wave, ζ is an Axicon angle, $i = \sqrt{-1}$, z is the propagation axis, t is time, and c is the speed of sound. It is seen from Eq. (1) that if $z \cos \zeta - ct = $ constant, Φ_{XBB_0} will be independent of z and t, i.e., the X wave is propagation-invariant.

The subsonic wave obtained by Donnelly[15] has the form of a sinc function

$$\Phi_{Sub}(\vec{r}, t) = \frac{\sin\left[\frac{\xi}{\sin^2 \zeta}\sqrt{(z - (c \cos \zeta)t)^2 + (r \sin \zeta)^2}\right]}{\frac{\xi}{\sin^2 \zeta}\sqrt{(z - (c \cos \zeta)t)^2 + (r \sin \zeta)^2}} e^{i\frac{\xi \cos \zeta}{\sin^2 \zeta}\left(z - \frac{c}{\cos \zeta}t\right)},$$ (2)

where the subscript "Sub" means subsonic (or subluminal), ξ and ζ are constants. The spectrum of $\Phi_{Sub}(\vec{r}, t)$ can be obtained from Eq. (2)[10]

$$\widetilde{\Phi}_{Sub}(\vec{r}, \omega) = \frac{\pi \sin^2 \zeta}{\xi c \cos \zeta} e^{i\left(\frac{\omega}{c} - \xi\right)\frac{1}{\cos \zeta}z} J_0\left(r \tan \zeta \sqrt{\left(\frac{\xi \cos \zeta}{\sin^2 \zeta}\right)^2 - \left|\frac{\omega}{c} - \frac{\xi}{\sin^2 \zeta}\right|^2}\right),$$ (3)

where $0 < \left|\frac{\omega}{c} - \frac{\xi}{\sin^2 \zeta}\right| < \frac{\xi \cos \zeta}{\sin^2 \zeta}$.

The propagation speed of the peak of the subsonic (or subluminal) wave in Eq. (2) is slower than the speed of sound (or light). This is opposite to the X wave whose peak propagates at a speed that is faster than the speed of sound (or light) (supersonic (or superluminal), see Eq. (1)).[2] It is seen that the frequency spectrum of the subsonic wave in Eq. (3) has limited bandwidth. The peak of the real or imaginary parts of the subsonic wave fluctuates as the wave propagates, which is a common feature of all localized waves (Eq. (2)).[10] This is because the localized waves always contain two propagation terms:

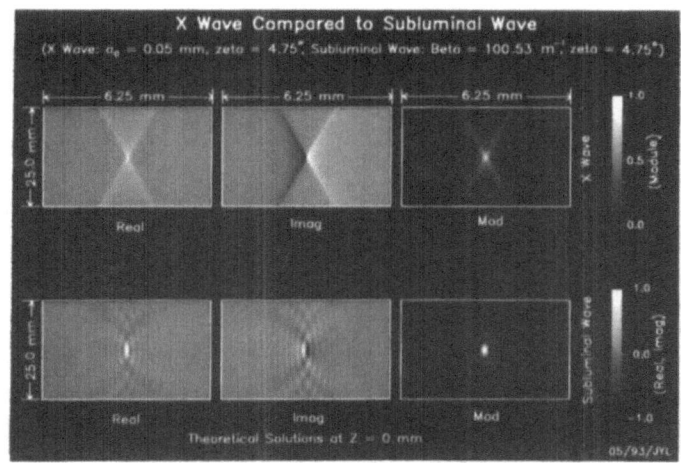

Figure 1. Theoretical X waves (top row) and subsonic (or subluminal) waves (bottom row) given by Eqs. (1) and (2) at the axial distance, $z = 0$. The horizontal direction represents time, the distance (6.25 mm) is calculated by assuming the speed of sound is 1500 m/s. The vertical direction is the radial direction and the images shown in the figure are a cross section through the wave axis. The dimension of each image is 6.25 mm × 25.0 mm. The columns from left to right correspond to real, imaginary, and modulus of the waves. Note that the scale bar of the modulus images is different from that of the real and imaginary images. The parameters for the X wave are $a_0 = 0.05mm$ and $\zeta = 4.75°$, and for the subsonic wave are $\xi = 100.53m^{-1}$ and $\zeta = 4.75°$.

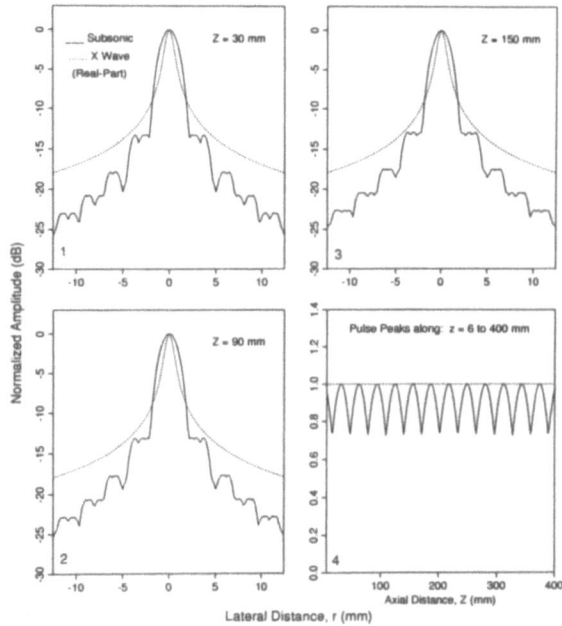

Figure 2. Line plots of the maximum envelope of A-lines of the real part of the X wave (dotted lines) and the subsonic wave (solid lines) shown in Fig. 1 at 3 axial distances: $z =$ (1) 30, (2) 90, and (3) 150 mm, versus the radial distance. Peak of the waves versus the propagation distance from 6 to 400 mm is shown in Panel (4). It is seen that the subsonic wave has lower sidelobes and the peak of the wave oscillates as it propagates. The waves are assumed to be produced with an infinite aperture and energy.

147

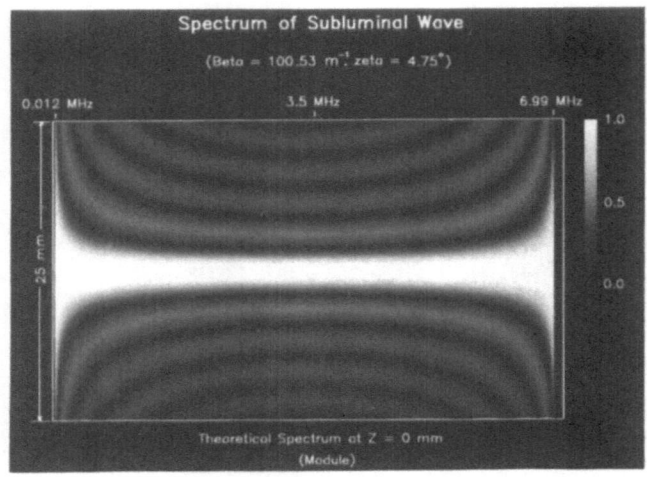

Figure 3. The modulus of the spectrum of the subsonic wave in Fig. 1 (see Eq. (3)) at the axial distance, $Uz = 0$. The horizontal direction represents frequency and the vertical direction represents the radial distance. With the parameters of the subsonic wave shown in Fig. 1, the frequency components are limited between 0.012 MHz and 6.99 MHz with a central frequency of 3.5 MHz.

$z - c_1 t$ and $z - c_2 t$, where c_1 and c_2 are different constants. For the subsonic wave (Eq. (2)), $c_1 = c \cos \zeta < c$ (subsonic) in the amplitude term and $c_2 = c / \cos \zeta > c$ (supersonic) in the phase term.

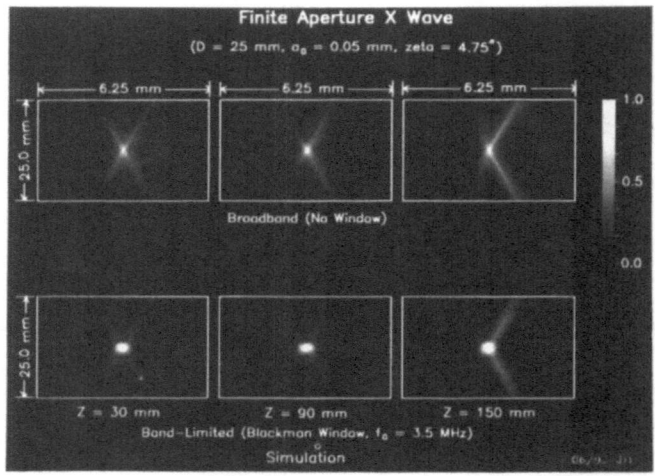

Figure 4. X waves produced at 3 axial distances ($z = 30$ (left column), 90 (middle column), and 150 mm (right column)) with a transducer of diameter of 25 mm. The images in the top row are produced with a transducer of infinite bandwidth, while the images in the bottom row are produced with a transducer of the Blackman window type of transfer function with a central frequency of 3.5 MHz and a -6-dB relative bandwidth of about 81%. The dimension of the images and the parameters of the waves are the same as those in Fig. 1.

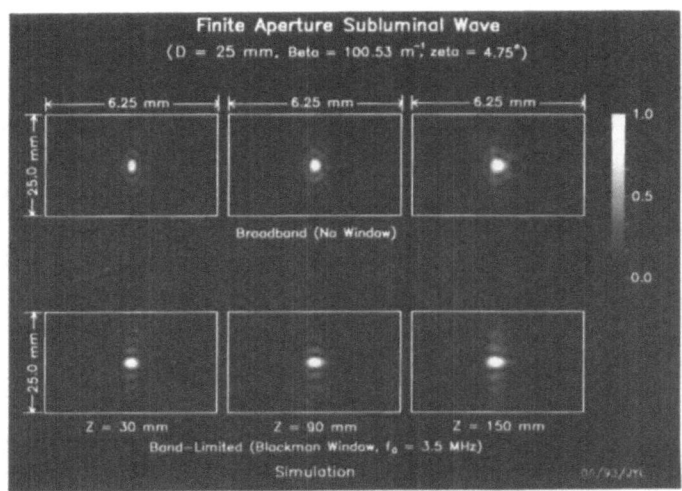

Figure 5. The subsonic (or subluminal) waves produced at 3 axial distances (z = 30 (left column), 90 (middle column), and 150 mm (right column)) with a transducer of diameter of 25 mm. This figure has the same format as that of Fig. 4 and the parameters of the waves are the same as those in Fig. 1.

RESULTS

A plot of Eqs. (1) and (2) at $z = 0$ is shown in Fig. 1. The parameters of the X wave and the subsonic wave are adjusted so that they have comparable mainlobe sizes. For the X wave, a_0 = 0.05 mm and ζ = 4.75°. And for the subsonic wave, ζ = 4.75° and ξ =

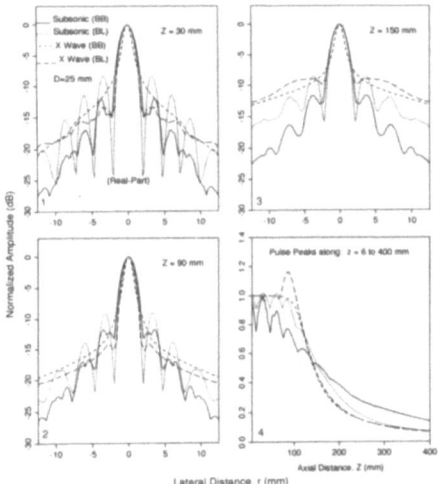

Figure 6. Line plots of the maximum envelope of A-lines of the real part of the X waves and the subsonic waves produced with a finite aperture transducer (25 mm in diameter) in Figs. 4 and 5, respectively, versus the radial distance. The plots are obtained at 3 axial distances: z = (1) 30, (2) 90, and (3) 150 mm. Peak of the waves versus the propagation distance from 6 to 400 mm is shown in Panel (4). "BB" and "BL" represent broadband and band-limited waves, respectively. Solid and dotted lines represent the broadband and the band-limited subsonic waves, respectively. Dashed and long dashed lines represent the broadband and the band-limited X wave, respectively.

149

100.53 m^{-1}. A line plot that corresponds to the real part of the waves in Fig. 1 is shown in Fig. 2. The difference of the sidelobes between the two waves is clearly seen. The sidelobes are obtained by plotting the maximum of each A-line (lines that are parallel to the wave axis (in horizontal direction of Fig. 1)) versus the radial distance of the beams. The oscillations of the peaks of the subsonic wave as it propagates are also obvious. The shape of the modulus of the subsonic wave in Fig. 1 is composed of concentric ellipses. That is why the peak of the subsonic wave propagates slower than the speed of sound.

The modulus of the spectrum of the subsonic wave is shown in Fig. 3. It has clear lower and upper cut off frequencies (see Eq. (3)). This is different from other localized waves[10] whose frequency components extend to infinity. The lower and upper cut off frequencies of the subsonic wave are given by $f_l = (1 - \cos \zeta) f_0$ and $f_h = (1 + \cos \zeta) f_0$, respectively, where $f_0 = \xi c / (2\pi \sin^2 \zeta)$ is the central frequency. The spectrum in Fig. 3 is obtained at $z = 0$ for various radial distances, r (in the vertical direction of the figure). The lower and upper cut off frequencies can be calculated with the parameters of the subsonic waves in Fig. 1: $f_l = 0.012$ MHz and $f_h = 6.99$ MHz (the central frequency is $f_0 = 3.5$ MHz). The relative bandwidth $((f_h - f_l)/f_0 = 2 \cos \zeta)$ of the spectrum is extremely large (about 199%).

In practical applications, waves must be produced with transducers of finite apertures and finite relative bandwidth. X waves (Fig. 4) produced by the transducer that has either a finite aperture (diameter $D = 25$ mm) or a finite aperture plus finite lower and upper cut off frequencies have been simulated using the Rayleigh-Sommerfeld diffraction formula.[18] The envelope of the real part of the X waves is displayed in Fig. 4. For the broadband X wave, a flat frequency response of the transducer is assumed. The lower and upper cut off frequencies for the band-limited X waves are 0 and 7.0 MHz, respectively, which correspond to the frequencies at zeroes of the Blackman window function.[17] The Blackman window has a central frequency of 3.5 MHz and its −6-dB relative bandwidth is about 81% of the central frequency. The simulated finite aperture subsonic wave is shown in Fig. 5 in the same format as Fig. 4. The parameters for the X wave and the subsonic wave are the same as those in Fig. 1.

Line plots of Figs. 4 and 5 are shown in Fig. 6 in the same format as Fig. 2. It is seen that sidelobes of the subsonic wave increase as the bandwidth decreases. For the X wave, sidelobes reduce with the bandwidth slightly.

Changes in the characteristics of the sidelobes of the subsonic wave with several Blackman window functions are shown in Fig. 7. These Blackman windows are shown in Fig. 8. It is noted that as the base bandwidth of the Blackman window function decreases from 14 MHz to 7 MHz, the sidelobes are increased dramatically (the base bandwidth is determined by the frequencies that correspond to the zeroes of a Blackman window function).

CONCLUSION

Sidelobes of limited diffraction beams are higher than those of conventional focused beams at their focuses. One way to reduce the sidelobes is to use localized waves that are non-spreading and propagate with only local deformations. However, to take advantage of the low sidelobes of localized waves, the transducers used to produce these waves must have a large relative bandwidth (at least 162% −6-dB relative bandwidth in the above example, see Fig. 7). Perhaps PVDF transducers, that have large bandwidth, could be used to produced low sidelobe localized waves.

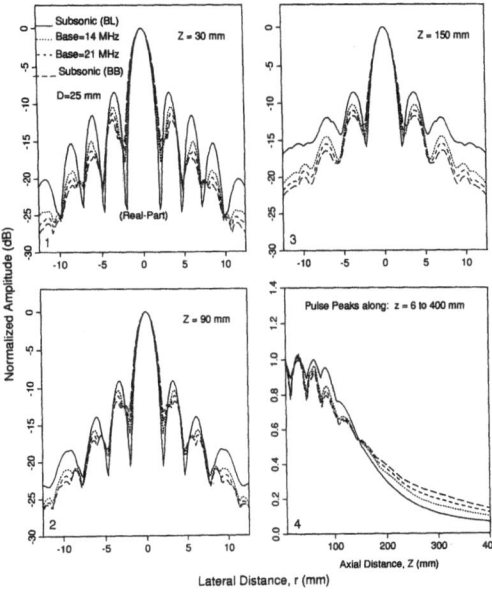

Figure 7. Line plots of the maximum envelope of A-lines of the real part of the subsonic waves produced with a finite aperture transducer (25 mm in diameter) versus the radial distance showing the change of sidelobes with transducer bandwidth. The figure has the same format as that of Fig. 6. "Subsonic (BL)" (solid lines) and "Subsonic (BB)" (long dashed lines) represent the subsonic waves obtained with the transducer that has a transfer function of the Blackman window type centered at 3.5 MHz, and are repeated from Fig. 6 for comparison. These plots correspond to the base bandwidths (the bandwidth between the two zeroes of the Blackman window, see Fig. 8) of 7.0 MHz and infinity respectively. Dotted and dashed lines represent the subsonic waves produced by the transducer of 14 MHz and 21 MHz base bandwidths, respectively (see Fig. 8).

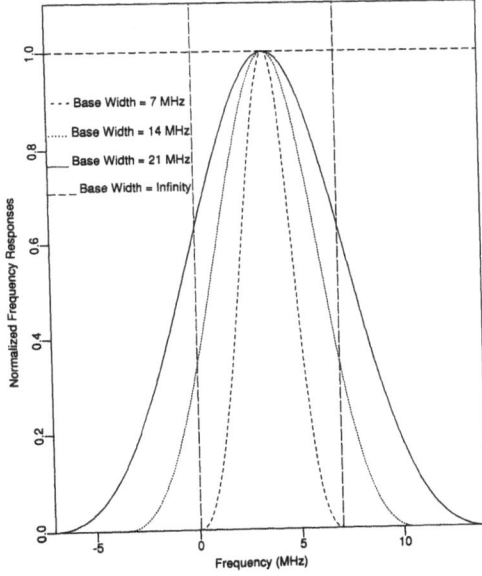

Figure 8. Examples of Blackman windows of different bandwidths: 7.0 MHz (dashed line), 14.0 MHz (dotted line), 21.0 MHz (solid line), and infinity (long dashed line). The two vertical dashed lines represent the lower and higher cut off frequencies of the subsonic waves calculated by using the parameters in Fig. 1. The -6-dB bandwidth of the Blackman window is about 41% of the base bandwidths above.

151

ACKNOWLEDGMENTS

The authors appreciate the secretarial assistance of Elaine C. Quarve. This work was supported in part by grants CA 43920 and CA 54212 from the National Institutes of Health.

REFERENCES

1. J. Durnin, Exact solutions for nondiffracting beams. I. The scalar theory, *J. Opt. Soc. Am.* 4(4):651–654 (1987).

2. Jian-yu Lu and J.F. Greenleaf, Nondiffracting X waves — exact solutions to free-space scalar wave equation and their finite aperture realizations, *IEEE Trans. Ultrason., Ferroelec, Freq. Cont.* 39(1):19-31 (Jan., 1992).

3. Jian-yu Lu and J.F. Greenleaf, Experimental verification of nondiffracting X waves, *IEEE Trans. Ultrason., Ferroelec., Freq. Contr.* 39(3):441-446 (May, 1992).

4. Jian-yu Lu and J.F. Greenleaf, Ultrasonic nondiffracting transducer for medical imaging, *IEEE Trans. Ultrason., Ferroelec., Freq. Contr.* 37(5):438-447 (Sept., 1990).

5. Jian-yu Lu and J.F. Greenleaf, Pulse-echo imaging using a nondiffracting beam transducer, *Ultrasound Med. Biol.* 17(3):265-281 (May, 1991).

6. Jian-yu Lu, and J.F. Greenleaf, Evaluation of a nondiffracting transducer for tissue characterization, [IEEE 1990 Ultrasonics Symposium, Honolulu, HI, Dec. 4–7, 1990], *IEEE 1990 Ultrason. Symp. Proc.* 90CH2938-9, 2:795-798 (1990).

7. Jian-yu Lu and J.F. Greenleaf, Producing deep depth of field and depth-independent resolution in NDE with limited diffraction beams, *Ultrason. Imag.* 15(2):134-149 (April, 1993).

8. R.W. Ziolkowski, Localized transmission of electromagnetic energy, *Phys. Rev. A.* 39(4):2005-2033 (Feb. 15, 1989).

9. J. Ojeda-Castaneda and A. Noyola-lglesias, Nondiffracting wavefields in grin and free-space, *Microwave and Optical Tech. Lett.* 3(12):430-433 (Dec. 12, 1990).

10. Jian-yu Lu and J.F. Greenleaf, Biomedical ultrasound beamforming, *Ultrasound Med. Biol.* (to be published in 1994).

11. J.N. Brittingham, Focus wave modes in homogeneous Maxwell's equations: transverse electric mode, *J. Appl. Phys.* 54(3):1179-1189 (1983).

12. R.W. Ziolkowski, Exact solutions of the wave equation with complex source locations, *J. Math. Phys.* 26(4):861-863 (April, 1985).

13. R.W. Ziolkowski, D.K. Lewis, and B.D. Cook, Evidence of localized wave transmission, *Phys. Rev. Lett.* 62(2):147-150 (Jan. 9, 1989).

14. R. Donnelly and R.W. Ziolkowski, A method for constructing solutions of homogeneous partial differential equations: localized waves, *Proc. R. Soc. Lond. A,* 437:673-692 (1992).

15. R. Donnelly and R.W. Ziolkowski, Designing localized waves, *Proc. R. Soc. Lond. A,* 440:541-565 (1993).

16. R. Donnelly, D. Power, G. Templeman and A. Whalen, Graphic simulation of superluminal acoustic localized wave pulses, *IEEE Trans. Ultrason., Ferroelec., Freq. Contr.* 41(1):7-12 (Jan., 1994).

17. A.V. Oppenheim and R.W. Schafer, *Digital signal processing.* Prentice-Hall, Inc., Englewood Cliffs, NJ, ch. 5 (1975).

18. J.W. Goodman, *Introduction to Fourier Optics.* McGraw-Hill, New York, chs. 2–4 (1968).

FERROELECTRIC POLYMER ARRAY SENSORS FOR ULTRASONIC IMAGING IN AIR

L. Capineri[1], A.S. Fiorillo[2] and S. Rocchi[3]

[1] Dipartimento Ingegneria Elettronica, Università degli Studi di Firenze,
Via S.Marta 3, 50139 Firenze, Italy
[2] DIIIE, Università di Salerno, Via Ponte don Melillo, 84084, Fisciano, Italy
[3] Istituto di Elettronica Università degli Studi di Perugia, S. Lucia Canetola
Perugia, Italy

INTRODUCTION

In this work we studied ferroelectric polymer films because they are suitable transducer materials for making air-borne ultrasonic transducers. They have several advantages as regards to other conventional piezoelectric materials, such as lightness, low cost and small dimensions when employed in robotic grasp-effectors for short range vision systems.

We built transducers with strips of PVDF film mounted with hemicylindrical shape. The underlying theory of this type of transducers has been already developed[1,6] and it shows that the resonance frequency depends on the inverse of the cylinder radius.

We used this feature in order to fabricate transducers in the frequency range 40-200 kHz, by developing a reliable fabrication technique for a single transducer and the preliminary results of the acousto-electric characterization are also reported.

Finally in order to achieve spatial resolution in the millimeter range we mounted this type of transducers in a linear array configuration.

The linear array of transducers allows the implementation of a broadband synthetic aperture imaging technique with multi-offset transducers that was developed for medical and non-destructive testing applications[2,3]. The broadband characteristics are obtained by combining the echo signals from two or more linear arrays of 16 elements each, working at different central frequencies. The multi-frequency and multi-offset synthetic aperture improves the lateral and the spatial resolution. By a digital beam former we simulated the response from a test-object consisting of a thin steel wire and the results have demonstrated a feasible imaging technique.

Acoustical Imaging, Vol. 21, Edited by
J.P. Jones, Plenum Press, New York, 1995

BASIC PRINCIPLES AND FABRICATION OF PVDF FILM TRANSDUCERS

Piezoelectric polymer films have suitable characteristics for making linear arrays of air ultrasonic transducers. The main advantages for their applications in ultrasonic short-range vision system are summarized below:

· low lost material,

· lightness,

· limited size (hemicylindrical shape),

· easy to fabricate in the frequency range 20 to 200 kHz,

· air matched acoustic impedance.

The good acoustic impedance matching of PVDF polymer with air is clear from the Table 1. The acoustic impedance of PVDF is close to the silicon rubber that is the common material used for matching PZT transducers with air.

Table 1. Acoustic impedance of different materials

Zac. air (at 20°C)	= 4.27 10^2 Kg m^{-2} s^{-1}
Zac. PVDF	= 3.82 10^6 Kg m^{-2} s^{-1}
Zac. PZT5	= 33.7 10^6 Kg m^{-2} s^{-1}
Zac. Silicone Rubber	= 1.07 10^6 Kg m^{-2} s^{-1}

The theory underlying this type of transducer, when mounted with hemicylindrical shape, was already developed[1,6] and here the main concepts on this type of resonator are outlined. We used a uniaxially stretched PVDF film vibrating in the length extensional mode along the stretching direction 1 in Figure 1. The PVDF film is metallised on both sides with thickness τ and width w (see Figure 1). The electric field along direction 3 is generated by the external voltage applied to the metallised surfaces. Here we used the PVDF film mounted with hemicylindrical shape. Due to this geometrical configuration, the resulting force of an elementary volume of PVDF works only in the radial direction and the integration over the film surface of all elementary sources, generates an ultrasonic wave in forward (concave side) and back (convex side) directions.

Fiorillo[1] has demonstrated the theory leading to the relationship between the resonance frequency f_r and the PVDF transducer characteristics:

$$f_r = 1/ 2\pi r \; (1/\rho_{PVDF} \; S^E_{11})^{1/2} \qquad (1)$$

where: r mean radius of cylinder, S^E_{11} tangential elastic compliance, ρ_{PVDF} average density of metallised film.

In equation (1), apart a constant factor depending on the material characteristics, the resonance frequency is proportional to the inverse radius of the cylinder. From this consideration emerges that transducers with different central frequency can be fabricated simply by adjusting the radius of the thin PVDF film. In Table 2 are shown the radius values for the frequency range of interest in our applications.

Table 2. Theoretical f_r values for different transducer radius r.

r[mm]	f_r[kHz]
3.88	61
2.88	83
1.38	173

The transducer sensitivity drops at higher frequencies because the size of the transducer becomes very small reducing also the radiating area equal to $\pi\, r\, w$, while the ratio r/λ_{air} remains constant and equal to 0.7 at any frequency.

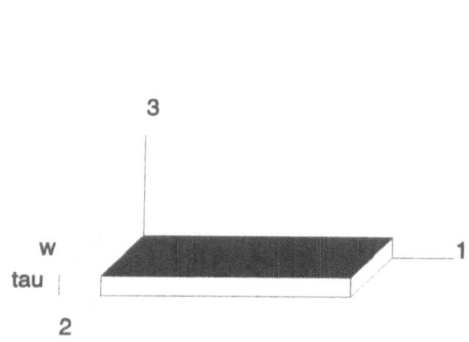

Figure 1 Ultrasonic resonator with uniaxially PVDF film stretched along direction 1.

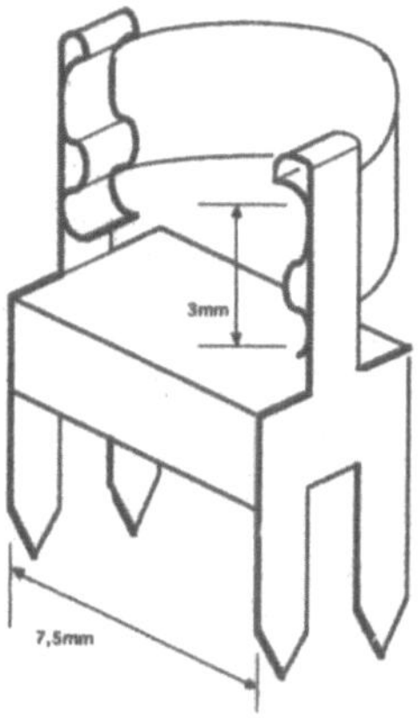

Figure 2. PVDF film transducer with hemicylindrical shape.

The transducer mechanical assembly is shown in Figure 2. A PVDF film metallised on both sides with gold, width w=3 mm and thickness τ=40 µm is used. The extremities of the film are bonded with silver loaded epoxy to copper pins pair. The copper pins are made by photolitography and the pins displacement is compatible with standard DIP integrated circuit sockets. The contact resistance between the film and the electrodes is reduced by folding the upper part of the pin onto the PVDF film and so the contact surface is doubled. In this assembly it is essential to keep parallel the two electrodes otherwise the radius is altered and the resonance frequency can result quite different than the theoretical value. Moreover the single transducer can be inserted in a 14 pin DIP socket together with the signal conditioning electronics.

TRANSDUCERS CHARACTERISATION

Prototypes of transducers at nominal central frequency 61 kHz were fabricated and characterised. An experimental setup was assembled for evaluating the response of two transducers from a plane reflector at distance 60 mm in air. The transmitting transducer was excited with a 150 V pulse by a Panametrics 5052PR pulser.

Though the broadband excitation technique is quite efficient in this case, low voltage narrow band signals can be considered with the advantage of a reduced electrical stress on the transducer contacts. As receiver we used a broadband amplifier with voltage gain 40 dB followed with a pass-band filter with bandwidth broader than the transducer one. An output signal of the order of 1 mV with S/N equal 15 dB was obtained. However, the S/N can be improved by replacing the broadband receiver with a tuned amplifier.

The amplitude spectrum of the received signal has a central frequency of 51 kHz that is lower than the nominal one (61 kHz) due to mechanical assembly tolerances. This value has been also confirmed with measurements of the transducer impedance phase. The actual resonance frequency value was confirmed with measurements of the transducer impedance phase. The bandwidth of these transducers at -3dB is 8.5 kHz and Q = 6. In the transducer far field a main lobe at - 6 dB of 50° with secondary lobes less than 30% of the main one was measured.

ULTRASOUND IMAGING IN AIR BY MULTI-OFFSET AND MULTI-FREQUENCY SAFT

In order to achieve spatial resolution in the millimeter range in air, we used this type of transducers in a linear array configuration. The linear array allows the implementation of a synthetic aperture focussing technique (SAFT) with multi-offset transducers. This imaging technique was developed in backscattering mode for medical[2] and non destructive testing applications[3] and it demonstrated some advantages in terms of lateral resolution and signal to noise ratio. Its ability in resolving inclined plane reflectors makes it suitable for the investigation of object contours.

Moreover the ultrasonic system bandwidth was increased by coherently summing the echo signals from two linear arrays with different resonance frequencies (64 and 83 kHz in our case).

Remanding the reader to pertinent literature for the basic theory of reflection SAFT imaging technique[5] in this section we explain briefly the main characteristics of the multi-offset SAFT in a double linear array configuration.

During the acquisition phase the single transmitting (TX) and receiving (RX) elements are used independently. Because the variable spatial offset of the TX-RX pairs, the object is investigated over a wide angle range. Let N the number of elements of the linear array. In the first step data are collected from all possible combinations of TX-RX pairs, then at the end of the acquisition process we have N^2 digitized radiofrequency signals.

In the second step the digitized ultrasonic signals are backprojected in the space domain by ellipsoidal scalar functions. The position of an elementary reflector is retrieved by considering that the time of flight (TOF), multiplied by the ultrasound air velocity, defines an ellipse for each transmitter-receiver pairs. Hence N^2 ellipsis are written in the computer image memory. Only in the position of elementary reflector the signal samples are coherently summed, resulting in a high pixel intensity. The radiofrequency image is finally detected for easy visualization.

The spatial resolution takes full advantage from large arrays and transducer elements with large main lobe. In fact is possible to demonstrate[3] that the theoretical lateral resolution δ_{lat} is given by the following expression:

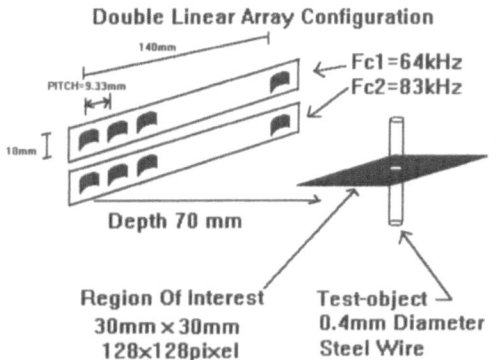

Double Linear Array Configuration

Figure 3 Multi-frequency imaging system with two 16 elements arrays.

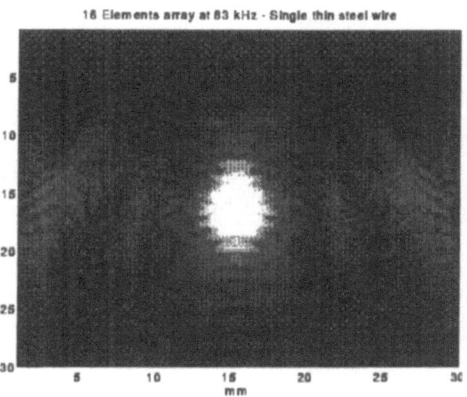

Figure 4 Point spread function with array at f_r 83 kHz.

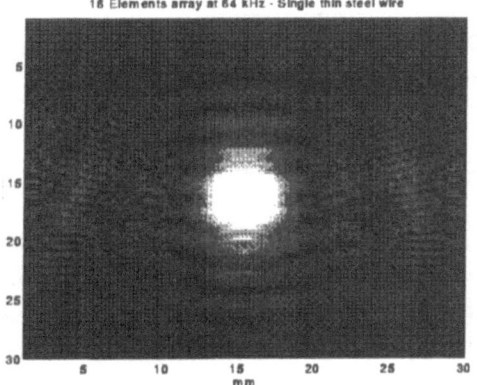

Figure 5.Point spread function with a linear array at f_r 64 kHz.

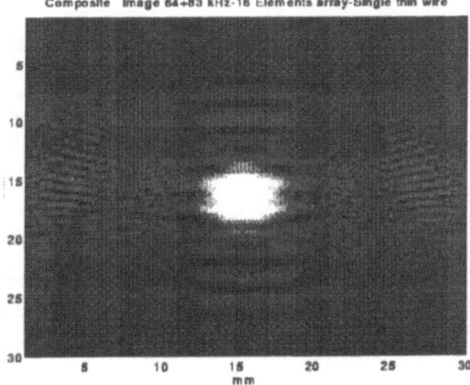

Figure 6 Point spread function with multi-frequency system (64 + 83 kHz).

157

$$\delta_{lat}(z) = \text{sinc}^2\left(k\,d\,z\,/\,2\,(x^2+z^2)^{1/2}\right) \tag{2}$$

where: x is depth, k wave number, d array length, z offset from array centre.

The proposed system consist of a double array configuration as shown in Figure 3. Each array has 16 elements with central frequency 64 and 83 kHz respectively. In the next section the results of computer simulations by a developed digital beamformer able to count for the real characteristics of the transducers and excitation signals, are presented.

RESULTS

A quantitative evaluation of the spatial resolution attainable with the proposed imaging system is possible by the reconstruction of the system point spread function. We simulated the ultrasonic response from a test-object consisting of a 0.4 mm diameter steel wire at depth 70 mm from the array. The diameter is lower than the acoustic wavelength so that the response is almost isotropic. Figure 4 and 5 illustrate the images obtained by processing the synthesized signals from single-frequency array systems respectively at 83 and 64 kHz. The images are 128x128 pixel and 30x30 mm. After the summation of the two data sets the results of the image reconstruction is shown in Figure 6. We can observe that the lateral resolution is unchanged but the axial resolution is improved in the multi-frequency image. Table 3 shows more quantitatively these results.

Table 3. Results of spatial resolution

Resonance frequency	83 kHz	64kHz	64+83kHz
Lat. Res. -6dB [mm]	6.34	6.34	6.5
Axi. Res. -6dB [mm]	4.42	5.0	4.0

These results show an improvement on the axial resolution but a deal of optimization remains in the choice of the transducer frequency. Ideally after the combination of two transducer bandwidth we would have a flat bandwidth but this was not completely achieved in our simulated system. The lateral cross sections displays some sidelobes caused by the coarse sampling of the array aperture at intervals greater than half wavelength.

REFERENCES

1. A.S. Fiorillo, Design and characterization of a PVDF Ultrasonic Range sensor, *IEEE Transactions on UFFC*, Vol. 39, N.6, pp. 688-692 (1992)
2. L. Capineri, G. Castellini, L. Masotti, S. Rocchi
   Computer simulated benchmarks of synthetic aperture techniques for vascular ultrasonography, *Micro and Microscopica Acta*, Vol.23, No.4, pp. 515-524, (1992)
3. L. Capineri, H.G. Tattersall, J.A.G. Temple, M.G. Silk, Time of flight diffraction tomography for NDT applications, *Ultrasonics*, Vol 30 No 5, pp. 275-288, (1992)
4. D.A. Berlincourt, D.R. Curran and H. Jaffe, *Piezoelectric and piezomagnetic materials*

and their function in transducers, W.P. Mason, Physical Acoustics, Principles and methods, Ed. New York, Academic Press, vol 1, part A, No. 3, pp. 169-270, (1964)

5. M. Soumekh, Surface imaging via wave equation inversion,
   Acoustical Imaging Vol. 16, pp. 383-393

6. A.S. Fiorillo, PVDF ultrasonic sensors for small objects location,
   Sensors & Actuators: A. Physical, vol 42, n. 1-3, pp. 406-409, (1994)

EXPERIMENTAL RESULTS ON PHASE ABERRATION CORRECTION FOR TWO-DIMENSIONAL CONFORMAL ARRAYS

Pai-Chi Li and Matthew O'Donnell

Electrical Engineering and Computer Science Department
University of Michigan
Ann Arbor, MI 48109-2122

INTRODUCTION

Very large, two-dimensional anisotropic arrays (also known as 1.5 dimensional arrays) have been proposed to enhance low contrast detectability for diagnostic ultrasound.[1] These arrays are undersampled in the non-scan direction to reduce channel number. They are not suitable for real-time three-dimensional imaging due to limited steering capability in elevation. Nevertheless, they can provide full three-dimensional focus. Moreover, if used with a fully adaptive front-end, they can correct for beamforming artifacts associated with a large aperture.

Clinical applications of large arrays have not been successful for the following reasons. First, if the aperture is large, a significant portion of the array is likely to be blocked by obstructions. Blocked elements receiving either no signals or strong reverberations can be easily detected using the receive amplitude distribution across the array. Such elements should not contribute to beamforming, and consequently should be turned off. Inoperable elements result in higher sidelobes in beam patterns and deteriorate contrast resolution. Furthermore, spatial resolution is also degraded if the total aperture size is reduced. To compensate for degradation due to blocked array elements, an object dependent algorithm has been proposed.[2] Details of this algorithm have been addressed by Li et al.[2,3]

Second, aberrations due to sound velocity inhomogeneities are also likely to produce severe beamforming artifacts. The sound velocity is usually assumed constant in tissue to calculate phasing patterns in forming acoustic beams from arrays. However, the velocity of acoustic waves in the body varies over a wide range. Therefore, ultrasound waves experience wavefront distortion which disrupts diffraction patterns and produces image artifacts. Conformality due to the contact between the aperture and the body is another problem in potential clinical applications of large arrays. Since the surface of the body is non-planar, a non-planar array must be used if adequate contact is to be maintained over a large area. The phase aberration problem is more pronounced if a larger array or a higher operating frequency is employed to increase resolution.

As shown in Figure 1, phase aberrations associated with large two-dimensional conformal arrays are produced by both sound velocity inhomogeneities in tissue and irregular geometries due to the flexibility of the array. Although small compared to the total array size,

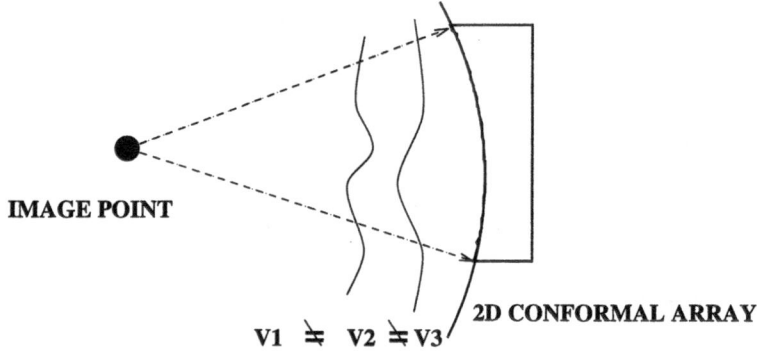

Figure 1. Phase aberrations with conformal arrays.

array conformality can be as large as several wavelengths. In this case, the loss in image quality due to the unknown array geometry is no longer only a few dB in contrast or several degrees in spatial resolution, it is possibly catastrophic.

Extensive research has been conducted on phase aberration correction.[4-9] One correlation based method has been developed and shown to be robust with no need of a beacon.[5-7] In the case of distributed scatterers, however, this algorithm has to be applied iteratively to compensate for the underestimation resulting from superposition of random scatterers. The main assumption behind this method is that the received signal in one channel can be approximated by a time delayed replica of the signal received by another channel. Therefore, phase aberrations can be found from the position of the maximum in the cross-correlation function of received signals between two array elements. Based on this method, a real-time aberration correction system using a baseband beam former has been described.[10]

Although this method is effective for small errors across a one-dimensional aperture, problems arise with these methods if applied to large two-dimensional arrays. Clearly, tissue inhomogeneities are not restricted only in the scan direction. With a large array length in the non-scan direction, it is obvious that proper estimation of phase aberrations must be performed two-dimensionally.

Phase aberration correction on a two-dimensional anisotropic array has been investigated.[1] Algorithms for two-dimensional phase aberration correction were developed by extending the one-dimensional, correlation based method. Two types of two-dimensional aberration correction architectures were proposed.

The first scheme is shown in Figure 2 where each array element can be correlated with four adjacent elements (except for boundary elements). In other words, the correlation path between two arbitrary array elements is not unique. Nevertheless, the measured phase error between any two elements should be independent of the correlation path according to the phase closure property. This redundant phase information provides several advantages. First, in the presence of inoperable elements, phase errors can still be unwrapped by finding a path circumventing these elements. Second, phase aliases can be detected using different correlation paths. It becomes more important if anisotropic arrays are used where element spacings in the non-scan plane can be several wavelengths. The full two-dimensional correlation structure, on the other hand, has several disadvantages. First, the number of correlators is approximately two times the number of channels. If the number of channels is large, this hardware complexity will be significant. Second, the full correlation architecture is not suitable for pipelined structures, i.e., it is difficult to implement in real-time.

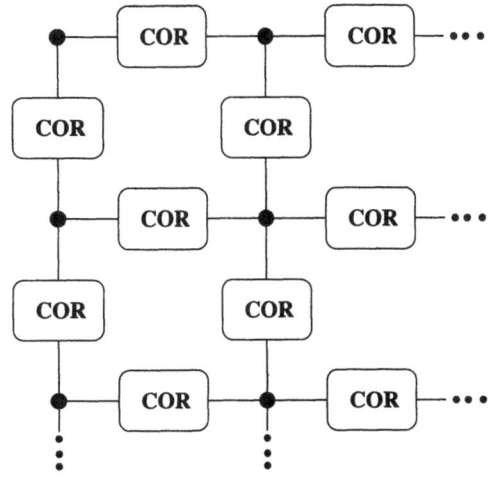

Figure 2. Full two-dimensional correlation structure.

An alternative (*row-sum*) structure is shown in Figure 3. In this case, correlation is

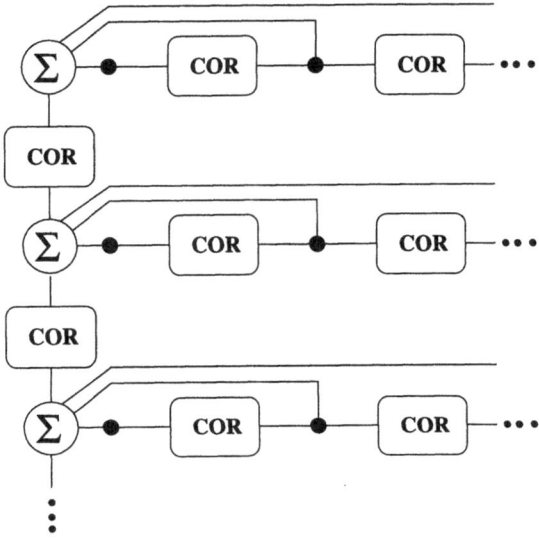

Figure 3. *Row-sum* correlation structure.

performed on neighboring elements on the same row in the scan direction. Correlation in the non-scan direction is only performed on the coherent sums of each row. In this way, the phase error is unwrapped along a unique path and the number of correlators required in this structure is about the same as the number of array channels. More importantly, this structure is suitable for pipelined architectures and real-time processing. The drawbacks are, first, aliases are more likely to occur and second, the effective center of a row with inoperable elements is moved.

Simulations showed that both methods produced high quality phase estimates. Even in the presence of inoperable elements, both architectures greatly improve image quality over that produced by smaller apertures. However, no experimental results were reported and only planar geometries were assumed. It is the goal of this paper to test the efficacy of the algorithms on estimating large time excursions across the aperture using real measurements. In addition, restoration of the image quality of adaptive imaging systems using two-dimensional conformal arrays will be further verified.

This paper is organized as follows. Experimental system configuration, two-dimensional array geometry and phantom design are described in the Method Section. Results on phase aberration correction using a two-dimensional conformal aperture are included in the Results Section. Finally, the paper concludes in the Discussion Section.

METHOD

Experiments were designed to study the degradation in image quality due to phase aberrations and to test the efficacy of the correlation based correction method. The system setup is shown in Figure 4. A 386-based microcomputer was used for measurement control.

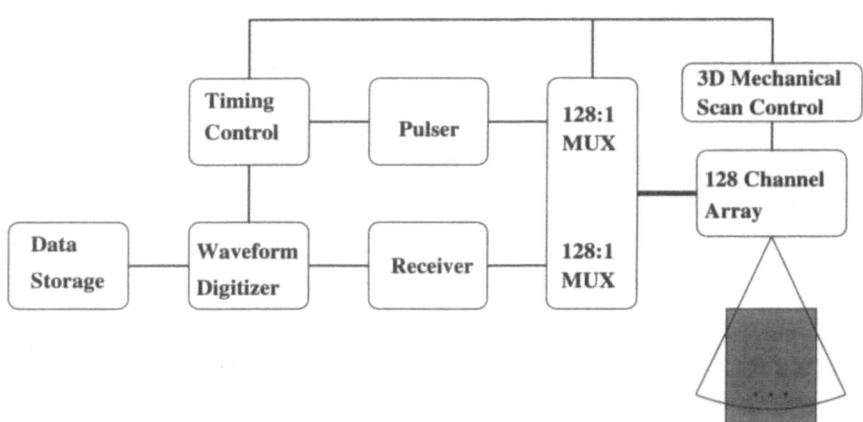

Figure 4. The measurement system.

A one-dimensional array was used to synthesize a two-dimensional, anisotropic array. The array operated at 3.5 MHz and had 128 elements with a 0.22 mm interelement spacing. To confine the transmission and reception of acoustic waves in the non-scan direction, an acoustic mask with an open aperture of only 2 mm was placed on the array. The mask was made of adhesive paper and provided a round-trip (transmit and receive) attenuation of about 30 dB. The three-dimensional mechanical scan system was used to move the one-dimensional array and synthesize arbitrary array geometries. The array was moved by 20 mm, i.e., there were 10 sub-apertures (10 scan steps) in the non-scan direction. During data acquisition, full transmit and receive records were obtained in the scan direction and a synthetic aperture approach was applied in the non-scan direction.

Using a reconfigurable, 256 channel multiplexer, this system permitted full transmission and reception from a 128 element array. For each sub-aperture, 128 × 128 radio frequency (rf) records were recorded with no time delays applied during data acquisition. Each record comprised 2048 rf samples at a 13.8889 MHz sampling rate. Signal averaging was applied to improve the electronic signal-to-noise ratio (SNR). Each rf record was accumulated during 16 identical firings with the same transmit and receive multiplexer positions (i.e., signal averaging of 16) and were digitized into 16 bit samples. Diverse simulations were performed separately and different images were displayed based on the same data set after data were transferred from a local storage device to a workstation.

A gelatin-based phantom was constructed to verify the full, three-dimensional focusing capability of large two-dimensional arrays as well as test the efficacy of two-dimensional phase aberration correction algorithms. As shown in Figure 5, three balls (glass beads) 0.5 mm in diameter were suspended in gelatin. The glass beads were used as three-dimensional

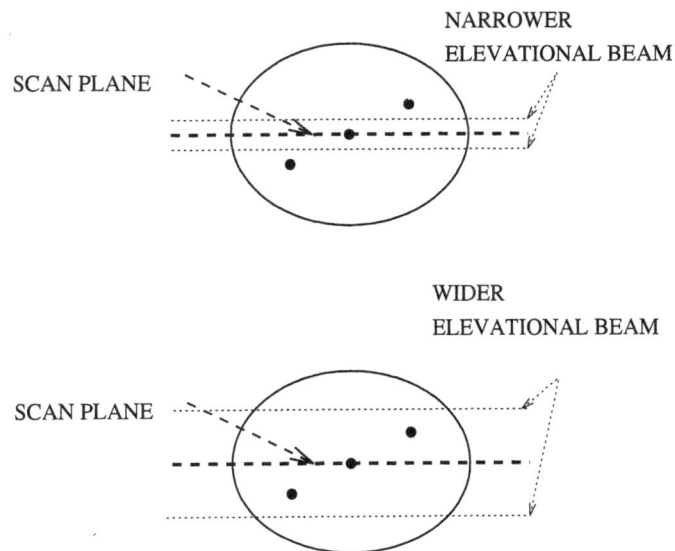

Figure 5. Top view of the gelatin-based phantom.

point targets placed 17 mm apart along the same line and 96 mm deep from the top of the phantom. The line containing the ball targets was purposely angled with respect to the scan plane such that only the center ball (ball 2) was within this plane. By varying the array length in the non-scan direction, the relative positions of the balls in the elevational beam also changed. Therefore, this phantom was used to show the reduction of slice thickness with large, two-dimensional arrays. The distance between outside balls (ball 1 and ball 3) and the scan plane was about 5.6 mm.

The array size used in this study was 2-3 cm in both dimensions. A partial cylinder, therefore, suitably approximates the lower order components of the array geometry conforming to a human body surface. Since an anisotropic array was used, it was more tolerant of deformation in the scan direction than in the non-scan direction. Therefore, the conformality of the array was modeled as follows. A phase pattern described by a third-order polynomial was applied along the scan direction. This polynomial approximated a 2 mm dip centered at array element 73 without an overall linear term. No low order phase patterns were applied between rows.

A two-dimensional, spatially low-passed, randomly generated error pattern was also produced to emulate both higher order components in the conformality and sound velocity inhomogeneities in the body. A contour plot scaled in radians of this random pattern is shown in Figure 6. The range is π at 3.5 MHz.

The emulated phase aberration pattern for sub-aperture 6 is shown in Figure 7. The top panel shows the spatially low-passed, randomly generated pattern and the middle panel shows the low order components. The overall pattern is shown in the bottom panel. Note that the low order components dominate the overall pattern and are expected to produce severe beamforming artifacts.

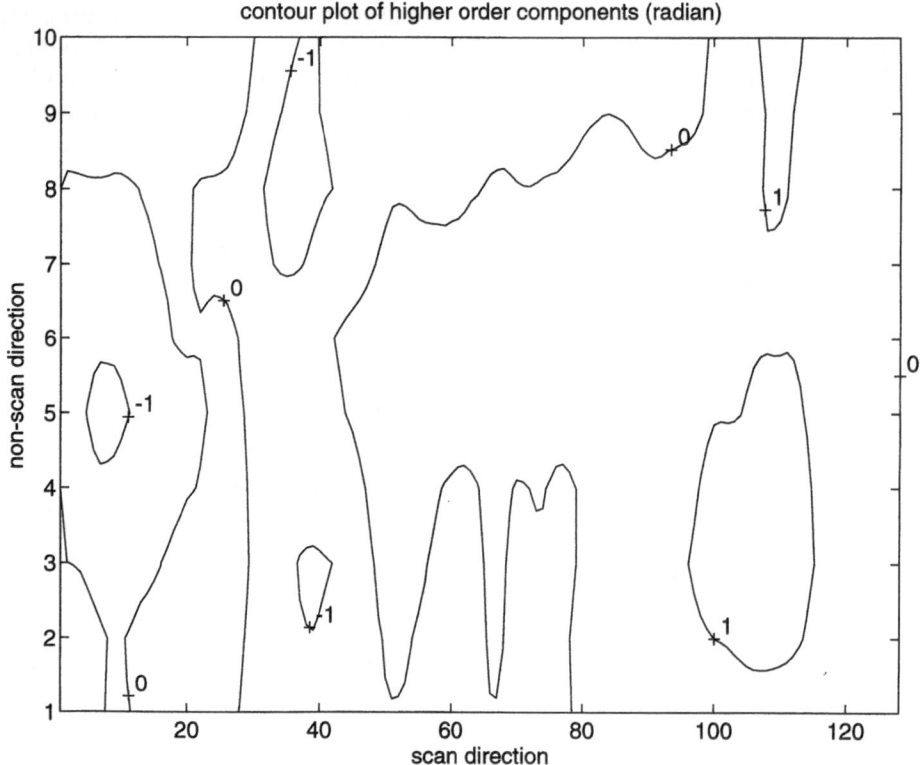

Figure 6. Contour plot of the emulated random phase pattern imposed on the two-dimensional array (in radians).

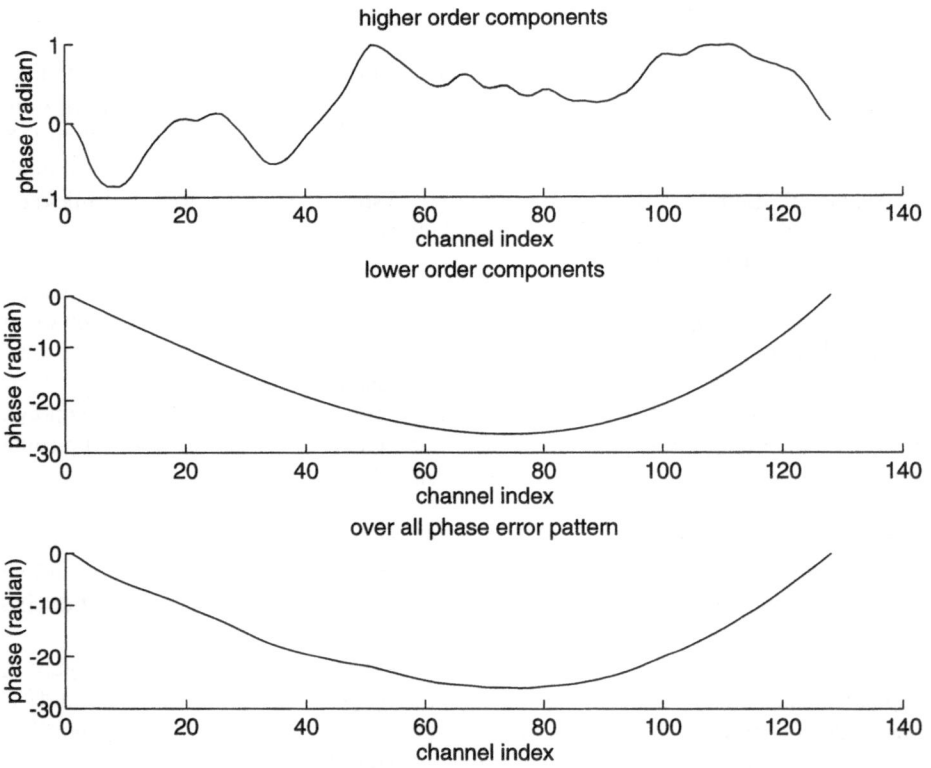

Figure 7. Aberration pattern for sub-aperture 6.

RESULTS

The one-dimensional array was used to synthesize a two-dimensional array. Geometric errors due to array movement occurred and were removed using the *row-sum* structure. The emulated aberration pattern was then applied to the synthetic two-dimensional array. For the center sub-apertures where the SNR was adequate, the *row-sum* architecture was used. Due to diffraction effects, on the other hand, outside sub-apertures exhibited poor SNR limiting the quality of phase aberration correction using only a single element. To solve this, a modified *row-sum* scheme was proposed. This scheme coherently groups N adjacent elements on the same row (e.g., elements n to $n + N - 1$) together to reduce noise levels. This signal is then correlated with the sum over the next N adjacent elements (e.g., elements $n + 1$ to $n + N$).

Figure 8 presents the contour plot of the residual phase error pattern. The residual

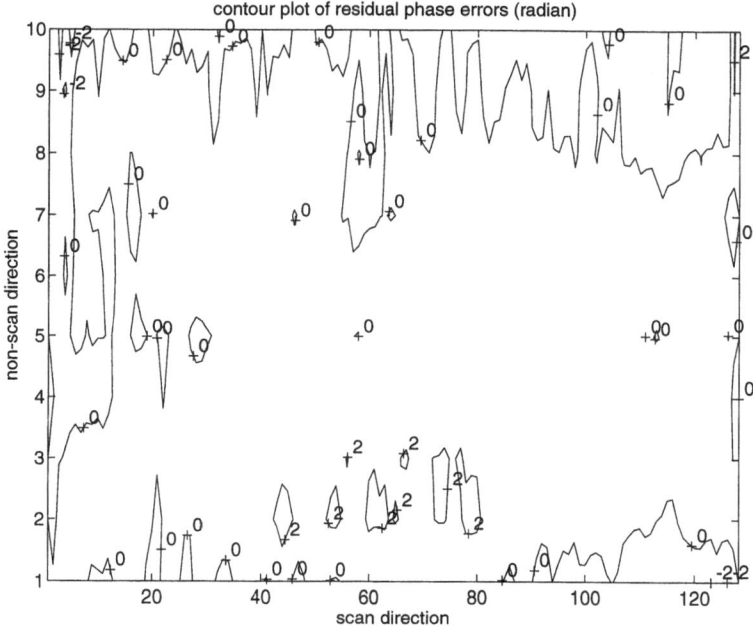

Figure 8. Contour plot of the residual phase error pattern (in radians).

pattern is very close to zero for the center sub-apertures. In other words, the *row-sum* scheme is adequate if the SNR is sufficiently good. Outside sub-apertures, however, have worse residual patterns with maxima approximating π even with the application of the grouping scheme. Nevertheless, the resultant image is much better than the uncorrected image and is comparable to the undistorted image. Indeed, even if no aberrations exist along the outside sub-apertures, the estimated aberration patterns are not close to zero due to poor SNR and are actually worse than the residual patterns shown in Figure 8.

Figure 9 shows the images with and without phase aberration correction. The left panel shows the uncorrected image and the right panel shows the corrected image. Both images are displayed over a 20 dB dynamic range. Note that the non-planar geometries due to array movement are already removed in both cases. Because of conformality, two strong lateral steering components produced a double image effect. On the other hand, due to the loss in signal level, artifacts resulting from system noise in the experimental setup appear in the image. Clearly, the elevational beam width has been restored after phase aberration correction. The corrected image is comparable to the undistorted image.

Figure 9. Images of the phantom without correction (left) and with correction (right).

As mentioned in Section 2, the balls were 96 mm from the top of the phantom. At this depth, the ball diameter was about 1/3 of the 6 dB lateral main beam width. Assuming a continuous-wave (CW) model, the mainlobe (i.e., the first zero crossing) ends where the sine of the beam angle is the wavelength divided by the total aperture size. At 3.5 MHz assuming a sound velocity of 1540 m/sec, the outside balls will be outside the mainlobe for any array larger than 7.5 mm in the non-scan direction. In other words, with a 20 mm length in the non-scan direction, gray levels of the outside balls represent sidelobe levels of the elevational beam at those directions.

For a n-element synthetic aperture, the average sidelobe level can be approximated as $-20 \log n$ dB relative to the mainlobe. In this experiment, the array had 10 elements in the non-scan direction and therefore the average sidelobe level in elevation was approximately -20 dB. The numbers from the experiment (-18 and -21.5 dB) were very close to the theoretical prediction. Certainly, small experimental errors existed.

DISCUSSION

Array conformality results in severe steering terms and produces a double image artifact, as shown in Figure 9. In other words, the target is moved from the desired spot by phase aberrations. A simple algorithm for estimating locations of regions of interest may be needed to improve the accuracy of phase aberration correction while applying the correction methods iteratively.

Aliases are likely to occur if phase aberrations are severe. In this paper, aliases can be easily detected using the first order difference of the estimated phase error pattern. Aliases produce spikes in the first order difference with magnitudes close to multiples of 2π if the first order difference of the true aberration pattern is relatively small. In general, however, no particular aberration patterns can be assumed and therefore more robust methods are needed to avoid aliases. This becomes more important if there is a large number of blocked elements.

A grouping scheme was used here and shown to be effective if the SNR is poor due to high electronic noise. For a real-time system, SNR is much higher since all array elements (1280 elements in the experiment) can contribute to beamforming at the same time. That is, because synthetic reconstruction is used to produce an image, the SNR for a single element is reduced by about 30 dB ($10 \log_{10}(1280)$) compared to a full real-time system assuming the same drive voltage. For a real-time system, therefore, simple *row-sum* structures without grouping may be adequate.

Point targets are used in this study to show the reduced slice thickness in elevation. In clinical applications, speckle is the primary scattering source. In this case, the synthetic aperture approach in elevation has to be modified because propagation paths between array elements must have sufficient overlap to obtain a high correlation for phase aberration correction. Future studies will address this issue.

Finally, combined with the blocked element compensation algorithm,[2,3] large, two-dimensional conformal arrays can dramatically improve image quality over current one-dimensional systems. Such improvements are only possible with a flexible, adaptive beam former permitting both real-time aberration correction and blocked element compensation.

REFERENCES

1. M. O'Donnell and P.-C. Li, "Aberration correction on a two-dimensional anisotropic phased array", *Proceedings of the 1991 IEEE Ultrasonics Symposium*, 91CH3079-1, pp. 1189-1193, 1991.
2. P.-C. Li, S.W. Flax, E.S. Ebbini and M. O'Donnell, "Blocked element compensation in phased array imaging", *IEEE Transactions on Ultrasonics, Ferroelectrics, and Frequency Control*, Vol. 40, No. 4, pp. 283-292, July, 1993.
3. P.-C. Li and M. O'Donnell, "Improved detectability with blocked element compensation", *accepted for publication in Ultrasonic Imaging (1994)*.
4. M. Hirama and T. Sato, "Imaging through an inhomogeneous layer by least-mean-square error fitting", *J. Acoust. Soc. Am.* Vol. 75, pp. 1142-1147, 1984.
5. S.W. Flax and M. O'Donnell, "Phase aberration correction using signals from point reflectors and diffuse scatterers: Basic principles", *IEEE Transactions on Ultrasonics, Ferroelectrics and Frequency Control*, Vol. 35, No. 6, pp. 758-767, Nov. 1988.
6. M. O'Donnell and S.W. Flax, "Phase aberration correction using signals from point reflectors and diffuse scatterers: Measurements", *IEEE Transactions on Ultrasonics, Ferroelectrics and Frequency Control*, Vol. 35, No. 6, pp. 768-774, Nov. 1988.
7. M. O'Donnell and S.W. Flax, "Phase aberration correction using signals from point reflectors and diffuse scatterers: Human studies", *Ultrasonic Imaging*, Vol. 10, pp. 1-11, 1988.
8. L. Nock, G.E. Trahey and S.W. Smith, "Phase aberration correction in medical ultrasound using speckle brightness as a quality factor", *J. Acoust. Soc. Am.*, Vol. 85, pp. 1819-1833, 1988.
9. D. Rachlin, "Direct estimation of aberrating delays in pulse-echo imaging systems", *J. Acoust. Soc. Am.*, Vol. 88, pp. 191-198, 1988.
10. M. O'Donnell and W.E. Engeler, "Correlation based aberration correction in the presence of inoperable elements", *IEEE Transactions on Ultrasonics, Ferroelectrics and Frequency Control*, Vol. 39, No. 6, pp. 700-707, Nov. 1992.

ACTIVE IMAGING ANALYSIS
VIA ELLIPSOIDAL PROJECTIONS

Forrest Anderson and Felix Morgan

Impulse Imaging Corporation
PO 1400, Bernalillo, NM 87004

ABSTRACT

Most active imaging systems, regardless of their beam forming approach, can be analyzed using ellipsoidal projections and ellipsoidal backprojections. Also, high quality, real-time 3D imaging is possible using ellipsoidal backprojection with defocused transmitted pulses. This paper shows that because conventional active imaging systems can be analyzed in terms of ellipsoidal analysis, ellipsoidal backprojection imaging machines can therefore be understood in terms of conventional imaging systems. Consequently, their point spread functions and sidelobe structures are similar. A basic example is given demonstrating this.

1. INTRODUCTION

Active imaging systems are typically analyzed in terms of their beam patterns or their point spread functions (the image of a point reflector.) Also, transmit time-space beam patterns from wideband arrays have been described and examples given [1]. Theoretical equations describing the two-way time-space beam patterns from the individual one-way transmit and receive time-space beam patterns have been developed [2]. The basis of this theory is the elementary solution (or Green's function) for active imaging which is an ellipsoid [3]. It has been shown that for the far-field, two-way, linear aperture beam pattern for monochromatic transmission, this analysis produces the expected sinc^2 function [2]. Most active imaging systems can be analyzed using ellipsoidal analysis (i.e. projections and backprojection using the ellipsoid. This analysis method has been described [2,3] and the techniques are reviewed and discussed further here.

Ellipsoidal projections [see note 1], and a special case--spherical projections, have been used to reconstruct 3D images in actual imaging systems [11, 12]. Ellipsoidal analysis is being used in the design of new ultrasound systems. Conventional, focused-pulse ultrasound machines can not provide real-time, 3D imaging. Because of the fixed, round-trip travel time in the body on the order of 200 μsec, there is not enough time in a 33 millisecond video frame to scan out a 3D volume of any practical size using focused pulses. Anderson [4,5] describes a practical, real-time, 3D ultrasound system which uses defocused transmit pulses and ellipsoidal backprojection. By illuminating the entire volume to be imaged with each and every pulse, the number of pulses required is dramatically reduced, overcoming the previous limitation for real-time 3D imaging. These systems are capable of providing high quality, real-time, 3D imaging [6]. The mathematical theory which underlies ellipsoidal backprojection imaging, integral geometry, [7], also is the basis for the Radon Transform used in CT and MRI.

This paper first reviews ellipsoidal analysis by describing the elementary solution, the ellipsoid, of active imaging and then applying this to continuous apertures. This elementary solution is also the point spread function of the 'elementary active imaging system'. A steered and focused array is then analyzed using ellipsoidal analysis. Next the theory of 2D synthetic focus imaging is reviewed, and an example system of this type is analyzed using ellipsoidal analysis, and its equivalence with the conventional steered and focused array is discussed. Finally an ellipsoidal backprojection imaging system is described and its equivalent steered and focused transmit beam conventional active imaging system is given.

{In general the point spread function can not be convolved with the true image to yield the actual imaging system image, but rather an inner product must be used [3]. Therefore, to emphasize this, henceforth the designation 'point image' will be used rather than the conventional 'point spread function'.}

2. REVIEW OF ELLIPSOIDAL ANALYSIS

In this section, the elementary active imaging system is described and it's point image is shown to be an ellipsoid. More complex active imaging systems can then be decomposed into a linear superposition of elementary active imaging systems. The point images of the complex active imaging system then is the linear superposition of all the elementary active imaging system point images, the linear superposition of ellipsoids.

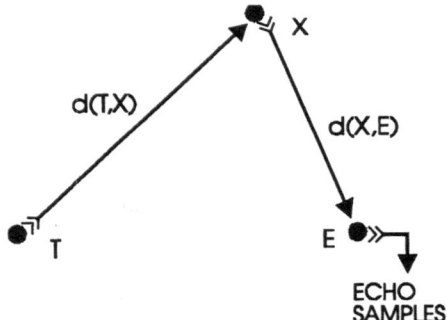

Figure 1. Elementary Active Imaging System Geometry.

2.1 The Elementary Active Imaging System

As shown in Figure 1, the elementary active imaging system geometry is a point transmitter emitting an impulse, a single reflecting point and a point receiver. The elementary solution for the wave equation is

172

$$\psi(r,t) = \frac{1}{4\pi r}\delta(r-ct)$$

$$with\ r = \sqrt{x^2+y^2+z^2}$$

(1)

This is a Dirac distribution with the sphere, r = ct, as its support; that is, it is an impulsive spherical wavelet propagating at a velocity, c. r is the distance from the source. (In the following the 1/r spreading effect will be neglected).

The steps leading to the point image follow: First the transmitter at $T = x_t, y_t, z_t$ transmits an impulsive spherical wavelet of the form $\delta(r-ct)$. The wavelet arrives at the reflecting point located at $P = x_p, y_p, z_p$ at time $t_{TP} = d(T,P)/c$. (In general, d(U,V) will denote the distance from U to V.) Second the reflecting point acts as a wavelet source of the form $\delta(r-c(t-d(T,P)/c))$. Third this wavelet arrives at the point receiver located at $E = x_e, y_e, z_e$ at time $t_{TE} = d(T,P)/c + d(P,E)/c$. The location of all points, X, which produce this same round-trip time-of-flight, is an ellipsoid with foci at T and E and with a major axis, 2a = d(T,P) + d(P,E).

For convenience, define the notation

$$d(T,P,E) \equiv d(T,P) + d(P,E)$$

(2)

Using this notation, the equation of the ellipsoid having foci at T and E, and having a major axis equal to 2a, becomes d(T,X,E) = 2a. The time waveform at the receiver is

$$\delta\left(t - \frac{d(T,P,E)}{c}\right) =$$

$$\delta\left(t - \frac{d(T,P)}{c}\right) * \delta\left(t - \frac{d(P,E)}{c}\right)$$

(3)

which is the convolution of the transmitted and reflected wavelets.

To derive the image of a single reflecting point, we first must define how to image a three dimensional volume with the elementary imaging system: We can range gate by sampling the echo sequence at a time equal to the round trip travel time to a selected point to be imaged. This time is computed by adding the propagation time from the transmitter to the point, t_{TP}, to the propagation time from the point back to the receiver, t_{PE}. By repeating this procedure, we can image any arbitrary point in the 3D volume, and therefore can obtain the point image of the elementary imaging system.

Although the range gating attempts to detect only the single reflecting point, the elementary

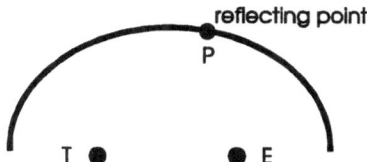

Figure 2. Elementary active Imaging System's Point Image.

imaging system will sense reflecting points lying anywhere on the surface of the ellipsoid having a major axis equal to d(T,P,E) and foci at T and E. Except for the actual reflecting point at P, all of these false points will be artifacts or sidelobe structure. The elementary active imaging system response for the image of a point X with reflectivity R(X), at time t, is given by

$$I(X, t) = \delta\left(t - \frac{d(T, X, E)}{c}\right) R(X) \tag{4}$$

Then the point image for a single unit reflecting point at P is found by setting t = d(T,P,E)/c, and we get

$$I(X) = \delta\left(\frac{d(T, P, E)}{c} - \frac{d(T, X, E)}{c}\right) \tag{5}$$

So the non zero points, X, in the image satisfy

$$d(T, X, E) = d(T, P, E) \tag{6}$$

and are those points having round trip travel time equal to the round trip travel time of the reflecting point. Thus the point image is an ellipsoid having the transmitter and receiver locations as foci, and containing the single reflecting point on its surface, as shown in figure 2.

For a transmitted wave of the form f(t) rather than a transmitted impulse, the response of the elementary active imaging system can be obtained by convolution with the impulsive response:

$$\begin{aligned} I(X, t) &= \delta\left(t - \frac{d(T, X, E)}{c}\right) * f(t) \\ &= f\left(t - \frac{d(T, X, E)}{c}\right) \end{aligned} \tag{7}$$

and the point image at t = d(T,P,E)/c is

$$I(X) = f\left(\frac{d(T, P, E)}{c} - \frac{d(T, X, E)}{c}\right) \tag{8}$$

2.2 Ellipsoidal Analysis of Continuous Apertures

The point image for a continuous aperture can be obtained by first decomposing the aperture into its constituent points or elemental areas. Then considering each possible combination of points in the aperture an elementary imaging system, and, finally, summing together the point images of each of these elementary imaging systems. The impulsively excited, continuous aperture's point image is the superposition of these ellipsoids.

For continuous apertures the impulsive point image becomes

$$I(X) = \iint_{E\ T} \delta\left(\frac{d(T, P, E)}{c} - \frac{d(T, X, E)}{c}\right) dT dE \tag{9}$$

The point image for a transmitted waveform f(t) rather than an impulse is

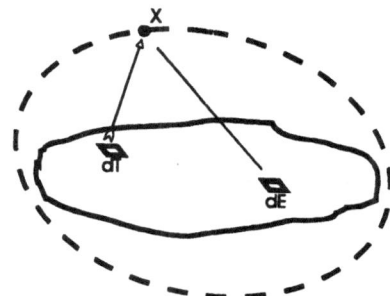

Figure 3. Ellipsoidal Analysis of a Continuous Aperture.

$$I(X) = \int_T \int_E f\left(\frac{d(T,P,E)}{c} - \frac{d(T,X,E)}{c}\right) dT dE \qquad (10)$$

and the geometry of the analysis is shown in figure 3.

2.3. Ellipsoidal Analysis of Discrete Apertures

For discrete transmitter and receiver apertures, $T \equiv T_1...T_m$ and $E \equiv E_1...E_n$, respectively, the integrals over the transmitter and receiver become summations:

$$I(X) = \sum_{i=1}^{m} \sum_{j=1}^{n} \delta\left(\frac{T_i,P,E_j}{c} - \frac{T_i,X,E_j}{c}\right) \qquad (11)$$

$$I(X) = \sum_{i=1}^{m} \sum_{j=1}^{n} f\left(\frac{T_i,P,E_j}{c} - \frac{T_i,X,E_j}{c}\right) \qquad (12)$$

3. EXAMPLE: ELLIPSOIDAL ANALYSIS OF A FOCUSED, STEERED ARRAY

Focusing a wideband array transmit beam for a point P in space, and steering the beam toward the same point, requires that the array elements be pulsed with appropriate delays such that the wavelets from each element arrive at P at the same time. So then the delays for the transmit elements will be $D_i = d(T_i,P)/c - k$, where k is the travel time associated with a transmitter element at the center of the array. If an impulse is transmitted the point P is 'illuminated' by

$$\sum_{i=1}^{m} \delta\left(t+D_i - \frac{d(T_i,P)}{c}\right) \qquad (13)$$

which is an impulse of 'weight m' occurring at t = k. If the transmitter is steered and focused for any other point in space, X, the illumination at P is

$$\sum_{i=1}^{m} \delta\left(t + \frac{d(T_i, P)}{c} - \frac{d(T_i, P)}{c} - k\right) = \sum_{i=1}^{m} \delta(t-k) \tag{14}$$

$$= \sum_{i=1}^{m} \delta\left(t + D_i - \frac{d(T_i, P)}{c}\right) \tag{15}$$

where $D_i = d(T_i, X) / c$. Then substituting for D_i yields

$$\sum_{i=1}^{m} \delta\left(t + \frac{d(T_i, X)}{c} - \frac{d(T_i, P)}{c} - k\right) \tag{16}$$

which is a series of impulses.

Focusing a wideband array receive beam for a point P in space, and steering the receive beam toward the same point, requires that the echoes from the array elements be delayed before being summed together in such a manner that the echoes detected by the array elements, resulting from a wavelet originating at P, will all be in aligned in time at the echo summer. So the delays for the receiver elements will be $D_j = d(E_j, P)/c - 1$, where 1 is the travel time associated with a receiver element at the array center. Then if P emits an impulse the receiver aperture output will be

$$\sum_{j=1}^{n} \delta\left(t + D_j - \frac{d(P, E_j)}{c}\right) \tag{17}$$

$$= \sum_{j=1}^{n} \delta\left(t + \frac{d(P, E_j)}{c} - \frac{d(P, E_j)}{c} - l\right) = \sum_{i=1}^{n} \delta(t-l) \tag{18}$$

which is an impulse of 'weight' m occurring at time $t = 1$.

If the receiver is steered and focused for any other point in space, X, and P emits an impulse, the receiver aperture output will be

$$\sum_{j=1}^{n} \delta\left(t + D_j - \frac{d(P, E_j)}{c}\right) \tag{19}$$

, which is a series of impulses:

$$\sum_{j=1}^{n} \delta\left(t + \frac{d(X, E_j)}{c} - \frac{d(P, E_j)}{c} - l\right) \tag{20}$$

Then a combined transmit element and receive element delay, D_{ij}, can be associated with each pair of transmit and receive elements in the array and $D_{ij} = d(T_i, P)/c + d(P, E_j)/c - k-l = d(T_i, P, E_j)/c - k-l$. So, if both the receiver and transmitter are steered and focused for P the receiver output is given by

$$I(P,t) = \sum_{i=1}^{m} \sum_{j=1}^{n} \delta\left(t + \frac{d(T_i,P,E_j)}{c} - \frac{d(T_i,P,E_j)}{c} - k - l\right)$$
$$= \sum_{i=1}^{m} \sum_{j=1}^{n} \delta\,(t-k-l) \tag{21}$$

which is an impulse of weight m times n occurring at t = k+l.

The response for any other point in space, given a single reflecting point at P is

$$I(X,t) = \sum_{i=1}^{m} \sum_{j=1}^{n} \delta\left(t + \frac{d(T_i,X,E_j)}{c} - \frac{d(T_i,P,E_j)}{c} - k - l\right) \tag{22}$$

Here, in the argument of delta, $-k+d(T_i,P)/c-l+d(P,E_j)/c$ is the combined transmit/receive steering and focusing delays for P and $-d(T_i,X,E_j)/c$ is the round trip travel time to some other point, X. Although the array is steered and focused for X, both for transmit and receive, P is illuminated by the 'sidelobe structure' of the transmitted pulse and the resulting echo from P is received by the sidelobe structure of the receive aperture. The resulting output is attributed to a nonexistent reflecting point at X.

Then the point image at time t = k+l is

$$I(X) = \sum_{i=1}^{m} \sum_{j=1}^{n} \delta\left(\frac{d(T_i,X,E_j)}{c} - \frac{d(T_i,P,E_j)}{c}\right) \tag{23}$$

Thus for a steered and focused four element array with a transmitted impulse, the

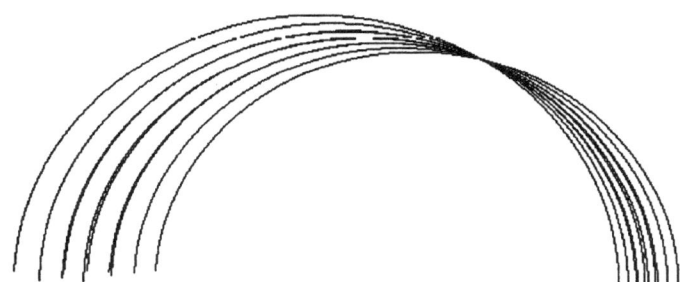

Figure 4. Point Image of a Steered and Focused Array.

point image is shown in Figure 4. This is the superposition of point images associated with the elementary imaging systems formed by all possible pairs of transmit and receive elements in the array. It consists of 4 circles and 6 ellipses (several nearly coincide in the figure). The ellipses have vaying major axes, the circles have different radii, and all intersect at the focal point. In the limit as the number of elements increases, this approaches the continuous aperture impulsive point image.

4. SYNTHETIC FOCUS IMAGING REVIEW

Synthetic focus (synthetic aperture) imaging [8, 9, 10] typically uses defocused transmitted pulses of 1-3 cycles of a damped sinusoid to form two dimensional images. Commonly this method is implemented with a linear array, where a single array element is pulsed and resulting echo time histories are collected at all of the array elements. This process is repeated until all of the array elements have been pulsed. The echo time histories are kept separate, so for example, for a 64 element linear array 4096 separate echo time histories $ETH_{ij}(n)$ are collected, where n is the sample number and i and j are the transmitter and receiver indexes.

Each echo sample in the echo time histories represents an elliptical projection of the object to be imaged. A transmitter element and a receiver element form an elementary imaging system except that a damped sinusoid rather than an impulse is used. Appropriate echo samples are then summed together to form the 2D image. Round trip travel time is the criteria used for echo sample selection to create image points:

$$I(X) = \sum_{i=1}^{m} \sum_{j=1}^{n} ETH_{ij}\left(\frac{d(T_i, X, E_j)}{c \Delta T}\right) \tag{24}$$

where, ΔT is the sample interval, ETH_{ij} is the echo time history set for receiver element j when transmitter element i is used, and the argument is rounded to the nearest integer.

If there is a single reflecting point at P, and if the transmitted pulse is an impulse, each echo time history will contain a single impulse at sample number $d(T_i, P, E_j)/c\Delta T$ and all of the rest of the echo samples will be zero. $ETH_{ij}(n)$ then can be replaced by $\delta(d(T_i, X, E_j)/c\Delta - d(T_i, P, E_j)/c\Delta)$ and we get

$$I(X) = \sum_{i=1}^{m} \sum_{j=1}^{n} \delta\left(\frac{d(T_i, X, E_j)}{c \Delta T} - \frac{d(T_i, P, E_j)}{c \Delta T}\right)$$
$$= \sum_{i=1}^{m} \sum_{j=1}^{n} \delta\left(\frac{d(T_i, X, E_j)}{c} - \frac{d(T_i, P, E_j)}{c}\right) \tag{25}$$

This last equation is the same as the equation for the point image for a conventional focused and steered array. Thus the previously analyzed conventional array produces the same point image as the synthetic focus system. If point images are obtained for a transmitted pulse other than an impulse via convolution, the images will still be the same and the sidelobe structures will be the same.

Figure 5 shows the point image for the 4 element synthetic focus imaging system with an impulsive transmission. Note that this figure is the same as Figure 4, the point image for a focused and steered conventional array. The synthetic focus imaging system achieves the equivalent of simultaneously having both dynamic receiver and transmitter aperture focusing--impossible to achieve in a conventional focused transmit beam imaging system.

The dynamic transmitter aperture is focused 'after the fact', hence the name "synthetic focus imaging." In a conventionally focused system, the wavelets from each transmitter element are physically summed together in space as they propagate. In a synthetic focus system the wavelets are essentially electronically summed after collecting the echoes, and the phasing, or delay, is adjusted for each image point.

5. SYNTHETIC FOCUS SYSTEM IN TERMS OF A CONVENTIONAL PHASED ARRAY

Because of the previous equivalency, the synthetic focus imaging system can be analyzed in terms of the appropriate, conventional focused and steered phased array system. If a conventional phased array is steered toward point P, and its transmitted pulse focused for P, and its receiver aperture focused for P, the amplitude of the image of the reflecting point at P will be exactly the same as the amplitude produced by a synthetic focus system. If this conventional steering and focusing process is repeated for each image point, the resultant 2D point image will be the same as the point image produced by a synthetic focus system. However, the conventional phased array must fire a separate pulse for each point in the 2D image, resulting in a large number of pulses and an impractically long time to produce the image.

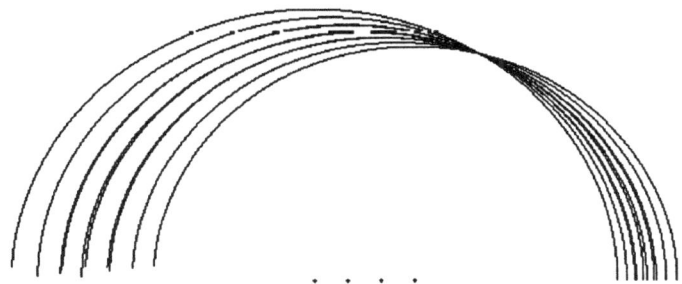

Figure 5. Synthetic Focus Point Image for Four Elements.

6. ELLIPSOIDAL BACKPROJECTION IMAGING REVIEW

Ellipsoidal backprojection imaging systems [4, 5, 6] acquire 3D images in a manner similar to acquisition of 2D images by synthetic focus systems. The point image of the simplest possible ellipsoidal backprojection system consists of a single ellipsoid having a single point transmitter and a single point receiver as foci and containing the single reflecting point on its surface. Thus, the simplest ellipsoidal backprojection system is the elementary active imaging system. A more complex ellipsoidal backprojection system consists of the linear superposition of a large number of elementary active imaging systems, and, assuming an impulsive transmission, its impulsive point image consists of many ellipsoids all intersecting at a common point--the location of the single reflecting point.

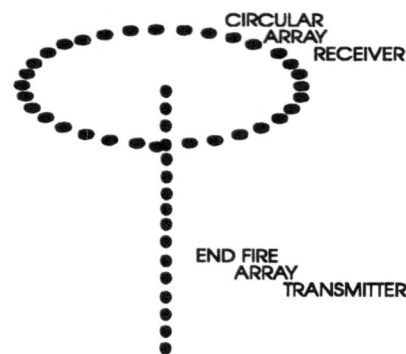

Figure 6. Ellipsoidal Backprojection Array Geometry.

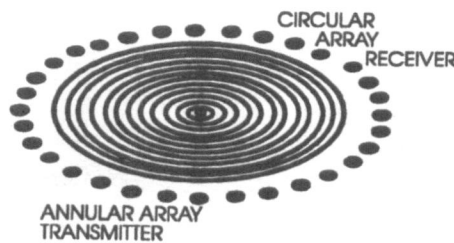

Figure 7. Impulse Imaging's Scan Head.

7. ANALYSIS OF ELLIPSOIDAL BACKPROJECTION SYSTEMS IN TERMS OF CONVENTIONAL SYSTEMS

So then an ellipsoidal backprojection imaging system can be decomposed into a number of elementary imaging systems, which can then be reassembled into an equivalent conventional focused beam imaging system for purposes of analysis. As an example, Figure 6 shows the transmitter and receiver geometry for Impulse Imaging's initial OB-GYN scan head. (The transmitter array is synthesized via a single annular array fired in a defocused mode, see reference 6, Patent no. 5,235,857, generating portions of spherical waves having various radii of curvature centered at virtual transmitter elements, see Figure 7) If the same transmitted pulse wave forms were used, and the waveshapes were kept intact through the beamforming and/or imaging processes, the ellipsoidal backprojection imaging machine can be viewed in terms of the equivalent conventional focused transmit imaging system, as shown in figure 8. And this equivalent imaging system can be used to predict resolution and sidelobe performance.

The ellipsoidal backprojection system in figure 6 is equivalent to the conventional focused transmit system of figure 8. The receiver aperture is a circular ring with 64

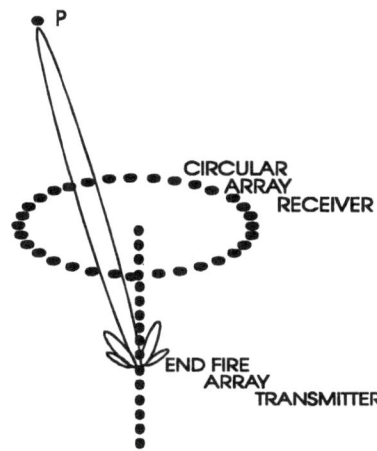

Figure 8. Equivalent Conventional Focused, Steered Array.

elements. The receiver is steered toward and focused for each point to be imaged, thus it is dynamically focused. The conventional array transmitter is an end fire array which is also steered toward and focused for each point to be imaged. Thus the transmitter must be pulsed a number of times equal to the number of points, or voxels, in the 3D image. This achieves the equivalent of dynamic transmit focusing.

In the actual ellipsoidal backprojection system, a filter to remove projection blurring and dynamic transmitter and receiver apodization are implemented. These have been neglected in this discussion.

8. CONCLUSION

The sidelobe structure of ellipsoidal backprojection imaging systems is the same as the equivalent conventional focused and steered imaging system. However, the conventional system must fire a separate focused and steered pulse for each point in space to be imaged to attain an equivalent image since both the receiver and transmitter need to be 'dynamically focused'. In general, the sidelobe structure of the ellipsoidal backprojection systems will be the same as that of conventional focused transmit systems.

9. REFERENCES

1. Anderson, F., Christensen, W., Fullerton, L., Kortegaard B., "Ultra-wideband Beamforming in Sparse Arrays", IEEE Proceedings-H, Antennas and Propagation, Vol. 138, No. 4, August 1991

2. Anderson, F., "Two Way Beam Patterns from Ultra-Wideband Arrays", SPIE vol. 1631, proc. of "Ultrawideband Radar", OE/LASE '92, Jan. 1992, Los Angeles, CA

3. Anderson, F., "Active Imaging Green's Function", Acoustical Imaging vol. 19, Plenum Press, NY

4. Anderson, F., "3D Real Time Ultrasonic Imaging Using Ellipsoidal Backprojection", SPIE vol. 1443, proc. of Medical Imaging V, Feb. '91, San Jose, CA

5. Anderson, F., "3D Ellipsoidal Backprojection Images From Large Arrays", SPIE vol. 1651, proc. Medical Imaging VI, Feb. '92, Newport Beach, CA

6. Anderson, F., United states Patents no.s 4,688,430, 4,706,499, 4,817,434, 5,005,418, 5.027,638, 5,090,245, 5,134,884, 5,235,857

7. Gelfand et. al., "Generalized Functions" vol 5, Academic Press, NY, 1964

8. S.A. Johnson, J.F. Greenleaf et al, "Digital Computer Study of a Real Time Collection, Post Processing Synthetic Focusing Ultrasound Cardiac Camera" Acoustical Holography vol. 6, 1975

9. P. Corl et al, "A Digital Synthetic Focus Acoustic Imaging System" Acoustical Imaging, vol. 8, 1980

10. S. Bennett et al, "A Real Time Synthetic Aperture Imaging System", Acoustical Imaging, vol. 10, 1982

11. S. Norton, M. Linzer, "Ultrasonic Reflectivity Imaging in Three Dimensions, Exact Inverse Scattering Solution for Plane, Cylindrical and Spherical Apertures", IEEE Trans. on Biomedical Engineering, BME-28, no. 2, pp 202-220, 1981

12. D. Miller, M. Oristaglio, and G. Beylkin, "A New Slant on Seismic Imaging: Migration and Integral Geometry", Geophysics, 52, no 7, pp 943-964, 1987

NOTES: 1. The term 'projection', as used in this paper, is the surface or line integral of the object to be imaged, eg., the surface integral over an ellipsoid. This is the common use of the term in engineering literature. However mathematicians require that a projection operator, P, satisfies P(P) = P. So to use mathematically correct terminology, the 'backprojected' 'projection' would be the projection.

NOVEL AIRBORNE ULTRASOUND TRANSDUCER

H. Dabirikhah and C.W. Turner

Department of Electronic and Electrical Engineering
King's College, University of London
Strand, London WC2R 2LS, U.K.

ABSTRACT

A novel transducer capable of transmitting and receiving airborne ultrasonic radiation at sub-mm wavelengths has been demonstrated experimentally in our laboratories to have significant advantages over alternative devices. The radiated beam profile is shown to depend directly on the characteristics of leaky Lamb waves propagating in immersed plates. The excitation of the antisymmetrical mode in the transducer structure generates significant longitudinal wave radiation into the adjacent fluid which is air. In this paper the basic operating principles and design considerations are presented. Experimental measurements are shown to confirm the results of numerical calculations. The maximum attainable operating frequency and bandwidth for this class of transducer is discussed and comparisons with the performance of other forms of airborne ultrasonic transducers are drawn.

INTRODUCTION

There are many applications which use high frequency ultrasonic airborne transducers such as robotic sensing, gas flow measurements and non-destructive testing. Despite this range of applications, the methods of generating and transmitting ultrasound in air are very limited and airborne ultrasonic transducers which can operate above 100 kHz are not yet readily available. In general there are three types of airborne transducers, which can be categorized as: i - composite transducer with a matching layer [1,2], ii - membrane transducer [3], iii - PVDF transducer [4,5]. In the first category because of the enormous mismatch of impedances between typical piezoelectrics and air, the most difficult task is to design and fabricate the matching layer. The second type of transducer uses electrostatic excitation to vibrate the membrane. This works well at low frequencies but it becomes extremely difficult to fabricate at higher frequencies. The third kind of transducer uses PVDF as the active element and various designs for using this piezoelectric polymer have been proposed. The operating frequency depends on the design and the reported frequencies range from 60 kHz to 800 kHz. In general the design and fabrication of each of these transducer types for higher frequencies becomes more difficult. Here we describe a novel airborne ultrasound transducer whose operation is based on the generation and propagation of leaky plate waves (Lamb modes). The ease of the design and fabrication on the one hand and high sensitivity, light weight and directionality on the other hand, give it an advantage over other designs. The experimental results obtained from the first prototype pair of this kind of transducer have been very satisfactory and will be reported in the following section.

THEORY AND NUMERICAL RESULTS

Plate waves or Lamb modes propagate along an ideal solid plate in vacuum without loss. They are dispersive waves but without any damping above their cut-off regions. When the plate is immersed in a fluid , no matter how tenuous the fluid is, some part of the energy of the wave travelling along the plate radiates or leaks into the surrounding fluid. The plate waves can be either symmetrical or antisymmetrical, depending on the particle motion with respect to the median plane of the plate. If the fluid density to solid density ratio is much smaller than unity, the spectrum of the leaky waves is very similar to the spectrum of Lamb modes for a free plate. The characteristic equations for the plate waves are very difficult to solve, but we have developed good tools using Muller's method for solving this type of transcendental equation. We have investigated the radiation to the adjacent fluid numerically [6] and some of these numerical results have been reported earlier for other applications [7,8]. The radiation occurs at a coincidence angle which is given by $\theta = \sin^{-1} V_F / V_{mode}$, where V_F is the velocity of longitudinal waves in the fluid and V_{mode} is the plate mode velocity. The characteristic equation can be written as [6]

$$\frac{\tanh(\frac{d}{2}\sqrt{K^2 - K_L^2})}{\tanh(\frac{d}{2}\sqrt{K^2 - K_S^2})} - \frac{4 K^2 \sqrt{K^2 - K_S^2} \sqrt{K^2 - K_L^2}}{(2K^2 - K_S^2)^2}$$

$$+ i \frac{\rho_F K_S^4 \sqrt{K^2 - K_L^2}}{\rho_s \sqrt{K_F^2 - K^2} (2K^2 - K_S^2)^2 \tanh(\frac{d}{2}\sqrt{K^2 - K_S^2})} = 0 \qquad (1)$$

for the antisymmetrical case, where K_F is the wave number of the acoustic wave in the liquid, and K, K_S, and K_L are the wave numbers of the plate modes, transverse (shear) waves and dilatational (longitudinal) waves in the plate material respectively. The liquid and solid mass densities are represented by ρ_F and ρ_S.

Solving equation (1) we have calculated the solutions for the complex wave numbers for flexural plate waves for immersed plates in air. Figure 1 and Figure 2 show the phase velocity and imaginary part of the wave number of the A_0 mode (zeroth antisymmetrical mode) for several materials in air. As is seen, when the plate material density gets smaller, damping which represents radiation increases and the excitation of the A_0 mode in a very thin plate or membrane produces significant radiation of longitudinal waves into air under certain conditions. Figure 3 shows $\sin \theta$ versus f.d (frequency times plate thickness) and the arrows show the values of f.d chosen as the operating points for two different frequencies.

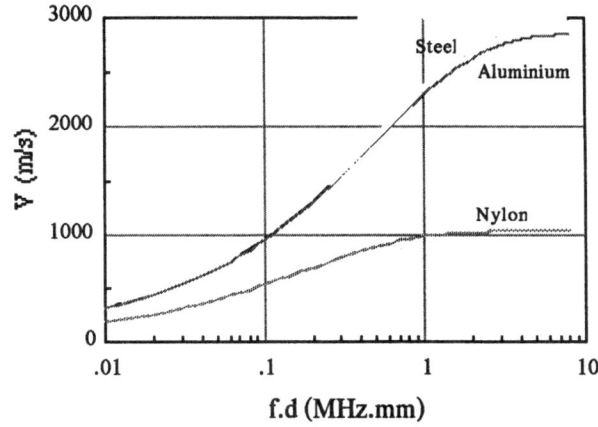

Figure 1. Phase velocity of A_0 mode for steel, aluminium and nylon plates in air.

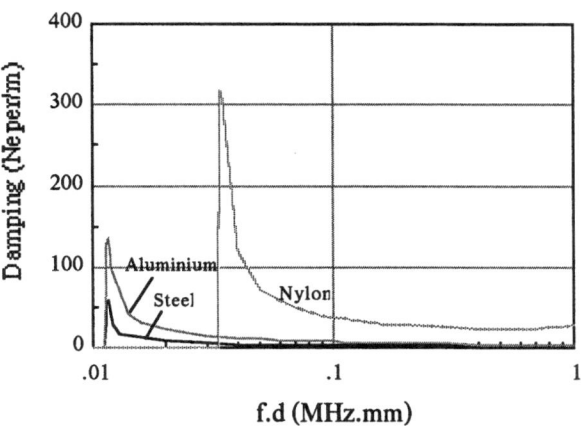

Figure 2. Damping of A_0 mode for steel, aluminium and nylon plates in air.

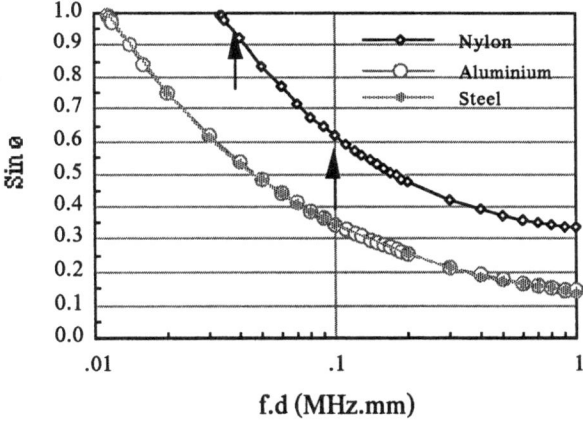

Figure 3. Sin (coincidence angle) of A_0 mode for steel, aluminium and nylon plates in air.

TRANSDUCER DESIGN

As is seen from the numerical results the A_0 mode is not cut-off even in very thin plates. The plate deformation in this mode is given in Figure 4 (a). This mode can be generated by the use of externally applied plane waves at oblique incidence at coincidence angle or by interdigital transducers positioned on the surface of the plate. Figure 4 (b) shows the schematic diagram of our transducer in which flexural modes are generated on a cone-shaped thin shell by the radial vibration of a piezoelectric ceramic disc glued to its circumference.

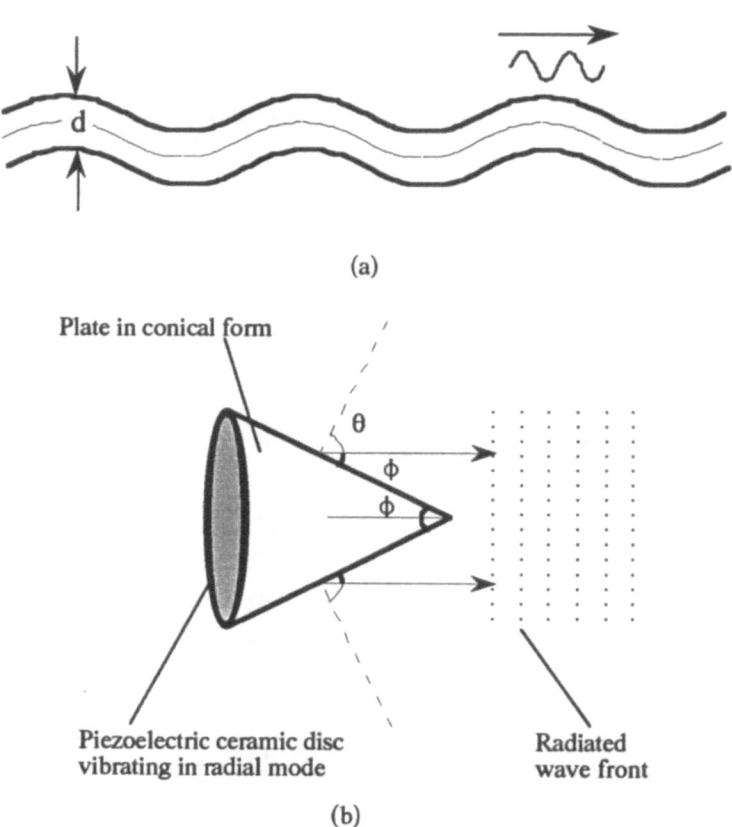

Figure 4. (a) Plate deformation for A_0 mode, (b) Schematic diagram of the conical shell transducer.

From numerical results it is seen that low density polymers are the most suitable materials for this type of transducer. We have used a nylon 66 cone with approximately 200 µm wall thickness which has an angle of about 15 degrees. The plate waves are excited by a piezoelectric ceramic disc with 10 mm diameter and 1 mm thickness. Its fundamental radial resonance frequency and harmonics were measured at 200 kHz, 530 kHz and 770 kHz. The radial motion of the disc is transferred to the circumference of the cone and excites flexural motion along the wall of the cone. The flexural wave leaks energy into air along its propagation path and because of the geometry of this transducer a relatively directional beam can be produced. Figure 5 is a photograph of the first prototype of this kind of transducer. In the following the experimental results using this pair of transducers are presented.

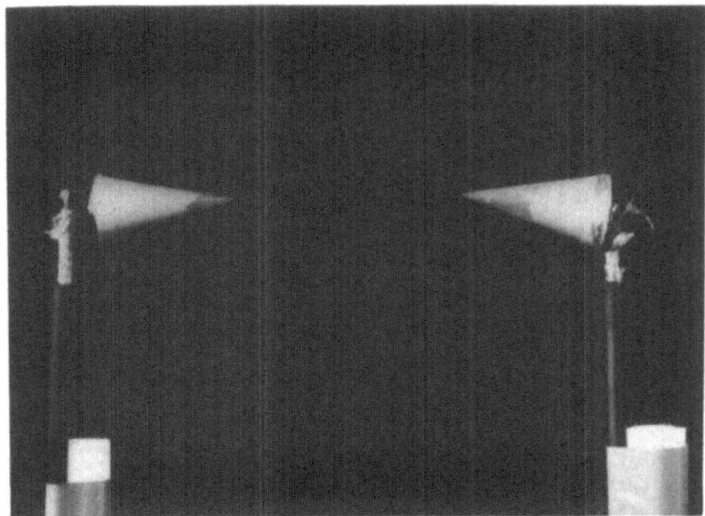

EXPERIMENTAL RESULTS

The prototype transducers described above were made for demonstration purposes without accurate or optimum choice of materials and angle of the cone. In fact they produced very satisfactory results which confirmed our design concepts and are presented in the following.

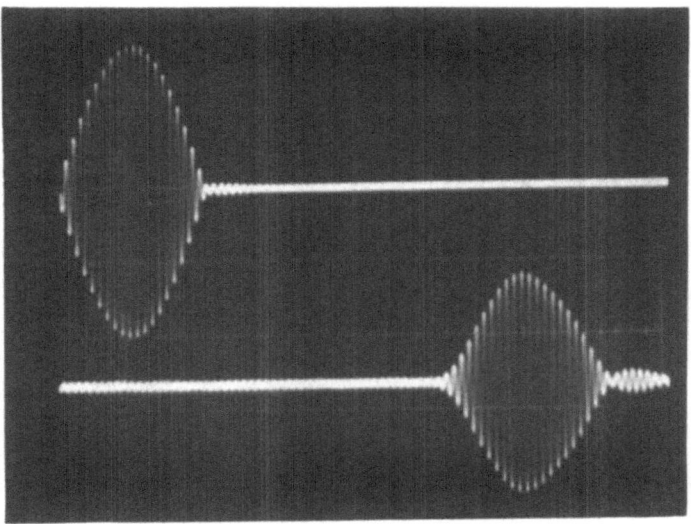

Figure 6. Oscilloscope photograph of the excitation and received pulses at 200 kHz center frequency.

Designating T_1 as the transmitter and T_2 as the receiver, we transmitted pulsed ultrasonic waves through air. The amplitude of the excitation pulses was as low as 0.2 volt. Figure 6 shows the oscilloscope photograph of a 10 V input pulse and the received output pulse when T_1 and T_2 were 20 cm apart. An amplification of 60dB was used for the received signal. The center frequency of the pulse was 200 kHz.

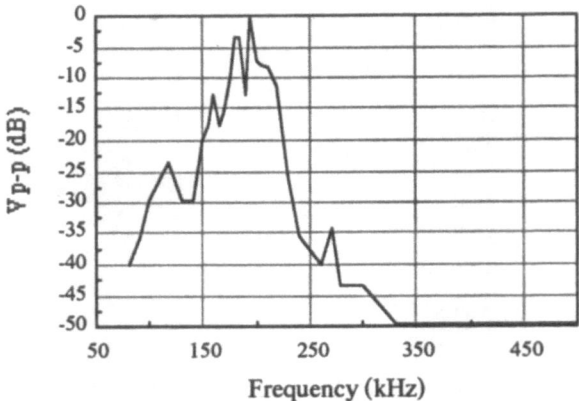

Figure 7. Frequency response of the unmatched transducer pair, normalized to peak value at resonance.

Figure 7 shows the frequency response for an unmatched pair of transducers. The absence of resonance peaks at higher frequencies is due to the fact that the two piezoelectric discs used are not precisely matched.

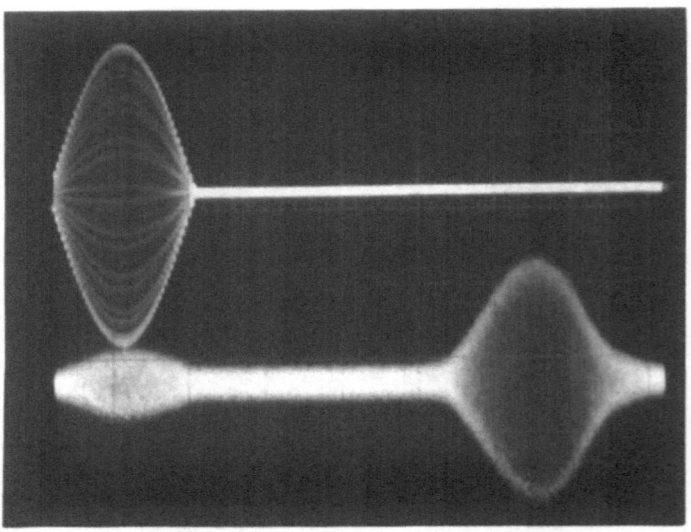

Figure 8. Oscilloscope photograph of the excitation and received pulses at 500 kHz center frequency.

Using a matched pair would allow us to transmit and receive higher frequencies using higher harmonics of the radial mode. In our experiments we used matching inductors to bring the second harmonic frequency of the piezoelectric discs closer to each other. In this way both of the transducers showed second harmonic resonance at 500 kHz. The experiment was repeated as for 200 kHz. We were able to transmit and receive ultrasonic pulses at 500 kHz. However we had to change the position of the receiver because of the change of the radiation angle predicted from the numerical results. Higher amplification for the received signal was also used since the amplitude of the resonance at higher harmonics is much smaller than the amplitude of the fundamental resonance. Figure 8 is an oscilloscope photograph of the input and output pulses with center frequency of 500 kHz.

The polar response of T_2 as a transmitter at 200 kHz using T_1 as a receiver is shown in Figures 8. The directivity of the beam evident from this figure is as we expected. Experiments showed that the received signal depends strongly on the orientation of the transmitter and the receiver cones with respect to each other.

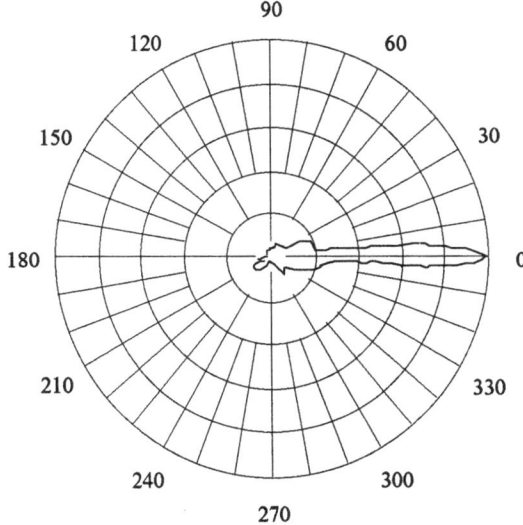

Figure 9. Radiation pattern of T_2 as a transmitter at 200 kHz, normalized to the peak value.

DISCUSSION

In this paper a novel ultrasonic transducer [9] has been introduced which operates through the use of leaky Lamb modes. The principal design features and underlying physical mechanism have been presented together with numerical and experimental results for a pair of prototype transducers. The measured radiation characteristic and frequency response agreed broadly with our theoretical prediction.

In comparison with other methods of generating high frequency ultrasound in air, this type of transducer is considered to have significant advantages in respect of its directionality and the possible scaling to frequencies above 1 MHz. Applications in acoustic microscopy and other types of non-contact acoustical imaging are seen to be feasible, given the possibility of producing either well-collimated or sharply focused beams.

Although the beam radiated from the prototype transducer described here is necessarily inhomogeneous, by careful control of the cone parameters, such as wall thickness and cone angle, in principle the beam can be made reasonably uniform. Other methods of excitation, using direct magnetostrictive and piezoelectric effects with conical foils of suitable materials are also being investigated for extending the frequency range of the transducer into the MHz region.

REFERENCES

1. M. I. Haller and B. T. Khuri-Yakub, 1-3 Composites for ultrasonic air transducers, *IEEE proceedings of ultrasonics symposium* 2, 937-939 (1992).
2. J. D. Fox, B. T. Khuri-Yakub, and G. S. Kino, High frequency acoustic wave measurements in air, *IEEE proceedings of ultrasonics symposium* 1, 581-584 (1983).
3. G. Curtis, A broadband polymeric foil transducer, *Ultrasonics* 12, 148-154 (1974).
4. A. S. Fiorillo, Design and characterization of a PVDF ultrasonic range sensor, *IEEE Trans. on ultrasonics, ferroelectrics, and frequency control* 39,688-692 (1992).
5. J. S. Schoenwald and J. F. Matin, PVF2 transducers for acoustic ranging and imaging in air, *IEEE proceedings of ultrasonics symposium* 1, 577-580 (1983).
6. H. Dabirikhah, Wave propagation in immersed membranes for acoustic device applications, *Ph. D. thesis,* University of London (1993).
7. H. Dabirikhah and C. W. Turner, Membrane acoustic beam deflector for imaging and monitoring applications, *Acoustical Imaging* 20, 395-402 (1993).
8. H. Dabirikhah and C. W. Turner, Anomalous behaviour of flexural waves in very thin immersed plates, *IEEE proceedings of ultrasonics symposium* 1, 313-317 (1992).
9. Patent applied for : No. GB 9306224.8

ACOUSTIC SENSORS USING LANGMUIR-BLODGETT FILMS

A. M. Kosevich [1], E. S. Syrkin[1], and M. V. Voinova[2]

1. B. I. Verkin Institute for Low Temperature Physics
   and Engineering of Ukrainian Academy of Sciences
   Lenin Ave. 47. 310164 Kharkov, Ukraine
2. Kharkov State University, Svobody Sq. 4
   310077 Kharkov, Ukraine

INTRODUCTION

The monolayer or multilayer assemblies constructed by means of Langmuir-Blodgett techniques (LB films) are the subject of the relatively new and multidisciplinary field of surface sciences [1,2]. These ultrathin ($d \sim 25\overset{\bullet}{A}$) films deposited on the solid substrate consist of amphiphile molecules which have both hydrophilic and hydrophobic properties [3]. Such unique molecular architecture allows to create different coupling between the substrate and the absorbed LB film. The weak coupling may be realized by wet chemical methods when the "hydrocarbon" side of LB film is deposited on hydrophobic substrate surface. The strong coupling corresponds to the chemical bound of LB polar head groups with hydrophilic substrate. These molecular "sandwiches" have been applied succesfully as functional material in modern sensors, biosensors, in particular, in surface acoustic wave devices [4,5]. In respect of LB film application in gas-sensing detectors [6], the quantitative analysis of acoustic changes in LB film, caused by gas absorbtion seems to be important.

The goal of this paper is the theoretical study of surface waves propagating in the system of solid substrate - LB film. We consider a molecular model of LB film differently coupled with the substrate. On the base of macroscopic dynamics of the solid surface with absorbed layer [7] , the dispersion equations for the surface acoustic waves have been analysed in different cases of LB film deposition. In particular, the change of surface waves velocity caused by the presence of LB film has been calculated in the case of acoustic waves with shear polarization as well as Rayleigh polarization.

THE MODEL OF LANGMUIR - BLODGETT FILM -SUBSTRATE "SANDWICH"

Let us consider the LB amphiphilic monolayer deposited on the solid substrate. Here a substrate may be described as an elastic half-space with absorbed molecular layer. To study an arbitrary coupling between the film and substrate surface, we first need to characterize the intralayer and interfacial forces.

Starting with the weak coupling, we consider the hydrophobicized substrate surface with deposited chain-to-chain hydrophobic side of LB monolayer (Fig. 1a). An opposite side of the film is hydrophilic part of the LB layer. In case of zwitterionic head groups of a film, they are packed into the lattice of parallel dipoles which are normal to the layer surface [8]. The hydrocarbon interior of "sandwich" consists of the chains with Van-der-Waals interaction. Then the case of weak coupling between the LB film and hydrophobicized substrate surface may be considered similar to bilayer membranes with weak interaction between two leaflets [3]. This assumption gives us a possibility to elavuate the interaction between the surface and hydrocarbon core of LB film.

The another situation is "face-to-face" deposition at which the polar side of LB film chemically bound with the hydrophilic substrate surface (Fig. 1b). In this case the LB film is extremely firmly bound (coupled) with the surface of substrate [2,3,4]. We suggest this deposition corresponds to the strong coupling between the leaflets of molecular sandwich.

Obviously, the results of our model are valid for the pile of weakly interactive LB layers, too. The free energy of such molecular sandwich in general form may be written as a sum of bulk (index"v") and surface (index"s") terms:

$$F = F_v + F_s \qquad (1)$$

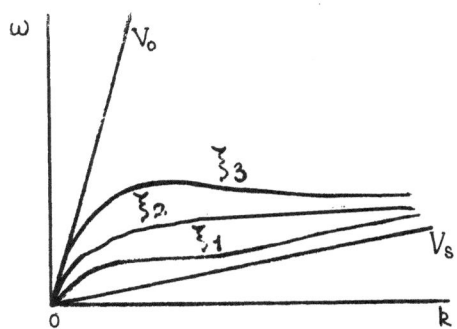

Fig. 1. Schematic depiction of Langmuir-Blodgett film-substrate system in case of (a) weaking coupling; (b) strong coupling between the substrate surface and LB monolayer.

More detail analysis of the surface part gives the density of elastic free energy in a form [7]:

$$F_s = 0.5 \, A_{ik}(u_i^s - u_i)(u_k^s - u_k) + 0.5 \, h_{\alpha\beta\gamma\delta} u_{\alpha\beta}^s u_{\gamma\delta}^s +$$

$$+ 0.5 \, g_{\alpha\beta} \frac{\partial u_i^s \, \partial u_i^s}{\partial x_\alpha \partial x_\beta}, \qquad (2)$$

where a new independent dynamical variable u^s is introduced. This variable corresponds to the elastic displacement of the film and differs from the substrate surface one. Here A_{ik}

is the force constants tensor put on in ref.7. The components of this tensor in our model are described the coupling between the substrate and LB film. Tensors $h_{\alpha\beta\gamma\delta}$ and $g_{\alpha\beta}$ characterize the elastic properties of the LB film and the remanent straines, respectively. The variation of the total free energy equilized to zero gives the bulk motion equations

$$\rho\,\frac{\partial^2 u_i}{\partial t^2} = \frac{\partial \sigma_{ik}}{\partial x_k} \qquad (3)$$

and correspondent boundary conditions at z=0:

$$\sigma_{iz} = A_{ik}(u_k^S - u_k) \rightarrow\rightarrow \qquad (4)$$

$$\sigma_{iz} = g_{\alpha\beta}\nabla_\alpha\nabla_\beta u_i^S + \delta_{i\beta}h_{\alpha\beta\gamma\delta}\nabla_\alpha u_{\gamma\delta}^S - \frac{\rho_s\partial^2 u_i^s}{\partial t^2} \qquad (5)$$

As in ref.7 we apply this method suggested for finding the dependence of wave frequency ω from the wave number **k**. In case of surface shear wave with horizontal polarization (SH - waves) this dispersion equation for the surface elastic wave interacted with LB film takes the form:

$$\varkappa_t\,(\rho_s\omega^2 - A_2 - h_{66}k^2) = \frac{A_2}{C_{44}}\,(h_{66}k^2 - \rho_s\omega^2) \ ,$$
$$\varkappa_t^2 = (k^2 - \rho\omega^2/C_{44}) \qquad (6)$$

where ρ_s and ρ are the surface mass of LB film and bulk density of the substrate respectively. The film elastic modulus $h_{66} = h_{xyxy}$ and the bulk modulus of solid subtrate $C_{44} = C_{xzxz}$, A_2 is the component of A_{ik} tensor which corresponds to the binding of LB film and substrate.

The dispersion equations for the Rayleigh polarization are [7]:

$$[(\rho\omega^2 - 2C_{44}k^2)^2 - 4C_{44}k^2\varkappa_t\varkappa_1][\rho_s\omega^2 - A_1\tilde{h}_{11}k^2]X$$

$$X(\rho_s\omega^2 - A_3 - g_1k^2) = A_1A_3(\rho_s\omega^2 - g_1k^2)(\rho_s\omega^2 - \tilde{h}_{11}k^2)X$$

$$X(k^2 - \varkappa_t\varkappa_1) + \rho\omega^2[A_1\varkappa_t(\rho_s\omega^2 - A_3 - g_1k^2)(\rho_s\omega^2 - \tilde{h}_{11}k^2) +$$

$$+ A_3\varkappa_1(\rho_s\omega^2 - A_1\tilde{h}_{11}k^2)] \qquad (7)$$

where the designations are introduced :

$$\varkappa_1^2 = (k^2 - \rho\omega^2 C_{11}), \quad \tilde{h}_{11} = h_{11} + g_1.$$

Let us analyse the dispersion relations (6),(7) for the cases of weak and strong coupling between the LB film and substrate surface.

SURFACE SHEAR WAVES WITH HORIZONTAL POLARIZATION

Let introduce following parameters:

$\delta = h_{66}/C_{44}$ the effective length of the interfacial elastic distorsion;

$l = \rho_s/\rho$ — the effective length of the interfacial mass distorsion;

$\xi = l\delta A_2/h_{66}$ — coupling parameter corresponds to the interaction between the solid substrate and LB film.

Using the following parametization

$$u = (k^2\delta^2 - \omega^2\delta l/V_s^2), \quad V_s^2 = \rho_s/h_{66},$$

the parametric solution of the dispersion equation (6) can be obtained:

$$\frac{\omega^2\delta^2}{V_0^2} = \frac{u^3 + p\xi u^2 + u(1+p\xi)}{(u+p\xi)(p-1)} \qquad (8)$$

$$k\delta = u^2 + \omega^2\delta^2/V_0^2, \quad V_0^2 = C_{44}/\rho. \quad p = \delta/l.$$

The relative change of wave velocity for shear surface wave is equal

$$\frac{\Delta V}{V} = 1 - \left[\frac{u^2 + p\xi u + p}{pu + p\xi(1 + pu)} \right]^{1/2} \tag{9}$$

For the limit strong coupling between the substrate and LB film the shift in wave velocity is a quadratic function of frequency:

$$\frac{\Delta V}{V} = 0.5 \frac{\omega^2}{V_0^2} (\delta-1)^2, \quad (\xi \to \infty) \tag{10}$$

In fact this limit case corresponds to Love wave polarization.

In Fig.2 the dispersion laws for the arbitraly coupling parameter values are shown.

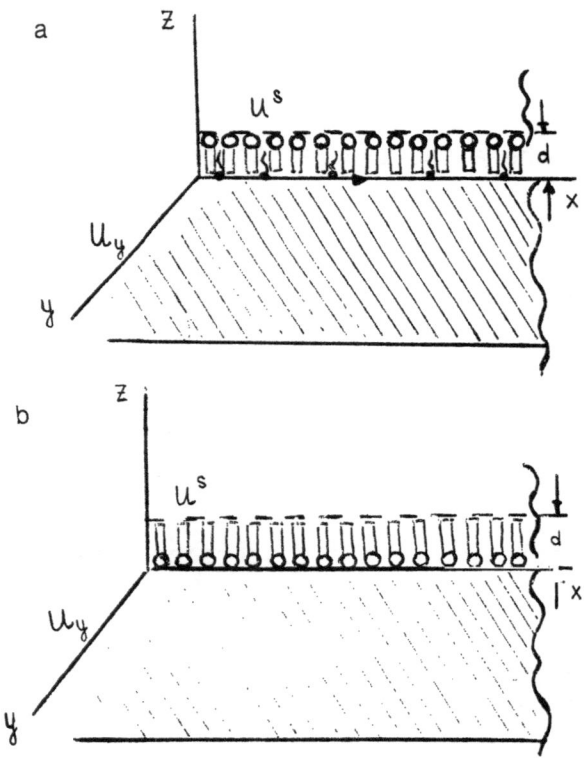

Fig.2. Schematic plot of dispersion laws of the surface acoustic shear waves for different coupling parameters $\xi: \xi_1 < \xi_2 < \xi_3$

In the opposite limit case of weak coupling $\xi \ll 1$ a new wave propagated in the film occurs. The velocity of this surface wave is determined as

$$V_s^2 = V_0^2 \delta / 1 = h_{66} / \rho_s \qquad (11)$$

From (11) it follows that the surface film wave velocity depends only on the inherent monolayer parameters - it's surface mass density and elastic modulus. On this wave frequency the resonance vibration [7] and the anomal absorption of the acoustic wave by LB film [9] will appear.

The results obtained in our model for the SH-waves allow to suggest a method of acoustical measurements of LB inherent parameters, ρ_s and h_{66}. From the analysis of weak coupling case (11) (Fig.1a deposition) it is obtained

$$h_{66} = V_s^2 \rho_s$$

and from the formula (10) for the strong coupling (Fig.1b deposition) it follows

$$\rho_s = \frac{\rho V_0}{\omega} \left[\frac{(1 - V_s^2 / V_0^2)}{V_0 / 2\Delta V} \right]^{1/2}$$

$$(12)$$

RAYLEIGH WAVES PROPAGATION

Let us analyse the dispersion equations for the Rayleigh polarization in case of limit strong coupling between LB film and solid elastic substrate when the condition of an acoustic contact is realized. In the interval of frequences $\rho \omega^2 \ll C_{44} k^2$ the reverse penetration length values are:

$$\varkappa_t \cong k \left(1 - \frac{\rho \omega^2}{2 C_{44} k^2} \right) , \qquad \varkappa_1 \cong k \left(1 - \frac{\rho \omega^2}{2 C_{11} k^2} \right)$$

The difference from the Rayleigh wave velocity V_R due to

the presence of monomolecular layer has the form:

$$V^2 = V_R^2 - \alpha V_R^2 ,$$

(13)

where

$$\alpha \cong \frac{k \left(\frac{V_R}{c_t}\right)^2 \left(\frac{\rho_s}{\rho} - \frac{h_{11}}{c_{44}}\right) \left(\sqrt{1 - \left(\frac{V_R}{c_t}\right)^2} + \frac{h_{11}}{g} \sqrt{1 - \left(\frac{V_R}{c_l}\right)^2} \right)}{2 \left[\frac{c_{44}/c_{11}}{\sqrt{1 - \left(\frac{V_R}{c_l}\right)^2}} + \frac{1}{\sqrt{1 - \left(\frac{V_R}{c_t}\right)^2}} + \left(\frac{V_R}{c_t}\right)^2 - 2 \right]}$$

(14)

When only a mass loading is taken into account, we obtaine the following estimation of the relative change in the velocity of the Rayleigh waves

$$\frac{\Delta V}{V_R} = k \frac{\rho_s}{\rho} \left(\frac{V_R}{c_t}\right)^2 .$$

(15)

The last formula agrees with the expression obtained in ref.1. Unlike the SH-waves (10) the relative shift of velocity due to the presence of LB film in the later case is a linear function of frequency.

The expresion (14) obtained here is more general case than in ref.[1] since it involves not only a mass loading but the elastic properties of LB film.

REFERENCES

1. Plesskii V.P., "Devices on surface acoustic waves included LB-films". Sov.Acoust.Journ. 37, N33 (1991), p.421.
2. Blinov L.M., "Langmuir Films". Uspekhi phizicheskikh nauk. 155, N3 (1988), p.443.
3. Gevod V.S., Ksenzhek O.S., and Reshetnyak I.L. " Artificial membrane structures and prospect for their practical application", Sov.Biol.membranes, v.5, N12 (1988), p.1237.
4. Ohnishi M., Ishimoto C. and Seto.J., "The biomimetic property of gas-sensitive films for oderant constructed by the Lan-

gmuir-Blodgett technigue". Thin Solid Films, 210/211 (1992) p.455.

5. Rapp M., Stranzel R., Schickfus M.V., Hunklinger S., Fuchs H., Schrepp W., Keller H., and Fleischmann B., "Gas detection in the Ppb-range with a high frequency, high sensitivity surface acoustic wave device". Thin Solid Films, 210/211, (1992), p.474.

6. Kurach T.N., Naumenko N.F. and Plesskii V.P., "Sensor "cut-off" of lithium niobate and quartz". Sov. Acoust Journ. 36, N3 (1990), p.561.

7. Kosevich Yu.A. and Syrkin E.S., "Macroscopic dynamics of a crystal surface with an adsorbed monolayer". Sov. Phys. Solid State, 31, N7 (1989), p.127.

8. Israelaechvili J.N., Mitchell D.J., and Ninham B.W., "Thermodynamics of self-organization of amphiphile assembles" Soc. Faraday Trans. II72 (1976), p.1525.

9. Kosevich Yu.A., "Anomalous acoustic wave absorption by a two-dimensional adsorbed layer on the surface of a solid". Sov. Phys Solid State, 33, N9 (1991), p.2597.

PHASED ARRAY TRANSESOPHAGEAL ENDOSCOPE FOR PEDIATRICS

J. E. Piel Jr.,[1] R. S. Lewandowski,[1] P. W. Lorraine,[1] L. S. Smith,[1] and D.J. Sahn[2]

[1]G.E. Corporate Research & Development
P. O. Box 8
Schenectady, NY 12301
[2]Oregon Health Sciences University
3181 S.W. Sam Jackson Park Road
Portland, Oregon 97201

INTRODUCTION

The use of ultrasonic phased array transducers in cardiac imaging had once been limited to the inspection of the heart through the chest wall or from under the lower ribs, looking up past the diaphragm. This method of cardiac imaging, transthoracic echocardiography, has limited views due to the acoustical interference from the ribs, lungs and other anatomical features between the transducer probe and the heart.

There has been increased interest in transesophageal echocardiography (TEE). This cardiac imaging method places the phased array transducer on the end of a long, flexible shaft (Figure 1) which is inserted into the esophagus of a patient. The esophagus, which is in intimate contact with the heart, provides a greatly improved acoustic window compared to the transthoracic approach. The crucial central area of the heart, including the left atrium, pulmonary veins, mitral valve area, subaortic area, and pulmonary arteries from a posterior approach, are all in close proximity to probes placed into the esophagus. TEE provides higher resolution for diagnostic echocardiology and has been used to monitor interventions such as catheter intervention and cardiac surgery. The TEE approach to cardiac imaging not only provides the ability to observe post-surgical recovery and to inspect for cardiac defects, but also can be used during heart surgery without the surgeon's direct involvement or distraction. The improved acoustic window of TEE also improves the phased array transducer's measurement of cardiac blood flow and pressure, both during and after surgery.

The advantages of TEE over the transthoracic method are many, but the obvious disadvantage is in the introduction of the endoscope into a vital passageway of the body. This requires the design of the endoscope's shaft and probe to be of a minimum profile, thereby facilitating the initial swallowing and manipulation of the endoscope. This small profile also minimizes any obstruction to the flow of air in the adjacent trachea.

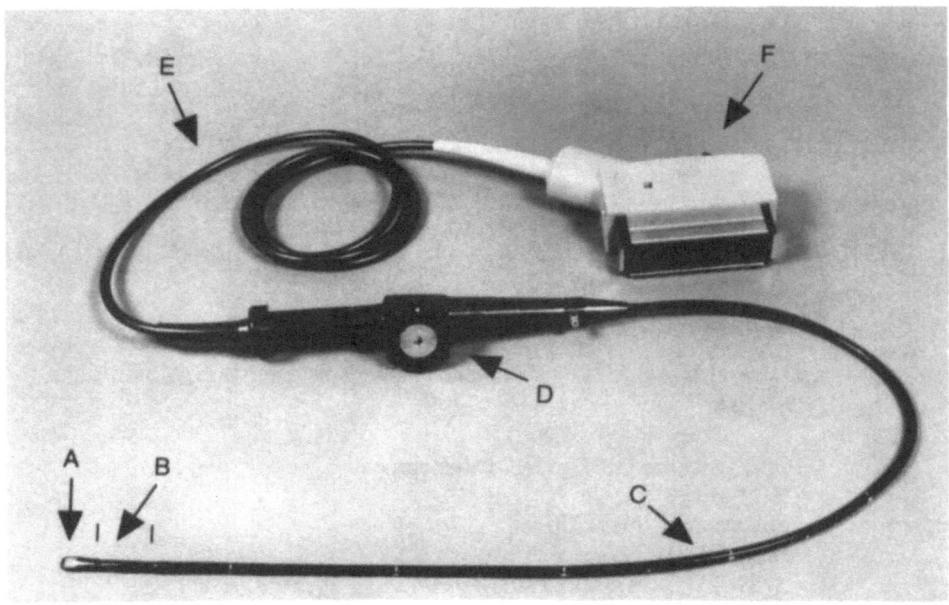

Figure 1. TEE Pediatric Endoscope: (A) Probe Tip, (B) Deflection Section, (C) Flexible Shaft, (D) Handle/Control Knob, (E) Console Cable, F) Console Connector

The design requirement for minimal probe and shaft diameters is constrained by the currently available technology for constructing the various parts of the endoscope. Within these constraints, endoscopes with shaft and probe tip diameters in the range of 10-15 mm have been built and are commercially available. These endoscopes provide excellent cardiac imaging in adult patients and also in older children.

PEDIATRIC TEE ENDOSCOPE DESIGN

The success of the adult-sized endoscopes, along with the increasing demand for early detection and treatment of congenital heart defects, has created a challenge to design and build an endoscope with the equivalent image quality of the adult endoscope but having the smaller physical dimensions needed for imaging neonatal patients (newborns). There are a few commercial endoscopes available for this segment of the pediatric population but they have compromised either the small physical dimensions or the image quality in their designs.

We began building an optimal pediatric TEE endoscope by defining the design criteria from clinical experience[1,2] involving numerous patients and the use of various existing pediatric endoscopes. These studies established three major requirements for building a pediatric TEE endoscope. The first requirement defines the physical dimensions of the flexible shaft and the probe tip needed to image newborns whose weight may be as low as 2 kg. A shaft diameter in the range of 6-7 mm and a probe tip whose widest dimension does not exceed 8 to 8.5 mm are needed for newborn imaging. Comparison of these dimensions to a typical adult-sized TEE endoscope can be seen in Figure 2. This shows a side by side view of an adult probe tip ~13 mm in width and our pediatric probe tip of ~8.5 mm in width. The shaft diameter of the adult endoscope in Figure 2 is 10 mm while the diameter of our pediatric endoscope is 6.2 mm.

The second requirement maintains the high element number of the transducer arrays used in adult endoscopes. This high element count produces images with good contrast and spatial resolution. Typical adult endoscopes with a single array architecture have phased array transducers with 64 elements in the array and therefore this is the element count used in our pediatric endoscope.

The third requirement maximizes the frequency at which the transducer will operate. Although signal attenuation in tissue is directly proportional to frequency, the shorter imaging distance needed for pediatrics allows the use of a higher frequency than in adult transducers. This higher frequency increases the spatial resolution thereby improving the ability to image the finer anatomical features of neonatal patients. This pediatric endoscope operated at 7.5 MHz rather than the 5 MHz usually used for adult endoscopes.

Building a TEE endoscope with these specifications would achieve the goal of producing a newborn-sized endoscope with excellent image quality, but the design problems introduced are numerous and they stretch the limits of available technology.

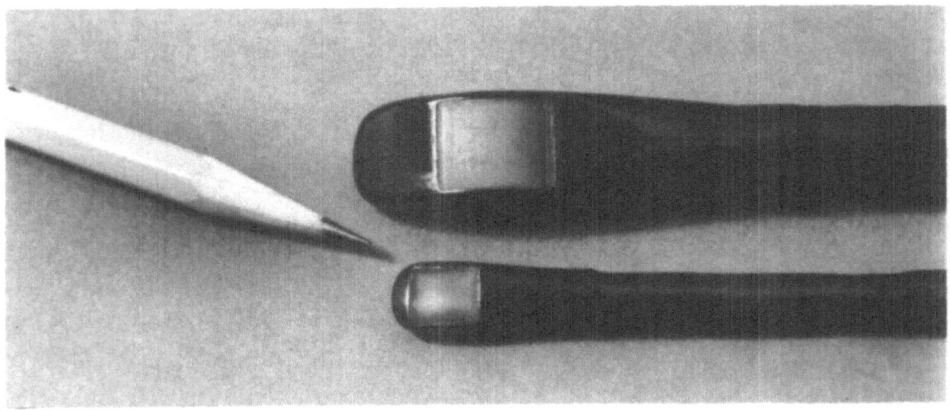

Figure 2. Comparative sizes of 64 element endoscopes: conventional 5 MHz adult (top) and the 7.5 MHz pediatric endoscope (bottom).

COAXIAL CABLE AND SHAFT ASSEMBLY

Reducing the diameter of the flexible shaft for the pediatric TEE endoscope yet maintaining the element count of the adult endoscope (64 elements) creates conflicting requirements. The free space available inside the shaft is decreased from a 5-7 mm diameter in the adult scope to only a 3 mm diameter for the 6.2 mm O.D. pediatric endoscope. Only 70% of the cross-sectional area of this free space can be filled by cables in order to maintain the flexibility of the shaft. These stringent space demands require state-of-the-art coaxial cable technology. This technology [Precision Interconnect, Portland,OR] provides individual 50 ohm coaxial cable with 44 AWG center conductor wire which is only 50 μm in diameter. Over this wire is a controlled thickness of dielectric material and a served shield made of strands of 52 AWG wire (20 μm diameter). These individual coaxial cables have a capacitance of approximately 30 pF per foot and have a finished outer diameter of less than 225 µm. Including several spare cables, a bundle of 69 of these miniature coaxial cables with a diameter of ~2.5 mm is fitted into the flexible shaft. This bundle just meets the design criteria for flexibility of the shaft.

LONGITUDINAL PLANE IMAGING

Although bi-plane imaging provides both transverse and longitudinal anatomical views, the longitudinal plane chosen here achieves the highest possible image quality and yet stays within the physical size constraints dictated by the size of the newborn patients. The half-wavelength pitch of the 7.5 MHz, 64 element array is ~100 µm and therefore the array aperture is ~7 mm (this includes four unused elements, two at each end of the array to reduce end effects). This aperture length was oriented along the axis of the endoscope shaft (Figure 3a). This helps determine the length of the probe tip but does not directly impact the critical width or thickness dimensions of the tip. The length of the individual elements of the array determines the elevation aperture. This aperture was chosen to be ~6 mm (approximately the diameter of the shaft) and is the major contributor to the width of the probe tip. This orientation of the array to the endoscope shaft also determines the orientation of the imaging plane to the heart and is referred to as longitudinal plane imaging (Figure 3b). If effective transgastric positioning is achievable, the rotation of the probe in the stomach effectively provides views covering the transverse imaging plane of the anatomic structures. Longitudinal plane imaging, therefore, offers high resolution imaging with minimal probe tip width.

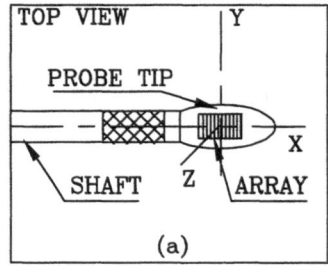 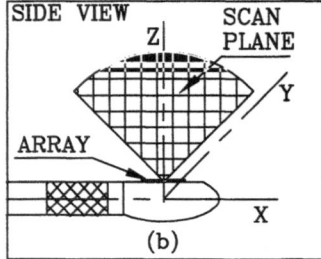

Figure 3a) Orientation of the transducer array elements to the endoscope shaft.
Figure 3b) Longitudinal plane imaging

ACOUSTIC DESIGN

The acoustic design of the transducer array (Figure 4) uses conventional methods for achieving wide bandwidth and short impulse response leading to good axial and lateral resolution.[3,4,5] The piezoelectric ceramic is matched to the body using one or more acoustic matching layers of an impedance intermediate between the ceramic and body. The thickness of these layers is carefully tailored to optimize performance. A backing material is used on the obverse side of the ceramic to absorb sound which leaks out of the back so that it does not subsequently reappear within the ceramic. This backing material is made from a highly attenuative material which minimizes its thickness and therefore the thickness of the probe tip is minimized. Special care was taken to isolate the elements of the array from each other, both electrically and mechanically, so that the imaging sector would not degrade at wide angles and so that steerable CW Doppler exams would be possible. The acoustic lens was precast from a low velocity, medical grade silicone material and fastened to the matching layer with medical grade adhesive. This lens sets the elevation aperture focus to ~2-3 cm, a distance determined optimal for pediatric echocardiology from clinical experience.

Sputtered metal on the matching layer side of the piezoelectric ceramic is used as the signal ground electrode and the metal on the opposite side of the ceramic is used as the signal electrode. In order to improve the quality of steerable CW Doppler exams, the signal ground electrode was divided at the middle of the array. During CW Doppler imaging mode, half of the array elements are used continuously as transmitters and the other half of the elements are used as receivers. This "split ground" feature isolates the electrical return paths of the higher voltage transmitter signals from the lower voltage receiver signals. These two separate grounds were maintained from the probe tip to the console connector.

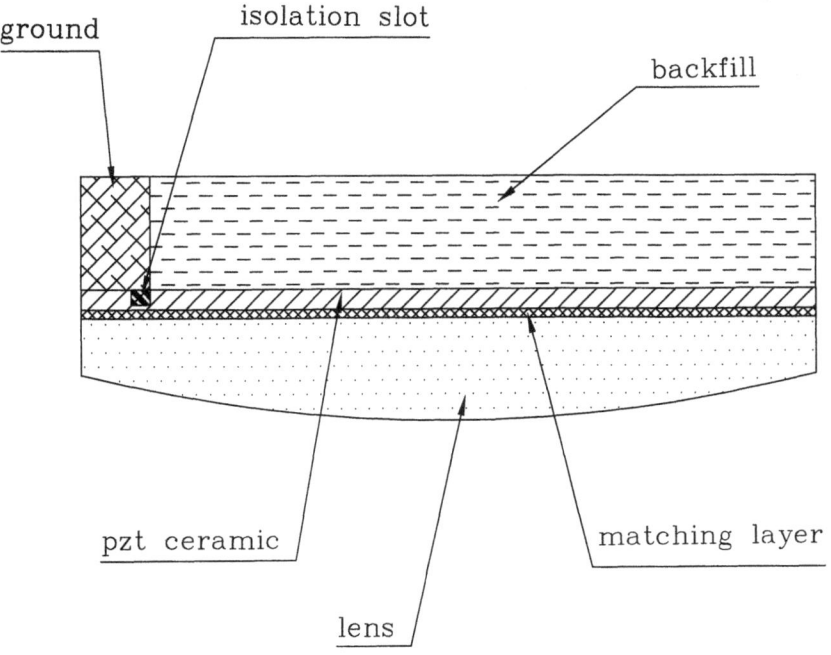

Figure 4. Cross-sectional drawing of phased array transducer.

INTERCONNECT TECHNOLOGY

Fabrication of ultrasonic phased arrays requires attaching individual signal leads to array elements of a piezoelectric ceramic. The high frequency, high element count requirements for pediatric TEE arrays have challenged the well established methods of lead connection, such as hand-soldering and wire bonding. The tight center-center element spacing (~100 μm) and the limited volume available to interface the coaxial cables to the element leads requires a sophisticated interconnect design.

We developed a laser-assisted soldering process[6] to address this interconnect problem. A photolithographic process is used to pattern the piezoelectric ceramic with a copper solder pad for each element of the array. Solder is then plated onto these copper pads. A flexible circuit board with a lead pitch corresponding to the solder pads (~100 μm c-c) is aligned with these pads. A CW Nd:YAG laser is focused onto each solder pad/lead assembly and is used to solder the lead to the ceramic. The laser steps from lead to lead, soldering each lead individually. The bond strength of the solder joints have been shown to be comparable to other solder methods.

The focused power of the laser heats a very localized area. This raises the solder to melting temperatures without excessive heating of the ceramic substrate, which can cause degradation of piezoelectric activity. This method also has the advantage of delivering the necessary soldering power without mechanical contact with the ceramic. With a mechanical method, such as a solder pencil or ultrasonic wire bonder head, the bonding tool is often wider than the desired pitch causing difficulty when connecting adjacent leads. Laser soldering's localized power also allows the use of flexible printed circuit board material to carry the signal leads away from the ceramic without damage to the flexible substrate. These flexible boards can be folded in such a way as to minimize the finished dimensions of the ultrasonic probe (Figure 5).

Figure 5.   Coaxial cable soldered to flexible circuit boards of transducer.

This interconnect technology required the development of special expertise in the production of extremely fine pitched flexible circuit boards with line/space pairs at approximately 100 micron pitch. Our approach begins with a relatively thin metal layer on the flexible dielectric substrate. Standard photolithographic techniques are used to pattern the traces and then additional metal is selectively plated to enhance the conductivity and improve reliability. Special care must be taken to assure metal ductility of the resulting structures for good flexibility. The extreme end of these traces fan out to a pitch of ~300 μm c-c to allow for attachment of coaxial cables. Four flexible circuit boards were used in order to maintain the fan out to less than the width of the array.

Each element's signal is connected through the flexible circuit board to the center conductor of one of the coaxial cables running through the shaft. This constitutes one of the most labor intensive operations in the assembly, simply because the wires must be hand soldered individually. The need to limit the volume occupied by this joint while maintaining good electrical isolation also complicates assembly. Additional ground planes are inserted between the individual signal boards to limit capacitive coupling among the elements.

ARRAY DICING

After the leads have been soldered to the piezoelectric ceramic, the backing material is applied to the signal side and the array assembly is prepared for dicing of the individual elements. A semiconductor dicing saw is used for this process. The dicing blade is carefully aligned to cut between the laser soldered leads. The kerf of the blade goes completely through the matching layer and the ceramic and also partially into the backing material. The backing material serves a dual purpose of attenuating unwanted reverberations and also holding together the diced elements of the array. The dicing procedure requires straight, uniform depth kerfs of minimum width. A width of ~25-30 µm is achievable, limited mostly by blade technology and cutting technique.

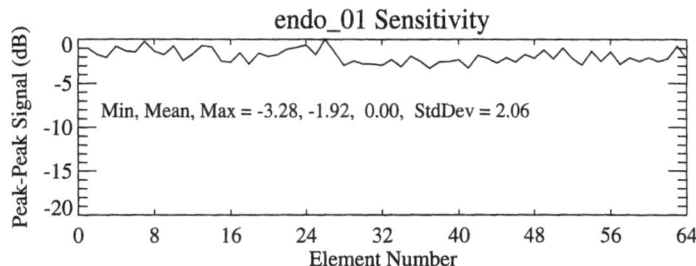

Figure 6.   Element sensitivity across array demonstrating good uniformity.

LENS, ENCAPSULATION AND TESTING

The diced array is fitted with an acoustic lens and the completed transducer is connected to the coaxial cables extending from the endoscope tip. The final assembly encapsulation of the transducer/cable assembly into the probe tip is performed in a special jig. The array is held by the lens face in the mold and a biocompatible epoxy covers the remaining structures and fixes the tip to the distal end of the deflection section. An RF shield is also fitted over the transducer structure prior to the encapsulation. Some hand crafted finishing work is required to give the best external appearance.

All probes have been qualified pneumatically and electrically before shipment to the clinical site. The endoscope shaft withstands an overpressure of four pounds/sq. in. without evidence of leakage. Electrical tests conform to standard industry(NEMA) and UL practice. High potential and leakage current tests were performed with the shaft and headshell immersed in saline. The devices withstood 1500 kVac, and had leakage current of less then 50 microamps.

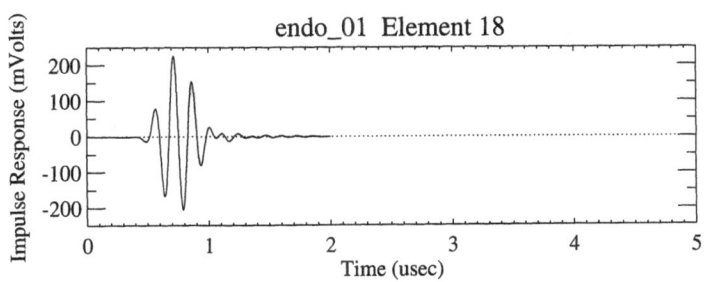

Figure 7.   Impulse response of single element.

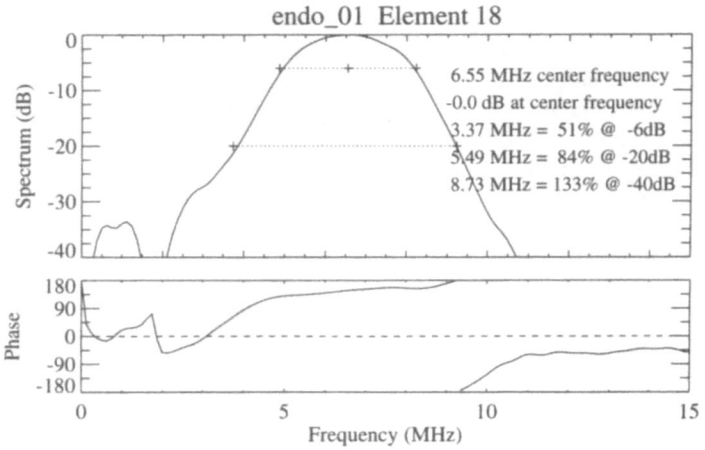

endo_01 Element 18

6.55 MHz center frequency
-0.0 dB at center frequency
3.37 MHz = 51% @ -6dB
5.49 MHz = 84% @ -20dB
8.73 MHz = 133% @ -40dB

Figure 8.   Spectrum of single array element.

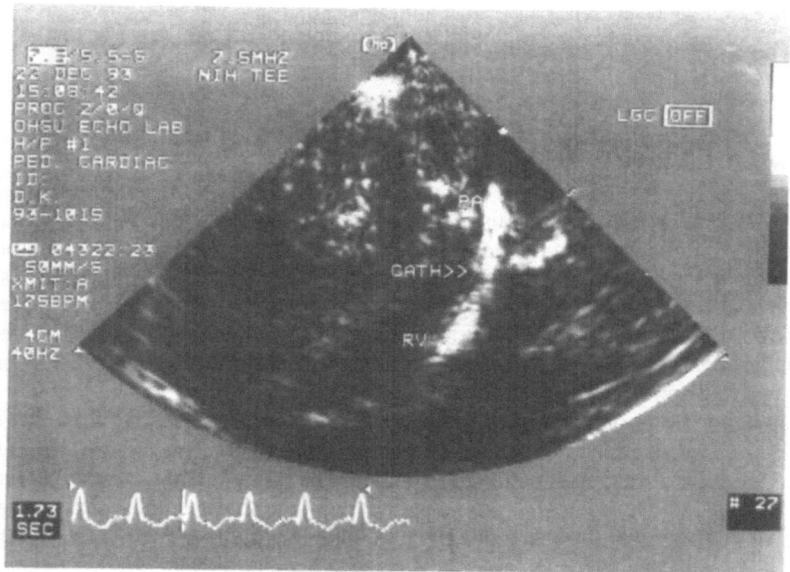

Figure 9.   7.5 MHz ultrasonic image of a pediatric heart. The endoscope provides anatomicguidance for balloon valvalopasty, a treatment for severe pulmonary stenosis. A catheter is clearly visible passing through the pulmonary valve.

RESULTS

Acoustic water tank measurements show that the arrays have excellent sensitivity, a short impulse response, good inter-element isolation, and good uniformity (Figures 6,7,8).

These endoscopes have been used in a variety of neonates with congenital heart disease. They have been especially useful in guiding therapeutic procedures performed with catheterization. In a particular example, the endoscopic images were used to position a balloon catheter for the enlargement of a severe pulmonary stenosis. The probes have yielded extremely high resolution images even in far field subgastric views. The probes function with very high sensitivity, not only for imaging, but for color Doppler, pulsed, and steerable CW Doppler, rarely requiring dropping the frequency for Doppler to 5.5 MHz, and often allowing full scanning capability at 7.5 MHz even out to 4 to 6 cm with extremely good near field resolution. Our probes have now been used in infants as small as 2100 g with extremely favorable results (Figure 9).

ACKNOWLEDGMENTS

We wish to thank John Smith of Schott Fiber Optics, Blake Merrick, Georgene Hornbuckle, and Jim Baker of Precision Interconnect. We would also like to thank our GE CR&D colleagues, Jim McMullen, Mike Fazzone, Marshall Jones, and Jerry Harrison for their superb technical assistance.

REFERENCES

1. D.J. Sahn. Invasive and non-invasive cardiovascular imaging in the child and fetus, Curr. Opin. Cardiol, 7:94-102,(1992).
2. D.J. Sahn. Recent and future technological advances for transesophageal echocardiology, Ch 12 in: "Transesophageal Echocardiology," H. Dietrich, ed., Mosby-Year Book, St. Louis, MO, (1992), pp 165-171.
3. C.S. DeSilets, J. D. Fraser, and G.S. Kino, The Design of Efficient Broadband Piezoelectric Transducers, IEEE Trans. Sonics and Ultrasonics, SU-25:115-125 (1978).
4. G.S.Kino, and C.S. DeSilets, Design of Slotted Transducer Arrays with Matched Backings, Ultrasonic Imaging 1:189-209 (1979).
5. T.L. Rhyne, and S. Panda, Design Implications of Using Transducer Noise Figure as the Basis of Sensitivity, 1993 Ultrasonic Symposium Proceedings.
6. M.G. Jones, J.T. Harrison, J.E. Piel, and L.S. Smith, Holder for Soldering a Flexible Circuit Board to a Substrate, U.S. Patent 5,148,962, Sep. 22, 1992.

CURVED ARRAY IMAGING FOR NAVIGATION IN AN UNKNOWN ENVIRONMENT

D.W. Purcell, and J.P. Huissoon

Department of Mechanical Engineering
University of Waterloo
Ontario
Canada, N2L 5E6

INTRODUCTION

Autonomous vehicles require the ability to sense and react to the environment in which they operate. This ability is necessary for all classes of autonomous vehicle operation, ranging from map-based navigation to roving within a completely unknown environment. Airborne ultrasound is a highly suitable sensing technique in many applications; for moderate vehicle travel speeds (up to about 1 m/s [3.6 km/hr]) and fairly close range detection (up to about 10 m), airborne ultrasound is probably the most cost-effective sensing technique available at present.

Airborne ultrasound has been used for obstacle detection and avoidance, vehicle self-localization and mapping an unknown environment [1-3]. In a static environment, where features that may be used to define potential routes remain unchanged, map-based navigation may be possible. In such applications, ultrasonic ranging may be used to maintain vehicle distance from defined corridor boundaries, as well as for self-localization. The latter is almost always required in a map-based guidance strategy, since the vehicle must somehow accommodate accumulated error in global distance measurements obtained through vehicle travel sensors (such as wheel encoders)[4]. However, unless the environment is truly static, transient equipment and other objects (especially mobile) that cannot be anticipated, require any navigation strategy to include intelligent obstacle detection and avoidance capabilities.

The navigation tasks outlined above require information in real time about the dynamic environment in which the vehicle is operating. While airborne ultrasound is an ideal sensing medium, its effective implementation has encountered difficulties due to data rate and bandwidth problems associated with commercially available transducers. Most guidance strategies have involved the use of Polaroid type discrete transducers that are either placed in a ring about the vehicle [5], or mechanically focused and oriented [6]. Using a ring of transducers, or a single stepped device, 360° scans of the environment may be performed. An effective autonomous vehicle has remained elusive however, as these strategies suffer respectively from limited angular resolution and response time.

A curved array transducer has been developed that offers significant advantages over discrete device transducer systems. The array can theoretically be manufactured with a near 360° field of view, eliminating the need for any mechanical scanning motion. A desired direction may be viewed by choosing a subset of elements facing that direction. Operating in a standard pulse-echo ranging mode, a narrow ultrasonic beam may be generated by appropriate phasing of multiple elements facing the desired scan direction. Phasing is reserved for beam shaping only, as the incremental spacing between elements is smaller than the minimum beam width; no benefit would be gained through phased beam steering. While the pulse echo technique is appropriate for certain navigation tasks, the array is also capable of utilizing more advanced imaging methods. A wide beam phased transmission, followed by multiple element reception, may be processed using beamforming or the synthetic aperture focusing technique (SAFT).

In any AGV navigation system, inefficient and over use of sensors must be avoided if the vehicle is to travel safely at acceptable speeds. Cumulative delays caused by heavy processing demands and the flight time of ultrasound can be limited through intelligent use of a sensor. The flexibility of the curved array permits imaging techniques to be chosen to best suit the navigation task at hand and thus increase efficiency.

Obstacle detection is conventionally performed by viewing the area of in front of the vehicle with a number of overlapping scans. The least system effort would be expended utilizing a single wide beam transmission, however if an obstacle is detected, more detailed information is required to navigate around it. The adaptability of the array is beneficial for this task of obstacle detection and avoidance. A wide beam transmission may be processed as a standard pulse-echo ranging signal until some obstruction is detected. More detailed information may then be obtained through beamforming or multiple narrow beam pulse-echo scans. A similar strategy could be applied to enable moving targets to be avoided. Regions around the vehicle where little activity has recently been registered may be viewed with wide angle pulse-echo ranging, while more interesting regions can be imaged in detail. Adaptive use of the array thus limits processing requirements for this navigation task.

The navigation task of self localization can also be implemented more efficiently utilizing the curved array. It is necessary to occasionally correct an AGV's position estimation due to dead reckoning errors that accumulate through the use of internal sensors such as wheel encoders. Naturally occurring environment features may be selected from a precomposed map and then searched for using ultrasonic sensors based on the current estimated position. New range and bearing information about an anticipated target can then be used to correct and update the AGV's position. Beamforming is particularly attractive for this purpose as the exact location of the natural beacon will not be known due to vehicle position error. A beamforming image may reveal the land mark with a single ultrasonic transmission, whereas multiple narrow beam scans would require more effort.

This "intelligent" use of the adaptive beam shaping capabilities of the curved array and its control processor, may greatly enhance the performance of autonomous vehicles over those using conventional sensors. The following sections describe the array control, and both experimental and theoretical results for its beam shaping and beamforming capabilities.

CURVED ARRAY DESIGN AND CONTROL

The prototype curved array ultrasound transducer is shown in Figure 1. This device consists of 31 equally spaced transducers formed around a 100 mm diameter perspex cylinder. The individual anodes of the array are etched onto a flexible printed

Figure 1. Curved Array Transducer   **Figure 2.** Array Electronics Board Location

circuit board. The board is then fastened around the body of the cylinder for a resulting angular spacing between elements of 3.75°. In between each element on this curved board are raised guard rails that serve to maintain an air gap between the anodes and a common cathode foil that is tensioned over them. The current implementation consists of 16 adjacent elements that can be driven for phase shaded transmission, and 4 equally spaced elements for echo reception. This arrangement permits the active array segment to view up to 60° of the environment.

Each array element is serviced by dedicated electronics to permit transmission and reception. These circuits are located immediately behind each element on the inside of the perspex cylinder. They are mounted on a curved circuit board positioned so that the elements are connected to their respective driving electronics via the shortest electrical path. This arrangement minimizes signal attenuation and cross-talk between adjacent elements. The position of the interior curved circuit board is shown in Figure 2.

An element is driven for transmission by a class B high voltage amplifier as shown in Figure 3. This circuit takes a TTL level signal (0-5 V) as an input and produces a low impedance high voltage output to drive the transmitting element (0-180 V). Components are chosen such that input TTL signals with frequencies in the order of 40-50 kHz are reproduced accurately on the array element. When the TTL input is inactive (0 V) this circuit provides the high impedance polarizing voltage that is necessary for the element to be used as a receiver of ultrasound.

The receive circuit is a very high impedance amplifier as shown in Figure 4. The array has driving and receiving circuits in place for 16 elements, however only four are currently being used to record return echoes. This circuit, which is located adjacent to each element, amplifies and transforms the impedance of the received AC echo signal. The amplified signal is then available for following stages where two further gains are applied before digitization and storage in a PC. Further data processing is performed in software in the PC.

The array can transmit using phase shading over sixteen elements symmetrically driven about the centre two. This phase shading is delivered to pairs of element amplifiers as eight independent TTL signals. These eight signals are stored in RAM as a look-up-table (LUT) which contains one cycle of phase information. Each bit of the RAM's output will drive a pair of element amplifiers with the wave form stored in the LUT. The schematic of this LUT is shown in Figure 5. A 12 MHz oscillator clocks a synchronous 8-bit pre-loadable counter to generate sequential LUT addresses. Each cycle of the counter

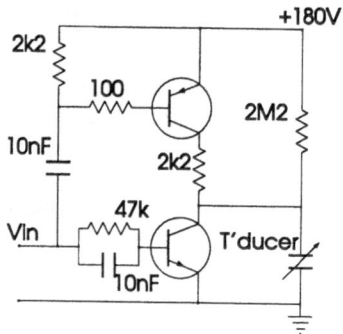

Figure 3. Transmission Amplifier Circuit for Each Array Element

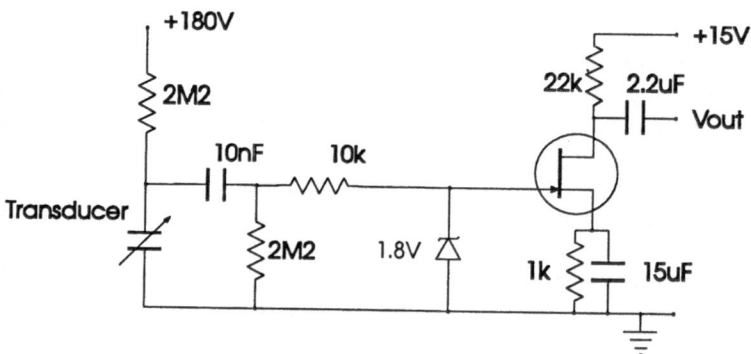

Figure 4. Reception Amplifier Circuit for Each Array Element

will clock through one transmit cycle of the LUT. The pre-load value of the counter will determine how many RAM locations are evaluated and hence the corresponding transmission frequency. The pre-load value is set via DIP switches. The data stored in the LUT correspond bit-wise to the logic levels applied to the transmitting amplifiers at any instant. With an oscillator frequency of 12 MHz and a counter pre-load value of

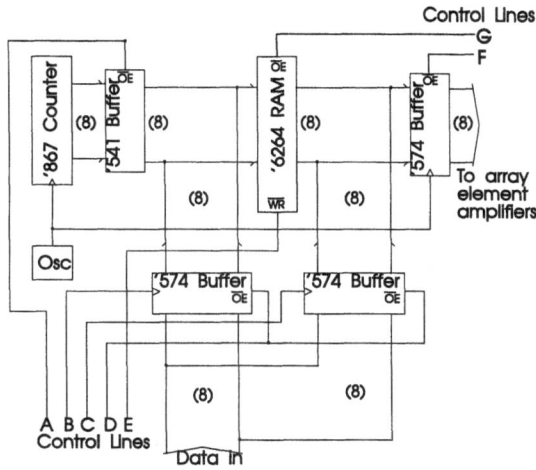

Figure 5. RAM Look-Up-Table for Phased Transmission of 16 Elements

255 the minimum phase resolution is obtained; 1.4° at a frequency of 46.875 kHz. An output latch ensures that all signals are synchronously applied to the amplifiers. The RAM system described here is capable of storing up to 16 phase LUT that may be selected to transmit different beam shapes as required by a navigation system. These LUT are generated with the knowledge of the desired relative phasing between elements and downloaded to the memory via a PC-AT. The relative phases are determined numerically using the techniques described in the following section.

BEAM PATTERN SIMULATION

The far-field sound pressure distribution P(R,θ), for a single rectangular element on the surface of a cylinder, in the plane normal to the axis of the cylinder and through the mid point of the element, is given as [7]:

$$p(R,\theta) = \frac{4\rho c A_0 h}{\pi^2} \frac{e^{j(kR-\omega t)}}{R} \cdot$$
$$\sum_{m=0}^{\infty} \frac{\sin m\alpha}{m\varepsilon_m} \frac{e^{-jm\pi/2}}{H_m^{(1)}(ka)} \cos m\theta \qquad (1)$$

To compute the beam pattern for the active array elements, 1440 station points (at 0.25° intervals) are first defined about the central axis of the transducer. The magnitude and phase of the far-field sound pressure for each array element are computed at each station point. The resulting beam pattern for any phase (and amplitude) shading can then easily be computed at the station points by simply superimposing the contributions of all active elements. Figure 6 shows the computed beam pattern for 16 array elements, all driven in phase, calculated using the above method. As would be expected, a fairly wide effective main lobe width is observed. On providing phase shading so that the contribution from each active array element is in phase along the central axis of the active array sector (a planar wave front), the beam pattern shown in Figure 7 results. This beam pattern is clearly more directional, although significant grating lobes are evident.

BEAM PATTERN OPTIMIZATION

The directivity of the beam pattern may be classified in a number of ways [8]. In this study, the primary criterion, main lobe width, was defined as the angle subtended between the first off-axis -6dB points. A second criterion, side-lobe rejection, was defined as the maximum off-axis sound pressure level beyond the main lobe limit. The ripple within the main lobe was used as a third criterion by which to judge the sound pressure distribution.

These three criteria were combined to generate a rating metric whereby the "quality" of various sound pressure distributions could be compared. Main lobe width was weighted as the square of the difference between the required and actual value. Side lobe rejection was referenced to a base value of -15dB, below which no penalty was incurred but above which the penalty was calculated as the square of the value over the reference level. Sound pressure level variation within the main lobe was rated by the difference between the greatest local maxima and minima (if any) about the on axis sound pressure level.

This beam pattern rating enables a minimization procedure to be numerically implemented. However, the results of any such procedure are somewhat subjective in that the relative weighting of the various criteria greatly influence the minimum converged to. Furthermore, the rating surface may well exhibit local minima, in which the minimization routine may become trapped. The results of each beam shape optimization were therefore further tuned using a graphical user interface. This enables incremental changes to be made to the phase shading profile via the keyboard, and immediately displays the effects of these changes on the screen. The shading profile can thus be manually tuned following the initial numerical estimate. While this is not a very scientific approach to beam shape optimization, it was found to be very effective in generating globally optimal beam shapes, as the following results indicate.

The beam pattern resulting from the in-phase transmission was tuned to give a 60° main lobe width as shown in Figure 8. While this does not differ greatly from that shown in Figure 6, the side lobe rejection has been increased by about 7dB, while the "effective" main lobe is better defined although at the expense of a slight (<2 dB) increase in main lobe ripple. However, the power distribution is now more concentrated within the main lobe, and this provides a significant improvement over the beam pattern of Figure 6.

Optimization of the focused transmission from the planar wave front shown in Figure 7, results in the beam shape shown in Figure 9. Some improvement is evident, primarily due to the transfer of power from the side lobes to the main lobe; this is also accompanied by a slight improvement in side lobe rejection although the main lobe width remains unchanged.

MEASURING BEAM SHAPE AND ANALYSIS

Experimental recording of the sound pressure distribution is implemented using a similar setup to that reported in [9]. Control of the test rig and data acquisition are performed by the PC. The curved array transducer is mounted on a turntable, the motion of which can be controlled by the PC, and the rotation of which can be measured accurately. A Polaroid transducer, located 1m from the array, is used to measure the emitted sound pressure from the array. The test procedure consists of slowly rotating the array and transmitting a 2ms tone burst every 1.1° of rotation, using the phase control hardware described above. 3.5ms after the start of the tone burst, the (amplified) signal from the detector is recorded for 0.5ms at a 1MHz sample rate, with 12 bit precision. Figure 10 shows a typical tone burst and the sampled section. Since the rotational speed of the transducer is such that successive tone bursts occur at 80ms intervals, this technique

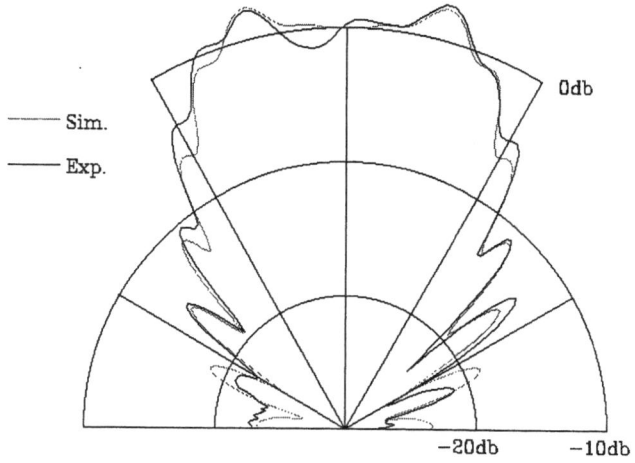

Figure 6. Experimental vs Simulated Data for Inphase Transmission

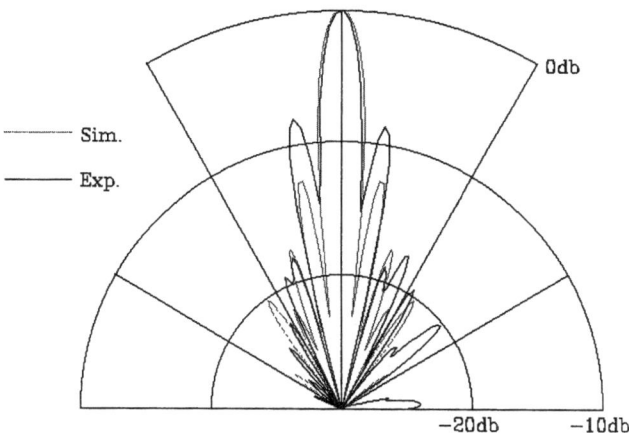

Figure 7. Experimental vs Simulated Data for a Planar Wavefront

ensures that the recorded signal is not corrupted by reflections; furthermore, the array rotates through less than 1.7 minutes of arc during the transmission, and the recorded signal thus accurately represents the sound pressure emitted by the array at that particular angle. The RMS value of the 500 recorded samples is computed following each transmission, and is stored with the angular position at which it was recorded. The recorded beam pattern is displayed on the screen following the completion of each test.

PULSE-ECHO DATA RECORDING AND ANALYSIS

Pulse-echo experiments are carried out by transmitting either a focused or wide beam of ultrasound and then recording the return echoes at four equally spaced array elements. Echo data is digitized from these four elements using an analog to digital

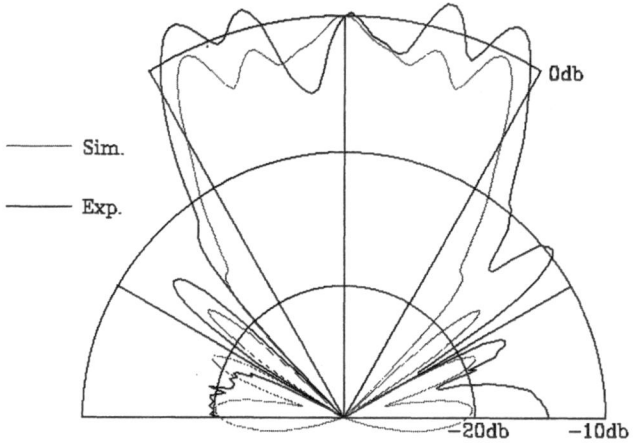

Figure 8. Experimental vs Simulated Data for 60° Beam

converter sampling at a frequency of 300kHz. In the following experiment, a specular reflector has been placed at ranges in the order of 1m in the array's view. Further displaced reflectors can be detected, but time gain amplification has not yet been implemented so ranges have been kept small. The transmitted beam shape is chosen depending on the desired navigation task. A narrow beam would be chosen for standard pulse-echo ranging scans in specific directions. In the experimental result that follows a wide 60° beam has been used.

The recorded echoes are then processed using a beamforming algorithm similar to that described in [10]. This algorithm is based on the coherent summation of data available from the receiving elements. The environment within the array's view is divided into a spatial grid for processing. At each point in this grid the appropriate data from each element is summed to produce an estimate of the certainty of a reflector's position. The data chosen for summation from each element is based on the time of flight from the array to the node under consideration and back to the element. At nodes where a reflector actually exists the data will tend to sum constructively to a larger value than at grid points where no reflector is present. This coherent summation of the recorded data for each node processed allows determination of the location of reflectors using a single broad beam transmission.

EXPERIMENTAL RESULTS

The phase shading profiles used to generate the simulated beam patterns shown in Figures 6 to 9, were programmed into the phase control circuit described in earlier. The recorded beam patterns for each are shown by the solid lines in the respective figures. It should be noted that all beam patterns have been normalized to the on-axis pressure. In each case it is evident that remarkable agreement exists between the angular location of features in the simulated and experimental beam patterns. From Figures 7 and 9, it is seen that the experimentally measured major grating lobe sound pressure level tends to be somewhat higher than that predicted. Also, the experimental beam patterns all exhibit a slight asymmetry, most evident to the left side of the central axis in Figures 6 and 8, and also in the significantly larger side lobes in the lower right quadrant. Delamination of the metallic film over a small area on the two array elements to the right side of the central axis is the likely cause of the first effect. Non-uniform foil tension, caused by the repair of a small tear in the foil close to the tension roller on the right side of the array is most likely responsible for the larger side lobes on that side. Nevertheless, the agreement between simulated and experimental results provides a high level of confidence in the numerical model. It is also possible to generate intermediate main lobe widths between those shown in Figures 8 and 9, although these are not presented here.

Pulse-echo experiments were carried out as described in the previous section using a transmission of 60° beam width. The received echoes were digitized at 300 kHz on four array elements as shown in Figure 11. The result of applying the beamforming algorithm to this data is shown in Figure 12. The x-y scales represent a grid that has been laid horizontally across the environment in front of the curved array. The z axis represents the certainty of a reflector's position in this grid. In this case the specular reflector placed to the forward right of the array has been accurately located using the beamforming technique. Figure 11 shows how the echo from this reflector returns to the nearest element (4) first and the furthest element last (1). Only at grid nodes close to where the reflector actually exists do these four waveforms add coherently to a large sum.

CONCLUSIONS

The benefits of a curved array for autonomous vehicle navigation have been discussed, including adaptive beam shaping and beamforming. Phase shading capabilities of the array have been demonstrated with several distinct experimental beam profiles, and good agreement has been shown with the numerical model. Beamforming results calculated from experimental data recorded at four array elements have been presented. The PC based phase control and data acquisition system implemented for the curved array has been described.

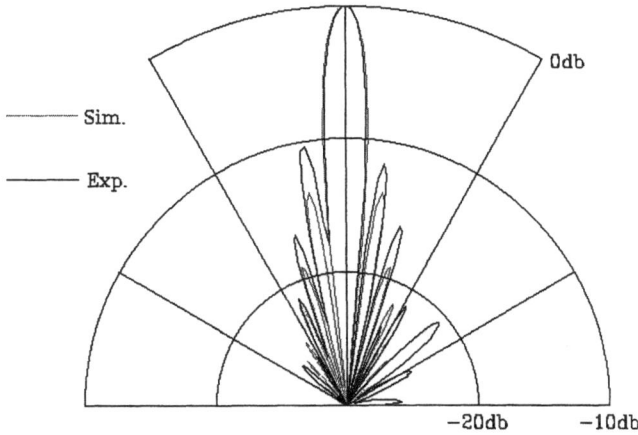

Figure 9. Experimental vs Simulated Data for Focused Transmission

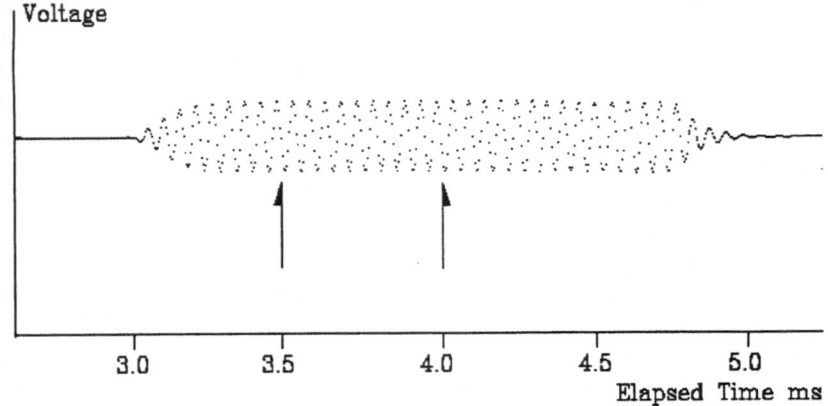

Figure 10. Typical Tone Burst

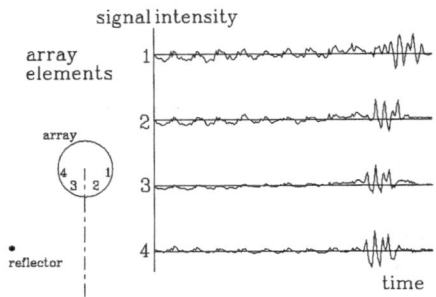

Figure 11. Return Echo Recorded at Four Elements

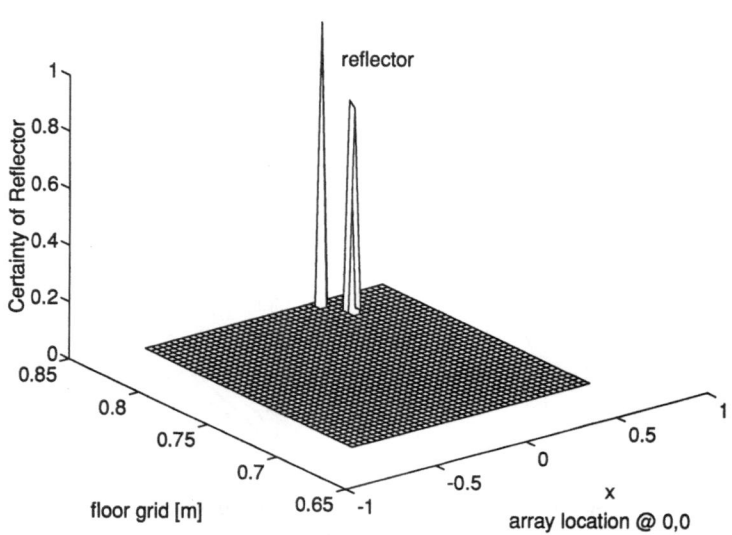

Figure 12. Beamforming for a Reflector Right of Centre

REFERENCES

[1]   Elfes, A., Sonar-Based Real-World   Mapping and Navigation, IEEE J Robot Automat (1987) RA-3 249-265.

[2]   Weisbin, C.R., de Saussure, G., Einstein, J.R., Pin, F.G., Autonomous Mobile Robot Navigation and Learning, , Computer (1989) June 29-35.

[3]   Everett, H.R., Gilbreath, G.A., A Supervised Autonomous Security Robot, Robotics (1988) 4 209-232.

[4]   Leonard, J.J., Durrant-Whyte, H.F.,   Mobile Robot Localization by Tracking Geometric Beacons, IEEE Trans Robotics Automat (1991) 7/3 376-382.

[5]   Holenstein, A.A., Badreddin, E., Collision Avoidance in a Behavior-based Mobile Robot Design, Proc 1991 IEEE Int Conf Robotics Automat (1991) April 898-903.

[6]   Bozma, O., Kuc, R., Building a Sonar Map in a Specular Environment Using a Single Mobile Sensor, IEEE Trans Pattern Analysis Machine Int (1991) 13/12 1260-1269.

[7]   Laird, D.T., Cohen, H., Directionality Patterns for Acoustic Radiation from a Source on a Rigid Cylinder, J Acoust Soc Am (1952) 24/1 46-49.

[8]   Huissoon,J.P., Multiobjective Design of Curved Array Sonar Transducers, Acoustical Imaging (1992) 19 231-236.

[9]   Huissoon, J.P., Moziar, D.M., Curved Ultrasonic Array Transducer for AGV Applications, Ultrasonics (1989) 27 221-225.

[10]   Purcell,D.W., Huissoon,J.P., A Curved Array Transducer for Autonomous Vehicle Navigation, Int Conf Acoustic Sensing Imaging (1993) March 32-37.

BIOPHYSICAL BASES OF ELASTICITY IMAGING

A.P. Sarvazyan,[1] A.R. Skovoroda,[2] S.Y. Emelianov,[2,3,4] J.B. Fowlkes,[3]
J.G. Pipe,[3] R.S. Adler,[3] R.B. Buxton,[5] and P.L. Carson[3]

[1] Department of Chemistry, Rutgers University, New Brunswick, NJ 08903
   Permanent Address: Institute for Theoretical and Experimental Biophysics
   Russian Academy of Sciences, Pushchino, 142292, Russia

[2] Institute of Mathematical Problems of Biology
   Russian Academy of Sciences, Pushchino, 142292, Russia

[3] Department of Radiology
   University of Michigan Medical Center
   Ann Arbor, Michigan 48109-0553

[4] Department of Electrical Engineering and Computer Science and
   Bioengineering Program, University of Michigan
   Ann Arbor, Michigan 48109-2125

[5] Department of Radiology
   University of California at San Diego
   San Diego, California 92103

INTRODUCTION

Elasticity imaging is based on two processes. The first is the evaluation of the mechanical response of a stressed tissue using imaging modalities, e.g. ultrasound, magnetic resonance imaging (MRI), computed tomography (CT) scans and Doppler ultrasound. The second step is depiction of the elastic properties of internal tissue structures by mathematical solution of the inverse mechanical problem. The evaluation of elastic properties of tissues has the potential for being an important diagnostic tool in the detection of cancer as well as other injuries and diseases. The success of breast self-examination in conjunction with mammography for detection and continuous monitoring of lesions has resulted in early diagnosis and institution of therapy. Self-examination is based on the manually palpable texture difference of the lesion relative to adjacent tissue and, as such, is limited to lesions located relatively near the skin surface and increased lesion hardness with respect to the surrounding tissue. Imaging of tissue "hardness" should allow more sensitive detection of abnormal structures deeper within tissue. Tissue hardness can actually be quantified in terms of the tissue elastic moduli and may provide good contrast between normal and abnormal tissues based on the large relative variation in shear (or Young's) elastic modulus.

All of the commonly used medical imaging modalities can provide useful information about various tissue properties. Ultrasound relies primarily on the heterogeneity of acoustical impedance, which is a combination of the bulk modulus and the tissue density, CT scans depict the spatial distribution of X-ray attenuation, basically electron density which generally correlates with the physical density distribution. The tissue properties which provide contrast in MRI are better described microscopically and this contrast is very sensitive to water and fat content. However, the tissue properties which are being measured by most conventional MRI techniques do not directly include the elasticity, which corresponds closely to the property of "hardness" obtained by the manual palpation. None of the above mentioned imaging techniques provide a direct quantitative measure of the tissue elastic properties related to the shear modulus.

The information which can be provided by developing an elasticity imaging technique is therefore independent, complementary to that from other imaging modalities and, as indicated above, relevant to medical applications. The ability to take advantage of the additional data provided by tissue elastic properties is dictated by the tissue properties themselves and so will not be subject to the inherent limitation of the "images" themselves. Each of these considerations in now examined.

The Elastic Properties of Tissues

The literature on mechanical properties of soft biological tissues is very limited. Available data concerns mainly the tissues having some sort of mechanical function, such as skin[1-7] and muscle[8-12] and little or nothing is reported on most other soft tissues. Changes in mechanical properties of tissues accompanying development of various lumps and cancerous lesions are discussed mainly in qualitative terms. Various authors describing mechanical properties of tissues use a number of common sense based terms such as "hardness", "stiffness", "elasticity", "compressibility", etc., in parallel with such rigorously defined terms as Young's modulus or shear modulus. Such a variety of terms used for characterizing mechanical state of tissue may result in ambiguity and cause misunderstanding. For most materials mechanical properties are usually among the most investigated but this is not the case for soft tissues. There are presently no comprehensive and adequate reviews on soft tissue mechanics, and therefore we present here a brief summary of qualitative and quantitative relationships on mechanical characteristics of soft tissues.

The major differences in mechanical properties between materials allow them to be separated into two groups: solids and fluids. As it will be shown below, soft tissues have features of both and are sometimes characterized as fluid-like (or liquid-like) solids. Mechanical properties of materials are expressed in terms of elastic moduli which are related to the ratio of a given stress to the resulting strain. One can obtain many elastic moduli of a solid material depending on the type of stress: bulk compression, torsion, shear stress, extension, etc. In a case of an isotropic linear elastic material the plurality of moduli can be expressed using only two independent parameters such as bulk compressional modulus, K, and shear modulus, G. Two other parameters[13] commonly used for characterizing mechanical properties of solid materials: Young's modulus, E, and Poisson's ratio, ν, (defined as the ratio of the transverse contraction of a bar to its elongation under a tensile stress), are related to K and G by the following expressions:

$$K = \frac{E}{3(1-2\nu)} \tag{1}$$

$$G = \frac{E}{2(1+\nu)} \ . \tag{2}$$

Formulas (1) and (2) are applicable to isotropic Hookean solids, but soft tissues are often far from being both isotropic and linear. Elastic moduli of tissues are, in general, not just constants and could depend on direction and scale of deformation, and also on time because of a wide range of structural relaxational processes in the stressed tissues. In such a heterogeneous media like soft tissues spatial variations in elastic moduli could be significant. To evaluate stress-strain relationship in tissues one could need, in general, many more that just two elastic moduli, even in the case of a linear approximation. Nevertheless, in the reviews on mechanical and acoustical properties of tissues[14-16] only bulk and shear elastic moduli, volume and shear viscosities and such related values as speed and attenuation of longitudinal and shear waves are presented. Limited data on mechanical anisotropy of soft tissues are available such as for the obviously anisotropic striated muscles where elastic moduli for the directions along and across fibers are given.[15] However, for most soft tissues such as liver, brain, kidney, spleen, lung etc., only the bulk and shear moduli are given.

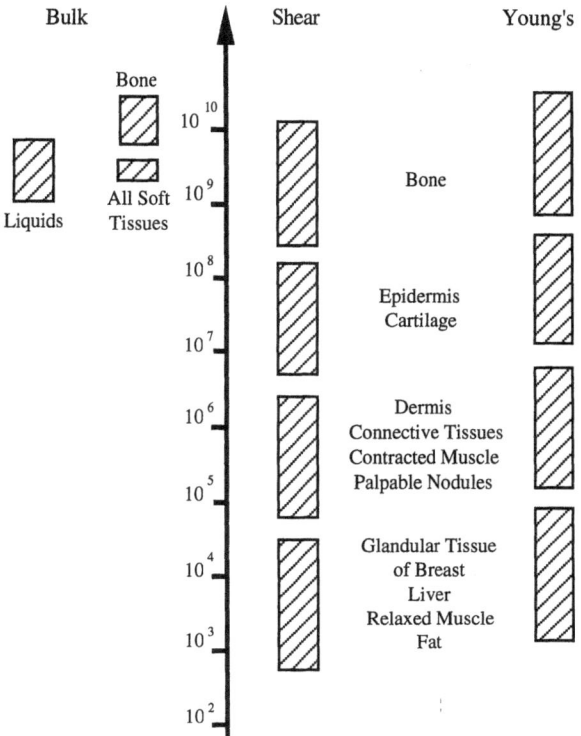

Figure 1. A summary of data from the literature concerning the variation of the shear modulus, the Young's modulus and bulk modulus for various materials and body tissues. Ranges for each of the moduli for a given tissue type are indicated as the shaded regions.

But in most cases this grossly simplified mechanical characterization is quite adequate, and exact and comprehensive mechanical description of real tissues which, in principle, could require many parameters is impractical and unnecessary. In the present paper we analyze mechanics of soft tissues, as a first approximation, using mainly the conventional language of just two bulk and shear elastic moduli, not considering in detail anisotropy, nonlinearity and heterogeneity of tissues. However, one should bear in mind that the level of appropriate simplification in characterizing a biomechanical system must be carefully estimated in the solution of any particular problem.

Of the elastic moduli K, G and E, the only mechanical parameter of soft tissues which has been widely investigated by many authors is bulk compressional modulus, K, because of two related reasons. First, spatial distribution of bulk modulus of tissues is of basic importance for the ultrasonic medical imaging and it has been a subject of intense study based on the wide spread use of acoustic imaging in tissues. Second, the bulk modulus can be easily estimated from the measurement of sound speed, V, and density, ρ, in tissues: $K \approx V^2 \rho$. Frequency dependence of the bulk modulus is small, therefore, the values of K obtained by ultrasonic measurements in the MHz region differ from the static value by no more than one percent. There are a number of excellent reviews on bulk elastic properties for nearly all human and animal soft tissues.[14,15] In contrast, not much is known about the other mechanical characteristics of tissue: G, E and V. In most cases only rough estimates have been made.

Figure 1 shows a scale of elastic moduli of tissues calculated from our own and literature data.[1-7,9,11,12,15,17-26] The ranges for elastic moduli are rough estimates calculated from very variegated and mixed data obtained by different authors using different techniques and in many

225

cases not confirmed independently. Most of these data are obtained by dynamic methods in the range of frequencies from 10^2 to 10^7 Hz and evaluation of static modulus requires extrapolation to zero frequency. Although the literature data on the frequency dependence of shear modulus of tissues is very limited, such an extrapolation can be done because the spectra of relaxation processes responsible for dispersion of shear modulus are to first approximation similar for all the soft tissues and continuously span the frequency range from nearly zero to up to hundreds of MHz. Figure 2 shows an example of such a frequency dependence of shear modulus of liver tissue, calculated from the data of Sarvazyan et. al.[26,27] and Madsen et. al.[21] Despite of the differences in the measurement techniques used, the experimental points are fit well by a smooth curve describing a system with a continuous spectrum of relaxation times typical for polymer-based materials.

Total error in the evaluation of shear or Young's moduli of tissues from the literature data may be large because of poorly grounded assumptions in extrapolating high frequency values to zero frequency, systematic errors of measurements by different and sometimes incompatible techniques, intrinsic variability of biological specimens, etc. But because tissues of the body may differ in their elastic moduli by a few orders of magnitude, even a large error in shear or Young's moduli does not change the general regularities observed in Fig. 1. The literature values for bulk, shear and Young's moduli presented on Fig. 1, are obtained based on the assumption of linear elastic media and should be considered effective means of the corresponding mechanical characteristics of the materials. At the same time, these data show the applicability of such an approach at least as a first order approximation in attempts to describe the mechanical properties of soft tissue for the diagnosis based on the significant variation in the shear elastic modulus.

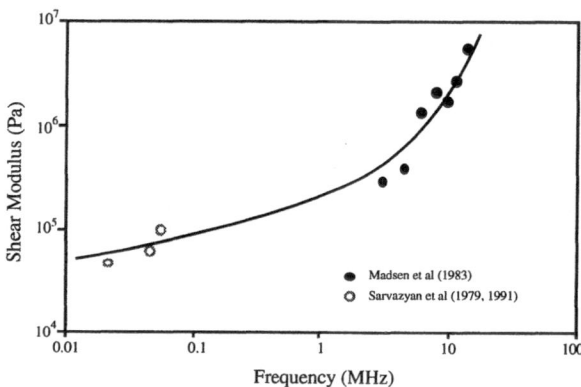

Figure 2. Frequency dependence of the shear modulus in liver tissue. Two sets of experimental data are taken from Madsen *et al.* (1983) and Sarvazyan *et al.* (1979 and 1991) for the higher and lower frequencies, respectively.

For the present consideration the most important point is a 1-2 orders of magnitude difference in shear or Young's elastic moduli of a tissue in different physiological states, such as, e.g., relaxed and contracted muscle, or a normal tissue and a pathologically transformed one, such as glandular tissue of breast and palpable nodules. As seen in Fig. 1, bulk moduli K for all soft tissues, liquids, as well as solids are roughly in the same range. That is why the speed of longitudinal (compressional) acoustic waves in all the liquid and solid materials is within about 1000-5000 m/s. The velocity of shear waves for solids is in the same range while in soft tissues the shear wave velocity is much lower and varies to a much greater extent (from 1 to over 100 m/s). A qualitative difference in the ratios of G/K for solids (G equals to a few tenth of K) and liquids (G/K=0) is an expression of the fact that a solid is able to conserve its shape in contrast to a fluid which conforms to the shape of a vessel. Although soft tissues have enough shear stiffness to conserve their shape to some extent, mechanically they are similar to liquids, i.e. G<<K (more precisely G=10^{-2}-10^{-6}K). The latter also means, according to Eqs. (1) and (2), that Poisson's ratio of soft tissues is in the range from 0.4900 to 0.4999

which is very close to the value ν=0.5 for liquids. For most solids Poisson's ratio is typically within 0.2 to 0.4.

Figure 3 is a graphical representation of Eqs. (1) and (2) and shows the ratios G/K and E/K versus Poisson's ratio. The choice of the ordinate in Fig. 3, where shear and Young's moduli are related to the value of bulk modulus, reflects the fact that the bulk modulus varies for different materials in much smaller range than other moduli. The shaded region indicates values of ν, G/K and E/K typical for solid materials, and the point in the vicinity of ν=0.5 corresponds to soft tissues. In the range near ν=0.5, as it follows from Eq. (2), Young's and shear moduli only differ by a factor of 3, i.e. E=3G. Therefore, these two moduli have the same information content and either of them can be used equally well for characterization of soft tissues.

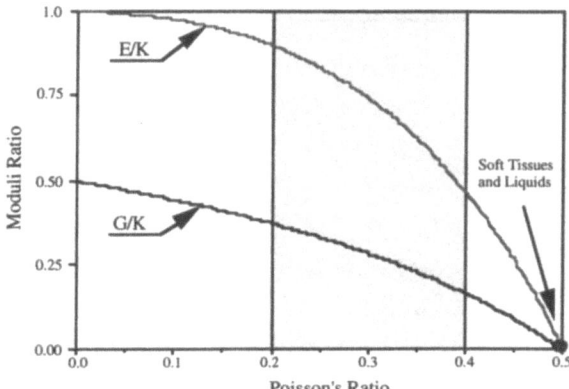

Figure 3. The ratios of E/K and G/K as a function of Poisson's ratio ν. G, K, and E are the shear, bulk, and Young's moduli, respectively. The shaded region corresponds to the range of ν for solids as generally known throughout the literature. The point in the vicinity of ν=0.5 corresponds to soft tissues.

Materials with ν close to 0.5, such as liquids, soft tissues and some rubbers, are often called "incompressible" materials. This term can be confusing because it has nothing to do with the low or high value of bulk compressional modulus itself. Soft tissues are called incompressible just because an external stress usually causes only a change in the shape of the stressed tissue, while the volume remains constant with a high degree of precision. For example, when one presses on a solid object: wood, glass, steel, etc., as with a piston, the resulting stress and stain patterns are determined by values of both bulk and shear elastic moduli because their values are comparable. But if a soft tissue is deformed the resulting stress and strain patterns will be completely defined by the shear elasticity only, regardless of the bulk modulus value for the tissue, and whether K is infinitely large or is only a few orders of magnitude higher than G. This can be well illustrated by the classical solution, given by Sadowsky,[28] when the circular piston with a radius R is acting on the semi-infinite elastic media having the elastic moduli G and K. In this solution the relationship between the force P applied to the piston and its vertical displacement W takes a form :

$$P = \frac{8GRW}{1 + \dfrac{G/K}{1+G/3K}} \quad ,$$

where the term containing the ratio of G/K vanishes for G<<K and then the relationship between W and P will be determined by the shear modulus G only.

Therefore, information on mechanical properties of tissues obtained by palpation can be fully described in terms of either the shear or Young's modulus and palpatory information is by no means related to bulk compressional modulus of the tissue. The latter point is very important when explaining that such a widely used imaging modality as ultrasonic echography which is based on the use of high frequency compressional waves does not yield any information directly related to palpatory "stiffness" of tissues. But it is only bulk, not shear, modulus which substantially dictates the impedance variation in tissues imaged by an ultrasonic device.

The palpatory detection of sufficiently large and not deeply situated cancerous lesions is based on the evaluation of rather distinct changes in the local tissue elastic properties. In principle, one could think of an acoustical imaging technique based on the use of shear waves to reveal internal tissue structures with differing shear elasticity, but this possibility is excluded by extremely high attenuation of shear waves in soft tissues.[12,21,24] Only a superficial soft tissues such as skin and the nearest underlying structures can be directly tested acoustically to visualize the pattern of distribution of shear elasticity[1,5,6,26,29] and evaluate various dermatitis and skin cancer.

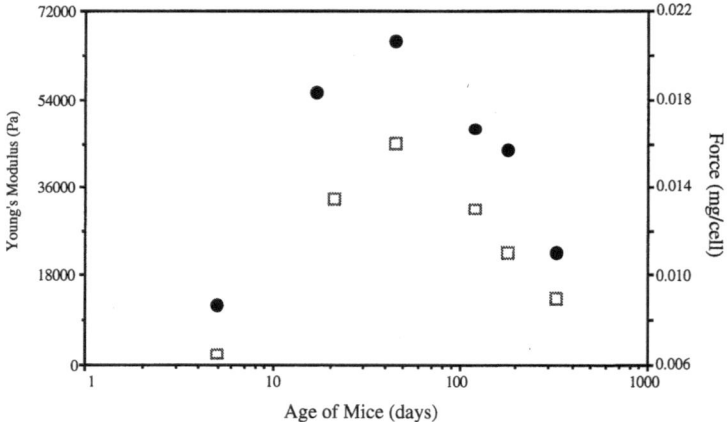

Figure 4. Measurements of the Young's modulus (indicated by circles) and intercellular separation force (indicated by squares) for liver tissue excised from mice of various ages. The force is that required to tear whole pieces of tissue (from the data of Malenkov and Asoyan, 1983).

Efforts for developing remote evaluation methods for shear elasticity of deeply situated internal tissues, i.e. elasticity imaging, are justified by the well established fact that development of cancerous lesions is commonly accompanied by significant changes in tissue "hardness". In addition, at least one series of experiments indicates that changes in elastic properties may even precede the presence of a distinct lesion and be <u>predictive</u> of cancer. Malenkov and Asoian[22] harvested the livers from mice of various ages and measured the Young's modulus along with the force required for the tissue to be torn along cellular boundaries. The results are shown in Fig. 4 where the squares correspond to the intercellular separation force, i.e. the force required to tear the tissue and the circles are the corresponding values of the Young's modulus. Note the close correlation between these two measurements. Previously, a measurement had been made in which mice were administered "a hepatic protector" in order to reduce the incidence of spontaneous hepatic carcinoma in a specially bred species of mice.[30] The percentage of animals which developed cancerous lesions was determined as a function of the intercellular separation force measured in the normal liver parenchyma, i.e. in regions with no cancer as determined by histology. The occurrence of cancer was well correlated with a decrease in the intercellular separation force. Furthermore in another study , the decrease in separation force actually appeared significantly in advance of cancer formation.[31] This is an example of a case when the start of the pathological process, such as cancer lesion formation in the mouse liver, may result in a significant decrease in elastic modulus. In other cases, a different dependence could possibly be observed and further study of changes in mechanical properties during the development of a lesion is needed. The point here is only that such correlation may exist and might be significant.

Therefore, prediction of this type of cancer might be possible based on a measurement of shear modulus. But the most obvious and perhaps largest application for elasticity imaging is in improving detection of cancerous lesions, which differ in mechanical properties from the surrounding tissue, while reducing the limitations of conventional palpatory diagnostics, such as the need for large and superficial nodules for detection. The second application area is monitoring cancer therapy by analyzing temporal changes in elasticity and geometry of a tumor.

The Use of Imaging Systems in the Measurement of Tissue Motion

There are two general approaches to obtaining elasticity images from standard imaging modalities. First, the elastic modulus can be determined for an internal tissue structure having boundaries which are detectable by any imaging modality. The deformation of these boundaries is measured under mechanical stress. The second approach is to evaluate the relative motion of tissue elements which are not separately resolvable by an imaging technique. Once again the reconstruction is based on a motion pattern induced by an applied stress. Therefore, in principle, any imaging modality can be used to obtain the motion/deformation pattern. There have been a number of experiments using ultrasonic techniques to measure elastic properties using these approaches.[11,25,32-44] Specific examples can be given.

The presence of hard lesions was detected using the measurement of Doppler shift signals for tissues being mechanically vibrated using an external source.[25,37] This technique produces qualitative (relative hardness) information about the tissue properties in two dimensions. A theoretical investigation based on similar concepts involving tissue vibration is a part of this work. Similarly, a number of other examples of tissue motion evaluation by Doppler measurements have been made.[11,38,45] Cross correlation techniques have also been applied to both raw and envelope-detected RF data for the determination of tissue motion. Tristam *et al.*[34] were able to discriminate normal liver parenchyma from hepatic metastasis using a multidimensional evaluation of Fourier coefficients associated with the cross correlation. O'Donnell *et al.*[46] have also developed Fourier-based algorithms for the tracking of the arterial wall motion which demonstrate a tissue motion variation in the vicinity of harder plaques. More recently the same algorithm has been used to evaluate the internal displacement and strain fields inside the gel-based phantoms.[47] The imaging of tissue elastic properties using cross correlation techniques has now developed to the point where 1-D evaluation of tissue motion has been effectively demonstrated to produce images of relative elasticity, so-called elastography.[42,44] The spatial variation in tissue elastic properties has also been evaluated using cardiac motion as a naturally occurring motion source in the prediction of fetal lung maturity[48,49] and in echocardiography.[41] In the former, M-mode images were analyzed to track the motion of specular reflectors during the cardiac cycle and the latter utilizes optical flow techniques.[50]

The measurement of tissue deformation using Magnetic Resonance Imaging (MRI) has not been directly addressed in the literature, although several techniques for measuring bulk motion exist.[51-56] Perhaps the most standard techniques involve the addition of a magnetic field gradient to standard imaging sequences. This "motion sensitive gradient" is the cause of an additional phase shift in the MR signal of a voxel which is proportional to the motion of that voxel in the direction of the applied gradient field. With simple assumptions about the motion (e.g. constant velocity), the degree of motion is easily extracted from the phase information and the sequence parameters. In order to avoid aliasing (i.e., phase shifts $> 2\pi$), the gradient strength must be chosen according to the velocity range being investigated. In addition, motion artifacts due to non-systematic motion are likely to be enhanced in these methods. Feinberg and Jakab[51,52] have demonstrated an extension of this technique designed to give the Fourier components of motion. Decorps and Bourgeois[53] have developed a different technique which correlates the actual MR signal intensity to motion. The nonlinear relationship between signal intensity and motion requires the setting of sequence parameters to target a specific range of motion. A recent method for measuring motion introduced by Zerhouni *et al.*[54] and extended by Axel and Dougherty[55] involves "tagging" tissue by selectively saturating the NMR signal available in that tissue. This saturation pulse can be applied in a variety of patterns which result in the appearance of radial or parallel lines or rectangular grids of low (saturated) signal intensity over an entire imaging slice. Tissue displacement after the application of this saturation pattern and before the acquisition of MRI data will result in a geometric distortion of that pattern, which can then be measured in the MR image. It is a specific implementation of this saturation pattern technique[56] which is used in one of the MRI experiments described here.[57]

Summary

As indicated above, the potential for producing elasticity images to detect cancer is great and should be investigated in a more rigorous fashion for at least three reasons: 1) the existence of imaging modalities to measure tissue motion, 2) the large variation in tissue elastic moduli, and 3) the known hardness associated with cancerous lesions and its possible correlation with shear properties of the tissue. The theoretical development and experimental demonstrations presented below is a result of work from several research groups interested in further developing this area.

THEORETICAL BASIS OF ELASTICITY IMAGING

A theoretical model has been used for predicting strain patterns in a stressed material with spatially varying elastic properties and for solving inverse problem of reconstructing mechanical structure of an object from the measurements of strain pattern.

The equations of the dynamic equilibrium of a body have the form:

$$\sum_{j=1}^{3} \frac{\partial \sigma_{ij}}{\partial x_j} + f_i = \rho \frac{\partial^2 u_i}{\partial t^2} , \qquad i=1,2,3 , \tag{3}$$

where f_i is the force per unit volume acting in the x_i direction, u_i is the particular displacement in the x_i direction, ρ is the density, t is the time and σ_{ij} are the components of the stress tensor.[58] Note, that in the static case the right hand side of Eq. (3) which is the inertial term is equal to zero. In the case of a linear elastic model of a compressible media, the stress-strain relations are described by the equations:

$$\sigma_{ij} = (K - \frac{2}{3}G)\Theta\delta_{ij} + 2G\varepsilon_{ij} , \tag{4}$$

where K and G are bulk and shear moduli, respectively, δ_{ij} is Kronecker's delta symbol, ε_{ij} and Θ are the components and the trace of strain tensor, respectively, defined in the linear approach as:

$$\varepsilon_{ij} = \frac{1}{2}\left(\frac{\partial u_i}{\partial x_j} + \frac{\partial u_j}{\partial x_i}\right) \tag{5}$$

$$\Theta = \sum_{i=1}^{3} \varepsilon_{ii} = \varepsilon_{11} + \varepsilon_{22} + \varepsilon_{33} , \tag{6}$$

where u_i are the particular displacements in the x_i direction.

In the case of the linear elastic model for an incompressible tissue and tissue-like materials with a Poisson's ratio very close to 0.5, the relations between stress and strain take the form:

$$\sigma_{ij} = p\delta_{ij} + 2G\varepsilon_{ij} , \tag{7}$$

where p is the static pressure. Incompressibility then yields the additional condition that

$$\varepsilon_{11} + \varepsilon_{22} + \varepsilon_{33} = 0 . \tag{8}$$

Generally the boundary conditions take the form:

$$\left(\sum_j \sigma_{ij} n_j - F_i\right) \delta(u_i - u_i^0) = 0 \quad , \tag{9}$$

where n_j is the jth component of a unit normal vector at the body surface, F_i is the force per unit area of the surface acting in the x_i direction and δ is a symbol of the variation. If there are any known external forces F_i applied to part of the surface of the body, the corresponding Eq. (9) is satisfied for this part of the surface by means of the first term. In contrast, if there are any known displacements u_i^0 of part of the body surface, this equation in (9) is satisfied there by means of the second term.

Equations (3-9) represent a boundary value problem for the system of first order partial differential equations and corresponds to the case of small deformations of the homogeneous isotropic and linearly elastic media. In the case of inhomogeneous media two different approaches are possible. If the heterogeneities are continuously distributed in the space (see Fig. 10b,c) the elastic moduli in the Eqs. (4) and (7) are continuous functions of the coordinates. If there are "clearly bounded" inclusions inside the phantoms (see Fig. 6a,b) the Eq. (3) must be satisfied for every homogeneous region (the region of base material in the phantom and regions of inclusions) where elastic moduli are constant. In addition, the standard conditions of continuity of stress and displacements on the boundary of each inclusion must be satisfied:

$$\left[\sum_j \sigma_{ij} n_j\right] = 0 \ , \quad [u_i] = 0 \ , \quad i,j = 1,2,3 \quad .$$

Here square parentheses denote the discontinuity of a corresponding term and the components of the normal vector on the boundary of the inclusion are denoted by n_j. In this paper both approaches are used. Note, that a multi-layered media is a particular case of inhomogeneous media, where "clearly bounded" inclusions are the layers themselves.

Generally, there are no analytical solutions for such boundary value problems even for this very simplified approach (Eqs. (3-9)) and, therefore, numerical methods have to be used. Substituting the expressions (4) or (7) for compressible or incompressible media, respectively, into the system of Eq. (3) and using Eq. (5), a closed system of second order partial differential equations for unknown displacements can be obtained. For the incompressible media the additional unknown is the pressure, and an additional Eq. (8) has to be taken into account. The method of finite differences was used to solve the boundary value problem in each particular example presented in this paper. Spatial derivatives of unknowns were approximated with a second order accuracy using the uniform rectangular grid, and well known numerical algorithms were used to solve the obtained discretized system.[59] Note that even in the static case, an iterative scheme must be used to obtain the solution when the changing size of the loaded boundary, i.e. the deformation-dependent boundary condition (see Fig. 5), is taken into account in a step-wise fashion.

The theoretical approach presented above is applicable with necessary changes for more complex biomechanical systems. In the case of large deformation the additional nonlinear terms in the expression (5) must be taken into account.[58] In the general case of anisotropic media the stress-strain relationship (4) should be changed for

$$\sigma_{ij} = \sum_{kl} A_{ijkl} \varepsilon_{kl} \ , \quad i,j,k,l = 1,2,3 \quad ,$$

where A_{ijkl} is the general tensor of elastic coefficients. For linear viscoelastic media the additional linear strain rate terms have to be taken into account in the stress-strain relationship. For the nonlinear media the exact form of this relation could be different depending on the nature of the object, but the general mathematical approach presented above remains the same differing only in the amount of numerical calculations.

In this paper we present some results for the application of the theory of elasticity to MRI experiments on remote evaluation of elasticity for internal structures. Intending to compare the experimental data for phantoms used in this work and the results of theoretical approach and to illustrate the main idea of method, we restrict our attention to the linear elastic model.

MATERIALS AND METHODS

Two MRI experiments were performed using cylindrical phantoms containing inclusions all of which had shear moduli which differed from that of the surrounding material. In the first experiment, the phantom was a disposable ultrasound standoff gel pad (85 mm in diameter and approximately 10 mm thick) in which three liquid-filled capsules (common Vitamin E capsules) had been inserted randomly in the pad through small incisions. This cylindrical phantom was placed in a non-ferrous clamp which allowed planar deformation of the phantom normal to its longitudinal central axis (as in Fig. 5) while inside a standard bird-cage coil designed for head imaging. The whole assembly was then placed into the bore of a 1.5 T whole body imaging MRI system (General Electric, Milwaukee) and was imaged with a standard spin echo sequence (TR = 400 ms, TE = 20 ms with 3 mm interleaved slices and a 256 X 256 matrix). The thin cylinder was constrained on the ends to simulate an infinitely long cylinder.

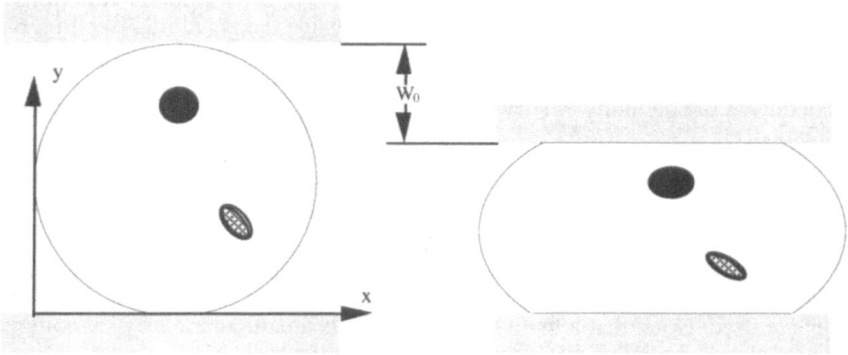

Figure 5. A depiction of the phantom before and after deformation during both MRI experiments. The x and y axes indicate the spatial dimension along which the displacement is measured. W_0 is the absolute displacement of the top plate from its original position.

The phantom used in the second set of MRI experiments, described in more details elsewhere,[57] was a silicone composite cylinder measuring 90 mm in length with an 85 mm diameter. Within the phantom were three smaller cylinders (9 mm dia.), referred to as inclusions, all having a shear modulus approximately 10 times greater than the background. Two of these cylinders were parallel to the central axis of the phantom and the other was positioned diagonally from one end of the phantom through approximately one-half of the length. The elastic properties of the phantom materials were evaluated on an MTS model 810 Materials Test System (MTS Systems Corp., Minneapolis, MN) using a constant compression-displacement test. The test was performed on a separate piece of the inclusion material and on the phantom as a whole where the contributions from the two inclusions parallel to the central axis of the phantom were eliminated. The moduli ratio between the inclusion and its surrounding material was estimated to be 10.5±1.5 for up to a 30% deformation of both materials.

For these studies, the phantom was placed in the hydraulic device consisting of two plates whose vertical separation could be changed quickly under remote manual control. The phantom/deformation device was placed in a bird-cage coil and these together in the bore of the 2-Tesla GE Omega MRI system. A uniform grid of coronal and saggital planes of NMR signal saturation was applied prior to an axial spin-echo image. The phantom was deformed vertically between application of the saturation planes and collection of the image data (similar to Fig. 5). The grid of saturation planes, applied using a SPAMM technique,[55,56] appears in the image as a grid of dark lines which has been predictably distorted by the surface deformation.[57] The resulting deformation field contains local perturbations at the locations of the hard inclusions.

The mathematical formulations corresponding to both MRI experiments are similar. In these cases, there are freely moving portions of the phantom's surface, where the external forces F_i are equal to zero, and for other surfaces the displacements are obtained in a step-wise fashion (see Eq. (9) and Fig. 5).

RESULTS AND DISCUSSION

MRI Experimental Results

Figures 6a and 6b are images from the first MRI experiment in which the phantom contained liquid-filled inclusions with unknown mechanical properties. Figure 6a is an image of the phantom prior to any external deformation. The inclusions appear as elliptical regions of mostly increased signal intensity. Figure 6b is an image of the phantom taken in the same fashion but with about 20% planar deformation along the vertical axis. The result is a change in the position and the eccentricity of the inclusions due to phantom deformation. One of the inclusions is no longer clearly visible in Fig. 6b and therefore not used in subsequent evaluations.

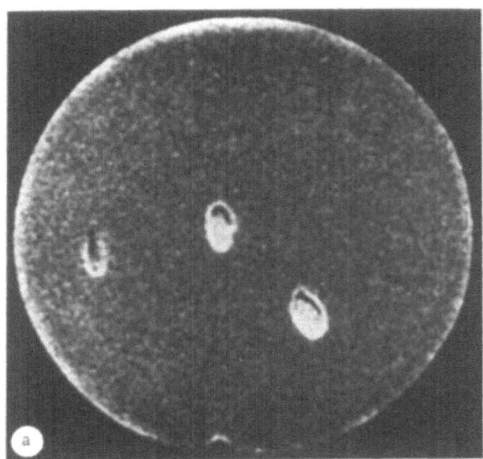 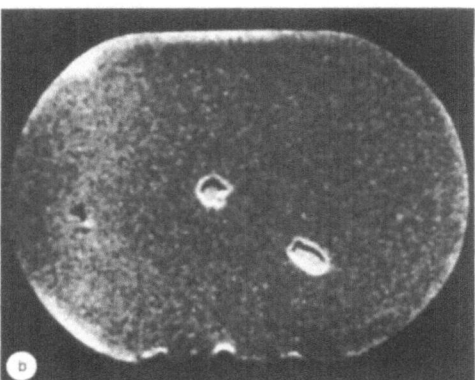

Figure 6. MRI spin echo images of a phantom composed of an ultrasound standoff gel pad containing three liquid-filled capsules. Images are taken before (a) and after (b) deformation of the phantom using plane along the top of the phantom. Experiments were performed at the University of California at San Diego.

The corresponding mathematical problem was solved by using a 3D model.[60] To simulate the phantom used in the first MRI experiment, where liquid-filled capsules were inserted in the standoff pad, the inclusions were modeled as elliptic bodies having thin rigid shells and soft internal portions. Detailed structure of a complex inclusion may not be visible experimentally, just as the thin hard shells of the inclusions are not visible in the MR images of Fig. 6. In this case, as a first approximation, one can describe the complex inclusions as homogeneous, i.e. Young's modulus is a constant inside the inclusion. Results of the mathematical simulation based on data taken from Fig. 6 show that the liquid-filled capsules with their complex structure provide local deformation effects similar to homogeneous inclusions having Young's moduli values two times smaller than the Young's modulus of the surrounding material, as described in the next section. In the second simulation, a more complex structure for an inclusion having a rigid shell and soft core was analyzed. The Young's modulus of the shell was assumed to be two times larger than the Young's modulus of the surrounding material and the value of the Young's modulus of the internal part of the inclusions was varied. By varying the value of the Young's modulus for the internal part of the inclusions, the theoretical results were adjusted to maximize correlation to the experimental measurements of Fig. 6b and thus the Young's modulus of the internal soft material is at least 5 times smaller than that of the surrounding material. This means that with the present resolution, the known zero value of the Young's modulus for liquids could not be estimated more precisely. However that estimation will be sufficient in most pathological conditions since the variation of the relative Young's modulus in the range of approximately 0.2 to 5 is important for early differentiation of disease states.

Figure 7 shows an example of the experimental results obtained in the second MRI experiment,[57] where the top plate of the deformation system (Fig. 5) was displaced 15 mm with respect to the bottom plate (an 18 % surface deformation). Two regions of low signal intensity appear corresponding to the position of the inclusions. The inclusions have an increased cross linking in the silicone which may reduce the spin mobility within those regions and cause the inclusions to appear as low signal intensity circles in the image. The low signal intensity also reduces the ability to detect any shift of saturation lines within the inclusion. With sufficient deformation the phantom, the strain/displacement of the material surrounding the inclusion would indicate the presence of the local change of shear properties.[57] It is important also to note that these inclusions cannot be detected by manual palpation.

Figure 7. MRI image of the 85 mm diameter phantom with two 9 mm diameter inclusions each with a shear modulus approximately 10 times larger than that of the surrounding material. The inclusions appear as the low signal intensity circles in the image. The phantom is deformed by 18 %. Experiments performed at the University of Michigan.

The corresponding mathematical problem was solved by using a 2D model. In this model, the phantom is assumed to extend infinitely in the longitudinal direction so that the only non-zero components of the displacement are in the cross-sectional plane of the phantom. The inclusions were simulated as homogeneous cylinders with Young's moduli equal to 10 times that of the surrounding material. The relative modulus values selected were based on the previously described compression-displacement test of the elastic moduli for the phantom and inclusion materials.

To quantitatively compare the experimental and theoretical results, the percent deformation was evaluated along horizontal and vertical lines through the center of each inclusion in the phantom subjected to 30 % surface deformation. The result for vertical line through the center of the inclusion located on the side of the phantom is shown in Fig. 8, where the presence of the inclusion is indicated by a sudden decrease in the deformation at the position of the inclusion. The experimental measurements are consistent with the theoretical predictions for the positions of the inclusions and the magnitude of the percent deformation generally matches the theory quite well. Similar results were obtained for smaller deformations of the phantom, but the data for 30% deformation of the phantom are presented here to illustrate that in some particular cases the linear mathematical model can be used for interpretation of experimental data even in the case of large deformation. In response to the localized reduction in deformation due to the harder inclusions, the deformation for other portions of the phantom would be increased over the 30% surface deformation of the phantom.

The evaluation along a line through the inclusion would be the equivalent to an ultrasound A-mode line measurement only with the reduced spatial resolution of 4 mm in the case of MRI. Further theoretical evaluations of detection sensitivity, deformation requirements, *etc.* will be given later.

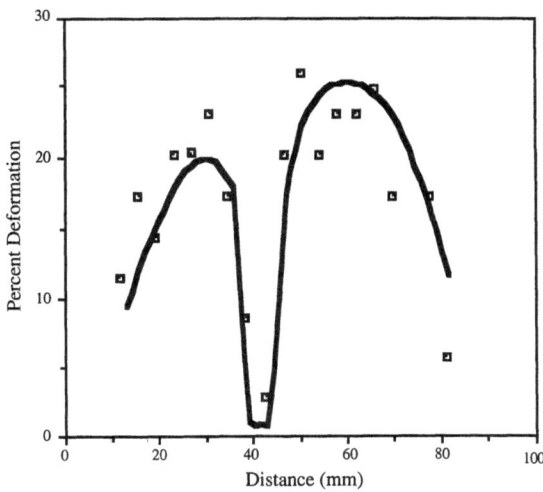

Figure 8. Comparison of the percent deformation measured experimentally (points) and predicted theoretically (solid line). The comparison includes the deformation measured parallel to a vertical line through the center of inclusion on the side of the phantom, and plotted as a function of the position along this line.

Extensions to the Theoretical Modeling

The experimental results of the presented MRI experiments can be understood better when one considers local displacements and strain patterns caused by an inclusion having a different shear (or Young's) modulus than the surrounding material. Figure 9a is a theoretical calculation demonstrating the movement of material around an inclusion (with Young's modulus E) in response to an overall 10% planar deformation along the vertical axis of the host material. The theoretical phantom, having the cylindrical inclusion located at the center of the phantom, is square shaped in two dimensions and infinitely long in the third dimension. The diameter of this cylindrical inclusion is 0.2 relative to the original size of the phantom. This phantom was vertically deformed by two planes, where slippage between the planes and the phantom boundaries was allowed. The cross-sectional view of the inclusion before deformation is shown as a circle. It is assumed that the deviation in material displacement is evaluated in the square part of the phantom which has an initial boundary G_0 around a circular inclusion. For this geometry, three cases are presented ($E=E_0$, $E>E_0$, $E<E_0$) where E_0 is the Young's modulus for the base material. The first case $E=E_0$ indicates homogeneity and the resulting deformation of the phantom causes the square to become rectangular with the boundary denoted as G_{hom}. If an inclusion ($E \neq E_0$) is placed in the center of the square G_0 prior to deformation, the shape of the deformed G_0 is modified. For example, the curve G_{hard} corresponds to the inclusion which is 3 times <u>harder</u> than the base material ($E>E_0$), and the curve G_{soft} corresponds to the inclusion which is 3 times <u>softer</u> ($E<E_0$).

To investigate the influence of the inclusion, let us consider the local characteristic of deformation

$$Q = \frac{d_x}{d_y} \quad ,$$

where d_x and d_y are the width and height of the deformed square as it is shown in the Fig. 9b. This characteristic is very sensitive to the presence of inclusions and can be used for the examination of the tissue. The actual value of Q depends on both the relative Young's modulus of inclusion and initial size of the square G_0 as shown in Fig. 9c-d. Here Q_0 is the value calculated for the homogeneous case, R is the radius of inclusion and the linear size of the square G_0 is greater than 2R. Effects shown on Fig. 9a-d are enhanced proportionally with the increasing of the range of deformation. This Q-value can be used for characterization of inclusions with complex shapes and spatial distributions of Young's modulus. For example, such a technique was used to analyze the theoretical displacements corresponding to the first MRI experiment where only the motion of the capsule boundary was measured. The motion detection alone is sufficient to characterize a lesion, whether or not it is of sufficient image contrast to be visually evident. In the phantom used in the first MRI experiment, where the Young's modulus of the internal soft material was estimated to be at least 5 times smaller than that of the surrounding material, the relative values Q/Q_0 for two identical inclusions located in different positions within the phantom were indeed found to be equal (having a value of $Q/Q_0=1.1$). This value corresponds to homogeneous inclusions with Young's moduli two times <u>smaller</u> than that of the surrounding material. The characteristic Q was introduced and investigated in detail by Skovoroda,[60,61] where it was shown that this characteristic can be used successfully both in 2D and 3D. As described above, the alterations in the displacement field due to the presence of an inclusion are distributed locally and the magnitude decreases as a function of the distance r from the inclusion as shown on Fig. 9d. Note that in practice there is no need to know the value Q_0 for the investigated object - in the presence of the local inhomogeneity the value Q will be sufficient to find and characterize this inhomogeneity.

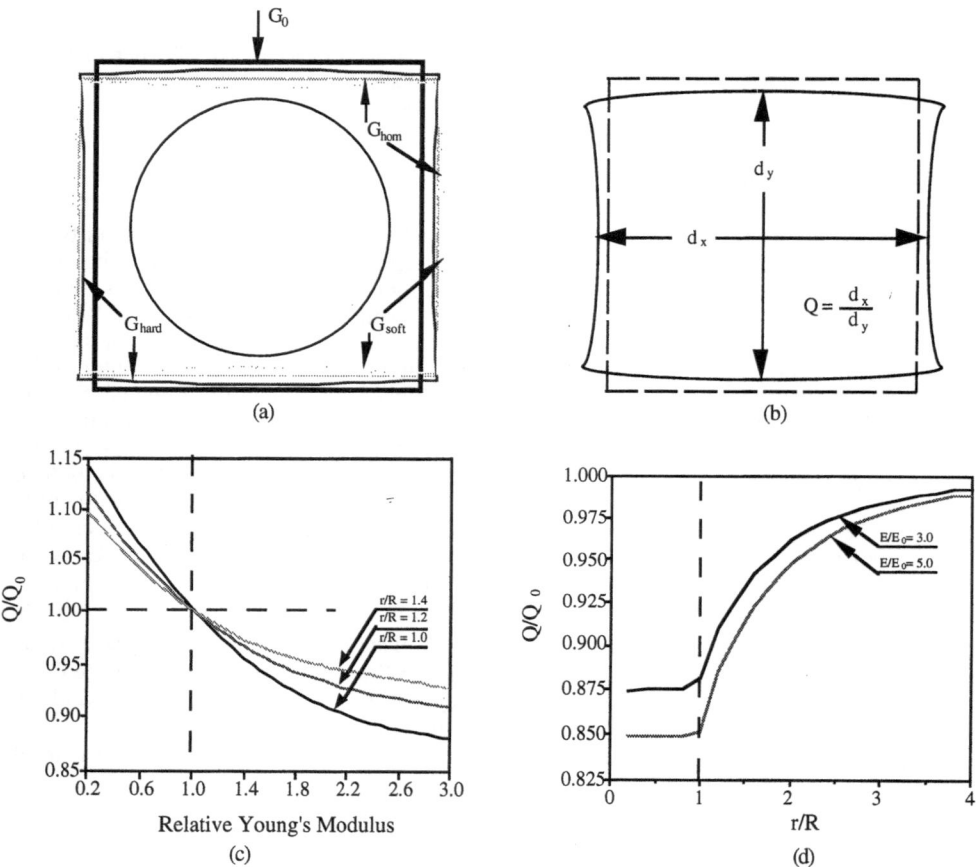

Figure 9. (a) Theoretical examination of the material motion for a block of material which has initial boundary G_0, is infinite in depth, and is contained within a larger phantom having a Young's modulus E_0. Within the block is an inclusion of radius R and a Young's modulus E. Three cases are presented for $E=E_0$ (G_{hom}), $E>E_0$ (G_{hard}), and $E<E_0$ (G_{soft}). Once the material is deformed, the boundary of the square is distorted in accordance with the relative Young's modulus value for the inclusion. (b) A schematic representation of a characterization of the distortion which defines the value $Q=d_x/d_y$. (c) The normalized value of Q plotted as a function of the relative Young's modulus for three normalized distances (r/R) from the center of the inclusion. (d) The normalized value of Q plotted as a function of the relative distance from the center of the inclusion for two inclusions of differing hardness. The dashed line refers to the position of the inclusion boundary.

The inverse problem should provide the distribution of the mechanical characteristics of the object and can be successfully solved based on complete information about the displacements of every particular point within the object. However, obtaining the needed 2D or 3D information may not be possible in practice. In this case, a general mathematical statement of the inverse problem cannot be solved, and additional constraints must be applied.

One possible way for correctly solving the inverse problem is to determine the class of monotonic functions based on solving the forward problem for variable relative hardness, displacements, etc. and use these functions to solve the inverse problem. Such a function, for example, is given in Fig. 9c. Indeed, if the size of the lesion is known, the relative Young's modulus can be uniquely reconstructed based on displacements measured using some imaging modality. In this scheme one uses some set of starting conditions corresponding to a reasonable estimate of the position and suspected hardness of a lesion, and solves the corresponding forward problems in an iterative fashion to maximize the correlation of the theoretical predictions and experimental data by changing these unknown values of the size, position and the Young's modulus of inclusion. The practical applicability of such techniques depends on the number of iterations needed to find the solution.

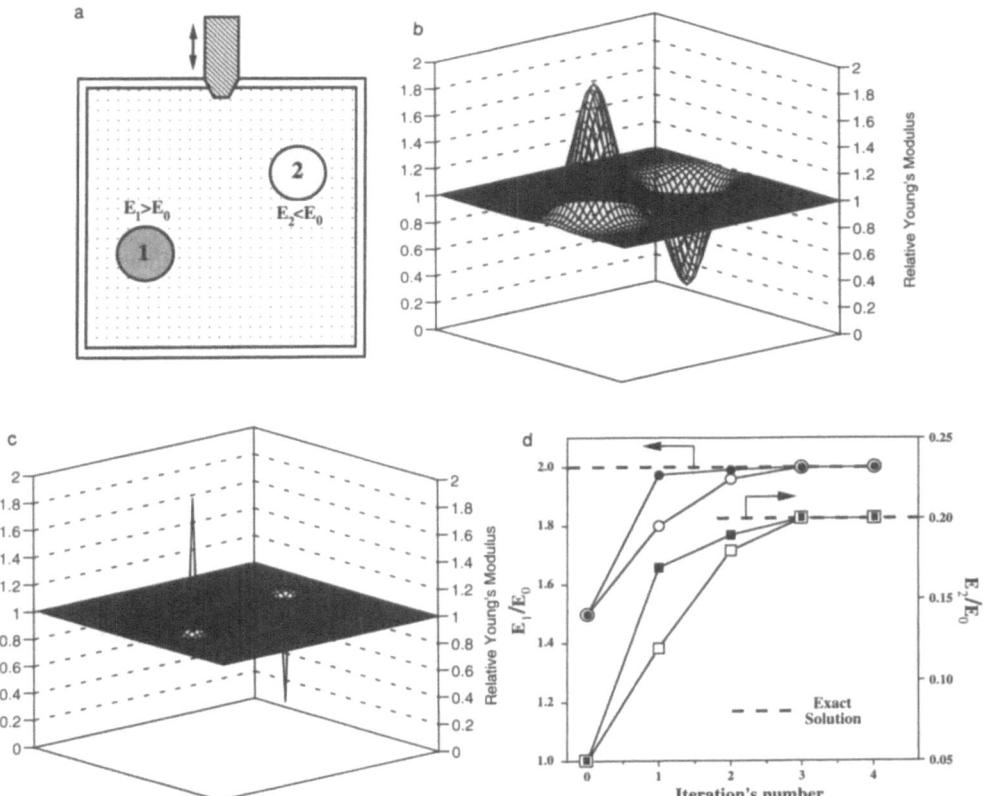

Figure 10. (a) Schematic for a theoretical calculation of relative motion within a square phantom containing two spatially distributed inhomogeneities with $E>E_0$ and $E<E_0$ where E_0 is the Young's modulus for the host material. The phantom is bounded rigidly on three of four sides and at all but the position of a mechanical driver on the fourth. (b) The spatial variation of the relative Young's modulus. (c) Same as (b) for a Young' modulus distribution where the inclusions are smaller. (d) An indication of the speed and convergence for the iterative scheme using the peak values for the Young's modulus of the inclusions as representative of the entire process. The open and closed symbols correspond to the inclusions of larger and smaller radii, respectively. Circles (squares) correspond to values for the inclusions having a Young's modulus two times larger (five times smaller) than the surrounding material.

Figure 10 shows the result of such an approach which was applied to a dynamic case. To obtain motion data for use in the investigation, the forward problem was performed on the theoretical phantom of Fig. 10a. Experimentally such motion data could be obtain by using Doppler ultrasound imaging.[37] The phantom is a slab of slightly compressible material bounded by a rigid frame on three complete sides and partially on the fourth. On this fourth side, a mechanical actuator is used to harmonically vibrate the phantom. Within the phantom are two spatially distributed inhomogeneities with differing hardnesses than the surrounding material, schematically shown on Fig. 10a. This example illustrates the applicability of the iterative technique in solving the inverse problem.

The Young's modulus distribution is described in Fig. 10b-c for two different sizes of the inclusions. Specifically, the two inclusions within the phantom have either a higher (inclusion 1) or lower (inclusion 2) Young's modulus than the surrounding material. Using an iterative inverse algorithm and an initial arbitrary guess at the value of E_1 and E_2, the motion information was used to reconstruct the relative Young's modulus distribution within the phantom. The minimum and maximum values of the Young's modulus were chosen to demonstrate the convergence rate of the iterative technique and the results appear in Fig. 10d. Dashed lines on this figure represent the exact values of the relative Young's moduli for inclusion 1 (E_1/E_0=2) and inclusion 2 (E_2/E_0=0.2). Note that the numerical solution obtains the exact solution within 4 iterations. This convergence rate is the same for other reasonable initial guesses at the value of the Young's modulus and in the case of the smaller size inclusions the convergence rate is higher.

SUMMARY

The large variation in tissue elastic moduli suggests elastic imaging may provide a sensitive and specific technique for cancer detection as well as potentially determining the susceptibility of tissues to the formation tumors. This large variation contrasts the small variation in the bulk modulus, the latter being used in conventional ultrasound imaging. Presently, tissue elasticity is not directly measured by any other imaging modalities commonly used in medicine. Even with a limited localization of a region differing in elastic properties, theoretical predictions indicate that the deformation of a tissue region surrounding a suspected lesion is sufficient for determining that the region contains abnormal tissue. For example, if a lesion only 3 times "harder" than the surrounding tissue was isolated to a region with a radius 40% larger than its own, Fig. 9c indicates that a 5% local deformation would be sufficient for lesion detection. Therefore, a 1 mm dia. lesion measured within an imaging grid of 1.4 mm should be detectable using the techniques described here and ultrasound for the deformation detection. Detection of lesions in this size range could greatly improve the early diagnosis of breast cancer. Finally, the discussed techniques are applicable to any imaging modality which can produce a static or quasi-static deformation profile of the tissue and so measure a local deviation in deformation with sufficient resolution.

ACKNOWLEDGMENT

The authors wish to recognize the assistance of Dr. Kevin Parker of the University of Rochester in discussions on tissue vibration analysis.

REFERENCES

1. Potts RO, Chrisman DA, and Buras EM, Jr. 1983, The dynamic mechanical properties of human skin in vivo, J Biomechanics **16**, 365-372.
2. Sarvazyan AP. 1983, Biophysical basis of ultrasonic medical diagnostics, In: Ultrasonic Diagnostic (Rus.), Institute of Applied Physics, Gorky, 80-94.
3. Madigosky WM, Lee GF, Haun J, Borkat F, and Kotaoka R. 1986, Acoustic surface wave measurement on live bottlenose dolphins, J Acoust Soc Am **79**, 153-159.
4. Dorogi PM, Dewitt GM, Stone BR, Buras EM, Jr. 1986, Viscoelastometry of skin in vivo using shear wave propagation, Bioeng Skin **2**, 59-70.
5. Pereira JM, Mansour JM, Davis BR 1990, Analysis of shear wave propagation in skin; application to an experimental procedure, J Biomechanics **23**(8), 745-751.

6. Pereira JM, Mansour JM, Davis BR 1991, The effects of layer properties on shear disturbance propagation in skin, J Biomechanics Engineering **113**, 30-35.

7. Vucelic D, Sarvazyan AP. 1989, Surface acoustic waves in medical diagnostics. Procs. 13th Intl. Cong. Acoust., Belgrade, **4**, 151-154.

8. von Gierke HE, Oestreicher HL, Franke EK, Parrack HO, and von Wittern WW. 1952, Physics of vibration in living tissues, J Appl Physiol **4** 886-900.

9. Pasechnik VL, Sarvazyan AP. 1969, On the possibility of examination of muscle contraction models by measuring the viscoelastic properties of the contracting muscle, Studia Biophys **13**, 143-150.

10. Fung YC. 1981, Biomechanics-mechanical properties of living tissues, Springer-Verlag; New York.

11. Krouskop TA, Dougherty DR, Levinson SF. 1987, A pulsed Doppler ultrasonic system for making non-invasive measurements of the mechanical properties of soft tissue, J Rehabil Res Dev **24**(2), 1-8

12. Kazakov VV, Klochkov BN, Chichagov PK. 1989, The study of dispersive characteristics of a wave on a human body. In: Methods of vibrational diagnostics of rheological properties of soft materials and biological tissues, Ed. V.A. Antonets, Institute of Applied Physics publications, Gorky.

13. Mase GE. 1970, Theory and problems of continuum mechanics, In Schaum's outline series, McGraw-Hill Book Company, New York.

14. Chivers RC, Parry RJ. 1978, Ultrasonic velocity and attenuation in mammalian tissues, J Acoust Soc Am **63**(3), 940-953.

15. Goss SA, Johnson RL and Dunn F. 1978, Comprehensive compilation of empirical ultrasonic properties of mammalian tissues, J Acoust Soc Am **64**, 423-457.

16. Duck FA. 1990, Physical properties of tissues. Academic Press.

17. Sarvazyan AP, Shnol SE, Pasechnic VL. 1969, Acoustical properties of gels and biological tissues in the low frequency sound fields. In: Properties and function of macromolecules and macromolecular systems Ed. G.M. Frank, Moscow, Nauka, 121-134.

18. Sarvazyan AP. 1969, Low velocity of sound in gels and biological tissues. PhD thesis, Pushchino, Institute of Biophysics, USSA Acad. Sci., 99.

19. Sarvazyan AP. 1975, Low frequency acoustical characteristics of biological tissues, Mechanics of Polymers **4**, 691-695.

20. Frizzell LA, Carstensen EL, Franke EK, Parrack HO, and von Wittern WW. 1952, Physics of vibration in living tissues, J Appl Physiol **4**, 886-900.

21. Madsen EL, Sathoff HJ, Zagzebski JA. 1983, Ultrasonic shear wave properties of soft tissues and tissuelike materials, J Acoust Soc Am **74**, 1346-1355.

22. Malenkov AG, Asoian KV. 1983, Correlation of acoustic characteristics and probable origin of a mouse liver tumor, Biofizika **28**(2), 326-329.

23. Pashovkin TN, Sarvazyan AP. 1989, Mechanical characteristics of soft biological tissue. In: Methods of vibrational diagnostic of rheological properties of soft materials and biological tissues. Ed. V.A. Antonents, Institute of Applied Physics publication, Gorky, 105.

24. Burke TM, Blankenberg TA, Sui AKQ, Blankenberg FG, Jensen HM. 1990, Preliminary results for shear wave speed of sound and attenuation coefficients from excised specimens of human breast tissue, Ultrasonic Imaging **12**, 99-118.

25. Parker KJ, Huang SR, Musulin RA, Lerner RM. 1990, Tissue reponse to mechanical vibrations for "sonoelasticity imaging", Ultrasound Med Biol **16**(3), 241-246

26. Sarvazyan AP, Skovoroda AR, Vucelic D. 1992, Utilization of surface acoustic waves and shear acoustic properties for imaging and tissue characterization. in Acoustic Imaging, Ermert, H, Harjes, HP, (eds) v.19, 463-467, Plenum Press, New York.

27. Sarvazyan AP, Klemin VI. 1979, Unpublished results.

28. Sadowsky M. 1928, Z Angew Math Mech **8**, 107.

29. Sarvazyan AP, Ponomarjev V, Vucelic D, Popovic G, Veksler A. 1990, Method and device for acoustic testing of elasticity of biological tissues. United States Patent #4,957,851, August 14, 1990.

30. Modjanova EA. 1974, Ontogenez **7**, 1022.

31. Modjanova EA, Malenkov AG. 1973, Alteration of properties of cell contacts during progression of hepatomas, Experimental Cell Research **76**(2), 305-314.

32. Dickinson RJ and Hill CR. 1982, Measurement of soft tissue motion using correlation between A-scans. Ultrasound Med Biol **8**, 263-271.

33. Tristam M, Barbosa DC, Cosgrove DO, Nassiri DK, Bamber JC, Hill CR. 1986, Ultrasonic study of *in vivo* kinetic characteristics of human tissue. Utrasound Med Biol **12**, 927-937.

34. Tristam M, Barbosa DC, Cosgrove DO, Bamber JC, Hill CR. 1988, Application of fourier analysis to clinical study of patterns of tissue movement. Ultrasound Med Biol **14**(8), 695-707.

35. Lerner RM and Parker KJ. 1987, Sono-elasticity in ultrasonic tissue characterization and echographic imaging. Procs. 7th Eur. Comm. Workshop, JM Thijssen, ed., October 1987, Nijmegen, The Netherlands.

36. Lerner RM, Parker KJ, Holen J, Gramiak R, Waag RC 1988, Sono-elasticity: Medical elasticity images derived from ultrasound signals in mechanically vibrated targets. Acoust Imaging **16**, 317-327.

37. Lerner RM, Huang SR, Parker KJ. 1990, "Sonoelasticity" images derived from ultrasound signals in mechanically vibrated tissues, Ultrasound Med Biol **16**(3), 231-239.

38. Yamakoshi Y, Sato J, Sato T. 1990, Ultrasonic imaging of the internal vibration of soft tissue under forced vibration, IEEE Trans Ultras.Ferro. Freq. Cont., **UFFC-37**, 45-53.

39. Ishihara K, Tanouchi J, Kitabatake A, Uematsu M, Masuyma T, Yoshida Y, Doi Y, Kondo H, Kamada T, Kishimoto S, Ogawa T, Yokozawa N, Mulai H, Kodama M. 1990, High speed digital subtraction echography: principle and preliminary application to arteriosclerosis, arythmia and blood flow visualization, Proceedings of 1990 IEEE Ultrasonic Symposium, **2**, 1473-1476.

40. Yamashita Y, and Kubota M, 1990, Tissue characterization from ultrasonic imaging of movement and deformation, Procs. of the 1990 Ultrasonics Symposium, **2**, 1371-1375.

41. Meunier J, Bertrand M, Mailloux G, Petitclerc R. 1988, Assessing local myocardial deformation from speckle tracking in echography, SPIE Proc. Medical Imaging II **914**, 20-29.

42. Ophir J, Cespedes I, Ponnekanti H, Yazdi Y, Li X. 1991, Elastography: a quantitative method for imaging the elasticity of biological tissues, Ultrasonic Imag. **13**, 111-134.

43. Parker KJ and Lerner RM. 1992, Sonoelasticity of Organs: Shear Waves Ring a Bell, J Ultrasound Med **11**(8), 387-392.

44. Ponnekanti H, Ophir J, Cespedes I. 1992, Axial stress distributions compressors in elastography: an analytical model," Ultrasound Med Biol **18**(8), 667-673.

45. Truong XT, Jarrett SR, Nguyen MC. 1978, A method for deriving viscoelastic modulus from transient pulse propagation, IEEE Trans Biomed Eng **24**(4), 382-385.

46. O'Donnell M, Skovoroda AR, Shapo BM. 1991, Measurement of arterial wall motion using fourier based speckle tracking algorithms, Procs. of the 1991 IEEE Ultrasonics Symposium, **2**, 1101-1104.

47. Yemelyanov SY, Skovoroda AR, Lubinski MA, Shapo BM and O'Donnell M. 1992, Ultrasound elasticity imaging using Fourier based speckle tracking algorithm, Procs. of the 1992 IEEE Ultrasonics Symposium, **2**, 1065-1068.

48. Adler RS, Rubin JM, Bland PH, Carson PL. 1989, Characterization of transmitted motion in fetal lung: Quantitative analysis, Med Phys **16**(3), 333-337

49. Adler RS, Rubin JM, Bland PH, Carson PL. 1990, Quantitative tissue motion analysis of digitized M-mode images: Gestational differences of fetal lung, Ultrasound Med Biol **16**(6), 561-569.

50. Horn KP, Schunck BG. 1981, Determining Optical Flow, Artificial Intelligence **17**, 185-203

51. Feinberg DA, Crooks LE, Sheldon P, Hoenninger J, Watts J, Arakawa M. 1985, Magnetic resonance imaging the velocity vector components of fluid flow, Magn Reson Med **2**(6), 555-566.

52. Feinberg DA, Jakab PD. 1990, Tissue perfusion in humans studied by Fourier velocity distribution, line scan, and echo-planar imaging, Magn Reson Med **16**(2), 280-93.

53. Decorps M and Gourgeois D. 1991, Very Slow Flow Imaging, Magn Reson Med **19**(2), 270.

54. Zerhouni EA, Parish DM, Rogers WJ, Yang A, and Shapiro EP. 1988, Human heart: tagging with MR imaging - a method for noninvasivee assessment of mycardial motion, Radiology **169**, 164-172.

55. Axel L, Dougherty L. 1988, Heart wall motion: improved method of spatial modulation of magnetization for MR imaging, Radiology **169**, 59-63.

56. Pipe JG, Boes JL, Chenevert TL. 1991, Method for measuring three-dimensional motion with tagged MR imaging, Radiology **181**, 591-595.

57. Fowlkes JB, Emelianov SY, Pipe JG, Skovoroda AR, Adler RS, Carson PL and Sarvazyan AP. 1994, The possibility of cancer detection based on remote MRI measurements of tissue elasticity, submitted for publication in Medical Physics.

58. Landau LD and Liftshitz EM, 1965, Theory of elasticity, Moscow, Nauka.

59. Samarskii AA, Nikolaev ES. 1978, Methods of the solution of the net equations, Nauka, Moscow.

60. Skovoroda AR. 1992, About the diagnosis of the local pathologies in the elastic medium (3D approach), Preprint, Pushchino Scientific Center of Russian Acad.Sci., Pushchino.

61. Skovoroda AR. 1992, About the diagnosis of the local pathologies in the elastic medium (2D approach), Preprint, Pushchino Scientific Center of Russian Acad.Sci., Pushchino.

RECONSTRUCTIVE ELASTICITY IMAGING

S.Y. Emelianov,[1,2] A.R. Skovoroda,[1] M.A. Lubinski,[2] and M. O'Donnell[2]

[1]Institute of Mathematical Problems of Biology,
Russian Academy of Sciences,
Pushchino, Russia 142292

[2]Electrical Engineering and Computer Science Department and
Bioengineering Program,
University of Michigan,
Ann Arbor, MI 48109-2122

INTRODUCTION

Changes in soft tissue elasticity are usually related to some abnormal, pathological process. Because the Young's modulus can differ by orders of magnitude between soft tissues,[1] there has been consistent interest in tissue elasticity. Unfortunately, no imaging modality, including ultrasound, nuclear magnetic resonance (MRI) and computed tomography (CT), can directly provide information about elasticity. Recently, several investigators[2-6] have used internal motion induced by external forces to monitor tissue mechanical properties. Although mechanical properties are ultimately linked to patterns of internal deformation, deformational geometry can greatly affect the pattern as well. Consequently, to uniquely image tissue elasticity, the Young's modulus must be **reconstructed** from estimates of internal displacement and strain.

Using real time ultrasound imaging devices to track tissue motion resulting from externally applied forces, it may be possible to reconstruct the Young's modulus from estimates of internal displacement and strain fields. Results to date, however, have been disappointing due to limitations in motion tracking and elasticity reconstruction techniques, i.e. limitations of traditional longitudinal speckle tracking algorithms for large absolute displacements, poor accuracy of lateral displacement measurements (due to lack of resolution), and simplistic elasticity reconstruction algorithms. In this paper, we present methods to overcome some of these issues.

In the theory section, the general theoretical approach to reconstruction based on the common model of a linear, elastic, isotropic, incompressible medium is presented. Practical methods to estimate the elastic modulus using limited experimental data are discussed in the next section. Measurements on a tissue equivalent phantom with a single hard inclusion are presented to demonstrate the accuracy and sensitivity of the displacement and strain imaging methods reported previously.[7,8] Methods to detect the boundary of an inclusion and actual reconstruction of the spatial distribution of Young's modulus are also presented in this section.[9,10] The paper concludes with a discussion of the results.

THEORY

Consider a three-dimensional (3-D) volume V of deformed media with $\mathbf{U}=(u_1,u_2,u_3)$ as a displacement vector in Cartesian coordinates $\mathbf{X}=(x_1,x_2,x_3)$. Volume V can be either the entire mechanical body, or the region of interest inside the object under study.

Forward Problem. The forward problem is formulated here as a boundary value problem satisfying the equations of continuum mechanics,[11-13] where the parameters describing the deformation of an elastic or viscoelastic body (i.e., spatial distribution of displacement vector U, strain, and stress) should be determined for a known Young's modulus distribution. The most general form of Newton's 2nd law describing the motion of a mechanical body under static deformation (i.e. the equilibrium condition) is

$$\sum_{j=1}^{3} \sigma_{ij,j} + f_i = 0 \quad , \quad i=1,2,3, \tag{1}$$

where σ_{ij} is one component of the 2nd ranked stress tensor, u_i is one component of displacement vector U and f_i is the body force per unit volume acting on the body in the x_i direction. In this equation, and the entire paper, the lower index after comma means differentiation with respect to the corresponding spatial coordinate. Equation (1) must be satisfied at every internal point of the body.

Assuming linear elasticity, the components of the stress tensor in an isotropic, continuous, compressible medium under static deformation are:

$$\sigma_{ij} = \lambda \Theta \delta_{ij} + 2\mu\varepsilon_{ij} \quad , \tag{2}$$

where

$$\Theta = \mathrm{div}\mathbf{U} = \varepsilon_{11} + \varepsilon_{22} + \varepsilon_{33} \quad , \tag{3}$$

is the trace of the strain tensor, δ_{ij} is the Kronecker delta symbol and ε_{ij} is one component of the 2nd ranked symmetric strain tensor, defined as

$$\varepsilon_{ij} = \frac{1}{2}\left(\frac{\partial u_i}{\partial x_j} + \frac{\partial u_j}{\partial x_i}\right) \quad . \tag{4}$$

In Equation (2) the parameters λ and μ are Lame coefficients, and, therefore, any statical deformed isotropic continuous mechanical body can be characterized by a **spatial distribution** of the elastic parameters, such as λ and μ. This model of a continuous mechanical body applies for soft tissue at a spatial scale sampled by diagnostic ultrasound (i.e., at a scale greater than or comparable to an ultrasound wavelength). Note that some tissues, such as muscle, are anisotropic,[13] necessitating a more general stress-strain relationship[11,12] than Eq. (2).

Since all soft tissue and tissue-like materials are incompressible[14] (i.e., the Poisson's ratio approaches 0.5 for most soft tissue), the general expressions of linear elasticity can be greatly simplified. If the mechanical body is incompressible, then

$$\Theta = \text{div}U = \varepsilon_{11} + \varepsilon_{22} + \varepsilon_{33} = u_{1,1} + u_{2,2} + u_{3,3} = 0 \quad , \tag{5}$$

which determines the volume change due to deformation. Similarly, if the material volume does not change, then the longitudinal Lame coefficient λ approaches infinity. Under these conditions, the stress-strain relation (2) for static deformation reduces to

$$\sigma_{ij} = P\delta_{ij} + 2\mu\varepsilon_{ij} \quad , \tag{6}$$

where P is the static internal pressure, defined[15] as

$$\lim_{\substack{\lambda/\mu \to \infty \\ \Theta \to 0}} (\lambda\Theta) = P \quad . \tag{7}$$

Therefore, deformation of an incompressible media can be completely characterized by a **single** material parameter, either shear modulus μ or Young's modulus E since they are simply proportional to each other, i.e. $E=3\mu$. That is, the Young's modulus completely describes the static elastic properties of soft tissue, where its value may vary widely between different types of soft tissue.

Finally, combining Eqs. (1, 4, 5, 6) and eliminating σ_{ij} and ε_{ij}, a closed set of coupled differential equations for unknown components of the displacement vector **U** and pressure $P(x_1,x_2,x_3)$ describing static deformation of a viscoelastic, incompressible material can be generated:

$$\frac{\partial P}{\partial x_1} + 2\frac{\partial}{\partial x_1}\left(\mu\frac{\partial u_1}{\partial x_1}\right) + \frac{\partial}{\partial x_2}\left(\mu\left(\frac{\partial u_1}{\partial x_2} + \frac{\partial u_2}{\partial x_1}\right)\right) + \frac{\partial}{\partial x_3}\left(\mu\left(\frac{\partial u_1}{\partial x_3} + \frac{\partial u_3}{\partial x_1}\right)\right) + f_1 = 0$$

$$\frac{\partial P}{\partial x_2} + \frac{\partial}{\partial x_1}\left(\mu\left(\frac{\partial u_1}{\partial x_2} + \frac{\partial u_2}{\partial x_1}\right)\right) + 2\frac{\partial}{\partial x_2}\left(\mu\frac{\partial u_2}{\partial x_2}\right) + \frac{\partial}{\partial x_3}\left(\mu\left(\frac{\partial u_2}{\partial x_3} + \frac{\partial u_3}{\partial x_2}\right)\right) + f_2 = 0 \tag{8}$$

$$\frac{\partial P}{\partial x_3} + \frac{\partial}{\partial x_1}\left(\mu\left(\frac{\partial u_1}{\partial x_3} + \frac{\partial u_3}{\partial x_1}\right)\right) + \frac{\partial}{\partial x_2}\left(\mu\left(\frac{\partial u_2}{\partial x_3} + \frac{\partial u_3}{\partial x_2}\right)\right) + 2\frac{\partial}{\partial x_3}\left(\mu\frac{\partial u_3}{\partial x_3}\right) + f_3 = 0$$

$$\frac{\partial u_1}{\partial x_1} + \frac{\partial u_2}{\partial x_2} + \frac{\partial u_3}{\partial x_3} = 0 \quad .$$

The first three equations are the equations of equilibrium for a mechanical body written in terms of the displacements, and the last equation is the condition of incompressibility. Note that any spatial variations in elasticity $\mu(x_1,x_2,x_3)$ are explicitly included in Eqs. (8).

The system of Eqs. (8) has an infinite number of solutions, where the unique solution is determined by the boundary conditions. The general statement of mechanical boundary conditions is:

$$(\sum_{j=1}^{3} \sigma_{ij} n_j - T_i)\delta(u_i - u_i^0) = 0 \quad , \quad i=1,2,3 \quad , \tag{9}$$

where n_j is the jth component of the unit normal vector at the body surface, T_i is the force per unit area at the surface acting in direction x_i and δ is a symbol of variation.[12] In general, the boundary conditions can be split, where external applied forces are specified for part of the boundary and displacements for other parts. For any given external force T_i applied to some part of the surface, the corresponding equation in (9) is satisfied by the first term with no restriction on the u_i component of the surface displacement. In contrast, if the displacement u_i^0 is given, the corresponding equation is satisfied by the second term with no additional restriction on the stress tensor. Note that the stress components in the boundary conditions (9) can be written in terms of the displacements and pressure using Eqs. (4) and (6).

Finally, by solving the system of Eqs. (8) with corresponding boundary conditions (9), the displacement field can be obtained and components of the strain tensor can be simply calculated from Eq. (4). Therefore, the system of Eqs. (8) with the boundary conditions (9) completely characterize the forward mechanical problem.

Inverse problem. The main goal of elasticity imaging is to reconstruct the elastic modulus of any desired tissue region using precise measurements of strain and displacement components. Based on the above analysis for incompressible media, the inverse mechanical problem can be formulated as a reconstruction of the spatial distribution of the Young's modulus $E(x_1,x_2,x_3)$ (or shear modulus $\mu(x_1,x_2,x_3)$). Assume that the Young's modulus E is an arbitrary function of position, i.e.

$$E(\mathbf{X})=E_0 k(\mathbf{X}) \quad , \tag{10}$$

where E_0=const and $k(\mathbf{X})$ is not generally a continuous function.

Elasticity reconstruction in an isotropic medium must accurately represent both continuous and discontinuous changes in the modulus. We focus first on local, **clearly bounded** inhomogeneities ("inclusions") having a boundary **G** and residing in tissue with otherwise smoothly varying mechanical characteristics.[9] For this type of inclusion the function $k(\mathbf{X})$ is not a continuous function, i.e. it has a discontinuity at the boundary **G**, and the inverse mechanical problem is formulated as Young's modulus detection at this boundary.

The stress continuity condition at the boundary **G** has the form:

$$\sum_{j=1}^{3}[\sigma_{ij}]n_j = 0 \quad , \quad i=1,2,3 \quad , \tag{11}$$

where the square parentheses denote discontinuity of the terms at the boundary points.

Substituting expression (11) into Eq. (6), combining the results and eliminating the unknown pressure P, the condition of stress continuity becomes:[9]

$$\Gamma\left(n_1 n_2 (\varepsilon_{11}^{ext} - \varepsilon_{22}^{ext}) + ((n_2)^2 - (n_1)^2)\varepsilon_{12}^{ext} + n_3(n_2\varepsilon_{13}^{ext} - n_1\varepsilon_{23}^{ext})\right) =$$
$$n_1 n_2 (\varepsilon_{11}^{int} - \varepsilon_{22}^{int}) + ((n_2)^2 - (n_1)^2)\varepsilon_{12}^{int} + n_3(n_2\varepsilon_{13}^{int} - n_1\varepsilon_{23}^{int})$$

$$\Gamma\left(n_1 n_3 (\varepsilon_{11}^{ext} - \varepsilon_{33}^{ext}) + ((n_3)^2 - (n_1)^2)\varepsilon_{13}^{ext} + n_2(n_3\varepsilon_{12}^{ext} - n_1\varepsilon_{23}^{ext})\right) =$$
$$n_1 n_3 (\varepsilon_{11}^{int} - \varepsilon_{33}^{int}) + ((n_3)^2 - (n_1)^2)\varepsilon_{13}^{int} + n_2(n_3\varepsilon_{12}^{int} - n_1\varepsilon_{23}^{int}) \quad . \tag{12}$$

Here superscripts "int" and "ext" refer to internal and external variables at the boundary points, and the relation of Young's modulus E^{ext}/E^{int} is denoted by Γ. In equations (12) a linear stress-strain relation (6) is assumed, but no assumption is made for the strain-displacement relation.[9]

If all components of the strain tensor are known, the first equation in (12) can be used to determine the value of Γ, i.e. reconstruct the Young's modulus ratio at the boundary, and the second one to estimate the accuracy of experimental strain measurements in the neighborhood of the boundary point. Also, if mechanical properties do not vary significantly inside the inclusion, the reconstruction is complete since the Young's modulus at the boundary is found. We also note that Eqs. (12) can be used for boundary detection, and such information derived from measured strain images can be extremely important in those cases where the inclusion cannot be directly visualized by the imaging system.

Not all inclusions, however, will exhibit a clear boundary and, therefore, reconstruction of the isotropic medium with **spatially continuous** changes in the Young's modulus is required. For this type of reconstruction the function $k(\mathbf{X})$ is assumed to be continuous with well defined spatial derivatives.

Using the equations defining the linear, incompressible elastic medium, the equilibrium condition (1) can be rewritten in the form:[9]

$$2\varepsilon_{12}(k_{,11}-k_{,22})+2(u_{2,2}-u_{1,1})k_{,12}+2\varepsilon_{23}k_{,13}-2\varepsilon_{13}k_{,23}+$$
$$(\Delta u_2+\omega_{12,1})k_{,1}-(\Delta u_1-\omega_{12,2})k_{,2}+\omega_{12,3}\,k_{,3}+(\Delta\omega_{12})k+F_{12}=0$$

$$2\varepsilon_{13}(k_{,11}-k_{,33})+2\varepsilon_{23}k_{,12}+2(u_{3,3}-u_{1,1})k_{,13}-2\varepsilon_{12}k_{,23}+ \qquad (13)$$
$$(\Delta u_3+\omega_{13,1})k_{,1}+\omega_{13,2}\,k_{,2}-(\Delta u_1-\omega_{13,3})k_{,3}+(\Delta\omega_{13})k+F_{13}=0$$

$$2\varepsilon_{23}(k_{,22}-k_{,33})+2\varepsilon_{13}k_{,12}-2\varepsilon_{12}k_{,13}+2(u_{3,3}-u_{2,2})k_{,23}+$$
$$\omega_{23,1}\,k_{,1}+(\Delta u_3+\omega_{23,2})k_{,2}-(\Delta u_2-\omega_{23,3})k_{,3}+(\Delta\omega_{23})k+F_{23}=0 \quad ,$$

where the pressure P was eliminated from this system of equations, and

$$\omega_{12}=u_{2,1}-u_{1,2}\,,\quad \omega_{13}=u_{3,1}-u_{1,3}\,,\quad \omega_{23}=u_{3,2}-u_{2,3}\,,\quad \Delta=\frac{\partial^2}{\partial x_1^2}+\frac{\partial^2}{\partial x_2^2}+\frac{\partial^2}{\partial x_3^2}\quad ,$$

$$F_{12}=(3/E_0)(f_{2,1}-f_{1,2})\,,\quad F_{13}=(3/E_0)(f_{3,1}-f_{1,3})\,,\quad F_{23}=(3/E_0)(f_{3,2}-f_{2,3})\,.$$

The unique solution of the system of Eqs. (13) is determined by the boundary conditions, i.e. the elastic modulus $k(\mathbf{X})$ must be specified at some boundary. Note also that analytical solution of Eqs. (13) is not generally possible for an arbitrary spatial distribution of the elastic modulus $k(\mathbf{X})$.

If all components of the displacement vector and strain tensor are known, then the reconstruction of the Young's modulus can be performed. However, current strain imaging systems based on ultrasonic speckle tracking are two-dimensional,[5,7,8] and, consequently, boundary detection of inclusions and reconstruction of the Young's modulus must be estimated based on displacement and strain data from a single imaging plane. Therefore, to reconstruct the elastic modulus using current ultrasonic equipment either some symmetries must be assumed in the distribution of Young's modulus, or the deformation pattern must be carefully controlled.[9,10] In particular, a specific type of deformation producing additional simplification to the general equations (12) and (13) is considered in the next section.

METHODS

Experiments were performed on a number of gel-based phantoms, with results reported here obtained on a particular phantom with a single hard inclusion inside. Originally, a homogeneous cylindrical gel 88 mm in diameter and 140 mm long was constructed from a 5.5 % by weight concentration of gelatin. Then a circular, longitudinal hole 30 mm in diameter was made in the center of the phantom. The hole was backfilled with the same gelatin concentration except for the central one third part of the hole, which was backfilled by a 12 % by weight concentration of gelatin producing a hard, 45 mm long, cylindrical inclusion in the center of an otherwise homogeneous phantom. The cross-sectional and longitudinal views of the phantom are shown schematically in Fig. 1. The Young's modulus of the inclusion was estimated to be 2.5 times larger than that of the surrounding material. In all phantom materials a small amount of polystyrene microspheres was added to act as ultrasonic scattering centers.

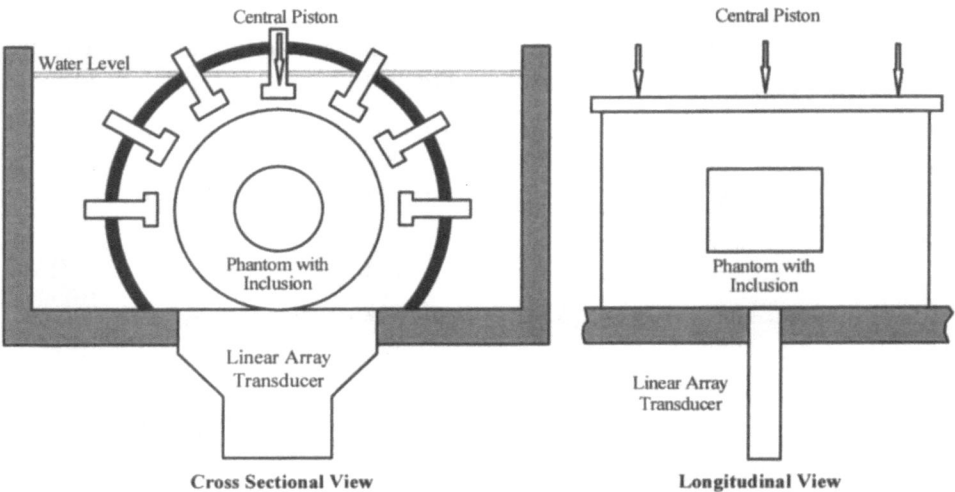

Figure 1. Schematic representation (cross-sectional and longitudinal views) of the experimental system and phantom geometry. The bottom of the cylindrical phantom, 88 mm in diameter and 140 mm long, contacts a 128 element linear transducer array used to image the cross-sectional central plane. A set of pistons providing deformation is on the top, where only the central piston was displaced by 5 mm in the present experiment.

The phantom was placed in a water tank with cylindrical axis perpendicular to the axis of a 3.5 MHz, 128 channel, 1-D transducer array attached to the bottom of the tank, as illustrated in Fig. 1. The phantom was centered so that the image plane approximated the central plane perpendicular to the longitudinal axis of the phantom. The tank was filled with water to provide contact between the array and phantom. In the present experiments, simple surface displacements were produced by a hydraulically driven piston located at the top of the phantom, where movement of this piston was controlled by measuring ultrasonic pulse arrival time differences to the central array elements. This piston was a 14 mm wide, rigid, rectangular block extending the entire length of the phantom.

As discussed previously, traditional speckle tracking breaks down for large (compared to an acoustical wavelength) displacements because of both out-of plane motion and decorrelation effects due to the finite strain magnitude.[7] To overcome these limitations, a large set of images were recorded with small relative displacements but significant total displacement from the beginning to end of the set. The results presented below used a set of 26 images over a total vertical piston displacement of 5 mm with 200 μm steps to quantitatively estimate the vertical displacement and longitudinal strain components. These images were computed by properly accumulating differential displacement and strain estimates between two neighboring images of this large set of images spanning the total deformational range.[7]

Based on the particular geometry of the experimental system and phantoms, Eqs. (12) and (13) can be simplified.[9,10] As reported previously,[8] in the area of the central vertical plane of the phantom ($x_3=0$), the deformation pattern produced by the current system will closely approximate a plane deformed state. For a plane deformed state, components u_1 and u_2 of the displacement vector $\mathbf{U}$ are functions only of x_1 and x_2, and $u_3=0$. With these conditions, the second equation of stress continuity condition (12) is automatically satisfied and the first equation leads to

$$\Gamma = \frac{n_1 n_2 (\varepsilon_{11}^{int} - \varepsilon_{22}^{int}) + ((n_2)^2 - (n_1)^2) \varepsilon_{12}^{int}}{n_1 n_2 (\varepsilon_{11}^{ext} - \varepsilon_{22}^{ext}) + ((n_2)^2 - (n_1)^2) \varepsilon_{12}^{ext}} , \qquad (14)$$

which should be satisfied along the entire boundary $\mathbf{G}$ of the inclusion contained in the imaging plane.[9] An additional simplification based on incompressibility is possible. For a plane deformed state, the condition of incompressibility (5) reduces to $\varepsilon_{11}+\varepsilon_{22}=0$, and, consequently, at any point along the boundary where $(n_1)^2=(n_2)^2$, Eq. (14) reduces to

$$\Gamma = \frac{\varepsilon_{22}^{int}}{\varepsilon_{22}^{ext}} . \qquad (15)$$

That is, at any point where the two unit normals to the inclusion's boundary are equal in magnitude, the Young's modulus can be accurately computed from a single component of the strain tensor. This is a key result since only a single component is accurately measured with current strain imaging systems.[5,7] Nevertheless, even if the accurate reconstruction of the Young's modulus based on the Eq. (15) is only possible at special points, the simple expression presented in this equation can be used to detect inclusion boundaries since there will be large changes in Γ near the boundary over a wide range in n_1 and n_2. This is also true even if the inclusion exhibits large but not discontinuous changes in the Young's modulus.

Equations (13) can also be simplified for a plane deformed state, reducing to the following:

$$(k_{,11} - k_{,22})(u_{1,2} + u_{2,1}) + 2k_{,12}(u_{2,2} - u_{1,1}) + \qquad (16)$$
$$2k_{,1}(u_{2,11} + u_{2,22}) - 2k_{,2}(u_{1,11} + u_{1,22}) + k(u_{2,111} + u_{2,221} - u_{1,112} - u_{1,222}) = 0 ,$$

where it was assumed that all f_i components of the body force vanish for the experimental setup used in the present study, and the condition of incompressibility $u_{1,1}+u_{2,2}=0$ was explicitly used.[9] Note, that only two-dimensional measurements are needed to reconstruct the spatial distribution of the Young's modulus in the plane $x_3=0$.

RESULTS

The first step in ultrasound elasticity imaging is the accurate estimation of all necessary displacement and strain components.[7,8] The measured vertical displacement distribution created by the central vertical piston is shown in Fig. 2a. In this figure the image covers an area of 100 by 100 mm and a quantitative gray scale is used, i.e., full white represents a 5 mm vertical displacement and full black corresponds to no displacement. The largest displacement magnitudes can be seen at the top of this image, where the piston provided the surface displacement, with smooth reduction to zero at the constrained bottom surface where the transducer was attached. Non-zero measured displacement outside the phantom is related to imaging artifacts and cannot be filtered out. Nevertheless, the quality of displacement images within the phantom is sufficient for quantitative strain imaging. Note that the presence of the inclusion is hardly noticeable on the displacement image.

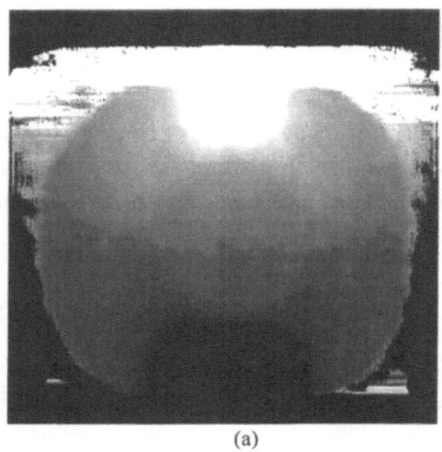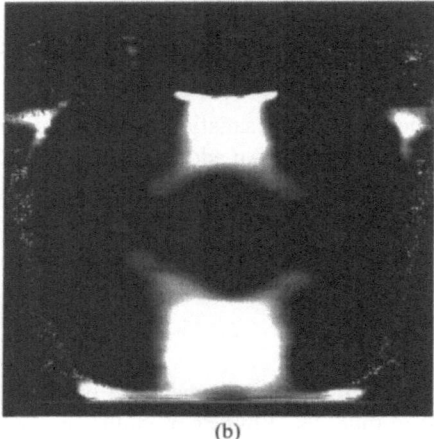

(a)   (b)

Figure 2. Measured image of a) vertical u_2 component of the displacement vector and b) vertical ε_{22} component of the strain tensor for the phantom with a single hard inclusion.

An image of the ε_{22} strain component displayed over the same 100 by 100 mm area is presented in Fig. 2b. Again, a quantitative gray scale is used, where the bright areas on the strain image represent regions with the highest strain magnitude. In this image the region of highest strain magnitude, starting at the top near the position of the piston and extending to the bottom where the transducer array was attached, can be easily seen, as well as the inclusion in the center of the phantom. The signal-to-noise ratio (SNR) of these images depends on position, but is estimated to be about 30 to 1 at the center of inclusion.[7] The SNR in this image is between one and two orders of magnitude greater than that in previously published studies.[5] Because the contact area of the phantom with the piston is not the same as that at the bottom of the tank, there is some difference in the strain distribution at the top and bottom of the phantom.

Based on these displacement and strain images, reconstruction of the Young's modulus proceeds as follows. First, inclusion boundaries must be identified. The result of boundary detection based on Eq. (15) is presented in Fig. 3. In Figure 3a, the spatial distribution of the parameter Γ is imaged using a quantitative gray scale. A step size of 0.4 mm was used for this image, and the gray scale was selected so that a Γ of 1.25 is pure white, a Γ of 0.75

is pure black and a Γ of 1.00 is mid-gray. Also, a normalization procedure was used for this image to minimize the effects of noise in low strain regions.[9,10] The boundaries of the inclusion can be clearly seen in Fig. 4a, but will be even more visible if a different display format, such as that of Fig. 3b, is used. In Figure 3b the parameter Λ, defined as $\Lambda=\Gamma-1$ for $\Gamma\geq1$ and $\Lambda=1/(\Gamma-1)$ for $\Gamma<1$, is displayed to highlight changes in Γ near inclusion boundaries. This image was threshed at the value of $\Lambda=0.06$, i.e., $\Lambda<0.06$ is black and $\Lambda\geq0.06$ is white. Clearly, the boundaries of the inclusion are identified on this image.

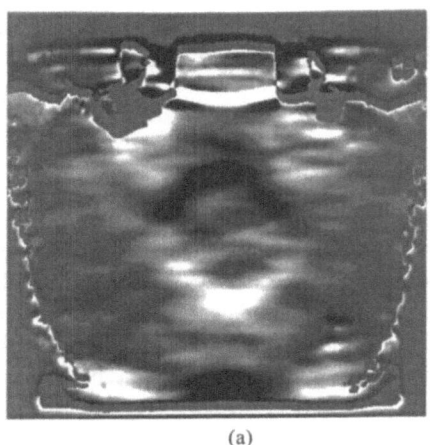 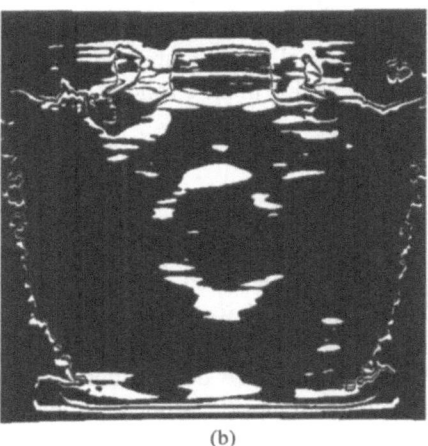

(a)   (b)

Figure 3. Spatial distribution of a) parameter Γ and b) threshed differential parameter Λ estimated in the phantom with hard inclusion.

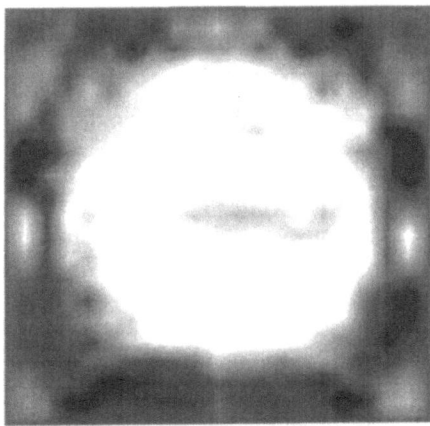

Figure 4. Reconstructed distribution of Young's modulus in the central plane of the phantom with hard inclusion. The Young's modulus distribution is displayed over a logarithmic gray scale.

The reconstruction of the Young's modulus using a specific program developed to solve Eq. (16) is presented in Fig. 4. It was assumed that the Young's modulus is uniform along the borders of the 40 by 40 mm square positioned at the center of the phantom. This square was chosen since it includes the boundaries detected in Fig. 3. The lateral u_1 component of the displacement vector was computed from the measured longitudinal u_2 displacement component assuming incompressibility and that u_1 was zero along a vertical central line of the phantom. Equation (16) was discretized with a 21 by 21 point grid (2 mm grid spacing), and all spatial derivatives were approximated by finite differences. Prior to reconstruction, the original displacement image (Fig. 2a) was spatially filtered to ensure that the model of continuous elasticity is well approximated.

The image in Fig. 4 represents the 40 by 40 mm region, where a logarithmic gray scale over the range $0.5<E/E_0<2.1$ is used. The gray scale was selected so that the relative Young's modulus of 1.0 is mid-gray, dark areas represent softer material and bright areas harder material. Even with the artifacts evident in Fig. 4, the presence of the hard inclusion is clearly visible in the elasticity image.

DISCUSSION

Reconstructive elasticity imaging is possible if internal displacement and strain fields are accurately measured with subsequent reconstruction of the elastic (Young's) modulus. In this report, we have shown that specific phase sensitive methods can be used to measure longitudinal displacement and strain images in the limit of large (compared to acoustic wavelength) surface deformation. Consequently, these measurements can be used for detection of inclusion boundaries and quantitative reconstruction of the Young's modulus distribution.

To fully characterize deformations of an elastic body, the complete displacement vector and strain tensor must be computed. In the case of a plane deformed state, only lateral and longitudinal components of the displacement vector must be measured since the other component vanishes. Consequently, only one longitudinal and one shear component of the strain tensor need be measured to represent the strain tensor for the plane deformed state.

In the present study, only one longitudinal component of the displacement vector was accurately measured in the imaging plane (Fig. 2a). However, the lateral u_1 displacement can be estimated directly from the measured longitudinal u_2 displacement using the principle of incompressibility, i.e., $u_{1,1}+u_{2,2}=0$. Using this condition, the lateral component was estimated assuming that it was zero along the central vertical line of the phantom. This is a good approximation for the experimental system and phantom used in the present study. In more general cases, however, this approximation may not hold and, therefore, accurate measurements of the lateral component of the displacement vector **at most** along any single vertical line are needed.

Several techniques have been proposed for sensitive cross-beam displacement estimation, but even these methods will produce significant error in the lateral displacement and, subsequently, shear components of the strain tensor.[16,17] By optimally deforming the surface to maximize displacement and strain along the ultrasound beam, error in lateral tracking will have minimal influence on strain images and subsequent elasticity reconstruction.

Boundaries of local discontinuous changes in the Young's modulus can be detected based on Eq. (15) where the plane deformed state is assumed. Moreover, even in cases where changes in Young's modulus are not discontinuous, the "boundary" of inclusions can be also detected with Eq. (15). The parameter Γ, however, will not represent the exact ratio of Young's moduli.

Once the boundary is determined and the ratio of Young's moduli along the boundary estimated, the actual reconstruction must be performed. Numerical solution of Eq. (16)

requires knowledge of the elastic modulus along some boundary, but not necessarily the boundary of the object, however. Using the inclusion boundaries determined by Eq. (15), the Young's modulus must be specified for any closed curve outside this inclusion. In the present studies, the modulus was specified along the edges of a 40 by 40 mm square (Fig. 4). In general, however, Goursat conditions are preferable, where the Young's modulus need be specified only along two intersecting characteristic curves.[9]

The quality of Young's modulus reconstruction is ultimately limited by the accuracy of measured displacements and strains. Previously, a method of tracking relatively large displacement was presented showing significant SNR improvement in both displacement and strain estimates.[7] Even if linear elasticity does not strictly hold for large scale deformations, Young's modulus reconstruction based on Eqs. (15) and (16) may be accurate.[9,10]

Finally, the results presented here suggest that quantitative reconstruction of the elastic modulus may be possible for complex objects such as the human body. Full reconstruction of the elastic properties of soft tissue without any assumptions should be based on reasonably good measurements of the complete 3-D spatial distribution of all necessary components of the displacement and strain. In the meantime, if limited measurements are available, then both a correct model of inhomogeneities and proper control of external deformation must be used for accurate reconstruction.

ACKNOWLEDGMENTS

Helpful discussions with Prof. Armen P. Sarvazyan are gratefully acknowledged. Partial support from the Office of Vice President of the University of Michigan and the National Institutes of Health under grant CA 54896 are gratefully acknowledged. M.A. Lubinski was supported under an NSF Graduate Research Fellowship. Finally, we thank Acuson and General Electric for assistance with this work.

REFERENCES

1. Sarvazyan AP, Pasechnik VI, Shnol SE, "Low speed of sound in gels and biological tissues," Biofizika **13**, pp. 587-594 (1968).
2. Krouskop TA, Dougherty DR, Levinson SF, "A pulsed Doppler ultrasonic system for making non-invasive measurements of the mechanical properties of soft tissue," J. Rehabil. Res. Dev. **24**(2), 1-8 (1987).
3. Lerner RM, Huang SR, Parker KJ, "Sonoelasticity" images derived from ultrasound signals in mechanically vibrated tissues," Ultrasound Med. Biol. **16**(3), 231-239 (1990).
4. Sarvazyan AP, Skovoroda AR, "The new approaches in ultrasonic visualization of cancers and their qualitative mechanical characterization for the differential diagnostics," Abstract of the All-Union Conference "The Actual Problems of the Cancer Ultrasonic Diagnostics", Moscow (1990).
5. Ophir J, Cespedes I, Ponnekanti H, Yazdi Y, Li X, "Elastography: a quantitative method for imaging the elasticity of biological tissues," Ultrasonic Imag. **13**, 111-134 (1991).
6. Fowlkes JB, Yemelyanov SY, Pipe JG, Carson PL, Adler RS, Sarvazyan AP, Skovoroda AR, "Possibility of cancer detection by means of measurement of elastic properties," Radiology. **185**(P), 206-207 (1992).
7. O'Donnell M, Skovoroda AR, Shapo BM and Emelianov SY, "Internal displacement and strain imaging using ultrasonic speckle tracking," to appear in the May, 1994 Issue of the IEEE Transactions on Ultrasonic Ferroelectrics and Frequency Control (1994).
8. Skovoroda AR, Emelianov SY, Lubinski MA, Sarvazyan AP and O'Donnell M, "Theoretical analysis and verification of ultrasound displacement and strain imaging," to appear in the May, 1994 Issue of the IEEE Transactions on Ultrasonic Ferroelectrics and Frequency Control (1994).

9. Skovoroda AR, "Inverse problems of the theory of elasticity in the diagnostics of soft tissue pathologies," Preprint, Pushchino Research Center of Russian Academy of Sciences, Pushchino (1992).

10. O'Donnell M, Emelianov SY, Skovoroda AR, Lubinski MA, Weitzel WF and Wiggins RC, "Quantitative elasticity imaging," Procs. of the 1993 IEEE Ultrasonics Symposium **93CH3301-9**, pp. 893-903 (1993).

11. Landau LD and Liftshitz EM, Theory of elasticity, Moscow, Nauka (1965).

12. Rabotnov YN, Mechanics of solid structures, Moscow, Nauka (1979).

13. Fung YC, Biomechanics-mechanical properties of living tissues, Springer-Verlag; New York (1981).

14. Sarvazyan AP, "Low frequency acoustical characteristics of biological tissues," Mechanics of Polymers **4**, 691-695 (1975).

15. Biot MA, Mechanics of Incremental Deformations, John Wiley & Sons, Inc., New York (1965).

16. Trahey GE, Allison JW and von Ramm OT, "Angle independent ultrasonic detection of blood flow," IEEE Trans. on Biomedical Eng. BME-**34**, pp. 965-967 (1987).

17. O'Donnell M, and Flax SW, "Phase aberration measurements in medical ultrasound: human studies," Ultrasonic Imaging 10, 1-11 (1988).

IMAGING SYSTEM OF PRECISE HARDNESS DISTRIBUTION IN SOFT TISSUE IN VIVO USING FORCED VIBRATION AND ULTRASONIC DETECTION

Katsunori Fujii*, Takuso Sato*, Keisuke Kameyama*, Toshikazu Inoue*,
Katsunori Yokoyama* and Koichi Kobayashi**

* Interdisciplinary Graduate School of Science and Engineering,
  Tokyo Institute of Technology , 4259 Nagatsuta, Midori-ku, Yokohama 227, Japan
** University of Tokyo, Japan

INTRODUCTION

This paper focuses on observation of biological tissue hardness (elasticity) by applying low frequency vibration to the tissue and detecting the vibration propagation characteristics with use of ultrasonic pulsed Doppler technique. It has been reported that the amplitude, phase and velocity of the propagating vibration can be mapped using ultrasonic Doppler method [1]. Traditionally, low frequency vibration propagation in soft tissues have been considered to be shear waves, and therefore tissue hardness have been formulated to be directly related to the propagating velocity of the vibration [2]. However, through experimental verification, it was found that vibration propagation is much affected by the vibrating conditions such as vibration frequency or the shape of the vibrator attachment. Here, a new formulation of vibration propagation in soft tissue which takes the vibrating conditions into consideration is introduced. Based on the derived formulation, a precise tissue hardness map estimation which is independent of the vibrating conditions is demonstrated. Additionally, a method to scan the vibration frequency and to map the average hardness estimated for multiple frequencies is used in order to reduce the effects of the developed standing waves which can be a serious obstacle for estimation of precise hardness maps. The method proved to be useful for reducing the false effects (ghosts) observed in hardness maps. The feasibility of the introduced methods were certified through observation of hardness for an agarose phantom with different elasticity and in vivo human thigh with various loads at the ankle.

FORMULATION OF LOW FREQUENCY VIBRATION PROPAGATION IN SOFT TISSUE AND ITS EXPERIMENTAL VERIFICATION

In visco-elastic media such as human tissue, the energy of the applied vibration of frequency lower than 1 kHz have been considered to be transferred to incompressive shear waves. In such a case, the governing dynamical equation is written as [2],

$$\rho \frac{\partial^2 u}{\partial t^2} = \mu_t \frac{\partial^2 u}{\partial z^2} + \eta_t \frac{\partial^3 u}{\partial t \partial z^2} \tag{1}$$

where μ_t, η_t and ρ are the medium's shear elasticity, shear visco-elasticity and density, respectively. Therefore, the wave velocity can be written as,

$$v_t = \sqrt{\frac{2\left(\mu_t^2 + \omega^2 \eta_t^2\right)}{\rho\left(\mu_t + \sqrt{\mu_t^2 + \omega^2 \eta_t^2}\right)}} . \tag{2}$$

In this case, Eq. (2), the velocity is affected only by the mechanical characteristics of the medium. In Fig. 1, the measured velocity of the wave propagating in an agarose phantom is compared with the theoretical velocity calculated from Eq. (2). It can be seen that the observation diverge from the theoretical values as the frequency tends lower and as the radius of the disk-shaped attachment of the vibrator increases (see Fig. 3 for measurement setup). As this tendency was reproducable and a similar velocity increase is reported in [3] for low frequency situations, a new formulation of low frequency vibration propagation was derived as follows.

For strict modelling, the vibration radiation is treated as propagation of incompressive waves. Let the attachment of the vibrator be contacted onto the surface of the measured object as depicted in Fig. 2. For convenience, it will be put that: 1) The vibration displacement (frequency f, amplitude a) is in the z-axis direction of Fig. 2. 2) The attachment of the vibrator is a disk of finite radius r. The following notations will be used hereafter to refer to the medium characteristics:

μ_l : longitudinal elasticity, η_l : longitudinal viscosity, μ_t : transverse (shear) elasticity,

η_t : transverse (shear) viscosity, ρ : density.

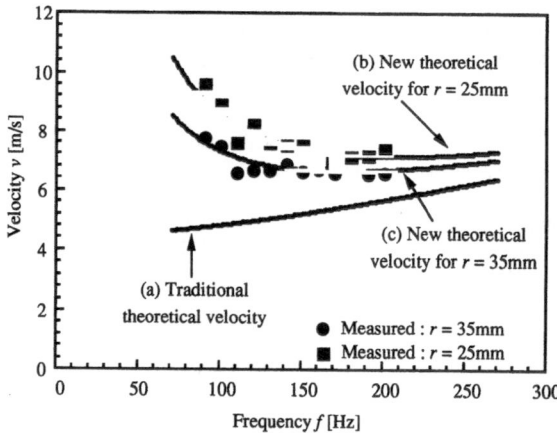

Fig. 1. Propagating velocity of the low frequency vibration. (a) Traditional formulation calculated from Eq. (2), New formulation calculated for two kinds of attachments ((b) $r=25$ and (c) 35mm) and the experimental results obtained in the setup shown in Fig. 3.

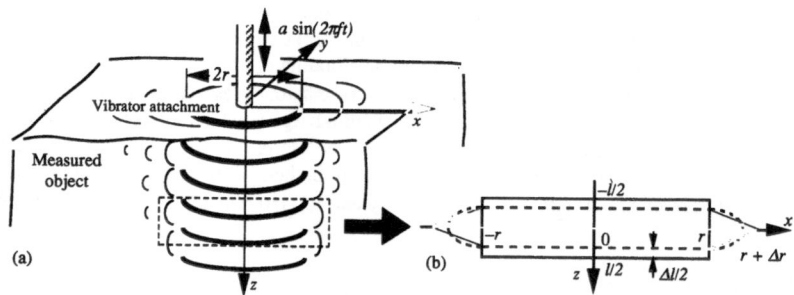

Fig. 2. (a) The low frequency vibration propagating in the soft medium as incompressive wave. (b) The model of half wavelength medium deformation.

The model of incompressive medium deformation caused by a wave of wavelength $2l$ radiated from a vibrator with attachment radius r is shown in Fig. 2. The cylindrical portion of the medium contoured with a solid line is squeezed by the wave into a shape drawn in a dashed line. The longitudinal strain occurring in this portion of the medium is,

$$\varepsilon = \Delta l \, / \, l \, . \tag{3}$$

The lateral deformation of semicircular sections can be approximated with triangles drawn in narrow lines in Fig. 2. Then, we can write according to incompressiveness of the medium,

$$\pi r^2 l = 2\pi \int_0^{l/2} \left(\frac{2\Delta r}{l - \Delta l} x + \Delta r + r \right)^2 dz \, . \tag{4}$$

By solving Eq. (4), we will have

$$\Delta r = \frac{\Delta l}{l - \Delta l} r \, . \tag{5}$$

254

Therefore, the relation between longitudinal strain ε and shear strain γ is,

$$\gamma = \frac{dr}{dz} \approx \frac{\Delta r}{\frac{l - \Delta l}{2}} = \frac{\frac{\Delta l}{l - \Delta l} r}{\frac{l - \Delta l}{2}} \approx \frac{2 \Delta l}{l^2} r = \frac{2r}{l} \varepsilon . \tag{6}$$

Using $2l = v/f$, Eq. (6) is rewritten as

$$\varepsilon = \frac{v}{4 r f} \gamma . \tag{7}$$

If $rf \gg v$, which is roughly the case when the vibration frequency is high or the radius of the vibrator attachment is large, $\gamma \gg \varepsilon$, and thus the shear wave dominates. On the contrary, longitudinal strain must be taken into account when r or f is small, and an equivalent shear elasticity μ can be written as

$$\mu = \mu_l + \frac{v}{4 r f} \mu_l . \tag{8}$$

Under dynamically varying stress, the mechanics of biological tissue can be modelled as elastic and viscous factors connected in parallel (Voigt model). Therefore, the effect of longitudinal viscosity will diminish as the vibration frequency f increases, and can be neglected. The dynamical equation of the propagation of low frequency vibration taking both longitudinal and shear strain is written as,

$$\rho \frac{\partial^2 u}{\partial t^2} = \left(\mu_l + \frac{v}{4 r f} \mu_l \right) \frac{\partial^2 u}{\partial z^2} + \eta_l \frac{\partial^3 u}{\partial t \partial z^2} . \tag{9}$$

From Eq. (9), the diffusing velocity is

$$v = \frac{2(a^2 + b^2)}{a \left(-s + \sqrt{\frac{x + \sqrt{x^2 + y^2}}{2}} \right) + b \sqrt{\frac{-x + \sqrt{x^2 + y^2}}{2}}} \tag{10}$$

where

$$a = \frac{\mu_l}{\rho}, \; b = \frac{\omega \eta_l}{\rho}, \; s = \frac{\pi \mu_l}{2 r \omega \rho}, \; x = s^2 + 4a \text{ and } y = 4b .$$

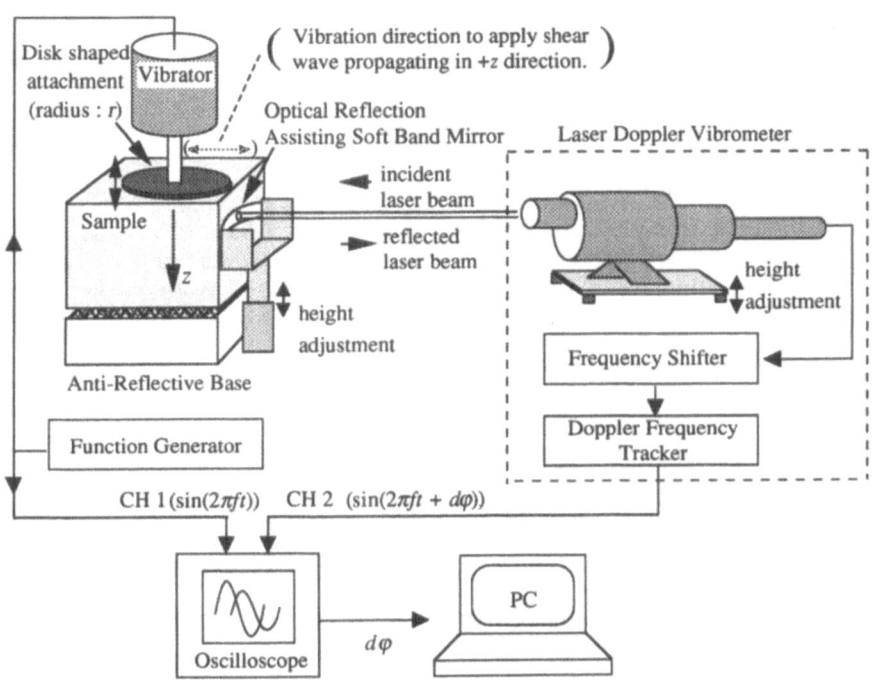

Fig. 3. Scheme of vibration propagating velocity measurement setup. The phase difference $d\varphi$ is measured for several points on the side of the sample by scanning the target of the laser beam along the z-axis, and the velocity is estimated by applying linear regression to the data.

The propagation velocity of the forced vibration was measured using a setup illustrated in Fig. 3. The sinusoidal vibration waveform generated by the function generator is amplified and drives the vibrator which is attached onto the measured object with a disk-shaped attachment. A base which inhibits the vibration reflection is placed under the measured object. The vibration propagating downwards is detected by a laser Doppler vibrometer. This vibrometer illuminates a point on the side of the measured object with a laser beam, and detects the scattered light. The surface displacement velocity is detected from the Doppler frequency shift of the scattered beam, and transferred to a electrical potential proportional to the movement velocity. In order to aid the beam reflectivity of the object surface, a flexible conveniently curved mirror (thin plastic band with aluminum film adhered) is attached to the object surface. The mirror will vibrate freely together with the object surface where the laser beam is targeted, thereby enabling to measure the movement of objects with various surface conditions. Both the detected movement and the signal from the function generator is sent to the synchroscope where the phase difference is measured and recorded with a personal computer.

The phase delay data were collected from several points on the side of the object along the z-axis, and by regression analysis of the data, an approximating line

$$z = A(d\varphi) + B \tag{11}$$

was estimated. Here, B can be neglected because $d\varphi = 0$ when $z = 0$. Using

$$\text{(Wavelength)} = 2l = v/f = A \cdot 2\pi \tag{12}$$

velocity v can be estimated as

$$v = 2\pi f A . \tag{13}$$

Next, we will refer to estimation of shear elasticity μ_t and shear viscosity η_t. Here, μ_t and η_t of the measured object was estimated in the following way. By changing the orientation of the vibrator of Fig. 3, pure shear wave was given to the top surface of the object. As the velocity of shear wave v_t is subject to Eq. (2), measuring v_t for two different frequencies $f = f_1$ and $f = f_2$ will make a system of equations for μ_t and η_t. Shear elasticity μ_t and shear viscosity η_t can be estimated by solving this system of equations. For the the relation between the longitudinal and shear elasticity,

$$\mu_l = 2(1+\sigma)\mu_t = 3\mu_t . \tag{14}$$

was used for the Poisson's ratio $\sigma = 1/2$ since we assume that the medium is incompressive.

By supplying these parameter values, the velocity in Eq. (10) was calculated. In Fig. 1, the results are shown together with experimentally obtained data for an agarose phantom. Velocity data were collected using two vibrator attachments ($r = 25mm$ and $r = 35mm$) and the curve derived from the formulation agrees with the obtained data very well. Especially, the velocity estimation in the low frequency region is improved compared with the traditional formulation assuming only the shear waves. The same experiment was carried out for a in vitro pig leg tissue and the measured velocities agreed with the theoretical values of the new formulation. From these experiments, the validity of the proposed formulation of the propagation of forced vibration in soft tissue has been proved.

HARDNESS MAPPING

Here, a practical approach of tissue characterization with forced low frequency vibration is discussed. Instead of the setup in Fig. 3, an enhanced ultrasonic scanner device is used. For detection of inner tissue displacement due to wave propagation, pulsed doppler technique is employed. By detecting the scattered pulse wave with Doppler frequency shift due to the movement of the scatterers, the tissue movement is estimated, thereby mapping the amplitude and the phase of the low frequency wave as reported in [1]. After estimating the velocity map from the phase map by using

$$v = 2\pi f / \alpha \tag{15}$$

where α is the local maximum gradient of the phase, hardness mapping is done as follows.

The propagation velocity of the forced vibrational wave v can be expressed as Eq. (10). Among the parameters in Eq. (10), medium density ρ, frequency f and attachment radius r are known. For μ_l, relation between μ_t of Eq. (14) is applicable, and can be replaced with $3\mu_t$. As for η_t, we will use an empirical relation found in our experimental results for agarose phantom, pig leg tissue and excised human uterine myoma such as,

$$\mu_t \ [N/m^2] \approx 1000[s^{-1}] \cdot \eta_t [Ns/m^2] , \tag{16}$$

for it is inconvenient to measure twice with different vibrational frequencies to obtain both μ_t and η_t. Using these conditions, the only unknown variable in Eq. (10) will be μ_t, and it can be obtained for each point in the observed region by numerical methods, thereby making the hardness (shear elasticity μ_t) map.

MEASUREMENT SYSTEM

The signal flow in the system consisting of an enhanced ultrasonic scanning device and a signal processor which was used for hardness mapping is shown in Fig. 4. The ultrasonic pulse (frequency 3.5 MHz, interval 250 ms) is generated by the 64 ultrasonic transducer/receiver array. The scattered wave received at the array is processed with the 90 degree phase shifter, multiplier and low-pass filter (f_{cut} = 500kHz), producing the complex Doppler signal. The spectral analysis of this signal is done in the digital signal processor, and using the spectral patterns of the Doppler signal, vibrational amplitude and phase are estimated. Calculation of velocity and hardness is processed in the computer where the obtained maps are displayed on the monitor. The time for measurement is about 90 seconds, and the hardness map can be obtained in another 90 seconds.

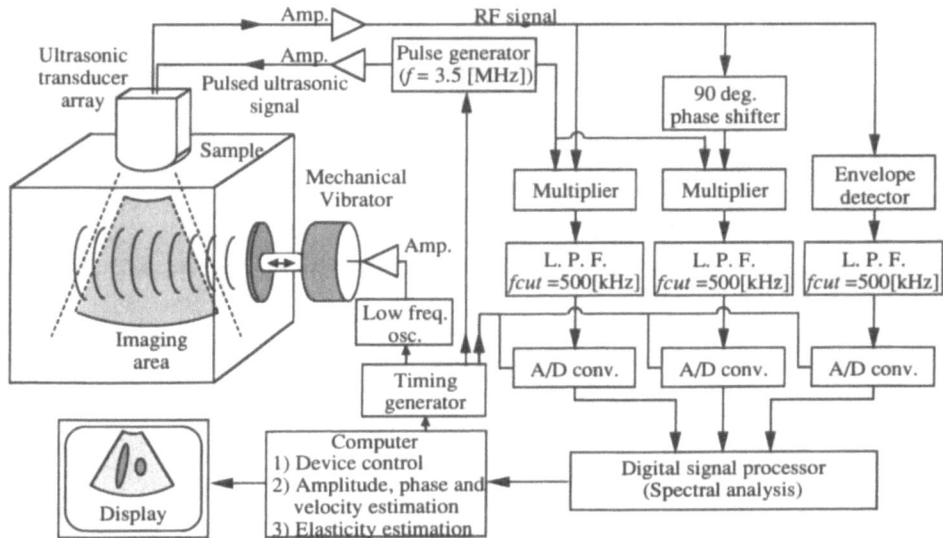

Fig. 4. Scheme of measurement and signal processing of the hardness mapping system

OBTAINED HARDNESS MAPS

Agarose phantom

The hardness of a prepared agarose phantom which consists of two kinds phantoms with different hardness was measured as shown in Fig. 5. On applying forced vibrations, existence of secondary waves such as standing waves are very common in observations of human body, where conditions meet for existence of various reflected waves. As the observed objects such as limb muscles and internal organs have approximately the same dimensional order with the vibration wavelength, and are contacted to adjacent organs with various boundary conditions, various reflected waves can exist. Presence of reflected waves and standing waves can cause false characterization of the tissue. However, complicated structures of the inner boundaries make it hard to anticipate and to eliminate false characterization caused by the reflected and standing waves.

Here, multiple hardness maps obtained by applying forced vibrations of different frequencies are averaged so that the effect of standing waves which tends to show false hardness features as seen in Fig. 5(a)~(d), can be oppressed. It is seen that the hardness difference is clearly mapped in the averaged hardness map.

In vivo human thigh

In Fig. 6, the measurement setup, the imaging area and the obtained results of *in vivo* human thigh hardness mapping with various loads at the ankle are shown. Vibration frequency scanning and averaging is also employed for these hardness maps. As it has been reported in a similar approach in [4], it can be seen that the increase of mean hardness is detected as the load at the ankle increases. Also, the lateral displacement of the active muscle is observed as the load is increased.

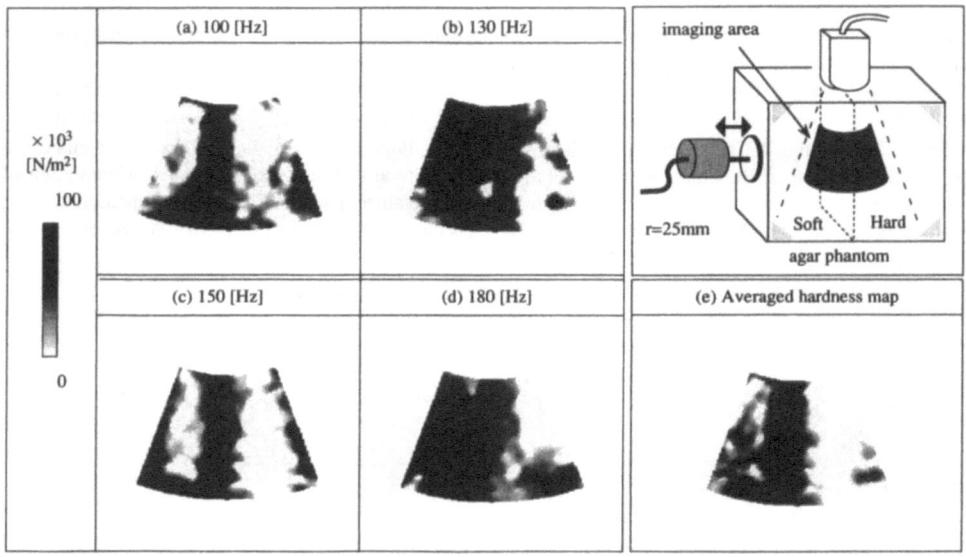

Fig. 5. Measurement setup of the agarose phantom hardness. (a)~(d) : The hardness maps obtained by each vibration frequency. (e) The averaged hardness map.

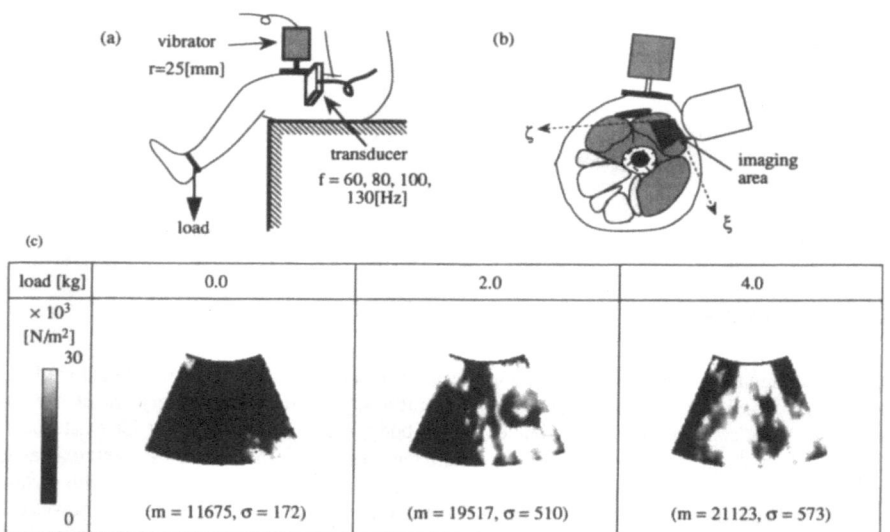

Fig. 6. (a) Measurement setup of the *in vivo* human thigh hardness. (b) The imaging area (cross section). (c) Hardness maps for various loads at the ankle.

REFERENCES

[1] Y. Yamakoshi, J. Sato and T. Sato "Ultrasonic Imaging of Internal Vibration of Soft Tissue Under Forced Vibration" IEEE Trans. UFFC, Vol. 37, No. 2, pp.45–53 (1990).

[2] H. L. Oestreicher, "Field and Impedance of an Oscillating Sphere in a Viscoelastic Medium with an Application to Biophysics" J. Acoust. Soc. Am., Vol. 23, No. 6, p. 707 (1951).

[3] V. Y. Kazakov and B. N. Klochkov, "Low Frequency Mechanical Properties of the Soft Tissue of the Human", Biophysics Vol. 34, No. 4, pp. 742–747 (1992).

[4] S. F. Levinson, M. Shinagawa, T. Sato, "Sonoelastic Evaluation of Dynamic Skeletal Muscle Elasticity", Ultrasonic Imaging Vol. 13, No. 2, p. 196 (1991).

SPECKLE DEFINITIONS

A. J. Healey and S. Leeman

Department of Medical Engineering and Physics
Kings College School of Medicine and Dentistry
Dulwich Hospital, East Dulwich Grove
London SE22-8PT, U.K.

INTRODUCTION

The coherent nature of the ultrasound imaging pulse allows for the possibility of interference effects and the appearance of the speckle artefact which occurs in virtually all medical ultrasound signals. Speckle is often regarded as noise, and as such has the potential to compromise many of the interpretations of ultrasound signals. This paper addresses the specific problem of defining and reducing the effects of speckle in ultrasound pulse-echo B-mode images. The reduction of speckle may be usefully regarded as a two stage procedure, consisting of (*i*) recognition — is the texture present in a clinical image segment the result of real (resolvable) structures or the speckle artefact?, and (*ii*) correction — what would the image look like in the absence of speckle?

If speckle is to be regarded as noise, say, then the difference between the actual signal and signal free of speckle (speckle-free), may be used to *define* the 'speckle' at a point in the signal. Any 'speckle reduction' technique may be evaluated in terms of its success in reducing the magnitude of the speckle, defined in this way, - either locally, or globally. Clearly, in order to implement such an evaluation, a precise specification of the speckle-free signal is of paramount importance. Remarkably, no universally agreed definition exists. Indeed, in the seminal paper on speckle in ultrasound images, by Burckhardt [1978], at least two definitions of the speckle-free signal are proposed, one based on an averaging procedure and another on maximum writing.

This problem is highlighted by consideration of a simple one dimensional convolutional model of rf A-line generation, see figure 1. Here an exact knowledge of the imaging pulse, scattering structure, and interaction between the two is known, and yet there is an ambiguity as to the exact nature of the speckle-free signal. The problem of defining speckle precisely is fundamental to all speckle reduction procedures in determining: (*i*) their operation and implementation; (*ii*) evaluating their effectiveness; (*iii*) to avoid the introduction of artefactual structure; and (*iv*) to minimise any loss of 'real' (diagnostically relevant) information.

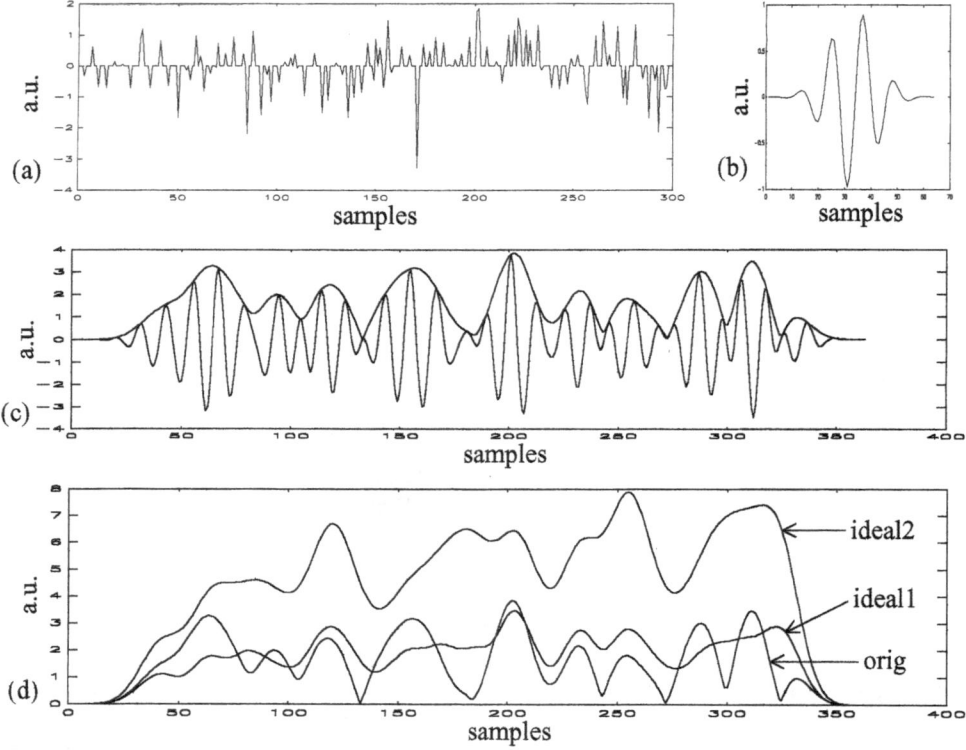

Figure 1. The speckle definition problem :- given the scatterer sequence in (a) and the pulse in (b) producing the rf A-line signal (c) what is the form of the signal devoid of any speckle component? ; (d) orig - original envelope, ideal1 & ideal2 - two attempts to define speckle given via Equations 4 & 5 respectively.

In this paper the problem of defining ultrasound speckle is analysed afresh. The conventional approaches towards speckle are defined in terms of a 'random phase' model. Frequency and angle compounding are analysed in terms of the simple convolutional model. The important problem of speckle reduction on a single image basis is considered, and a fundamental uniqueness problem is highlighted. A method of overcoming this uniqueness problem is proposed in terms of an 'effective' scatterer set. This forms the basis of an alternative approach towards defining and correcting speckle on a single image basis.

THE ROUGH VOLUME MODEL

Ultrasound speckle may be described via the consideration of a 'rough volume' model (based on the rough surface model in laser optics), i.e. one comprised of many point scatterers per resolution cell. The large number of scatterers provides the opportunity for coherent interference between overlapping echoes, sometimes being predominantly destructive in nature, and sometimes predominantly constructive. The resultant variable nature of the signal is associated with the speckle artefact, and the latter is hence regarded as having its origins in interference effects. At each time instant, the signal may be regarded as a phasor sum, with each component in that sum being a contribution from the many overlapping echoes at that point. The description of such a signal structure is analogous to that of the well-known random walk problem, and, in this way, the statistical properties of the backscattered echoes may be derived.

If the amplitude and phase of the jth contributing component are denoted by P_j and φ_j, respectively, then the (complex) value of the random phasor sum, at any temporal instant in the signal, may be expressed as:

$$A \exp(i\Theta) = \sum_{j=1}^{n} P_j \exp(i\varphi_j) \qquad (1)$$

The term 'fully developed' speckle is given to the situation where φ_j is uniformly distributed between 0 and 2π, and where the central limit theorem may be applied to the components P_j. In fact, the precise condition imposed is that the P_j be normally distributed. In a more general case, P_j may conform to any statistics, it being sufficient that n be large enough for the central limit theorem to be applicable. However, it should be noted that these conditions are sufficient in order to recover Burckhardt's results — they are not necessary in order to recover the characteristic (and definitive) statistical features of 'fully developed speckle', which is, in practice, recognised as indicated below.

In order to devise an algorithm for correct speckle suppression, a definition of the speckle-free signal is required which encompasses both the 'fully' and 'partially developed' cases. Indeed, the latter may, in some instances, be the more interesting problem, as some potentially resolvable structure exists in the image, which may be masked by the presence of speckle. In terms of Burkhardt's approach, partially developed speckle may be understood to refer to a situation for which: φ_j is non-uniform between 0 and 2π; the central limit theorem may not be applied; n is small; or n is random. The obvious question arises as to which conditions produce *no* speckle component in the signal. In the following section, two possibilities (as proposed by Burkhardt [1978] for fully developed speckle only) are extended to the general case: one in which an average (interference) signal level is taken; and, alternatively, when the signal is constrained to take on its maximum possible level. This more generalised approach towards speckle is also couched in terms of a random phase model.

THE RANDOM PHASE MODEL

The theory will be developed for the case that the (analytic) rf A-line signal, $s(t)$, is obtained as the result of a convolution between a strictly bandlimited and analytic pulse, $p(t) = b(t)\exp(i\omega t)$ (where b is the pulse shape and ω is the centre frequency), and a wide-band 'scatterer sequence' comprised of a finite number, N, of point scatterers with (complex) amplitudes, a_j. Expressing the scatterer sequence as $r(t) = \sum_{j=1}^{N} a_j \delta(t - t_j)$, it follows that the echo signal may be written

$$s(t) = \int_{-\infty}^{\infty} p(\tau)r(t-\tau)d\tau \equiv p(t) \otimes r(t) \qquad (2)$$

Different realisations of $s(t)$ may then obtained via the incorporation of a random phase component, $\exp(-i\Phi_j)$, associated with each of the a_j in $r(t)$: such a procedure essentially defines the random phase model. As the magnitudes and positions of the scattering structures remain constant from realisation to realisation, differences in the signal envelopes may be attributed to a change in the speckle noise. All possible realisations, $s_j(t)$, together define a set of random signals, S, with each member encoding the same 'structure', (a_j, t_j), but exhibiting a different speckle-corrupted envelope, $|s_j(t)|$.

Since the image displays the signal envelope, one approach [Burkhardt, 1978] is to define the speckle-free signal as the expectation value of the envelopes of the realisations in S. For computational convenience, however, it is more appropriate to define the speckle-free signal as the RMS average of the signal envelopes, viz. $\sqrt{\Xi\{|s_j|^2\}}$, where $\Xi\{..\}$ denotes the expectation value. The square of the envelope is then given by

$$|s(t)|^2 = s(t)s^*(t)$$

$$= \int_{-\infty}^{\infty} p(\tau)\sum_{j=1}^{N} [a_j\delta_j(t-t_j-\tau)\exp(-i\Phi_j)]d\tau \int_{-\infty}^{\infty} p^*(\tau')\sum_{k=1}^{N} [a_k^*\delta(t-t_k-\tau')\exp(i\Phi_k)]d\tau'$$

$$= \sum_{j=1}^{N} [a_j p(t-t_j)\exp(-i\Phi_j)] \sum_{k=1}^{N} [a_k^* p^*(t-t_k)\exp(i\Phi_k)] \tag{3}$$

If Φ_j is uniformly distributed between 0 and 2π, then, because

$$\Xi\{\exp(-i\Phi_j)\exp(i\Phi_k)\} = \delta_{jk}$$

the speckle-free signal is given by

$$s_R(t) = \sqrt{\Xi\{|s(t)|^2\}} = \left\{\sum_{j=1}^{N} |a_j|^2 |p(t-t_j)|^2\right\}^{\frac{1}{2}} = \left\{|p(t)|^2 \otimes |r(t)|^2\right\}^{\frac{1}{2}} \tag{4}$$

Constructive and destructive interference may be defined as the respective positive and negative differences between the envelopes of a speckle free signal and a particular realisation. Note that the speckle-free signal is formulated by removing the possibility of interference between convolution components. The speckle definition implied in Eq. 4 is general in that it applies to partially- as well as fully- developed speckle. Moreover, it conforms to the usual condition for incoherent superposition, viz. $E = \sum_i E_i$, where E is the total energy, at a temporal point in the signal and E_i is the contributing energy from the ith component.

An alternative definition for the speckle-free signal was also suggested by [Burckhardt, 1978], whereby the maximum value of the set, S, determines the envelope value of the desired output. Hence, the speckle-free signal, $s_M(t)$, is defined as l.u.b.$\left\{\sum_{j=1}^{n} P_j\exp(i\varphi_j)\right\}$, and is given by

$$s_M(t) = |p(t)| \otimes |r(t)| \tag{5}$$

Note that this definition for the speckle-free signal is also a general one. Even though it is not directly equivalent to the previous definition, $s_R(t)$, it is similarly based on a convolution between appropriate functions of the pulse and structure, thereby removing the possibility of interference between convolution components. The maximum writing approach, as embodied in Eq. 5, has a strong point in its favour, inasmuch as it makes fewer demands on the structure of the original signal, in order to arrive at the desired result.

Equations 4 and 5 provide two distinct definitions for the speckle-free signal, and it is important to appreciate that they lead, in general, to different end-points: the 'true' structures (without speckle) defined by Eqs 4 and 5 are not necessarily identical (Fig. 1). But they do both encode (resolution limited) information about the parameters which define the structure, viz. $\{a_j\}$ and $\{t_j\}$. This highlights the critical need for precise and consistent definitions if attempts are to be made to assess speckle correction methods in a meaningful way.

The above arguments show that the 'correct' (speckle-reduced) envelope signal is not known, even for the simplified 1D convolutional model: consequently, there is some ambiguity about the way in which speckle reduction procedures should be evaluated. Indeed, even the fundamental assumption (incorporated implicitly in a large number of single image speckle-reduction techniques) that, for fully developed speckle, the speckle free signal is a uniform grey level is not necessarily compatible with the definitions provided by Eqs. 4 and 5 [Healey and Leeman, 1993].

COMPOUNDING TECHNIQUES

In practice the random phase component is generated via the variation of one or more imaging parameter(s) in order to obtain (perhaps only partially) uncorrelated speckle realisations. Common methods for producing image sets are variation of: frequency of transmitted pulse; view angle; and receiver aperture, among others. The choice of parameter(s) may affect the distribution of the random phase components, Φ_j, which may no longer be uniform over the range 0 to 2π.

Frequency compounding

This technique produces an image set via the variation of the centre frequency of the imaging pulse. Hence, $|s(t)|^2$ becomes

$$\int_{-\infty}^{\infty} b(\tau)\exp(i\omega\tau)\sum_{j=1}^{N}[a_j\delta_j(t-t_j-\tau)]d\tau \int_{-\infty}^{\infty} b^*(\tau')\exp(-i\omega\tau')\sum_{k=1}^{N}[a_k^*\delta(t-t_k-\tau')]d\tau'$$

Averaging over all frequencies results in

$$\lim_{L\to\infty}\frac{1}{2L}\int_{-L}^{L}\sum_{j=1}^{N}[a_jb(t-t_j)\exp(i\omega(t-t_j))]\sum_{k=1}^{N}[a_k^*b^*(t-t_k)\exp(-i\omega(t-t_k))]d\omega \qquad (6)$$

Since $\lim_{L\to\infty}\frac{1}{2L}\int_{-L}^{L}\exp(i\omega t_j)\exp(-i\omega t_k)d\omega = \delta(t_j-t_k)$, frequency compounding and the random phase model possess equivalent speckle-free image definitions (in terms of the simple 2D convolutional model).

Angle compounding

This approach produces an image set via the variation of view angle (and hence the angle of incidence of the illuminating pulse). Angle compounding requires a large acoustic window and thus its applicability is limited to relatively few clinical sites, such as the breast. The speckle-free image associated with angle compounding, $s_A(\mathbf{r})$, is not equivalent to that of frequency compounding, in the random phase model. This may be demonstrated by considering a structure component comprising two scatterers, of amplitudes a_1 and a_2, and located at $\mathbf{r}_1 = 0$, and $\mathbf{r}_2 = \Delta\hat{\mathbf{x}}$. The 2D (complex) image is given by

$$s_2(\mathbf{r}) = a_1 b(\mathbf{r})\exp(-i\mathbf{k}\bullet\mathbf{r}) + a_2 b(\mathbf{r}-\Delta\hat{\mathbf{x}})\exp(-i\mathbf{k}\bullet\{\mathbf{r}-\Delta\hat{\mathbf{x}}\}). \qquad (7)$$

Thus,

$$\begin{aligned}
s_A(\mathbf{r}) \equiv \sqrt{\Xi\{|s_2(\mathbf{r})|^2\}} = [&|a_1|^2 b^2(\mathbf{r}) + |a_2|^2 b^2(\mathbf{r}-\Delta\hat{\mathbf{x}}) \\
&+ 2.Re\{a_1 a_2\}.b(\mathbf{r})b(\mathbf{r}-\Delta\hat{\mathbf{x}}).J_0(k\Delta)]^{1/2}
\end{aligned} \qquad (8)$$

where the third term, containing the zero-order Bessel function, J_0, may be regarded as an interference term. Note that the presence of the interference term causes the speckle free signal to depend on $k \equiv |\mathbf{k}| = \omega/c$. On the other hand, for pulses of the type considered here, the definitions in Eqs. 4 and 5 imply that if an object is imaged with pulses of two different centre frequencies, but identical envelopes, then the speckle-free signals are identical. This has the troublesome implication that if an object is imaged twice, with pulses of two centre frequencies (which have identical envelopes), say at 2 and 20 MHz the speckle free signals associated with these two images are identical. However, it may seem reasonable to expect some difference in these images which is not attributable to speckle. Angle compounding does not suffer this inconsistency, but contains interference terms, and hence does not conform to the usual incoherent image definition. Hence the speckle-free signals provided by angle and frequency compounding are not without problem.

THE SINGLE IMAGE PROBLEM

Commonly, the acquisition of a full image set in order to effect speckle reduction is not practicable and the speckle reduction options become limited to those techniques which are applicable with a single image (for a brief overview of methods, see Bamber [1992]). Conventional techniques regard speckle as (usually multiplicative) noise with some predictable statistics, and a number of techniques exist which essentially trade some other factor of image quality, such as resolution, for speckle reduction. The merits of particular approaches may be considered in terms of the type and efficiency of trade-off involved. A level of operator interaction may even be introduced to establish the exact level of trade-off desired.

Frequency diversity

This technique (Magnin et al, 1982) attempts a degree of frequency compounding by splitting the received rf signal into a number of overlapping frequency bands and compounding the corresponding envelope signals. It represents an attempt to construct the frequency compounding speckle-free signal as defined by Eq. 4. An extension of the method is also possible in two-, and even in three-, dimensions.

Adaptive filtering

As implemented, these techniques tend to rely on statistical parameters to recognise the local presence of speckle. For the adaptive unsharp mask filter, correction of the envelope signal is based on the *assumption* that, for fully developed speckle, the 'correct' output is

264

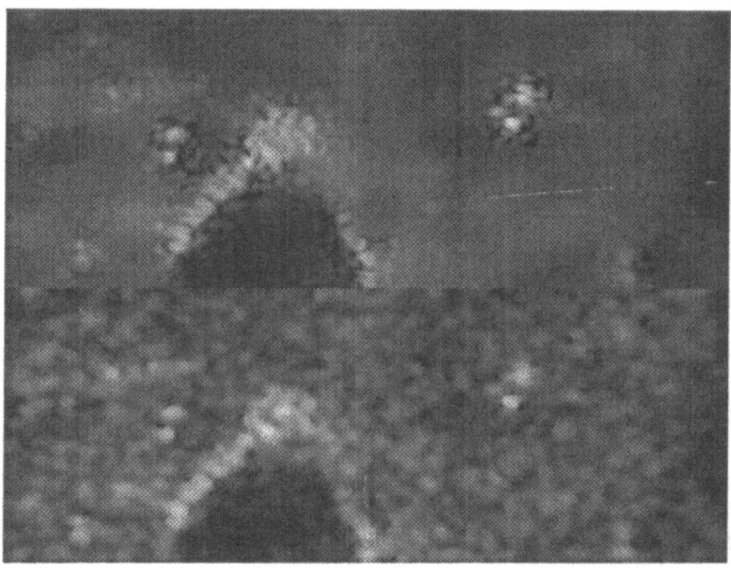

Figure 2. Results of adaptive unsharp mask filtering applied to *in vivo* data, with a kernel size of (above) 35x35 pixels, and (below) 7x7 pixels. Which image more accurately depicts the 'true' structure, and is closer to the 'correct' speckle-free signal?

the local mean value. This procedure is justified by its advocates on the basis that the only statistically relevant parameter associated with a fully developed speckle region is the mean level. In a situation of recognised partially developed speckle, the signal is smoothed less agressively than for fully developed speckle, and in a situation of no speckle, the signal is left unaltered. A number of problems arise as to the exact nature of the implementation of these techniques, since a number of parameters, such as the filter kernel shape and size, need to be fixed. Their exact values are not explicitly predicted by theory, but semi-quantitative methods to fix them have been suggested [Healey and Leeman, 1993]. Different values of these parameters can lead to non-trivial differences in the filter outputs — an example being given in Fig. 2.

One consistency requirement for such two stage techniques is that the quantity used for speckle recognition may be employed to judge the effectiveness of the reduction process, i.e. provide an evaluation index. Surprisingly, the image features commonly used for recognition, such as SNR and the second moment M2, can actually show an *increase* after the image has been processed [Healey and Leeman, 1993]. Such problems demonstrate the practical importance of the need for precise speckle definitions.

THE UNIQUENESS PROBLEM

The strictly bandlimited nature of the rf signal, which, in the convolutional model, is enforced by the imaging pulse, implies that many different structures can give rise to *identical* rf A-line signals. In terms of the simple 1D convolutional model the bandlimited imaging pulse may be regarded as providing a Fourier domain 'window' (between ω_l and ω_h respectively) onto the wideband structure spectrum [Hutchins and Leeman,

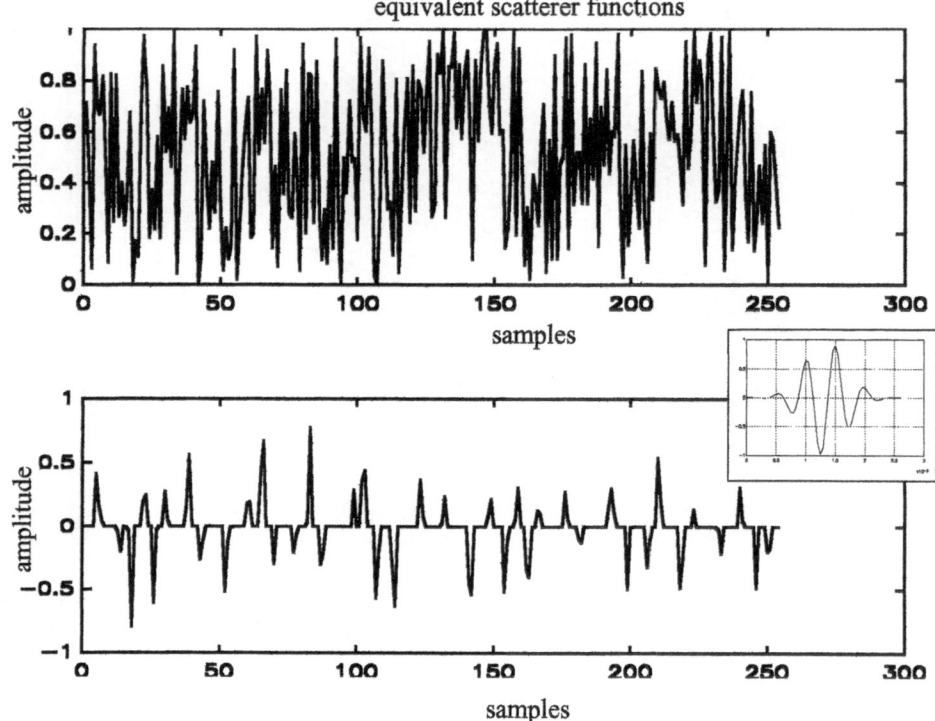

Figure 3. Two equivalent scatterer sequences that produce *identical* rf A-line signals when convolved with the imaging pulse (insert).

1982]. An infinite set of structures may be produced that are identical over the finite bandwidth of the imaging pulse, but not so outside this band. Each member of such a set produces an identical rf A-line signal when convolved with the strictly bandlimited imaging pulse. A fundamental question arises as to which specification of the structure to marry with the definition for the speckle-free signal. The uniqueness problem is more prominent for speckle reduction techniques that operate on a single image basis, since multi-image techniques have the potential to (partially) resolve some of this ambiguity.

Effective scatterers

In order to overcome the structure uniqueness problem in a consistent manner it is useful to introduce the notion of the 'effective' scatterers associated with a particular rf A-line signal. These may be loosely regarded as the minimum 'strength' of scattering required to produce the rf A-line signal. More formally, we propose that, for the finite length, discrete, simple convolutional model

$$s[n] = p[n] \otimes r[n] \quad ; \quad n = 0, 1, ..., N-1 \tag{9}$$

(where $\otimes$ denotes circular convolution) the effective scatterer sequence, $r_e[n]$, may be calculated from the parameter set θ_m^*, which is obtained from a L_1-minimisation of the function

$$\varepsilon(\theta_m) = \sum_{n=0}^{N-1} |r_b[n] + r_{ob}[n, \theta_m]| \tag{10}$$

266

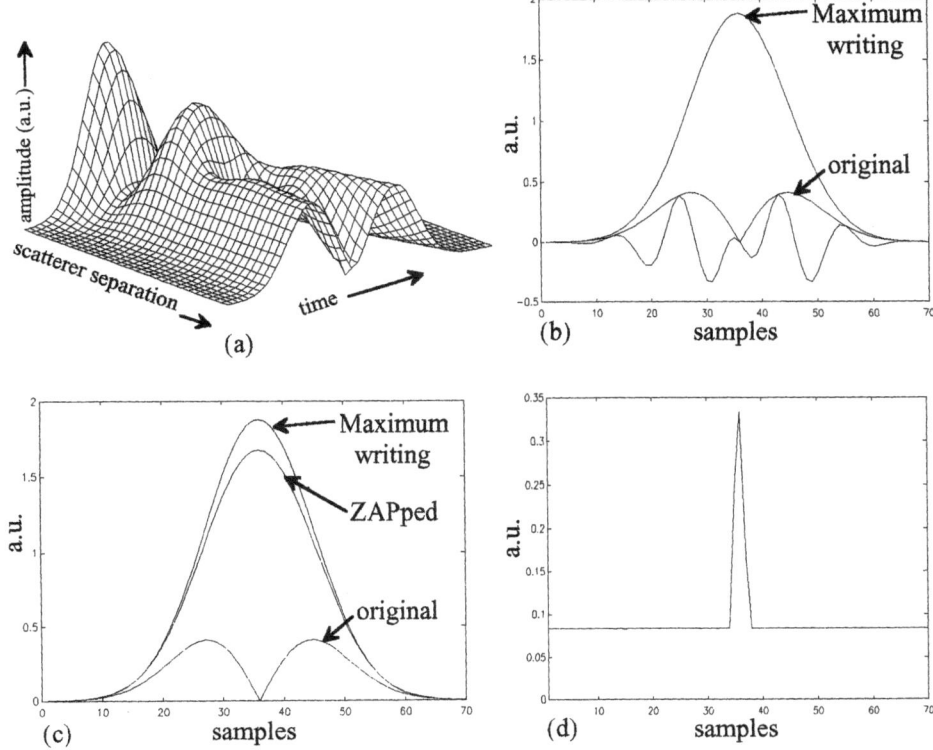

Figure 4. (a) envelope of sliding two scatterer simulation; (b) rf A-line signal at a separation of 1/2 wavelength; (c) results of ZAPping; (d) instantaneous frequency associated with the rf A-line signal in (b).

where $p[n]$ represents the analytic pulse; $r_b[n]$ the ideal bandpass ($\omega_l < \omega < \omega_h$) component of $r[n]$; and $r_{ob}[n, \theta_m]$ the ideal notch ($-\pi < \omega < \omega_l$; $\omega_h < \omega < \pi$) component of $r[n]$. The (finite) parameter set $\theta_m = \{\theta_n | n = 1, 2, ..., m\}$ has elements which are the discrete Fourier transform coefficients associated with the frequency ranges $-\pi < \omega < \omega_l$ and $\omega_h < \omega < \pi$.

Thus, for any particular image, the many random scatterers per resolution cell required to generate a fully developed speckle pattern have an associated effective scatterer distribution that consists of a much smaller number of elements (Fig. 3). The effective scatterer concept, defined as above, resolves the ambiguity problem by selecting the minimum 'strength' scattering structure consistent with the measured data. Note that the above procedure for determining the effective scatterers is applicable only in the context of a spatially invariant, noise-free, one dimensional convolutional model.

A NEW DEFINITION FOR SPECKLE

The effective scatterer concept makes it possible (within the model restrictions) to devise an unambiguous maximum writing speckle reduction scheme that can be applied on a single image basis. The speckle-free signal is defined in terms of Eq. 5, but with $r(t)$ replaced by the effective scatterer sequence. The consideration of a simple two-scatterer model [Healey et al, 1991] reveals that a close approximation to the speckle-free signal,

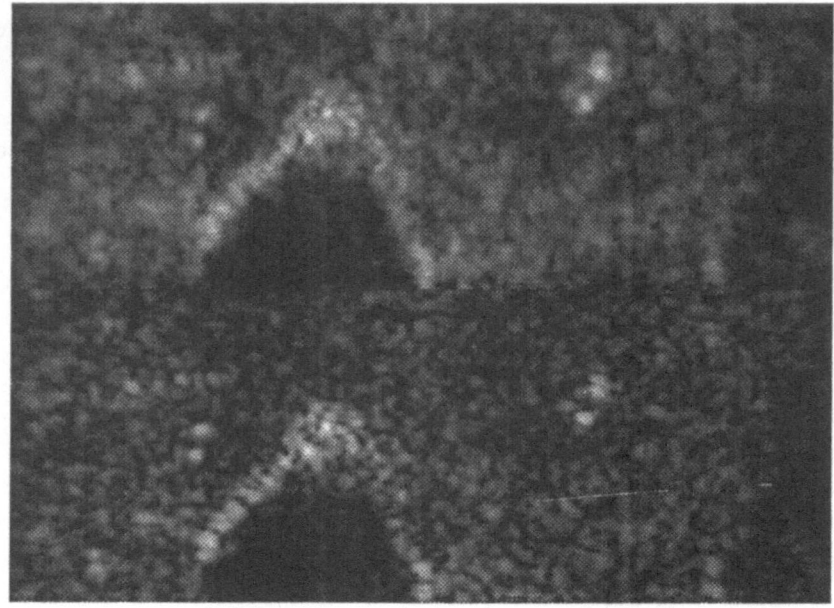

Figure 5. Top - zero adjusted image; bottom - original image.

defined in this way, may be obtained *without* explicit calculation of either the parameter set, θ_m, or the pulse $p[n]$.

In order to highlight the features of the new speckle definition it is informative to consider its operation in the simplest interference situation. Consider, therefore, the situation where two unit scatterers are imaged while the separation distance between them is varied (Fig. 4). When the separation is equal to a multiple of the pulse wavelength, λ, total constructive interference occurs. The resultant signal contains no speckle component according to Eq. 5, but possesses substantial speckle according to Eq. 4. At a separation distance of $\lambda/2$, severe destructive interference occurs, and a large speckle component is introduced, according to both Eqs. 4 and 5. An instantaneous frequency excursion occurs at the centre of destructive interference, and is readily detected. Zero Adjustment Processing ('ZAPping') [Healey *et al*, 1991] attempts to correct for the signal loss by inserting the maximum writing envelope. Note that the correction stage always adds a positive envelope correction. Zero Adjustment Processing appears to be the only algorithm able to form an envelope correction in such an additive way.

Results

An example of the correction is shown in Fig. 5, for a segment of a clinical image (liver). An important feature of the new approach is that no inherent resolution trade-off is demanded by the correction procedure, which is based on the 'effective scatterer definition' for speckle. One consistency requirement is not maintained (as with other single image techniques) in that the effective scatterer set changes with view angle and hence the speckle-free signal changes with view direction.

CONCLUSIONS

In order to effect some form of correction for speckle in pulse-echo B-mode images, a precise definition of the artefact is required. Surprisingly, a number of conflicting definitions exist in the literature. Single image techniques suffer a uniqueness problem, which leads to conventional reduction procedures being based on statistical considerations. A deterministic single image speckle reduction technique has been proposed which overcomes the uniqueness problem by the introduction of an effective scatterer set, and does not suffer from the usual resolution trade-off. Methods exist to implement this approach efficiently with rf data.

ACKNOWLEDGEMENTS

The Scientific and Engineering Research Council, and the Wellcome Trust are gratefully acknowledged for their financial support.

REFERENCES

Bamber J. C., 1992, Speckle Reduction, in : *Advances in Ultrasonic Techniques and Instrumentation*, ed. P. N. T. Wells, Churchill Livingstone, New York.

Burckhardt C. B. 1978, Speckle in Ultrasound B-mode Scans, *IEEE Trans. Sonics and Ultrasonics* **SU-25,** 1-6.

Magnin P. A., von Ramm O. T., and Thurstone F. L., 1982, Frequency compounding for speckle contrast reduction in phased array images, *Ultrasonic Imaging* **4**, 267-81.

Healey A. J., Leeman S. and Forsberg F., 1991, Turning off speckle, *Acoustical Imaging*, eds Ermert and Harjes, **19**, 433-436.

Healey A. J. and Leeman S., 1991, Zeroing Out Speckle, *Proc. 6th World Congress in Ultrasound, WFUMB*.

Healey A. J. and Leeman S., 1993, Speckle Reduction Methods in Ultrasound Pulse-echo Imaging, *IEE International Conference on Acoustical Sensing and Imaging*, **369**, 68-76.

Hutchins L. and Leeman S., 1982, Pulse and Impulse response in human tissues, *Acoustical Imaging* **12** eds. Ash E A and Hill C R, pp 459-67.

ALGEBRAIC APPROACH TO ULTRASONIC B-MODE IMAGING FOR RESOLUTION ENHANCEMENT

Xiao-Liang Xu, Hehong Zou, and James F. Greenleaf

Biodynamics Research Unit
Department of Physiology and Biophysics
Mayo Clinic and Foundation
Rochester, MN 55905

ABSTRACT

In this paper we investigate an iterative algebraic algorithm for resolution enhancement in ultrasonic B-mode imaging. The algorithm, called iterative back reconstruction (IBR), considers the measured signal as the initial "scattering image", and in each iteration it forms an updated image by adding a correction term to the old one. The correction term is calculated from a system model, processing operators, and the old image. In addition to fast convergence, IBR can be applied either to the measured *rf*-signal or to the detected envelope (either logarithmic-compressed or uncompressed) and can also consider the 2-dimensional spatially variant nature of the point spread function. In our preliminary simulations, it is shown that with only three iterations the IBR algorithm effectively enhances the resolution or the dynamic range without boosting image noise.

INTRODUCTION

B-mode ultrasound, a pulse-echo imaging technique, is a widely used technique in medical diagnostics to produce cross-sectional images of biological structures especially soft-tissues within the body. Although their quality has been improved significantly since the last decade, the B-mode ultrasonic images suffer from low resolution (especially in the lateral direction) and low contrast.

Many digital signal processing techniques have recently been applied to improving the image quality including the deconvolution technique via Wiener filtering (e.g., [1-7]). The deconvolution technique via Wiener filtering assumes the measured signal to be the convolution of the reflection coefficient describing the object and the point spread function describing the shape of the ultrasonic pulse. It has been shown by a number of papers (e.g., [1,2,3,7]) that under certain assumptions and conditions this technique achieves resolution enhancement without increasing noise. However, its performance is affected or limited by the following factors: 1) the point spread function is typically spatially variant, 2) signal and noise spectra are usually unknown, 3) envelope-detection of the *rf*-signal and the logarithmic compression are not linear operations.

This paper investigates the algebraic approach to ultrasonic B-mode imaging for resolution enhancement. Algebraic imaging techniques provide images by solving a set of equations modeling the measured signal. Many iterative algebraic algorithms have been developed in the digital signal processing field for image restoration and tomographic reconstruction. Some of them can be applied to ultrasonic imaging with slight modifications. An advantage of the algebraic approach when applied to ultrasonic images is its ability to model a spatially variant point spread function and nonlinearity in the imaging system. If the system can be modeled accurately, the images obtained by an algebraic algorithm possesses high resolution and may have low noise if the iterating stops within a small number of iterations and/or if an appropriate regularization scheme is employed.

We particularly focus on an iterative algebraic algorithm called the iterative back reconstruction (IBR) algorithm. The IBR method was originally developed for computed tomography, and it was shown that it improved the quality of tomographic images in terms of image resolution and noise.[9] The method has certain attractive features: it is easy to implement, it converges quickly and it is flexible. In next sections, the IBR algorithm is described, results are shown in comparison with regular B-mode images, and then we discuss the results.

BACKGROUND

In this paper we use x to denote the lateral axis and z the axial axis. Under certain assumptions the received *rf*-signal is a linear function of the reflection coefficient of the object, and it can be expressed as a 2-D (2–dimensional) integral of the reflection coefficient function and a point spread function (PSF), given by

$$B_{rf}(x,z) = \int \int h(x,x',z,z')r(x',z')\, dx'\, dz' + n(x,z), \tag{1}$$

where $B_{rf}(x, z)$ is the received *rf*-signal, $r(x, z)$ is the reflection coefficient of object at (x, z), $n(x, z)$ is noise, and $h(x, x', z, z')$ is the PSF which models the linear spread function from the image space to the *rf*-signal space. In eq. (1) we have converted the time variable to z. The PSF $h(\cdot)$ depends on many factors including the transmitted pulse, transducer beam, and medium. $h(\cdot)$ is in general spatially variant, especially for conventional transducers. If z is limited to a small range around the focal plane of the beam, $h(\cdot)$ may be considered to be approximately spatially invariant, and thus eq. (1) is reduced to a 2–D convolution of the reflection coefficient $r(x, z)$ and the PSF $h(x, z)$, that is,

$$B_{rf}(x,z) = \int \int h(x-x',z-z')r(x',z')\, dx'\, dz'. \tag{2}$$

Many researchers (e.g., [2,4,6]) further simplified eq. (2) by modeling lateral and axial convolution separately or even by modeling only the lateral convolution.

The final image that represents $r(x, z)$ is formed by applying envelope-detection to the *rf*-signal $B_{rf}(x, z)$. Because of the finite aperture of the transducer and the band-limited transmitted pulse, the obtained ultrasonic images are blurred, suffering from low resolution (especially in lateral axis) and low contrast.

Classical deconvolution is one way to improve the image quality especially for resolution enhancement. In theory, deconvolution is only valid for the *rf*-signal. It is shown by a number of investigators that deconvolution based on Wiener filtering achieves resolution enhancement without increasing noise by performing deconvolution on the *rf*-signal and limiting z within a small range to make $h(\cdot)$ to be linear and spatially invariant. Recently, devolution has been applied to the envelope of the *rf*-signal or the logarithmically compressed envelope, but little success has been achieved.[4,6] In general, the performance of deconvolution for ultrasonic imaging is affected by one or more of the following factors: 1) the point spread function is typically spatially variant, 2) signal and noise spectra are usually unknown, and 3) envelope-detection of the *rf*-signal and the logarithmic compression are not linear operations.

This paper considers ultrasonic imaging as an estimation problem using a computer. Therefore we form the discrete version of eq. (1) by replacing the integral by summation, giving

$$\mathbf{b}_{rf} = \mathbf{Hr} + \mathbf{n}, \tag{3}$$

where $\mathbf{b}_{rf}$ denotes the column vector of the *rf*-signal formed by stacking samples from each A-line, $\mathbf{r}$ denotes the column vector of reflection coefficients formed in a manner consistent with $\mathbf{b}_{rf}$, $\mathbf{n}$ is the noise vector, $\mathbf{H}$ is a transition matrix whose components are determined by the PSF $h(\cdot)$. (We use bold lowercase to denote column vectors and bold uppercase to denote matrices.) Note that $\mathbf{H}$ can model 2–D spatial variations in $h(\cdot)$.

Let $\mathbf{b}_{en}$ denote the envelope of $\mathbf{b}_{rf}$, and it can be expressed as

$$\mathbf{b}_{en} = f_1(\mathbf{b}_{rf}), \tag{4}$$

where $\mathbf{b}_{en}$ represents the envelope of $\mathbf{b}_{rf}$, and f_1 represents the envelope detection, which is a nonlinear operation. In the conventional B-mode imaging system, $\mathbf{b}_{en}$ is the image representing the absolute value of the reflection coefficient of the object, i.e., the estimate of $|\mathbf{r}|$.

In commercial machines, the final image is often displayed in the form of a logarithmically compressed envelope of the *rf*-signal, The compressed envelope is given by

$$\mathbf{b}_{co} = f_2(\mathbf{b}_{en}), \tag{5}$$

where $\mathbf{b}_{co}$ is the compressed envelope, and f_2 represents the compression operation, which is also nonlinear.

METHOD

Consider ultrasonic imaging as a generic estimation problem. This paper addresses a method to estimate the reflection coefficient image $|\mathbf{r}|$ from the available data vector $\mathbf{b}$, where $\mathbf{b}$ can be the *rf*-signal, its envelope, or the logarithmically compressed envelope. $\mathbf{r}$ and $\mathbf{b}$ satisfy the following equation,

$$\mathbf{b} = f(\mathbf{r}) + \mathbf{n}, \tag{6}$$

where $\mathbf{n}$ is a noise vector and $f(\cdot)$ is a function representing one or more operations of $\mathbf{H}$, $f_1(\cdot)$ and $f_2(\cdot)$, depending on the form of $\mathbf{b}$. For example, if $\mathbf{b} = \mathbf{b}_{rf}$, then $f(\mathbf{r}) = \mathbf{Hr}$;

273

if $\mathbf{b} = \mathbf{b}_{en}$, then $f(\mathbf{r}) = f_1(\mathbf{Hr})$; if $\mathbf{b} = \mathbf{b}_{co}$, then $f(\mathbf{r}) = f_2(f_1(\mathbf{Hr}))$. We assume $\mathbf{H}$ and therefore $f(\cdot)$ can be accurately formed. The least square estimate of $\mathbf{r}$ can be obtained by minimizing the L-2 norm of the difference between $\mathbf{b}$ and $f(\mathbf{r})$, i.e.,

$$\min_{r} \| \mathbf{b} - f(\mathbf{r}) \|^2 . \tag{7}$$

There are many iterative algorithms to solve eq. (7), including the conjugate gradient method, steepest decent algorithm and the downhill simplex method.[8] If f is a linear operation, other simpler algorithms such as ART, SIRT and Landweber iteration can be used.[8,9] Unfortunately, for the problem of ultrasonic imaging these algorithms are either extremely complicated or computationally very expensive.

In this paper we investigate a specific iterative algorithm for resolution enhancement that has features of simplicity, relatively inexpensive computation and fast convergence. This algorithm was originally developed for tomography and termed "iterative filtered projection" (IFBP).[10] With necessary modification for the problem given by eq. (6), the algorithm takes the form

$$\widehat{\mathbf{r}}^{(k+1)} = \widehat{\mathbf{r}}^{(k)} + \eta \left(\mathbf{b} - f \left(\widehat{\mathbf{r}}^{(k)} \right) \right), \tag{8}$$

where η is a relaxation parameter and $\widehat{\mathbf{r}}^{(k)}$ is the estimate of $\mathbf{r}$ from the kth iteration. As an iterative algorithm for image restoration and reconstruction, the IFBP behaves like other iterative methods: as the number of iterations increases, the image resolution increases but the image noise also increases. It is often necessary to stop the iteration procedure early or apply a regularization scheme to each iteration or the final image to achieve image resolution enhancement without boosting image noise.

To avoid confusion we term eq. (8) as the iterative back reconstruction (IBR) algorithm in this paper. The IBR algorithm can be applied to the rf-signal, its envelope or the compressed envelope. For the rf-signal, the IBR iteration scheme becomes

$$\widehat{\mathbf{r}}_{rf}^{(k+1)} = \widehat{\mathbf{r}}_{rf}^{(k)} + \eta \left(\mathbf{b}_{rf} - \mathbf{H}\widehat{\mathbf{r}}_{rf}^{(k)} \right). \tag{9}$$

The final estimate of the reflection coefficient is given by

$$\widehat{\mathbf{r}} = C_{rf} \{ f_1 \left(\widehat{\mathbf{r}}_{rf} \right) \}, \tag{10}$$

where C_{rf} is an operator which includes a low pass filter and a regularization scheme (if necessary). Here $\widehat{\mathbf{r}}$ is an estimate of $|\mathbf{r}|$. If the IBR is applied to the detected envelope, the iteration becomes

$$\widehat{\mathbf{r}}_{en}^{(k+1)} = \widehat{\mathbf{r}}_{en}^{(k)} + \eta \left(\mathbf{b}_{en} - f_1 \left(\mathbf{H}\widehat{\mathbf{r}}_{en}^{(k)} \right) \right). \tag{11}$$

The final image is then given by

$$\widehat{\mathbf{r}} = C_{en} \{ \widehat{\mathbf{r}}_{en} \}, \tag{12}$$

where C_{en} is an operator that includes a non-negative constraint and a regularization scheme (if necessary). In the case, $\widehat{\mathbf{r}}$ is again an estimate of $|\mathbf{r}|$.

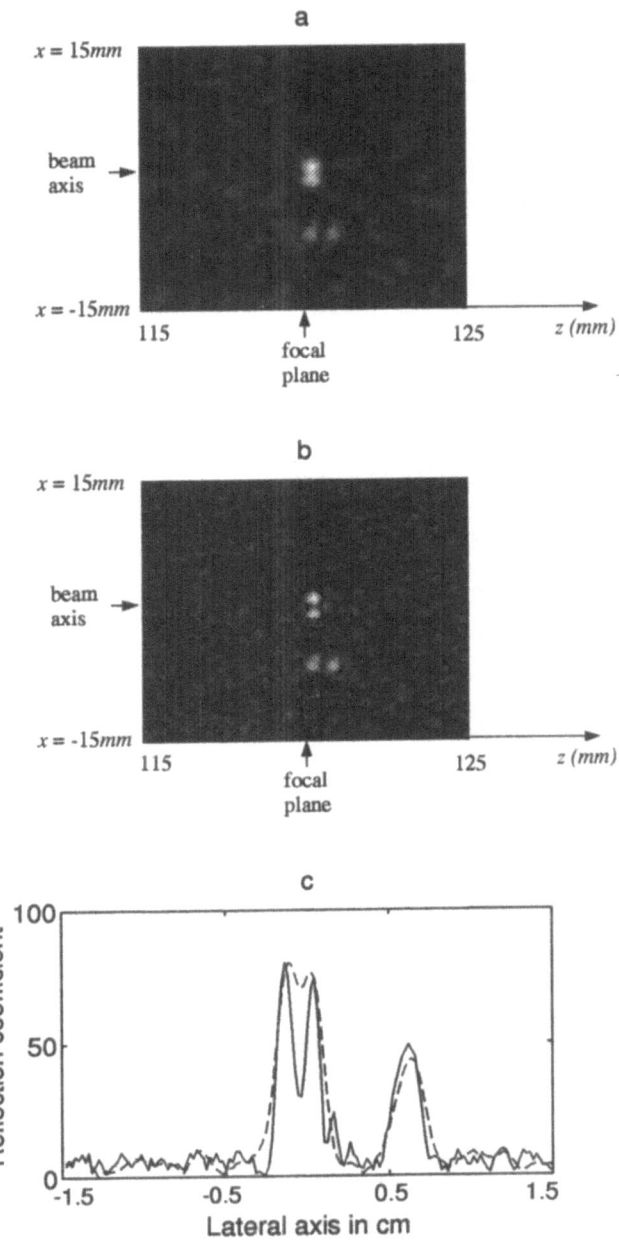

Figure 1 Images and lateral profiles from B-mode and the IBR applied to the *rf*-signal. (a) original uncompressed B-mode image, (b) the IBR image with three iterations and without regularization, (c) lateral profiles across the two closely spaced scatterers at the focal line of the two images (the solid-line for the IBR and the dash-line for regular B-mode).

For the logarithmically compressed envelope, we can first decompress the compressed envelope and then apply eqs. (11) and (12). But we can also apply the IBR algorithm directly to the compressed signal. In a manner similar to eqs. (11) and (12), we obtain an iteration scheme for the compressed envelope, which is given by

$$\widehat{\mathbf{r}}_{co}^{(k+1)} = \widehat{\mathbf{r}}_{co}^{(k)} + \eta\Big(\mathbf{b}_{co} - f_2\Big(f_1\Big(\mathbf{H}\widehat{\mathbf{r}}_{co}^{(k)}\Big)\Big)\Big). \tag{13}$$

The final image is given by

$$\widehat{\mathbf{r}} = C_{co}\{\widehat{\mathbf{r}}_{co}\}, \tag{14}$$

where C_{co} is a regularization operator. Here $\widehat{\mathbf{r}}$ is an estimate of $f_2(f_1(|\mathbf{r}|))$.

RESULTS

We have tested the IBR algorithm in a number of simulations using different types of transducers including Gaussian unfocused or focused transducers and a Bessel transducer[11] and using different forms of data including the rf-signal, its envelope and the logarithmically compressed envelope. Here we present only simulations that used a Gaussian focused transducer and applied the IBR iteration to the rf-signal, its envelope and the logarithmically compressed envelope, respectively. We compare the images obtained for the IBR algorithms with the regular B-mode images; all IBR images were obtained by performing only three iterations.

For all simulations, the electrical pulse was in the form of

$$f(t) = e^{-t^2/t_o^2} \sin\left(2\pi f_c t\right), \tag{15}$$

where $t_o = 0.4\,\mu s$ and $f_c = 3.0\,MHz$. The sound speed was assumed to be $c = 1.5\,mm/\mu s$. The sampling rate was $f_s = 16\,MHz$. We assume that a 2-D scatterer phantom was placed in attenuation free water; the phantom contained strong coherent scatterers that were closely spaced and weak diffusion scatterers and noise in the background. The diameter of the Gaussian focused transducer was 50mm, the FWHM of the Gaussian weighting on the transducer surface was 25mm, the focal distance was 120mm, and the FWHM at the focal plane was 2.1 mm. The phantom was placed in the focal region. We focused on a region that was 30mm wide and 10mm deep, and centered in the intersection of the focal plane and the axial axis. Thus, the region of interest was* $x \in (-15mm, 15mm)$, and $z \in (115mm, 125mm)$.

The results are demonstrated in Figs. 1, 2 and 3, where (a) and (b) are the regular B-mode image and the IBR image (the horizontal axis is the beam axis and the vertical axis is the lateral axis), and (c) contains lateral profiles of these two images. In particular, Fig. 1 shows the image of the IBR applied to the rf-signal in comparison with the image of the original uncompressed envelope; Fig. 2 shows results of the IBR applied to the envelope image in comparison with the original envelope image; Fig. 3 shows results of the IBR applied to the logarithmically compressed envelope in comparison with the original compressed envelope image. No regularization scheme was used in Figs. 1 or 2, while a sieve regularization scheme[9] (3×3 points smoothing) was used in Fig. 3. It is clearly

---

* In this paper we assumed that the origin of the coordinate was the center point on the transducer surface.

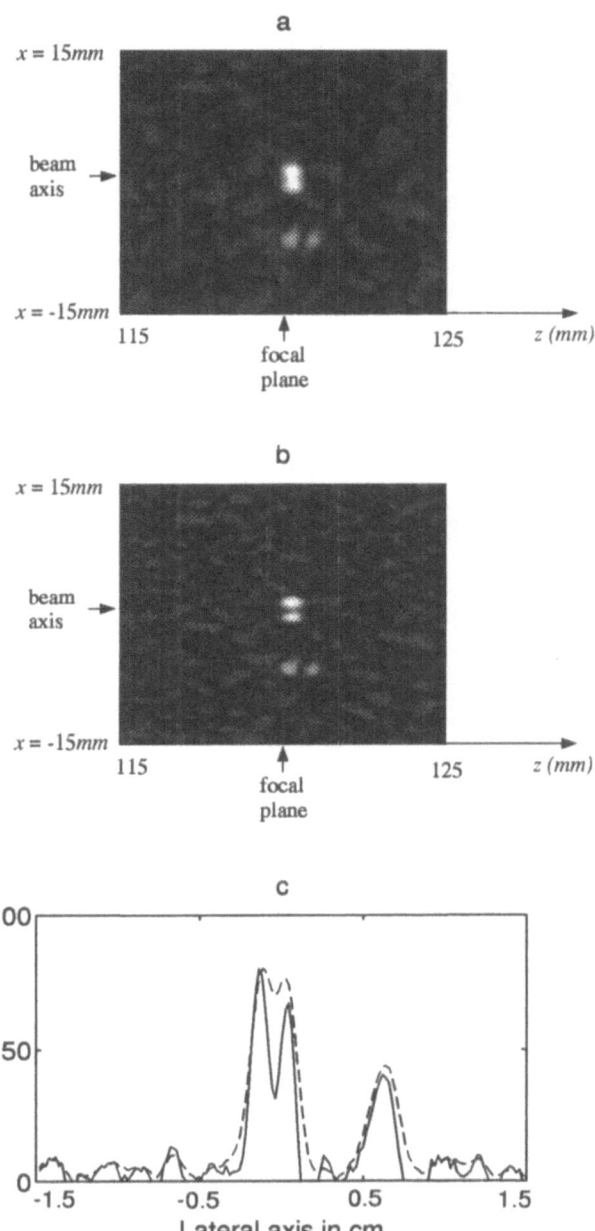

Figure 2 Images and lateral profiles from B-mode and the IBR applied to the envelope of the *rf*-signal. (a) original uncompressed B-mode image, (b) the IBR image with three iterations and without regularization, (c) lateral profiles across the two closely spaced scatterers of the two images (the solid-line for the IBR and the dash-line for regular B-mode).

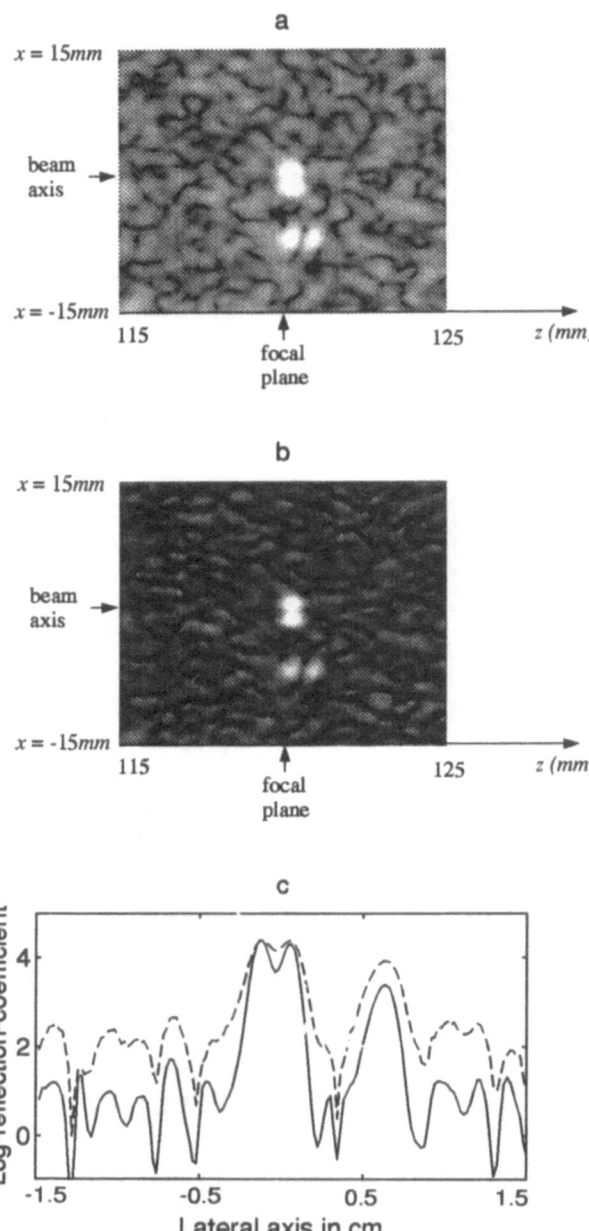

Figure 3 Images and lateral profiles from B-mode and the IBR applied to the logarithmically compressed envelope of the *rf*-signal. (a) original logarithmically compressed B-mode image, (b) the IBR image with three iterations and with a sieve regularization, (c) lateral profiles across the two closely spaced scatterers of the two images (the solid-line for the IBR and dash-line for regular B-mode).

shown that the IBR achieves resolution enhancement in Figs. 1 and 2 without boosting noise. Although resolution enhancement is not seen intuitively in Fig. 3, which was due to the regularization, we observed that the original IBR image has a better dynamic range than the B-mode image, where the dynamic range is defined as the ratio of the object peak to the root-mean-square value of the background.

DISCUSSION

From the simulations we found that the IBR algorithm applied to the envelope performs best. When it is applied to the *rf*-signal, its performance is more sensitive to phase error and noise. When it is applied to the logarithmically compressed envelope, the problem becomes more ill-conditioned, and therefore it converges more slowly and requires a regularization scheme.

Although it is assumed that the PSF $h(\cdot)$ is known or can be accurately estimated, we observed that the IBR algorithm still performed well, especially if applied to the envelope of the *rf*-signal, when an approximate of the PSF was used, e.g., the PSF of a specific point was used for an area (30*mm* wide and 10*mm* deep) in the neighborhood of this point.

CONCLUSION

We have studied the iterative back reconstruction (IBR) for resolution and contrast enhancement in B-mode imaging. The simulation results qualitatively show that IBR achieves resolution enhancement and/or dynamic range improvement without obvious increase in image noise.

In the future we will test the IBR algorithm on real data, perform quantitative evaluation of its performance, and study its robustness against different uncertain factors.

Although this approach to ultrasonic imaging requires extensive computation which makes real-time imaging impractical, we believe that because of its unique features it may have a future as computer technology advances.

ACKNOWLEDGMENT

This work was supported by CA 43920 from NCI. The authors thank Dr. Jian-yu Lu for his assistance in generating simulated Bessel acoustic beams and their *rf*-signals and are grateful for the secretarial assistance of Elaine C. Quarve.

REFERENCES

[1] E.E. Hundt and E.A. Trautenberg, "Digital processing of ultrasonic data by deconvolution, " *IEEE Trans. Sonics Ultrason.* SU-27:249 (1980).

[2] W. Vollmann, "Resolution enhancement of ultrasonic B-scan images by deconvolution," *IEEE Trans. Sonics Ultrason.* SU-29:78 (1982).

[3] C.N. Liu, M. Fatemi, and R.C. Waag, "Digital processing for improvement of ultrasonic abdominal images," *IEEE Trans. Med. Imag.* MI-2:66 (1983).

[4] H. Schomberg, W. Vollmann, and G. Mahnke, "Lateral inverse filtering of ultrasonic B-scan images," *Ultrasonic Imaging* 5:38 (1983).

[5]   G. Demoment, R. Reynaud, and A. Herment, "Range resolution improvement by a fast deconvolution method," *Ultrasonic Imaging* 6:435 (1984).

[6]   D.E. Robinson and M. Wing, "Lateral Deconvolution of ultrasonic beams," *Ultrasonic Imaging* 6:1 (1984).

[7]   J.A. Jensen, "Deconvolution of ultrasound images," *Ultrasonic Imaging* 14:1 (1992).

[8]   W.H. Press, B.P. Flannery, S.A. Teukolsky, and W. T. Vettering, *Numerical Recipes in C*, Cambridge University Press, Cambridge (1988).

[9]   X-L. Xu, J-S. Liow, and S.C. Strother, "Iterative algebraic image reconstruction algorithms for emission computed tomography: A unified framework and its application to PET," *Medical Physics* 20:1675 (1993).

[10] X-L. Xu, J-S. Liow, and S.C. Strother, "A novel iterative algorithm for emission computed tomography," *Proc. 1993 IEEE International Conf. on Acoustics, Speech and Signal Processing* (1993).

[11] J-y. Lu and J.F. Greenleaf, "Ultrasonic nondiffracting transducer for medical imaging," *IEEE Trans. Ultrason. Ferroelec. Freq. Contr.* UFFC-37:438 (1990).

PULSE CODING IN MEDICAL ULTRASOUND: SOME POSSIBLE APPLICATIONS

N.A.H.K. Rao, Dong-Li Yang, R. Raman, P. Chandraroy
and S. Venkatraman

Center for Imaging Science
Rochester Institute of Technology
Rochester, NY 14623

ABSTRACT

Pulse coding and pulse compression techniques have been used extensively in radar, sonar, non-destructive testing and seismic imaging[1-4]. Its application in medical imaging has been sparse and few[5-8]. Possible reasons could be, (i) limitations on time-bandwidth product[5,7] due to transducers operating in 1 to 10 MHz range with only 60% fractional bandwidth, (ii) frequency dependent attenuation in soft tissue[5] and (iii) a need to keep the sidelobe levels that result from pulse compression processing below certain levels[7], for improved contrast resolution. With significant improvements in processing hardware and software, real time implementation does not appear to be a significant problem[7,10]. In this paper we have examined two possible applications of pulse coding techniques. The possibility of increasing the effective bandwidth of a transducer (and hence the time-bandwidth product) through pre-enhancement of the drive signal has been evaluated with simulation studies. The increasing penalty on the sidelobe levels is also examined. As a second application, the flexibility to change the center frequency f_0 and bandwidth Δf of the coded signal has been exploited to study experimentally the sub-resolution scattering microstructure of a scattering object.

INTRODUCTION

Some preliminary work in recent years has shown that pulse coding techniques might be of some value in medical applications. We have used linear frequency modulation (FM) in this study which is just one form of pulse coding among many others[8,10]. Instead of a short pulse (few μs), a longer duration (10-30 μs) frequency modulated pulse is launched for interrogation (Fig 1(a)). This pulse may be further amplitude modulated or if the frequency sweep is across the entire bandwidth of the transducer, the transducer frequency response itself will impose its own modulation[9] as shown in Fig 1(b). The time autocorrelation of the FM pulse in Fig. 1(b) is shown in Fig. 1(c) and is referred to as the compressed pulse. The pulse width (computed on the envelope of the compressed pulse) depends only on the "effective" bandwidth of the FM pulse and not on its long pulse duration. However, the ratio of the <u>peak intensity</u> of the compressed pulse to the <u>peak intensity</u> of the FM pulse, called the signal to noise ratio (improvement) depends on the product of the time duration and the "effective" bandwidth of the FM pulse. This factor is referred to as the time-bandwidth product of the system and is responsible for its improved performance over conventional short pulse systems under peak power and/or noise limited conditions[1,5]. It appears that in medical imaging time-bandwidth product on the order of 10 to 30 may be possible[5,7].

PULSE CODING WITH PRE-ENHANCEMENT

Generally the 6dB bandwidth of the medical transducers is about 50-60% of its center frequency and their frequency response can be assumed to be approximately Gaussian. In order to improve the axial resolution (i.e. decrease the pulse width after pulse compression processing), in coded systems the frequency sweep bandwidth has to be increased. This is ultimately limited by the transducer bandwidth[9], and therefore can be increased further only by operating with transducers of higher center frequency, consequently increasing the signal attenuation in soft tissue and decreasing the penetration depth[5]. Therefore, it is useful to investigate possibilities of increasing the effective bandwidth of the low frequency transducers by pre-enhancement of the signal driving the transducer. Inverse filtering or signal deconvolution applied after the backscattered signal has been received is also a possible alternative but is expected to be limited by the system noise.

In this paper, the pre-enhancement method has been evaluated through simulation. We consider a transducer with center frequency of 3.5 MH$_z$ with a Gaussian frequency response (amplitude of the transfer function) and a 6dB bandwidth of 1.75 MH$_z$. We assume that the phase of the transfer function is either zero or is approximately linear with frequency and departure from linearity are small. The transducer response is considered as a linear shift invariant system. Therefore, the received FM pulse after two way transmission through the transducer system can be calculated by multiplying the Fourier transform of the input FM pulse (drive signal) with the two way transducer transfer function, and then inverse Fourier transforming the product. Following equation was used for the input FM pulse:

$$P(t) = B(t).e^{+\frac{(t-T_o/2)^2}{D^2}} \; Sin\left[2\pi\left\{(f_o - \Delta f/2).t + (\Delta f/T_o^2)t^2\right\}\right]: \; o \le t \le To \qquad (1)$$

where To = 12.56 μs was the time duration of the FM pulse and D was used to control the degree of pre-enhancement. f_0 and Δf define the center frequency and the sweep bandwidth of the input FM pulse. B(t) was a 12th order Butterworth window function used to reduce the ringing artifact in the fast fourier transform (FFT) computations[11]. B(t) was equal to 1 for about 80% of the time duration To of the pulse.

Results

Fig. 1 (a), (b) and (c) corresponds to the case of almost no pre-enhancement (D = 28 μs). The f_0 of the FM pulse was matched to the transducer center frequency of 3.5 MH$_z$ and Δf = 5 MH$_z$ was used in Eqn. (1). Fig. 1 (a) and (b) show the input and corresponding output respectively from the transducer and Fig 1(c) shows the autocorrelation of the output FM pulse in Fig. l(b). This corresponds to the case[9] where we have in fact swept the frequency across the entire bandwidth of the transducer and no further improvement in compressed pulse width or the time bandwidth product can be achieved by further increasing Δf. Amplitude pre-enhancement can, however, produce some additional improvement as evidenced from Fig 1(d), (e) and (f). D = 2.35 μs was used to produce pre-enhancement shown in Fig. 1(d). After the pulse in Fig. 1(d) has gone through the transducer response the output pulse is shown in Fig 1(e) along with its autocorrelation in Fig. 1(f). The output FM pulse looks more or less equalized in amplitude at different frequencies and has an "effective" bandwidth higher than the 6dB bandwidth of the transducer.

Due to increase in effective bandwidth, the peak amplitude of the compressed pulse in Fig. 1(f) is higher than that in Fig. 1(c) but the pulse width, on the other hand is smaller. We have quantitatively evaluated this improvement and the results are shown in Fig. 2 and 3. Fig. 2 shows the relative increase in the peak amplitude due to increase in

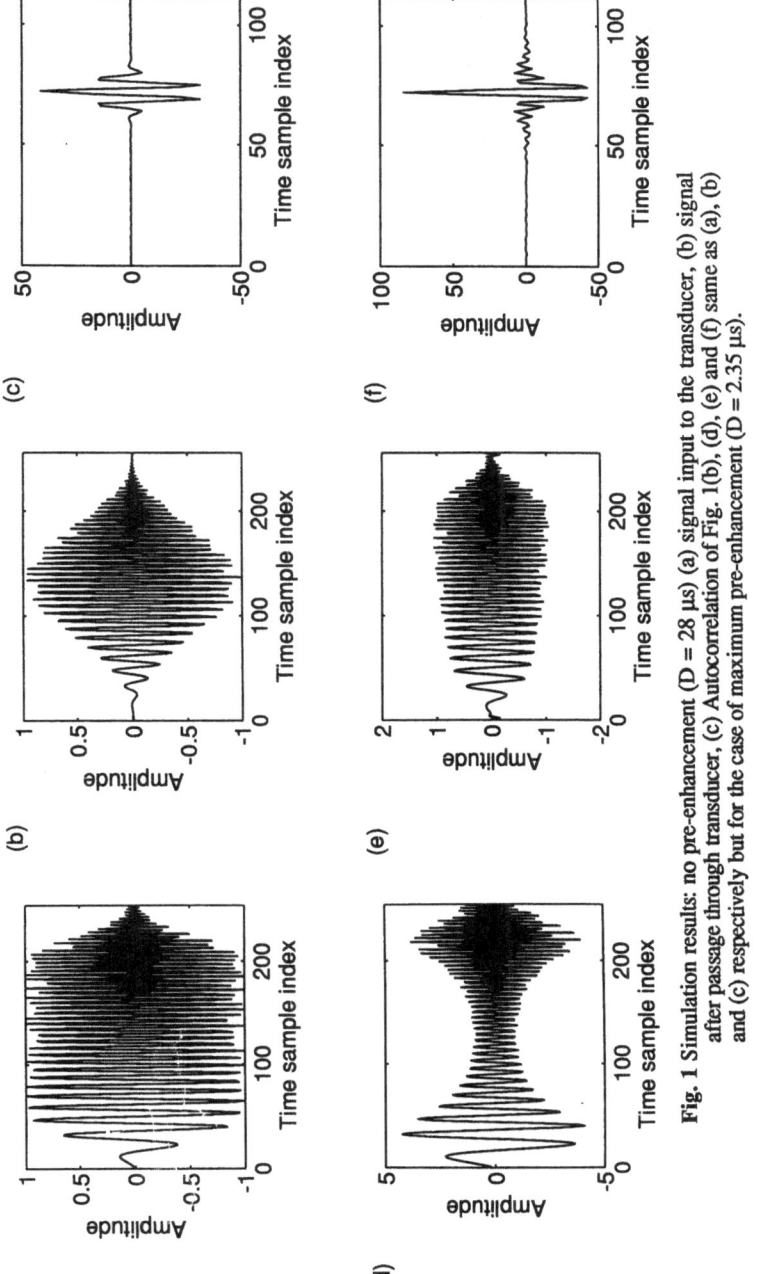

Fig. 1 Simulation results: no pre-enhancement ($D = 28$ µs) (a) signal input to the transducer, (b) signal after passage through transducer, (c) Autocorrelation of Fig. 1(b), (d), (e) and (f) same as (a), (b) and (c) respectively but for the case of maximum pre-enhancement ($D = 2.35$ µs).

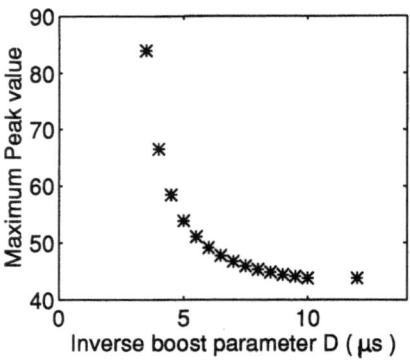

Fig. 2 (relative) change in peak amplitude of the compressed pulse as a function of pre-enhancement factor D.

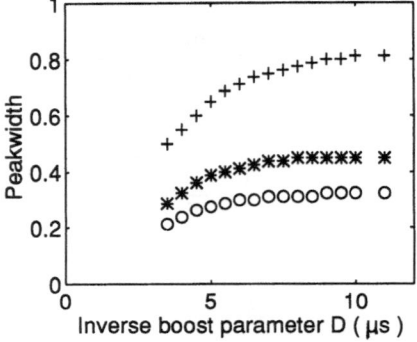

Fig. 3 change in pulse width of the envelope detected compressed pulse as a function of D. (○3dB, * 6dB and + 20db)

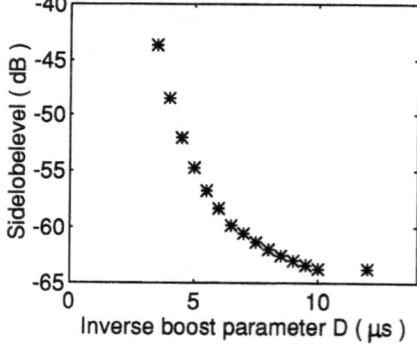

Fig. 4 Highest sidelobe level as a function of D.

the time-bandwidth product as a function of the parameter D that controls the degree of pre-enhancement. D = 28 μs corresponds to approximately no pre-enhancement and the point farthest to the right corresponds to this case. Fig. 3 shows corresponding decrease in the 3dB, 6dB and 20dB pulse width of the envelope detected signal. Envelope detection was performed using Hilbert transform of the compressed pulse. From the simulation results, it appears that 60-70% improvement in resolution should be possible. However, this improvement comes with a penalty in terms of sidelobe levels. Fig. 4 shows the highest measured sidelobe level as a function of D. These side lobe levels were measured on the envelope of the compressed pulse data plotted on a dB scale as shown in Fig. 5. We observe gradual increase in sidelobe levels from 64dB to about 43dB as the degree of pre-enhancement is increased. Speckle or clutter signal in medical ultrasound is usually below 30dB compared to a specular signal. Therefore sidelobe levels below 30-40dB should be maintained so that we do not compromise on contrast resolution. The simulation results indicate that this should be possible.

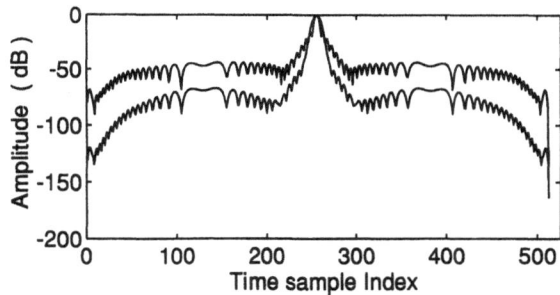

Fig. 5 Compressed pulse values after envelope detection plotted on a dB scale to show sidelobe levels. Upper curve is for maximum pre-enhancement (D - 2.35 μs) and lower is for no pre-enhancement (D = 28 μs).

SCATTERING STRUCTURE CHARACTERIZATION

A prototype pulse coding and correlation processing method described in the previous section was designed and implemented in a laboratory system. We briefly describe the system here but the details can be found elsewhere[12,13]. An Analogic Corporation arbitrary waveform generator was programmed to generate linearly swept frequency modulated pulses:

$$P_{in}(t) = \tfrac{1}{2}\left[1 - \cos 2\pi\left(\tfrac{t}{T_o}\right)\right]\left[\operatorname{Sin}2\pi\left(f_s t + kt^2\right)\right]; \quad o \le t \le T_o \,(\mu s)$$

(2)

$$\text{where } f_s = (f_0 - 1.0141\,\Delta f + 0.0708) \text{ and } k = (\Delta f - 0.068)/(0.928\,T_0)$$

where T_0 is the time duration of the pulse, f_0 and Δf are the center frequency and 6dB amplitude spectrum bandwidth in MH_z respectively. The last two parameters can be set independently. The first term is the hanning weighting applied to the pulse amplitude. The argument of the sinusoid has two terms, the first one defines the starting frequency and the second term essentially represents the sweep constant k of the FM signal. The constants of the equation were determined empirically so as to match the expected spectrum of the weighted pulse at several different values of f_0 (ranging from 2 to 4 MH_z) and Δf with T_0 set at 20 μs. Programming of the FM pulse in this form was motivated by the need to probe the tissue microstructure with well defined FM pulses with different f_0 and Δf. The generator uses 2000 points to calculate the function and outputs it at 100 MH$_z$. A panametric unfocused piezoelectric circular disk transducer was used to transmit the FM pulse into the medium. The transducer had a diameter of 1.27 cm. It had a center frequency of 2.4 MH$_z$ and a 6dB bandwidth of about 1.4 MH$_z$. The backscattered RF signal, after amplification, was digitized by an Analogic Corporation DATA 6000 at 50 MH$_z$ with 8 bits resolution. 1024 data points covering about 1.5 cm depth (20 μs) were stored in the memory for every line. The FM pulse used to drive the transducer was also digitized and stored for crosscorrelation processing.

For every f_0, Δf was varied from 0.5 MH$_z$ to 1.5 MH$_z$ in steps of 0.1 MH$_z$; f_0 was changed from 1.6 MH$_z$ to 2.8 MH$_z$ also in steps of 0.1 MH$_z$. In all, there were 143 different combinations of f_0 and Δf. Experiments were done in a water tank. The subresolution scattering medium was a water-filled sponge with cell size much smaller than the resolution of the imaging system. For pulse compression processing, the backscattered RF signal were crosscorrelated with the corresponding FM pulse used to drive the transducer. The region of interest selected in the far field, was located 9 cm from the transducer.

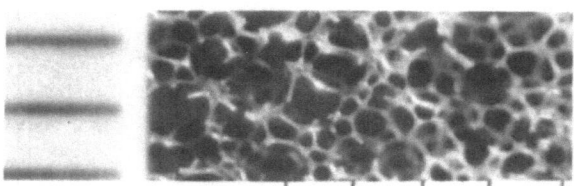

Fig. 6 Photograph of the sponge surface

Fig. 6 is a photograph of the surface of the sponge. The distance between the lines on the scale is 1 mm. The pore sizes ranged in value from 0.3 mm to 0.9 mm. The axial resolution measured in separate experiments, ranged from 1 mm to 2.5 mm for various different bandwidths. The transducer beam patterns were also measured and the 6dB beam widths ranged from 5 mm to 10 mm. Clearly in most of the cases, we have a large number of sponge interfaces perpendicular to the beam within the resolution cell volume[14]. However, the scattering structure is not regular. There is probably a mean spacing 'a' for the interfaces with some associated standard deviation. Therefore in the next section we consider its implication on the measured backscattered signal and develop a procedure to analyze the data.

Theoretical Considerations

Considering that the acoustic reflection is a linear time-invariant process and the fact that pulse compression processing is also linear, the composite RF signal backscattered from a random medium, after pulse compression processing, can be expressed[15] as:

$$r(t) = \sum_{n=1}^{N} a_n B(r_n) \tilde{A}(t-t_n) e^{i2\pi f_0(t-t_n)} \qquad (3)$$

We use analytic signal notation here with the understanding that the experimental signal is considered to be the real part of $r(t)$. The composite signal is considered to be a linear sum of several weighted and time shifted compressed pulses $\tilde{A}(t)$, similar to the one in Fig. 1(c). a_n represents the reflection coefficient and $t_n \approx 2 z_n/c$ the two way travel time to the nth interface, with c being the speed of ultrasound and z_n its distance from the transducer. $B(r_n)$ is due to the beam profile of the circular disk transducer. Assuming circular symmetry for the profile, $B(r_n)$ is a function of r_n only, the perpendicular distance of the nth interface from the central beam axis. Mean spacing smaller than axial resolution implies that $\tau = t_n - t_{n-1}$ on the average is small compared to the 6dB pulse width of $\tilde{A}(t)$. Therefore Eqn. (3) represents an interference pattern set up by several overlapping compressed pulse wavelets. The degree of overlap is controlled by the pulse width of $\tilde{A}(t)$, which in turn is controlled by the bandwidth Δf of the FM pulse echoes. The constructive and destructive nature of the interference is controlled by the center frequency f_0. Also note that N is the total "effective" number of reflecting interfaces seen by the ultrasound beam within a distance equivalent to a pulse width at time t, which is essentailly determined by the pore size and the resolution cell volume of the system[14].

For a sufficiently narrow bandwidth Δf, Eqn. (3) represents a random walk problem in the complex plane, with r(t) being a vector sum over N number of phasors. A detailed discussion of the statistics of the stochastic signal r(t) under various different conditions is given by Jakeman and Tough[15]. In the limit $N \to \infty$, and uniformly distributed phase $\Phi_n = 2\pi f_0 t_n$, r(t) becomes zero mean circular complex Gaussian process with Kurtosis $K = <r^4>/<r^2>^2 = 3$ where $<\cdot>$ stands for statistical average. The intensity I(t) = envelope square of the stochastic process r(t) has also been studied extensively, and the normalized second moment $<I^2>/<I>^2$ approaches a value of 2 in the same limit[14]. Departure from this limiting behavior can occur when (i) N becomes small[15,16]; (ii) underlying statistics of $\{a_n\}$ change; (iii) underlying statistics of $\{\Phi_n\}$ change[17,18]. In this paper we have examined the third type of behavior. We effectively change $\{\Phi_n\}$ by changing the center frequency f_0 of the interrogating FM pulse (in the narrow bandwidth case of $\Delta f = 0.5$ MH$_z$). When the Bragg's condition is satisfied, i.e. $<\Phi_n> = 2\pi f_0 <t_n> = 2\pi<2 z_n>/\lambda = n.2\pi$, where n = integer, we expect a buildup of a constant phasor in the random walk problem, causing a departure from Gaussian behavior[15]. Although the statistics of the envelope of the stochastic signal r(t) has been investigated[18] under condition (iii), it is not well known as to what happens to the non-Gaussian statistics of the RF signal r(t). In particular, how does Kurtosis K depart from its limiting value of 3? To find an answer to this question, we essentially repeated the theoretical simulation of Tuthill et. al.[18] A one dimensional periodic array of scatterers with spacing = 0.48 mm was considered in addition to high density random scatterers. The mean power calculated from the amplitude of the periodic scatterers was four times the mean power of the random scatterers. Sound speed was assumed to be 1520 m/sec. A Gaussian modulated sinusoidal function was used to generate the interrogating short pulse $\tilde{A}(t)$ with a 6dB pulse width of 0.95 μs. Therefore the periodic spacing was below the system axial resolution. The center frequency f_0 of the interrogating pulse was varied over a wide range and the Kurtosis K was computed from the simulated RF signal (convolution of pulse with the array of scatterers). Fig. 7 shows K as a function of the wavelength $\lambda_0 = c/f_0$ of the interrogating pulse. The wavelength has been normalized by the periodic scatterer spacing. K decreases significantly from its Gaussian limit value of 3 whenever (scatterer spacing)/λ_0 = n/2, where n = integer. Corresponding to each minimum there is an adjacent maximum at (scatter spacing)/λ_0 = i/4, where i = odd integer.

Results

Guided by the theoretical consideration of the last section, we now take up the analysis of our experimental data. Kurtosis K was calculated from the RF signal after pulse compression step for every combination of f_0 and Δf. In Fig. 8 we show a plot of K as a function of f_0 (with solid circle points) for a fixed $\Delta f = 0.5$ MH$_z$. Kurtosis shows two minima at $f_0 \approx 2.7$ MH$_z$ and 2 MH$_z$, and a maximum at f_0 approximately 2·3 MH$_z$. At the minima, the Kurtosis value drops below 3, similar to the simulation results. The

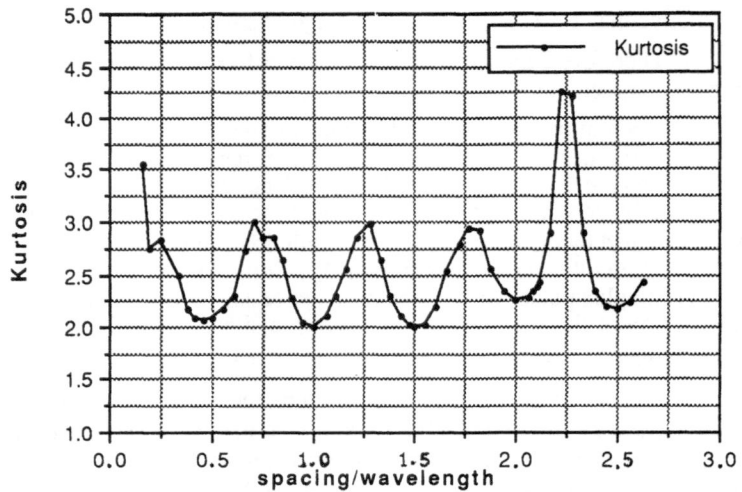

Fig. 7 Simulation results: change in Kurtosis K as a function of wavelength $\lambda_0 = f_0$. The periodic spacing was held fixed but f_0 was varied.

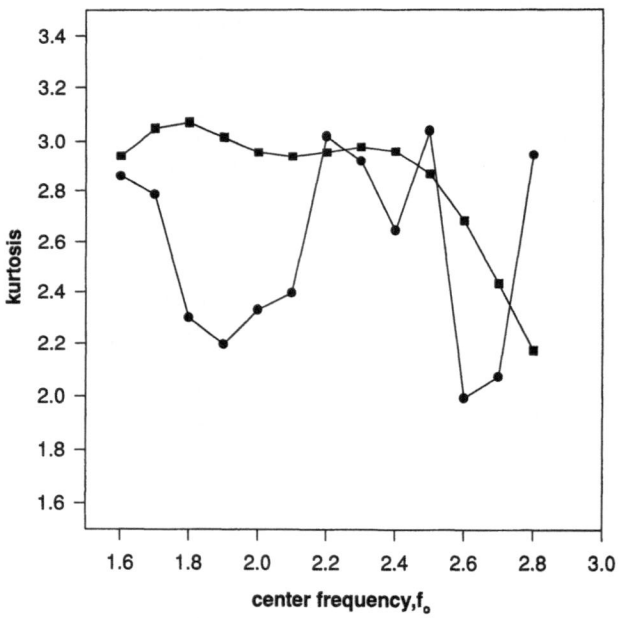

Fig. 8 Experimental results on sponge: (·) change in Kurtosis K as a function of f_0, the center frequency of chirp. Bandwidth of the chirl was held fixed at $\Delta f = 0.5$,MHz. (■) Kurtosis vs f_0 obtained from split spectrum analysis of the wide bandwidth chirp data ($f_0 = 2.2$ MHz, $\Delta f = 1.5$ MHz) from identical region of the sponge.

first pair of maximum and minimum could be associated with an average pore size of about 0.5 mm. The second minimum is perhaps due to another dominant spacing around 0.76 mm.

It may be argued that the same underlying stochastic information is also contained in the backscattered signal obtained with a wide band FM pulse. One may be able to extract this information by split spectrum processing[19] of the wide band signal. We have examined this hypothesis by analyzing our wide band RF signal (after pulse compression) obtained in response to the FM pulse with $f_0 = 2.2$ MH_z and $\Delta f = 1.5$ MH_z. The region of interest was identical. This signal was decomposed into a set of 13 narrow bandwidth signals with a bank of 13 narrow bandwidth filters defined in the frequency domain. The bank of filters were Gaussian in shape with 6dB bandwidth of 0.5 MH_z and centered at 13 frequencies ranging from $f_0 = 1.6$ MH_z to 1.8 MH_z in steps of 0.1 MH_z. Kurtosis K was calculated on each one of these 13 filtered sub band signal in the time domain and the results are shown in Fig. 8 as square points. We do not see the fine structure (maxima/minima) due to the periodicity of the interrogated medium. The phase sensitive information is lost in this type of processing.

CONCLUSIONS

Two possible applications of pulse coding technique have been investigated in this research. Linear frequency modulation (FM or chirp), one form of pulse coding was considered. Pre-enhancement of the FM signal was considered with an objective of improving the bandwidth of the signal coming out of the transducer. Simulation results indicate that 60-70% improvement should be possible without paying significant penalty on the increased side lobe levels.

A second application involved the experimental investigation of the possibility of extracting scattering structure information from narrow bandwidth multispectral backscattered data collected when tailored FM pulses are used to probe the scattering medium. The scattering structure (water filled sponge) had some degree of order (periodicity), but on a scale that was below the resolution capability of the transducer. It is believed that such structures exist in soft tissue and their quantification may prove to be of value in tissue characterization problem. Our results demonstrate that Kurtosis, calculated from the backscattered RF signal, shows a series of maxima and minima at specific center frequencies f_0 of the chirp signal. These frequencies can be related to the "mean" spacial period of the scattering microstructure. Although such experiments with tailored short pulse may also be possible, the advantage of the chirp lies in its superior signal to noise ratio.

ACKNOWLEDGMENT

This work was supported in part by a grant from the Whitaker Foundation.

REFERENCES

1. J.R. Klauder, A.C. Price, S. Darlington and W.J. Albertshim. "The theory and design of chirp radars," *Bell System Tech. J.,* 39:745-809 (1960).
2. M.D. Parent and T.R. O'Brien, "Linear-swept FM (chirp) sonar sea floor imaging system," *Sea Technology*, June:49 (1993).
3. F.K. Lam, "Microcomputer-based digital pulse compression system for ultrasonic NDT," *Ultrasonics* 25:166 (1987).
4. J.C. Fowler and K.H. Waters, "Deep crusted reflections recording using 'vibrosis' methods - A feasibility study," *Geophysics*, 40:399 (1975).
5. N.A.H.K. Rao, "Investigation of a pulse compression technique for medical ultrasound: A simulation study," *J. Medical & Biological Engg. & Computing*, to be published in March 1994.
6. N. Rao and S. Mehra, "Medical Ultrasound Imaging Using Pulse Compression," *Electronic Letters*, 29:649-651 (1993).

7. M. O'Donnel, "Coded excitation system for improving the penetration of real-time phased-array imaging systems," *IEEE Trans. Ultrason. Ferro. Freq. Control*, **39**:341-350 (1993).

8. J.Y. Chapelon, "Pseudo-random correlation imaging and system characterization," Chap. 6 in *Progress in Medical Imaging*, Ed. by V.L. Newhouse, Spring-Verlag 1988.

9. M. Pollakowski, H. Ermert, L. vonBernus and T. Schmedit, "The optimal bandwidth of chirp signal in ultrasonic application," *Ultrasonics*, **31**:418-420 (1993).

10. G. Hayward and Y. Gorfu, "A digital hardware correlation system for fast ultrasonic data acquisition in peak power limited applications," *IEEE Ultrason. Ferro. Elect.*, **35**:800 (1988).

11. A.V. Oppenheim and R.W. Schafer, "Discrete-time signal processing," Prentice Hall, Englewood Cliffs (1989).

12. N.A.H.K. Rao and W.M. Aubry, "Evaluation of a pulse compression technique for ultrasound speckle reduction," *Electronics Letters*, **29**:1900 (1993).

13. N. Rao and S. Mehra, "Experimental point spread function of FM pulse imaging scheme," *1991 IEEE Ultrasonic Symposium Proceedings* 2:1249 (1991).

14. N.A.H.K. Rao, "Simulation study of changes in ultrasound speckle statistics with the system point spread function," *J. Acoust. Soc. Am.*, **95**:1161 (1994).

15. E. Jakeman and R.J.A. Tough, "Non-Gaussian models for the statistics of scattered waves," *Advances in Physics* **37**:471 (1988).

16. N. Rao and H. Zhu, "Modeling ultrasound speckle formation and its dependence on imaging system's response," *SPIE Proc. on Medical Imaging V: Image Physics* **1443**:81 (1991).

17. L. Weng. J.M. Reid, P.M. Shankar, K. Soetanto, and X. Lu, "Nonuniform phase distribution in ultrasound speckle analysis - Part I: Background and experimental demonstration," *IEEE Trans. Ultrason. Ferroelec. Freq. Contr.*, **39**:352-359 and 360-365 (1992).

18. R.A. Tuthill, R.H. Sperry, and K.J. Parker, "Deviations from Rayleigh statistics in ultrasonic speckle, *Ultrasonic Imaging* **10**:81-89 (1988).

19. J.D. Aussel, "Split-spectrum processing with finite impulse response filters of constant frequency-to-bandwidth ratio," *Ultrasonics* **28**:229 (1990).

A NONLINEAR THEORY FOR DIFFRACTION TOMOGRAPHY - THE FRACTAL DIFFRACTION

W.S.Gan

Acoustical Services Pte Ltd
29 Telok Ayer Street
Singapore 0104
Republic Of Singapore

INTRODUCTION

So far theories for diffraction tomography are linearized. In view that diagnostic medical ultrasound systems and other systems in underwater acoustics and seismic applications produce nonlinear effects and nonlinearity problems are gaining much attention during the past decade and linearity works have reached maturity, we propose a nonlinear theory for diffraction tomography. During the past few years there has been much activities in applying fractal analysis to medical images resulting in the enhancement of edges as fractal dimension is very sensitive to edge analysis. This confirms the fractal structure of the human tissue empirically and phenomenologically. Fractal structure also occurs for geological structures and complex materials and underwater acoustics (turbulence). Our research proposes a fractal structure for the human tissue and complex materials and underwater acoustics (turbulence) and also to incorporate in diffraction and wave nature for the ultrasound propagation in a fractal structure. To describe the wave nature, we use the KZK equation and with parabolic approximation. The purpose is to obtain the scattered wavefield in a fractal medium. This is the forward problem.

SOLUTION OF THE KZK EQUATION.

One dimensional solutions to the nonlinear wave equation, such as those of Blackstock[1] can provide some useful information for medical ultrasound systems but are unable to reproduce the fine detail and phase variations seen in "real" pressure fields. Highly diffracted and focused pressure fields require more rigorous treatment. Smith and Beyer[2] commented on the "lack of appropriate theoretical analysis" when they published nonlinear measurements on a focused acoustic source operated at 2.3 MHZ. One of the most significant theoretical advances came in 1969 when Zabolotskaya and Khokhlov[3] published a solution of the nonlinear wave equation for a confined sound beam in which it was assumed that " the shape of the wave varies, slowly both along the beam and transversely to it". In 1971 Kuznetsov[4] extended their treatment to include absorption and the resulting equation is now widely known as the KZK equation. He also obtained solution which is known as the parabolic approximation to the nonlinear wave equation and is equivalent to the paraxial approximation used in optics. The KZK equation accounts for diffraction, absorption and nonlinearity and is valid for circular apertures that are many wavelengths in diameter and will accept arbitrary source conditions.

General Wave Equation

The KZK equation is a nonlinear wave equation for a scalar potential Φ, in consideration of the dynamics of a viscous heat conducting fluid. It is correct to the second order with terms for diffraction, absorption and nonlinearity:

$$\frac{\partial^2 \Phi}{\partial t^2} - c^2 \nabla^2 \Phi = \frac{\partial}{\partial t}[2\alpha ck^2 \nabla^2 \Phi + (\nabla\Phi)^2 + \frac{B}{2A}\frac{1}{c^2}(\frac{\partial\Phi}{\partial t})^2] \tag{1}$$

where c= speed of sound, α= absorption coefficient, k= wave number.
The left-hand side of eq.(1) is the three dimensional linear Helmholtz wave equation. Of the three terms on the right-hand side, the first term is a linear term and accounts for absorption, the second term is due to convective nonlinearity in the equation of state.

Parabolic Approximation

Kuznetsov also showed that the eq. (1) could be simplified by approximation, in the case of a quasi-plane wave field and the Laplacian (∇^2) can be replaced by the transverse Laplacian (∇_1^2). A circular aperture that is many wavelengths in diameter (i.e. ka is large) falls in this category since most of the energy is confined to a beam in the axial direction. This is known as the parabolic (or paraxial) approximation and is equivalent to the Fresnel approximation that is sometimes used in the diffraction integral for near-field calculations. Kuznetsov's parabolic approximation can be expressed in a normalised form:

$$[4\frac{\partial^2}{\partial\tau\partial\sigma} - \nabla_1^2 - 4\alpha R_o\frac{\partial^3}{\partial\tau^3}]\overline{P} = 2\frac{R_o}{l_D}\frac{\partial^2}{\partial\tau^2}\overline{P}^2 \tag{2}$$

where $\overline{P} = (p/p_o)$ is the acoustic pressure normalised by the source pressure and $\tau = (\omega t - kz)$ is the retarded time, i.e. includes a phase term for a plane wave travelling in the z direction,
R_0 = Rayleigh distance= $ka^2/2$ and l_D = shock distance. In this equation σ is the axial coordinate normalised by the Rayleigh distance and ξ is the radial coordinate normalised by the aperture radius i.e.

$$\sigma = \frac{2z}{ka^2} \quad \text{and} \quad \xi = r/a$$

A trial solution was then assumed in the form of a Fourier series (for the time waveform) with amplitude and phase that were functions of the spatial coordinates, i.e.

292

$$P(\sigma, \xi, \tau) = \sum_{n=1}^{\infty} q_n(\sigma, \xi, \tau) \sin(n\tau + \psi_n(\sigma, \xi, \tau))$$

or

$$P(\sigma, \xi, \tau) = \sum_{n=1}^{\infty} g_n(\sigma, \xi, \tau) \sin(n\tau) + h_n(\sigma, \xi, \tau) \cos(n\tau) \qquad (3)$$

where $g_n = q_n \cos\psi_n$, $h_n = q_n \sin\psi_n$ and n is the harmonic number with n=1 representing the fundamental frequency. Substituting the trial solution (3) in eq. (2) and collecting terms in $\sin(n\tau)$ and $\cos(n\tau)$ gives a set of coupled differential equations for g_n and h_n :

$$\frac{\partial g_n}{\partial \sigma} = -n^2 \alpha R_0 g_n + \frac{1}{4n} \nabla_1^2 h_n + \frac{nR_0}{2l_0} [\frac{1}{2} \sum_{k=1}^{n-1}(g_k g_{n-k} - h_k h_{n-k}) - \sum_{p=n+1}^{\infty}(g_{p-n}g_p + h_{p-n}h_p)] \quad (4)$$

$$\frac{\partial h_n}{\partial \sigma} = -n^2 \alpha R_0 h_n + \frac{1}{4n} \nabla_1^2 g_n + \frac{nR_0}{2l_D} [\frac{1}{2} \sum_{k=1}^{n-1}(h_k g_{n-k} - g_k h_{n-k}) + \sum_{p=n+1}^{\infty}(h_{p-n}g_p - g_{p-n}h_p)] \quad (5)$$

Eqs (4) and (5) form the basis of the numerical solution which can be implemented in a FORTRAN program.

FRACTAL STRUCTURE AS A DIFFRACTION MEDIUM

Sound propagation is related to the elastic properties of the medium. Most solids have tensorial elasticity and can support transverse as well as longitudinal sound waves. Hence we need to investigate the nature of their elasticity first. To attack this problem, we formulate fractals as a growth problem and we use the Diffusion Limited Aggregation (DLA) model, a growth model. For DLA, we need to calculate the growth probabilities. Here we focus on multi-fractality's relation to transport properties of the fractal medium.

Let P(r,t) = probability of finding the random walker on sites at a fixed distance r from the starting point. The probability P(r, t) to find the walker at l at time t is a Gaussian,

$$P(l,t) = P(o,t) \exp(-\frac{l^2}{4Dt}) \qquad (6)$$

where D = fractal dimension. The moments of the probability density $\langle P^q(r,t)\rangle$ can be written as a convolution integral:

$$\langle P^q(r,t)\rangle = \int_0^\infty Q(r/l) P^q(l,t) dr \qquad (7)$$

where Q(r/l) = probability of finding the sites separated by a chemical distance l and Euclidean distance r. The chemical distance is the shortest path between two sites on the cluster. In the general case, the qth moment $\langle P^q(r,t)\rangle$ can be written as

$$\langle P^q(r,t)\rangle = \frac{1}{N_r} \sum_{i=1}^{N_r} P_i^q(r,t) \qquad (8)$$

where the sum is over all N_r sites located a distance r from the origin (N_r may include many configurations or a single configuration with a very large number of cluster sites). The sum equation (8) can be separated into sums over different l values (N_m values of l_m):

$$\left\langle P^q(r,t)\right\rangle = \frac{1}{N_r}\{\sum_{i=1}^{N_1} P_i^q(l,t)+\sum_{i=1}^{N_2} P_i^q(l_2,t)+...\}$$

$$= \frac{1}{N_r}\sum_m N_m \times \frac{1}{N_m}\sum_{i=1}^{N_m} P_i^q(l_m,t) \quad = \frac{1}{N_r}\sum N_m\left\langle P^q(l_m,t)\right\rangle \tag{9}$$

This covers all the scattering points within the fractal medium.

In this problem we assume that the random walker starts at the origin 0 and after t time steps can be found at r[x], with very different probabilities at different sites.

For the scattering of sound by a fractal medium one needs to treat all sites of the fractal as starting points and the various parameters like sound velocity, attenuation coefficients etc. have to be modified for fractals. Fractal media are characterized by not having a very characteristic length scale and they have a very inhomogeneous density distribution. We can therefore expect to find very different physical properties in materials with fractal structure compared to the ordinary solids. Furthermore, real fractals are disordered and highly irregular. In some sense they can be regarded as ideally disordered materials. In ordinary diffraction tomography theory, we consider only scattering by one point and ignoring the object size-First Born Approximation. Now we consider object size as consisting of several scattering points.

For the diffraction of sound wave by a fractal medium, awe need to consider all sites of the fractal as scattering points. We call this type of diffraction "fractal diffraction". We start with nondiffracting case using Xray as an example. We calculate the Fourier transform of the pair correlation function g(r),

$$S(q) = 1 + \frac{N}{V}\int_v |g(r)-1|e^{i\underline{q}\cdot\underline{r}}d\underline{r} \tag{10}$$

where $g(r) = G(r)/\rho$ for r > 0 and by definition, $\frac{N}{V}=\rho$.

This gives the power spectral density (PSD) or intensity of scattered Xray beams. The full pair correlation functions is given by

$$G(r) = \frac{1}{V\rho}\int_v <\rho(r')\rho(r'+r)>dr' \tag{11}$$

where $\rho=<\rho(r)>$ is the mean density, p(r) is the local density and the brackets mean ensemble average is then approximately given by

$$G(r) \cong \frac{1}{N}\sum \theta(r')\theta(r+r') \tag{12}$$

The function $\theta(r)$ takes the value 0 or 1 depending on whether a given pixel is occupied or not and the sum runs over all the pixels. For a fractal system, the resulting correlation function should follow a power law of the form:

$$G(r) \sim (r/a)^{-A} \tag{13}$$

where A is related to the fractal dimension d_f and the embedding space dimension d by

$$d_f = d - A.$$

WAVE SCATTERING MODIFIED BY THE FRACTAL MEDIUM

The expression for scattered acoustic pressure wavefield is modified by the correlation coefficient which contains the fractal dimension of the medium.

We have obtained scattered wavefield amplitude fluctuation as

$$P(\sigma, \omega, \omega t - kz) = \sum_{n=1}^{\infty} q_n(\sigma, \zeta, \omega t - kz) \sin[n(\omega t - kz) + \varphi_n(\sigma, \zeta, \omega t - kz)] \qquad (14)$$

For diffraction of sound wave by a fractal medium. We need to consider all sites of the fractal as scattering points. For this reason, we have chosen our correlation coefficient as eq(9). By substituting eqn(9) into eq.(14), then the autocorrelation function for the amplitude fluctuation is given by the following formula:

$$\overline{P_1(t)P_2(t)} = \int_0^{R_1} \int_0^{R_2} \iiint_{-\infty}^{+\infty} P_1(\sigma, \omega, \omega t_1 - kz)P_2(\sigma, \omega, \omega t_2 - kz)$$

$$< \rho^q(r, t) > d\sigma_1 d\sigma_2 dz_1 dz_2 d\omega_1 d\omega_2 \qquad (15)$$

where the coordinates of the receivers are $(R_1, 0, 0)$ and $(R_2, 0, 0)$. So far we consider the forward problem.

The intensity of scattered wave = Fourier transform of autocorrelation function =

$$\int_{-\infty}^{\infty} \overline{P_1(t)P_2(t)} e^{-j2\pi ft} dt \qquad (16)$$

where f = frequency. Equation (16) is a frequency domain representation. It shows that both nondiffracting and diffracting cases involve correlation function. What is difference is reflected in the correlation function which involves scattered wave amplitude.

INVERSE PROBLEM

This is to obtain information of the medium from the scattered wave at the receiver. When a wave propagates in a medium with random inhomogeneities, fluctuations of the characteristics of the wave-field due to the superposition of the scattered waves and the primary wave are observed. There must be dependence between the fluctuations of the characteristics of the wavefield and the fluctuations of refractive index. The inverse problem is usually not unique and cannot be solved without additional assumptions. However, if we make reasonable assumptions about the form of the correlation coefficient of the refractive index, then by measuring the scattered field fluctuations we can determine the mean value of the refractive index fluctuations.

PROPOSED EXPERIMENTAL SET UP

The diffraction tomography scheme will be used. The wave nature and diffraction effect have been incorporated in eq.(15). A typical diffraction tomography experiment set up is shown in Fig (1)

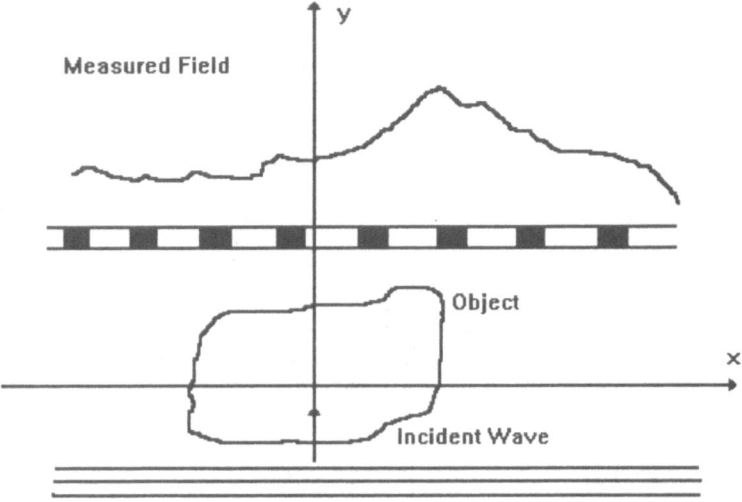

Fig 1 A Typical Diffraction Tomography Experimental Scheme

Here the incident wave is used to illustrate the object and the scattered field is measured on the far side of the object. This is transmission tomography. There are two mathematical algorithms for inverting the scattered data to estimate the object's complex refractive index. The two algorithms can be considered as interpolation in the frequency domain and interpolation in the space domain. Now we will describe the frequency domain algorithm assuming plane wave illumination. There are two schemes for frequency domain interpolation. The more conventional approach is polynomial based and assumes that the data near each grid point can be approximated by polynomials. A second approach is known as the unified frequency domain (UFR) and interpolates data in the frequency domain by assuming that the space domain reconstruction should be spatially limited.

We first describe the polynomial interpolation. In order to discuss the frequency domain interpolation between a circular arc grid on which the data are generated by diffraction tomography and a rectangular grid suitable for image reconstruction, we must first select parameters for representing each grid and then write down the relationship between two sets of parameters. The space domain algorithm is the backpropagation algorithm[5] which is computationally more intensive. We have decided to use the frequency domain algorithm. We have chosen the frequency space domain algorithm because our expression for the autocorrelation function for the amplitude fluctuation in equation (15) is given in the frequency domain. By measuring the field fluctuations as given by equation (15), we can determine the value of the refractive index fluctuations.

The main program in the reverse process is to compute a two dimensional estimate of the object function or the refractive index based on the scattered data. The input to this program is the complex amplitude of the scattered field along the receiver line generated by the main program for the forward process. This routine uses the frequency domain interpolation algorithm in the reconstruction process. With these algorithms, the Fourier transform of the object along the arc, $Q(\omega, \phi)$ is obtained. (ω = angular frequency, ϕ projection angle). Once the Q value corresponding to the NxN grid is found, an inverse Fourier transform is performed to obtain the object function. To find the refractive index, the equation below is used :

$$n\delta = [+ - \sqrt{O+1} - 1] \tag{17}$$

where O = object function and $n\delta$ = refractive index. Lastly, the output data is output as an NxN array of binary single precision floating point numbers to a file.

So far the interpolation algorithm we have chosen is only valid for weak scattering. We propose to incorporate a primary wave corrector[6] for moderately scattering. We use post-compensation, i.e. we will make an image in the conventional way, then correct it by using a proper corrector. This correction effect will be represented by Φ_n, the internal primary-wave distribution. If $\Phi_n = 1$, then there is no primary wave scattering. With primary wave scattering correction, the reconstructed image will be equation (15) multiplied by

$$\frac{1}{\frac{1}{2\pi}\int_{-\pi}^{\pi} \Psi_n(\xi_0, \eta_0, \theta)d\theta}$$

where θ is the incident angle, ψ_n=internal primary wave distribution and (ξ_0, η_0) = position coordinate.

The algorithm for reconstructing the refractive index from diffracted projections is briefly as follows:
(1)   Evaluate Eq(15) and these are the input data.
(2)   Fourier transform each projection.
(3)   Estimate the 2-dimensional Fourier transform of the refractive index from the transformed projections.
(4)   Perform a 2-dimensional inverse Fourier transform to get an estimate of the refractive index.

At each step of this procedure, signal processing theory suggests a number of procedures to improve the reconstruction
These include:
(a)   Zero padding the projection data to reduce the effects of interperiod interference. This also increases the resolution in the frequency domain and should make interpolation easier.
(b)   Apply a Hamming window to the projection data to smooth out the data at the ends of the receiver.
(c)   Multiply the two dimensional Fourier transform of the object by a Low Pass Filter (LPF) (a Hamming window in this case) to reduce the effects of high frequency noise.

An important part of the reconstruction process is filtering the projection data. For efficiency reasons the filter is implemented with an FFT algorithm but these algorithms do not perform an aperiodic convolution like that used in linear systems theory. Instead a filter implemented with FFT's performs circular convolution. We are now in the process of testing the interpolation algorithm.

CONCLUSIONS

Complex fractal structure is a nonlinear medium. The scattering of sound wave by such a fractal medium will form chaotic diffracted sound beams which in turn will form chaotic fractal images, a term coined by us to describe the type of images formed. For our case, these images will be in terms of refractive index fluctuations. The advantage of chaotic fractal images is high sensitivity to the change in initial parameters. So any slight change in abnormality in the human tissue will be detected. Fractal diffraction tomography is a new field of study and cannot be handled by conventional diffraction theory such as Kirckhoff diffraction theory or the pertubation theory.

REFERENCES

1. D.T.Blackstock Generalised burgers equation for plane waves, J.A.S.A. 77:2050(1985).
2. C.W.Smith and R.T.Beyer, Ultrasonic radiation field of a focusing spherical source at Finite Amplitudes, J.A.S.A. 46:806 (1969).
3. E.A.Zabolotskaya & R.V. Khokhlov,Quasi-plane waves in the nonlinear acoustics if confined beams, Sov.Phys Acoust. 15:35 (1969).
4. V.P. Kuznetsov, Equations of nonlinear acoustics, Sov. Phys. Acoust. 16:467 (1971).
5. A.J.Devaney, A filtered backpropagation algorithm for diffraction tomography, Ultrasonic Imaging, 4:336(1982).
6. A.Cai, Y.Nakagawa, G.Wade and M.Yoneyama, "Imaging the Acoustic Nonlinear Parameter with Diffraction Tomography" Acoustical Imaging , H. Shimizu, N. Chubachi and J. Kushibiki, ed., Plenum Press, New York and London, 17:273(1988).

NONLINEAR PARAMETER TOMOGRAPHY BY
SECONDARY SOUND WAVES

X.F.Gong D.Zhang and P.Day

Institute of Acoustics, Modern Acoustics Lab.
Nanjing University, Nanjing, China

Introduction

Acoustic nonlinear parameter can be defined by the ratio of coefficients of quadratic term to linear term in the Taylor expansion of pressure-density relationship for a medium, and is a very basic parameter in nonlinear acoustics. In recent ten years much attention has been focused on the investigation of acoustic nonlinear parameter in biological media due to the occurrence of nonlinear phenomena, such as wave distortion, harmonic generation, acoustic saturation etc., in the range of biomedical frequencies and intensities. The existence of these nonlinear ultrasonic effects can accelerate the sound losses and limit the amount of sound power deliverable to the media. Thus, the wide applications of ultrasound in medical diagnosis and therapy have increased the importance and necessity to study the nonlinear properties, especially the nonlinearity parameter B/A in biological media. Moreover, many works indicate that the nonlinear parameter B/A is evidently dependent on the structural hierarchy of specimens and is sensitive to the pathological state of tissues. Therefore, this parameter consists of dynamical information of the materials and is expected to become a new parameter in tissue characterization and imaging.

Three experimental methods of B/A determination have been proposed in our laboratory: finite amplitude insert-substitution method, improved thermodynamic method (phase shift method) and method of parametric array. Recently, we proposed a method based on second harmonic wave detection to image the nonlinearity parameter by using FAIS method. In this paper we present a nonlinearity parameter tomography by using acoustic parametric array. The principle of this method is to generate a secondary wave with sum or difference frequency due to the nonlinear interaction of two primary waves. Using conventional CT technique the pressure amplitude of difference frequency sound wave will be received as projection data and then the nonlinear parameter B/A tomography can be reconstructed by filter-back projection algorithm. Some computer simulation and experimental images for liquid model are presented in this paper.

Theoretical analyze

1. It is well known that the Burgers' equation may be used to describe the nonlinear propagation of a finite amplitude wave in fluids as follows:

$$\frac{\partial W}{\partial \sigma} - W\frac{\partial W}{\partial y} = 0 \tag{1}$$

where $W = u/u_o$, $\sigma = x/l$, $y = \omega\tau$, $\tau = t - x/c_o$, u and u_o-instantaneous and peak values of particles velocity. $l = 1/(\beta M k)$-discontinuity distance, $\beta = 1 + B/(2A)$, B/A-nonlinear parameter, $M = u_o/c_o$, $k = \omega/c_o$, c_o-sound velocity. For the boundary condition of a two-frequency source with ω_1 and ω_2 at $x = 0$,

$$u(0, t) = u_{01}\sin(\omega_1 t + \phi_1) + u_{20}\sin(\omega_2 t + \phi_2)$$

F.H. Fenlon gave an exact solution of Burgers' equation in a series form for the case of a losses medium. If $p_1(0) > > p_2(0)$, $\omega_2 > \omega_1$, which correspond to the interaction of a low frequency pumping wave(ω_1, p_1) and a high frequency probing wave(ω_2, p_2), then from Fenlon's solution the sound pressure amplitude of sum and difference frequency components produced by nonlinear interaction of these two waves can be written by:

$$p_{\omega_2 \pm \omega_1} = \frac{2p_2}{\sigma_2 \pm \varepsilon\sigma_1} J_1(\sigma_2 \pm \varepsilon\sigma_1) J_1[(\mu+1)\sigma_1] \tag{2}$$

where $\mu = \omega_2/\omega_1$, $\varepsilon = u_{o2}/u_{o1}$. When the argument of Bessel function is small, Eq.(2) can be simplified and the secondary wave sound pressure amplitude after propagation a distance L is expressed by:

$$p_s(L) \approx \frac{\omega_s}{2} p_1 p_2 \int_0^L \frac{\beta(x)}{\rho(x) c^3(x)} dx = \frac{\omega_s}{2} p_1 p_2 \int_0^L \beta'(x) dx \tag{3}$$

Where $\omega_s = \omega_2 - \omega_1$, ρ-density of material. $\beta'(x) = \beta(x)/(\rho(x)c^3(x))$.

2. CT imaging by comparative method

When the transmitter and receiver with distance L are immersed in degassed distilled water where $(B/A)_o$ is known, then from (3) we obtain:

$$p_{s0}(L) = \frac{\omega_s L}{2} p_1(0) p_2(0) \beta'_0 \tag{4}$$

where $\beta'_o = \beta_o/(\rho_o c_o^3)$, $2\beta_o = 2 + (B/A)_o$. Now, if the specimen with unknown $(B/A)_x$ is inserted in the degassed distilled water, then at same distance L the secondary wave sound pressure amplitude with difference frequency can be expressed:

$$P_{sx}(L) = \frac{\omega_s}{2} P_1(0) P_2(0) \int_0^L \beta_x'(x) \, dx \qquad (5)$$

From equation (4) and (5), we obtain

$$\frac{P_{sx}(L)}{P_{so}(L)} = \frac{1}{L\beta_0'} \int_0^L \beta_x'(x) \, dx \qquad (6)$$

where $\beta'_x(x) = \beta_x(x)/\rho(x)c^3(x)$, $2\beta_x(x) = 2 + (B/A)_x$. Eq.(6) shows that $p_{sx}(L)/p_{so}(L)$ is the Radon transform of $\beta'_x(x)$. Therefore, using formula (6) and CT imaging method the nonlinearity tomography can be reconstructed. In order to get a quite good image the modified SL filter function is used for reconstruction. For a lossy medium a modified correction matrix is necessary to compensate the attenuation. However, in our case the liquid sample with less absorption is used in two dimensional imaging, so for simplicity the correction matrix will be ignored.

Experiment and Results

1.Experimental arrangement

   The block diagram of experiment is shown in Fig.1

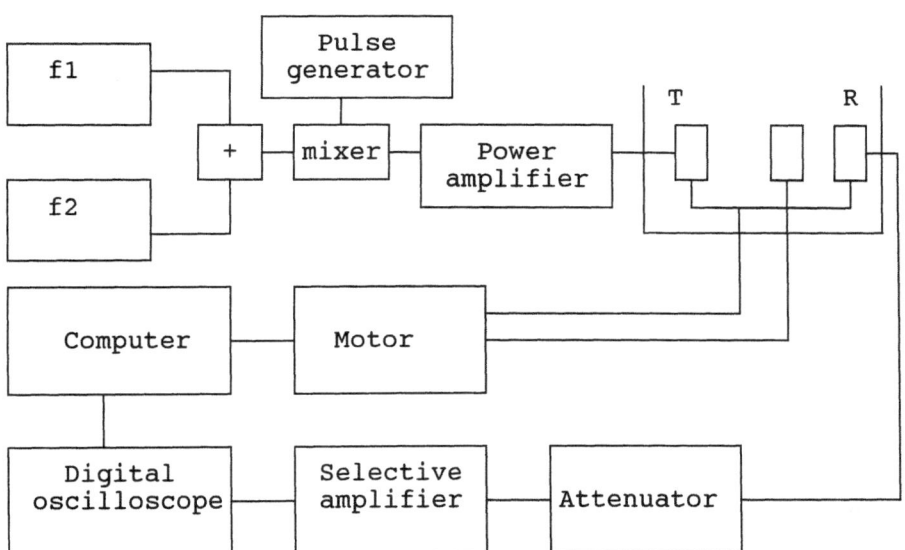

Figure 1. The block diagram of experiment system

Signal generators f_1 and f_2 are the pump wave source and probe wave source respectively. The signal after adder is modulated by pulse generator using a mixer. The source transducer

T is a PZT-ceramic disk with resonance frequency 1MHz. The transducer transmits both the pump signal (1MHz) and probe signal (at near the 5th overtone of the fundamental frequency of transmitter) and is driven by broadband power amplifier. The PZT receiver with resonance frequency 4MHz is used to detect the difference frequency signal. The mechanical system and signal processing system are used to sample the received signal and to reconstruct the image. The whole system is operated automatically by using the motors and HP-IB interface. The image of B/A is visualized on the TVGA monitor.

2. Some experimental results

(1) From Eq(4) amplitude of difference frequency component is proportional to the primary waves sound pressure. Fig.2 shows this dependence of measured difference frequency signal on pumping wave amplitude for the water and ethanol at distance L=4cm.

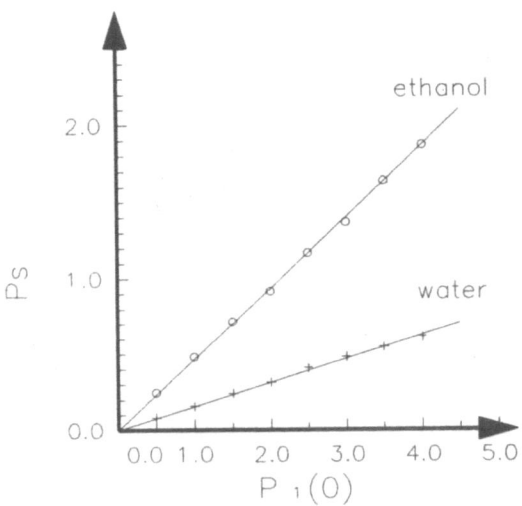

Figure 2. Dependence of measured difference frequency component on pump wave amplitude for degassed distilled water and ethanol at L=4cm

(2) One-dimensional image of B/A

Some liquid samples with different nonlinear parameter B/A filled in sample box are immersed in water as shown in Fig3. The transmitter and receiver are scanned simultaneously in same direction. The amplitudes of difference frequency sound wave are measured in samples and in water respectively. Then, using comparative method the one dimensional B/A image will be obtained and is shown in Fig.4 for ethanol(s_1),30% dextrose aqueous solution (s_2) and ethylene glycol(s_3).

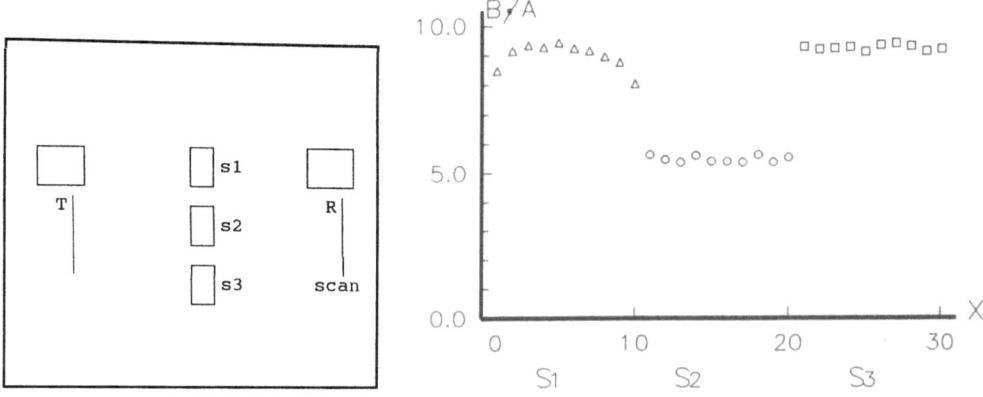

Figure 3. Illustration of
one-dimensional scanning

Figure 4. One dimensional image

(3) Nonlinear parameter tomography

Fig.5 gives a two layer liquid model (water-ethanol). For computer simulation of B/A tomography we use modi-SL filter function to reconstruct the B/A image as shown in Fig.6. Fig.7 is the experimental results using block diagram as Fig.1.

Conclusion Remark

On the basis of Burgers' Equation and Fenlon's theory we present a nonlinearity parameter imaging method by secondary sound waves. The computer simulation and experimental

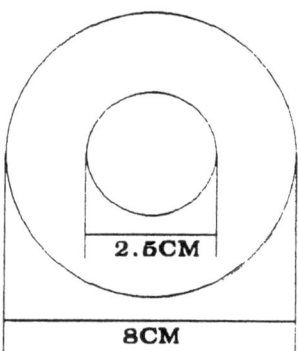

Figure 5. Two liquid layer model. inside - ethanol,
outside - water

 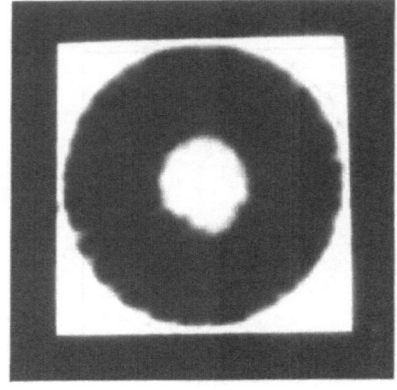

Figure 6.computer simulation of B/A image   **Figure 7.**Experimental result of B/A image

result of B/A images for liquid are obtained with some success. However, we have not enough experimental data to compare this method with the method of second harmonic detection. Therefore, further study this method with lower difference frequency sound wave for biological tissues is in progress.

This work is supported by National natural Science Foundation of China.

References:

1.X.F.Gong, Z.M.Zhu, T.Shi and J.H.Huang, "Determination of acoustic nonlinearity parameter B/A in biological media using FAIS and ITD methods", *J.Acoust.Soc.Am.*, 86,1,(1989)

2.Y.Nakagawa etal. *Proc. IEEE 1984 Ultrasonic Symposium* 637(1984)

3.Y.Nakagawa, W.Hou, A.Cai, N.Arnold and G.Wade, "Nonlinear parameter imaging with finite amplitude sound waves", *Ultrasonic Symposium,* 901(1986)

4.X.F.Gong, "Nonlinear parameter in tissue characterization and imaging", *Abstract of 20th International Symposium on Acoustical Imaging,* 1992

5.D.Zhang, X.F.Gong and S.G.Ye, "Nonlinear parameter Imaging", *Abstract of 20th International Symposium in Acoustical Imaging,* 1992

6.F.Dunn, J.Zhang and L.A.Frizzell, "Composition and structural dependence of B/A for biological media", *Proc. of 12th ISNA,* 385(1990)

7.X.F.Gong, Z.M.Zhu, B.Fan and D.Zhang, "Parametric effect and nonlinearity parameter in biological media", *Proc. of 12th ISNA,*397(1990)

LOW FREQUENCY ACOUSTICAL IMAGING

USING A VIBRATING TIP

Fabrice Sthal, Bernard Cretin

Laboratoire de Physique et Métrologie des Oscillateurs
associé à l'Université de Franche-Comté-Besançon
32 avenue de l'Observatoire - 25044 Besançon Cedex - France

ABSTRACT

In conventional acoustic microscopy, the Abbee barrier limits the resolving power of imaging systems. The use of near-field enables super-resolution. Thus, the spatial resolution is related to the acoustic source size and not to the acoustic wavelength. This paper shows that near-field acoustical images can be obtained by pressing a vibrating tip onto the sample surface. Such a technique allows high resolution and high signal to noise ratio. Presented images show subsurface defects or crystalline orientation inside metals and semiconductors at a few tens kilohertz. Resolution range is 1-40 microns depending on the tip radius and usual signal to noise ratio is about 90 dB. The hertzian model points out that the observation depth is restricted to a few tip diameters as indicated by images of variable size subsurface structure.

INTRODUCTION

During the last years, many authors have reported near-field techniques showing acoustic super-resolution. In all cases, a small tip (such as AFM tip or truncated cone[1,6]) or a pinhole[7] is used to generate an acoustic field very close to the sample surface. The source diameter Φ is smaller than the acoustic wavelength λ_a (typically $\Phi \leq \lambda_a/10$). Corresponding spatial resolution is about Φ and is not given with the Rayleigh criterion.

In our setup, we have maximized the acoustical coupling by pressing a vibrating tip onto the sample surface. Generated stress is detected in transmission mode with a PZT transducer. Amplitude and phase images of the scanned sample lead to topographic and elastic information of the microdeformation volume.

This paper shows that a first theoretical approach can be based on the hertzian theory of contact since the sample size is smaller than the acoustic wavelength that is used to investigate the material. Then, the setup principle is described and the electromechanical transfer function of a new bimorph excitation device is presented. Obtained images point out that the penetration depth is restricted to a few tip radius and is not related to excitation frequency in the used range. Moreover, a 1 μm spatial resolution is shown at low frequency (typically, the resolution is better than $\lambda_a/10\,000$) and demonstrates the significance of nearfield methods for NDE.

MODEL OF STATIC DISPLACEMENT

In our microscope, a contact pressure between tip and sample is required for high acoustic coupling. Static displacement and stress can be calculated by using the hertzian theory. That theory is restricted to static and non adhesive contact between two perfectly elastic solids. Assuming the tip is a sphere of radius R and the sample a cylinder bounded with a plane surface, the contact radius is expressed as :

$$a = \sqrt[3]{(k_1 + k_2)\frac{3RF}{4}}$$

where F is the bearing force ; $k_i = [(1-v_i^2)]/(\pi E_i)$, v_i is the Poisson ratio and E_i the Young's modulus ($i = 1$ for sample and $i = 2$ for tip) ; a defines the acoustic source radius for the contact vibrating tip. In practice, vibration frequency is low and the sample size is very small compared with the acoustic wavelength. So, a static model gives an approximate behaviour of the sample in the microdeformation volume that is imaged. If the origin is the center of the contact area, the stress distribution is hemispherical. Using cylindrical coordinates yields to simple expressions of stress for[8, 9] $r = 0$:

$$T_{zz} = p_o \frac{1}{1 + z^2 / a^2}$$

$$T_{rr} = T_{\theta\theta} = p_o \left\{ (1+v)\left[1 - z/a\, atan\,(a/z) \right] - \frac{1}{2}\frac{1}{1 + z^2/a^2} \right\}$$

$$T_{rz} = \frac{1}{2}\, |T_{rr} - T_{zz}|$$

where p_o is the stress T_{zz} at $r = 0$ and $z = 0$ (maximum stress between tip and sample)

The stresses are shown in Fig. 1 versus adimensional coordinate z/a, at $r = 0$. Only the shearing stress is maximum inside the volume. T_{zz} rapidly decreases ; an approximate observation depth may be given by the z value z_{obs} corresponding to T_{zz} ($z = 0$) / 10 ; $z_{obs} = 3a$. Thus, the observation depth and apparent penetration depth are related to the tip radius and the contact pressure.

This model shows the possibility of sample characterization with tip static pressure and stress or displacement measurement. Experimentally, the static displacement is typically in a 1 - 10 nm range and the measure is perturbed by thermal drift. So, the best way to detect the microdeformation consists in low frequency excitation and detection.

MICROSCOPE SETUP

Microscope principle

The microscope is illustrated in Fig. 2. A diamond or sapphire tip is linked to the exciting flexure mode transducer by a cantilever beam. A CW generator sets the vibrations of the tip that is pressed onto the sample surface. Micro-deformations induce strain inside the sample. The strain is detected in transmission mode with a 2 cm long PZT rod that is glued onto the opposite face of the sample. Transducer electrical signal is averaged with a double-phase lock-in amplifier. Amplitude and phase information is stored and displayed by the computer. Three-stage piezoelectric translation units enable sample scanning and bearing force adjustment.

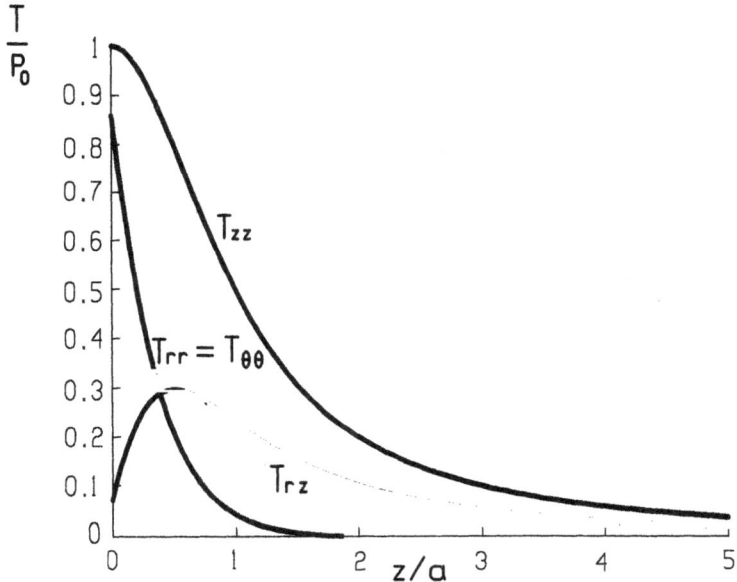

Figure 1. Normalized stresses versus adimensional coordinate z/a (Poisson ratio : v = 0.36)

Improved excitation device

Excitation head may be simplified by using a bimorph beam (piezoelectric layer deposited onto a cantilever beam). The beam can be composed of different materials such as metals, quartz or silicon. However, single-cristal silicon remains the basic material due to its outstanding mechanical properties and the knowledge of etching processes. Electromechanical coupling is obtained with ZnO or AℓN film, but AℓN is more favourable at low frequencies due to its high resistivity.

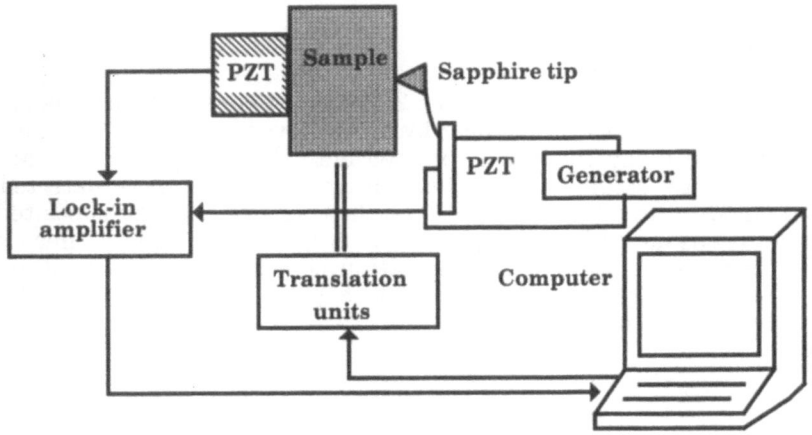

Figure 2. Setup of the acoustic microscope using a vibrating tip

An example of realization is shown in Fig. 3. The single arm bimorph cantilever is composed of a 30 μm thick silicon beam covered with a 2 μm thick AℓN film. The diamond tip has been glued onto the AℓN film surface with a silver paste. Presented tip radius is 2 μm and the complete cantilever stiffness is about 10 N/m. Electrodes are the doped silicon beam and a thin layer of silver paste deposited on AℓN film.

Figure 3. View of a diamond tip glued onto a single arm bimorph beam (AℓN on silicon)

EXPERIMENTAL RESULTS

Electromechanical transfer function of a bimorph beam

The presented bimorph beam has been excited with a programmable generator and the resulting bending has been measured for each frequency with a high resolution interferometric laser probe.[10] Displacement amplitude and phase were averaged with a lock-in amplifier.

Figure 4 shows the electromechanical transfer function of the cantilever (silicon beam size : $\ell \times w \times t = 3\,500 \times 500 \times 30 \, \mu m$). Experimental Q-factor is a few tens whereas its value is $\sim 1\,000$ for the simple silicon beam : damping is related to the presence of silver paste and tip.

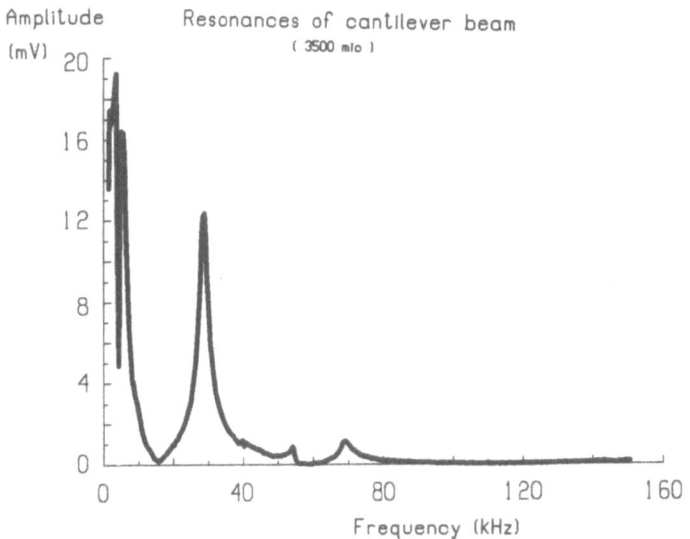

Figure 4. Electromechanical transfer function (tip displacement amplitude versus frequency) of a bimorph beam loaded with a diamond tip.

Imaging of solid samples

Before imaging, the global transfer function (including the sample) of the microscope is plotted and the excitation frequency is set to a resonance peak.[11]

First presented sample is a silicon wafer (360 μm thickness, crystalline orientation [100]). Parallel grooves have been etched on one face, the opposite face remaining polished. Figure 5 is a cut of the sample that was coupled to the support with an ultrasonic gel.

Scanned surface is the plane face of the sample where the grooves are optically invisible. Figure 6 shows the amplitude image obtained with a 40 μm radius tip at 31.6 kHz frequency. Subsurface grooves appear as parallel white stripes. Estimated observation depth is 130 μm, i.e. close to the value suggested by the hertzian model.

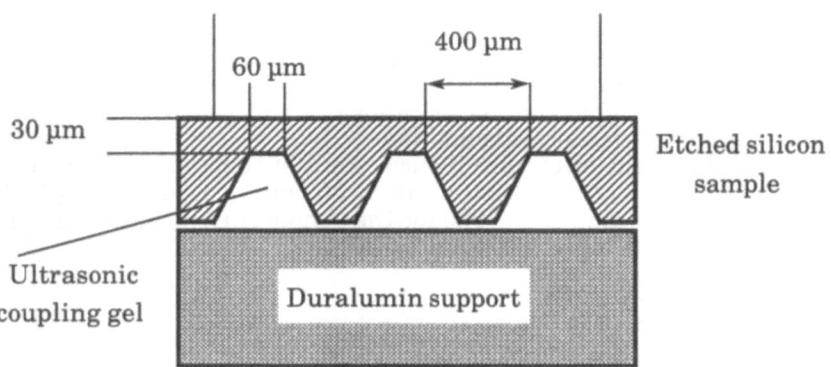

Figure 5. Geometry of the etched silicon sample.

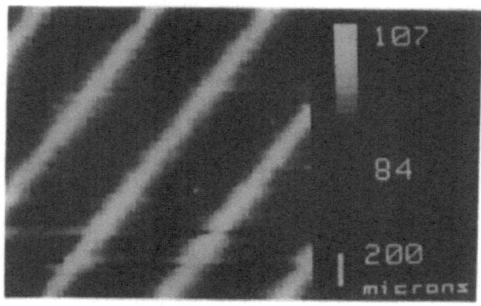

Figure 6. Amplitude image of the subsurface pattern obtained at 31.6 kHz with a 40 µm tip.

Second presented sample is a polycristalline stainless steel. Grains size is in the 30 - 50 µm range. The sample surface has been mechanically polished before imaging. In figure 7, the amplitude image results from a scanning with a 2 µm radius tip at 73.6 kHz. Estimated spatial resolution is ~ 1 µm (smaller than λ_a / 10 000). Image contrast is related to the grain orientation.

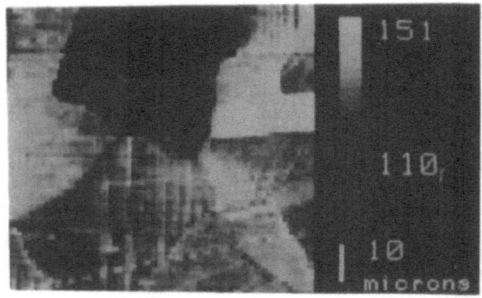

Figure 7. Amplitude image of stainless steel sample obtained with a 2 µm radius tip

CONCLUSION

The acoustic microscope using a vibrating tip enables subsurface imaging at low frequency. The present resolution is in a 1 - 50 µm range, and depends on the tip radius and bearing force. The classical hertzian model shows that the observation depth is a few tip radius, as confirmed with experimental images. Images contrast is related to the microdeformation volume properties and to surface topography.

The presented microscope may become an alternative to high resolution acoustic microscopes that operate at very high frequencies. Indeed, strong attenuation is observed for high frequency acoustic waves (attenuation typically depends on f^2 for most fluids used as coupling medium). Potential application fields of this technique are in material science : all hard and quasiflat objects may be investigated. The microscope enables grain characterization, hardness evaluation and subsurface defects imaging. So it could become a new industrial nondestructive evaluation method.

ACKNOWLEDGEMENTS

The authors would like to thank CETEHOR, Besançon and LTSI, Saint-Etienne, France for their help in bimorph cantilever realization.

REFERENCES

1. W. Dürr, D.A. Sinclair, E. A. Ash, A high resolution acoustic probe, *Electron. Lett.* 21:805 (1980).
2. J. K. Zienuk, A. Latuszek, Ultrasonic pin scanning microscope. A new approach to ultrasonic microscopy, Proceedings IEEE Ultrasonics Symposium, Williamsburg, B.R. Mc Avoy, ed., Pittsburgh, pp. 1037-1039 (1986).
3. B. T. Khuri-Yakub, S. Akamine, B. Hadimioglu, H. Yamada and C.F. Quate, Near field acoustic microscopy, SPIE, 1556, *Scanning Microscopy Instrumentation*, 30 (1991).
4. K. Takata, T. Hasegawa, S. Hosaka, S. Hosoki, T. Komoda, Tunneling acoustic microscope, *Appl. Phys. Lett.* 55:1718 (1989).
5. P. Günther, U. Ch. Fisher, K. Dransfeld, Scanning nearfield acoustic microscopy, *Appl. Phys. B*, 48:89 (1989).
6. A. Kulik, J. Attal, G. Gremaud, Nearfield scanning microscopy, Acoustical Imaging, Nanjing, China, Plenum Press (1992).
7. B.T. Khuri-Yakub, C. Cinbis, C. H. Chou, P. A. Reinholdsten, Nearfield scanning acoustic microscope Proc. IEEE Ultrasonic Symposium, pp. 805-807, (1989).
8. M. T. Huber, Zur theorie der Berührung fester elastischer Körper, *Ann. Physick*, 14:153 (1904).
9. K. L. Johnson, Contact Mechanics, Cambridge University Press, (1987).
10. B. Cretin, W. X. Xie, S. Wang, D. Hauden, Heterodyne interferometers : practical limitations and improvements, *Optics Communications*, 65(3):157 (1988).
11. B. Cretin, F. Sthal, Scanning microdeformation microscopy, *Appl. Phys. Lett.*, 62(8):829 (1993).

A NEW FIELD MEASUREMENT TECHNIQUE FOR ULTRASONIC COMPUTED TOMOGRAPHY

M.Betts, A.Healey, and S.Leeman

Dept. of Medical Engineering and Physics
King's College School of Medicine and Dentistry
Dulwich Hospital, London SE22 8PT, UK

INTRODUCTION

There has been considerable interest in computed tomography with ultrasound (CUT), since the successful introduction of computed tomography with X-rays. Compared to X-ray methods, CUT suffers from a number of field propagation artefacts, that more severely compromise the accuracy of the projection data. These are: refraction and diffraction [Greenleaf, 1982], multipath transmission [Crawford and Kak, 1982], and reflection [Dines and Kak, 1979]. The simpler inversion procedures usually make a number of simplifying assumptions regarding ultrasonic propagation, such as straight path propagation, and diffraction free measurements, and these may significantly degrade the final reconstructed image. The field measurement technique, that we propose, reduces the effects of these artefacts, in comparison to those made with small point hydrophones [Greenleaf and Johnson, 1978], or with transducers of similar characteristics to the transmitting transducer [Glover and Sharp, 1977].

The main purpose of this paper is to explore some of the potential advantages of field measurements made with a large aperture receiving device, of the type that has been demonstrated to be diffraction insensitive [Leeman et al, 1988], and which has a number of potentially advantageous properties for tomographic imaging [Leeman et al, 1991]. A particular property of the hydrophone used in these experiments is its very sharp directivity. This allows measurement of the direction of propagation at which energy emerges from an object after refraction, and thus promises the eventual possibility of a more rapid, and unambiguous, first order correction for refraction artefacts.

Quantities that have been measured to obtain CUT images are the velocity, [Greenleaf et al, 1975], attenuation [Greenleaf, 1974], attenuation slope [Kak and Dines, 1978], and the nonlinearity parameter [Sato et al, 1986]. The large aperture hydrophone, in principle at least, has advantages for all these implementations of CUT, but since velocity reconstructions appears to be the most accurate application of standard field measurement techniques, only "Time of Flight" (TOF) measurements are explored here. Data are acquired for refraction compromised measurements of "fast" and "slow" circularly symmetric phantoms. Reconstructed cross-sections are shown, in order to demonstrate some of the advantages of the large aperture hydrophone technique over that based on the use of point hydrophones.

TIME OF FLIGHT MEASUREMENTS

The simplest geometry for TOF data collection comprises of two transducers, positioned opposite one another, in the standard "line of sight" path. The transmitting transducer emits a pulse, which travels through an object, to be received by a second receiving transducer. The projection data are collected by translating the transducers, or the phantom, and repeating this for many different angles of view. This is not the only scanning geometry but is the one most commonly used [Greenleaf et al 1975; Dines and Kak 1979; Sato et al, 1986].

The transmitted pulse is normally measured using a small point hydrophone [Greenleaf and Johnson, 1978], or a receiver similar to the transmitting transducer [Glover and Sharp, 1977]. We use a large aperture hydrophone, which is constructed from a uniformly sensitive planar PVDF membrane [Leeman et al, 1988], large enough to effectively intercept all of the field. In our experiments its diameter is 50 mm, which appears to be just acceptable for the 19 mm diameter, focused, transmitting transducer.

The two more serious problems compromising TOF measurements are: refraction, which alters the direction of the transmitted energy, and multipath transmission, whereby the transmitted energy reaches the receiving transducer by more than one, non-equivalent, pathway. Only these more serious classes of error will be examined here.

The output from the large aperture hydrophone has a relatively sharp maximum when the normal to its plane is aligned orthogonally with the propagation direction of an incident field. This allows the accurate and rapid assessment of the acoustic axis of a transmitting transducer [Costa et al, 1988]. Thus, in TOF measurements the propagation direction of emergent refracted pulse can be detected by angling the hydrophone and searching for a maximum in the output signal.

Fig. (1.a.) depicts a pulse received through water only; Fig. (1.b.) shows a pulse received after transmission through the phantom, with no large aperture hydrophone or phantom angulation; and Fig. (1.c.) shows a pulse received after transmission through the phantom, with the large aperture hydrophone angled to receive the refracted pulse. The pulse shown in Fig. (1.c.) required the hydrophone to be angled at 2.4° away from direction for the maximal output of a water path signal. In the experimental set-up, the hydrophone is stepped in angular increments of 0.6°, so it is not possible for this procedure to align the hydrophone perfectly orthogonal to the incident field. This represents the accuracy to which the direction of a refracted pulse pathway can be established, relative to the water path. A geometric correction is applied to the large aperture hydrophone, compensate for changes in TOF when measured at the same angulation in a water-only path. This is carried out to correct for the TOF difference with angulation of the hydrophone.

The TOF for a point hydrophone is estimated by establishing the instant when the envelope of the received signal exceeds a fixed threshold value [Crawford Kak; 1982], which set to 4 bits, for an 8 bit digitizer. On the other hand, the TOF for the large aperture hydrophone measurements was estimated by cross correlating the received signal against that for a water-only path. Cross correlation is more suitable for the angle-adjusted measurements made with the large aperture hydrophone, as the received signal, after transmission through the phantom, closely resembles the water-only signal, a circumstance not always evident with

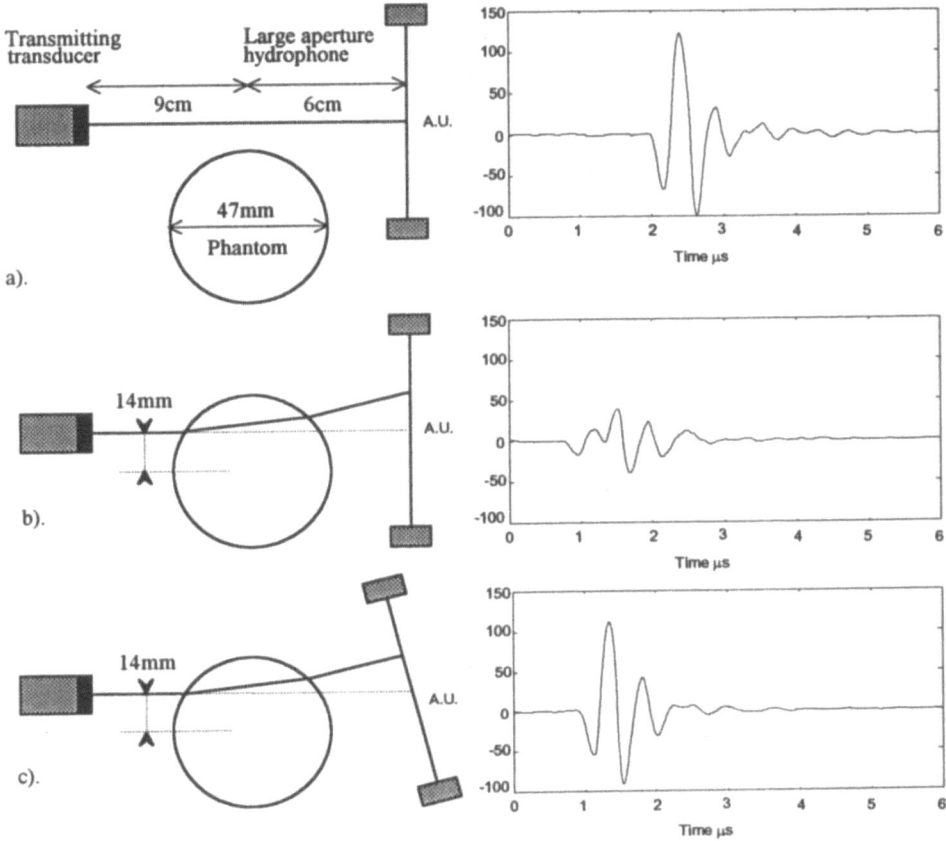

Figure 1. Large aperture hydrophone measurements, a). water-only path only, with hydrophone angled for maximum amplitude signal. b). received after transmission through phantom with no hydrophone angulation. c). pulse received through phantom with hydrophone angled by 2.4 degrees.

point hydrophone measurements. The "direction" of an emergent, refracted pulse is established by the angle for which the large aperture hydrophone signal, $s(t)$, exhibits a maximum "energy", viz. $\int s(t)^2 dt$. In the measurements reported here, all hydrophone signals were recorded by a fully automated technique, without any operator intervention whatsoever. While such an approach is automated, it implies that accuracy of measurements may be considerably improved if the criteria for automated detection were to be tuned more finely.

EXPERIMENTAL MEASUREMENTS

The data acquisition system consists of a Gould GD4050 digital oscilloscope connected to an IBM PC compatible computer via an IEEE 488 interface bus. The signals were acquired at 100 MHz sampling rate, to a nominal 8 bits accuracy and transferred to the PC for off-line processing. The transmitting transducer was a commercial Phillips 19 mm

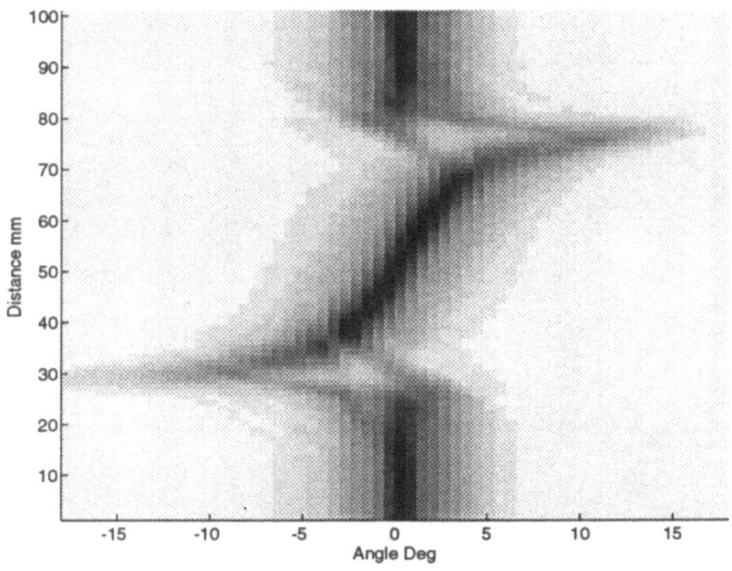

Figure 2. Plot of energy received vs. hydrophone angle for the saline phantom.

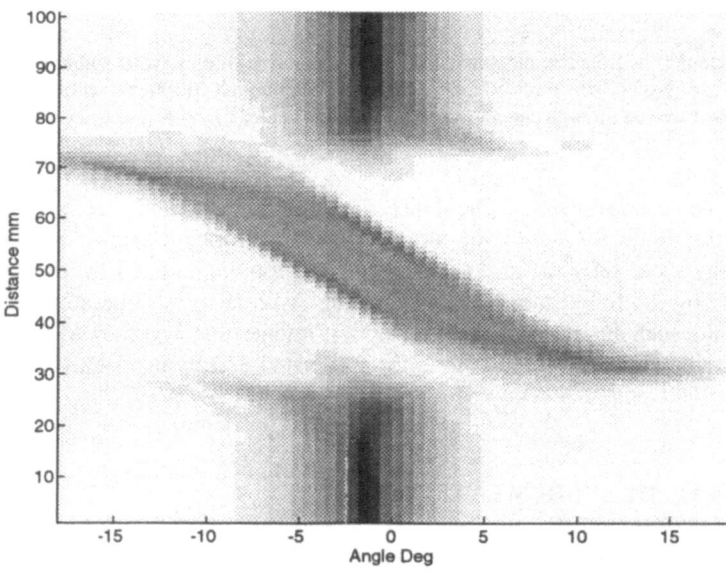

Figure 3. Plot of energy received vs. hydrophone angle for the water/alcohol phantom.

316

diameter, 9 cm focus, 2.25 MHz transducer. The phantom was constructed from a thin latex rubber tube, of 47mm diameter. The phantom is translated a total of 101 mm in increments of 1 mm. The point hydrophone was a 1 mm^2 PZT crystal, while the large aperture hydrophone was a 50 mm diameter, planar PVDF membrane of 25μ thickness

Fig. 2 shows refraction measurements for a phantom filled with saline, which exhibits a velocity that is 4% faster than the surrounding medium. The plot shows energy received vs. hydrophone angle. The horizontal lines show energy received for a particular phantom location. Starting at the top left hand corner, the hydrophone is angled at -18 degrees with respect to the maximum output for water path, the hydrophone is then stepped in increments of 0.6 degrees through to +18 degrees, by the use of computer controlled stepper motors. For water-path transmission, at 0 and 100 mm, the angular dependence of the received energy essentially depicts the directivity pattern of the transmitting transducer [Costa et al, 1988]. The edge of the phantom is intercepted at approximately 30 mm, with the energy refracted strongly away from the central position, and a large loss in signal amplitude due to reflection. The deviations of the refracted energy then decreases as the centre of the phantom is approached at approximately 55 mm, then increases again as the bottom of the cylinder is approached. The water-path only signal is regained, at approximately 80 mm.

The refraction measurement for a "slow" cylinder is shown in Fig. 3. The solution used is a water/isopropyl alcohol mixture, with a velocity 8% slower than the surrounding distilled water. The depicted pattern is, to first approximation, a mirror image of that measured for the "fast" cylinder, allowing for the velocity difference. This is, the expected behaviour for the "slow" cylinder. The actual velocity of the solutions used for these experiments was measured independently by a substitution method.

RECONSTRUCTED PROFILES

Reconstructions were carried out with a standard filtered-backprojection algorithm, on the assumption of perfect rotational symmetry for the phantom

A cross-sectional profile through the centre of the reconstructed image of a saline cylinder is shown in Fig. 4. For·point hydrophone measurements the profile exhibits an overestimation of the diameter, in agreement with other published results [Greenleaf, 1982], and an underestimation of the velocity. Large aperture hydrophone measurements are better at recovering the diameter of the cylinder, and approach closer to the correct velocity.

Reconstructed profiles for the alcohol cylinder are shown in Fig. 5. The point hydrophone profile has a large overestimate of the velocity, particularly at the edges of the cylinder, and an underestimation of the diameter of the cylinder. The large aperture hydrophone measurement has a better estimate of the diameter of the cylinder, but with an overestimation of the velocity particularly at the edges, although not as great as the error with point hydrophone measurements.

An improvement in the recovered velocity values would be expected if the reconstruction, for both measurement techniques, incorporated the "true" propagation paths, rather than the straight ray paths assumed for the filtered backprojection algorithm.

CONCLUSIONS

Experimental results have been shown that contrast point hydrophone measurements, against a new large aperture hydrophone measurement for use in ultrasonic TOF tomography. The initial results show a quantitative improvement for the measurements made

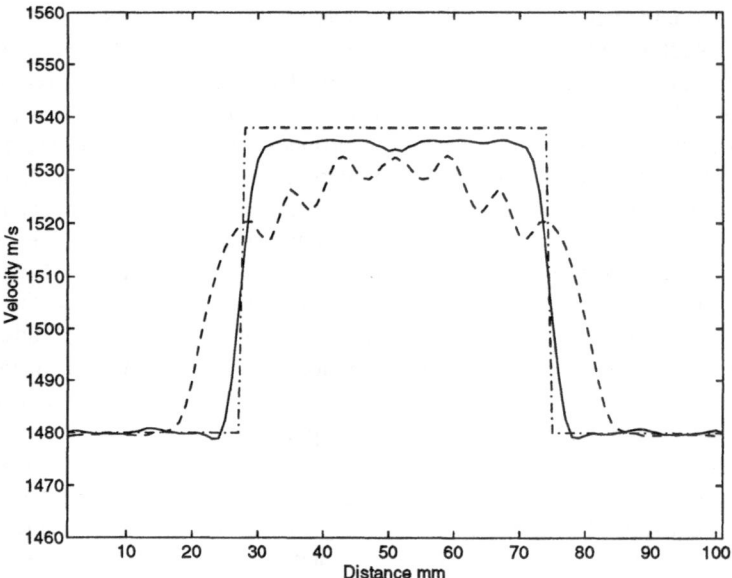

Figure 4. Reconstructed velocity profiles for the saline phantom. -.- Ideal, - Large aperture hydrophone, -- Point Hydrophone.

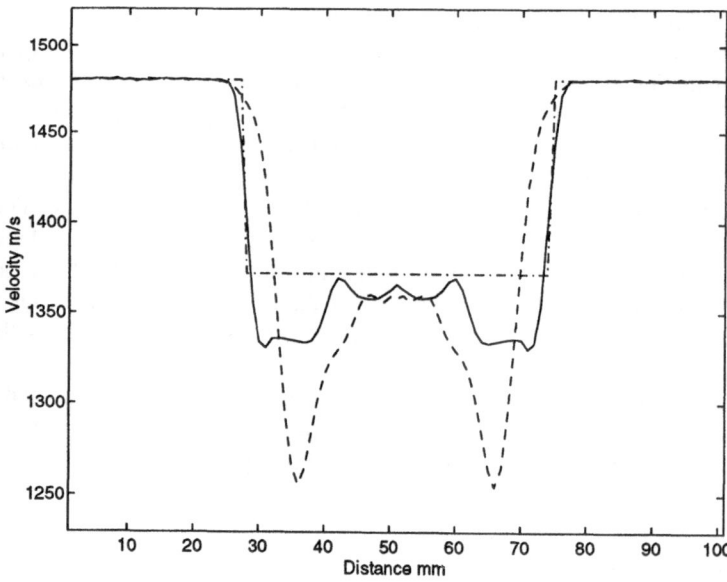

Figure 5. Reconstructed velocity profiles for the water/alcohol phantom. -.- Ideal, - Large aperture hydrophone, -- Point Hydrophone.

with the large aperture hydrophone, for two simple "fast" and "slow" circularly symmetric phantoms, reconstructed using the filtered backprojection algorithm. An advantage of the technique is that the acoustic axis of emergent refracted pulses can be identified, which may be useful as an input to a refraction correction technique that estimates, the true propagation paths. Work is in progress to investigate more realistic imaging situations.

ACKNOWLEDGEMENTS

The Scientific and Engineering Research Council, and the Welcome trust are gratefully acknowledged for their financial support.

REFERENCES

Costa. E. T, Leeman. S, Hoddinott. J. C, "A New Approach Towards Measurement of Three-Dimensional Pulses: Linear and Non-linear fields", Physics in Medical Ultrasound II., Eds: Evans. D. H, and Martin. K, IPSM, York U.K., pp. 87-93, 1988.

Crawford. C. R, Kak. A. C, "Multipath Artefact Corrections in Ultrasonic Transmission Tomography", Ultrasonic Imaging, Vol. 4, pp. 234-266, 1982.

Dines. K. A, Kak. A. C, "Ultrasonic Attenuation Tomography of Soft Tissues", Ultrasonic Imaging, Vol. 1(1), pp.16-33, 1979.

Glover. G. H, Sharp. J. C, "Reconstruction of Ultrasound Propagation Speed Distributions in Soft Tissue: Time of Flight Tomography", IEEE Tran. Son. Ultrason., Vol. 24, No. 4, pp. 229-234. July 1977.

Greenleaf. J. F, Johnson. S. A, Lee. S. L, Herman. G. T, Wood. E. H, "Algebraic Reconstruction of Spatial Distributions of Acoustic Absorption Within Tissue from their Two Dimensional Acoustic Projections", Acoustical Holography, Vol. 5, pp. 591-603, 1974.

Greenleaf. J. F, Johnson. S. A, Samoya. W. F, Duck. F. A, "Algebraic Reconstruction of Spatial Distributions of Acoustic Velocities in Tissue from their Time of Flight Profiles", Acoustical Holography, Vol. 6, pp 71-90, 1975.

Greenleaf. J. F, Johnson. S. A, "Measurement of Spatial Distribution of Refractive Index in Tissues by Ultrasonic Computer Assisted Tomography", Ultrasound in Med. & Biol., Vol. 3, pp. 327-339, 1978.

Greenleaf. J. F, "Effects of Diffraction and Refraction on Computer-Assisted Tomography with Ultrasound", Proc. Soc. Photo. Opt., pp. 223-239, 1982.

Kak. A. C, Dines. K. A, "Signal Processing of Broadband Pulsed Ultrasound Measurement of Attenuation of Soft Biological Tissues", IEEE Trans. Biomed. Eng. BME-25, 321-344, 1978.

Leeman. S, Costa. E. T, Healey. A. H, "Diffraction Artefacts and their Removal", Acoustical Imaging, Vol. 17, pp. 403-411, 1988.

Leeman. S, Costa. E. T, Healey. A. J, "Reconstruction Imaging Without Artefacts", Acoustical Imaging, Vol. 19, pp. 23-27, 1991.

Sato. T, Yamakoshi. Y, Nakamura. T, "Non-linear Tissue Imaging", Proc. IEEE Ultrasonic Symposium, pp. 889-900, 1986.

LOW BIT-RATE IMAGING IN SCANNING TOMOGRAPHIC ACOUSTIC MICROSCOPY

S. Davis Kent and Hua Lee

Department of Electrical and Computer Engineering
University of California, Santa Barbara
Santa Barbara, California 93106–9560

INTRODUCTION

Acoustic microscopy has been successfully employed in many areas, such as medical imaging and non-destructive evaluation. Using acoustic waves, the technique is able to obtain important information of the specimen under evaluation that is not available or that is more difficult to obtain with other means. Properties such as the acoustic velocity and attenuation in biological tissue, for example, is readily obtained with acoustic microscopes. This information can be interpreted by pathologists to identify the condition of tissues. In the area of non-destructive evaluation, the differences in attenuation of flaws such as voids and the surrounding materials is easily identified using acoustic waves. In order to expand the usefulness of acoustic microscopes, holographic and tomographic techniques have been employed which have the effect of improving system resolution, but at the expense of increased computational complexity and increased bit rates.[1, 2]

In the optimization of imaging systems, such as acoustic microscopy, we are constrained by limits on computation time, data rate or bandwidth, data acquisition periods, and data storage. In this paper, we study a method for reducing the bit rates associated with the Scanning Tomographic Acoustic Microscope (STAM). The goal of this paper is to reduce the amount of data required to perform tomographic reconstructions without sacrificing system resolution.

PROBLEM FORMULATION

In the STAM system, high frequency plane waves are used to illuminate a three-dimensional specimen as shown in Fig. (1).[3] The incident acoustic wave becomes modulated by the specimen's structure and propagates to the detection plane where a mirrored coverslip deflects in response to the acoustic wave. A laser beam scans out a raster pattern over the surface of the coverslip and is reflected to a knife-edge and photodetector where the angular modulation of the light beam caused by the acoustically induced ripple is converted into an intensity variation. This intensity signal is passed through a quadrature demodulator which provides both the in-phase (I) and the quadrature-phase (Q) components of the detected wave field. Detecting both the I and Q components allows retention of both magnitude and phase components of the wave field that has propagated through the specimen. Knowledge of the phase is crucial since it allows us to backpropagate, or focus, the received wave field to the desired depth, Δz, by application of the backpropagation filter of Eq. (1)

$$H(f_x, f_y; \Delta z) = \begin{cases} \exp\left\{-j2\pi\Delta z \sqrt{1/\lambda^2 - f_x^2 - f_y^2}\right\}, & f_x^2 + f_y^2 < 1/\lambda^2 \\ 0, & otherwise \end{cases} \quad (1)$$

where f_x and f_y are the spatial frequency indices in the x and y directions respectively, and λ is the wavelength of insonification.[4, 5, 6] In holographic reconstructions, demodulation of the resulting wave field yields the final image. In tomographic reconstructions several backpropagated wave fields are combined to form a mean-squared error estimate of the transmission characteristic of a plane in the specimen. All images presented in this paper are formed in this manner.

In a STAM system currently under development, the I and Q channels of the quadrature demodulator are sampled in a 256 x 256 grid using eight bits per pixel per channel. For a single projection this totals 128 kB of storage space per projection. In multiple angle tomography where projections may be recorded at steps of 1° of rotation of the incident wave field, for example, the total amount of storage required is 45 MB. For noise reduction, multiple frequency projections may also be obtained, which can increase the memory storage requirements significantly. Therefore, there is sufficient motivation for identifying memory-efficient methods for quantizing the received wave fields while maintaining image quality. With low bit-rate data representation, memory-efficient storage requirements can be achieved.

In order to evaluate the performance of bit allocation schemes three criteria are used for a quantitative analysis: entropy, contrast, and the mean-squared error of the reconstructed image. These criteria are also related to a qualitative evaluation of image quality.

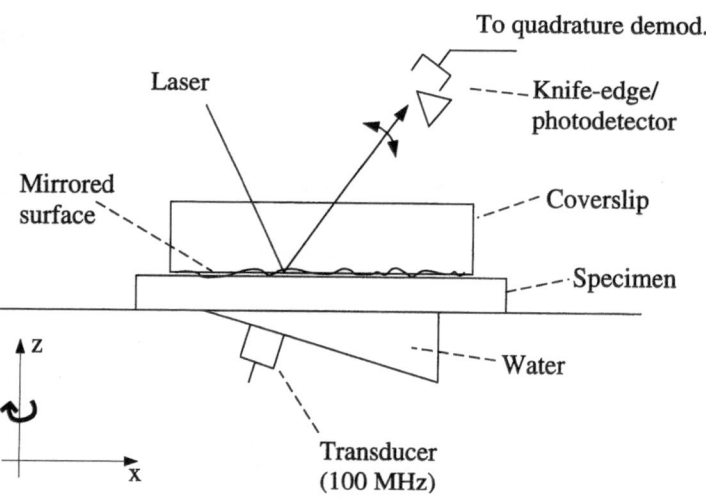

Figure 1. Geometry of Scanning Tomographic Acoustic Microscope Stage.

Contrast

The definition of contrast used in this paper is

$$C(I) = \frac{E\{I^2\} - E\{I\}^2}{E\{I\}} \qquad (2)$$

where I is the intensity of the reconstructed wave field and $E\{\cdot\}$ is the expectation operator evaluated over some region of the image. As defined, contrast is a normalized variance of the intensity, which is useful for measuring the quality of an image since the human eye is sensitive to spatial variations in intensity and since contrast in an image facilitates discerning structures. By using the contrast measure we can determine bit allocations such that contrast and dynamic range are optimized.[7]

Entropy

In defining the entropy of an image, it is convenient to treat each pixel in the image as an event. The probability associated with the jth pixel or event is considered to be proportional to its intensity. Thus, the probability is defined as

$$p_j = I_j / \sum_{i \in \gamma} I_i \qquad (3)$$

where γ is the neighborhood of pixels around pixel j. Entropy, which is a non-decreasing measure of disordered energy in a system, is defined by

$$H(I) = -\sum_{j \in \gamma} p_j \log p_j. \qquad (4)$$

The entropy measure has been used in acoustical imaging to perform such tasks as the estimation of imaging parameters such as structure depth and wave propagation velocity. Used in this way, entropy measures the amount of blurring or spreading of the reconstruction of the specimen. Similarly, here we assume that the most accurate reconstruction of the specimen will exhibit the greatest structure or order. Therefore, we seek a bit allocation strategy that minimizes the entropy of the reconstructed image.[8, 9]

Mean-Squared Error

As a quantitative measure of the accuracy of a reconstruction, the mean-squared error (MSE) criterion of Eq. (5) is used.

$$MSE = \frac{1}{X_n Y_n} \sum_{i=1}^{X_n} \sum_{j=1}^{Y_n} (x_{ij} - \tilde{x}_{ij})^* (x_{ij} - \tilde{x}_{ij}) \qquad (5)$$

In Eq. (5), X_n and Y_n are the dimensions of the image, x_{ij} is the $(i,j)^{th}$ pixel of the original reconstruction, and $\tilde{x}_{ij}$ is the $(i,j)^{th}$ pixel of the reconstruction using a reduced number of bits to represent the received wave field. This measure assumes that the original reconstruction yields the best possible reconstruction.

The data used in this paper are obtained from an older STAM system equipped with only a 256 x 240 6 bit frame grabber. For a single projection, two frames must be captured to obtain both the I and Q channels. Therefore, the data is quantized to 64 levels for the in-phase and 64 levels for the quadrature-phase. This poses an upper limit of 4096 total levels on the quantizer resolution.

The goal of this study is to reduce the amount of memory required to represent the data while maintaining low MSE and entropy, high contrast, and a perceptually good image. This may be performed by a process known as resampling or requantization. One method is to reduce directly the number of levels used to quantize the I and Q components. However, since $I = A \cos \phi$ and $Q = A \sin \phi$, where A is the amplitude of the signal and ϕ is the phase, if we assume A and ϕ are statistically independent, the statistics of the I and Q components are identical to each other. This assumption has been found to be valid when the object being imaged can be modelled as a collection of point sources, none being dominant.[10] Therefore, a quantization scheme should allocate quantization levels equally to both the I and Q channels. This fact removes flexibility in assigning quantization levels to represent the data.

An alternative is to convert the I and Q data to polar form. The polar version has some important advantages over the rectangular resampling. Most important is the ability to address the issue of accurate phase representation. In holographic and tomographic reconstructions of acoustical subsurface images, research has determined that the phase plays a significant role in image quality. On one extreme, knowledge of only the magnitude prohibits the possibility of subsurface imaging. On the other extreme, however, tomographic reconstruction is possible with only the phase data.[9, 11, 12]

Experimental Setup

In order to determine an efficient allocation scheme, multiple angle projections of a two layer specimen are taken at rotation angles of 0, 90, 180, and 270 degrees, clockwise about the z-axis. For multiple frequency studies, projections with insonifying wave field frequencies of 98, 100, 104, and 105 MHz are also available. These received wave fields are converted to magnitude and phase data files. Each file is then requantized, allocating a certain number of levels to the magnitude and a certain number of levels to the phase. These magnitude and phase files are recombined to form the requantized wave fields. Holographic and tomographic reconstructions are then performed and the contrast, entropy, and MSE criteria are evaluated.

Quantizer Design

The quality of the reconstruction using a small number of quantization levels is highly sensitive to the quantizer design. In order to requantize the magnitude and phase, a two step quantizer design process is implemented:[13]

    I.   Initial quantizer design by a splitting algorithm, and
    II. Quantizer refinement via the Lloyd II iterative algorithm.

This procedure is chosen for its simplicity and its ability to generate accurate quantizers tailored to the input data.

The splitting algorithm of Step I begins by determining the proper quantization output level necessary to minimize the MSE of a 1 level quantizer. Without making other changes to the quantizer, the algorithm then determines where a second output level should be placed in order to achieve the largest improvement in MSE over the 1 level quantizer. With the addition of this output level, the partitions that map the input values to the closest output level are determined. This process of adding output levels and computing partitions continues until a quantizer with the desired number of levels is formed. This algorithm has the attractive feature that the MSE always decreases with each iteration. To provide for a better initial quantizer, it is possible to insert a refinement stage, such as a Lloyd algorithm step, after each iteration. This has the benefit of shortening the total quantizer design time when a quantizer with a large number of levels is being generated.

The second quantizer step is to refine the initial quantizer provided by step I. A modified Lloyd II algorithm is employed which iteratively improves the quantizer design using the actual data to be quantized. This algorithm modifies an existing N-level quantizer to produce an N-level quantizer with smaller MSE. This is done in two steps. First, the partitions are adjusted so that the input data are mapped to the nearest output level. Second, the output levels are adjusted to minimize the MSE of the quantizer for these new partitions. If the improvement in MSE of the new quantizer is small, the algorithm terminates, otherwise this process is repeats. This design procedure generates a non-uniform quantizer whose output closely represents the data file. Once the data is requantized, the quantizer may be discarded or it may be used as the initial quantizer for quantizing the next data file.

RESULTS

Tomographic Imaging (Multiple Angle)

Each multiple angle projection of the two-layer specimen of Fig. (2) is requantized in magnitude and phase, using up to 16 levels for each. The requantized wave fields are used to form a tomographic reconstruction of the bottom layer, as described earlier. Each reconstruction is evaluated using the entropy, contrast, and mean-squared error criteria, as well as a visual comparison of the resampled images and the reconstruction from the original data.

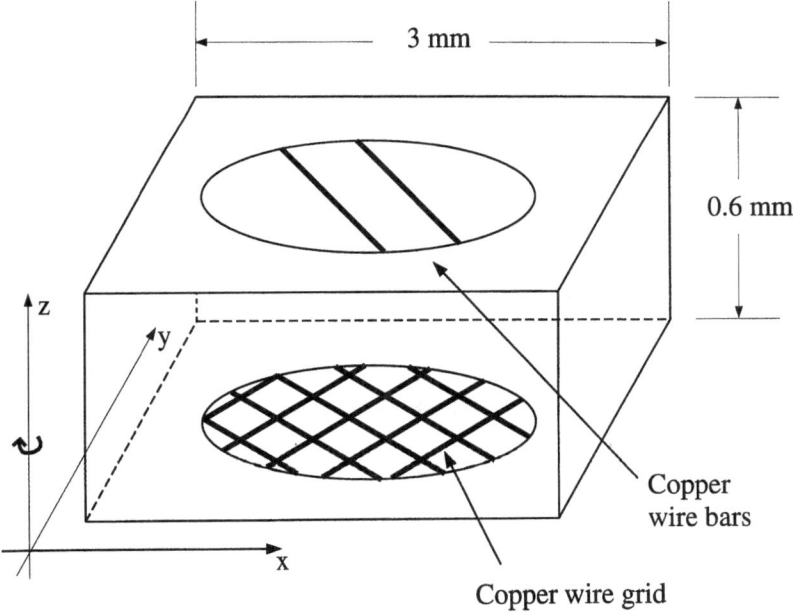

Figure 2. Illustration (not to scale) of the 3 mm square wire bar and grid specimen used for all experimental results. The specimen is illuminated from below, and the propagated wave field is detected 0.3 mm above the top plane.

Figs. (3) and (4) depict the contrast and the MSE of the magnitude, respectively, for the multiple angle case as a function of levels allocated to the magnitude and phase. In these figures, the dashed curves represent the possible allocation of bits to the magnitude and phase for a constant number of total bits. The contours of the contrast plot are evenly spaced, as shown. The contours of the MSE graph are spaced at 1 dB intervals, where the decibel values are computed via the equation $db = 10 \log A/A_{max}$ where A is the value of the MSE and A_{max} is the maximum value of the MSE.

In order to arrive at a desirable allocation of bits, these contour plots are studied. The most striking feature of these graphs is that for any number of levels of phase, the incremental improvement in MSE or contrast with increasing levels of magnitude quickly diminishes, indicating that excessive levels attributed to the magnitude is unnecessary. For example, if we consider the case where we represent the data using a total of 6 bits, we see that allotting 4 bits to the phase and 2 bits to the magnitude is superior to an alternative of 2 bits to the phase and 4 bits to the magnitude. From these graphs it is evident that for a fixed number of bits it is preferable to allocate finer representation to the phase than to the magnitude. To determine a suitable bit allocation strategy, a 1 dB criteria is used to indicate how few levels to assign to the magnitude and still stay within 1 dB of the minimum MSE for a given number of phase levels. The 1 dB criteria is chosen because a visual inspection of the reconstructions reveals little noticeable difference in image quality when the MSE increases by 1 dB. Using this criteria, we see that for a large number of phase levels, we need fewer than half that number of magnitude levels. For convenience in data handling, a design goal for data compression is to use no more than 8 bits (1 byte) per pixel without sacrificing image quality. With this constraint, it is found that representing the magnitude with $L_M = 10$ levels and the phase with $L_P = 25$ levels is desirable. It is interesting to note that although this choice uses only 250 of the 256 levels possible with 8 bits, the results are superior by 1.5 dB in MSE of the magnitude to the equal allocation case $L_M = L_P = 16$, which uses all possible levels. A visual, qualitative comparison of the bit-rate reduced reconstruction to the original reconstruction reveals little difference, with only a slightly noisier image and no apparent loss of resolution. The results of the entropy measure are not presented, however, the results are consistent with the MSE and contrast measures.

Tomographic Imaging (Multiple Frequency, Conventional Quantization)

Two methods of requantizing the multiple frequency data are studied. The first is identical to that used for the multiple angle data. The second method takes advantage of the redundancy in magnitude between multiple frequency projections. Doing so allows us to represent the magnitude with fewer levels than quantizing each projection separately.

Requantizing each projection separately yields results similar to the multiple angle case, as seen in the contrast and MSE plots of Figs. (5) and (6), respectively. The evaluations based on contrast and entropy indicate that near-maximum contrast is achieved for as few levels as $L_M = 7$ and $L_P > 10$, with the plot showing steady contrast for $L_P > 14$, thus employing 7 bits. Again, approximately twice as many levels should be allocated to the phase than to the magnitude. If 8 bits are used in the quantization, using $L_M = 10$ and $L_P = 25$ yields an improvement of 1.3 dB in MSE of magnitude and 0.34 dB in MSE of phase over the case of $L_M = L_P = 16$.

Tomographic Imaging (Multiple Frequency, Redundancy Quantization)

When performing multiple frequency tomography, there is a significant redundancy in the magnitude between projections. If this redundancy can be removed, then it is possible to perform more efficient quantization. An effective procedure to accomplish this goal is to quantize the difference of the magnitudes of two projections. Given the large amount of correlation between projections the difference will have a smaller variance, allowing for improved quantization.[13] There is very little apparent correlation in the phase and, therefore, it is better to quantize the phase of each projection independently of the others. The method employed in this study is to quantize the magnitude difference between each file and a reference file. Doing so allows simple decoding of the stored image, requiring only the summation of the difference file and the reference. An alternative, which is rejected as being cumbersome and prone to propagation of noise, is to quantize and store the difference of successive frequency projections.

The contrast and MSE plots of Figs. (7) and (8) show a substantial gain in performance of over the conventional quantization case. For example, when we choose the preferable allocation of $L_M = 10$ and $L_P = 25$, the MSE of the reconstruction is improved by about 1.2 dB in magnitude over the case when redundancy is not considered, and each projection is requantized separately. While the gains of redundancy quantization are substantial, qualitatively there is no significant visual difference between the two methods.

CONCLUSION

In this paper we study low bit rate imaging as it applies to the Scanning Tomographic Acoustic Microscope. A principal goal is to provide guidelines for allocating quantization levels to the magnitude and phase. The need for this flexibility is indicated in the density of the contours of the contrast and MSE plots. We also provide a mechanism for estimating future system performance when storage requirements dictate specific bit rates.

For an existing STAM system, we have found that 8 bits of information, as opposed to 12 bits for the existing system, is sufficient for accurately representing the received wave fields when performing holographic, multiple-angle, or multiple-frequency reconstructions. This represents a 33% memory savings. The importance of the phase information has been emphasized and proper bit allocation to the phase has been shown to allow high resolution image quality at reduced bit rates.

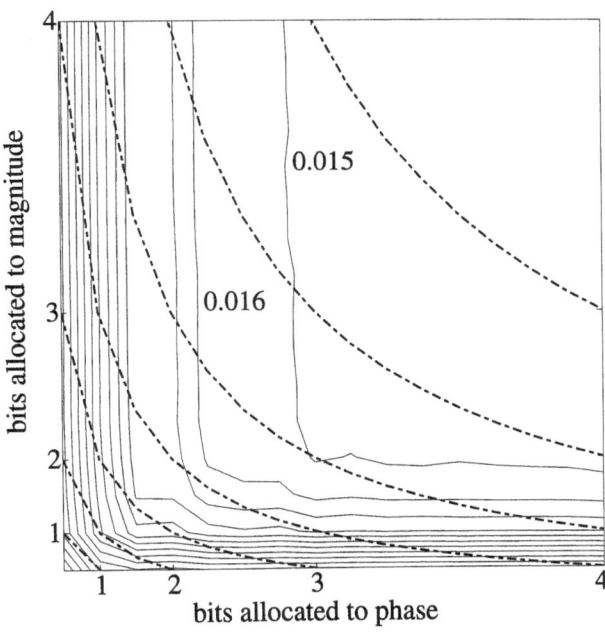

Figure 3. Contour plot of contrast as a function of bits assigned to the magnitude and the phase for a multiple angle tomographic reconstruction. Dashed lines are lines of constant total bits used in representation.

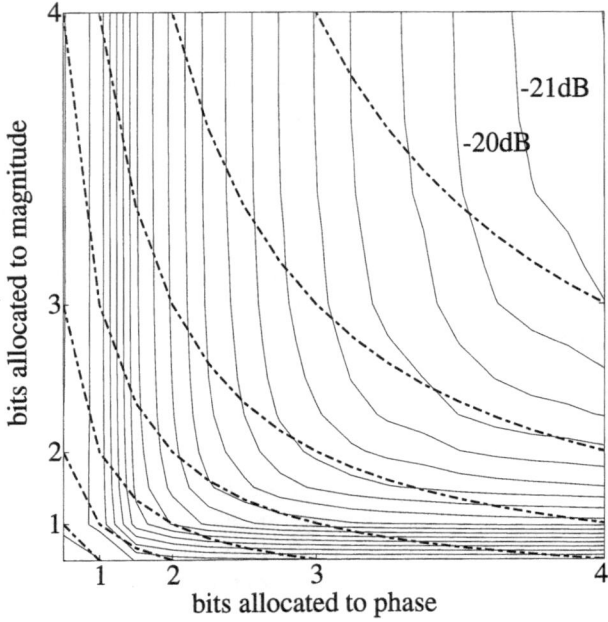

Figure 4. Contour plot of MSE of the magnitude as a function of bits allocated to the magnitude and the phase of a multiple angle tomographic reconstruction. Contours at 1 dB intervals. Dashed lines are lines of constant total bits used in representation.

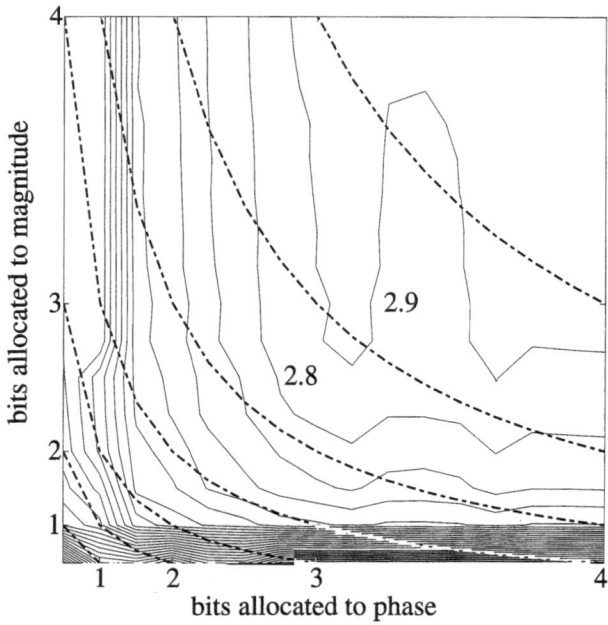

Figure 5. Contour plot of contrast as a function of bits assigned to the magnitude and the phase for a multiple frequency tomographic reconstruction using conventional quantization. Dashed lines are lines of constant total bits used in representation.

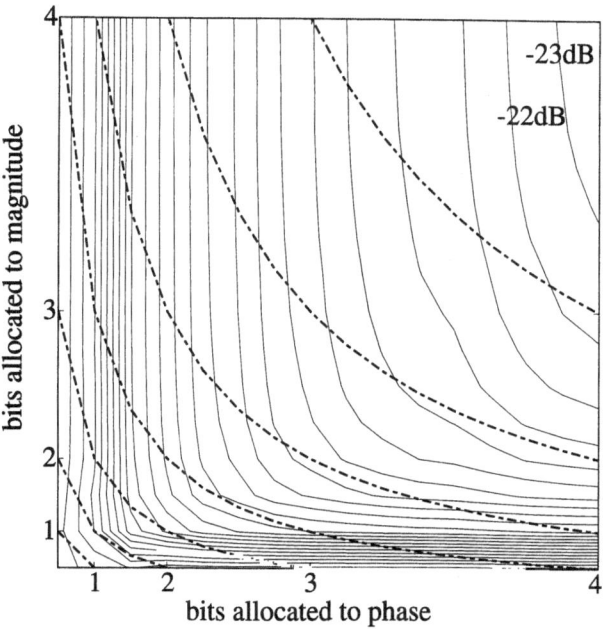

Figure 6. Contour plot of MSE of the magnitude as a function of bits allocated to the magnitude and the phase of a multiple frequency tomographic reconstruction using conventional quantization. Contours at 1 dB intervals. Dashed lines are lines of constant total bits used in representation.

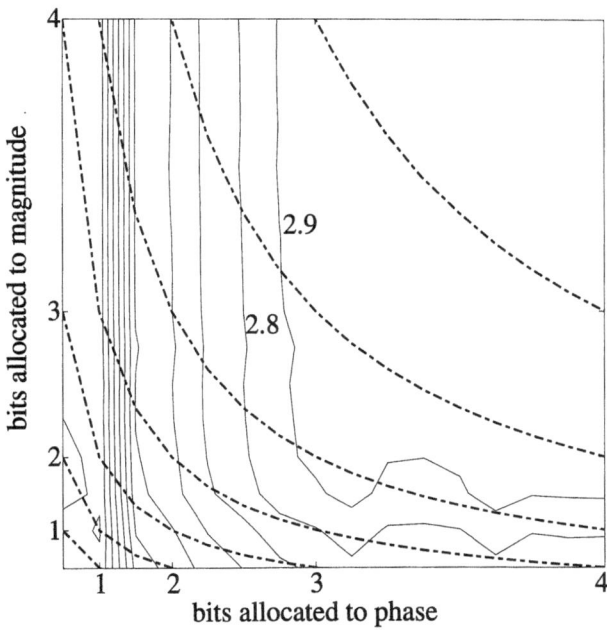

Figure 7. Contour plot of contrast as a function of bits assigned to the magnitude and the phase for a multiple frequency tomographic reconstruction using redundancy quantization. Dashed lines are lines of constant total bits used in representation.

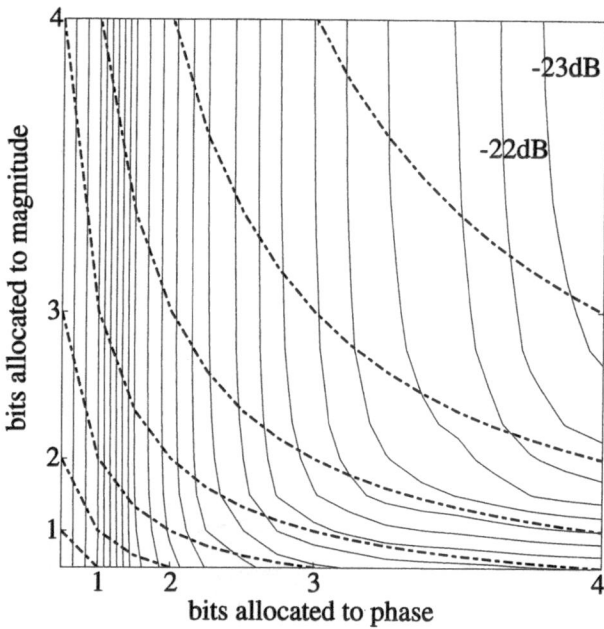

Figure 8. Contour plot of MSE of the magnitude as a function of bits allocated to the magnitude and the phase of a multiple frequency tomographic reconstruction using redundancy quantization. Contours at 1 dB intervals. Dashed lines are lines of constant total bits used in representation.

ACKNOWLDEGEMENT

This research is supported by the National Science Foundation under Grant No. MSS-9020556.

REFERENCES

1. J. F. Havlice and J. C. Taenzer. Medical ultrasonic imaging: An overview of principles and instrumentation. *Proceedings of the IEEE*, 67(4):620–641, Apr. 1979.

2. L. W. Kessler and D. E. Yuhas. Acoustic microscopy — 1979. *Proceedings of the IEEE*, 67(4):526–536, Apr. 1979.

3. H. Lee and C. Ricci. Modification of the scanning laser acoustic microscope for holographic and tomographic imaging. *Applied Physics Letters*, 49(20):1336–1338, Nov. 1986.

4. Z. C. Lin, H. Lee, and G. Wade. Back-and-forth propagation for diffraction tomography. *IEEE Transactions on Sonics and Ultrasonics*, SU-31:626–634, Mar. 1984.

5. R. Y. Chiao. *Signal Processing and Image Reconstruction for Scanning Tomographic Acoustic Microscopy*. PhD thesis, University of Illinois, Urbana-Champaign, July 1990.

6. J. W. Goodman. *Introduction to Fourier Optics*. McGraw-Hill Book Company, New York, 1968.

7. J. S. Lim. *Two-dimensional Signal and Image Processing*. Prentice-Hall, New Jersey, 1990.

8. A. Papoulis. *Probability, Random Variable, and Stochastic Processes*, chapter 15. McGraw-Hill, New York, 2nd edition, 1984.

9. B. L. Douglas, S. D. Kent, and H. Lee. Depth parameter estimation and the use of phase information in tomographic acoustic microscopy. *Proceedings of 1992 IEEE Ultrasonics Symposium*, 1992.

10. B. D. Steinberg. A theory of the effect of hard limiting and other distortions upon the quality of microwave images. *IEEE Transactions on Acoustics, Speech, and Signal Processing*, ASSP-35:1462–1472, Oct. 1987.

11. H. Lee and G. Wade. Evaluating quantization error in phase-only holograms. *IEEE Transactions on Sonics and Ultrasonics*, SU-29(5):251–254, Sept. 1982.

12. H. Lee and J.-H. Chuang. Performance evaluation of phase-only technique for high-resolution holographic imaging. In L. W. Kessler, editor, *Acoustical Imaging*, volume 16, pages 227–236. Plenum Press, New York, 1988.

13. A. Gersho and R. M. Gray. *Vector Quantization and Signal Compression*. Kluwer Academic Publishers, Boston, 1992.

SINGLE-CHANNEL DIGITAL DEMODULATION
OF ULTRASOUND BANDPASS SIGNALS FOR FLOW-IMAGING APPLICATIONS

L.Bessi, F.Guidi, F.Gucci, C.Atzeni and P.Tortoli

Electronic Engineering Department - University of Florence
via S.Marta, 3 - 50139 Firenze, Italy

INTRODUCTION

Operation of ultrasound Pulsed Wave (PW) flowmeters is based on the emission of a burst of acoustic energy at a rate PRF[1]. Typically, the Radio-Frequency (RF) received echoes are synchronously demodulated on two quadrature channels, so that, for each depth of interest, a couple (I and Q) of samples are selected. The continuous Doppler signal can be reconstructed through Low-Pass Filters (LPF's) which are also useful to minimize the noise bandwidth. Finally, suitable High-Pass Filters (HPF's) can be used to reject the unavoidable high-level low-frequency components ("clutter"), originated from vessel wall slow movements. For a correct detection of Doppler frequencies, careful amplitude and phase match between the two channels must be ensured.

A possible alternative to quadrature demodulation schemes is represented by "digital demodulation", which involves direct sampling of the received RF signal[2]. The approach is attractive because it significantly reduces the amount of needed hardware, by avoiding, in particular, any distorsion due to demodulators or channel mismatch. The first problem to be faced is represented by the possible overlap of adjacent spectral images originated by sampling. This can be overcome by reconstructing the complex band-pass signal[3], i.e. by sampling the received signal, two times for each depth of interest, with an interval of one-fourth the period of the transmitted carrier[4]. Again, the timing requirements between two subsequent sampling pulses are here critical, and call for the use of very high speed sampler and Analog to Digital Converter (ADC).

We have investigated the possibility of limiting acquisition to a single RF sample, taken in correspondence of the depth of interest, for each pulse repetition interval. It can be shown that, even if the complex bandpass signal is not reconstructed, no information is lost if the PRF is related to the frequency of the transmitted burst, f_0, through the relationship:

$$PRF = \frac{4f_o}{(2K + 1)} \qquad (1)$$

(with K being a suitable integer number[5]). This in fact corresponds to transmitting pulses with a phase which is changed by 90° from one pulse repetition interval to the next (see Fig.1). This phase shift impressed to the transmitted burst is directly transferred to the echoes received in subsequent intervals: by sampling these phase-shifted echoes at the PRF rate, a PRF/4 frequency shift is inherently impressed to the Doppler spectrum. Digital demodulation is thus performed in such a way that low Doppler frequencies corresponding to fixed or slowly moving targets are moved around PRF/4, and possible "clutter" must be eliminated through a band-reject filter in place of the HPF used in conventional systems.

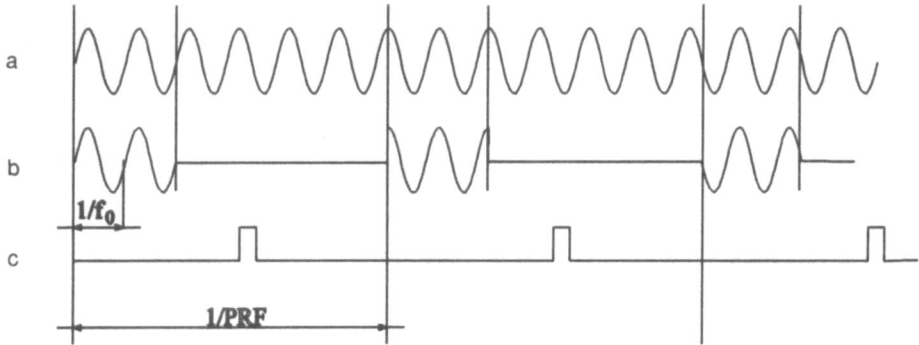

Figure 1. Timing of the single channel digital demodulator:
a) RF carrier b) Transmitted pulses c) Receiver sampling gates

Discrimination of flow direction is still possible, since flow approaching the ultrasound transducer yields frequencies between PRF/4 and PRF/2, while receding flow yields frequencies in the 0-PRF/4 range. The maximum Doppler shift frequencies must be limited within ±PRF/4, because the spectrum of the "real" signal sampled at the PRF rate is symmetrical around the origin of the frequency axis, with images spaced PRF apart. The analysable bandwidth is therefore one half with respect to that analysable in quadrature systems (i.e., one sampling channel is exchanged with a reduction of the analysable bandwidth).

In the next section, an implementation of the proposed technique is presented. The possibility of filtering out unwanted components, directly in the frequency domain, is discussed. A solution to the clutter rejection problem, which is critical in flow-imaging systems, is also presented and experimentally demonstrated.

EXPERIMENTAL SYSTEM

We have built an ultrasound PW system implementing the RF sampling technique discussed above. The system operates with a 32 MHz master clock, and analog electronics includes only transmitter and receiver sections around an 8 MHz ultrasound probe. The PRF is obtained by division of the 32 MHz system clock through programmable digital counters. In the experiments shown here, the PRF was fixed at approximately 20000 Hz, corresponding to 32 MHz divided by 1601 (i.e., according to eq.(1), K was here chosen equal to 800). Sampling is achieved by means of a low-jitter Sample/Hold (AD9100 by Analog Devices) while a low-cost ADC has been chosen with 12 bit resolution to accomodate the wide dynamic range due to the possible simultaneous presence of clutter and low-level useful signal.

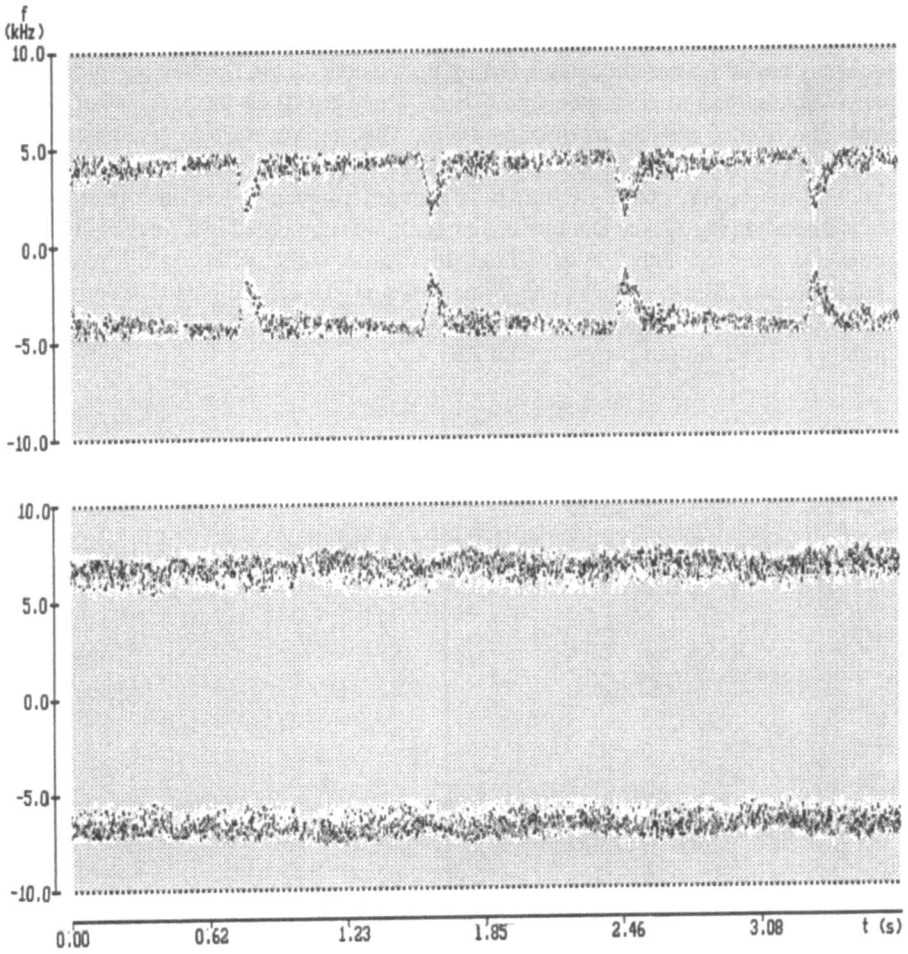

Figure 2. Full spectrograms obtained by RF sampling, at 20 kHz rate, human carotid artery (top) and jugular vein (bottom). Clutter was here removed through frequency domain filtering.

Frequency analysis is performed by means of a Fast Fourier Transform (FFT) algorithm implemented on the TMS320C25 Digital Signal Processor (DSP). Fig.2 shows the results obtained by analysing samples from human carotid artery and jugular vein, displayed in the spectrogram format typically used in clinical applications (with time on horizontal axis, frequency on vertical axis and the brightness of each pixel proportional to the detected instantaneous power spectral density). These results confirm that the spectrum is symmetrical around the origin of the frequency axis and therefore FFT analysis could be limited to the computation of spectral points corresponding to the positive frequencies. Apart from the expected PRF/4 frequency offset, these spectrograms are perfectly equivalent to those obtained by means of conventional approaches. In particular, bidirectional information contained in the RF signal is maintained, as evidenced by the different frequency ranges occupied by spectrograms related to contradirected flows such as those in carotid artery and jugular vein.

These results have been obtained by operating the needed band-reject filtering directly in the frequency domain, after imposing a Hamming window on the Doppler samples at the FFT input. Weighting is particularly beneficial here[5], since it restricts the presence of clutter components to a limited number of frequency bins. This improvement is demonstrated in Fig.3 where spectra derived from analysis of a carotid artery in correspondence of the systolic peak of the cardiac cycle, are shown without any frequency-domain filtering. If no weighting is imposed (Fig.3a) the Doppler components are submerged in sidelobes of clutter components. On the other hand, by weighting the signal, clutter effects are limited to a narrow range around PRF/4, and Doppler components at -15 dB are clearly identified. In the latter case, it is also evident that clutter can be easily filtered out by setting to zero a limited number of FFT output points around PRF/4.

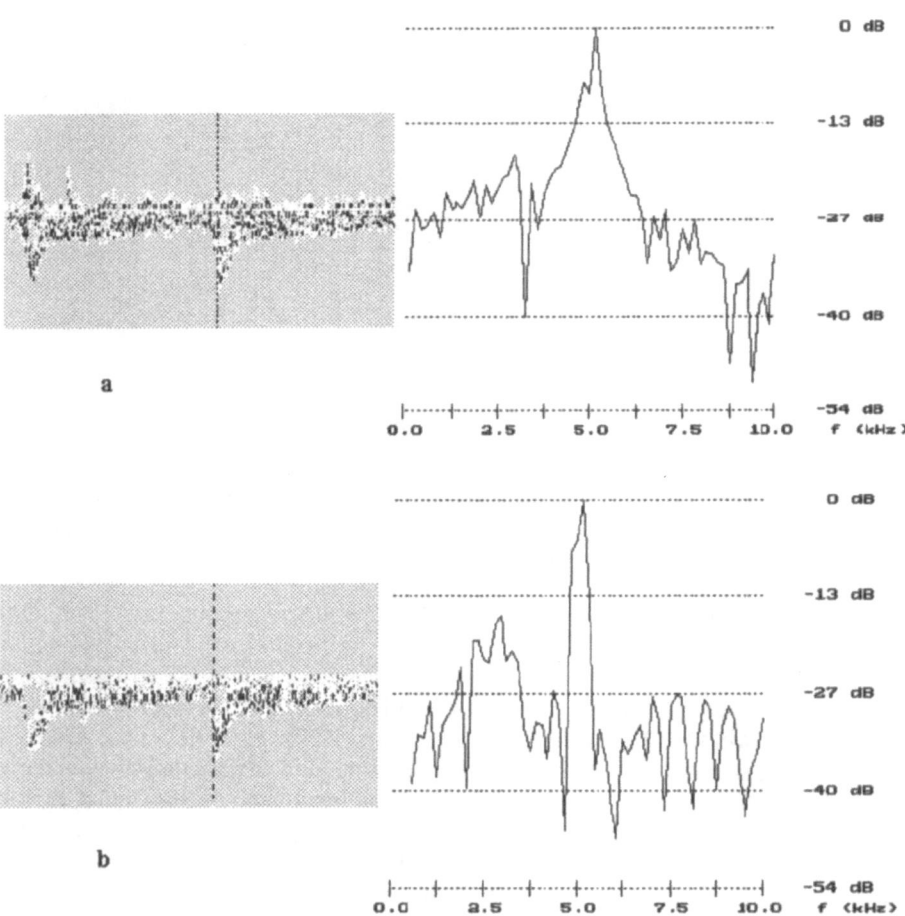

Figure 3. Examples of unfiltered spectrograms (left) obtained from a carotid artery. Dotted lines indicate the time position of the instantaneous spectra shown on the right. Rectangular and Hamming windows have been imposed on the Doppler signal in a) and b) respectively.

DISCUSSION

Direct sampling of a PW Doppler signal at a PRF rate can be performed without loss of information, provided that the PRF is related to the carrier frequency f_0 according to the equation (1). When compared to conventional PW systems based on quadrature demodulation, the technique limits noticeably the complexity of electronic circuits, at the expenses of a reduction in the analysable bandwidth. The above mentioned PRF/4 frequency offset seems not to constitute a problem, at least in cases where spectral analysis is employed for Doppler frequency detection.

The need of digitizing a signal where clutter is still superimposed to the useful components involves the use of an ADC capable of operating over a wide dynamic range. This problem is the same encountered in two dimensional flow-imaging systems, where the "continuous" Doppler signal from a specific depth cannot be reconstructed, since a number of samples originated from different depths must be subsequently digitized for each transmitted pulse. In both cases the noise bandwidth is equal to the RF bandwidth which, in turn, is conditioned by the system requirements in terms of range resolution.

In this possible application of the proposed technique, the PRF should still comply with equation (1), but the actual A/D conversion rate could be related to the range cell spacing, only. The main difference with respect to a conventional quadrature approach[4] is that in the latter case a couple of samples must be taken for each depth of interest. Since these samples must be spaced $1/4f_0$ apart, very sophisticated high-resolution ADC's are needed for appropriately sampling[6] ultrasound signals with f_0 in the usual range (3-10 MHz).

REFERENCES

1. D.H.Evans, W.N.McDicken, R.Skidmore, J.P.Woodcock, "Doppler Ultrasound - Physics, Instrumentation and Clinical Application", John Wiley & Sons, Chichester (1989).

2. F.Forsberg, M.Ø.Jørgensen, Sampling technique for an ultrasound Doppler system, *Medical & Biological Eng. & Computing.* 207:210 (1989).

3. A.V.Oppenheim, R.W.Schafer, "Discrete-time Signal Processing", Prentice Hall, New York (1989).

4. J.E.Powers, D.J.Phillips, M.A.Brandestini & R.A.Siegelman, Ultrasound phased array delay lines based on quadrature sampling techniques, *IEEE Trans. Sonics Ultrason.*, SU-27, 287:294 (1980).

5. P.Tortoli, L.Bessi, F. Guidi, Bidirectional Doppler signal analysis based on a single RF sampling channel, *IEEE Transactions on Ultrasonics, Ferroelectrics, and Frequency Control*, VOL.41, N.1 (1994).

6. A.P.G.Hoeks, T.G.J.Arts, P.J.Brands, R.S.Reneman, Comparison of the performance of the RF cross correlation and Doppler autocorrelation technique to estimate the mean velocity of simulated ultrasound signals, *Ultrasound in Medicine and Biology*, VOL.19, N.9, 727:740 (1993).

ACOUSTICAL IMAGING USING AN OPTIMAL COMBINATION OF SIGNAL PREFILTERING AND PULSE COMPRESSION

Helmut Ermert,[1] Martin Pollakowski,[1] Christian Passmann,[1] and Ludwig von Bernus[2]

[1]Ruhr-Universität Bochum, Institut für Hochfrequenztechnik
D-44780 Bochum, Germany
[2]Siemens AG, KWU S411
D-91050 Erlangen, Germany

INTRODUCTION

Resolution in acoustical imaging is characterized by the spectral properties of the transducer used in an imaging system, with the center frequency determining the lateral resolution (together with the aperture size and the object distance) and the bandwidth determining the axial resolution. As opposed to other diagnostic imaging methods (like X-ray imaging) the resolution of acoustical imaging systems usually needs to be optimized. Resolution can be increased by using high frequency transducers with a large bandwidth. But there are limitations in approaching higher frequencies and higher bandwidths due to the attenuation of the medium which has to be imaged or which is surrounding the objects of interest. This attenuation is limiting the allowable object distance as well as the (axial) size of the object areas.

Inverse filtering is a method to improve the axial resolution of an acoustical imaging system[1]. Pulse compression[2] known from RADAR leads to an enlargement of the system range. Some basic investigations concerning a suitable combination of both techniques have been carried out and successfully applied to ultrasonic imaging for nondestructive testing and medical diagnostics.

SYSTEM CONSIDERATIONS FOR ULTRASONIC IMAGING

An ultrasonic imaging system operating in the pulse-echo mode can be considered as a linear time-invariant transmission system, with input and output signals $x(t) \Leftrightarrow X(f)$ and $y(t) \Leftrightarrow Y(f)$, respectively, and $h(t) \Leftrightarrow H(f)$ the impulse response and the system transfer function, respectively, as far as stationary reflectors are representing the object. If this object is assumed to be an ideal plane reflector like in Figure 1 the system transfer function defined as $Y(f) = H(f)X(f)$ can be written as

$$H(f) = H_{tr}(f)H_{elec}(f)H_{med}(f)H_{refl}(f)e^{-j2\pi f\tau} \tag{1}$$

with τ the 2-way time delay of the acoustic signal and H_{tr}, H_{elec}, H_{med}, and H_{refl} being the transfer functions of the transducer, the electronics (including duplexer), the medium and the reflector, respectively. This transfer function also characterizes the imaging capability of a system in a regular operation mode with multiple scatterers of different size and different range locations. Linearity can be assumed as long as nonlinear acoustical transmission through the medium can be neglected (which usually is correct for imaging applications using low signal levels) and if the nonlinear duplexer is designed to work linearly within both the transmission mode and the receiving mode. Neglecting frequency dependencies of the medium's attenuation the magnitude of the overall system transfer function is mainly determined by the transducer so that $|H(f)| = |H_{tr}(f)|$ can be assumed. Optimization of the axial resolution and some compensation of disadvantageous phase and amplitude characteristics of the transfer function can be obtained using suitable filtering procedures.

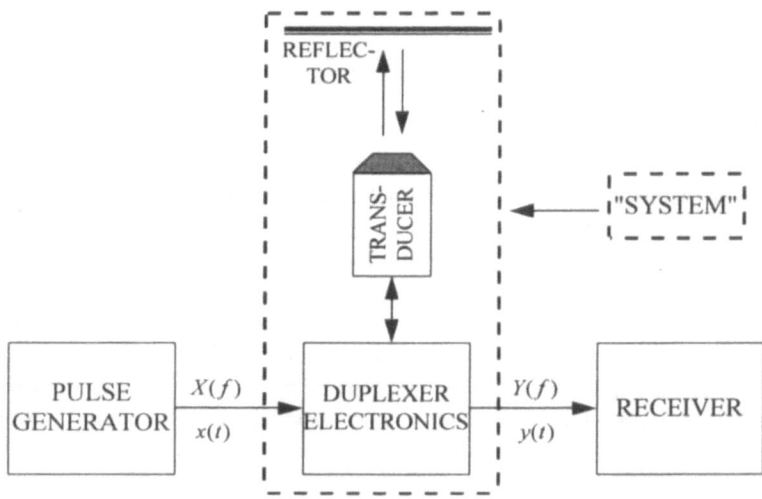

Figure 1. Ultrasonic imaging system

FILTERING CONCEPTS FOR ULTRASONIC IMAGING SYSTEMS

If a filter with a transfer function $F(f)$ is used for echo signal processing according to Figure 2 the resulting output signal $Z(f)$ is determined by

$$Z(f) = F(f)Y(f) = F(f)H(f)X(f). \tag{2}$$

An inverse filter which uses a filter transfer function defined by

$$F(f) = Y^{-1}(f) = H^{-1}(f) \tag{3}$$

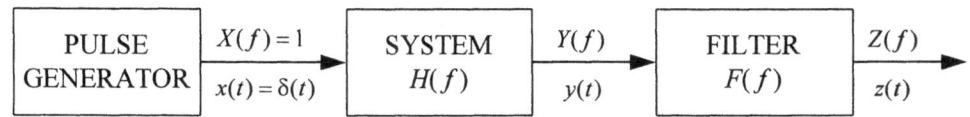

Figure 2. Postfiltering

leads to an output signal $Z(f) = 1$, which corresponds to $z(t) = \delta(t)$ under ideal broadband conditions, while the matched filter

$$F(f) = Y^*(f) = H^*(f) \tag{4}$$

produces the output signal $Z(f) = |H(f)|^2$. The inverse filter leads to short output signals representing the optimum of axial resolution. But it causes problems in the presence of noise, and because of the band-limitation of the transfer function the inverse filter can only be used in combination with a suitable spectral window $W(f)$. It also causes problems in the presence of noise. The matched filter is known to lead to an optimization of the signal-to-noise-ratio and represents an operation mode which utilizes the optimal range of the imaging system, but it does not have optimal axial resolution properties. The Wiener filter can be understood as a more general one which includes the inverse filter as well as the matched filter depending on the noise conditions.

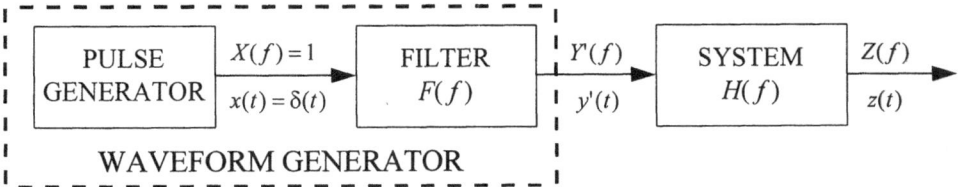

Figure 3. Prefiltering

Because of the linearity of the whole system the sequence of the system components can be changed like shown in Figure 3. The filter in front of the system combined with a pulse generator can be represented by a waveform generator which produces prefiltered signals with the subsequent system working as an inverse filter

$$Y'(f) = H^{-1}(f)W(f) \Rightarrow Z(f) = W(f) \tag{5}$$

or as an matched filter

$$Y'(f) = H^*(f) \Rightarrow Z(f) = |H(f)|^2, \tag{6}$$

respectively. By splitting the filter, for example into an amplitude filter and a phase filter at the front and at the end of the imaging system, respectively, an additional concept including a combination of prefiltering and postfiltering can be realized as shown in Figure 4. Postfiltering is a trivial necessity in the case of pulse-compression (usually with an allpass network as the compressing filter). Prefiltering may have some advantages by allowing special spectral weighting and amplitude shaping of the input signal in order to obtain an efficient utili-

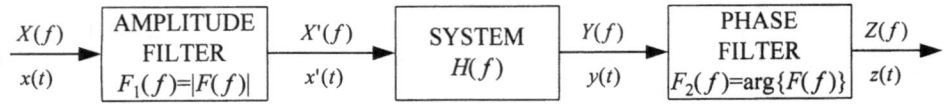

Figure 4. Splitted filters: prefiltering & postfiltering

zation of the transducer bandwidth and of the dynamic range of the whole system. There-fore, both can be used as an optimal combination in acoustical imaging systems.

PULSE COMPRESSION IN ACOUSTICAL IMAGING SYSTEMS

The range of an acoustical imaging system can be increased by using transmitter sig-nals with higher energy. Signal energy can be increased by increasing the amplitude of a pulse signal and maintaining it's bandwidth B. This may have disadvantages because of limitations in the available transmitter power or in the power capabilities of an transmitter power amplifier and of the transducer. An additional disadvantage may occur due to the nonlinear properties of the medium. An alternative way of increasing the signal energy is to enlarge the signal duration T while holding the transmitter signal amplitude on a constant level. This can only be achieved without bandwidth losses by using special signals charac-terized by a large time-bandwidth product BT. Echo signals with $TB >> 1$ can be com-pressed with respect to their duration in a suitable filter (all-pass circuit) leading to a com-pressed pulse with a duration t_d and a time-bandwidth product $t_d B \approx 1$. As the bandwidth B will not be reduced by this filtering procedure, t_d is much smaller according to the compres-sion ratio is $T / t_d \approx TB$, and the amplitude gain is $AG \approx \sqrt{BT}$. Three categories of signals can be used: (1) *frequency modulated pulses ("CHIRP"-signals)* like linear frequency modulated chirps (LFM), non-linear frequency modulated chirps (NLFM), and signals with non-steady modulation (frequency hopping etc.), *(2) pseudo-random signals* like Barker Codes, M-Sequences, and Golay-Codes, and *(3) noise signals*. Because of the relatively nar-row bandwidth of ultrasonic imaging systems due to the frequency response of transducers CHIRP-type signals turn out to be the most suitable ones for application[3] rather than pseudo random or noise signals which only can be utilized advantageously in case of larger band-width systems only. There is another limitation of the maximum pulse compression ratio and the amplitude gain which can be obtained in ultrasonic imaging systems. Compared to RADAR, in diagnostic imaging systems working at ultrasonic frequencies the distances be-tween the transducer and the object are relatively small. As the transmitted signal and the returning echo signal may not overlap in the transducer and the electronics the maximum signal duration T is determined by 2 times the distance divided by the speed of sound.

Pulse compression significantly enhances the signal-to-noise-ratio. Inverse filtering improves the axial resolution but it has disadvantages in the presence of noise. Therefore, a combination of both improving range and axial resolution promises to cancel out both dis-advantages. A major problem is, that for inverse filtering a system transfer function with a non-rectangular amplitude spectrum (e.g. Gauss- or $\cos^2$-shape) needs an input signal with an amplitude spectrum having an inverted envelope. Using a linear frequency modulated chirp (LFM) the envelope of this signal also has to be an inverted one with high amplitudes at the beginning and also at the end of the signal. Such a signal does not utilize the available power of the transmitter electronics efficiently. A chirp signal with a rectangular envelope and a constant power level well matched to the system hardware would be advantageous. A rectangular envelope is possible by using a chirp with non linear frequency modulation (NLFM). In a NLFM-chirp with a constant amplitude the weighting of the spectral components can be determined by a proper choice of the modulation function $f(t)$. NLFM-chirps can be used for stationary (non moving) objects only[4]. As opposed to RADAR, this applies to a wide range of diagnostic ultrasonic imaging (as far as liquid flow analysis using DOPPLER techniques is not considered).

NLFM chirp signals can be designed using a numerical approach[3,5]. As they have a constant amplitude they can be generated approximately by frequency modulated square wave generators[5,6]. This is a much simpler and cheaper alternative to a digital waveform generator. An additional filter for a subsequent low-pass filtering in order to suppress higher order harmonics of the square waves is not necessary because of the filter behavior of the transducer with it's limited bandwidth.

BASIC EXPERIMENTS

The effect of inverse filtering combined with NLFM pulse compression has been demonstrated experimentally in a set up for non-destructive testing which was equipped with a broadband PVdF transducer[3]. The 6 dB transducer bandwidth was limited by a lower frequency $f_L = 1{,}37$ MHz and an upper one $f_U = 3{,}93$ MHz. The transducer was located on a steel-block with a hemispherical back wall and adjusted for a maximum longitudinal wave echo from this wall. 3 different window functions $W(f)$ based on the expression

$$W(f) = k + (1-k)\left(\sin\left(\pi\,\frac{f - f_1}{f_2 - f_1}\right)\right)^2 \tag{7}$$

have been used with parameters presented in Table 1. These window functions and the system transfer function $H(f)$ are presented in Figure 5. Corresponding to these windows 3

Table 1. Parameters of window functions

window	k	f_1/MHz	f_2/MHz	shape
$W_1(f)$	0	0.1	5.3	without offset
$W_2(f)$	0.25	0.1	5.3	with offset
$W_3(f)$	1	0.1	5.3	rectangular spectrum

different NLFM chirp signals with a duration of 40 μs were used. The plots of their instantaneous frequencies $f(t)$, of the magnitudes of their spectra $|X(f)|$, and of the signal functions $x(t)$, are presented in Figures 6 to 8. The amplitude spectra of the resulting echoes in Figure 9 clearly represent the corresponding window functions. In Figure 10, the amplitudes of the compressed echo signals have been normalized with respect to their individual maxima.. This makes it easier to compare the lengths and sidelobes of these signals. Table 2 summarizes the results and includes also a simple LFM-chirp compression experiment as well as a combination of matched filter and NLFM chirp compression[3,5,7]. It can easily be seen that the inverse filtering concept can lead to significant improvements with respect to the axial resolution without major losses of echo amplitudes.

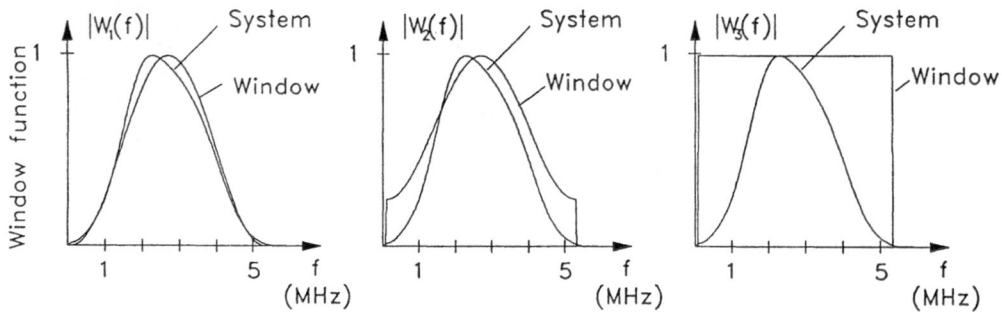

Figure 5. Window functions for pulse compression and inverse filtering

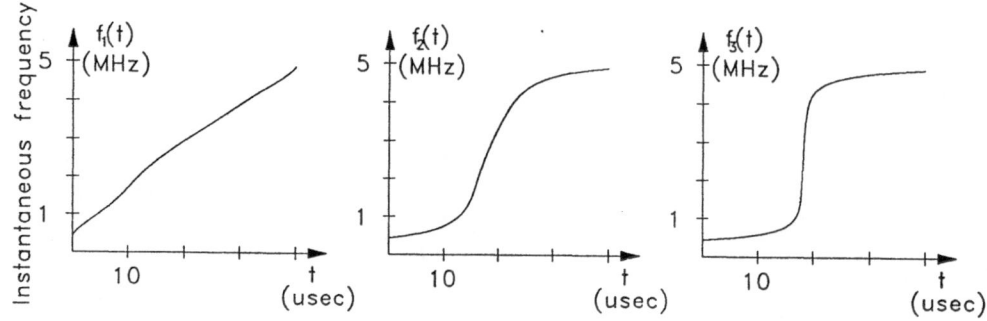

Figure 6. Instantaneous frequencies $f(t)$ of the transmitted chirp signals

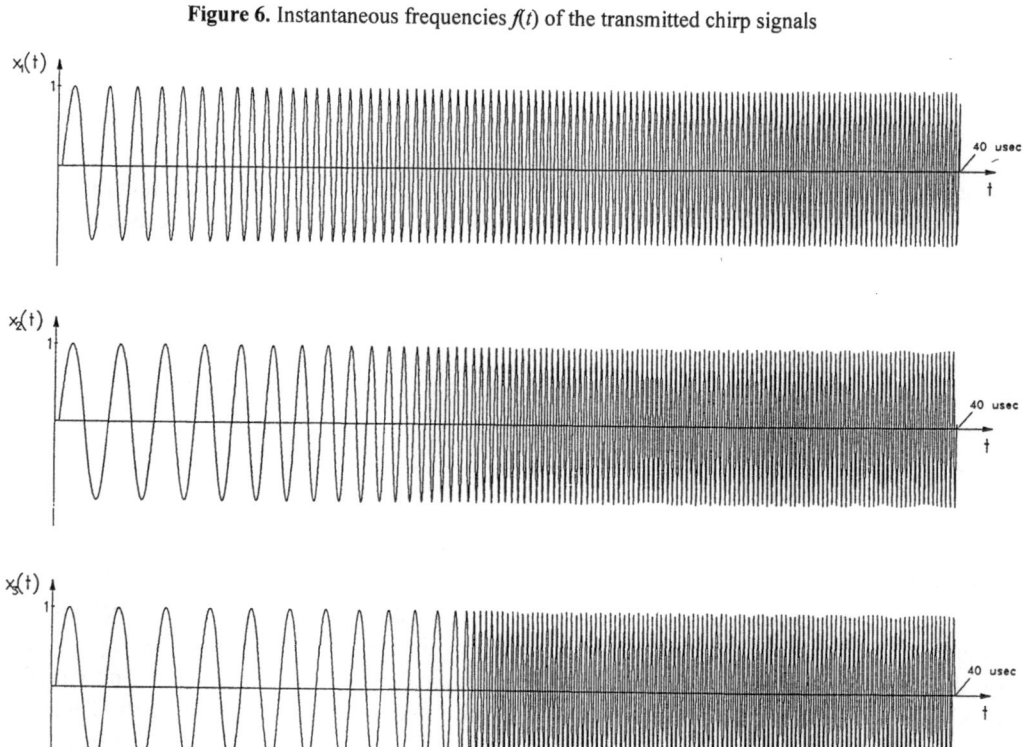

Figure 7. Transmitted chirp signals $x(t)$, time domain representation, chirp duration 40 μs

342

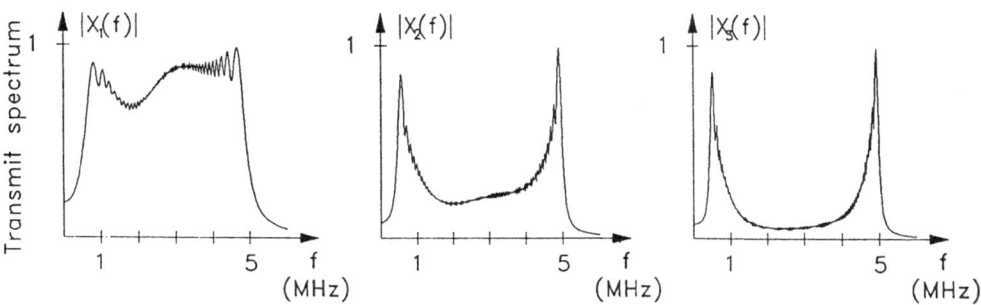

Figure 8. Magnitude spectra of the transmitted chirp signals

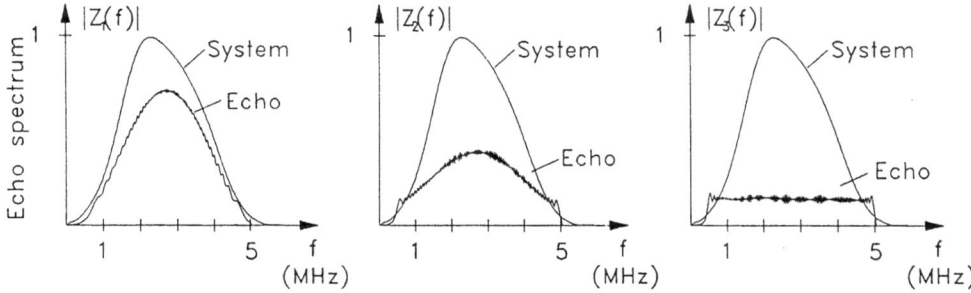

Figure 9. Spectra of the compressed echoes

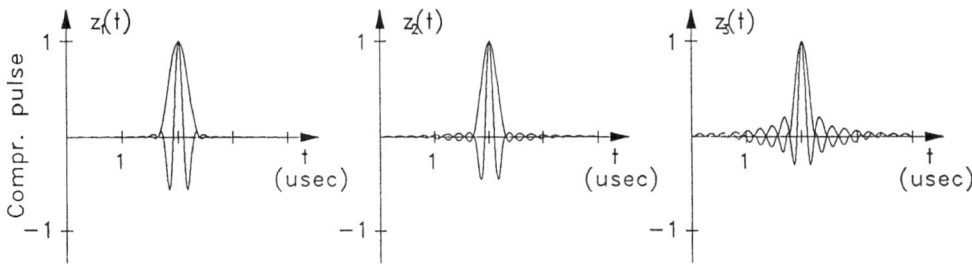

Figure 10. Compressed echoes $z(t)$, time domain representation

Table 2. Comparison of pulse compression results

filter type	amplitude maxima (related to matched filter)	pulse durations (between 50% amplitude points)
linear	0,98	480 ns
matched	1,00	510 ns
inverse (k=0)	0,88	350 ns
inverse (k=0,25)	0,57	330 ns
inverse (k=1)	0,31	270 ns

EXPERIMENTAL IMAGING RESULTS

Pulse compression can be obtained by analog filters at the input of the echo receivers as well as by digital filtering of digitized echoes. Analog filters do not lead to any limitations in real time imaging systems and they also do not produce additional noise due to signal digitization. On the other hand, digital filtering is more flexible and can also be realized for real-time applications using special hardware components. Imaging experiments in the 1 to 10 MHz range using digital pulse compression for medical imaging (e.g.[8]) and for non-destructive testing (e.g.[3]) have been reported previously. Pulse compression is of special advantage for imaging applications at high frequencies with associated high attenuation in the acoustical transmission media. In dermatological imaging the transducers usually do not touch the skin directly. The ultrasound has to travel between the transducer surface and the transducer focus, which is located inside the region of interest, through water with a path length suitable for the utilization of pulse compression. Results of pulse compression in a 50 MHz imaging system using analog compression filters and LFM chirps have already been published[9]. Recently, pulse compression has also been applied in a 150 MHz system[10]. The block diagram is presented in Figure 11. Digital prefiltering using a waveform generator (8 bit, 500 MHz D/A-conversion rate) combined with digital NLFM chirp compression and

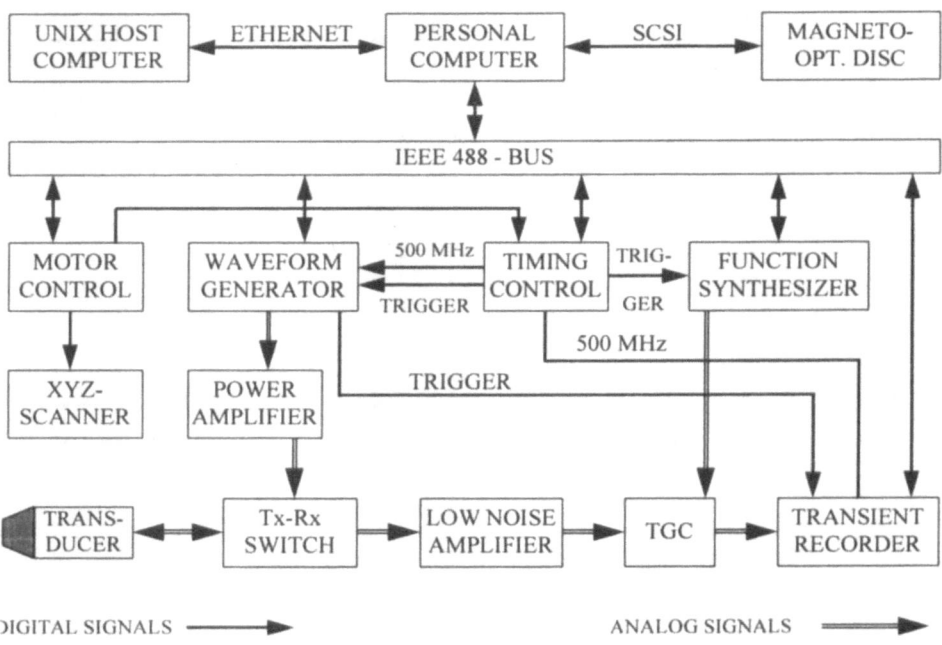

Figure 11. Block diagram of an high frequency dermatologic imaging system

inverse filtering has been applied. The waveform generator output signal is amplified by an 250 W power amplifier. The duplexer ("Tx-Rx-switch") has 3 dB attenuation in the transmission mode and 84 dB isolation. After low-noise preamplification (30 dB) the echo signals are digitized using a transient recorder with 500 MHz sampling rate, 8 bit resolution and a 0.5 Mbytes memory. The analog components of the system cover a frequency range from 20 to 250 MHz with an overall dynamic range of 117 dB. A broadband transducer

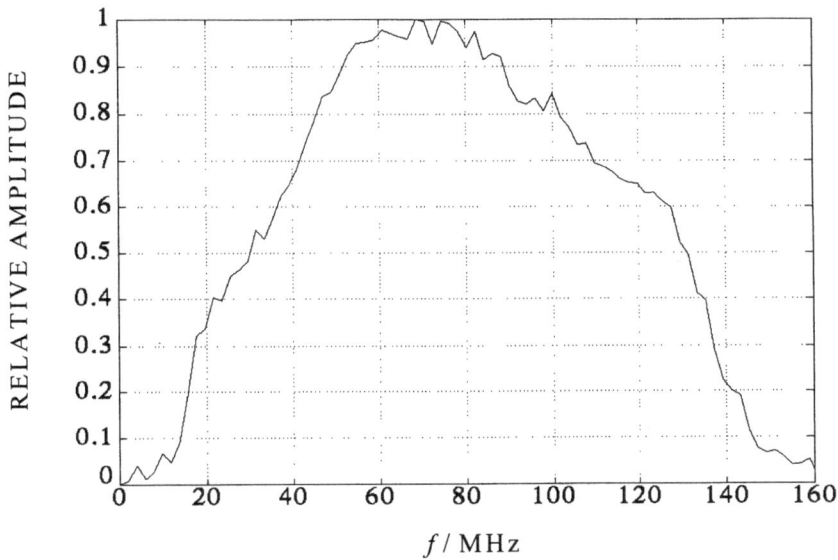

Figure 12. Frequency response of a focused ceramic transducer (ULTRAN)

(ceramic ULTRAN) with-12 dB lower and upper frequency limits of $f_L = 18$ Mhz and $f_u = 138$ Mhz, respectively, a focal length of 4.3 mm and a F-number of 1.34 has been used. The frequency response of this transducer is presented in Figure 12. The system data lead to a lateral resolutjion $\delta_{lat} = 27$ μm and an axial resolution $\delta_{ax} = 8,5$ μm with a maximum penetraion dept in skin of about 3 mm. Narrow bean transducers ("pencil beam") suitable for conventional B-scan imaging cause a significant discrepancy between axial and lateral resolution[9]. Therefore, using a focused transducer, a modified B-scan technique combined with an additional axial scan ("Z-scan") was applied which is illustrated in Figure 12. With

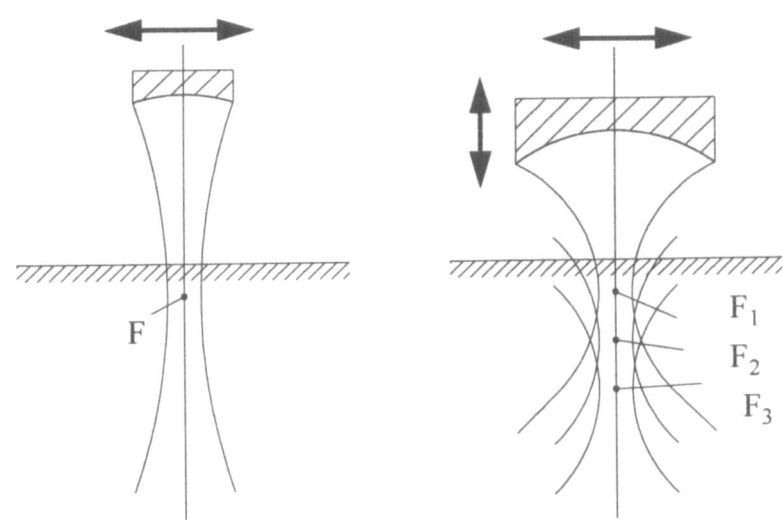

Figure 13. Comparison: B-scan (left) versus combined B-scan & Z-scan (right); F= focal areas

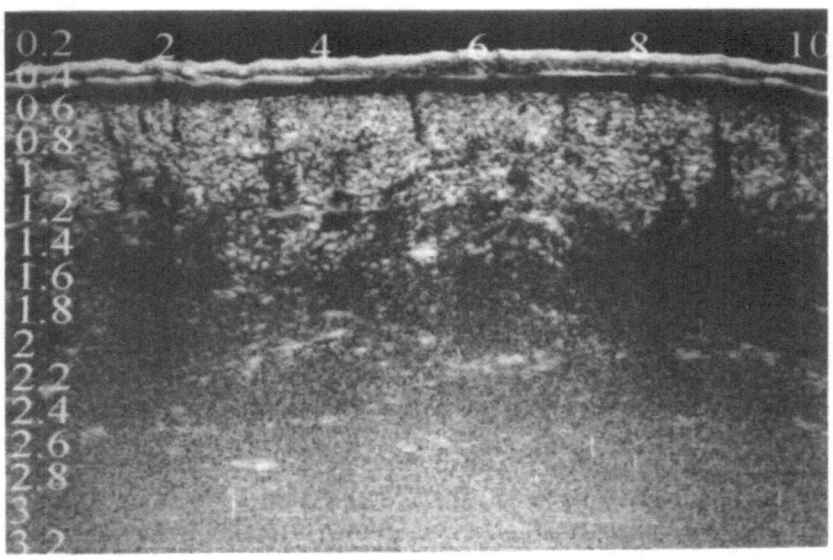

Figure 14. *In-vivo* image of a normal human skin obtained without pulse compression (numbers: mm)

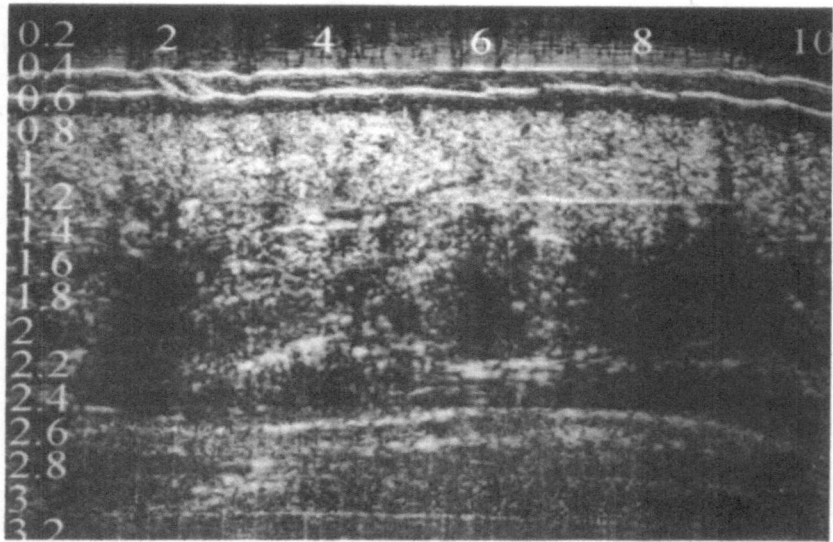

Figure 15. Image (corresponding to Figure 14) obtained with pulse compression

respect to different axial positions of the focus in the skin up to 32 B-scans of subsequent layers representing different depths can be obtained and combined to a complete cross sectional image of the skin. The maximum number of A-scans per layer is 1024. Using 512 A-scans the scanning time for each layer is 2.5 s. The maximum scanning area is 10.24 mm (lateral: scanning direction) x 3.2 mm (axial: depth). *In-vivo* images of a normal skin without pulse-compression (Figure 14) and with digital NLFM-chirp compression combined with inverse filtering (Figure 15) shall be compared. The chirp signal duration was 0.5 µs, the length of the digital filter was 512 points x 2 ns. These images consist of 8 depth layers obtained by the B-scan / Z-scan technique. The amplitude gain obtained by pulse compression was 12 dB. The corresponding improvement of signal-to-noise ratio and of imaging range is demonstrated in Figure 15. It would be possible to apply smaller compression ratios to the upper layers in order to prevent non-linear effects and inadequate echo signal brightness in the upper area of the skin (Figure 15).

CONCLUSIONS

It has been shown that for ultrasonic echo-mode imaging transmitter signal prefiltering and subsequent echo signal postfiltering representing a combination of inverse filter and pulse compression is a powerful concept for an optimization of the axial resolution and the imaging range of the system. Chirp signals with a non linear frequency modulation (NLFM) allow an efficient utilization of the transducer bandwidth and of the available power of the transmitter electronics. Basic experiments using a NDE imaging set-up in the 1 to 5 MHz range demonstrated the improvement of axial resolution. Additionally, an application of the NLFM-chirp compression technique associated with an inverse filter in the 150 MHz range for diagnostic *in-vivo* imaging in dermatology verifies a significant extension of the penetration depth and of the resulting imaging range of the system.

REFERENCES

1. J. Schmolke, D. Hiller, H. Ermert, J.O. Schaefer, and G. Gräbner, Generation of optimal input signals for ultrasound pulse-echo systems, *1982 IEEE Ultrason. Symp. Proc.*, 929-934.
2. D.K. Barton, "Radars Volume 3: Pulse Compression," Artech House, Dedham, MA. (1975)
3. M. Pollakowski, "The application of pulse-compression technique to ultrasonic non-destructive testing," Ph.D. Thesis, Ruhr-Universität Bochum, Shaker Verlag, Aachen, Germany (1993)
4. R. S. Berkowitz, "Modern Radar," John Wiley, New York (1965).
5. M. Pollakowski, H. Ermert, Chirp Signal Matching and Signal Power Optimization in Pulse-Echo Mode Ultrasonic Nondestructive Testing, *IEEE Trans. Ultrason. Ferroelec. Freq. Contr.*, 41 (1994), in press.
6. M. O'Donnell, Coded Excitation System for Improving the Penetration of Real-Time Phased-Array Imaging Systems, *IEEE Trans. Ultrason. Ferroelec. Freq. Contr.*, 39:341-351 (1992).
7. M. Pollakowski, H. Ermert, L. von Bernus, and T. Schmeidl, The optimum bandwidth of chirp signals in ultrasonic applications, *Ultrasonics* 31:417-420 (1993).
8. H. Bressmer, U. Faust, P. Schwarzer, Die Anwendung der Pulskompression in Ultraschall-Echo-Systemen, *Biomedizinische Technik*, 34 (Ergänzungsband):144-145 (1989).
9 A. Höß, H. Ermert, S. el Gammal, P. Altmeyer, Signal processing in high-frequency broadband imaging systems for dermatologic applications, in: "Acoustical Imaging 19," H. Ermert and H.-P. Harjes, eds., 243-249, Plenum Press, New York (1992).
10. C. Passmann, H. Ermert, 150 MHz *in-vivo* ultrasound of the skin: imaging techniques and signal processing procedures targeting homogeneous resolution, *1994 IEEE Ultrasonics Symposium Proceedings,* (to appear).

SPLATTING AND SPLINES IN 3D MEDICAL ULTRASOUND IMAGING

Timothy J. Pitt[1], Leonard A. Ferrari[2], Andy Healey[3],
R. Anthony Reynolds[3], Keith N. Humphries[1]

[1]Medical Physics Department
[3]Department of Radiology
Hammersmith Hospital
London, W12 OHS

[2]ECE Department
University of California at Irvine

[3]Dept. of Medical Eng. and Physics
Kings College School of Medicine
London SE22

INTRODUCTION

There are several potential benefits of three dimensional, or volume ultrasound scanning over existing planar, or two dimensional techniques, including, for example, accurate volume measurements, structural imaging of organs and vessels, and the ability to view planes through the (volume) image at orientations that are difficult or impossible to scan. However, despite the great increase in availability and power of digital electronic systems that have given rise to the potential of three dimensional ultrasound, there is still a significantly large amount of data that needs to be processed from such a volume scan. In principle, scanning a cubic volume of sides of 10 cm with a sample spacing of 1 mm alone yields a million data values, which potentially can be generated, using pulse echo ultrasound, within 1.3 seconds. This in itself is not so demanding, however this data is typically not in a regular, orthogonally sampled form. Existing experimental transducer systems employ either a linear or curvilinear array to produce two dimensional (slice) images, which are either mechanically driven to fixed, known positions by a rotation round an axis within the imaging plane, or by allowing the probe to be steered by an operator while the probe's position is measured by some suitable device. Both of these approaches lead to data which must be interpolated before it can be displayed using conventional surface or volume rendering algorithms, which assume simple, orthogonally sampled data sets with cubic voxels. This three dimensional interpolation is a computationally demanding task. Where the probe is mechanically driven, much of the calculation work can be done *a priori*, and the interpolation can be implemented as a series of look-up tables. With a hand steered system however, no assumptions can be made concerning the positions of the sampled data, so the interpolations for each scan are unique. The demands of the clinician are no less stringent. The main demand is that an image is formed, if not in real time, at least within minutes of the scan being performed, to ensure that the scan has been correctly performed, in case the patient needs to be recalled. At the same time this function must be fulfilled at a reasonable financial cost.

This paper aims to consider how these demanding tasks may be realised with existing, modest hardware demands (that is, without resorting to using powerful and expensive computer systems).

FORWARD MAPPING ALGORITHMS

The acquisition of pulse-echo ultrasound data is, by its very nature a serial process, that is the data values arrive one at a time, and the data for the whole volume takes some finite time to accumulate. With existing, steered planar transducer three dimensional acquisition systems, this takes of the order of ten seconds. This time is not likely to be significantly improved upon without the development of a two dimensional array transducer, since it is limited by the speed at which a conventional probe is scanned across the skin surface, whilst retaining good acoustical coupling to give a reasonable image of the subject. Ten seconds is certainly not an insignificant time in computing terms, and it can be used effectively. If, ultimately, a real-time system is desired, then all the necessary processing for each sample value must be achieved before the next data value is received. To this end, the development of a forward mapping algorithm is paramount. The limitation of many interpolating algorithms, is that they are backward mapping or must solve an inverse problem. They require a knowledge of several neighbouring (in many cases, all) data values to make an estimate of the image 'function' value for each voxel position in the output image. A cubic spline interpolation, for example, requires the solution of a matrix, for which all the input data values must be known, to provide the weighting values for the basis functions. If such an algorithm were implemented, it would require that the object was fully scanned before interpolation and display could even begin. It would be very difficult to make such an algorithm efficient enough to be interactive.

A purely forward mapping (or "splatting") algorithm, in contrast, requires only one incoming data point at a time to perform the interpolating calculations for that data point. The energy for each sampled value is 'splatted' into the output volume space, according to some kernel function. By accumulating the results from each input value, the final image will be fully interpolated immediately after the final data point has been sampled and processed. This algorithm is an adaptation of that developed for performing volume rendering - see Westover (1990). This type of algorithm can take advantage of the scanning time to perform the required calculations. It is also easier to perform this type of interpolation with arbitrarily positioned data, since it is very difficult to sort the data in such a way as to make a backward mapping algorithm efficient. The main disadvantage of this type of algorithm is that to perform the interpolation correctly, the sample spacing should be known, which is not typically the case where hand steered scanning is used in the acquisition system.

The kernel function, of course, determines the style of interpolation that is to be used. Ideally a sinc function could be used, although it's simple truncation leads to some small errors, and it is time consuming to calculate. Nearest neighbour, linear or (approximate) sinc function interpolations are quite feasible, since the kernel function contains only a single sample value term. The next section describes an efficient algorithm for the implementation of Hermite interpolation, although the general principle may be applied to other interpolating functions.

HERMITIAN INTERPOLATION

We examine the case of a one-dimensional curve, since the extension to images (surfaces) is rather straightforward. The interpolating function, $Y(\bar{u})$, can be written as

$$Y_i(\bar{u}) = h_{00}(\bar{u})P_i + h_{10}(\bar{u})P_{i+1} + h_{01}(\bar{u})D_i + h_{11}(\bar{u})D_{i+1} \tag{1}$$

where $\{P_i\}_{i=0}^{N-1}$ are the N data points which are to be interpolated and D_i are computed from the central differences of P_i. That is

$$D_i = \frac{1}{2}(P_{i+1} - P_{i-1}) \tag{2}$$

The matrix form of (1) is given by

$$Y(\bar{u}) = (1 \quad \bar{u} \quad \bar{u}^2 \quad \bar{u}^3)\begin{pmatrix} 1 & 0 & 0 & 0 \\ 0 & 0 & 1 & 0 \\ -3 & 3 & -2 & -1 \\ 2 & -2 & 1 & 1 \end{pmatrix}\begin{pmatrix} P_i \\ P_{i+1} \\ D_i \\ D_{i+1} \end{pmatrix} \tag{3}$$

Primary Function   Secondary Function

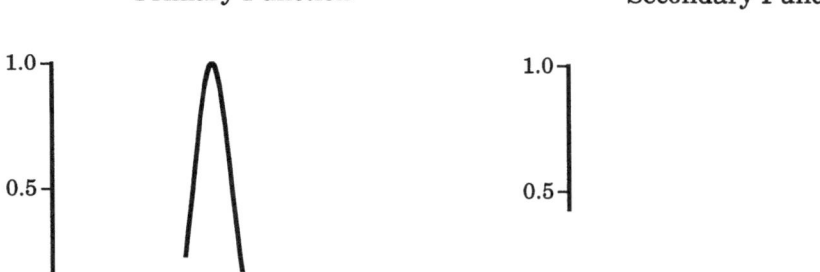

Figure 1   The primary and secondary functions which are convoluted with the data to effect Hermite interpolation. h_0 is a combination of h_{00} and h_{10}, h_1 is h_{01} and h_{11}.

Derivative-Summation Implementation

The functions h_0 and h_1 shown in Figure 1 are multiple knot splines, derived from the combination of the functions h_{00} and h_{10}, and h_{01} and h_{11} respectively. The first, second, third and fourth derivatives of each function are given in Figure 2, where the discontinuities in the second and third derivatives are evident.

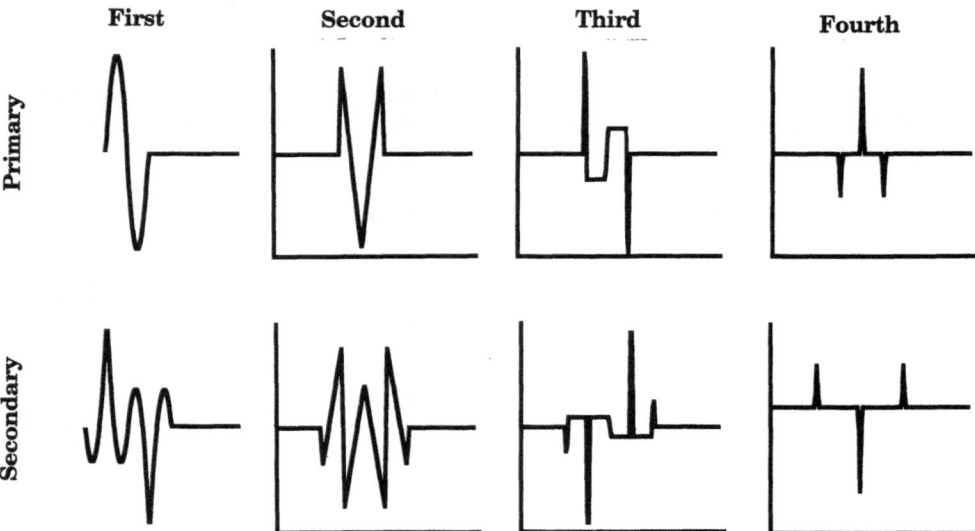

Figure 2 Top line:- first to fourth derivatives of the h0 function. Bottom line:- first to fourth derivatives of the combined h1 function.

Suppose we wish to create M-1 new data points between successive data points P_i and P_{i+1}. Then we can compute the discrete samples of Y, viz., $Y(n)$, using linear convolution operations. That is,

$$Y(n) = \sum_i Y_i(n)$$

$$= P(Mn) * h_0(n) + D(Mn) * h_1(n) \tag{4}$$

where $P(Mn)$ and $D(Mn)$ represent the original sequences P_i and D_i filled with M-1 zeros between each of the original samples.

Equation 4 can be implemented extremely efficiently, using only simple shifts and integer addition operations. We start with samples of the fourth derivative of $Y(\bar{u})$:

$$Y^{(4)}(\bar{u})\big|_{\bar{u}=\frac{n}{N}} = P(Mn|N) * h_0^{(4)}(\bar{u})\big|_{\bar{u}=\frac{n}{N}} + D(Mn|N) * h_1^{(4)}(\bar{u})\big|_{\bar{u}=\frac{n}{N}}$$

$$= P(Mn) * h_0^{(4)}(n) + D(Mn) * h_1^{(4)}(n) \tag{5}$$

Now, since $h_0^{(4)}$ and $h_1^{(4)}$ contain only a sparse set of Dirac functions, the implementation of the Equation 5 is simple. Recursive summation can be used to compute $Y^{(3)}(\bar{u})|_{u=n/N}$ from the results of Equation 5. The technique is explained in detail in Saukan, Silbermann and Ferrari (1994). A similar process is used to obtain $Y^{(2)}(\bar{u})$, $Y^{(1)}(\bar{u})$ and finally $Y(\bar{u})|_{u=n/N}$. The impulse response of the system and the results of these computations for the example of a cosine function are shown in Figure 3. The results are identical to those obtained using the matrix form expressed in Equation 3.

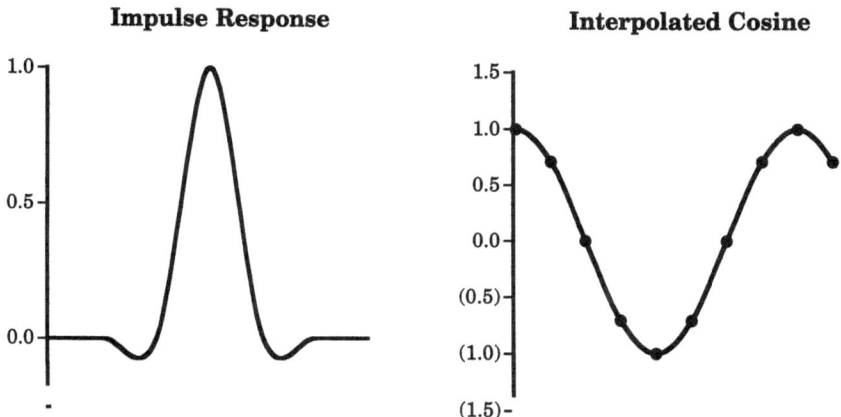

Figure 3   Left:- impulse response of the interpolation. Right:- Sampled cosine, interpolated by a factor of eight.

TEST DATA

We acquired some ultrasound volume data at Hammersmith Hospital using a PC based acquisition system. This digitises the video signal from a conventional scanner, and uses an electromagnetic system to measure the position and orientation of the ultrasound probe for each video image captured.

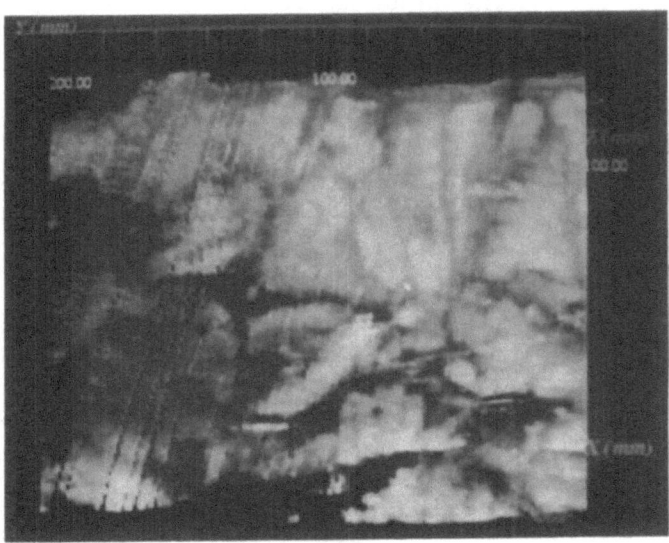

Figure 4   Volume rendered image from a nearest neighbour reconstruction of the data set, using 0.5 mm voxels. The foetal head is on the lower left of the image, with the right arm centre front, and the umbilical cord on the right towards the back.

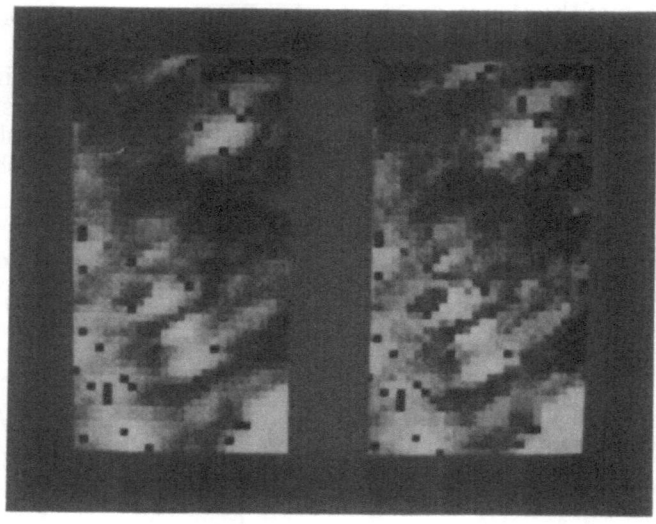

Figure 5 Expanded details of reformatted slices through the nearest neighbour (left) and linear (right) interpolated data sets, showing a cross section of the umbilical cord.

The subject was a 21 week gestation foetus (*in vivo*), which was scanned by hand, with a single slow sweep lasting thirteen seconds. 170 slice images were recorded, 184 x 174 pixels, 92 x 87 millimetres. The distance between slices was approximately 2 millimetres. This data set was reconstructed using a forward mapping algorithm, implementing nearest neighbour, linear, and Hermitian interpolations, to give cubic voxel data sets, that were volume rendered using a commercially available software package. Voxel sizes of 1, 0.5, and 0.25 mm were used to reconstruct sub-regions of the subject. Volume renderings and two dimensional slices through the output data set were considered subjectively, in terms of the computational complexity, and potential clinical use.

Figure 6 Details of reformatted slices through the linear (left) and Hermite (right) interpolated data sets that have 0.25 mm voxels.

Volume rendered images from the nearest neighbour, linear and Hermite interpolations were qualitatively the same, which may be expected considering the subsequent processing the data incurs in providing the rendered images. Reformatted slices through interpolated volume data sets show some difference in detail however. Figure 5 shows that linear interpolation with 0.5 mm voxels produces a more acceptable image than nearest neighbour. Oblique slices through linear and Hermite interpolated data sets with voxels of size 0.25 mm, show some difference in detail (Figure 6) with the linear interpolation appearing to have a greater smoothing effect than the Hermite interpolation, even though the voxel resolution is comparable with that of the ultrasound system.

DISCUSSION

As would be expected, there are some qualitative improvements in the output images when higher order interpolations are used, but the differences are only seen where the output pixel sizes are of a corresponding size. In general it is unlikely that such small voxel sizes would be used in practice, although this will be dependant on the precise clinical application. There is some potential advantage, however, in using a higher order interpolation, if it can be made efficient, in that coarser sampling can be used, offsetting the computational complexity against the amount of data that needs to be processed.

The application of the above algorithm for interpolating the regularly sampled volume from the arbitrary ultrasound image slices would consist of two very computationally simple phases. First the incoming samples would be serially 'splatted' according to a three dimensional derivative kernel, consisting only of a number of Dirac functions. This would then be followed by a post acquisition phase of recursive summations. In general only a few interpolated values (those which lie at the centres of voxels) are required from each input value, so some modification of the algorithm could be developed to take advantage of this. By making the mode of data acquisition more predictable, i.e. by covering the region to be scanned in a single sweep, processing time could be saved by starting each recursive summation stage, a soon as the previous stage has completed any required region. Since the algorithm requires only simple integer arithmetic to be performed, a highly parallel, and cost effective, hardware system can easily be envisaged, which may provide three dimensional images, if not in real time, then with only a short delay.

REFERENCES

P. Saukan, M. Silbermann, L. Ferrari, Curve and surface generation and refinement based on a high speed derivative algorithm, *CVGIP: Graph.Models and Im. Proc.* 56(1):94 (1994)

L. Westover, Footprint evaluation for volume rendering, *Computer Graphics*, 24(4):367(1990)

SPATIAL FILTERING OF PARTIALLY COHERENT ACOUSTICAL IMAGES

A. I. Khil'ko

Department of Hydrophysics and Hydroacoustics
Institute of Applied Physics,
46 Uljanov Str., 603600 Nizhny Novgorod, Russia

INTRODUCTION

In different practical applications of the acoustical vision systems the illumination fields are partially coherent. From the one hand the random inhomogeneities of the medium and motion of the elements of the vision system destroy of the sound coherence. From the other hand the special methods have been developed for the construction of the noncoherent illumination of vision objects, for the suppression of speckle-noise of images [1]. In any case we have to investigate the partially coherent (PACO) sound scattering by objects. The problem of acoustical image construction in natural media can also be build on this base. The theoretical estimation and experimental data show [1,2] that the possibilities of the spatial filtering of the images lose for the incoherent illumination systems. It is interesting to investigate , for example, the cases of the oceanic vision, when the observed inhomogeneities are smooth and forward scattering - major part of sound and the vision is realized in presense of rough surfaces. The application of PACO sound for the construction of the acoustical images gives us the possibilities of decreasing the coherence noise and spatial filtration of images.

In present investigation the problem of spatial filtration of acoustical partially coherent hydroacoustical images is developed when the spatial distribution of the scatterers is constructed by horizontal array which is situated in layered waveguides. The theoretical analysis of the low-frequency hydroacoustical images is accompanied by experiments which have been made in laboratory conditions.

DIFFRACTION OF PACO-FIELDS IN LAYERED WAVEGUIDES

Coherent field. For the simplification, we assume that the field of the source S illuminate a large (in the wave scales) absolutely ridges scatterer σ (See Figure 1). It is possible to represent the diffracted fields for the far region of the scatterer in waveguides as a sum of the waveguide waves by using of the Green's function of the nondisturbed

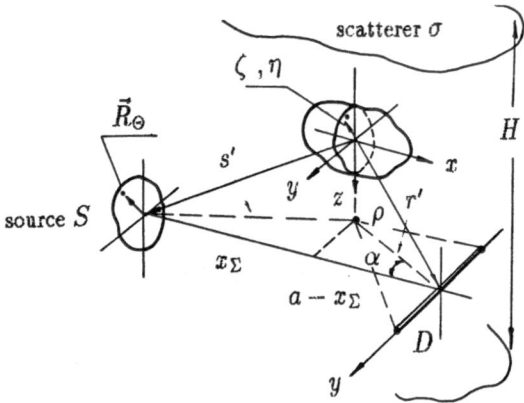

Figure 1. The disposition of the source S, the scatterer σ and region of observation array D in the waveguide

waveguide G:

$$G\left(x_s, y_s, z_s; x, y, z\right) = \sum_{n=1}^{N} \varphi_n(z_s)\varphi_n(z)\exp[i(h_n|\vec{r}_s - \vec{r}| - \frac{\pi}{4})](h_n|\vec{r}_s - \vec{r}|)^{-1/2} \quad (1)$$

where $\vec{R} = (\vec{r}, z) = (x, y, z)$, φ_n and h_n are the eigenfunctions and eigenvalues of the waveguide respectively, N is the entire number of the propagating waveguide modes. Let us assume that it the conditions of the small-angle approximation is satisfied. In this case the diffracted field from the Green's theorem using the Kirhgoff-hypothesis can be deduced. For the potential of displacement speed we have: $\Psi_s = \Psi - \Psi_0$:

$$\Psi_s = \sum_{m=1}^{M} \varphi_m(z)\exp[i(h_m S' - \frac{\pi}{4})](h_m s')^{-1/2} \sum_{n=1}^{N} \varphi_n(z_s)\exp[i(h_n r' - \frac{\pi}{4})](h_n r')^{-1/2} S_{nm},$$

$$(2)$$

where Ψ_0 is illumination field, $s' = (x_\Sigma^2 + y_s^2)^{1/2}$, $r' = ((a - x_\Sigma)^2 + y^2)^{1/2}$ (See Figure 1), x_Σ is displacement of the scatterer, a is distance between the source and the obsrvation region and M is a number of the diffracted waveguide modes. The matrix of the scattering of the waveguide modes S_{nm} is determined by form of the scatterer σ and waveguide characteristics:

$$S_{nm} = T_{nm}L_{nm} = ih_n \int_\zeta T(\zeta)\varphi_m(\zeta + z_\Sigma)\varphi_n(\zeta + z_\Sigma)d\zeta \int_\eta L(\eta)e^{i[\frac{h_n y_s}{s'} + \frac{h_m y}{r'}]\eta}d\eta, \quad (3)$$

where the form of the shadow-generated line $\sigma(\zeta, \eta)$ is approximately represented as a combination of two functions $\sigma(\zeta, \eta) \simeq T(\zeta)L(\eta)$.

As it follows from (2) and (3), the short-wave diffraction in waveguides contains the transformation of the waveguide modes on vertical direction (matrix elements T_{nm}) and similar to the diffraction in horizontal plane, of infinite medium which is described by matrix L_{nm}. The resulting field is constructed as a sum of all diffracted waveguide modes, each of them is formed by all illuminating modes [5].

PACO fields diffraction in the waveguides. Let us assume that quazimonochromatic illumination source S_Θ is described by a correlation function K_Θ:

$$K_\Theta = K_\Theta(\vec{R}_{\Theta 1}, \vec{R}_{\Theta 2}, \omega_0) =$$

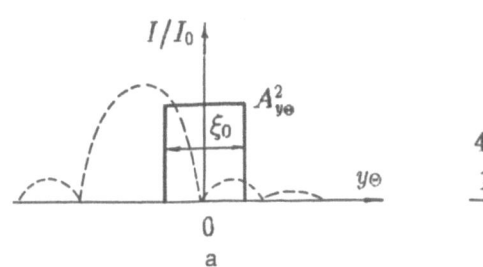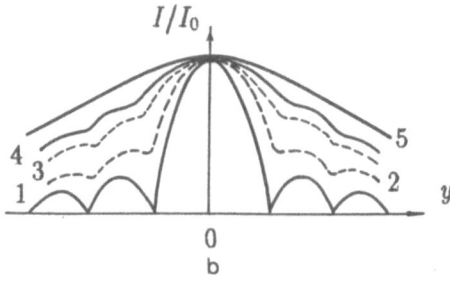

Figure 2. The graphical interpretation of the convolution in (6) - (a), and intensity variations of the signal for the different sizes of the source ξ_0 (b), $(1 - \xi_0 = \frac{D}{10}; 2 - \xi_0 = \frac{D}{2},$, $3 - \xi_0 = D, 4 - \xi_0 = 2D, 5 - \xi_0 = 5D)$.

$$< \rho(\vec{R}_{\Theta 1}, \omega_0)\rho^*(\vec{R}_{\Theta 2}, \omega_0) >= A_0^2 A_{z\Theta}^2(z_{\Theta 1})A_{r\Theta}^2(\vec{r}_{\Theta 1})\delta(z_{\Theta 1} - z_{\Theta 2})\delta(|\vec{r}_{\Theta 1} - \vec{r}_{\Theta 2}|)e^{-i2\pi f_0 \tau}, \tag{4}$$

where $A_0 =$ is a constant, $A_{z\Theta}$ is source spatial distribution function, δ is a Dirac function, τ is temporal delay. The PACO diffracted field in the waveguides can be represented as a sum of the diffracted structures for all elementary point-sources waiting by correlation function (4). Using expressions (2-4), the formula for the spatial coherence function can be obtained : $K_{12}^g = \sum_{n\mu} \varphi_m(z_1)\varphi_\mu^*(z_2) < b_m^g b_\mu^{g*} > $ ($b_{n,m}^g$ are the radience cofficients of the modes):

$$K_{12}^g = - \sum_{nm\nu\mu} A_0^2\varphi_\nu\varphi_\mu^* \int_{z_\Theta} A_z^2(z_\Theta)T_{nm}T_{\nu\mu}^* h_n h_\nu \varphi_n(z_0)\varphi_\nu^*(z_\Theta)dz_\Theta$$

$$\int_{r_\Theta} A_r^2(r_\Theta)L_{nm}L_{\nu\mu}^*[h_m h_\mu |\vec{r}_\Theta - \vec{r}_1||\vec{r}_\Theta - \vec{r}_2|]^{-1/2}e^{i[h_m|\vec{r}_\Theta - \vec{r}_1| - h_\mu|\vec{r}_\Theta - \vec{r}_2|]}dr_\Theta \tag{5}$$

This expression connects the sptial coherence function of the diffracted field with the form and displacement of the scatterer. On the contrary to the formula for the infinite medium the expression (5) describes the effects of the interference of the waveguide modes (n, m, ν, μ - indexes). It can be shown that multimode propagation complicate the diffraction patterns when scatterer sizes and spatial delays considerably exceed the scales of the mode interference.

If $A_{z\Theta} << \Delta_z$ where Δ_z is a scale of the changability of the waveguide field for vertical direction , and if $\frac{\eta^2}{x_\Sigma} << 1$, $\frac{y^2}{x_\Sigma} << 1$ we have from (5):

$$K_{12}^g = \sum_{nm\nu\mu} L_{nm}^{\nu\mu} \Phi_{nm}^{\nu\mu} \int_{y_\Theta} A_{y\Theta}^2 sinc([ph_n + qy_1 h_m]\eta_0)sinc^*([ph_\nu + qy_2 h_\mu]\eta_0)dy_\Theta, \tag{6}$$

where $p = \frac{y_\Theta + y'}{x_\Sigma}$, $q = \frac{1}{a - x_\Sigma}$, $L_{nm}^{\nu\mu} = -A_0^2\varphi_n(z_\Theta)\varphi_\nu^*(z_\Theta)\varphi_m(z)\varphi_\mu^*(z)h_n h_\nu T_{nm}^{\nu\mu}\eta_0^2$, $\Phi_{nm}^{\nu\mu} = [x_\Sigma(a - x_\Sigma)h_n h_m h_\nu h_\mu]^{1/2} \exp[i(x_\Sigma(h_n - h_\nu) + (a - x_\Sigma)(h_m - h_\mu))]$, $sinc(x) = \frac{\sin x}{x}$ and η_0 is the scatterer size for the horizontal direction. The expression (6) shows that the spatial coherence of the diffracted field in the waveguide can be presented as a convolution of the function $A_{y\Theta}^2$ which describs source and Fourie-transformation of the horizontal distribution of the scatterer (projection of the scatterer to the plane of the source) for the illuminating and diffracted waveguide modes. The Figure 2a shows the structure

359

of the integral over y_Θ in (6) when $y = y'$, $n = \nu$, $m = \mu$, where the solid line is the function $A^2_{y_\Theta}$ and dash line is a function $sinc^2([p+qy]\eta_0 h_n)$ which is determined by the width of the function $\Delta_\eta \simeq \frac{2\pi x_\Sigma}{\eta_0 h_n}$ and its displacment $\frac{y}{x_{eq}}$, where $x_{eq} = x_\Sigma(a - x_\Sigma)a^{-1}$. The distribution of intensity in observation plane is determined by joint region of these functions and is averaged for the increase of the source size (see Figure 2b).

Spatial filtration of diffracted PACO fields. Let us discuss the work of the horizontally distributed (along y axis) array, constructing the images of the scatterer in the waveguides. The aperture function of the array is determined by the expression: $M^\Sigma_m(z,y) = M_m(y)M'_m(z)\exp[i(h_m y\sin\alpha - \frac{h_m y^2}{\rho})]$, where $M_m(y)$ and $M'(z)$ describe the array construction and α and ρ determine the wave front slope and focus of array, respectively. In this case the image constructed by array can be presented as a convolution of the diffracted field (6) and aperture function $M^\Sigma_m(z,y)$. For the simplification let we assume that it the mode selection (for example, by angle or reaching time of the propagating pulse gating) has been made. Thus, for the image constructed by array we have:

$$< |P_\Sigma(\alpha,\rho)|^2 >= \int_{y_\Theta} A^2_{y_\Theta} S'^2_{mm} | \int_\eta L(\eta)$$

$$\int_D \exp\left(-i[p+qy]h_m\eta + h_m y^2[\frac{q}{2} - \frac{1}{\rho}] - h_m y\alpha\right)M_m(y)dyd\eta|^2 dy_\Theta, \qquad (7)$$

where $S'^2_{mm} = A^2_0 h^2_m \varphi^2_m(z_\Sigma)\varphi^2_m(z_\Theta)d^2 H^{-2}h^2_m x_\Sigma(a - x_\Sigma)$ and H is a depth of the waveguide. The procedure of the image construction is realized by focusing to all points of the observation region and summing the signals of all array sensors. The work of this imaging system is characterized by contrast transfer function (CTF) $K(\alpha,\rho,u)$, where u is a spatial frequency for y direction. For the deduction of the function K we substituate to (7) the observed object in a form:$L(\eta) = \exp(i\eta u) + \exp(-i\eta u)$. In this case, using the CTF determination, we obtain the expression for the CTF:

$$K(\alpha,\rho,u) = \int_{y_\Theta} A^2_{y_\Theta} M_m[l(ph_m - u)]M^*_m[l(ph_m + u)]\exp\left(i[\epsilon h_m l^2 2h_m p + 2\alpha h_m l]u\right)dy_\Theta,$$

$$(8)$$

where : $l = \frac{a-x_\Sigma}{h_m}$,$\epsilon = \frac{1}{2(a-x_\Sigma)} - \frac{1}{\rho}$ is a focusing parameter of imaging system. Deduced expression for the CTF describes the image construction in the waveguide (for the single mode approximation) for the partially coherent illumination. For the asymptotical cases when $\xi_0 << \frac{l}{D}$ and $\xi_0 >> \frac{l}{\Delta_{min}}$ (Δ_{min} is the distance between sensors) the equation (8) describes the well-known (in optics [1]) coherent and noncoherent imaging systems.

The Figure 3a shows the results of the calculation of the CTF structure transformation for spatial frequency and for the different sizes of the source, when the filter of the low frequency (dark field- filtration) is used. The calculation has shown, that, when $\xi_0 < D$ the possibilities of the elimination of the direct field background have kept to it is important for the reconstruction of the week oceanic inhomogeneities. The results of the background elimination for computer simulated images are presented in Figure 3b.

SOME EXPERIMENTAL RESULTS

For the estimation of the theoretical results the experiments in ultrasonic region (with frequency 140 kHz) and in optical fields (as the simple model from point of view of the modelling of the partially coherent illumination) have been carreed out.

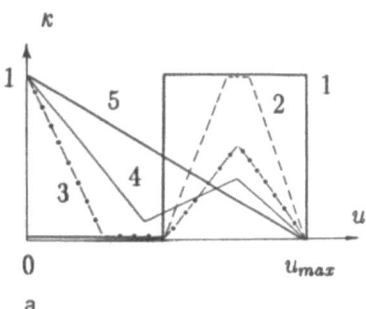
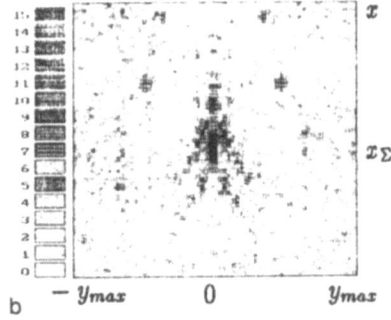

Figure 3. The distribution of modulus of CTF for the different source sizes (sizes of the source in Fiugre 2 and Figure **3** are the same) (a) and computer simulated coherent image of the scatterer after filtration for the low spatial frequency (b)

Ultrasonic imaging. The synthetic aperture with lense 28 cm, which can be constructed by motion of the point receiver, was used for the imaging of the scatterer spatial distribution (see figure 4a). The quasimonochromatic pulse with time-duration 300 μs was used for the reverberation decrease by time-gating. The isovelosity water layer which depth was equal to 3 cm was used as a simple model of the layered oceanic waveguide. The distance between the source and observtion region was equal to 44.6 cm was used.

A vertical steel cylinder of diameter 0.25 cm was situated in the middle of the distance between source and receiving array. Figures 4b,c show the measured data images, which are constructed accordanly with the algorithm (7) in the case when source was coherent. Figure 4b presents image without spatial filtration. We can only see the image of the source distorted by waveguide mode interference (see vertical black and white lines). The results of the low spatial frequency filtration (dark field method) are shown in Figure 4c, where the disposition of the cylinder is well determined. It can be note that the spatial filter based on the utilization of the waveguide mode interference was used.

Optical modelling. For the comparison of the theoretical results of the theoretical analysis of the spatial filtration of the PACO images with experimental data was built the special laboratory facilities which gave the possibilities to construct the partially coherent images in optical fields [2,3]. The major part of the experimental system is the source of light with switched over spatial coherence (see Figure 5).

The utilization of these facilities gave the opportuniaty of modelling the high frequency ultrasonic imaging system but this approach can't be used for the investigation of multimode imaging. The block-scheme of the arrangement includes the optical lense that is a analogy of the acoustical array (2 in Figure 5) sources with variable coherence: the thermal source with variable sizes (elements 1,2,3,4) and laser combined with random diffuser (method using the same facilities for the radiation of the noncoherent ultrasound was recently discussed in acoustics[6]). Figure 5[c] shows the results of the CTF measurements for the spatial low frequency filtration (dark field method) for the different sizes of the source. It give us the possibilities to estimate effeciency of the partially coherence images filtration. The example of this filtration is showed on Figure 5d, where images of letters after low frequency filtration are presented for the different sizes of source.

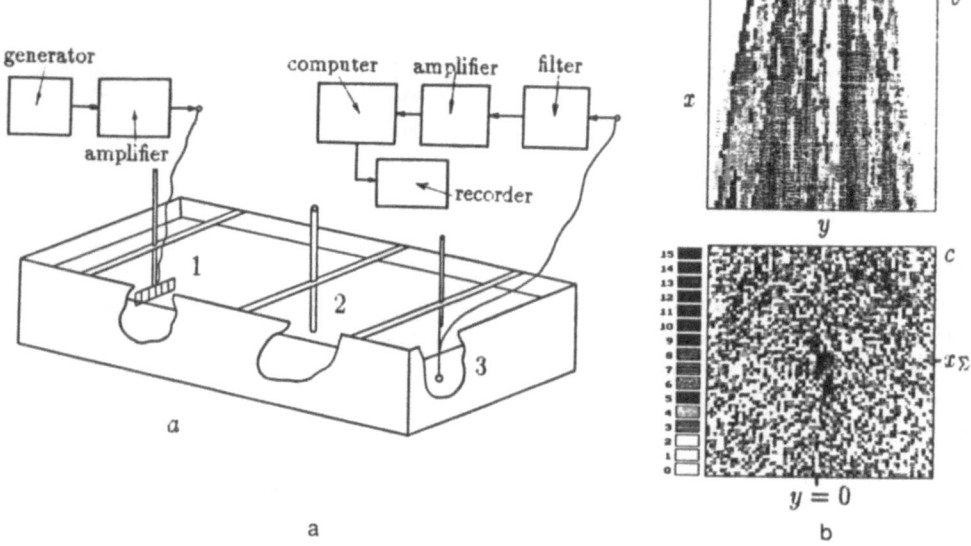

Figure 4. The experimental ultrasonic imaging system (a) (1 — source, 2— scatterer, 3 — receiving array); b,c — the coherent images before and after "dark-field filtring", respectevely.

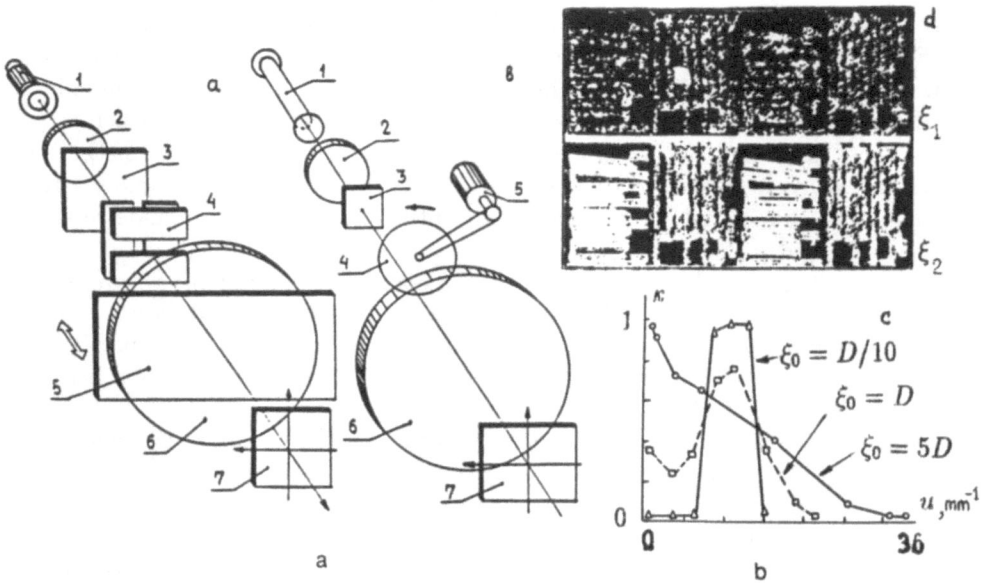

Figure 5. The optical source with tunable coherence. The case (a)- scheme with thermal source (1-incoherent lamp, 2-microlens, 3- frequency filter, 4- optical slits, 5-frosted screen, 6- lens, input plane), the case (b)-scheme with laser source (1-laser, 3-spatial filter, 4-rotation diffusor, 5-engine); d — the images of letters in the case of the different sizes of source, for $\xi_1 = \frac{D}{10}$ and $\xi_2 = \frac{D}{4}$, respectevely.

CONCLUSIONS

The aim of presented investigations is the development of the acoustical vision system in natural layered waveguides, such as the ocean and atmosphere, where it is necessary to observe the smooth inhomogeneities immersed into the random medium. The utilization of the different approximations which are founded on the experimental data allows to obtain the ample simple expressions for the physical analysis of the peculiarities of the spatial filtration of the partially coherent images in waveguides, for example, for the estimation of the effeciency of the dark field method. The theoretical calculations and experimental measurements show that effeciency of the low frequency filtration decreases on the thirteen percents when the coherency equal to a half of the aperture of the receiving array. The wave mode interference plas a great role for the image construction in the waveguides because it leads to the multiplication of the resulting image when it can not be made a special transformations of in the algorithm of the image reconstruction on the basis of a prior information about of the characteristics of waveguides. It is clear that the more complicated experimental analysis of the peculiarities of the PACO fields filtration should use the special ultrasonic system with sound source with variable spatial coherence like that which was described earlier [6].

REFERENCES

1. J. W. Goodman, "Introduction to Fourie Optics, Mcg-Hill Book Com., NY (1968).
2. V. A. Zverev, A. V. Shisharin and A. I. Kihl'ko, The application of the sources of incoherent white light in scheme of spatial filtering of images", *Avtometria* 2:108 (1978).
3. A. I. Khil'ko, Partially coherent image reconstruction, in Proc. of SPIE Symposium, Orlando, Vol. pp. (1993).
4. L. M. Brekhovskikh and Y. P. Lysanov, "The Theoretical Fundament of the Ocean Acoustics", Gidrometeoizdat, Moscow (1982).
5. N. V. Gorskaya, S. M. Gorskiy ,V. A. Zverev, et al., Short-wave diffraction in a multimode layered waveguides, *Sov. Phys. Acoust.* 34(1):29 (1988).
6. T. Sato, S. Wadaka, J. Ishii and T. Sunada, A new ultrasonic imaging system by using a rotating random phase disk and power spectral and third order correlation analysis, *Acoustical Holography* 7:167 (1978).

EXPERIMENTAL IMAGE-RECONSTRUCTION OF DIFFRACTION TOMOGRAPHY WITH DATA OBTAINED BY COMPOUND-SCANNING A PAIR OF TRANSDUCERS

Keinosuke Nagai, Tomoki Yokoyama and Koichi Mizutani

Institute of Applied Physics, University of Tsukuba
Tsukuba Ibaraki, 305 Japan

INTRODUCTION

We show the experimentally reconstructed images of the diffraction tomography. The procedure was proposed in the last symposium[1] and it has an advantage that the image is easily obtained in the experiment.

The acoustic wavelength used in medical diagnostics and the non-destructive inspection is usually three orders longer than the optical wavelength. The information of the acoustic images is, therefore, three orders in quantity less than that of the optical images. In order to use efficiently rather poor information and to obtain clear images the acoustic imaging method should compensate diffraction which degrades the images seriously. This is the motivation by which the diffraction tomography has been developed.

A large number of papers concerning to the diffraction tomography have been published. Most of them, however, treat it only theoretically. Few show the experimental results. The main reason seems for us that the clear image is not so easily obtained in the experiment. There are two ways to lluminate objects in the diffraction tomography. One uses a plane wave and the other a fan beam[2] . It is difficult to generate an ideal plane wave with the transducer of a finite extent. Fan beam is excited more easily. The image reconstruction is, however, still difficult. Fourier transform on a certain circle in the Fourier domain is usually calculated from the data obtained in

Acoustical Imaging, Vol 21, Edited by
J.P. Jones, Plenum Press, New York, 1995

the diffraction tomography[2,3] . If the image is reconstructed by the inverse Fourier transform using FFT (fast Fourier transform), the interpolation of data is required on the lattice points from those on the circle. This introduce much error which severely degrades the image. The interpolation free procedure has been proposed[3] but very time-consuming. Because it cannot use FFT.

Our proposed method[1] adopts the fan beam illumination which is easily generated with the transducer attached to a mask with a narrow slit. The Fourier transform of the object on a line rather than on the circle is computed from the data acquired on a scanning line . We describe the interpolation free implementation in this paper which can use FFT and is accomplished in shorter time. The images of phantoms are experimentally reconstructed and displayed. The resolution of these images are roughly estimated less than 1mm when the wavelength used is 0.75mm.

DATA ACQUISITION

The geometry is shown in Fig.1 where an object is embedded in homogeneous medium. The situation is assumed constant in the direction perpendicular to this plane to be considered two dimensional problem. The extension of the discussion in this paper to the three dimension is simple. Consider Cartesian coordinates (x, y) on the plane. The object is put about the origin. Its density and compressibility are $\rho(x, y)$ and $\kappa(x, y)$, respectively. Those of the homogeneous medium are ρ_0 and κ_0 . A transmitter is set to face a receiver. The coordinates of the transmitter and the receiver are (x_t, Y_t) and $(x_t, -Y_t)$, respectively.

Sinusoidal wave, which is replaced with tone burst in the experiment is shot by the transmitter. The scattered wave $p_s(x_t)$ from the object is received by the receiver which is represented by the following equation.

$$
\begin{aligned}
p_s(x_t) \quad = \quad & \iint_{S_o} \{\gamma_c(x, y)k^2 g_w(x_t, -Y_t|x, y) p_w(x, y) \\
& + \gamma_d(x, y) \nabla g_w(x_t, -Y_t|x, y) \cdot \nabla p_w(x, y)\} dx dy.
\end{aligned}
\tag{1}
$$

where S_o arbitrary closed surface including the object, $p_w(x, y)$ acoustic pressure and k wavenumber. $g_w(\cdot)$ is the Green's function:

$$
g_w(x_t, -Y_t|x, y) = \frac{i}{4} H_o^{(2)} \{k\sqrt{(x - x_t)^2 + (y + Y_t)^2}\},
$$

where $H_o^{(2)}(\cdot)$: the 0th order Hankel function of the 2nd kind.

γ_c and γ_d represent the acoustic characteristics of the object and are given by the following equations

$$
\gamma_c(x, y) = \frac{\kappa(x, y) - \kappa_o}{\kappa_o},
\tag{2}
$$

$$
\gamma_d(x, y) = \frac{\rho(x, y) - \rho_o}{\rho(x, y)}.
$$

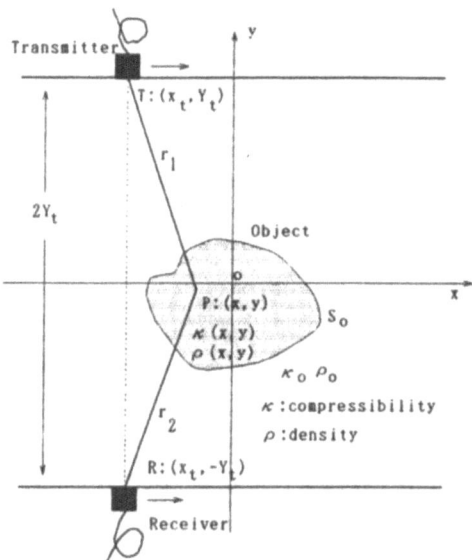

Figure 1. Geometry for data-acquisition. An object is set in the homogeneous medium and between a transmitter and a receiver. Tone burst wave is shot from the transmitter, scattered by the object and caught by the receiver. The wave is then quadrature-detected and recorded as a datum of a complex number. The transmitter and the receiver are simultaneously scanned on each line parallel to a certain line (the x-axis in this figure). Data are acquired at a proper step in a appropriate range on the line. Then the line is rotated at a step angle and the data acquisition is repeated. This procedure is continued till the rotation angle becomes totally 180 °.

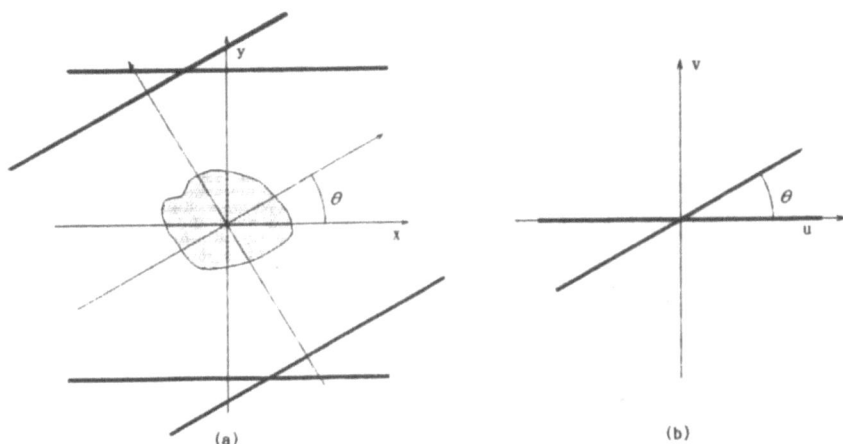

Figure 2. Relationship between (a) the scanning line of transducers and (b) the line on which the Fourier transform of the object is known. The Fourier transform of the object is obtained on the u-axis in the domain from data acquired on the scanning line of the x-axis. The Fourier transform on the line at the angle θ from the u-axis is then calculated from data obtained on the line at θ from the x-axis. If we rotate the angle over 180° and acquire data on the line at proper step angle, we can know the whole Fourier transform of the object from data thus obtained.

Noting that γ_c and γ_d are zero outside the object the integral in eq.(1) is meaningful only inside the object.

We apply Born approximation to eq.(1). That is, $p_w(x, y)$ in the object is assumed to be the incident wave $p_i(x, y)$. We use the transmitter with narrow aperture for the incident wave to be considered as cylindrical wave:

$$p_w(x, y) \simeq A\frac{i}{4}H_o^{(2)}\{k\sqrt{(x - x_t)^2 + (y - Y_t)^2}\},$$

$$A : \text{constant.}$$

We further assume the density of the object is almost same as that of the surrounding medium: $\gamma_d \simeq 0$. And when the distance between the transmitter and the receiver is much more longer than the size of the object and the range of the scanning: $Y_t >> |x|, |y|, |x_t|$, we can apply the paraxial approximation. Then eq.(1) reduce to

$$p_s(x_t) \simeq C \iint_{S_o} \{\hat{\gamma}_c(x, y) \exp\{-ik\frac{(x - x_t)^2}{Y_t}\}dxdy, \tag{3}$$

where C is a constant and

$$\hat{\gamma}_c(x, y) = \gamma_c(x, y) \exp\{-k\frac{y^2}{Y_t}\}. \tag{4}$$

IMAGE-RECONSTRUCTION

$\gamma_c(x, y)$ is the function of the compressibility of the object as shown in eq.(2). Then we regard $\gamma_c(x, y)$ as the 'object'. The amplitude $|\gamma_c|$ is displayed as the image, though γ_c is a complex number. We can, therefore, use $\hat{\gamma}_c(x, y)$ in place of $\gamma_c(x, y)$ as the object, because from eq.(4)

$$|\hat{\gamma}_c(x, y)| = |\gamma_c(x, y)|.$$

'Reconstruction' means calculating $\gamma_c(x, y)$ from input data $p_s(x_t)$ using the relationship of eq.(3). As eq.(3) is the convolution it can be readily resolved by the use of Fourier transform.

Let $P_s(u)$ be the one dimensional Fourier transform of $p_s(x_t)$, that is

$$P_s(u) = \int_{-\infty}^{\infty} p_s(x) \exp(-iux)dx, \tag{5}$$

$H(u)$ be the transform of $\exp\{-ik\frac{y^2}{Y_t}\}$,

$$H(u) = \int_{-\infty}^{\infty} \exp\{-ik\frac{y^2}{Y_t}\} \exp(-iux)dx, \tag{6}$$

and $\hat{\Gamma}_c(u, v)$ be the two dimensional Fourier transform of $\hat{\gamma}_c(x, y)$

$$\hat{\Gamma}_c(u, v) = \iint_{-\infty}^{\infty} \hat{\gamma}_c(x, y) \exp\{-i(ux + vy)\}dxdy. \tag{7}$$

Substituting eqs.(5) to (7) in Fourier transforms of the both side of eq.(4) yields

$$P_s(u) = C\hat{\Gamma}_c(u, 0) \cdot H(u). \tag{8}$$

Equation (8) means that we can know the value on the u-axis of the two dimensional Fourier transform $\hat{\Gamma}_c(u, 0)$ of the object $\hat{\gamma}_c(x, y)$ from the data $p_s(x)$ acquired on the x-axis. This process is shown in Fig.2. Similarly we can know the Fourier transform on the line at angle θ from the u-axis from the data obtained on the line at θ from the x-axis.

$$P_s(U) = C\hat{\Gamma}_{c\theta}(U, 0) \cdot H(U). \tag{9}$$

where $U\cos\theta = u, U\sin\theta = v$.

We can know therefore the whole Fourier transform $\hat{\Gamma}_c(u, v)$ by the rotating the scanning lines at proper angle, acquiring the data on the line and repeating this for rotating angle of totally 180 degrees. Once we know the whole Fourier transform we can reconstruct the image by the inverse transform,

$$\hat{\gamma}_c(x, y) = \frac{1}{(2\pi)^2} \iint_{-\infty}^{\infty} \hat{\Gamma}_c(u, v) \exp\{i(ux + vy)\} du dv. \tag{10}$$

In the previous paper[1] we interpolated the data on the lattice in the (u, v)-plane from those on the radial lines to calculate the inverse Fourier transform by using FFT. The interpolation introduce inevitable errors which degrade the image severely. We show, however, in the following that we are able to achieve interpolation free reconstruction,

$$\hat{\gamma}_c(x, y) = \frac{1}{4\pi^2 C} \int_0^\pi \int_{-\infty}^{\infty} \frac{P_s(U)}{H(U)} \exp(iUX) |U| dU d\theta. \tag{11}$$

We can expect the clear image reconstructed by using this equation.

EXPERIMENT

The experiments have been done under the condition shown in Table I. The first object is a vinyl tube. Its outer diameter and inner one are 18 mm and 14 mm, respectively. The data obtained on the certain line are shown in Fig.3. The reconstructed image is shown in Fig.4. The shape of the vinyl tube is well reconstructed which is recognized not to be perfectly cylindrical. The second object is cut sausage. Figure 5 shows the image. The brightness corresponds to the compressibility of the composite materials such as fat, muscle and meat.

Considering the thickness of the tube in Fig.4 and the grain size of the sausage in Fig.5 the resolution of these images is roughly estimated to be less than 1 mm.

CONCLUSIONS

The procedure of the diffraction tomography is proposed. It has the advantage that the image could be easily reconstructed in the experiment. The images are reconstructed and shown in this paper. The resolution of these images is roughly estimated less than 1 mm when the wave length is 0.75 mm.

Table I. Parameters for the experiment

Frequency of the ultrasonic wave f	2.0 MHz
(wave length λ)	(0.75 mm)
Step of the linear scanning $\triangle x_t$	0.5 mm
Length of the sampling region L	64 mm
Step of the rotation angle $\triangle\theta$	6°
Distance between transmitter and the receiver $2Y_t$	280 mm

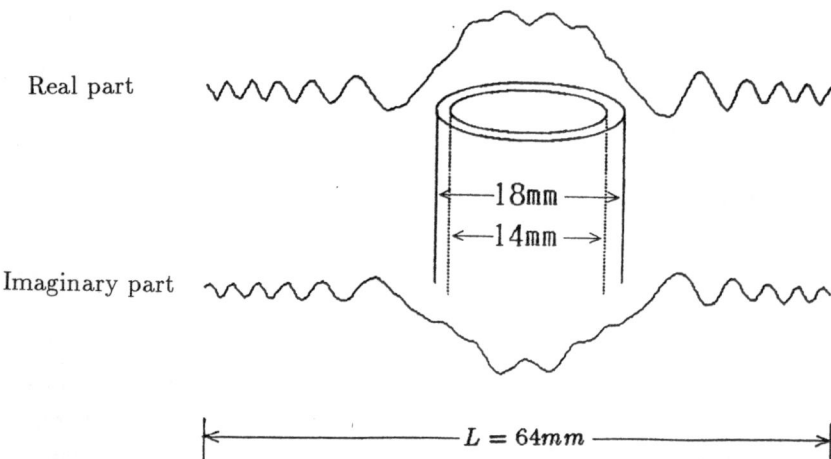

Figure 3. Example of data acquired on a line. The object is vinyl tube with 18mm outer diameter and 14mm inner diameter. The complex data of real and imaginary parts are the result from the quadrature detection of the received wave. The data are displayed at the relative position along the horizontal line corresponding to that of the object.

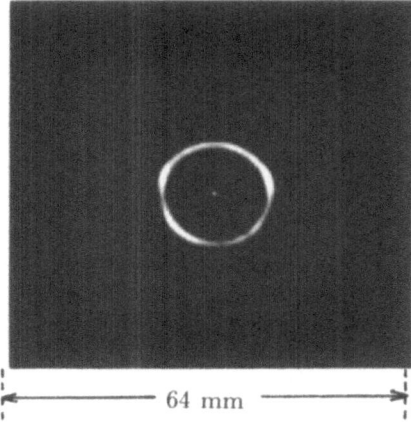

64 mm

Figure 4. The reconstructed image of the vinyl tube. It is reconstructed from data acquired on whole scanning lines over 180°, the part of which is shown in Fig.3. We can see the tube is not perfectly cylindrical.

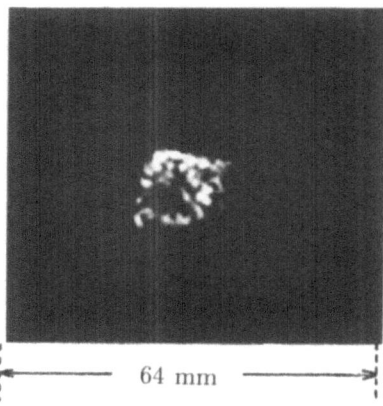

64 mm

Figure 5. The reconstructed image of the cut sausage. The compressibility of the composite are represented as the brightness of the image. The resolution of the image is roughly estimated less than 1mm.

We will adopt the transducer array to acquire data more speedily and for this procedure to be applicable to the practical system.

REFERENCES

1. K.Nagai, T.Yokoyama and K.Mizutani,A method of diffraction tomography : Compound scanning of transmitter and receiver, in: "Acoustical Imaging vol.20" Y.Wei and B.Gu ed., pp.619-623, Plenum Press, New York 1993.

2. A.J.Devaney and G.Beylkin, Diffraction tomography using arbitrary transmitter and receiver surface, Ultrasonic Imaging 6, 181-193, (1984).

3. D.Nahamoo, S.X.Pan and A.C.Kak, Synthetic aperture diffraction tomography and its interpolation free computer implementation, IEEE Trans. on Sonics and Ultrasonics, SU-31, 4, 218-229, (1984).

TOWARDS MORE INFORMATIVE PULSE-ECHO IMAGES

OF THIN MEMBRANES

A. J. Healey[1], S. Leeman[1], Marie Restori[2] and Jacquetta Allaway[1]

[1]Department of Medical Engineering and Physics
Kings College School of Medicine and Dentistry
London SE22 8PT, U.K.
[2]Moorfields Eye Hospital, London, U.K.

INTRODUCTION

The problem under consideration is that of measurement of the thickness of 'thin' (of the order of the wavelength of the imaging pulse) membrane-like layers, such as the human retina, using ultrasound pulse-echo data. An obvious approach to this problem is to record the location of a feature of the signal, such as an envelope maximum, associated with the front and rear faces of the membrane. The time separation of these two features then provides an indication of the (temporal) thickness dimension of the structure. A modification, (which may be advantageous for e.g. noisy signals) is to examine features of the cross-correlation between the imaging pulse and the rf A-line data segment of interest. However, a number of effects compromise such methods, a major problem being that of interference (see Fig. 1).

Another problem exists if echoes are present from structures adjacent to the membrane. This may lead to problematic interference effects and may also obscure the precise location of the membrane echo signal, leading to an ambiguity in exact location of the data extraction window. The proposed technique attempts to solve such ambiguities and to correct for interference effects when estimating membrane dimensions.

MODEL

The algorithm for producing separation estimates may usefully be described with the aid of a simple discrete two-scatterer convolutional model. The resulting rf A-line signal, $rf[n]$, is then comprised of a function representing the 'structure' component, $r[n]$, convolved with a function, $p[n]$, which represents the imaging pulse. The pulse is assumed to remain spatially invariant over the short data segments under consideration. Hence,

$$rf[n] = r[n] \otimes p[n] \qquad (1)$$

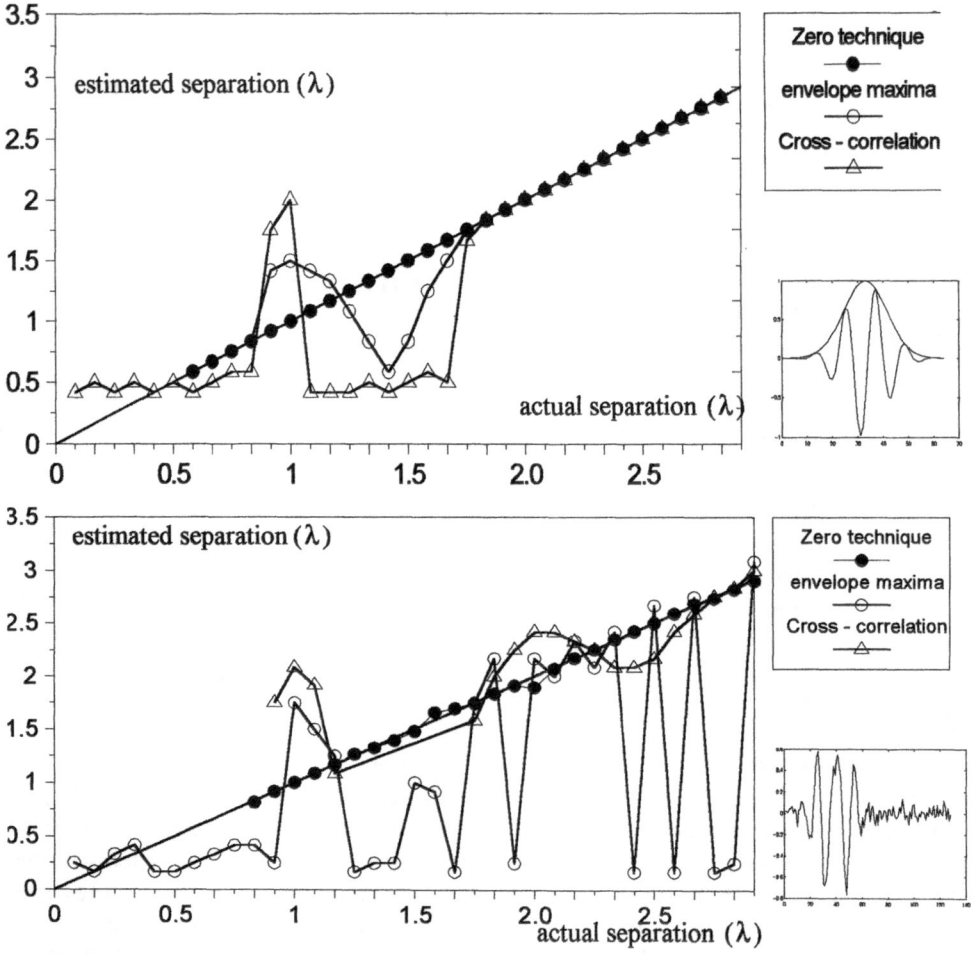

Figure 1. Simulation results based on a simple 1D convolutional model. Note that for pulse-echo signals the actual spacings will be half those indicated. (Top) Separation estimates in the absence of noise; pulse used in simulations shown as insert, (Bottom) Separation estimates in the presence of additive noise, typical noisy signal shown as insert.

The reflectivities of the front and rear faces of the membrane-like structure, embedded in a uniform medium, may be modelled as two discrete delta functions, of opposite sign, $r[n] = \alpha_0 \delta[n-d] - \alpha_1 \delta[n-d-n_0]$. The z-transform of the structure component may be expressed as

$$R(z) = \sum_{n=-\infty}^{\infty} \{\alpha_0 \delta[n-d] - \alpha_1 \delta[n-d-n_0]\} z^{-n} \qquad (2)$$

The roots of $R(z)$ are given by $z = \left(\frac{\alpha_1}{\alpha_0}\right)^{\frac{1}{n_0}}$, and can be seen to be located in the complex z-plane on a circle with centre at the origin, with radius $\left|\frac{\alpha_1}{\alpha_0}\right|$. The phase angle of the nth root is given by

$$\left\{Arg(\frac{\alpha_1}{\alpha_0}) + 2\pi n\right\}/n_0 \qquad (3)$$

A smoothness criterion may be associated with the energy density spectrum of the imaging pulse, in that it is relatively 'slowly varying' or 'smooth' over the pulse bandwidth.

In the present context, this implies that a zero-free region, close to the unit circle and the carrier frequency, may be associated with the pulse energy density spectrum. Convolution in the spatial domain is equivalent to multiplication in the inverse (spatial frequency) domain and hence the zero set of the z-transform of $rf[n]$, is comprised of the addition of the zero sets of the components of the convolution (plus a scaling factor). Therefore, any zero found close to both the unit circle and the carrier frequency, may be attributed to the structure component, i.e. it is a contribution from $R(z)$. A consideration of Eq. 3 provides a relationship between the phase angle of any structure zeros and the separation distance of the two interfaces in the spatial domain. As the two-scatterer model assumes a structure comprising two delta functions with a relative (scattering) phase difference of π, a zero of $R(z)$ will be fixed at $\left|\frac{\alpha_1}{\alpha_0}\right|$. Hence, the phase difference required may be obtained from the identification of two (or more) adjacent zeros of $R(z)$, or from only a single zero located at $\left|\frac{\alpha_1}{\alpha_0}\right| \exp\left(i\frac{2\pi}{n_0}\right)$, if it may be unambiguously identified. For ultrasound pulse-echo data the time separation thus obtained may be converted into a (spatial) thickness if the velocity of ultrasound is known in the membrane-like layer.

An important consideration in the performance of the algorithm is its behaviour with respect to an additive noise component, and the effects of windowing to extract the local signal segment corresponding to the structure of interest: these topics are analysed in the following sections.

NOISE

The effect of a distortion of the (finite number of) signal samples $rf[n]$, due to some form of noise and/or windowing procedure is to alter the positions of the zero locations in the complex z-plane. Of particular interest is the change in location of the zero(s) associated with the thickness estimate. Because of the finite data length of the segment $rf[n]$, its z-transform may be expressed as the polynomial $X(z) = b_0 - \sum_{k=1}^{N} b_k z^{-k}$. Setting $b_0 = 1$ for convenience, (as achieved via a scaling factor) we obtain

$$X(z) = 1 - \sum_{k=-1}^{N} b_k z^{-k} = \prod_{j=1}^{N} (1 - z_j z^{-1}) \tag{4}$$

By following an approach similar to that first used by Kaiser in a rather different context (Kaiser, 1966), the sensitivity of the ith zero to a change in the kth coefficient (due to the presence of a noise term) may be expressed as

$$\frac{\partial z_i}{\partial b_k} = \frac{z_i^{N-k}}{\prod_{j=1, j \neq i}^{N} (z_i - z_j)} \tag{5}$$

Hence the accumulated shift in the location of the ith zero z_i, to $z_i + \Delta z_i$, due to a noise term changing the b_k coefficients to $b_k + \Delta b_k$, is given by

$$\Delta z_i = \sum_{k=1}^{N} \frac{\partial z_i}{\partial b_k} \Delta b_k, \quad i = 1, 2, ..., N \tag{6}$$

The result shown in Eq. 5 follows directly from the observation that

$$\left(\frac{\partial X(z)}{\partial z_i}\right)_{z=z_i}\frac{\partial z_i}{\partial b_k} = \left(\frac{\partial X(z)}{\partial b_k}\right)_{z=z_i} \tag{7}$$

Eq. 7 provides a measure of the sensitivity of the ith zero to a change in the kth coefficient, and may be substituted into Eq. 6 to provide a measure of the total change in the ith zero due to small variations in the coefficients b_k (data samples). Thus, the inverse of the LHS of Eq. 5 may be interpreted as providing a measure of the relative stability of the zero z_i to a change in the b_k coefficients in the spatial domain.

An examination of the denominator of Eq. 5 reveals that if the zeros z_j are close to or clustered around z_i (small values of $(z_i - z_j)$) then z_i will be relatively sensitive to changes in coefficient value, the opposite being true for relatively 'isolated' zeros. Eq. 6 thereby forms the basis for the identification of the dominant (viz., most stable) structure zero of $R(z)$, as well as providing a measure of the relative confidence of the separation estimate.

WINDOWING

A local reduction in signal amplitude due to destructive interference produces a corresponding deviation in the magnitude of the instantaneous frequency (I.F.). Hence I.F. may be used to deterministically identify the precise location of destructive interference effects. If destructive interference arises between echoes from the front and rear faces of the membrane, then an I.F. maxima will occur. This may be used to locate the centre of the local data extraction window. A Hanning window of approx. 1.5 pulse lengths was used (in order to encapsulate the local interference effects) for the separation estimates shown in Figs 2 and 3. The separation estimate is relatively insensitive to the length of data window.

ALGORITHM

The algorithm consists of the following stages :
i   Calculate the discrete instantaneous frequency (D.I.F.) signal.
ii Detect maxima in the magnitude of the D.I.F.
iii Threshold D.I.F. maxima, and produce an image (see Fig. 4). For the data in Fig. 4 the membrane signal was selected by the continuous nature of the D.I.F. across the image (between adjacent A-lines). If no such features are present the object is imaged with a different centre frequency pulse or, alternatively, filtered and processing is returned to step i.
iv Centre the spatial domain window on the selected D.I.F. maxima, in order to select the local signal segment corresponding to the membrane.
v   Calculate the z-domain zeros of the extracted data segment.
vi Calculate stability of the z-domain zero set via Eq. 5.
vii Use the most stable zeros to calculate the separation estimate via a consideration of Eq. 3. Note that the most stable zero is assumed to provide an optimal estimate, as it is assumed to be the least affected by any noise component and the windowing procedure.

EXPERIMENTAL RESULTS

Fig 4 shows an image obtained of a Mylar film of thickness 0.346 ± 0.001 mm sandwiched between two layers of polyurethane foam. The transducer was a broadband Philips 19mm diameter, focused, commercial transducer. The data acquisition system

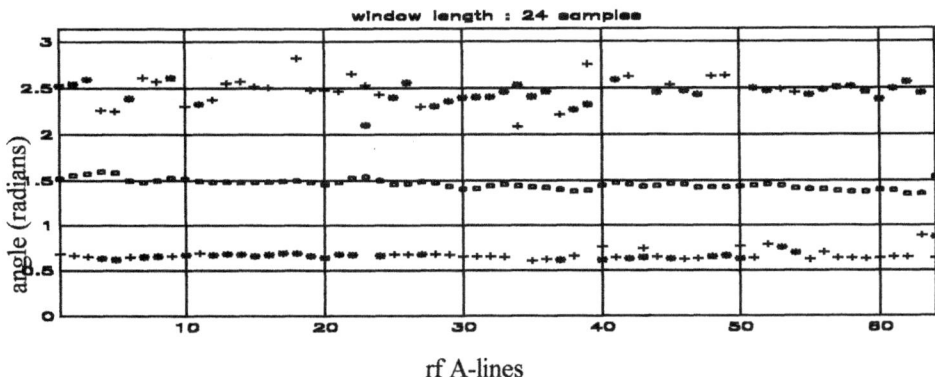

Figure 2. Plot of the phase angle associated with the z-plane zeros for the three least sensitive zeros, 'o', '+', and '*' respectively. The separation estimate is based on the least sensitive zero - as this is the least likely to be adversely affected by the windowing procedure.

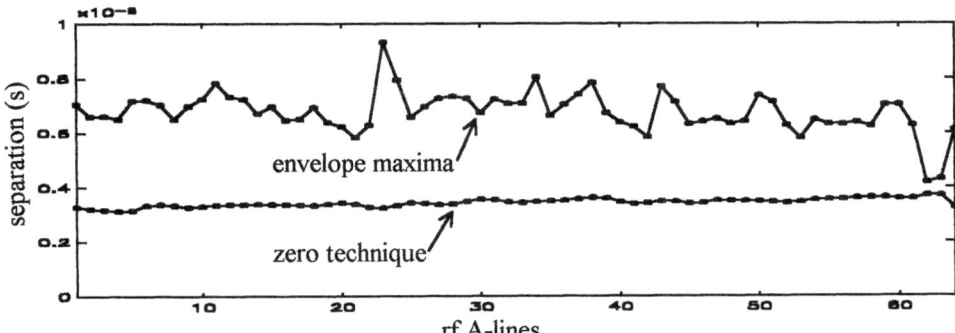

Figure 3. Time separation estimates for the front and rear membrane faces. The estimate based on envelope maxima has a bias of the order of 100%. The 'true' time separation is expected to be close to 0.3 microseconds, and is in agreement with the estimate produced via the new algorithm.

consists of a Gould GD4050 digital oscilloscope connected to an IBM PC compatible computer via an IEEE 488 interface bus. The signals were acquired at a 20 MHz sampling rate to a nominal 8 bits and transferred to the PC for off-line processing. Fig 5 shows a typical windowed data segment (b), and its associated zero set (a). The three most stable zeros associated with each rf A-line for the image in Fig. 4 are shown in Fig. 2. The time separation estimate provided by the most stable zero is shown in Fig. 3. In order to determine the 'true' distance separation a measurement of the velocity of ultrasound in Mylar (via a substitution method) provided an estimate close to 0.3 microseconds.

CONCLUSIONS

A signal processing technique has been presented which is capable of resolving the (sub-wavelength) thickness of membrane-like structures using pulse-echo ultrasound signals. If the technique is applied in conjunction with B-Mode imaging an angle correction may be implemented in order to correct the 'line-of-sight' thickness seen in an individual rf A-line basis. A consideration of the sensitivity of the z-plane zeros demonstrates that the relative sensitivity of the 'structure' zero, used in the thickness estimate, is potentially very stable to noise and windowing procedures, because of its location close to the unit circle

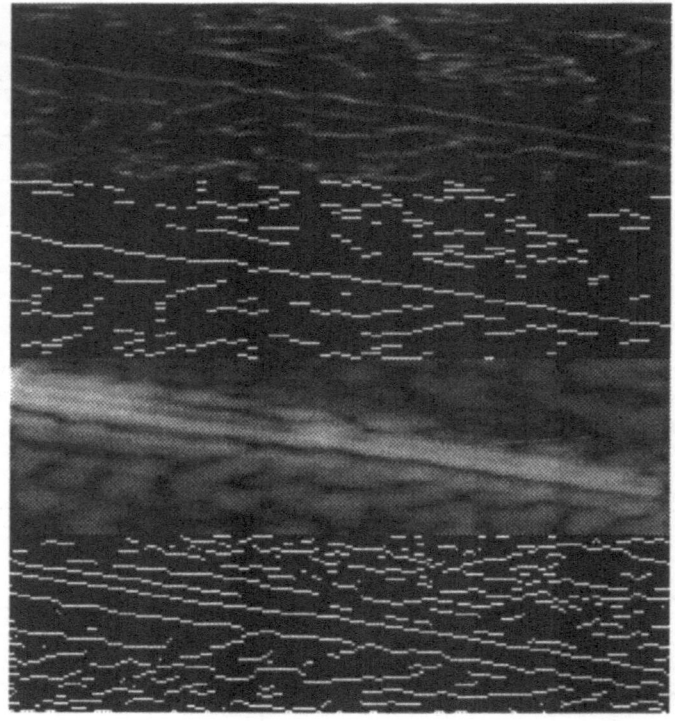

Figure 4: From top to bottom : instantaneous frequency image; threshold instantaneous frequency maxima, cutoff 0.5 MHz; corresponding envelope image; envelope maxima image.

and the carrier frequency of the imaging pulse. An important feature of the algorithm is, hence, that it is inherently relatively stable to noise. Another feature of the technique is that when a structure zero is absent, a membrane thickness below the resolution limit of the algorithm is indicated, rather than the provision of a biased estimate. The algorithm is currently capable of resolving membrane thicknesses below one half wavelength.

ACKNOWLEDGEMENTS

The Scientific and Engineering Research Council, and the Wellcome Trust are gratefully acknowledged for their financial support.

REFERENCES

Kaiser J. F., 1966, Digital filters, Chap. 7, *in*: "System Analysis by Digital Computer", F.F. Kuo and J.F. Kaiser, eds., John Wiley and Sons, New York.

A NEW CONSIDERATION OF DIFFRACTION COMPUTED TOMOGRAPHY FOR BREAST IMAGING: STUDIES IN PHANTOMS AND PATIENTS

Michael P. André[1,3], Peter J. Martin[2,3], Gregory P. Otto[2],
Linda K. Olson[1], Todd K. Barrett[2], Brett A. Spivey[2],
and Douglas A. Palmer[2]

[1]Department of Radiology, University of California
San Diego, CA 92093-9114
[2]Thermotrex Corporation, San Diego, CA
[3]Radiology Service, Veterans Affairs Medical Center, San Diego, CA

INTRODUCTION

Previous medical applications of ultrasound computed tomography showed promise of increased resolution over conventional pulse-echo imaging and exhibited potential for quantitative tissue characterization.[1-4] This research was limited by significant technical difficulties due in part to inadequate beam sampling and, in some cases, the assumptions associated with the straight-line ray-optical approach to pulse propagation. Nonetheless, experimental systems were constructed and credible clinical results were obtained in the breast where the range of tissue properties is less than in other parts of the body. Theoretical research has continued but few experimental systems have been built and there has been limited new work in tissue.[5,6]

In the past two years, we developed a human research system that uses diffraction tomography methods and that shows promise of resolving some of the earlier difficulties. This system was developed to test the feasibility of imaging the breast with these methods. Diffraction tomography involves illuminating an object with ultrasound and measuring a set of scattered wave data around the object. Image reconstruction computes the internal properties of the object by considering the wave equation with diffraction effects. The reader is referred to two review articles for a summary of the various approaches to diffraction tomography.[7,8]

Our approach differs from other work in that we employ a unique combination of features to facilitate data acquisition and reconstruction for breast imaging. For example, we: a) apply lower frequencies (≤ 1 MHz) to reduce phase aberration and reconstruction artifacts, b) utilize continuous-wave transmission to improve measurement accuracy and resolution, c) formally utilize cylindrical geometry with a cylindrical array and with concentric cylindrical wavefronts, d) employ a dense array of 1024 transducers with λ/2 spacing for high spatial sampling, e) acquire and average data at ten discrete frequencies to minimize artifacts, f) incorporate phase aberration correction in the image reconstruction, and g) apply a sophisticated parallel data acquisition and processing system to reduce imaging time to three seconds.[9]

Although not acceptable as a breast screening modality, high resolution conventional ultrasound continues to be explored as an adjunctive tool in the management of solid breast masses. When expertly applied in conjunction with mammography, ultrasound may greatly increase the accuracy of diagnosis and reduce the need for biopsy. However, present ultrasound technology has several limitations: it is difficult to perform, it is not sufficiently reproducible, it images only a very small portion of tissue at a time, and it lacks a global image of the breast for visual comparison. Each of these issues is addressed by our approach to diffraction tomography.

This paper describes the development of our experimental system, shows the performance of the method through simulation and phantom measurements, and presents preliminary results in an asymptomatic volunteer subject.

MATERIALS AND METHODS

Experimental System

Figure 1 shows a diagram of the cylindrical transducer array. Each element in turn acts as a transmitter while the remaining transducers record the signal emanating from the object. 1024 separate transmissions are made around the object producing a total of 1,048,576 transmit/receive pairs. The process is repeated for ten discrete frequencies from 687-1250 kHz. The large 20-cm ring of transducers illuminates the tissue via a heated coupling bath (Figure 2). The acoustic properties and/or temperature of the bath is adjusted to enhance penetration through the skin. The array is controlled through a high-speed 16-channel multiplexing network (Thermotrex Corporation, San Diego, CA) that completes the acquisition process in three seconds. Low acoustic intensities are employed.

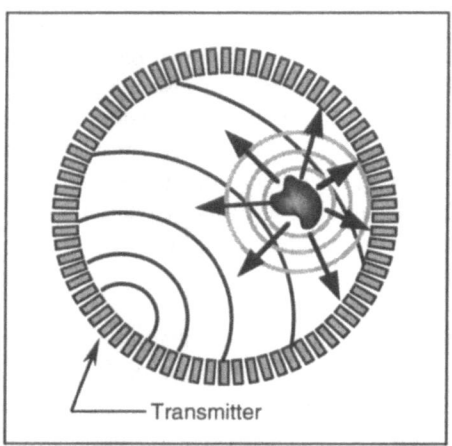

Figure 1. Diagram of the ring of transducers which sequentially transmit continuous-wave sound while the remaining transducers record the scattered signal. The process is repeated for ten frequencies, 687-1250 kHz.

Figure 2. 1024 transducers are contained in a 20-cm bath of coupling fluid and are controlled by a 16-channel multiplexer. The entire acquisition takes three seconds.

The experimental array was mounted on a modified stereotactic breast biopsy table contributed by Lorad Medical Systems, Inc. (Danbury, CT). The 20-cm cylindrical transducer array of 1024 elements are mounted below the table opening inside the water bath. The data acquisition and processing electronics are housed in the transducer head and the computer console. The patient lies prone on the table with the breast in the dependent position suspended in the water bath. The transducer elements are long in the elevation direction to ensure cylindrical wavefronts. This produces a slice thickness of 8.5 mm in water.

Detailed sampling of the wavefront at ten discrete frequencies permits image reconstruction by novel methods. The resulting image is a 1024x1024 matrix of the scattering potential of the medium dependent on speed of sound and attenuation, as described in the next section.

Image Reconstruction

The complex scattering potential, $S(\vec{x})$, may be derived

$$S(\vec{x}) = k^2(\vec{x}) - k_0^2 - \rho(\vec{x})^{\frac{1}{2}} \nabla^2 \rho(\vec{x})^{-\frac{1}{2}} \tag{1}$$

where the complex wave number is defined as

$$k(\vec{x}) = \frac{\omega}{c(\vec{x})} + i\alpha(\vec{x}) \tag{2}$$

and $c(\vec{x})$ = speed of sound, $\alpha(\vec{x})$ = attenuation coefficient, $\rho(\vec{x})$ = density, $\omega_0 = 2\pi f_0$.

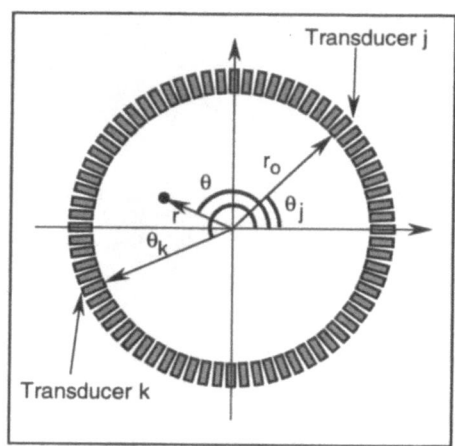

Figure 3. Coordinate system for reconstruction. Object is illuminated by a cylindrical wavefront from transducer k. Incident and scattered fields are measured at transducer j.

We describe the complex amplitude m_{jk} (j,k=1,2,...1024) of the scattered acoustic wave measured at transducer j due to insonification of the medium by transducer k (Figure 3) by

$$m_{jk} = \sum_{p,q=-\infty}^{\infty} \int d\vec{x}\, H_p'(k_0 r_0)\, H_q'(k_0 r_0)\, J_p(k_0 r)\, J_q(k_0 r)\, e^{ip(\theta-\theta_j)+iq(\theta-\theta_k)}\, S(\vec{x}) \tag{3}$$

In Equation (3), J_n and H_n' are the Bessel function and first derivative of the Hankel function, respectively. H_n' is indicative of a dipole (cos θ) radiation beam pattern and was confirmed for our transducer elements by direct measurement with a needle-point hydrophone. This first-order approach to reconstruction utilizes the Born approximation to the wave equation.

The homogenous water bath data, or incident wave, is subtracted from the measured object data which contains the incident plus scattered field.

$$m_{jk} = m_{jk}^{object} - m_{jk}^{incident} \tag{4}$$

We can invert equation 3 to obtain

$$\tilde{S}(k_x, k_y) = \tilde{S}(-2k_0\, \sin\bar{\theta}\, \cos\frac{\Delta\theta}{2},\, 2k_0\, \cos\bar{\theta}\, \cos\frac{\Delta\theta}{2}) = \mathscr{F}_2[\frac{\mathscr{F}_2[(m_{jk})_{pq}]}{H_p'\, H_q'}]_{\alpha\beta} \tag{5}$$

where $\tilde{S}$ is the discrete two-dimensional Fourier transform of the true scattering potential of the medium, $\mathscr{F}_2[\ ]$ represents the discrete two-dimensional Fourier transform operation and

$$\overline{\theta} = \frac{\theta_\alpha + \theta_\beta}{2} \quad , \quad \Delta\theta = \theta_\alpha - \theta_\beta \tag{6}$$

The discrete inverse two-dimensional Fourier transform, $\mathscr{F}_2^{-1}[\ ]$ of equation (5) above recovers $S(\vec{x})$

$$S(\vec{x}) = \mathscr{F}_2^{-1}[\ \tilde{S}(k_x, k_y)\ ] \tag{7}$$

$S(\vec{x})$ is now defined in terms of a large symmetrical set of measured complex values, m_{jk}, known positions of the transducers, and tabulated Hankel functions. This algorithm produces an image or map of $S(\vec{x})$ with a spatial bandwidth of $2k_0$ which is equivalent to a diffraction limit of $\lambda/4$ which is 0.37 mm at 1 MHz in water. The matrix size is 1024x1024 over a 20-cm field of view yielding a pixel size that was measured to be 0.23 mm.

The solution to Equation (1) is proportional to Equation (7) and is expressed in terms of $k^2(\vec{x})$, where the real component represents sound speed and the imaginary component represents attenuation.

Lower operating frequencies reduce phase aberration but also reduce image resolution. The Born Approximation ignores higher order terms in Equation 1 and requires that only small phase shifts occur in the wavefront across the field. For breast imaging, speed of sound variations are $\approx\pm8\%$. The phase aberration error due to the neglected terms in the approximate wave equation can be shown to increase with [frequency]$^{3/2}$ for a simulated breast. The diffraction limit of $2k_0$ means that resolution decreases at lower frequencies. Our approach to image reconstruction is to reduce phase aberration such that the Born approximation is valid while achieving adequate resolution.[10] We apply a combination of methods including lower frequencies and iterative reconstruction which employs phase shift corrections.

RESULTS

Two 0.5-inch diameter cylinders separated by 1.25 inches with +2% and +4% sound speed relative to the water bath were reconstructed from simulated scatter data at 0.5 MHz by the method described above. The real, imaginary and magnitude components are shown in Figure 4 in the left, center and right, respectively, with the +4% cylinder located at the top. X and y pixel profiles are plotted through the center of each image. The reconstructed values of $S(\vec{x})$ differ by approximately a factor of two and the edges are sharp. These results support the proportional relationship between the object and the image.

Reconstructions of $S(\vec{x})$ were made for the same 0.5-inch cylinders in which the sound speed was varied from -12% to +12% relative to the water bath. The values are shown in Figure 5. In the Born approximation, the real component of $S(\vec{x})$ should be linear with sound speed while the imaginary component should be zero (no attenuation for this specific case). In objects which exceed the range of validity of the Born approximation, phase aberration occurs which results in mixing of the real and imaginary components. For

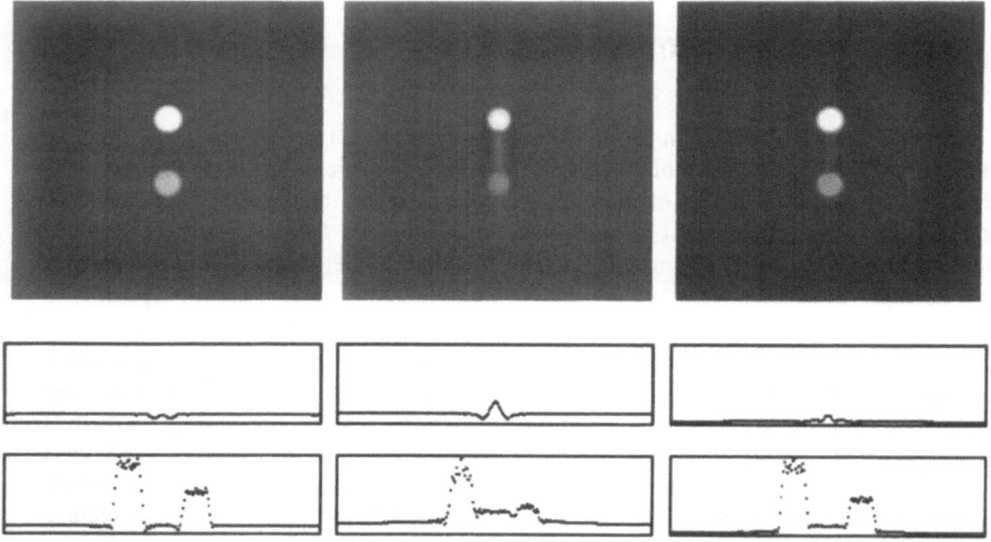

Figure 4. Two 0.5-inch diameter cylinders separated by 1.25 inches with +2% and +4% sound speed were reconstructed from simulated scatter data and are shown with x and y pixel profiles drawn through the center of the image. The real, imaginary and magnitude images are shown in the left, center and right, respectively.

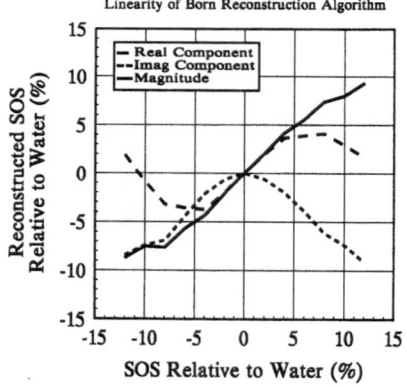

Figure 5. Reconstructed values of simulated $S(\vec{x})$ for 0.5-inch cylinders with -12 to +12% relative sound speed. For this simulation, computing the magnitude of the complex components shows reasonable linearity over the entire sound speed range.

this simulation, computing the magnitude of the complex components shows reasonable linearity over the entire sound speed range.

An experiment with the transducer array was conducted to compare results with the simulations. Four thin latex rubber tubes (~0.5 inch diameter) containing solutions of ethanol and saline were prepared to produce a range of sound speeds of -5.7, 0.0, +3.0, and +6.4% relative to water. These solutions had insignificant attenuation differences. Reconstructed values of $S(\vec{x})$ are shown in Figure 6 with x and y pixel profiles. The real, imaginary and magnitude images are shown in the left, center and right, respectively. The image of the neutral water-filled cylinder (real) is shown inset in which several small air bubbles are visible next to the latex. The tubes were not perfectly cylindrical which is reproduced well in the images.

The sound speed of each of the solutions above were independently measured in a controlled water tank with a pulsed single-element transducer (Ultran, Pittsburgh, PA). The mean pixel values for the real component in Figure 6 are plotted versus measured sound speed in Figure 7 and show reasonable linearity.

Three different reconstruction algorithms were compared for a simulation of a large cylinder (3-inch diameter) with two 0.5-inch voids immersed in water. The cylinder has +2% sound speed relative to water while the voids have the same sound speed as water. Significant phase aberration would be expected in this object.

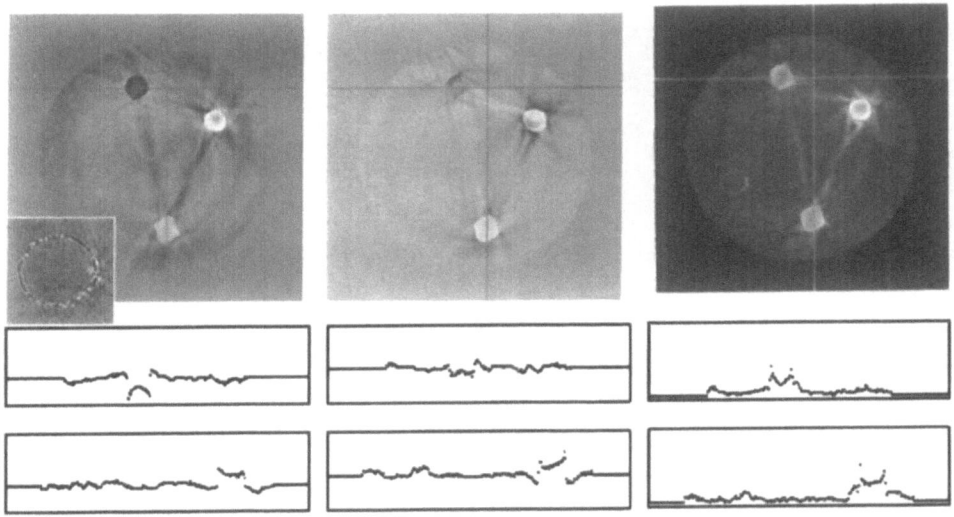

Figure 6. Rubber tubes containing solutions of ethanol and saline (-5.7, 0.0, +3.0, and +6.4%) were imaged. The real, imaginary and magnitude images are shown in the left, center and right, respectively. The image of the neutral water-filled cylinder (real) is shown inset in which several small air bubbles are visible next to the thin latex wall.

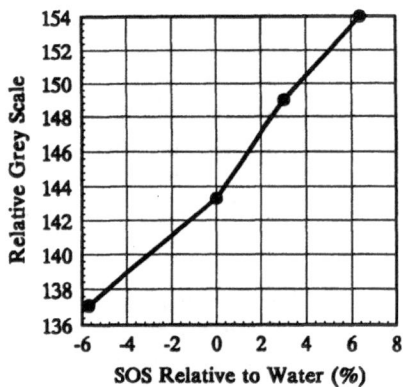

Figure 7. The mean pixel values in each tube of Figure 6 (real) are plotted versus sound speed showing reasonable linearity.

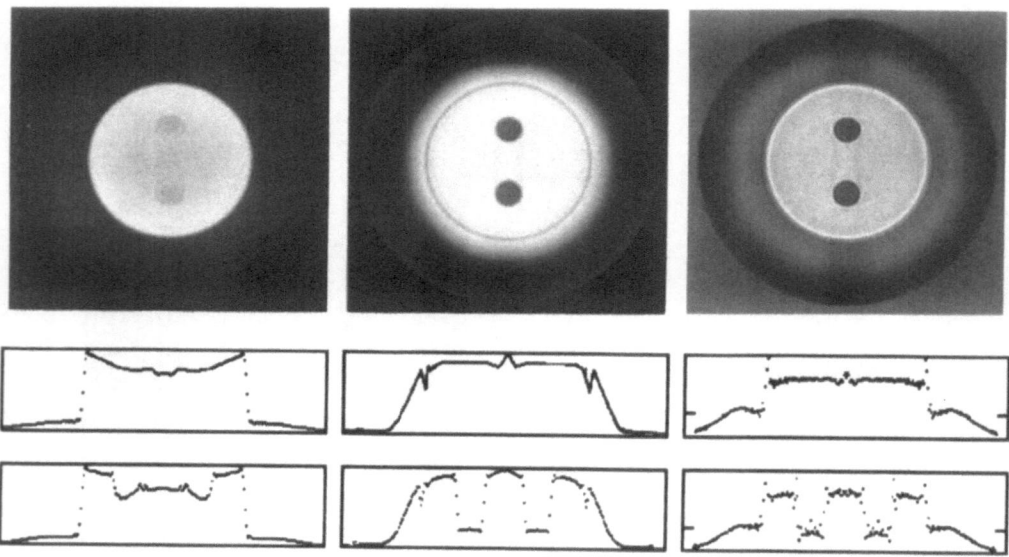

Figure 8. Three different reconstruction algorithms are compared for a simulation of a 3-inch cylinder with two 0.5-inch voids in water. The cylinder has +2% relative sound speed, the voids have the same sound speed as water. The left image employs no phase aberration correction but averages five discrete frequencies. The center and right images employ increasingly complex iterative aberration correction schemes.

The first reconstruction algorithm acquires scatter data at five discrete frequencies from 687-1250 kHz. Separate backpropagation images are formed and then summed which seems to reduce some artifacts. The left-hand image in Figure 8 was reconstructed by this algorithm. Note the non-uniformity or "cupping" from the edge to the center of the cylinder which is especially apparent in the x and y pixel profiles below the image. This approach is computationally very efficient and requires only two to three minutes on an 80486 personal computer.

The algorithm used to reconstruct the center image in Figure 8 requires several steps.

a. Scatter data are acquired at ten discrete frequencies from 687-1250 kHz. The multi-frequency backpropagation image data is used to form synthetic pulse beams. Synthetic beams from all transducers sequentially focus at each point in the medium and then propagate back to all transducers.

b. The differential transit time to each point in the object and the attenuation of the pulse peak value is calculated relative to a uniform water bath. Sound speed and attenuation maps are calculated from this data which results in low-spatial frequency images.

c. Each single-frequency backpropagation image is corrected for phase aberration using the transit time information. This results in high spatial frequency maps.

d. The images of step (c) are summed to reduce artifacts.

e. A "composite" image is produced by blending the high spatial frequency image of step (d) with the low spatial frequency transit time map of step (b).

The resultant image shows substantial reduction of the non-uniformity evident in the left-hand image of Figure 8. The image contrast of the water-filled voids is also greatly increased. Blurring of the edge of the cylinder seems to have increased as well, however. This reconstruction requires more than 60 minutes on an i860 RISC processor.

The right-hand image in Figure 8 was reconstructed by an advanced phase aberration correction scheme that maintains the high spatial frequency image data. This method utilizes a transit time map that represents the weighted-average arrival time of the synthesized pulse from the second algorithm above. Image sharpness and uniformity are improved in this approach.

Each of these reconstruction algorithms is under experimental study including experiments with human tissue. A 35 year-old asymptomatic female volunteer was recruited for study on the experimental array. The subject's mammogram showed a smaller breast with dense glandular tissue and no abnormalities. A set of seven reconstructed slices imaged with the system are presented in Figure 9. Images for the left breast are shown in coronal format sequentially from the chest wall to the nipple and viewed as though facing the patient. The subject's head is towards the top, lateral aspect at the right, medial at the left. The human images show good depiction of glandular structures, skin and subdural fat. The water bath appears as a uniform grey region surrounding the breast and the transducer ring is located at the edge of the field of view (black border). Only a portion of the image field of view is presented in each slice.

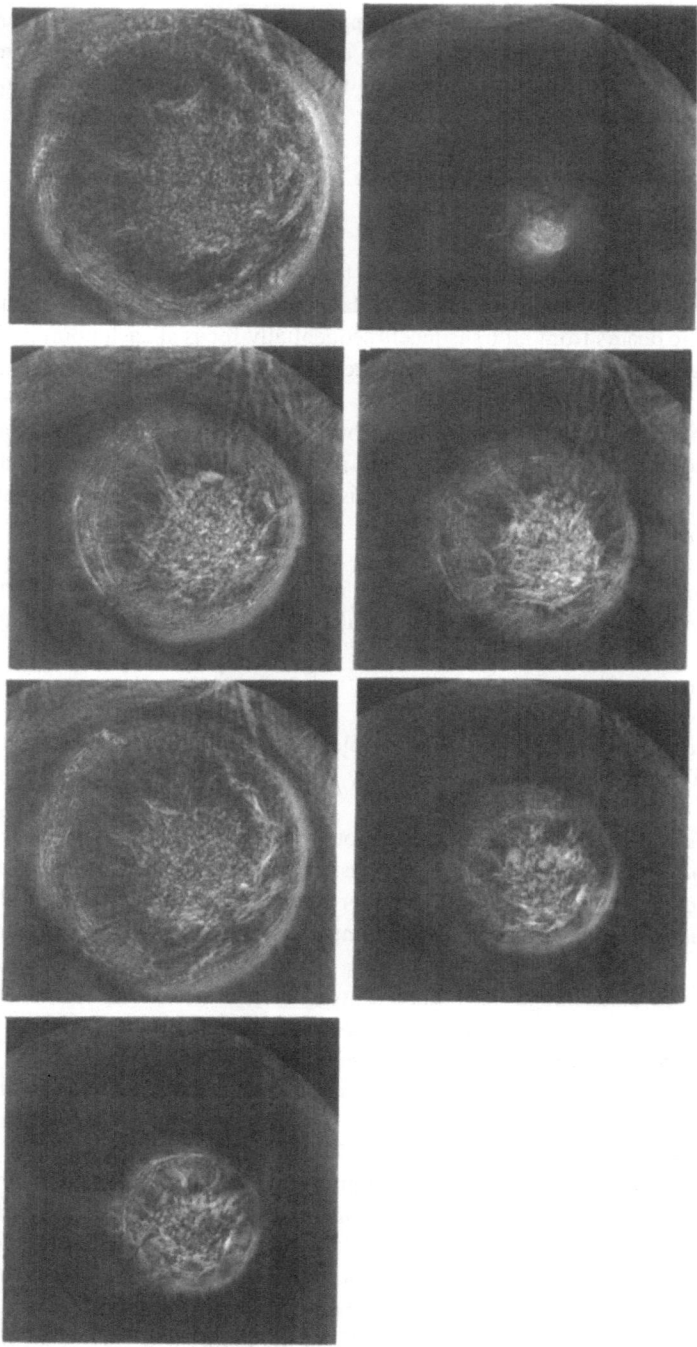

Figure 9. 35 year-old asymptomatic female volunteer with a smaller-sized dense breast. Images for the left breast are shown in coronal format sequentially from the chest wall to the nipple.

CONCLUSIONS

Computer simulations and test phantoms were studied to assess feasibility of our approach to diffraction tomography. Image quality was greatly improved and phase aberration artifacts were significantly reduced by using a novel multi-frequency iterative algorithm which adjusts for phase aberration. Low acoustic intensities are employed.

Several asymptomatic and symptomatic women have been successfully imaged by this technique and the results show promise for future development.[11]

Although less successful in previous studies, this new approach to diffraction tomography works well in phantoms and is able to accommodate large-scale refraction. By using multiple lower frequencies and a high-density cylindrical transducer array, this new technique provides good image quality and minimal artifacts in a unique large-scale coronal format.

REFERENCES

1. J.F. Greenleaf, S.A. Johnson, W.F. Samayoa, F.A. Duck, Algebraic reconstruction of spatial distributions of acoustic velocities in tissue from their time-of-flight profiles, *in:* "Acoustic Holography," N. Booth, ed., Plenum, New York, (1975).

2. P.L. Carson, C.R. Meyer, A.L. Scherzinger, T.V. Oughton, "Breast imaging in coronal planes with simultaneous pulse-echo and transmission ultrasound," *Science* 214:1141-1143 (1981).

3. G.H. Glover, J. C. Sharp, "Reconstruction of ultrasound propagation speed distributions in soft tissue: Time-of-flight tomography", *IEEE. Trans. Sonics Ultrasonics* SU-24, 229-234 (1977).

4. D. Hiller, H. Ermert, "Ultrasound computerized tomography using transmission and reflection mode: Application to medical diagnosis," *Acoustical Imaging* 12, 553-564 (1982).

5. M. Kaveh, R. Mueller, R. Rylander, T. Coulter, M. Soumekh, "Experimental results in ultrasonic diffraction tomography," *Acoustic Imaging* 9, 433-450 (1979).

6. A. Devaney, "A filtered backpropagation algorithm for diffraction tomography," *Ultrasonic Imaging* 4, 336-350 (1982).

7. H. Jones, "Recent activity in ultrasonic tomography," *Ultrasonics* 31(5), 353-360 (1993).

8. R.K. Mueller, M. Kaveh, G. Wade, "Reconstructive tomography and applications to ultrasonics," *Proceedings of the IEEE* 67(4), 567-587 (1979).

9. P.J. Martin, M.P. André, B.A. Spivey, D.A. Palmer, "Computed tomography using multi-hologram reconstruction of low-frequency ultrasound," *J Ultrasound Med* 12, S26 (1993).

10. G.P. Otto, P.J. Martin, M.P. André, B.A. Spivey, T. Barrett, D.A. Palmer, "Phantom studies of phase aberration correction methods in diffraction tomography," *Medical Physics* 21(6), 998 (1994).

11. M.P. André, P.J Martin, L.K. Olson, T.K. Barrett, G.P. Otto, B.A. Spivey, "Preliminary studies of breast imaging with sonic computed tomography," *Amer J Roentgenol* 162(3), 114 (1994).

SIMULTANEOUS IMAGING SYSTEM OF THREE KINDS OF PARAMETERS OF NONLINEARITY FOR MEDICAL DIAGNOSIS

Seiya Hasegawa, Katsuhiko Hayashi, Takuso Sato, and Keisuke Kameyama

Interdisciplinary Graduate School of Science and Engineering
Tokyo Institute of Technology, 4259 Nagatsuta, Midori-ku Yokohama 227
Japan

INTRODUCTION

This paper discusses an imaging system for biological tissue characterization for medical diagnosis. This system uses ultrasonic wave (5MHz) for probing and pump wave (350kHz) for pressure perturbation (~1atm). In addition to the most widely used pulsed echo image (linear reflectivity), on this system, we suggest three kinds of tissue characteristic parameters which are obtained by an interaction of the tissue and the pump wave as listed in the following.
1. Probe phase shift due to sound velocity dependency to pressure.
   (Phase Shift Parameter)
2. Probe phase distortion caused by microstructural positional movability.
3. Reflectivity change caused by microstructural change in orientation due to pumping.

For stable data acquisition, heartbeat synchronized detection to remove effect of tissue movement due to heartbeats has been developed. Moreover, to obtain precise image of the phase shift parameter, a new estimation method has been proposed. As the system is intended for clinical use, a portable measurement system for easy access to *in vivo* subject has been developed.

Images of *in vivo* human liver were observed, and information fusion of the multiple images for efficient diagnosis has been attempted.

PRINCIPLE

Principle of measurement of the three kinds of nonlinearities generated by the pumping wave is shown in Fig. 1. Two coaxial concave transducers generate the pumping wave (350kHz) and the probing wave (5MHz). Then, pumping wave transmitted to the observing region interacts with the tissue. Using reflected signals of the probing waves superimposed on three states of pumping waves (steps 1 - 3), each nonlinearity parameters of the tissue is estimated.

The difference of phase changes of the reflected probing wave at step 1 and step 2 is given by, [1]

$$\Delta\varphi(z) = \Delta\varphi_+(z) - \Delta\varphi_-(z)$$
$$= \omega_0 \int_{z_0}^{z} N'(\xi)\{\Delta p_+(\xi) + \Delta p_-(\xi)\}d\xi + \Delta\varphi_D(z) \tag{1}$$

where ω_0 is angular frequency of the probing wave, $\Delta p_\pm$ is static pressure generated by the pumping waves, N' is phase shift parameter which is related to dependence of sound velocity upon pressure, and $\Delta\varphi_D$ is phase distortion which corresponds to dynamic change of relative position of microstructure of the tissue. Separating

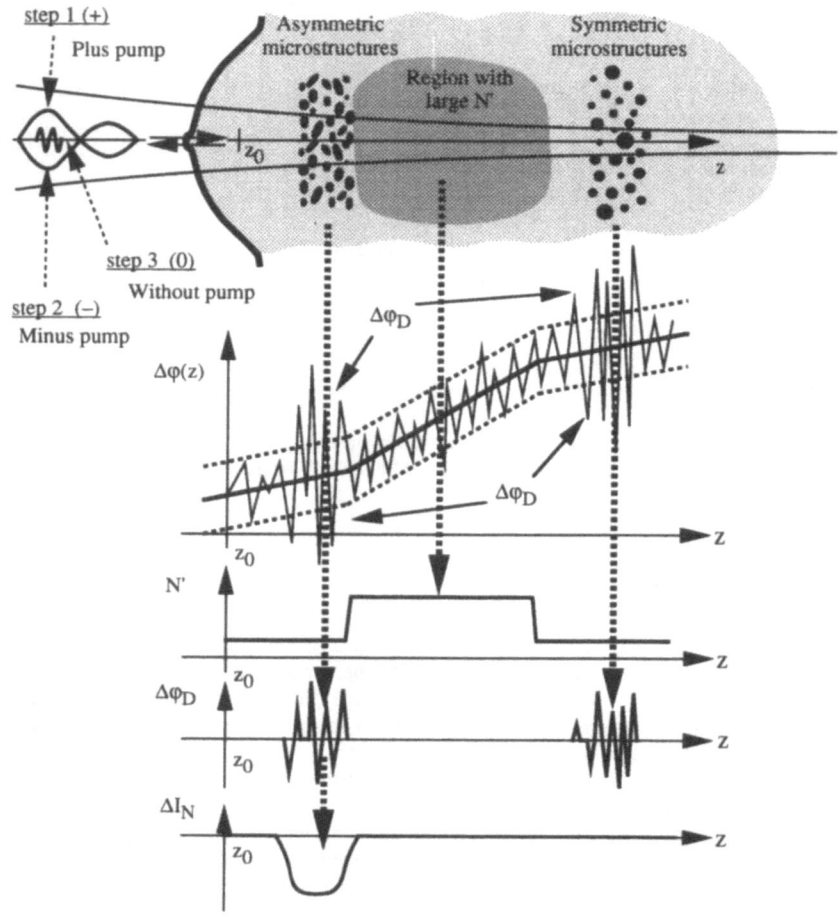

Fig. 1. Measurement principle of the three kinds of nonlinearities generated by tissue - pump interaction.

the terms of Eq.(1), these two parameters of nonlinearity are obtained.

Next, normalizing the difference of intensity changes of reflected the probing wave at steps 1, 2 and 3, another nonlinearity parameter is given as

$$\Delta I_N(z) = \frac{I_{with\ pump}(z) - I_{without\ pump}(z)}{I_{without\ pump}(z)}. \qquad (2)$$

This ΔI_N is a normalized reflectivity change parameter which is considered to be corresponding to the changes of orientations of the tissue microstructures caused by the dynamic change of pressure due to pumping.

Phase shift parameter N' is estimated from the average spatial inclination of the actual probe phase shift $\Delta \varphi$ after the ASC (Adaptive Sensitivity Compensation) procedure [1] which reduces the phase shift distortion due to the medium structure. Even after the ASC procedure, a noticeable phase distortion $\Delta \varphi_D$ will prevail (as in Fig. 1) in the region with symmetric microstructures, because the energy of pumping waves is mainly consumed for relative positional change of the microstructure. In the region with asymmetric microstructures, smaller $\Delta \varphi_D$ and large ΔI_N is observed since pumping energy is used for both relative positional and orientational change of the microstructures.

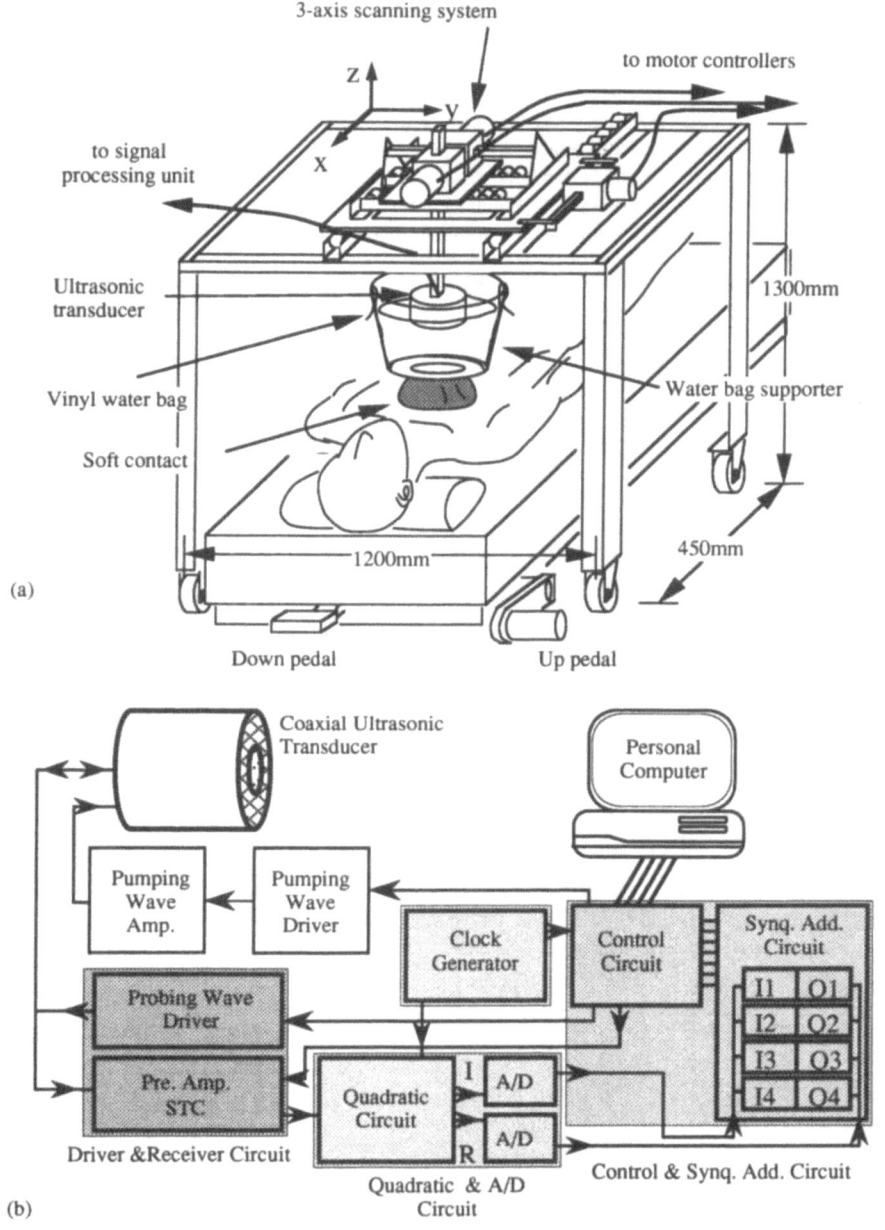

Fig. 2. The constructed measurement system. (a) Accessing system to human body. (b) Block diagram of the signal processing system.

SYSTEM CONSTRUCTION

Scheme of the constructed measuring system is shown in Fig. 2. In order to access to human body, a 3-axis scanning system and a height variable bed that can easily adjust the measuring position are used. The soft contact illustrated in Fig. 2(a) easily fits to the *in vivo* subject's body shape and enables measurement without giving any significant shape deformation or a static pressure deviation over the observed region.

Fig. 2(b) is a block diagram of signal processing system. All circuits are controlled by a controller circuit connected to a personal computer. The pumping wave driver and the probing wave driver drives the transducers according to the instructions from the controller. The results of the interaction between the pumping wave and the tissue is detected by the reflected probing wave received by the same inner transducer which is also used for ultrasonic generation. The phase of reflected signal is detected by a conventional quadratic phase detector. A/D converted phase signal is led to synchronous adder, and read by the computer after the addition for a certain number of times.

To remove the effect of movement due to heartbeat for stable data acquisition, heartbeat synchronized line detection system is developed. Since *in vivo* human tissue, especially the abdominal region, has a large displacement due to heartbeating during the measurement, images obtained by using ordinary mechanical scanning method has many horizontal stripes. To remove this disturbance, detecting heartbeat signal lead by two electrodes attached on the subject's right hand and body, line scan synchronized to the signal is done . In the image obtained by heartbeat synchronized detection, no stripes due to tissue movement were observed.

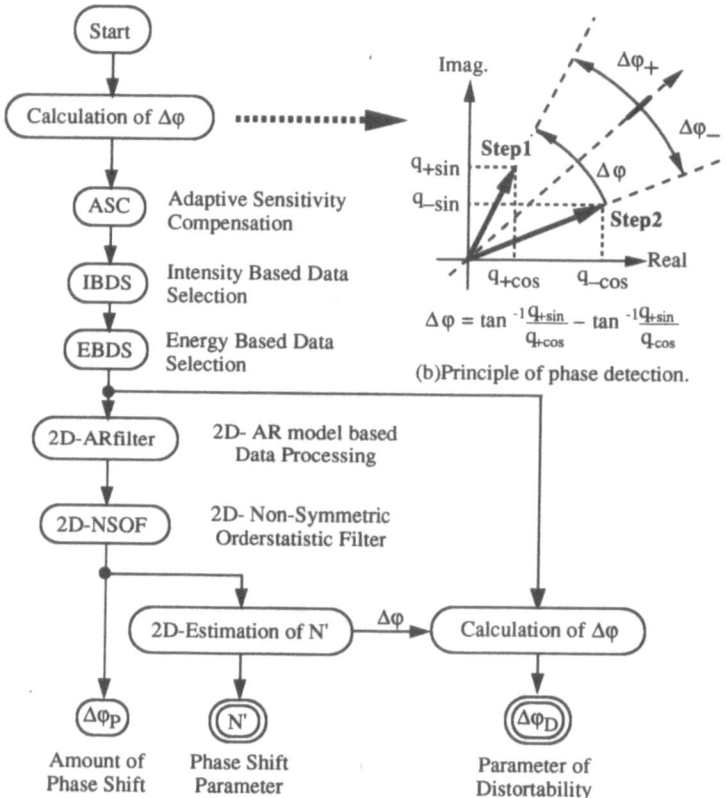

Fig. 3. Signal processing flow.

SIGNAL PROCESSING

Obtained phase signals of the probing waves are processed according to flow chart in Fig. 3. The details of ASC (Adaptive Sensitivity Compensation), IBDS (Intensity Based Data Selection), EBDS (Energy Based Data Selection), and NSOF (Nonsymmetric Order Statistics Filtering) are explained in [1]. Here, a new method for phase shift parameter N' estimation for precise N' mapping is detailed.

In Fig. 4, the newly proposed least variance method is shown. When the measuring object has two different N' regions (smaller N' in the subcutaneous region), observed phase shift typically becomes as shown in Fig. 4(a). In the case when the parameter N' of the point of interest in Fig. 4(a) is estimated, the inclination of Δφ for all local window region containing the particular point is calculated by linear fitting. If a local window region contains the point where N' changes (Fig. 4(c)), the Δφ data with in the region has a large variance, and otherwise, the variance will be relatively small (Fig. 4(b)). Therefore, if the spatial inclination of Δφ calculated from the data within a region window which gives minimum variance is used as an estimation of the local N', the boundary of regions of different N' is detected exactly. After estimation of N', the distortability parameter Δφ_D is estimated by reconstructing the phase shift from the estimated distribution of N', and assessing the discrepancies of the phase shift Δφ after ASC and the reconstructed phase.

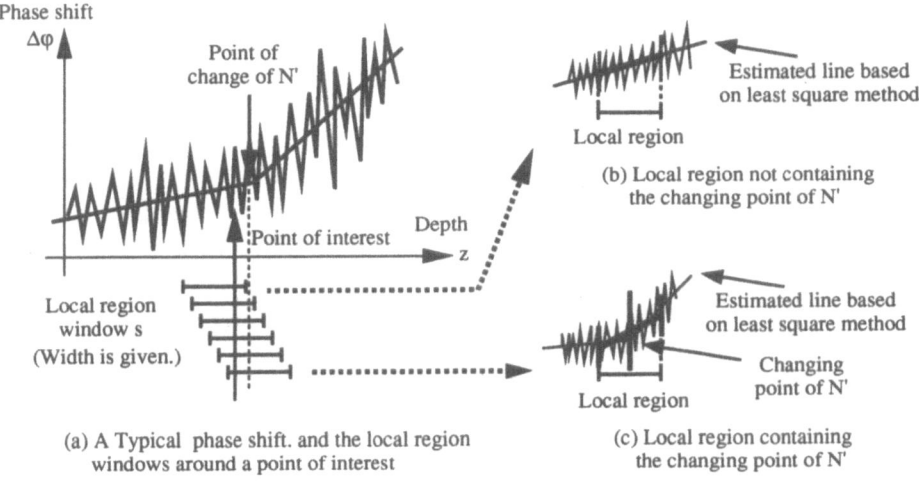

Fig. 4. Least variance method for N' estimation.

RESULTS

Observed images of *in vivo* human liver is shown as Fig. 5. The images have been observed by measuring with a medical doctor, and the tissue structure observed region is schematically shown in Fig. 5(b). In phase shift parameter N' map shown in Fig. 5(d), the fatty subcutaneous layer has a larger value than the inner tissue which corresponds to liver. In reflectivity changing parameter ΔI_N shown in Fig. 5(e) and distortability parameter $\Delta\varphi_D$ shown in Fig. 5(f), since the inner tissue has larger values compared with the subcutaneous, it is estimated that the inner liver layer consists of microstructures easily moved by the pumping pressure.

The tissue classification map based on the tissue microstructure model is also attempted for another scan over the liver of the same person. The effects given by the pumping wave to asymmetrically shaped microstructures are mainly change of orientation observed as reduction of reflectivity. Thus, large ΔI_N is observed in the region with a lot of asymmetric microstructures. In the case of symmetric microstructure, the change of relative position of microstructures is dominant and its effect is observed as large $\Delta\varphi_D$. On this account, combining the images of the two parameters, classification map of the tissue microstructure is obtained as Fig. 6.

CONCLUSIONS

A portable ultrasonic imaging system which jointly uses the pumping wave and the probing wave for detection of the tissue's nonlinear acoustical characteristics has been constructed. Using the system, the *in vivo* images of three nonlinear parameters were observed and their interpretations based on the features in these multiple images are also derived. Information fusion of these images based on the mechanism of the generation of each parameters has been attempted as a preparation for a much effective use of the maps in medical diagnosis.

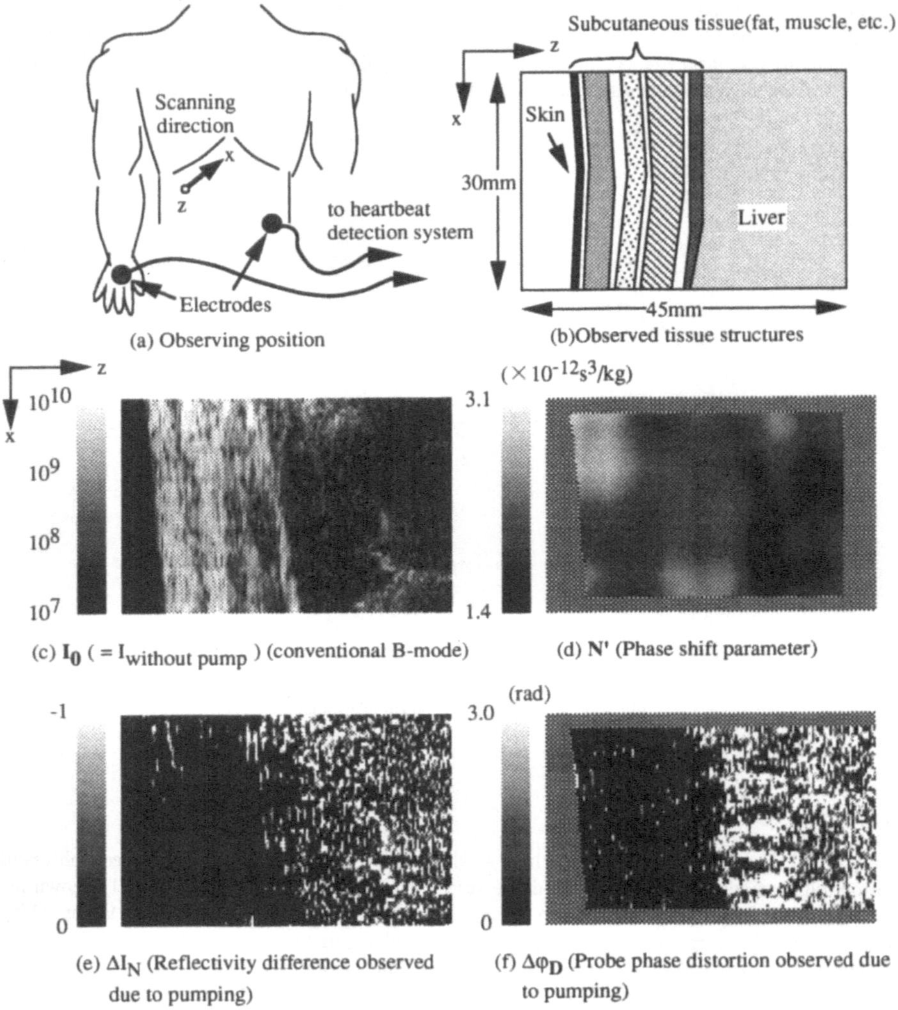

Subcutaneous tissue(fat, muscle, etc.)

Scanning
direction

to heartbeat
detection system

Electrodes

Skin

30mm

Liver

45mm

(a) Observing position

(b)Observed tissue structures

$(\times 10^{-12}s^3/kg)$

10^{10}

10^9

10^8

10^7

3.1

1.4

(c) I_0 ($= I_{without\ pump}$) (conventional B-mode)

(d) N' (Phase shift parameter)

-1

0

(rad)

3.0

0

(e) ΔI_N (Reflectivity difference observed
due to pumping)

(f) $\Delta\varphi_D$ (Probe phase distortion observed due
to pumping)

Fig. 5. Observed multiple image for *in vivo* human liver.

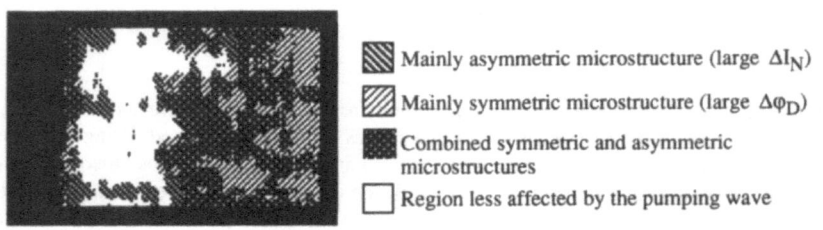

Mainly asymmetric microstructure (large ΔI_N)

Mainly symmetric microstructure (large $\Delta\varphi_D$)

Combined symmetric and asymmetric
microstructures

Region less affected by the pumping wave

Fig. 6. Tissue classification map as example of information fusion

REFERENCES

[1] T. Sato, E. Mori, K. Endo, Y. Yamakoshi and M. Sase "A Few Effective Signal Processings For Reflection-type Imaging of Nonlinear Parameter N of Soft Tissues" Acoustical Imaging Vol. 19 (H. Ermert and H. P. Harjes Eds.), pp. 363 - 368, Plenum Press, (1992).

[2] T. Sato, K. Yamashita, H. Ninoyu, K. Y. Jhang and Y. Kosugi "Imaging of Acousitical Nonlinear Parameters and Its Medical and Industiral Applications: A Viewpoing as Generalized Percussion" Acousical Imaging Vol. 20 (Y. Wei and B. Gu Eds.), pp. 9 - 18, Plenum Press, (1993).

3-D MOTION IMAGE RECONSTRUCTION USING VOLUME RENDERING

TECHNIQUE FROM ULTRASOUND ECHOGRAPHIC IMAGES

Tsuyoshi Yamamoto, Taisei Mikami†, Jun'ichi Teranishi†,
Akira Kitabatake† and Yoshinao Aoki‡

Computing Center, Hokkaido University
†Medical scool, Hokkaido University
‡Dept. of Engineering, Hokkaido University, Sapporo Japan

INTRODUCTION

Recently, use of 3D and 4D volume visualization from X-ray CT and MRI images has increased in many fields[1]. Realtime image generation is one of the advantage of acoustical method to MRI and X-ray CT techniques. Many ultrasound echographic devices can visualize sliced sections of moving organs in realtime. By gathering image slices to cover certain volume space and time interval, a space-time volume dataset or 4D volume dataset can be constructed. Volume visualization techniques can be applied to the dataset to create photo-realistic images and it is also possible to make a movie by assembling them.

Nelson et. al. [2] show the possibility of three dimensional reconstruction from ultrasound images, where still organs such as bones, liver and embryo are considered as objects. They employed volume rendering technique to visualize three dimensional structure of object and the result shows potentiality of ultrasound images as a three dimensional scanner.

One of the advantage of ultrasound imaging is its realtime nature and potentiality of imaging dynamic object.

In this paper, we propose a technique to visualize dynamics of human heart beat from multiple longitudinal images obtained by the step-wise rotation of transesophageal echocardiographic images. Since ultrasound image frames are recorded with ECG chart, it become possible to pick up volume data for specific phase of heart beat by setting delay time from the reference point of ECG waveform. Volume data for specific delay time is visualized by volume rendering technique and then assembled for animation.

4D ULTRASOUND DATA ACQUISITION

Currently, ultrasound image data are acquired as planar images using commercially available grey-scale and color-flow Doppler imaging equipment. 4D volume data or time varying 3D volume data of periodically moving organ is acquired by collecting movie segments of the space by changing position or angle of transducer. In the case of heart beat visualization, electronic cardiograph or ECG can be used for acquisition to the same phase of cardiac cycle. Figure 1 illustrates the coordinate system of 3D volume data acquisition for this experiment.

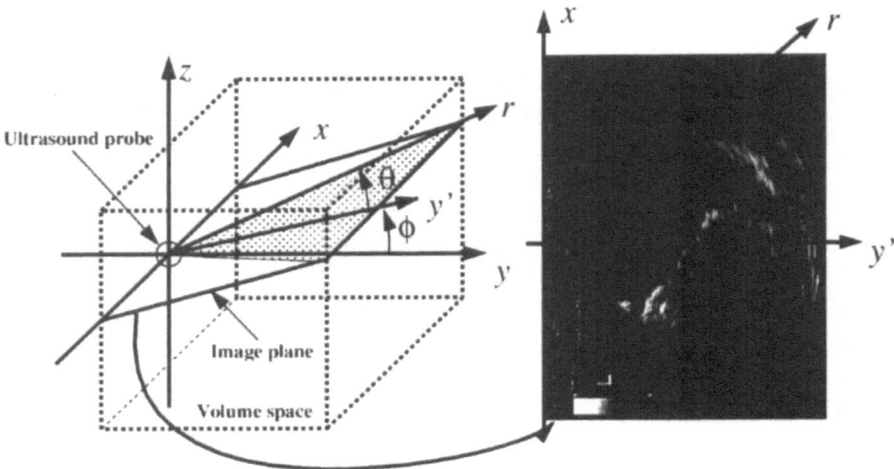

Figure 1: Coordinate system of 3D volume data acquisition using ultrasound echogram

As in the figure, 3D volume space is scanned by changing longitudinal angle or ϕ and the primary 3D volume data for time phase t is represented as $S_1(t, \phi, x, y')$ using cylindrical coordinate system.

However, most volume rendering algorithms assume source data on a 3D lattice and resampling processing is required. Let $\hat{S}_2(t, x, y, z)$ be the resampled volume data in 3D lattice coordinate, then the relation of S_1 and $\hat{S}_2$ is as follows;

$$
\begin{aligned}
\hat{S}_2(t, x, y, z) &= S_1(t, \phi, x, y') \\
y' &= \sqrt{y^2 + z^2} \\
\phi &= \tan^{-1}(z/y).
\end{aligned} \tag{1}
$$

VOLUME RAY TRACING

Each frame of heart beat animation is created by rendering corresponding 3D volume data by volume ray tracing method[3].

Volume ray tracing conducts an image order traversal of the image plane pixels, finding a color and opacity for each pixel. A ray is fired from each pixel through the data volume and continues in a straight line until found terminal condition or the ray exits the rear of the volume. The most general terminal condition is so called isosurface detection where ray tracer searches a point in volume at which ultrasound echo strength equals to the pre-defined threshold value. If tracer found the isosurface, the shading calculation is performed. In the calculation, surface normal vector $\vec{n}$ is nessesary and it can be substituted by gradient vector at the point.

$$
\begin{aligned}
g_x &= \hat{S}_2(t, x(p_0) + 0.5, y(p_0), z(p_0)) - \hat{S}_2(t, x(p_0) - 0.5, y(p_0), z(p_0)) \\
g_y &= \hat{S}_2(t, x(p_0), y(p_0) + 0.5, z(p_0)) - \hat{S}_2(t, x(p_0), y(p_0) - 0.5, z(p_0)) \\
g_z &= \hat{S}_2(t, x(p_0), y(p_0), z(p_0) + 0.5) - \hat{S}_2(t, x(p_0), y(p_0), z(p_0) - 0.5) \\
\vec{n} &= \vec{g}/|\vec{g}|
\end{aligned}
\tag{2}
$$

The brightness of the pixel is computed by Phong's shading model[4] using illumination vector $\vec{l}$, normal vector $\vec{n}$, and viewing vector $\vec{v}$. The brightness of the pixel or I is written as

$$
\begin{aligned}
I_0 &= max(\vec{l} \cdot \vec{n}, 0) \\
I_1 &= max((2\vec{n}(\vec{n} \cdot \vec{v}) - \vec{v}) \cdot \vec{l}, 0) \\
I &= k_0 I_0 + k_1 I_1^\alpha
\end{aligned}
\tag{3}
$$

where k_0 and k_1 are the irregular and specular reflection factors respectively. α is the parameter that control highlight focus and it may have the value of $4 < \alpha < 100$. In this shading model, shadow caused by he interference of other surface is not taken account.

Volume ray tracing is CPU intensive processing and it took approx. 450 sec. for one frame of image reconstruction.

Figure 2-(A) shows an example of primary image recorded by Aloka SSD870 using 5MHz bi-plane ultrasound transducer array. Figure 2-(B) is an reconstructed images of mitral valve viewed from the cranial direction through the opened left atrium. The primary images are recorded by Aloka SSD870 using 5MHz bi-plane ultrasound transducer array.

Volume ray tracing is CPU intensive processing and our implementation took approx. 450 sec. for one frame of image reconstruction using IRIS4D/30 workstation (30MIPS, 4.2MFLOPS). Since cardiac cycle of human heart beat is approximately 0.9 sec., we must perform 27 frames of reconstruction for animation. As for as using small workstation, CPU time for the processing will exceed 3 hours.

EXPERIMENTAL SYSTEM

Figure 3 illustrates the configuration of experimental visualization system. Ultrasound movie segments are recorded in VCR at the bedside and the movie is copied into optical disk system that can access arbitrary frame randomly.

Each movie segments corresponds to some angle ϕ and it last at least one cardiac cycle. The starting point of each segment can be defined by using ECG waveform recorded in the movie.

The flowchart of proposed visualization process is illustrated in figure 4. 4D visualization or animation of 3D volume space is created by repeating 3D volume visualization for different time phase of cardiac cycle and assembling the images resulted on optical disk.

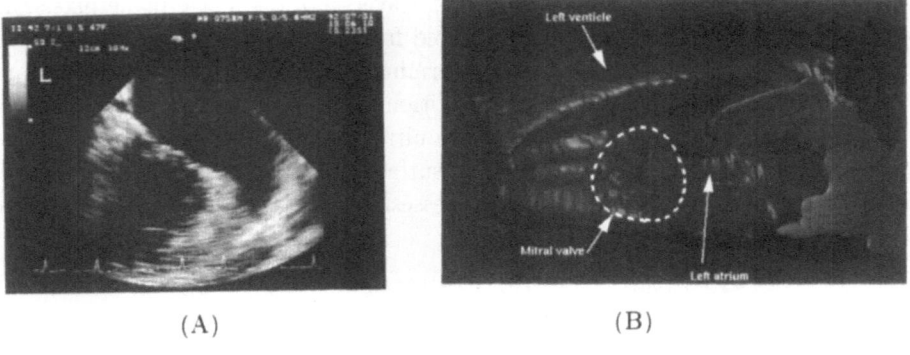

(A)   (B)

Figure 2. Primary ultrasound image(A) and reconstructed 3D images (B)

CONCLUSION

We have introduced experimental result of 3D motion image reconstruction of human heart beat from ultrasound images. By using volume rendering technique, shaded 3D image can be generated without making corresponding geometric model.

So far, computation power is the major restriction for practical applications of this technique, however, rapid increasing of processing power will free this restriction. Another problem is the quality of ultrasound image.

In proposed method, The accuracy of 3D volume reconstructed is depends on the accuracy of primary ultrasound images. However, many conventional ultrasound imaging techniques have not not enough geometric accuracy because it is affected by the speed of sound in specimen. Poor SNR is also problem and earlier result was suffered from artifacts caused by noise components. It is clear that higher SNR and less artifacts is important for 3D and 4D image reconstruction.

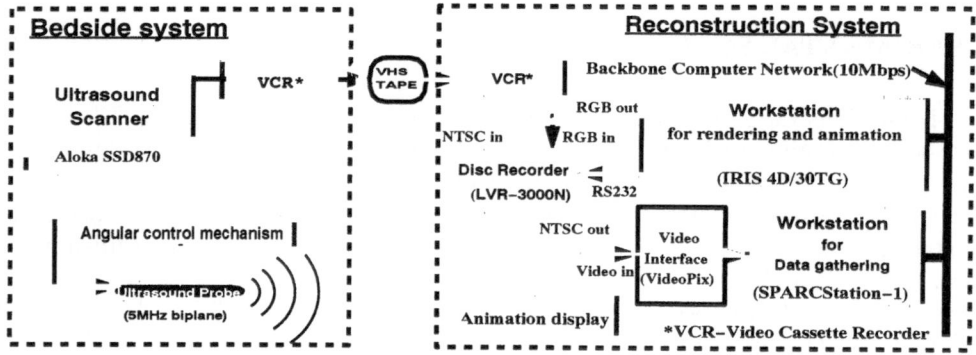

Figure 3. Block diagram of proposed experimental visualization system.

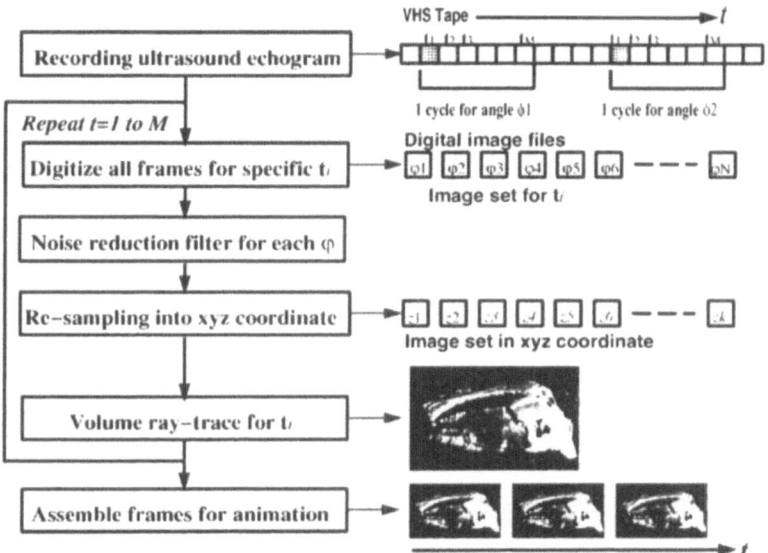

Figure 4. Flow chart for 3D motion image reconstruction from ultrasound echographic images

References

[1] Yamamoto, T.: "Heart Beat", The SIGGRAPH93 Visual Proceedings, pp. 72 (1993)

[2] Nelson R.T. and Elvins T.T.:"Visualization of 3D Ultrasound Data", IEEE CG&A Vol.13, No.6, pp.50-57 (1993)

[3] Kajiya, J.T. and Von Herzen, B. P.:"Ray Tracing Volume Densities", Computer Graphics Vol.18, No. 3,(SIGGRAPH'84 Conference Proceedings), pp.165-173(1984)

[4] Bui-Tuong, P.:"Illumination for Computer-Generated Pictures", Communications of the ACM, Vol.18, No.6, pp.311-317(1975)

A COMPARISON OF ALGORITHMS FOR THE DETECTION

OF STENOTIC VESSELS

Katherine Ferrara

Riverside Research Institute
330 West 42nd Street
New York, NY 10036

INTRODUCTION

In order to develop a sensitive detection scheme for minor stenoses, the received acoustic signal from regions within a stenosis and distal to a stenosis has been modeled and evaluated using experimental data. In comparison with a parabolic profile, analysis of the signal returned from vessel regions distal to the stenosis has shown that the correlated signal interval increases near the vessel wall and decreases in the center of the vessel.[1] With the goal of spatially mapping these changes to locate the source of the flow disturbance, two indicators of the correlated signal interval are presented and evaluated for known experimental flow conditions. The first parameter is the normalized likelihood function, evaluated at the maximum likelihood velocity, with coherent summation over a sequence of 8 ultrasonic pulses. The second parameter is the magnitude of the correlation at a lag of one pulse interval divided by the signal power. The experimental results show that the use of a parameter which coherently sums the signal correlation over a set of sequential pulses and lag values provides a sensitive indication of changes in a flow profile, and may be used to develop a spatial map of the flow conditions. This analysis differs from previous work in the transmission of a train of short, wideband pulses, and the resulting increased spatial resolution of the parameters.

ESTIMATION STRATEGIES

Normalized Likelihood Function

The wideband maximum likelihood estimate (WMLE) for a discrete velocity value v, corresponding to a two way travel time d, is given below by equation 1.[2] The presence of v within the signal envelope indicates the change in the time required for the signal to reach a group of red blood cells moving with respect to the transducer, while the presence of v within the exponential term indicates the frequency shift produced by this motion. Derived using the slowly fluctuating point target approximation, the WMLE involves the sum over the pulse train indexed by k, and the integral over time, t. The integrand is the product of the complex envelope of the received signal, an estimate of the complex envelope of the received signal denoted by $s_0'(\,\cdot\,)$, and the Doppler shift. In (1), the asterisk denotes complex conjugation, and the remaining notation is defined below.

Table 1-Notation

c	acoustic propagation velocity
$l(v)$	likelihood of velocity v
$nl(v)$	normalized likelihood of velocity v
$r'(t)$	complex envelope of the received signal
T	period of the transmitted pulse
$s_0'(t)$	deterministic portion of the received complex signal envelope
v	axial velocity of the estimator
w_C	center frequency of the transmitted signal (radians/second)

$$l(v) = \left| \sum_k \int_{-\infty}^{\infty} r'(t)s_0'^*(t-d-kT[1+2v/c])\exp[j2w_c vt/c]dt \right|^2 \tag{1}$$

In order to show the relationship between the expected value of the likelihood function and the expected signal correlation, the expected value of (1) is evaluated, in (2):

$$E\{l(v)\} = E\left\{ \left| \sum_k \int_{-\infty}^{\infty} r'(t)s_0'^*(t-d-kT[1+2v/c])\exp[j2w_c vt/c]dt \right|^2 \right\} \tag{2}$$

Expanding equation 2 and approximating the phase shift as a constant for each pulse, we obtain (3):

$$E\{l(v)\} = \sum_k \sum_m \exp[j2w_c vT(k-m)/c] \int_{-\infty}^{\infty} \int_{-\infty}^{\infty} E\{r'(t)r'^*(u)\}$$
$$\cdot s_0'^*(t-d-kT[1+2v/c])s_0'(u-d-mT[1+2v/c])dtdu \tag{3}$$

The expected value of the likelihood function can be interpreted as the summation over pulse indices k and m of weighted estimates of the signal correlation. Within the integral is the product of the signal correlation and temporal windows $s_0'(t)$ $s_0'^*(u)$, shifted to track a group of moving scatterers. This is integrated over the temporal window and summed over pulse indices. Thus, as the magnitude of the signal correlation decreases in the presence of a velocity spread target, the magnitude of the integrand decreases. When the expected likelihood is evaluated at the maximum likelihood velocity for a group of scatterers with a uniform velocity, the complex exponential term will cancel the phase of the correlation. For a velocity spread target, the expected likelihood decreases rapidly due to the incoherent summation of phases within the signal correlation. [1]-[3]

With the goal of developing an angle dependent indicator of the flow conditions within the vessel, a normalized form of the likelihood function is introduced. The normalization factor contains the summed signal power from a group of scatterers which move in the axial direction over a set of 8 pulses. The ratio of the likelihood and normalization factor is given by (4), and will be referred to as the normalized likelihood function (NLF) in the remaining sections.

$$nl(v) = \frac{\left| \sum_k \int_{-\infty}^{\infty} r'(t) s_0'^* (t-d-kT[1+2v/c]) \exp[j2w_c vt/c] dt \right|^2}{\sum_k \left| \int_{-\infty}^{\infty} r'(t) s_0'^* (t-d-kT[1+2v/c]) dt \right|^2} \qquad (4)$$

Normalized Lag One Correlation (NLOC)

Kasai[4] discussed a narrowband mean velocity estimation structure for use in color flow mapping, which will be referred to as the autocorrelator. The phase of the autocorrelation at a lag of one transmitted period is estimated and used in an inverse tangent calculation of the estimated mean Doppler shift. The estimated mean velocity, v_{mean}, of the scattering medium is then determined by scaling the estimated Doppler shift by several factors, including the expected center frequency of the returned signal.

In addition, Kasai proposed a turbulence measure given by the ratio of the magnitude of the correlation at a lag of one pulse period, divided by the signal power. This measure will be referred to as the normalized lag one correlation NLOC and will be compared with the performance of the normalized likelihood function.[4]

PREDICTED FLOW CONDITIONS

The flow profile through a 50% stenotic vessel, has been modeled using Fluent, a finite element flow modeling software package. The predicted flow profiles for 30%, 50%, and 80% stenoses were reported in [1], and the predicted flow profile for a 50% stenosis is summarized in this section. The vessel has maximum diameter of 0.9525 cm and a minimum diameter of 50% of the maximum. It was designed to have a symmetrical contoured surface with a taper which narrows with an approximate 2.5 to 1 ratio of change in the axial and radial dimensions, respectively. In the next section, experimental results will be presented for two flow rates, 5 mL/s and 15 mL/s. These conditions produce a parabolic profile at the input to the stenotic region with a peak velocity of 14 cm/s and 42 cm/s, respectively. At a beam vessel angle of 45 degrees the corresponding peak axial velocity are 9.9 cm/s and 29.6 cm/s.

Compared with laminar parabolic flow at the same flow rate, the velocity profiles distal to the stenosis show decreases in axial velocity and velocity spread within sample volumes near the vessel wall, and increases in axial velocity and velocity spread in the center of the vessel. The region which displays a decrease in the axial velocity becomes larger as the degree of stenosis increases. The peak axial velocity occurs at the center of the vessel near the end of the stenosis; it is predicted to be 31.8 cm/s at the 5 mL/s flow rate and 82 cm/s at the 15 mL/s flow rate. The axial velocity is smaller than that expected for a parabolic profile for a distance from the vessel wall less than 0.35 cm at a 5 mL/s flow rate and 0.3 cm at a 15 mL/s flow rate. Reverse flow components are predicted for both the 5 mL/s and 15 mL/s flow rates with a velocity of 1.59 cm/s and 7.1 cm/s, respectively. In each case, the rapid radial change in the magnitude of velocity produces a significant velocity gradient.

Distal to the stenosis, the radial velocity gradient decreases near the wall in comparison with the velocity gradient expected for a parabolic profile, particularly at the lower flow rate of 5 mL/s. In the center of the vessel, the radial velocity gradient is increased. For laminar flow, Fluent predicts the gradient to be largest near the wall. For the 50% stenosis the radial velocity gradient near the center of the vessel is significantly increased. Its peak is shifted to a point 0.36 cm from the vessel wall at the 5 mL/s flow rate and 0.33 cm from the wall at the 15 mL/s flow rate, while the gradient near the wall remains small.

EXPERIMENTAL RESULTS

In this section, the normalized likelihood function (NLF), and the normalized lag one correlation (NLOC) at a lag of one pulse period are evaluated using experimental data acquired using a vessel with a 50% stenosis and a calibrated flow pump. These functions are evaluated for data acquired at the center of the stenosis and a location distal to the stenosis. This analysis will show that the coherent processing of the signal from eight pulses using the NLF produces a parameter which sharply decreases when the sample volume contains scatterers with a large lateral velocity (across the beam) or a large range of axial (along the beam centerline) velocities. This parameter also is shown to increase near the vessel wall distal to a stenosis, particularly with the low volume flow rate appropriate to diastolic flow conditions. The increase in the normalized likelihood function results from the decrease in the lateral velocity and axial velocity gradient distal to the stenosis. Therefore, such a measure could be used to map regions with an increased velocity within the stenosis, particularly during systole, and with a decreased velocity near the wall distal to the stenosis, particularly during diastole.

Parameters of the experimental system

As previously detailed in [1],[3], a Philips ultrasound system has been interfaced to a LeCroy 8828 8 bit digitizer operating in burst mode, which samples the data immediately following wideband amplification and a bandpass filter centered at the carrier frequency. The transducer excitation approximates an impulse. The central clock of the ultrasound system was used to derive the pulse repetition control signal sent to the transducer and digitizer. Other parameters are summarized below:

Transducer center frequency	7.5 MHz
Transducer lateral beam width at 1.5 cm depth	0.75 mm
Focal Depth	2 cm
Focal Length	1 cm
Sampling rate	200 MHz
Pulse Repetition Rate	8012.8 Hz
Receiver bandwidth	6 MHz

The flow is produced by a UHDC flow system (London, Ontario). As in [1],[3], the blood mimicking material consists of Orgasol 3501 ex D at a 45% volume concentration suspended in emulsified oil, water and detergent. Orgasol 3501 ex D, is a powdered nylon particle, with a mean particle size of 9 μm. The density of the resulting solution is 1027 kg/m^3, the dynamic viscosity of the solution is 0.004 kg/(m·sec) and the speed of sound is 1545 m/s.

Evaluation of the Normalized Likelihood and Normalized Lag One Correlation

The evaluation of the NLF and NLOC is presented in this section for the experimental conditions detailed above, with a 50% stenotic vessel phantom and two flow conditions. The evaluation of these parameters is presented first for the signal from the minimum diameter within the stenosis, and next for a location downstream from the stenosis. In each case, the velocity profile estimated using the wideband likelihood estimator (WMLE) with 8 pulses and the magnitude as estimated by the NLF evaluated at the maximum likelihood velocity are compared with the profile as estimated by the autocorrelator and the magnitude as estimated by the NLOC. In the following figures, the velocity and normalized magnitude are plotted as a function of position along the transducer beam (angled at 45 degrees to the vessel), with position indexed in 3.86 μm increments, corresponding to 5 ns increments. The velocity profile is then compared to a likelihood threshold of 0.2 and a normalized power threshold of 0.2. The power is normalized by the peak wall filtered power along a single line of sight. For regions in which either the normalized likelihood or the normalized power fall below the threshold, the velocity is replaced with 0 cm/s.

Figures 1-4 first present the experimental results for the data acquired within the stenosis. In this region the velocity profile is predicted to be symmetric; the magnitude of the likelihood and correlation should decrease at the higher flow rate, and show a maximum in the center of the vessel if the profile is parabolic.[1]

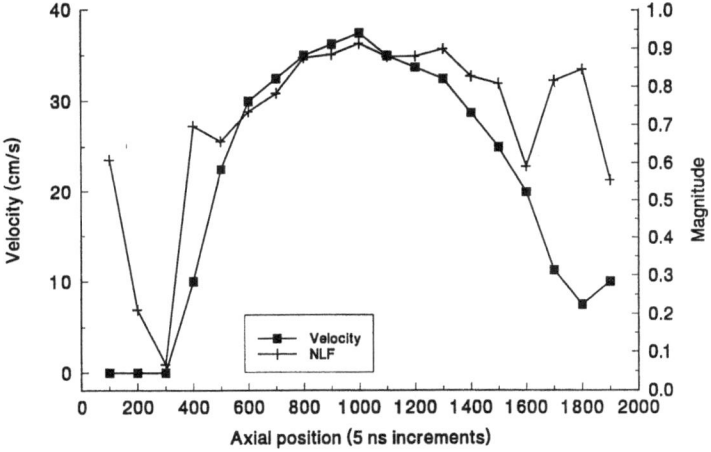

Figure 1. Velocity profile as estimated by the WMLE and magnitude of the NLF for a 5 mL/s volume flow rate, within the stenosis.

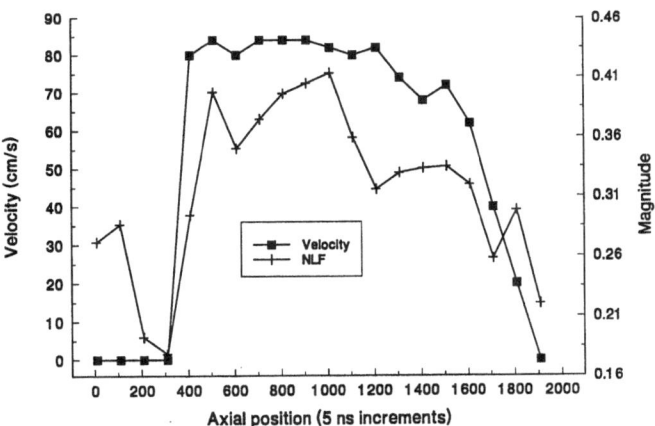

Figure 2. Velocity profile as estimated by the WMLE and magnitude of the NLF for a 15 mL/s volume flow rate, within the stenosis.

The peak velocity in the center of the vessel was predicted to be 31 cm/s and 82 cm/s for the 5 and 15 mL/s flow rates. The estimated peak velocities using the WMLE are 36 and 83 cm/s. The estimated velocities using the autocorrelator are 29 cm/s and 52 cm/s. Note that the transmitted signal has a 50% fractional bandwidth. Using the autocorrelator averaged over eight pulses at this velocity, a significant bias is produced, due to the incompatibility of the wideband signal and narrowband estimation technique.

For the 5 mL/s flow rate, the peak of the NLF has a magnitude of 0.89 and occurs in the center of the vessel. At this position in the vessel, the flow profile is nearly parabolic and therefore the minimum velocity gradient (minimum shear) is predicted to occur in the center of the vessel. The peak magnitude of the NLF drops to 0.4 at the 15 mL/s flow rate, with the peak in the center of the vessel and a slightly asymmetrical tapered profile near the vessel walls. The NLOC is nearly uniform for the 5 mL/s flow rate, with an amplitude ranging between 0.9 and 1. At the 15 mL/s volume flow rate, the peak magnitude of the NLOC decreases to 0.73, and the profile is asymmetrical.

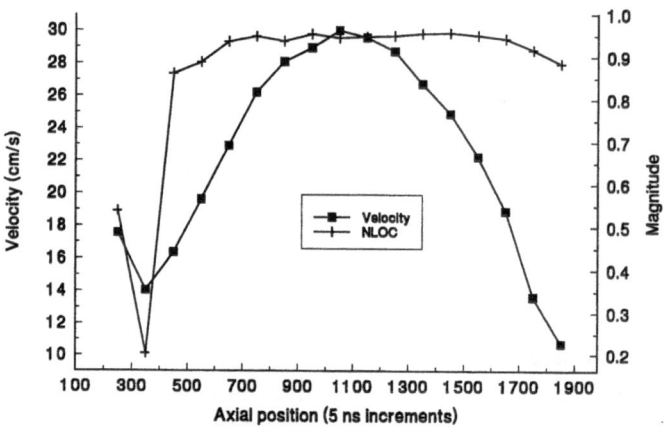

Figure 3. Velocity profile as estimated by the autocorrelator and magnitude of the NLOC for a 5 mL/s volume flow rate, within the stenosis.

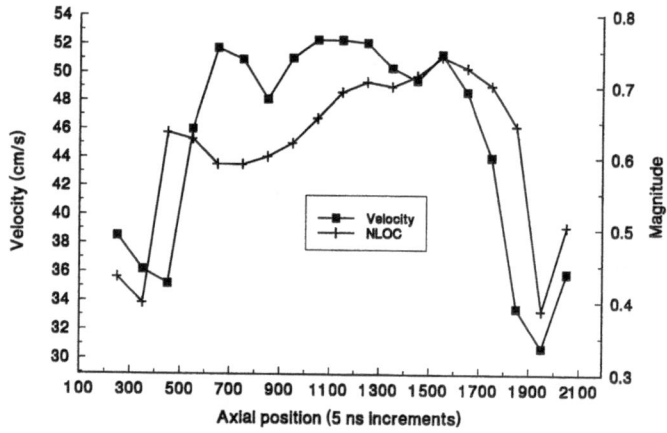

Figure 4. Velocity profile as estimated by the autocorrelator and magnitude of the NLOC for a 5 mL/s volume flow rate, within the stenosis.

Figures 5-8 present the experimental results for the data acquired distal to the stenosis. In this region it is anticipated that the profile is not symmetric due to the 45 degree angle between the beam and vessel. Applying the flow conditions predicted by Fluent shows that the beam will encounter the region of reverse flow at the top of the vessel, and cross a region in which the profile has recovered near the second vessel wall, at a downstream distance of 0.925 cm   The magnitudes of the likelihood and correlation functions should be asymmetrical due to these flow conditions, with a large magnitude at the top of the vessel and a decrease in the center of the vessel; this asymmetry should be particularly evident at the 15 mL/s flow rate.   The peak velocity in the center of the vessel was predicted to be 28 cm/s and 78 cm/s for the 5 and 15 mL/s flow rates, with a peak reversed velocity component of 1.59 and 7.1 cm/s at the 5 and 15 mL/s flow rates.   The estimated peak velocities using the WMLE are 32 and 90 cm/s. The estimated peak velocities using the autocorrelator are 28 cm/s and 55 cm/s. The reverse flow components are accurately mapped using the WMLE, with velocities of 2 and 7 cm/s at the 5 and 15 mL/s flow rates. The reverse velocity is estimated as 3 and 4 cm/s using the autocorrelator.

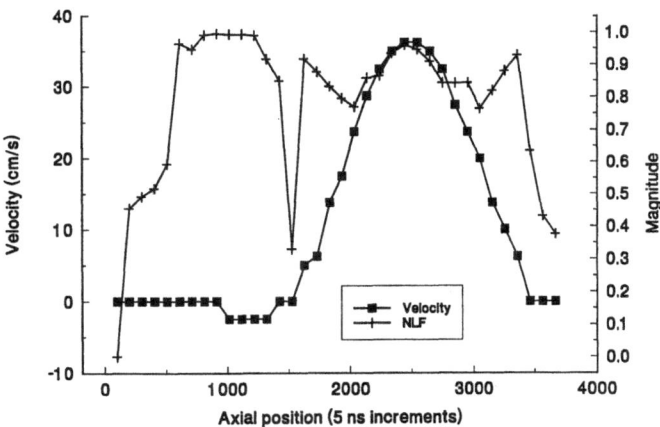

Figure 5. Velocity profile as estimated by the WMLE and magnitude of the NLF for a 5 mL/s volume flow rate, distal to the stenosis.

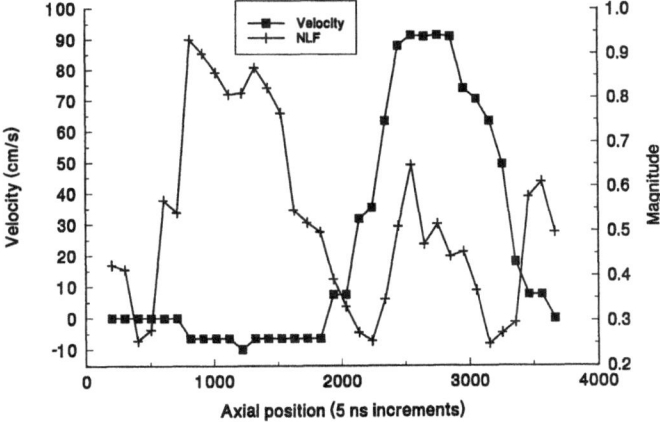

Figure 6. Velocity profile as estimated by the WMLE and magnitude of the NLF for a 15 mL/s volume flow rate, distal to the stenosis.

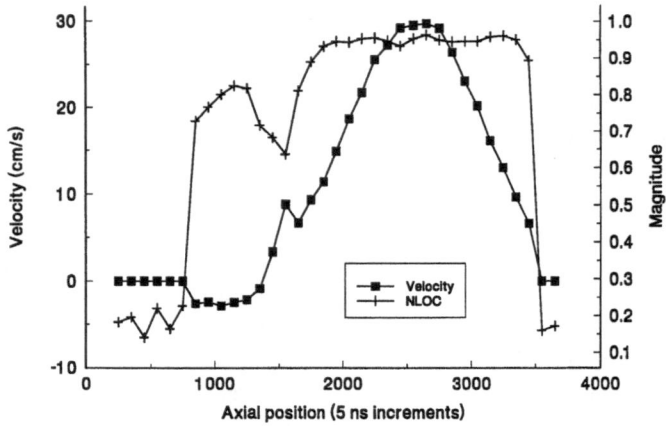

Figure 7. Velocity profile as estimated by the autocorrelator and magnitude of the NLOC for a 15 mL/s volume flow rate, distal to the stenosis.

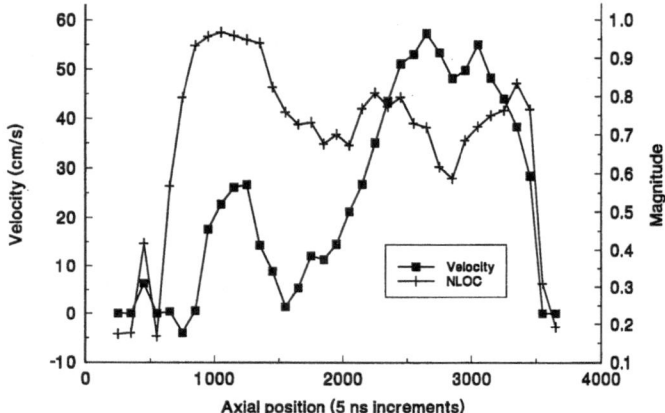

Figure 8. Velocity profile as estimated by the autocorrelator and magnitude of the NLOC for a 15 mL/s volume flow rate, distal to the stenosis.

Distal to the stenosis and near the vessel wall, the low mean velocity and very small shear rate which occur in the region of flow reversal produce a signal which is highly correlated over the eight pulse processing interval. Examination of Figure 5 shows that the NLF is asymmetric with a peak magnitude of 0.97 near the vessel wall, and lower values in high shear regions. In the region between the forward and reversed flow, the NLF is nearly equal to zero. The magnitude of the NLF increases to a second peak in the center of the vessel, in a region with a small shear rate near the peak of the parabolic velocity profile. At the 15 mL/s flow rate shown in Figure 6, the NLF is also asymmetric with a peak of 0.93 near the first vessel wall, dropping to zero between the regions of forward and reversed flow, and returning to a peak of 0.65 in the center of the vessel.

The NLOC, as shown in Figure 7, does not accurately reflect the spatial variations in the correlated signal interval or shear rate across the vessel for the 5 mL/s flow rate. In Figure 8 at the 15 mL/s flow rate, the correlation peak decreases to 0.8 in the center of the vessel, and increases to 0.95 in the region of flow reversal. This parameter does not accurately map regions of large shear, but in some instances does decrease when the sample volume contains flow with a large lateral velocity component.

CONCLUSION

With the goal of spatially mapping correlated signal intervals, two indicators of the correlated signal interval have been evaluated for known experimental flow conditions. The experimental results show that the NLF, which coherently sums the signal correlation over a set of pulses and lag values, provides a sensitive indication of shear rate and lateral velocity, and may be used to develop spatial maps of flow conditions. The NLOC inconsistently decreases with the lateral velocity, and cannot be used to map regions of significant shear.

ACKNOWLEDGMENTS

This research was supported by National Science Foundation grant BCS-9108940 and National Institutes of Health grant R15 HL48273.

REFERENCES

[1]   K.W. Ferrara and V.R. Algazi. "A Theoretical and Experimental Analysis of the Received Signal From Disturbed Blood Flow," *IEEE Transactions on Ultrasonics, Ferroelectrics and Frequency Control*, March 1994.

[2]   K.W. Ferrara and V.R. Algazi. "A New Wideband Spread Target Maximum Likelihood Estimator For Blood Velocity Estimation Part One-Theory," *IEEE Transactions on Ultrasonics, Ferroelectrics and Frequency Control*, January 1991, 1-16.

[3]   K.W. Ferrara and V.R. Algazi. "A Statistical Analysis of the Received Signal from Blood During Laminar Flow," IEEE Transactions on Ultrasonics, Ferroelectrics and Frequency Control, March 1994.

[4]   C. Kasai, K. Namekawa, A. Koyano, and R. Omoto, "Real-Time Two-Dimensional Blood Flow Imaging Using an Autocorrelation Technique," IEEE Transactions on Sonics and Ultrasonics, Vol. SU-32, No. 3, May 1985.

STENOSES CHARACTERIZATION WITH ULTRASOUND

O. Bonnefous

Laboratoires d'Electronique Philips, 22 avenue Descartes, BP 15
94453 Limeil Brévannes Cedex (France)

INTRODUCTION

The detection and evaluation of stenoses are important public health issues. After a morphologic detection of the stenosis is performed through ultrasound or angiography, the physicians must decide whether a surgical intervention or a recanalization is needed. The latter is used only if the stenosis reduces the arterial lumen by 70 % or less, and the first solution is decided when the vessel is nearly occluded. Today, the medical diagnoses are performed according to criteria, such as percentage of reduction of the lumen area or "pressure gradient", the latter being estimated from the Doppler blood velocity measurement inside the stenosis. Although these approaches are clinically validated, we think that they are only makeshifts, mainly because of the subjectivity introduced in the criteria interpretation. Indeed, these criteria rely on the modelization of the blood pressure gradient across the stenosis.

According to the simplified Bernouilli equation : $V^2 = k\Delta P$ (where V is the blood velocity, ΔP the pressure gradient and k a constant), the pressure gradient across the stenosis is derived from the blood velocity measured using ultrasonic techniques (Doppler, CVI[1]). However, this modelization is not satisfactory because the Bernouilli equation does not apply to the propagation of the blood flow in the arterial network. The physicians are thus lead to classify the stenoses qualitatively as "significantly haemodynamic" or "not significantly haemodynamic" rather than to quantify the alteration of the vessels.

Beside this subjective classification, an apparent contradiction is observed between the increase of the blood velocity inside a stenosis due to the lumen reduction, and its dramatic decrease when the lumen reduces to a thrombosis (artery nearly occluded).

We propose here a new investigation direction which, first, explains this apparent contradiction, and, second, leads to a new quantitative approach for the stenosis characterization. The practical means proposed for the observation of the stenoses and the related measurements are described through in vivo acquisitions from normal and pathological cases.

ANALYSIS OF THE STENOSIS DYNAMICS

In the case of a stenosis, the arterial wall is badly injured, which causes a strong disturbance of its function. The sane artery behaves like a propagation line because of its intrinsic elasticity. When plaques or stenoses spread out, this elasticity is altered and the propagation of the pressure wave is disrupted accordingly. Our goal is to evaluate the consequences of the wall disease on its propagative performances thanks to an analysis of the arterial dilation during the cardiac cycle. In a first order, the arterial tissue being elastic, there is a linear relation between the diameter and the pressure. This consideration justifies the external measurement of the arterial diameter rather than the invasive one of arterial pressure.

Blood flow and pressure (or now the arterial diameter) are conjugated parameters which propagate at the same propagation velocity. The two equations describing the pressure/flow relations in an elastic tube (with no friction) are the following :
- blood mass conservation :

$$S \frac{du}{dx} + \frac{dS}{dt} = 0 \tag{1}$$

- fundamental dynamic equation :

$$\rho \frac{du}{dt} + \frac{dP}{dx} = 0 \tag{2}$$

where S is the arteria section area, u the blood velocity, P the blood pressure, x the coordinates along the vessel axis, t the time, and ρ the volumetric blood density. The elasticity characteristic of the artery is expressed through the compliance C defined by :

$$C = \frac{dS}{dP} \tag{3}$$

corresponding to the ability of the artery to change its size when the pressure varies. As it is well known[2], the propagation velocity c is directly related to the compliance in the following manner :

$$C = \frac{S}{\rho c^2} \tag{4}$$

It is then easily understood that an increasing rigidity accelerates the propagation while it reduces the arterial dilation amplitude. Combining Eq. (3), (4) and (1), we get a new relation :

$$\rho c^2 \frac{du}{dx} + \frac{dP}{dt} = 0 \tag{5}$$

Substituting the parameter P in Eq. (5) and (2), and considering that locally the elasticity and consequently the propagation velocity c is disturbed by a plaque or a stenosis, we finally obtain :

$$\frac{du^2}{dt^2} - 2c \frac{dc}{dx} \frac{du}{dx} - c^2 \frac{du^2}{dx^2} = 0 \tag{6}$$

which is the very well known differential equation of an exponential horn. Of course, the perturbation is considered here as a constant inside the stenosis as a first approximation : dc/dx = cte. The resolution of the expression (6) exhibits a cut-off pulsation $\omega_c = dc/dx$. Thus, two types of solutions appear for waves of pulsation ω :

a) Damped stationary waves if $\omega < \omega_c$:

$$u = e^{-mx} [A\, e^{a(\omega)x} + B\, e^{-a(\omega)x}]\, e^{j\omega t}$$

A and B being two constants, and :

$$m = \frac{\omega_c}{c}\,,\; a(\omega) = m\sqrt{1 - (\frac{\omega}{\omega_c})^2}$$

b) Damped propagative waves if $\omega > \omega_c$:

$$u = e^{-mx} [A\, e^{j\omega(t - x/c(\omega))} + B\, e^{j\omega(t + x/c(\omega))}]$$

with :

$$m = \frac{\omega_c}{c}\,,\; c(\omega) = \frac{c}{\sqrt{1 - (\frac{\omega_c}{\omega})^2}}$$

We can extrapolate from this analysis that, as soon as the elastic perturbation is high enough to create a cut-off pulsation ω_c which is comparable with the cardiac one, the latter no longer propagates through the stenosis. Moreover, when the pulsation is higher than the ω_c limit, the propagation velocity depends on the pulsation which corresponds to a dispersive propagation. In the case of the stationary waves, their damping also depends on their pulsation. The higher the frequency, the higher the damping. Figure 1 depicts these various effects : figure 1a is the plot of the normalized propagation velocity $(c(\omega)/c)$ for propagative waves, and figure 1b is the plot of the damping 1-a $(\omega)/m$ as a function of the normalized pulsation ω/ω_c.

Simple numerical calculations help in evaluating the effects. Let us suppose a relative axial propagation velocity change of m = ω_c/c = $(dc/dx)/c$ = 1 %/mm, a soft perturbation. For a typical propagation velocity (carotid artery) c = 6 m/s, we find ω_c = $2\pi f_c$ = 60 rd/s and f_c = 10 Hz ! That means that, in this case, all the frequencies under 10 Hz are trapped in the stenosis. The wave associated to f_c = 10 Hz is damped by a damping coefficient m = 10^{-2} mm^{-1}. A 1 cm long plaque produces then a damping of 0.90.

For a serious perturbation m = 50 %/mm and a higher velocity propagation c = 10 m/s (old rigid arteries subjected to plaques), f_c = 5 kHz ! We can then consider that there is no more propagation through the plaque and that all the significant frequencies are damped. The attenuation coefficient takes the following form : m-a(ω) = m $\omega/2\omega_c$, creating a strong deformation of the waveform crossing the stenosis.

METHODS AND MEASUREMENTS

We use the motion of the arterial wall to visualize and evaluate the perturbations in the propagation of the arterial waves. As a matter of fact, the arterial dilation related to the pressure through the compliance can be estimated non invasively. However, this implies a scanning procedure which must be compatible with the time constraints involved by the propagation .

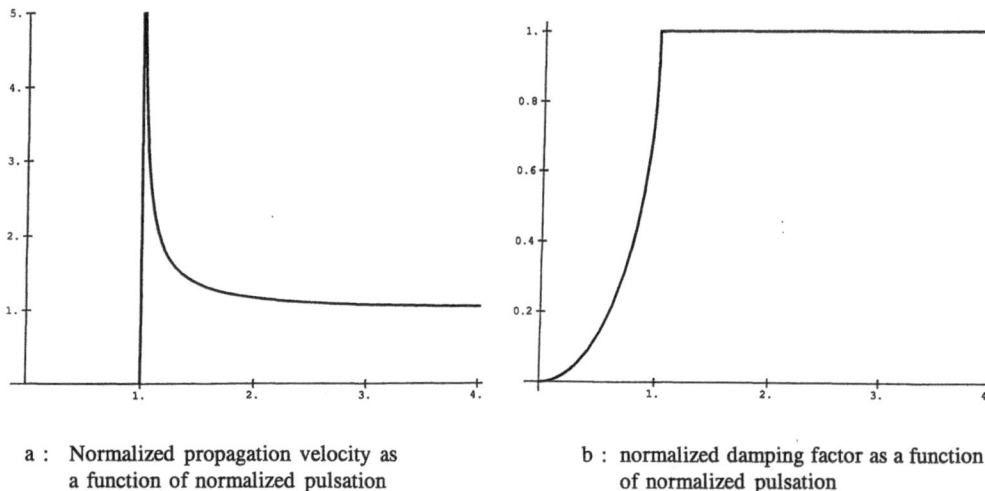

a : Normalized propagation velocity as
    a function of normalized pulsation

b : normalized damping factor as a function
    of normalized pulsation

Figure 1 : Propagation characteristics through a stenosis

The basic measurement principle is the time domain correlation technique already used with CVI[3], which is an echo traking process. The time sequence is the following : a plane including the axis of the artery is scanned and recorded at a high frame rate. That means that a unique ultrasonic firing is used for each image line. The cross correlation procedure is performed between the recurrent lines of the successive images. This produces 2D velocity image sequences. These sequences are then time integrated to get displacement images. After identification of the posterior and anterior arterial walls, the displacement averages on the wall thicknesses are extracted and combined to get the arterial dilations along the artery as a function of time. Figure 2 shows the arterial dilation from a sane artery : figure 2a presents the dilation of many artery diameters measured along a 3 cm long artery part where all the curves are presented superimposed, figure 2b shows the same curves, seen perpendicularly to the presentation of figure 2a and located along the artery axis.

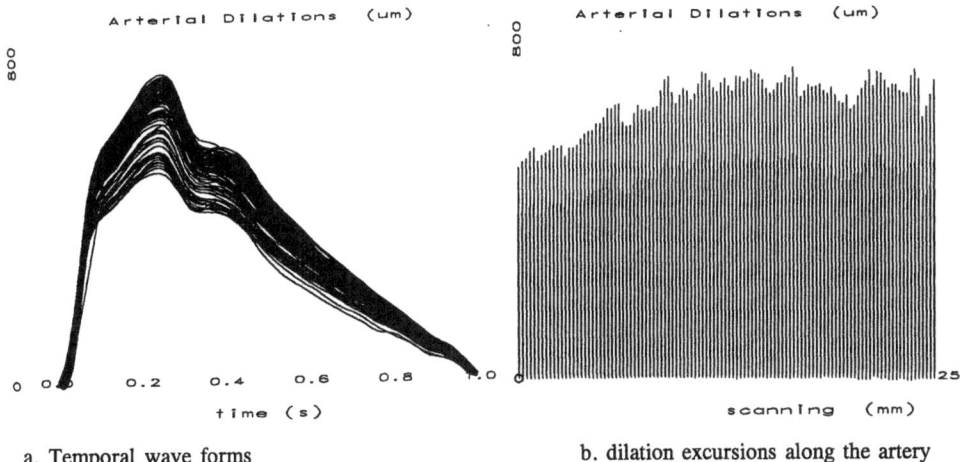

a. Temporal wave forms

b. dilation excursions along the artery

Figure 2 : Arterial dilations : normal common carotid artery

Figure 3 gives the same information but for an artery with a small plaque, which is judged as non pathologic by physicians. The spectacular results presented in figures 3a and 3b show a strange behaviour not described in the clinical files likely because it was impossible to evidence it : in fact, the artery appears to contract itself, rather than dilate immediately downstream the plaque. The contraction is about 200 μm since the upstream dilation is about 500 μm.

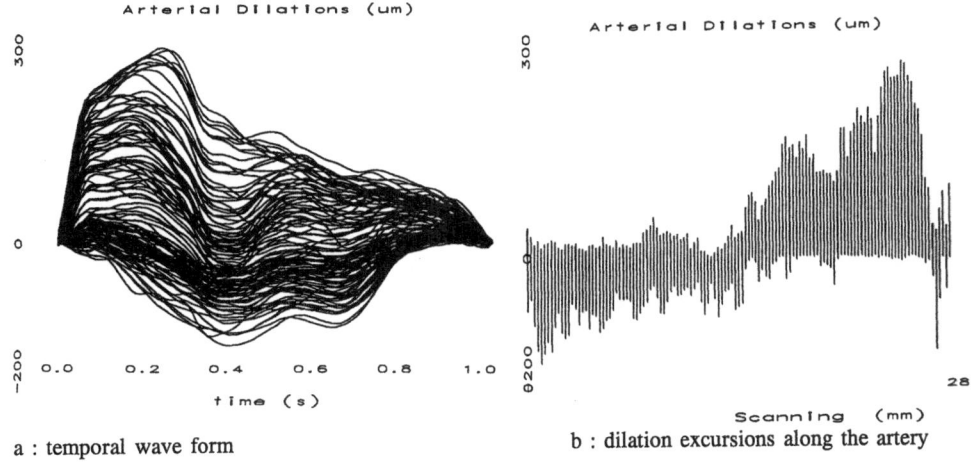

a : temporal wave form

b : dilation excursions along the artery

Figure 3 : Arterial dilations : small plaque in the internal carotid artery

A more detailed analysis, compatible with the previous theoretical presentation, is a harmonic decomposition. This is illustrated in the figure 4 series where colour images in the plane (t,x) show respectively the full displacement (figure 4a), harmonic 1 (figure 4b), harmonic 2 (figure 4c), harmonic 3 (figure 4d), and harmonic 4 (figure 4e). The red line superimposed corresponds to the null displacement points, allowing the identification of positive and negative zones. Observing the various perturbations in the propagation of these frequencies, it is clear that the first harmonic -corresponding to the cardiac

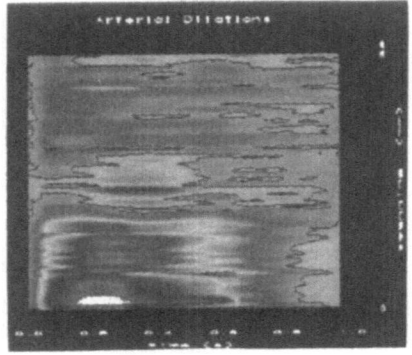

a : Full displacement

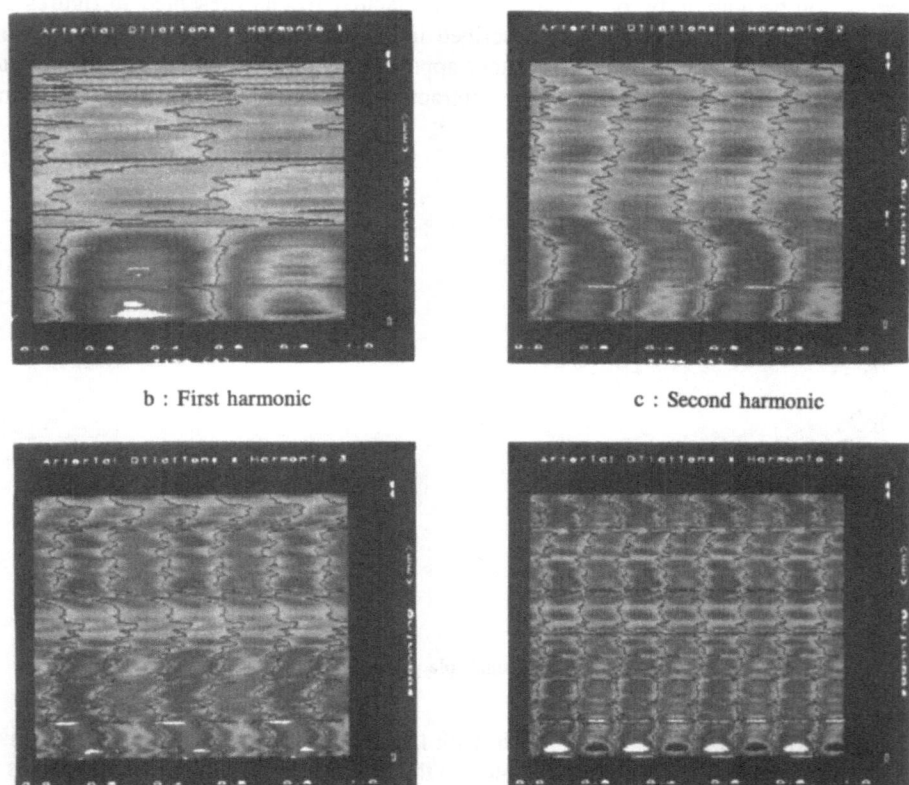

b : First harmonic

c : Second harmonic

d : Third harmonic

e : Fourth harmonic

Figure 4 : Harmonic decomposition of the arterial dilation through the plaque

frequency- does not propagate through the plaque : a sign inversion is evidenced along the x axis, characterizing stationary waves. The higher frequencies succeed in progressing through the plaque, even though the displacements are attenuated. In this case, the cut-off frequency is between the 1st and the 2nd harmonic, more precisely between 1 and 2 Hz. The local perturbation can then be evaluated to only $dc/dx = 10 \text{ s}^{-1} = 10^{-2}$ (m/s)/mm, which seems to be negligible and in agreement with the diagnosis of the angiologist.

CONCLUSIONS AND PERSPECTIVES

These first results are very encouraging. More than the evaluation of the arterial dilation which is still relevant and a descriptive information, we can already propose a true quantitative characterization of the local arterial injury.

However, this new approach needs an important campaign of clinical tests with different cases of arterial lesions. The validation of this incipient technique requires the confrontation with the state-of-the-art ones. Through our technico-clinical association ACTUA (Applications Cliniques des Techniques Ultrasonographiques Avancées), we have established a collaboration program with the "Urgences Neuro-Vasculaires" Service from the Salpêtrière Hospital in Paris which should be of great help and interest.

BIBLIOGRAPHY

1. O. Bonnefous, "Time domain colour flow imaging : methods and benefits compared to Doppler ", Acoustical Imaging, 19, 301-309 (1992)
2. W.W. Nichols, M.F. O'Rourke, "Wave propagation in an elastic tube", McDonald's Blood Flow in Arteries, 3rd Ed. p. 85, W.W. Nichols and M.F. O'Rourke. Edward Arnold
3. O. Bonnefous, P. Pesqué, "Time domain formulation of pulse Doppler ultrasound and blood velocity estimation by cross-correlation", Ultrasonic Imag. 8, 73-85 (1986)

INTRAVASCULAR ULTRASOUND AND SCANNING ACOUSTIC MICROSCOPY EVALUATION OF AORTIC WALL

Yoshifumi Saijo[1], Hidehiko Sasaki[1], Hiroaki Okawai[1], Motonao Tanaka[1], and Floyd Dunn[2]

[1]Department of Medical Engineering and Cardiology,
 Institute of Development, Aging and Cancer, Tohoku University.
 4-1 Seiryo-machi, Aoba-ku, Sendai 980, Japan
[2]Bioacoustics Research Laboratory and Computer Engineering,
 University of Illinois.
 1406 West Green Street, Urbana, IL 61801, USA

INTRODUCTION

Intravascular ultrasound (IVUS) imaging is now widely used to evaluate aortic and coronary diseases in clinical cardiology to provide tissue character information of the arterial wall not detectable by angiography[1-6]. While angioscopic evaluation of arterial tissue has also been developing, it only provides information of surface morphology[7-8].

Assessment of physical properties of the arterial wall by ultrasonic method is important in percutaneous transluminal coronary angioplasty and in percutaneous transluminal angioplasty because the angioplastic procedures produce artificial stress on the arterial wall. The low amplitude ultrasonic waves produce no stress, and can provide both morphological and physical details of the arterial wall.

In many studies, the muscular arteries, such as the coronary artery, have exhibited a three-layered appearance, and the elastic arteries, such as aorta and the peripheral arteries have exhibited a single-layer hypo-echoic appearance. These findings were considered to be related to the histological structure of artery. However, the acoustic properties of the layers have not as yet been evaluated.

IVUS and histological evaluation of excised arteries have been studied[9]. As the state of the artery is clearly different in the in vivo and in the in vitro states, the present study was undertaken.

The purpose is to evaluate the three-layered appearance of the arterial wall in IVUS imaging by comparing in vivo IVUS results and those obtained in vitro scanning acoustic microscopy employing an animal model.

METHODS

Three adult goats weighing 45–55 kg were anesthetized with intravenous thiopental sodium (2.5 mg/kg) and intravenous ketamin sodium (5.0 mg/kg). They were placed on a volume limited respirator following tracheal intubation. The anesthesia was maintained with nitrous oxide anesthesia under mechanical ventilation.

An 8F IVUS transducer was inserted through an 8F carotid arterial sheath, and the carotid artery, aorta and left ventricle were thus observed and recorded on a VHS videorecorder. The observed regions of the aorta were marked by silk threads, in the adventitia and the aorta was excised after the observation and specimen sacrifice.

Aortic specimens were formalin–fixed, paraffin embedded and sectioned approximately 10 μm in thickness with a microtome. The specimens were mounted on glass slides. They were not covered by cover slips while measurements were made with the SAM system. The paraffin was removed from the mounted sections, by the graded alcohol method, just before the ultrasonic measurement.

The ALOCA SSD–550 IVUS apparatus was employed in the study. Radial mechanical scanning was used, the center frequency was 20 MHz, and the outer diameter of the transducer was 2.4 mm.

Two–dimensional distributions of the attenuation constant and the sound speed can be obtained with the SAM system[10-11]. Figure 1 is a block diagram of the SAM system. The ultrasonic frequency is variable over the range of 100 to 200 MHz and the beam width at the focal volume ranges from 5 μm (at 200 MHz) to 10 μm (at 100 MHz). The mechanical scanner is so arranged that the ultrasonic beam is transmitted for every 4 μm interval over a 2 mm width. The number of sampling points is 480 in one scanning line and 480 × 480 points make one frame within 8 sec.

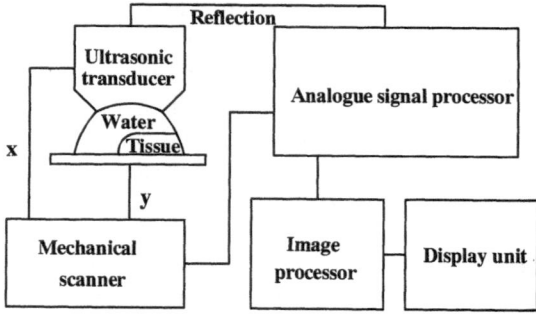

Figure 1. Block diagram of the scanning acoustic microscope (SAM) system.

The original signals for imaging produced in the analogue signal processor can be displayed on the CRT screen directly. However, the original signals are not sufficiently accurate values of attenuation constant or sound speed due to the lack of uniformity in thickness of the sectioned specimens. The flow chart showing the processing steps for correcting the attenuation constant and the sound speed is shown in Figure 2. The image processor stores the original image signals in 16 frames; eleven frames of amplitude images in 10 MHz steps in the range 100 to 200 MHz, and five frames of phase images in 10 MHz steps in the range 100 to 140 MHz. The thickness of the specimen is determined from the frequency dependent characteristics of the amplitude and the phase of the received signals obtained at the same position for each of the sixteen frames[11].

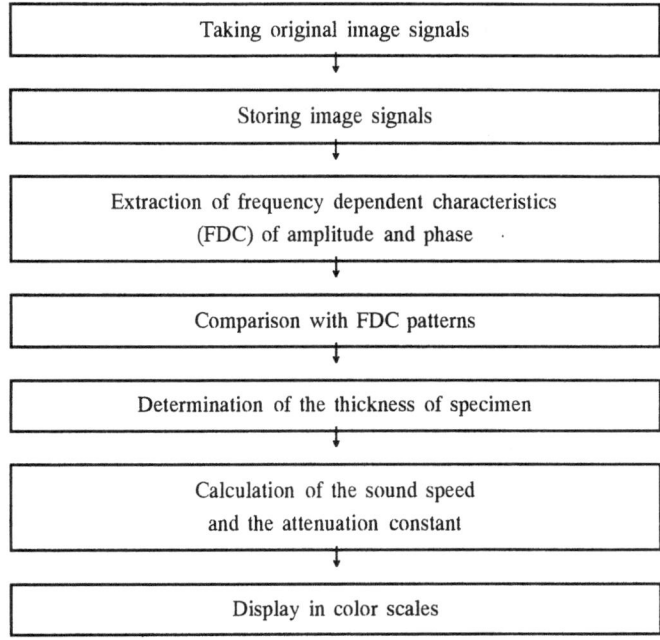

| Taking original image signals |
| Storing image signals |
| Extraction of frequency dependent characteristics (FDC) of amplitude and phase |
| Comparison with FDC patterns |
| Determination of the thickness of specimen |
| Calculation of the sound speed and the attenuation constant |
| Display in color scales |

Figure 2. Flow chart of the processing steps for determining the ultrasonic attenuation constant and the sound speed.

The attenuation constant and the sound speed are calculated by the computer using the equations;

$$A = \frac{L}{fd}$$

$$C = \frac{1}{\dfrac{1}{Cw} - \dfrac{\phi}{2\pi fd}}$$

where A is the attenuation constant, L is the magnitude of the reflected ultrasound, f is the frequency used, d is the thickness of the specimen, C is the sound speed in the specimen, Cw is the sound speed in the coupling medium, and ϕ is the phase shift.

A linear frequency dependence of the attenuation in the ultrasonic frequency range of 100–200 MHz, was obtained from consideration of the interference of the sound wave components. The data of attenuation constant and sound speed thus obtained were converted into color signals which corresponded to the volume of the amplitude and phase, and were displayed on a color monitor as two–dimensional distribution patterns. Table 1 shows the relationship between color coded scales and values of attenuation constant and sound speed. Distilled water was used for the coupling medium, which was maintained between 20 and 22 ℃ during the measurement procedure. A neighboring section of the SAM specimen was stained with Elastica–Masson's trichrome stain, and used for optical microscopy.

Table 1. Relationship between color coded scales and values of sound speed and attenuation constant.

Attenuation constant (dB/mm/MHz)	Color	Sound speed (m/s)
1.9 ~	Red	1690 ~
1.7 ~ 1.9	Magenta	1670 ~ 1690
1.5 ~ 1.7	Orange	1650 ~ 1670
1.3 ~ 1.5	Brown	1630 ~ 1650
1.1 ~ 1.3	Yellow	1610 ~ 1630
0.9 ~ 1.1	Green	1590 ~ 1610
0.7 ~ 0.9	Olive green	1570 ~ 1590
0.5 ~ 0.7	Cyan	1550 ~ 1570
0.3 ~ 0.5	Royal blue	1530 ~ 1550
0.1 ~ 0.3	Blue	1510 ~ 1530
~ 0.1	Black	~ 1510

RESULTS

Figure 3 shows the IVUS image of the goat aortic wall. The aortic wall exhibits a mixture of elastic arteries and a single–layered appearance is shown in this figure as reported in the past studies. However, a three–layered appearance is shown in Figure 4, comprising an inner bright echo zone, a central diffuse echo zone, and an outer bright echo zone. In the study of the goat aortic wall, the IVUS images of the arterial wall showed different appearances in the different regions.

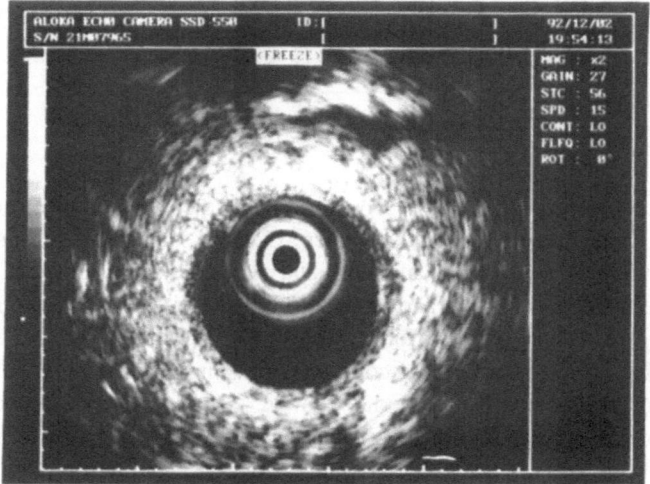

Figure 3. IVUS image of the goat aortic wall demonstrating a single–layered appearance.

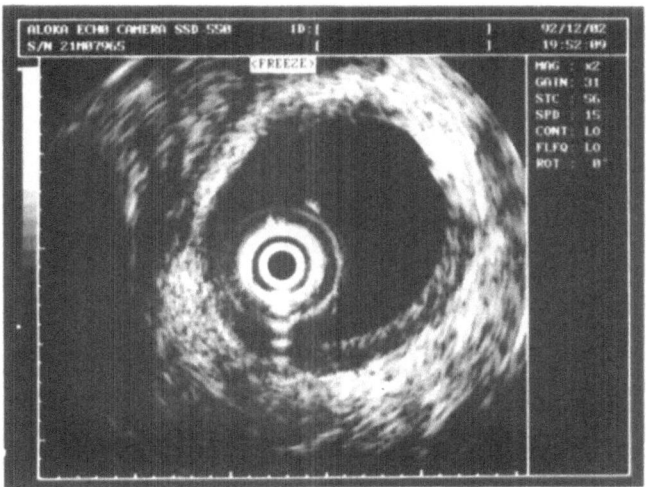

Figure 4. IVUS image of the goat aortic wall demonstrating a three-layered appearance.

The goat aortic wall is seen to consist of three parts, viz., the intima comprised of a single-layered endothelium and inner elastic membrane, the media comprised of elastic fibers and smooth muscle, and the adventitia comprised of sparse collagen fibers in the optical microscopic imaging. Figure 5 shows that the histological structure of the goat aortic wall is not identical to that of the human aortic wall. The elastic fibers are observed throughout the entire layer and are mixed with smooth muscle fibers in the human aortic media whereas the elastic fibers and the smooth muscle fibers are largely separated in the goat aortic media.

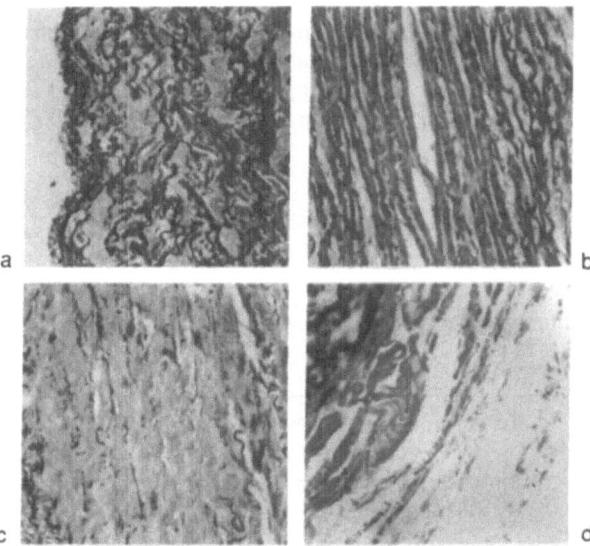

Figure 5. Optical microscopic image of the goat aortic wall ($\times 400$, Elastica–Masson's stain). a: intima, b: elastic fiber in the media, c: smooth muscle of the media, d: adventitia.

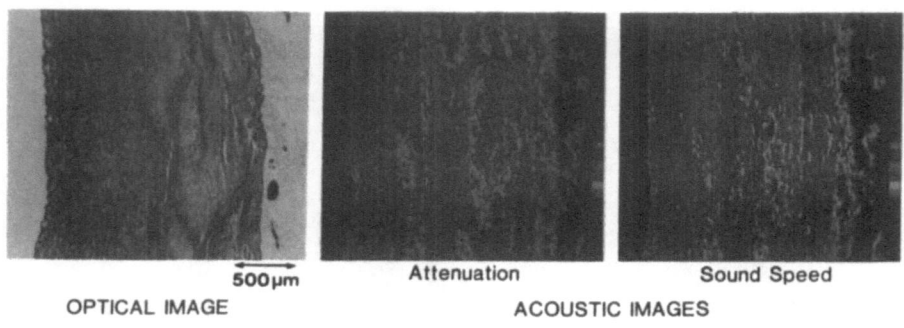

OPTICAL IMAGE   Attenuation   Sound Speed
                                        ACOUSTIC IMAGES

Figure 6. Optical microscopic image ($\times$40, Elastica–Masson's stain) and the acoustic microscope images (left: attenuation, right: sound speed) of the goat aortic wall.

Figure 6 is the optical microscope image ($\times$40, Elastica–Masson's stain) and the acoustic microscope images (left: attenuation, right: sound speed) of the goat aortic wall. The values of attenuation constant and sound speed are represented in color coded scales (Table 1).

The acoustic properties of the four tissue elements, viz., intima, elastic fiber of the media, smooth muscle of the media, and adventitia, were determined.

The region of interest (ROI) for acoustic microscopy was determined from optical microscopic observations. The size of the ROI was 200×200 μm in the optical microscopy and 50×50 sampling points in the acoustic microscopy. For each specimen, ultrasonic attenuation constant and sound speed in the tissue elements were obtained from an average values in the ROI. Two different regions in each goat aorta were observed, and the mean and standard deviation of the values of attenuation constant and sound speed were determined from six images.

Table 2 and Figure 7 show the data obtained. The values of attenuation constant and sound speed are 0.9 ± 0.3 dB/mm/MHz and 1590 ± 44 m/s in the intima, 1.4 ± 0.3 dB/mm/MHz and 1682 ± 61 m/s in the elastic fibers in the media, 0.9 ± 0.1 dB/mm/MHz and 1622 ± 13 m/s in the smooth muscle of the media, 0.5 ± 0.2 dB/mm/MHz and 1525 ± 20 m/s in the adventitia, respectively.

Table 2. Values of attenuation constant and sound speed in the tissue elements of goat aortic wall.

	Attenuation constant (dB/mm/MHz)	Sound speed (m/s)
Intima	0.9 ± 0.3	1590 ± 44
Media	1.1 ± 0.3	1636 ± 43
Elastic fiber	1.4 ± 0.3	1682 ± 61
Smooth muscle	0.9 ± 0.1	1622 ± 13
Adventitia	0.5 ± 0.2	1525 ± 20

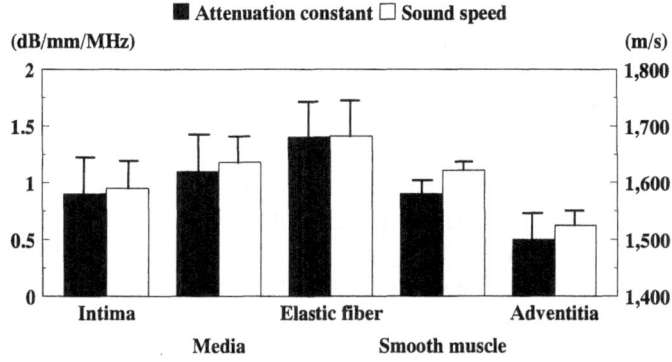

Figure 7. Graph showing the values of attenuation constant and sound speed in tissue elements of the goat aortic wall.

DISCUSSION

The ultrasonic frequency used in the IVUS study is 15–30 MHz, and the frequency used in clinical transthoracic and transesophagial echocardiography is 2.5–5.0 MHz. Precision observations of the tissue specimens are available in the IVUS study because the ultrasonic wave length is about 1/10 that of transthoracic and transesophagial echocardiography. In the present study, the ultrasonic frequency was 20 MHz and its wave length is about $80\,\mu$m, so that the IVUS apparatus provided for axial resolution of $200\,\mu$m, thus the layered structures thicker than $200\,\mu$m could be readily detected.

St. Goar et al, reported that the thickness of the intima of normal coronary arteries in young adults was about $120\,\mu$m. They also reported that the strong echo in the intima could not be detected and that the coronary arteries in young adults showed a single–layered appearance in IVUS imaging[12].

Fitzgerald et al, reported that the thickness of the intima was significantly thinner in the cases that showed a single–layered appearance of the coronary artery than in the cases that showed a three–layered appearance. They supported their results by the relationship between the thickness of the intima and the axial resolution of IVUS[13].

These reports suggested that the thickness of the layer was important in the assessment of association of acoustical properties and IVUS imaging[14].

The thickness of the intima in the present study, was found to be about $20\,\mu$m, that of the elastic fiber was $100\,\mu$m, that of the smooth muscle in the media was $500\,\mu$m, and that of the adventitia was $300\,\mu$m. Thus, smooth muscle and the adventitia were considered to be detectable by the IVUS methodology. However, as smooth muscle coexists with elastic fiber in the media, the echo signals from the smooth muscle only could not be detected easily by the IVUS method.

Thus, the inner bright echo in this study is considered to originate from another source.

In its simplest form, the specific acoustic impedance Z is expressed as

$$Z = \rho c$$

where ρ is the material density, and c is the sound speed in that material.

The dB level of the relative intensity of the reflected ultrasound at the interface of a pair of fluid–like media materials is
where I_r is intensity of the reflected ultrasound component in medium a and I_i is the

$$dB = 10\log\frac{I_r}{I_i} = 10\log\frac{(Z_a - Z_b)^2}{(Z_a + Z_b)^2}$$

incident component in medium a, Z_a is the specific acoustic impedance of material a, and Z_b is the specific acoustic impedance of material b.

In the present study, the values of the sound speed were determined using the scanning acoustic microscope, but the values of density were not measured. As the value of density ranges 1.05 g/cm^3 to 1.09 g/cm^3 in biological soft tissues[15], the ideal reflected ultrasonic intensity can be obtained from the difference of sound speeds between a pair of neighboring tissue elements as follows.

$$dB = 10\log\frac{(C_a - C_b)^2}{(C_a + C_b)^2}$$

where C_a is the sound speed in material a, and C_b is the sound speed in material b.

Figure 8 shows the schematic illustration of the aortic wall. The left figure shows the model in which blood, intima, media and adventitia are considered. The relative intensity of the reflections are − 38.0 dB between blood and intima, −36.9 dB between intima and media, and −29.1 dB between media and adventitia. However, the thickness of the intima is too small to be detected by the IVUS, so the actual reflection may be occurring between blood and intima−media complex, and between intima−media complex and adventitia. The right figure in Figure 8 indicates that the relative intensities are −31.4 dB between blood and intima−media complex and −29.1 dB between intima−media complex and adventitia.

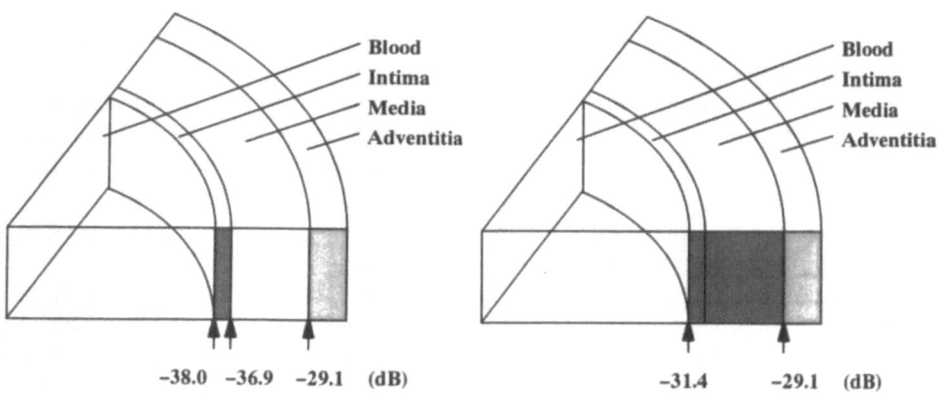

Figure 8. Schematic illustration of the goat aortic wall.

Although quantitative analysis of the echo signal was not done, the relatively strong echos in the inner zone and the outer zone are considered to be produced at the interface between blood and the intima−media complex and the interface between the intima−media complex and adventitia, respectively.

Both single−layered appearances and three−layered appearances of the aortic wall were observed by IVUS in this study. The three−layered appearance observed herein is not same as that for in coronary artery. One reason for this difference is that the histological structure of the aortic wall in the goat aorta is different from human aorta. In the human aorta, the media exhibits an homogeneous complex of elastic fiber and smooth muscle, while the goat aorta is inhomogeneous, nearly separated by elastic fiber and smooth muscle. Both single-layered and three-layered appearances were observed in the human aorta in the transformation region from elastic artery to muscular artery.

Another reason for this phenomenon is that the position of IVUS transducer may not have been positioned in the center of the aorta. Improper reflection may have occurred when the ultrasonic beam is incident from a different direction.

CONCLUSION

The acoustic properties of the tissue elements in goat aortic wall were determined as the microscopic level in this study. The data suggest that the strong internal echo is produced at the interface between blood and vessel wall and that the strong external echo is reflected ultrasound occurring between intima-media complex and adventitia. Measurement of the acoustic properties of diseased states such as calcification, fibrosis and atheroma[16] will provide valuable clinical information by the IVUS evaluation methodology.

REFERENCES

1. Tobis JM, Mallery JA, Gessert J, Griffith J, Mahon D, Bessen M, Moriuchi M, McLeay L, McRae M, Henry WL, Intravascular ultrasound cross-sectional arterial imaging before and after balloon angioplasty in vitro, *Circulation* 80:873 (1989).
2. Nissen SE, Grines CL, Gurley JC, Sublett K, Haynie D, Booth CDDC, DeMaria AN, Application of a new phased-array ultrasound imaging catheter in the assessment of vascular dimensions: in vivo comparison to cineangiography, *Circulation* 81:660 (1990).
3. Potkin BN, Bartorelli AL, Gessert JM, Neville RF, Almagor Y, Roberts WC, Leon MB, Coronary artery imaging with intravascular high-frequency ultrasound, *Circulation* 81:1575 (1990).
4. Tobis JM, Mallery J, Mahon D, Lehmann K, Griffith J, Gessert J, Moriuchi M, McRae M, Dwyer ML, Henry WL, Intravascular ultrasound imaging of human coronary arteries in vivo, *Circulation* 83:913 (1991).
5. Anderson TJ, Meredith IT, Uehata A, Mudge GH, Selwyn AP, Ganz P, Yeung AC, Functional significance of intimal thickening as detected by intravascular ultrasound early and late after cardiac transplantation, *Circulation* 88:1093 (1993).
6. Pinto FJ, St Goar FG, Gao SZ, Chenzbraun AC, Fischell TA, Alderman EL, Schroeder JS, Popp RL, Immediate and one-year safety of intracoronary ultrasonic imaging: Evaluation with serial quantitative angiography, *Circulation* 88:1709 (1993).
7. Mizuno K, Miyamoto A, Satomura K, Kurita A, Arai T, Sakurada M, Yanagida S, Nakamura H, Angioscopic coronary macromorphology in patients with acute coronary disorders, *Lancet* 337:809 (1991).
8. Mizuno K, Satomura K, Miyamoto A, Arakawa K, Shibuya T, Arai T, Kurita A, Nakamura H, Ambrose JA, Angioscopic evaluation of coronary-artery thrombi in acute coronary syndromes, *N Engl J Med* 326:287 (1992).
9. Siegel RJ, Ariani M, Fishbein MC, Chae JS, Park JC, Maurer G, Forrester JS, Histopathologic validation of angioscopy and intravascular ultrasound, *Circulation* 84:109 (1991).
10. Okawai H, Tanaka M, Dunn F, Chubachi N, Honda K, Qualitative display of acoustic properties of the biological tissue elements, *Acoustical Imaging* 17:193 (1988).

11. Saijo Y, Tanaka M, Okawai H, Dunn F, The ultrasonic properties of gastric cancer tissues obtained with a scanning acoustic microscope system, *Ultrasound in Med and Biol* 17:709 (1991).

12. St Goar FG, Pinto FJ, Alderman EL, Fitzgerald PJ, Stinson EB, Billingham ME, Popp RL, Detection of coronary atherosclerosis in young adult heart using intravascular ultrasound, *Circulation* 86:756 (1992).

13. Fitzgerald PJ, St Goar FG, Connolly AJ, Pinto FJ, Billingham ME, Popp RL, Yock PG, Intravascular ultrasound imaging of coronary arteries, Is three layers the norm? *Circulation* 86:154 (1992).

14. Pignoli P, Tremoli E, Poli A, Oreste P, Paoletti R, Intimal plus medial thickness of the arterial wall: a direct measurement with ultrasound imaging, *Circulation* 74:1399 (1986).

15. Hikichi H, Tanaka M, Ultrasono-cardiotomographic evaluation of histological changes in myocardial infarction, *Jpn Heart J* 22:287 (1981).

16. Barzilai B, Saffitz JE, Miller JG, Sobel BE, Quantitative ultrasonic characterization of the nature of atherosclerotic plaques in human aorta, *Circ Res* 60:459 (1987).

DETERMINATION OF ELASTICITY DISTRIBUTION IN TISSUE FROM SPATIO-TEMPORAL CHANGES IN ULTRASOUND SIGNALS

F. Kallel[1], M. Bertrand[1,2], J. Ophir[3] and I. Céspedes[3]

[1]Institut de génie biomédical, École Polytechnique,
[2]Institut de Cardiologie de Montréal
[3]University of Texas Medical School at Houston

INTRODUCTION

A few years ago, Lerner et al. (Lerner, 1988) and Krouskop et al. (Krouskop, 1987) proposed to visualize the velocity distribution in a tissue undergoing an external low frequency (100-1000 Hz) vibrational excitation. A color flow Doppler system was used to display the tissue motion superimposed on its B-scan image. This method was named *sono-elasticity*. With this approach it is now possible to infer parameters such as local tissue stiffness and viscosity (Yamakoshi, 1990). To do this, the amplitude, as well as the phase of the tissue vibration, has to be estimated.

Currently, attention is being drawn to processing spatio-temporal changes in echographic signals (RF or envelope detected) to observe tissue or fluid motion, and also to infer from this certain mechanical parameters describing the bio-fluid or the biomaterial; such parameters could be contraction patterns of a skeletal muscle or of a region of the myocardium or shear rate in blood flow (Meunier, 1989; Bertrand, 1989; and Shehada, 1992). Recently, tissue motion estimation has been proposed as a method to determine and then map the elastic properties of soft tissue (Ophir, 1991). The method consists of applying to it an external small quasi-static compression (about 1%) and uses the RF A-lines before and after compression to estimate the local axial motions with a correlation technique. These motion estimates which represent the axial displacement field of the tissue lead to the axial strain field which is visualized as a gray level image. The image has been named *elastogram* and the imaging technique *elastography*.

Under the assumption of a constant stress field, the elastogram could be interpreted as a relative measure of elasticity distribution, the strain being large in compliant (i.e. soft) tissue, and small in a rigid (hard) one. At this time, the potential of the method has been demonstrated *in vitro* using a variety of test objects. Strain imaging is also being investigated by others (O'Donnell et al., 1993) using computer modeling and experiments on phantom made of gel and excised tissue specimens.

Elastography has also been used *in vivo* to study certain breast pathologies (Céspedes, 1993a). Figure 1 presents a clinical example to illustrate this. It shows a conventional sonogram of a breast tumor compared to its elastogram. As can be seen here, the elastogram in (b) shows a much better contrast for the tumor, i.e. in this particular case, the tumor appears to be made of a tissue whose elastic properties differ significantly from its surrounding. In elastography, the current convention is to display tissue hardness as black and tissue softness as white. Therefore in this elastogram, the tumor can be identified as a soft tissue surrounded by a hard black capsule, which is indicative of a carcinoma with a necrotic core. This indeed was proven by pathologic examination of the excised tissue but could not be foreseen on the sonogram.

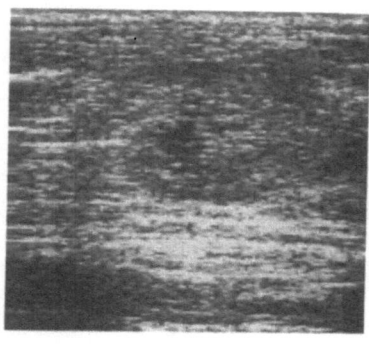 

(a)   (b)

Figure 1. Clinical example showing the potential of elastography in detecting and identifying breast cancer. The region of interest is 40x60 mm and the compression 0.7mm. (a) Sonogram showing a large tumor and a slightly hypoechoic area. (b) Corresponding elastogram showing the soft (white) necrotic tumor core surrounded by a hard (black) capsule. These images were obtained at the University of Texas Medical School (Ophir et al.).

As currently implemented, the elastogram is in fact an axial strain image; using Hook's law, it shows elasticity distribution only when the lateral strain is small and the stress field is known (Ophir, 1991). However, in practice the lateral strain would often be near 50% of the axial strain and the stress field would be unknown since it is conditioned by the elasticity distribution itself. Therefore, it is only in simple cases that the elastogram can be easily interpreted as a quantitative measure of local elastic properties. Surely, in complex real life situation, quantitative elasticity imaging would require an approach that can go beyond strain imaging.

In this paper, we propose to compute the elasticity distribution within the framework of inverse problem solving. Here, a forward problem formulation is used to predict the tissue motion for a known elasticity distribution and known boundary conditions. In this case, the inverse problem consists in predicting the elasticity distribution given observed tissue motion and known boundary conditions. In essence this is achieved by comparing the observed motion to the one predicted by solving the direct problem, and then adjusting the direct-problem elasticity distribution so that the observed and predicted motions agree.

FORWARD PROBLEM MODELS IN ELASTOGRAPHY

Figure 2 gives a comprehensive forward problem formulation for quasi-static conditions, i.e. the tissue has reached static equilibrium after the compressive step had been applied. The model brings together both the tissue acoustic and elastic properties in order to form the ultrasound signal from which the tissue displacement will be estimated. For the acoustic properties, a pre-compressed tissue is modeled as an acoustic impedance continuum. It is an arrangement of scatterers of known acoustic properties, geometry, density and distribution embedded in a supporting matrix. In the model, the elastic properties are carried through the supporting matrix. This matrix obeys the governing elasticity equations, i.e. matrix compression result in a displacement field which is specific to the elasticity distribution and compression patterns. This displacement field is impressed on the tissue acoustic impedance distribution. The resulting geometrically modified acoustic impedance is then used by an image formation model to produce the ultrasound scan at a given stage of tissue compression.

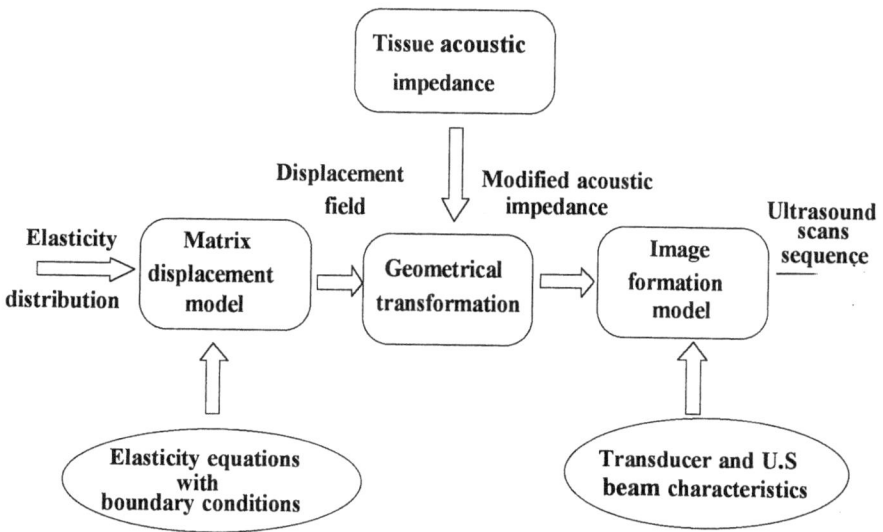

Figure 2. Forward problem in elastography: from elasticity distribution to ultrasound image sequence.

In this paper, the forward problem serves two purposes. First it is used to provide the simulated data that are required for this study; second it serves as part of inverse problem solving, by providing tissue displacement data which are to be compared against observed tissue motion.

Tissue matrix displacement model

For the purpose of describing its deformation in response to an external force, a tissue can be modeled as an isotropic continuous elastic medium that constitutes a matrix for embedded scatterers. For this case the tensorial stress/strain relationship is given by (Saada 1989):

$$\sigma_{ij} = 2\mu e_{ij} + \lambda \delta_{ij} e_{nn} \tag{1}$$

where λ, μ are the Lamé's constants, e_{ij} is one component of the strain tensor, σ_{ij} is one component of the stress tensor and δ_{ij} is the Kronecker delta, defined as:

$$\delta_{ij} = \begin{cases} 1 & \text{if } i = j \\ 0 & \text{if } i \neq j \end{cases} \tag{2}$$

Under static equilibrium we have:

$$\frac{\partial \sigma_{ji}}{\partial x_j} + F_i = 0 \tag{3}$$

where F_i are the body forces which could be associated with gravity; these are neglected assuming the density is constant. The relationship between the strain tensor and the displacement tensor (u_i) is given by:

$$e_{ij} = \frac{1}{2}\left(\frac{\partial u_i}{\partial x_j} + \frac{\partial u_j}{\partial x_i}\right) \tag{4}$$

Substituting equation 1 and 4 in equation 3 we obtain for a homogeneous medium:

$$\mu \nabla^2 u_i + (\lambda + \mu)\frac{\partial e_{nn}}{\partial x_i} + F_i = 0 \tag{5}$$

where ∇^2 is the Laplace operator. For an inhomogeneous medium this equation is more complex. A simple form can however be found under the assumption of a plane elasticity problem for which out of plane motion is neglected. In that case the elasticity equations reduce to (Reddy, 1984):

$$\begin{cases} -\dfrac{\partial}{\partial x_1}\left(c_{11}\dfrac{\partial u_1}{\partial x_1} + c_{12}\dfrac{\partial u_2}{\partial x_2}\right) - c_{33}\dfrac{\partial}{\partial x_2}\left(\dfrac{\partial u_1}{\partial x_2} + \dfrac{\partial u_2}{\partial x_1}\right) = f_1 \\[2ex] -c_{33}\dfrac{\partial}{\partial x_1}\left(\dfrac{\partial u_1}{\partial x_2} + \dfrac{\partial u_2}{\partial x_1}\right) - \dfrac{\partial}{\partial x_2}\left(c_{12}\dfrac{\partial u_1}{\partial x_1} + c_{22}\dfrac{\partial u_2}{\partial x_2}\right) = f_2 \end{cases} \tag{6}$$

where $c_{ij} = hC_{ij}$, in which h is the tissue transverse dimension and C_{ij} are given by:

$$C_{11} = C_{22} = \frac{E(1-v)}{(1+v)(1-2v)}; \quad C_{12} = C_{21} = \frac{Ev}{(1+v)(1-2v)} = \lambda; \quad C_{33} = \frac{E}{2(1+v)} = \mu \tag{7}$$

and E and v being respectively the Young modulus and the Poisson's ratio. In theory, this model is not suitable for incompressible material (Poisson's ratio=0.5) since C_{11}, C_{22} and C_{12} and C_{21} become infinite. In practice, Poisson's ratio is set below but close to 0.5, relaxing the soft tissue incompressibility condition and thus stabilizing the solution.

In this paper the above partial differential equations are solved for a given set of boundary conditions using a finite elements method (Reddy, 1984). This leads to the displacement field used to deform the acoustic impedance distribution of the tissue.

Image formation model

The image formation model we used for our computer simulation is that of a linear and space invariant system characterized by an image impulse response or a point spread function (PSF) (Meunier, 1993). In this model the RF image $i(x,y)$ is given by a convolutional integral $\otimes$ as:

$$i(x, y) = h(x, y) \otimes z(x, y) \qquad (8)$$

where $h(x,y)$ is the PSF modeled here by a modulated Gaussian function :

$$h(x, y) = \exp-\left(\frac{x^2}{2\sigma_x^2} + \frac{y^2}{2\sigma_y^2}\right)\cos 2\pi f_0 y \qquad (9)$$

where σ_x is a lateral beam-width parameter, σ_y is a pulse duration parameter; and f_0 is the spatial frequency (cycles/m) which is proportional to the transducer frequency f_{tr} ($f_0=2f_{tr}/c$, c is the ultrasound velocity). $z(x,y)$ is the tissue acoustic impedance distribution modeled here by a Gaussian random field.

INVERSE PROBLEM MODELS FOR ELASTOGRAPHY

Figure 3 identifies the functions currently involved in establishing the elasticity distribution of a tissue using pre- and post-compression echographic signals. First, the axial displacement field is estimated using the RF images before and after tissue compression. In conventional elastography the axial gradient of the estimated displacement field (i.e. the axial strain) is used as a relative measure of the tissue elasticity distribution assuming the stress field is constant. Due to boundary conditions this is not the case even when the medium is homogeneous. Hence in practice, effects of stress field decay with distance from boundaries would be corrected by computing axial strain/stress ratio, assuming the stress is equal to that of a homogeneous medium (Ophir, 1991; Ponnekanti, 1992; Ponnekanti, 1994). This assumption is however less valid near an inclusion where stress concentration effects are known to occur.

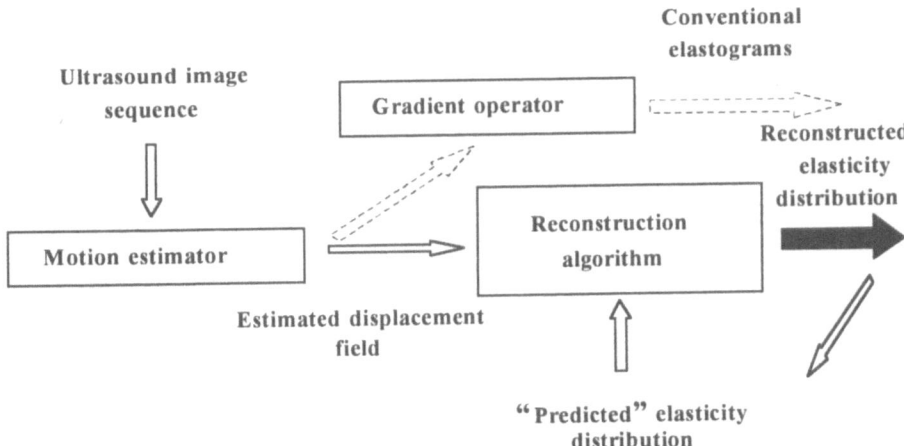

Figure 3. The inverse problem in elastography: from ultrasound image sequence to elasticity distribution.

In this work we propose to use a reconstruction algorithm that computes the elasticity distribution whose displacement field approaches the estimated displacement field. This reconstruction method is based on a linear perturbation method and is discussed in the next section.

Solving the inverse problem

Inverse problem solving is a very active field of research in electromagnetic, optics, or geophysics (Ghosh Roy, 1991). In the biomedical field it has been extensively studied in bioelectricity to determine the potential distribution on the surface of the heart from a limited number of peripheral potential measurements. But it is relatively new in the field of continuum mechanics (Maniatty, 1989; Gao, 1992), and it remains untackled in the field of biomechanics to which elastography belongs, at least for some of its fundamental aspects.

As an initial condition in our inverse problem solving, we assume a homogeneous tissue with an elastic distribution E subjected to a compression; this compression produces a displacement field $\vec{U} = (u,v)$. Now for a tissue that contains an elastic inhomogeneity ΔE the compression will result in a displacement field $\vec{U} + \Delta\vec{U}$. The difference between these two displacement fields ($\Delta\vec{U}$) is labeled the *displacement field perturbation*. Now if the elastic inhomogeneity is small, in both amplitude and size, we may assume a linear relationship between the displacement field perturbation and the elasticity perturbation (inhomogeneity) distribution. For an elasticity distribution sampled at N points and a displacement field sampled at M points, this is given by:

$$\Delta U = S\Delta E + n \tag{10}$$

where ΔU, the samples of the displacement field perturbation, is an Mx1 vector, S (to be described later) is an MxN matrix, ΔE, the samples of the elasticity inhomogeneity distribution, is an Nx1 vector and n, a noise term, is an Mx1 vector. Thus, equation 10 defines a linear system of M equations of N unknowns, these being the samples of the elasticity perturbation distribution ΔE. ΔU is the set of the data to be used for the elasticity reconstruction and is defined as:

$$\Delta U = U_i - U_h \tag{11}$$

where U_h is the displacement field sample vector for a homogeneous medium subjected to the problem's boundary conditions, and U_i is the displacement field sample vector for the presumed inhomogeneous medium. Notice here that the set of data ΔU used for the reconstruction could consist of both the axial and the lateral displacement field components or only one of them. In this work, only the axial component of the displacement field will be used for the reconstruction.

In equation 10, the matrix S is called the sensitivity matrix. Each column of this matrix is analogous to a system impulse response (i.e. samples of a Green's function). In this system the input pulse is a local small elastic inhomogeneity and the output response is the displacement field perturbation vector. Practically, this sensitivity matrix is constructed column-wise using the forward problem model for a given set of boundary conditions and a single inhomogeneity.

The classical least mean squared solution for the linear system given by equation 10, is the one that minimizes the squared error between the estimated displacement field perturbation and theoretical displacement field perturbation vectors:

$$\Delta\widetilde{E} = \arg\ \min_{\Delta E}\left\{\|\Delta U - S\Delta E\|^2\right\} \tag{12}$$

The solution which minimizes the least-squared error between the observed displacement perturbations (ΔU) and the forward model ($S\Delta E$), is given by:

$$\Delta \widetilde{E} = \left(S^T S\right)^{-1} S^T \Delta U \tag{13}$$

where superscript T stands for matrix transpose. When $S^T S$ is ill-conditioned the solution is unreliable; for conditioning a regularization scheme is used, and the solution may take the form:

$$\Delta \widetilde{E} = \left(S^T S + \lambda R\right)^{-1} S^T \Delta U \tag{14}$$

where R is a symmetric positive definite matrix and λ is a regularization parameter. Defining matrix R and the regularization parameter λ that best suit a given problem remains a complex matter, and there are many ways to do so as discussed in Galatsanos (Galatsanos, 1992). Generally, this would stem from the *a priori* knowledge about the solution sought and from the characteristics of the noise in the measurements (Titterington, 1985; Demoment, 1989; Nashed, 1981).

Simulation example

In this section, we present a typical simulation example illustrating the proposed reconstruction algorithm. Figure 4a shows the finite element mesh used to solve the 2-D elasticity equation. Figure 4b is a gray level image of the elasticity distribution for an inhomogeneous medium with a hard inclusion. This inclusion is 2.5 times harder than its surrounding.

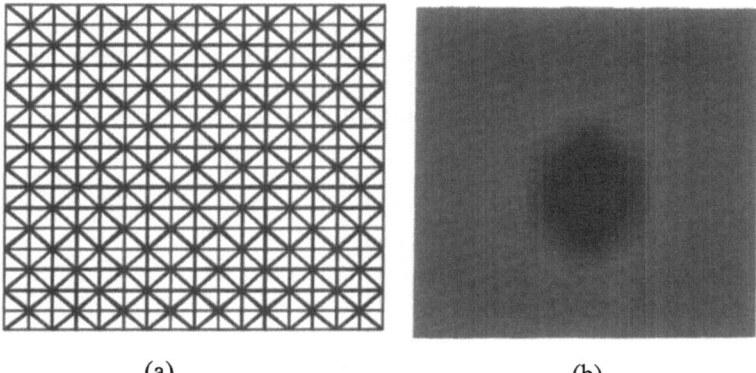

(a)   (b)

Figure 4. Finite element mesh (a) and elasticity distribution (b) used for simulation. The tissue elasticity distribution contains a hard inclusion in the center. The inclusion is 2.5 times harder than its surrounding area.

Figure 5 shows three contour plots of the axial displacement field. The first axial displacement field (Fig. 5a) is for a homogeneous medium and 1% uniform compression. The second axial displacement field (Fig. 5b) is for the inhomogeneous medium of elasticity distribution shown in figure 4b. The last displacement field is also for the inhomogeneous medium but now computed from RF correlation (i.e. using the simulated image formation). Figure 5b and 5c clearly show the perturbation of the axial displacement field due to the hard inclusion. In figure 5c the axial displacement field computed from RF data appears as wrinkled curves indicating that the estimate is quite noisy. This displacement noise originates from speckle decorrelation and as discussed in Kallel (1993), it is related to the imaging system characteristics (σ_x, σ_y and f_0 in equation 9), to the local tissue strain and to the signal processing method used for strain estimation (correlation apodization window, window overlap area). It cannot be ignored in reconstructing the elasticity distribution, and would in fact be taken into account by the regularization matrix we have seen before.

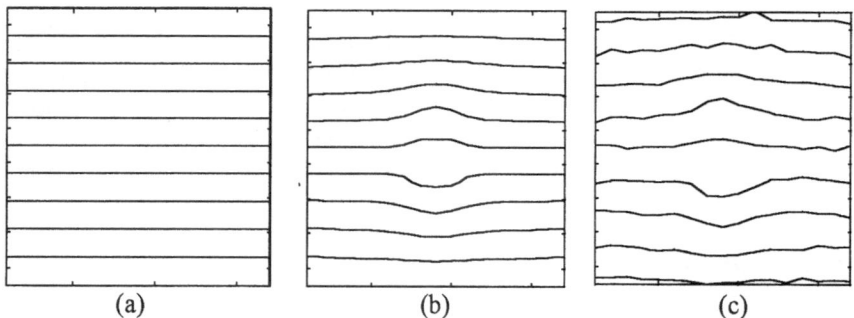

(a)   (b)   (c)

Figure 5. Contour plots of the axial displacement field. (a) Axial displacement field for a homogeneous medium. (b) Axial displacement field for an inhomogeneous medium. (c) Axial displacement field computed from simulated RF data of the inhomogeneous tissue before and after compression.

Figure 6a shows an ideal conventional elastogram; it is the axial derivative of the axial displacement field shown in figure 5b. Figure 6b shows the corresponding axial strain field estimated from the simulated RF data (axial derivative of the axial displacement field of figure 5c). The elastogram now appears very noisy, much more than what could be anticipated from the axial displacement field of figure 5c; this results from the fact that strain estimates are provided here through a derivative operation which therefore enhances the high frequency components of the noise. This effect could be reduced by pre-processing the RF signals prior to correlation (Céspedes; 1993b) but it would still remain quite significant.

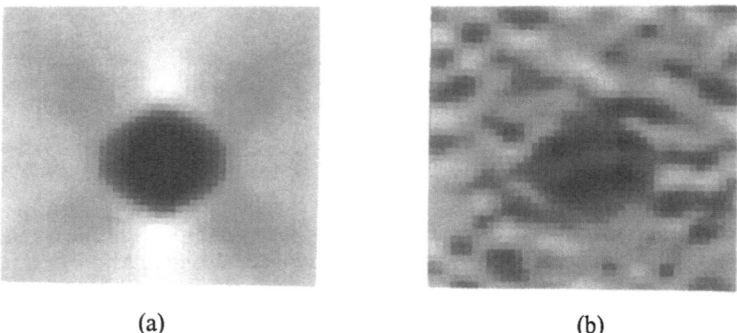

(a)   (b)

Figure 6. Current elastogram. (a) Ideal conventional elastogram. (b) Elastogram estimated from RF correlation.

Figure 7a and 7b show the reconstructed elasticity distributions corresponding to the data of figure 5b and 5c respectively; they were obtained using equation 14 where the identity matrix was chosen for matrix R and λ was set heuristically. In both elasticity images, the maximum inclusion/background elasticity ratio was found to be 2.1, which compares well to the theoretical value (2.5) used in the simulation. However in figure 7a, stress concentration effects (seen here as bright area below and above the inclusion) are still present but are in fact smaller than those of the conventional elastogram. In figure 7b, the elasticity image is less noisy then its elastography counterparts (figure 6b) and therefore the inclusion appears to have a better contrast.

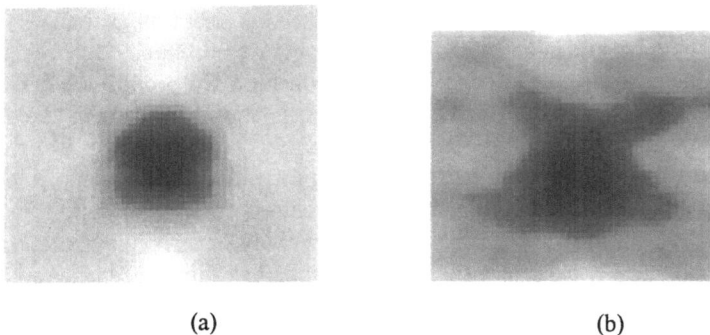

(a)   (b)

Figure 7. Reconstructed elasticity distribution using the perturbation method. (a) Reconstructed elasticity from the theoretical displacement field. (b) Reconstructed elasticity from estimated displacement field.

Application to clinical data and discussion

We have also processed the clinical data obtained from sonographic examinations conducted at the University of Texas Medical School on a volunteer breast cancer patient following the procedure described by Céspedes et al. (Céspedes, 1993a). Preliminary results of our elasticity reconstruction are shown in figure 8c.

(a)   (b)   (c)

Figure 8. Clinical example of elastogram and reconstructed elasticity compared to sonogram. The region of interest is 40x60 mm, and the compression 0.7 mm. (a) Sonogram of a breast section. (b) Elastogram of the breast section. (c) Reconstructed elasticity distribution of the breast section obtained using our perturbation method. Contour lines are shown to delineate elastic inhomogeneities

On both the elastogram and the elasticity image, the tumor shows as a soft region, surrounded by three hard "nodules". However, these seems delineated differently in the elasticity image; in particular the soft (white) region appears more circular; in addition the hardest nodule has a higher contrast in the elasticity image.

There are several factors that affects the quality and the appearance of our elasticity image. As with elastography, there is obviously the quality of the displacement estimates (noise, resolution, sample rate). In addition however there is also the regularization scheme used (matrix R and parameter λ), which in turn need to be specified properly. Here we have used a very simple heuristic approach, setting R to unity, and adjusting λ so the image "looks" correct. This is perhaps acceptable as a starting point in this research, but a more rigorous approach is needed for reliable quantitative elasticity imaging. As indicated above, this should stem from the characteristics of the displacement noise (distribution, variance and auto correlation); these could be studied using the forward model we presented.

The forward problem is also an integral part of the method we propose to determine the elasticity distribution: it presently serves to establish the displacement field for the homogenous medium and to determine its perturbation in the presence of isolated elastic inhomogeneities (i.e. the sensitivity matrix). The way the forward problem is implemented , i.e. the finite element method used, the geometry of the elements, their distribution and size, all affects the accuracy of the displacement field/sensitivity matrix computation. As a consequence, there are also "noise" terms in these, which in turns influences the regularization scheme and the elasticity image.

Finally, one may question the validity of the linear hypothesis in the perturbation model itself. In this model we implicitly assume a principle of superposition: the displacement perturbation produced by two adjacent inhomogeneity elements would simply be the sum of the perturbations produced by the same, but isolated elements (this is similar to a Born approximation). Indeed this would be true for a very light inhomogeneity, which is probably not the case here since we have simulated an inhomogeneous/homogeneous elasticity ratio of 2.5. This would explain that in figure 7a, the effects of stress field concentration partly remain in the elasticity reconstruction as soft area above and below the hard inclusion. Interestingly, this effects would also occur if the inclusion were soft: hard region artifacts would also appear above and below the inclusion. Obviously, this problem needs now to be solved. The approach we are now studying consists in re-iterating the perturbation model, using as the reference elasticity distribution not that of the homogenous medium, but rather a distribution that has been updated using the initial perturbation model.

ACKNOWLEGEMENTS

This work was supported by the National Science Engineering Research Council, the Quebec Ministry of Education, and the Fond de Recherche de l'Institut de Cardiologie de Montréal. The work performed at the University of Texas Medical School at Houston was supported by NIH grants CA38515 and CA60520, and by a grant from Diasonics Ultrasound, Milpitas, California.

REFERENCES

Bertrand, M., Meunier, J. Doucet, M. and Ferland, G., Ultrasonic biomechanical strain gauge based on speckle tracking, 1989 IEEE Ultras. Symp. Proc., pp. 859-863.

Céspedes, I. Ophir, J., Ponnekanti, H., and Maklad, N.F., Elastography: elasticity imaging using ultrasound with application to muscle and breast *in vivo*, Ultrasonic Imaging, Vol. 15, pp. 73-88, 1993a.

Céspedes, I. and Ophir, J., Reduction of image noise in elastography, Ultrasonic Imaging, Vol. 15, pp. 89-102, 1993b.

Demoment, G., Image reconstruction and restoration: Overview of common estimation structures and problems, IEEE Trans. Acoust., Speech, Signal Processing, Vol. 37, pp. 2024-2036, 1989.

Galatsanos, S. and Katsaggelos, A., Methods for choosing the regularization parameter and estimating the noise variance in image restoration and their relation, IEEE Trans. Image Processing, Vol. 1, pp. 322-336, 1992.

Gao, Z., Mura, T., Non-elastic strains in solids: an inverse characterization from measured boundary data, Int. Journal Eng. Sci., Vol. 30 (1), pp. 55-68, 1992.

Ghosh Roy, D.N., " Methods of Inverse Problems in Physics, " CRC press, 1986.

Kallel, F. and Bertrand, M., A note on strain estimation using correlation techniques, 1993 IEEE Ultras. Symp. Proc., pp. 883-887.

Krouskop, T.A., Dougherty, D.R. and Levinson, S.F., A pulsed doppler ultrasonic system for making non-invasive measurements of the mechanical properties of soft tissues , J. Rehabil. Res. and Dev., Vol. 24, pp. 1-8, 1987.

Lerner, R.M., Huang, S.R. and Parker, K.J., Sono-elasticity: medical elasticity images derived from ultrasound signals in mechanically vibrated targets, Acoustical Imaging, Vol. 16, pp. 317-327, 1988.

Maniatty, A. Zabras, N., Stelson, K., Finite element analysis of some inverse elasticity problems, Journal of Engineering Mechanics, Vol. 115 (6), pp. 1303-1317, 1989.

Meunier, J., Analyse dynamique des textures d'échographies bidimensionnelles du myocarde, Ph.D dissertation, Ecole Polytechnique Montreal, 1989.

Meunier, J., Bertrand, M., Echographic image mean gray level change with tissue dynamics: a system-based model study, Submitted to IEEE Trans. on Med. Imaging, (1993).

Nashed, M. Z., Operator-theoretic and computational approaches to ill-posed problems with applications to antenna theory, IEEE Trans. Antennas Propag., Vol. AP-29, pp. 220-231, 1981.

O'Donnell M., Yemelianov, S.Y., Skovoroda, A.R., Lubinski, M.A,, Weitzel, W.F. and Wiggins, R.C., Quantitative elasticity imaging, 1993 IEEE Ultras. Symp. Proc., pp. 893-903.

Ophir, J., Céspedes, I., Ponnekanti, H, Yazdi, Y., Li, X., Elastography: a quantitative method for imaging the elasticity of biological tissues, Ultrasonic Imaging, Vol. 14, pp. 111-134, 1991.

Ponnekanti, H., Ophir, J. and Céspedes, I., Axial stress between coaxial compressors in elastography: an analytical model, Ultras. Med. and Biol., Vol. 18 (8), pp. 667-673, 1992.

Ponnekanti, H., Ophir, J. and Céspedes, I., Ultrasonic imaging of the stress distribution in elastic media due to an external compressor, Ultras. Med. and Biol., Vol. 20 (1), pp. 27-33, 1994.

Reddy, J.N., "An Introduction to the Finite Element Method," McGraw-Hill, Inc., 1984.

Saada, S., "Elasticity, Theory and Applications, " Pergamon Press, New York, 1989.

Shehada, R.E.N., Cobbold, R.S.C and Bascom, P.A.J, Ultrasound methods for investigating the non-Newtonian characteristics of whole blood, IEEE Trans. on UFFC, Vol. 41 (1), pp. 96-104, 1994.

Titterington, D., General Structures of regularization procedures in image reconstruction, Astron. Astrophys., Vol. 144, pp. 381-387, 1985.

Yamakoshi Y., Sato, T., Ultrasonic imaging of internal vibration of soft tissue under forced vibration, IEEE Trans. on UFFC, Vol. 17 (2), pp. 1371-1375, 1990.

A NOVEL PULSE-ECHO ATTENUATION IMAGING TECHNIQUE

Sidney Leeman[1], Andrew J. Healey[1], and Leonard Ferrari[2]

[1]Department of Medical Engineering and Physics
King's College School of Medicine and Dentistry
London SE22 8PT, U.K.
[2]Department of Electrical and Computer Engineering
University of California Irvine
Irvine, CA 92717

INTRODUCTION

Frequency-dependent attenuation is one of the most marked features of wide-band ultrasound pulse propagation in human soft tissues. Specification of the functional dependence of the attenuation on frequency (in the diagnostic frequency range) is, to a large degree, a matter of choice — the most favoured being a simple power law, with non-integral index. In that case. there are two parameters which characterise the attenuation, and the values of both of these (or their progressive changes with disease) are widely regarded as conveying some information about the pathological state of the tissue.

In the clinic, virtually all ultrasound information about a patient is acquired via pulse-echo scanning, and only very imprecise knowledge of the attenuation can be conveyed by the displayed image. On the other hand, there has always been a strong hope that digital processing of the backscattered rf signal should lead to good quantitative assessment of the attenuation. Given the bandwidth of the pulses employed, and the techniques which have been developed, it turns out that it is the attenuation gradient ($d\alpha/df$) which is the more robust parameter. In practice, however, averaging over large regions of interest appears to be mandatory, if reasonably consistent, and hopefully accurate, values are to be obtained.

Another, perhaps more appealing, approach is to attempt to image the attenuation, or some feature derived from it, rather than to obtain a single, global, average value over a large region of interest. Given the posited value of the attenuation for tissue characterisation, it is to be expected that such 'attenuation images' would convey a great deal of diagnostically valuable information. One approach towards attenuation imaging is provided by computerised ultrasound tomography, but its demand for a very large acoustic window and a wide range of transmission pathways, suggests that it has limited clinical potential. A more pragmatic solution would be to seek a method for pulse-echo attenuation imaging: only a modest resolution need be the aim, since the lack of fine detail should, in principle at least, be more than offset by the enhanced diagnostic value.

PULSE-ECHO ATTENUATION IMAGING

Methods for estimating the attenuation from backscattered rf signals generally proceed from a dissection of the echo sequence into a number of contiguous or overlapping data segments. The rf A-line data are assumed to be piece-wise stationary, with a convolutional structure holding within each segment. Formally, this means that the amplitude spectrum, S, of the segment that is located at distance x, from the transducer, can be expressed as:

$$S(f;x) = P(f;x) \bullet T(f;x) \tag{1}$$

where f denotes frequency, P denotes the amplitude spectrum of the propagating pulse, and T is the amplitude spectrum of the tissue (back-) scattering function. It is further assumed that the pulse in the segment at distance x is an appropriately attenuation-modified form of the pulse in some reference segment ($x_0 < x$). It is convenient to make the usual assumption of linear frequency dependence for the attenuation coefficient, viz.:

$$\alpha(f;x) = \alpha_0(x)f \tag{2}$$

where $\alpha(f;x)$ expresses the frequency dependence of the attenuation, at the location x in the inhomogeneous medium, and $\alpha_0(x)$ is a location-dependent, but frequency-independent, function, which represents the local value of the attenuation gradient.

The amplitude spectrum of the pulse at some location, x, in the attenuating medium may now be expressed as

$$P(f;x) = P\left(f;x_0\right) \bullet exp\left\{-\int_{x_0}^{x} \alpha_0(x')dx'.f\right\} \tag{3}$$

Some algebraic manipulation shows that the gradient of the mean frequency of the pulse spectrum is given by

$$d\bar{f}/dx \equiv \tfrac{d}{dx}\int fP^2(f;x)\,df = -2\,\alpha_0(x)\int f^2.P^2(f;x)\,df \equiv -2\alpha_0(x).\overline{f^2} \tag{4}$$

Thus, provided that $\bar{f}$ and $\overline{f^2}$ can be estimated from the pulse-echo data, the variations in the local attenuation slope may be depicted as a function of depth (i.e., 'imaged'). It is clear that such an approach towards attenuation slope imaging is entirely equivalent to the problem of pulse estimation from back-scattered echo data.

In practice, entities such as $\bar{f}$ have been estimated by calculating the mean frequency of a short data segment, S(f;x). The estimates prove to be very noisy, because of the corrupting influence of the tissue scattering function, and a great deal of averaging, over relatively large tissue regions, may be required in order to arrive at convergent results. Such an approach is valid only on the assumption that the underlying statistics of the scattering function are essentially constant over the whole region of interest. It is therefore hardly surprising that attenuation slope imaging has not, up to now, been considered a realistic option. A number of techniques exist which attempt to recognise those segments which experience a severe noise distortion, and once identified, distorted segments may then be discarded. One approach examines the singular value decomposition of the discrete Wigner distribution in order to assess the extent of interference effects [Marinovic and Smith, 1985], and others have been based on the behaviour of instantaneous frequency [Hoddinott *et al*, 1987].

The approach advocated here departs from the notion that the fluctuations in $\bar{f}$ may be dampened only by averaging, or discarded if they may be recognised: instead, it is proposed that the spectrum of a single data segment may be *corrected* to reduce the dominant source of error in the mean frequency estimate. In this way, the possibility of a new type of informative ultrasound pulse-echo image is suggested, albeit at a resolution that is lower than that achievable in B-mode imaging.

PULSE ESTIMATION FROM BACKSCATTERED DATA

Two methods, in particular, have been used to estimate the pulse spectrum from backscattered data: spectral smoothing and homomorphic filtering [Leeman *et al*, 1984]. Both techniques originate in the observation that the data spectrum indicated in Equ. 1 comprises two components of very different behaviour. The pulse component, P, is expected to be smooth, bandlimited, and single-peaked; while the tissue component, T, is expected, in general, to exhibit a relatively wideband, rapidly fluctuating, random character.

Spectral smoothing attempts to remove the influence of the tissue component by, essentially, 'smoothing' its fluctuations to its presumed constant (over the pulse frequency range, at least) mean value. This is achieved in one of two ways: either by averaging the amplitude spectra of a large number of equivalent data segments; or by first filtering, and then averaging, the spectra of a (hopefully) relatively small number of equivalent data segments. Homomorphic filtering transforms the data to a domain in which the pulse component is separated from the tissue component, and the former is isolated by an appropriate filter in the new domain: P is then estimated by an inverse transformation back to the Fourier domain. Another filtering procedure is the so-called 'spectral smoothing' technique, which attempts to isolate the pulse component by filtering the data segment's autocorrelation function. In practice both filtering methods require a certain amount of averaging to attain convergent results, and are hence unsuitable for attenuation imaging. The need for such averaging — presumably to remove some random component — serves to underline the inability of these techniques to isolate the desired pulse component completely from the influence of the tissue scattering structures. Moreover, the prospects for achieving the separation by conventional deconvolution techniques appear to be equally poor. If it is remembered that the backscattered signal is composed of many overlapping echoes, it becomes clear that the dominant source of 'noise' in pulse estimation stems from interference effects in the backscattered echo field. Indeed, computer simulations suggest that corruption in mean frequency may typically be of the order of tens of percent — an order of magnitude greater than the effect being measured!

Attenuation is not the only effect which modifies the propagating pulse spectrum. Diffraction effects are often considered to be a major source of artefact in pulse-echo attenuation estimates, particularly via short-time Fourier techniques. However, the demonstration of significant spectral modification in water-tank experiments does not necessarily imply that the same holds for backscattered signals from tissues: indeed, theoretical arguments suggest that diffraction artefacts (in relation to backscattering mean frequency estimates) may be somewhat reduced under such circumstances. Computer simulations and careful experiments with tissue phantoms have led to similar conclusions, even though some researchers have advocated the use of semi-empirical 'diffraction corrections'. Our own experiments have also tended to confirm that diffractive influences on mean frequency estimates are very small: consequently, there is no attempt to correct for them in the work reported below. It should also be borne in mind that that diffraction artefacts may be considerably reduced in water tank transmission experiments, where they are very noticeable, by resorting to appropriate field measurement and/or pulse shaping techniques. Even if diffraction artefacts were to be evident in the backscattered signals, it appears likely that their impact on mean frequency estimation is swamped by the corrupting influence of the interference effects noted above.

INTERFERENCE ARTEFACTS AND COMPLEX FREQUENCY ZEROS

The salient features of interference artefacts may be demonstrated in the context of a simple two-scatterer model. Consider the pulse-echo sequence generated by an incident pulse, with Fourier spectrum $P(f)$, incident on two scatterers, of amplitude 1 and r, located a distance z apart. In the Fourier domain, the pulse-echo sequence may be expressed as:

$$S(f) = P(f) + r.exp(i2zf).P(f) \qquad (5)$$

so that

$$|S(f)| = |1 + r.exp(i2zf)|.|P(f)| \qquad (6)$$

If the scatterers have equal strengths, r = 1, and Equ. 6 indicates the possibility that (depending on the precise values taken by z, r, and the bandwidth of the pulse spectrum) the echo spectrum may become zero at some particular, *real*, frequency value(s). If the appropriate conditions are met, the resultant deep gash(es) in the spectrum — generated by the destructive interference between the echoes from the scatterers — will, in general, severely compromise the hypothesis that the mean frequency of the propagating pulse may be adequately approximated by the mean frequency of the pulse-echo sequence.

In practice, with *in vivo* pulse-echo sequences, it is unlikely that data segments will have originated from only two scatterers, and the validity of the above model may be questioned. But if the two scatterers in the model are interpreted as "effective scatterers" [Healey and Leeman, 1994], then the above considerations will apply to the majority of data segments. In any case, the general features of the two-scatterer model remain valid for multi-scatterer data: the departures of the echo spectrum from the pulse spectrum may be ascribed to the effects of zeros in the analytic continuation, into the complex frequency domain, of the former. The largest deviations are caused by zeros which lie close to the (real) frequency band of the pulse, and these perturbations may be reduced by effecting a transformation which removes the offending zeros to more distant locations.

In practice the validity of the simple two scatterer model for *in-vivo* pulse-echo data sequences is in question, and will be applicable only to specific locations in the signal. Therefore a reliable method for selecting signal segments for which the model is a good approximation to the underlying data is essential to the success of the technique. If the two scatterers in the model are interpreted as "effective scatterers" [Healey and Leeman, 1994], then regions where the model applies may be unambiguously identified by locating maxima in the excursions of the signal's instantaneous frequency. The departures of the data spectrum from the pulse spectrum may then be ascribed to the effects of zeros in the analytic continuation (in the complex frequency domain) of the former. Large deviations in the parameter of interest, f̄, are caused by zeros which lie close to the (real) frequency axis, within the band of the pulse. These perturbations may then be reduced by relocating the complex zeros to more distant locations.

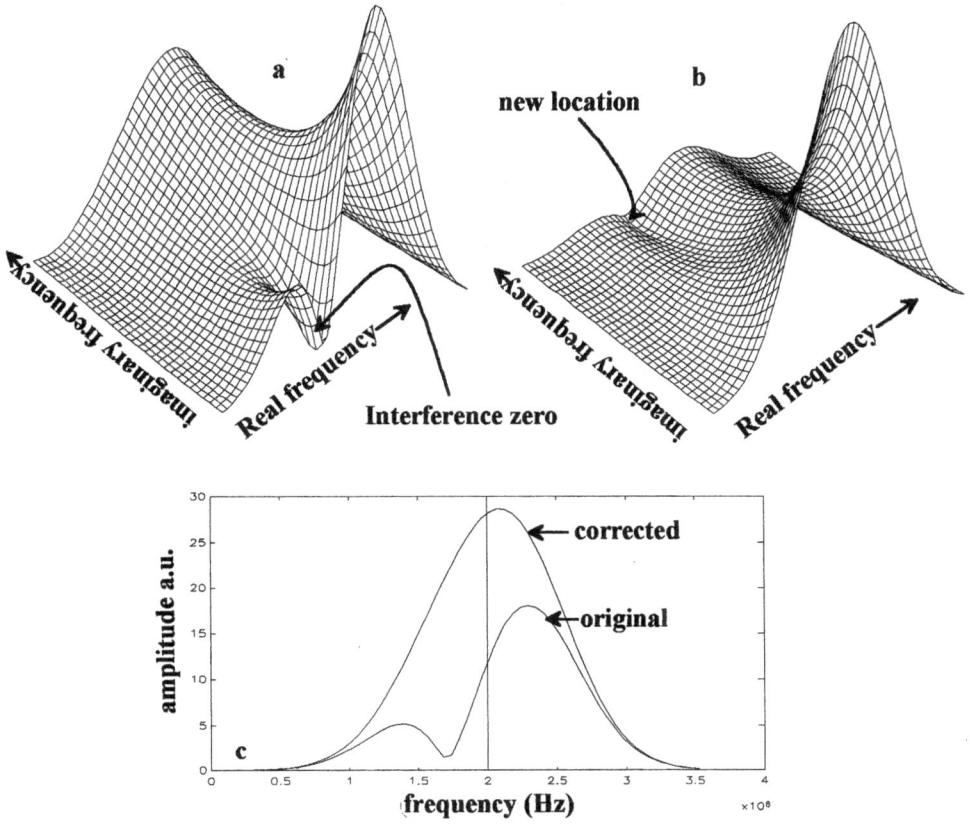

Figure 1. (a) Complex frequency representation of corrupted data segment. (b) The data in (a) after complex zero relocation. (c) Effect on the energy density spectra — pulse, original and corrected data mean frequencies are 2.00, 2.28, and 2.06 MHz (to 3 s.f.) respectively

Fig. 1 illustrates some of the above points. A gaussian pulse is convoluted with a multi-scatterer sequence. The analytic continuation of the amplitude spectrum of the echo data is shown, and the effects of the complex frequency zeros are evident. The transformed spectrum, after manipulation of only one zero, approximates much more closely to that of the pulse. In this particular example, the deviation of the mean frequency of the data spectrum from that of the pulse was reduced from 13.8% to 2.5% by the zero relocation. Note that the only *a priori* information being utilised is the 'smoothness' assumption associated with the imaging pulse and the wide-band nature of the scatterer sequence.

METHOD FOR MEAN FREQUENCY ESTIMATION

The rf A-line data are separated into non-overlapping, Hanning-windowed segments, approximately 1.5 pulse lengths in extent. The choice of such relatively short segments is dictated by the desire to maintain an acceptably high resolution for the resultant attenuation image. Potentially corrupted data segments are identified by large, spike-like, instantaneous frequency excursions: only those selected in this way are subjected to further processing to reduce the interference artefacts. Data segments are windowed symmetrically about the instantaneous frequency 'spike'. With *in vivo* signals, as well as those from realistic tissue phantoms, such selected segments are adequately dense to cover the entire pulse-echo sequence, without significant gaps.

In practice, it has not proved possible to unambiguously identify all the interference zeros, and only the 'dominant zero' (*viz*: the one with the strongest corrupting influence on the mean frequency) has been processed. Since this zero is such a marked feature of the signal, it may be identified as being the least sensitive to small changes in the original data. If the time domain data samples are denoted by $[a_k]$, and the locations of the complex zeros by $[f_k]$, then a 'stability measure', σ_i, for the ith zero may be defined as

$$\sigma_i \equiv \left| 1/\Delta f_i \right| \quad \text{with} \quad \Delta f_i = \sum_{k=1}^{N} \frac{\partial f_k}{\partial a_k} \Delta a_k \quad \text{and} \quad \frac{\partial f_k}{\partial a_k} = \frac{f_i^{N-k}}{\prod_{j=1, \neq i}^{N} (f_i - f_j)} \tag{7}$$

A first-order correction, as applied in the images shown below, corrects for the effect of only the dominant zero. Clearly, higher order corrections may be applied in an obvious way. Once the dominant zero has been identified, its influence is reduced by displacing it to a location remote to the real frequency axis. The translation is along the direction of the imaginary frequency axis (thereby maintaining the real component of the zero) for an optimal distance, whose magnitude has been decided on the basis of simulations with more than 15000 data segments. Once the zero translation has been carried out, $\bar{f}$ is calculated from the corrected spectrum in each segment.

ATTENUATION-WEIGHTED IMAGES

'Attenuation' (actually, $\frac{\partial \bar{f}}{\partial x}$ -) images are constructed from the corrected spectral mean frequencies. The data is uniformly resampled, utilising a cubic spline interpolation method, prior to calculating the discrete derivative, which is established by the central finite difference method. Some smoothing is applied to reduce the differentiation noise. Since $\bar{f^2}$ is expected to vary relatively slowly, it was thought reasonable to avoid the extra complexity of incorporating it, as per Equ. 4, into the attenuation imaging algorithm. The resultant attenuation maps are relatively free from the noise ('speckle') artefacts which intrude [Seggie, *et al.*, 1987] if uncorrected spectra are used.

When applied to *in vivo* data, the procedure produces images that are difficult to interpret, so it is thought appropriate to combine the attenuation map with the more familiar B-mode information in a single image. Consequently, the $\partial \bar{f}/\partial x$ - images are used modulate the corresponding (logarithmically compressed) B-mode display, as shown in Fig. 2 for a typical normal liver scan. In this case, an attenuation modulation factor that varied between 0.7 and 1.3 was chosen. Such an attenuation-weighted representation maintains the good resolution of the B-mode image, and appears to enhance the contrast, and even detectability, of some of the structures. Note, also, that the visual impact of B-mode 'shadowing' and 'bright-up' artefacts is considerably reduced. While considerable caution should be exercised, at this early stage, in ascribing real significance to the changes visible in the *in vivo* attenuation-weighted images, results are sufficiently encouraging to justify further development of the approach.

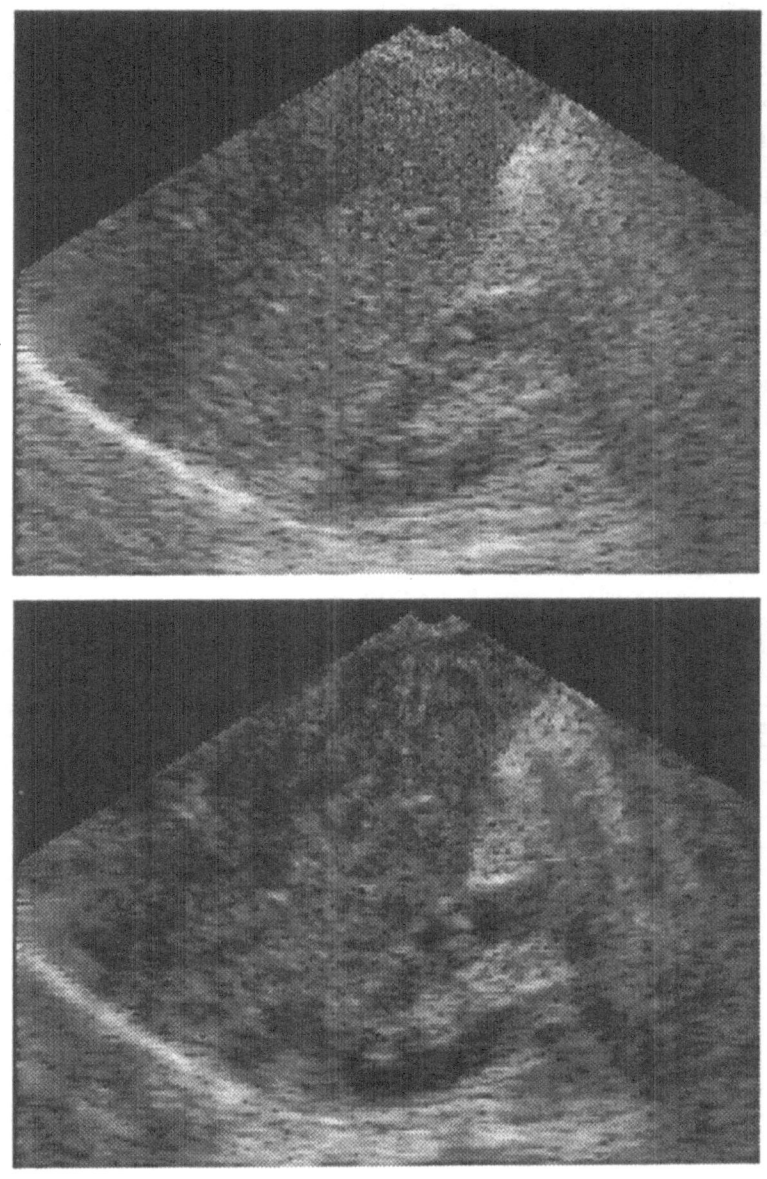

Figure 2. (Top) Original B-mode image, and (bottom) attenuation-weighted image.

CONCLUSIONS

It is apparent that interference artefacts constitute the dominant factor corrupting the spectral features of pulse-echo data from tissues. However, potentially corrupted (short) data segments may be identified by time-domain methods (instantaneous frequency 'spikes'), and the spectral distortions may be corrected by a novel type of filtering in the complex frequency domain. Only first-order corrections have been applied in this investigation, but the extension to higher order is straightforward, in principle. At present, the method has been used to produce only $\partial f/\partial x$ - images, at reasonable resolution.

When dealing with *in vivo* data, it is thought most appropriate to combine the $\partial \bar{f}/\partial x$ - information with the more familiar B-mode image, within a single display. Such attenuation-weighted B-mode images show sufficient promise for the method to be developed further.

ACKNOWLEDGEMENTS

The SERC and The Wellcome Trust are thanked for their support. *In vivo* data were supplied by J.A. Jensen of the Danish Technical University (Lyngby).

REFERENCES

Healey, A.J. and Leeman, S., 1994, Speckle definitions, See this volume.

Hoddinott, J.C., Seggie, D.A., Leeman, S. and Costa, E.T., 1987, Attenuation Mapping for imaging and characterisation, *in*: "Ultrasonics International 87", 37, Butterworths, Guildford.

Leeman, S., Ferrari, L.A., Jones, J.P., and Fink, M., 1984, Perspectives on attenuation estimation from pulse-echo signals, *IEEE Trans. Sonics and Ultrasonics*, 31/4:352.

Marinovic, N.M., and Smith, W.A., 1985, Suppression of noise to extract the intrinsic frequency variation from an ultrasonic echo, *in*: "Proc. 1985 IEEE Ultrasonics Symposium", IEEE, New York.

Seggie, D.A., Hoddinott, J.C., Leeman, S. and Costa, E.T., 1987, Mapping ultrasound pulse-echo non-stationarity, *Proc. S.P.I.E.*, 768:241.

DESIGN OF A TISSUE CHARACTERIZATION SYSTEM FOR DIAGNOSING CARDIAC REJECTION AFTER HEART TRANSPLANTATION BY MEANS OF TEXTURE ANALYSIS

E. Lieback, I. Hardouin, R. Koch, and R. Hetzer

German Heart Institute Berlin
13353 Berlin, Germany

INTRODUCTION

The aim of our study was to perform an objective evaluation of ultrasonic images in order to help physicians diagnose cardiac rejection. This diagnosis is difficult to make due to the possible lack of clinical symptoms. Ultrasonic imaging is based on reflection and diffusion of ultrasonic impulses through the target organ. The echo structure is highly related to the acoustical properties of insonified tissue. Because of disturbing factors such as noise, speckles, medium and frequency dependent attenuation, and multiple echoes, ultrasonic images are very different from optical ones. They are strongly textured and, therefore, can be described with textural features.

ULTRASONIC IMAGING OF TRANSPLANTED HEART

Selection of Heart Images

Using an ultrasonic imaging system, the heart can be viewed with very different angles. The visualisation angle is chosen according to the needs of the physician and the assumed disease. In the case of cardiac rejection, distinctive changes in acoustical properties are noted on images of the myocardium. For this reason a short axis view (fig. 1) is preferred in order to describe the myocardium with textural features.

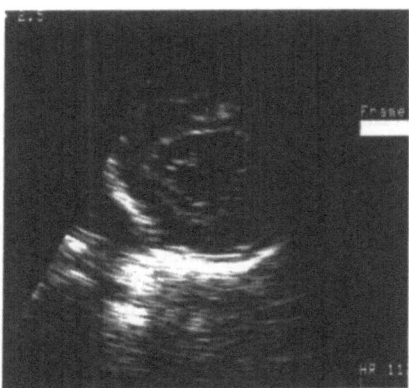

Figure 1. Short axis ultrasonic image of the heart. This viewing angle gives a clear view of myocardium. Changes in its acoustical properties are used for diagnosing cardiac rejection.

Selecting the Region Of Interest

It is well known that in ultrasonic imaging the signal–to–noise ratio has relatively low values which decrease even more with growing depth. Additionally, the geometric properties of the beam (during which time the heart is insonified by a mechanical sector scanner) imply that spatial resolution decreases significantly with depth. For these reasons, we prefer to select Region Of Interest in the upper part of the echocardiograms.

When selecting an R.O.I. in the myocardium, we avoid the borders where strong reflections normally occur. This is very important because cardiac rejection causes changes in heart muscle contractility which leads to stronger echoes. Simply summarised : bright regions in the R.O.I. are a sign of rejection (fig. 2b) while relative homogeneity in myocardial appearance (fig. 2a) indicate a well–accepted transplantation.

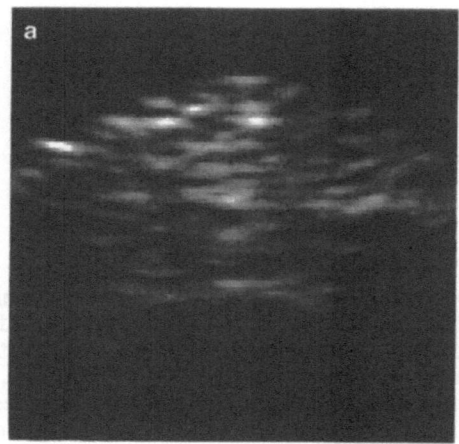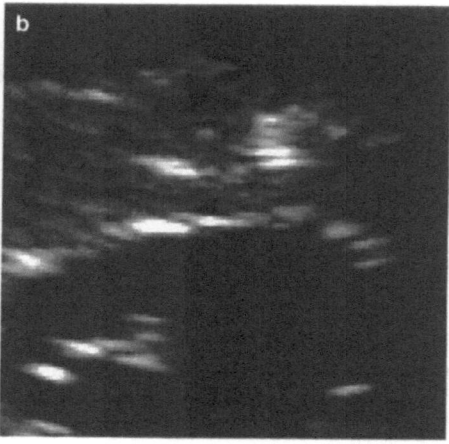

Figure 2. Regions of Interest set in echocardiograms.

a) Image of myocardium: No rejection.
The muscle looks relatively homogenous. Brighter regions correspond to myocardial borders.

b) Image of myocardium: Rejection.
This image shows brighter regions in the middle of the muscle.

TEXTURE ANALYSIS

What Is Texture ?

In the past, many definitions of texture were proposed without one becoming more accepted than the others. The lack of information about that which the human eye perceives while looking at texture can explain the difficulty in clearly defining texture. Texture is often described with qualitative words such as *fine, coarse, raw,* corresponding to the visual impression. Texture and visual perception are strongly correlated. We found a definition of texture, which satisfies us in Smolarz[1] :
A texture is an image region for which a window of minimal size can be defined, the visualisation of which leads to the same perception for every translation inside of this region.
or summarised :
The perception of a texture is translation invariant inside a region of a unique texture.

There are two dimensions by which texture can be described (Haralick[2]) : a structural one, at which *texel* (pattern) are defined and a statistical dimension, at which the spatial arrangement, dependence, and interaction between texels is described. The second dimension allows differentiateing between deterministic (analytically describable texel arrangement : often textures of manufactured products) and stochastic (described with random variables : natural texture as grass, wood etc.). In the case of our echocardiograms, the texel is reduced to the pixel and their spatial arrangement is randomly distributed, so that the texture has to be described as stochastic one.

Stochastic textures can be modelled (Gibbs or Markov random fields) or described using features. We chose the second alternative.

Possible Texture Features

Different properties of texture can be represented using features. They are, for example, the statistical distribution of gray levels in the image (histogram), their spatial distribution (co-occurrence and run-length matrices between others), or the spatial frequencies present in the texture (using a Fourier transformation).

Histogram analysis : A histogram is computed from the gray values existing in the image and its statistical properties are given by features like mean, standard deviation, and statistical moments of higher order.

Co-occurrence matrix : A co-occurrence matrix is related to a displacement vector. It is a square matrix from size N^2, N being the gray levels resolution. The term $C(i,j)$ in such a matrix corresponds to the number of times a pixel with gray level i was at the start point of the displacement vector while a pixel with gray level j was at the end point (for more precision see Haralick[3]). Here again, the matrix is described by its statistical moments (regularly fourteen).

Run-length matrix : A run-length is a succession of neighbouring pixels having the same gray level. A run-length matrix is defined in a certain direction and its size equals the gray level resolution by the maximal length measured. Galloway[4] proposed five features for the matrix description.

Fourier transform : A two–dimensional Fourier transform allows determining the spatial frequencies of the image and their directions. The transform itself is described using energy values in sectors (direction information) and rings (frequency information).

Features Selected for Recognition Of Cardiac Rejection

A preceding study (Lieback[5]) showed that 51 texture parameters were sufficient for making the distinction between "No rejection" and "Rejection". Those parameters are features of the histogram analysis, features of six different co-occurrence matrices (horizontal and vertical displacement of one, two and four pixels), features of the horizontal and vertical run-length matrix and energy features from the Fourier transform.

These 51 textural features are computed inside overlapping windows of size 25x25 pixels covering the set R.O.I. That means that an image is generally described by about six features vectors (51-dimensional).

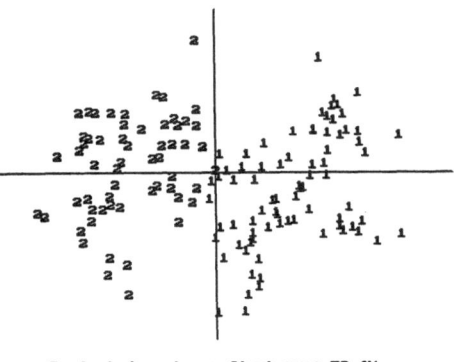

Projected variance first axe: 73.4%
Projected variance second axe: 7.8%

Figure 3. Representation of the clusters in the plane of the two first principal axes.
Points marked with "1" belongs to the class "No rejection" while points marked with "2" belongs to "Rejection". The two classes appear to be separated even if not distant.

A Principal Components Analysis allows the visualisation of the points cluster without loosing too much information while reducing the dimension (here from 51 to 2). On figure 3 we can see that the two classes can be distinguished from another although they are not very distant.

CLASSIFICATION AND DECISION METHODS

The domain of Statistical Pattern Recognition offers some classification methods which has been applied in our study. Each R.O.I. is now described by some points in a 51 dimensional space and should be recognised as a membership of one of the two cluster "Rejection"/"No Rejection". Two different classifiers were tested : a statistical and a neural.

Otherwise, the variations of norm of the feature vectors in this space can relate the evolution of the patients' state. This represents the third decision methods proposed.

Statistical Classifiers

Assumed that the feature vectors are normally distributed, a statistical decision rule allows the attribution of a new vector to one of several classes. Each class is assumed to have a multivariate Gaussian distribution, the parameter of which are estimated from a learning set. The latter was elaborated by a human trainer (supervised learning). A class is now summarised as a mean vector and a covariance matrix.

The probability that a vector belongs to a class is given by p_k:

$$p(v) = \frac{1}{\sqrt{(2\pi)^n \det(K_k)}} e^{-\frac{1}{2}(v-m_k)^T K_k^{-1}(v-m_k)} \tag{1}$$

m_k: mean vector of the kth class
K_k: covariance matrix of the kth class

The vector to be classified is affected to the class k for which the probability is the highest, if its value is greater than a given threshold. If assumed that the covariance matrices of all classes are the same, the class affectation corresponds to the minimal value of the Mahalanobis distance:

$$(v-m_k)^T K_k^{-1}(v-m_k) \tag{2}$$

This way, computing time can be reduced. However we did not take this simplifying assumption because of the restricted gain of time (computers are today efficient enough!).

Neural Networks Used as Classifier

Today, the trend in computer science is to produce more intelligent computers than faster ones, i.e. computers which can make decisions in given situations and are capable of learning. In other words: computers which imitate the functions of the human brain. The typical computer can calculate much faster than man but it is not able to learn. As Ritter[6] wrote:

It is remarkable that today computers can realise tasks which are very difficult for the humans but they miss quotidian tasks which humans very easily do.

In order to reach this topic, neural networks were developed. Such a network is composed from several neurones which are associated together so that they exchange information simultaneously. An artificial network works like a human one: if the input value greater than a threshold is then the output value becomes "1" (else "0"). Each neurone belongs to a layer in the network. The number of layers is not theoretically prescribed but generally two layers exists: the input layer, which receive the information from outside and the output layer with give information to outside. The number of layers between the latter is empirically chosen according to the experience of the user.

456

A neural network can learn if the fact that each information have a different weight in the decision is taken in account. This is done by weighting differently each link between the neurones. During the learning operation, these weights are iteratively updated until the desired output is obtained. The updating of the weights occurs according to a learning rule which depends on the type of the neural network.

The choice of the suitable type depends on the concrete application, as Schöneburg[7] remarked:

Concrete classifications of neural networks according to their suitability for given task are in fact not already known. Today assumptions like "type X (for example the Perceptron) is not suitable for task Z (solution of the XOR-problem)" exist. The question, which network for which problem can be answered only in exception cases.

Like a statistical classifier a neural one learn from examples but in contrary, no assumption is made concerning the distribution of the examples, what is an advantage of neural classifier.

For classification purpose, a back propagation network is commonly chosen. Such a network learns from its faults, which are used to update the weights (from this, its name). It can only be used if an a priori knowledge of the classes exists. For the classification of echocardiograms this knowledge corresponds to the diagnosis of each image of the learning set.

Multidimensional Distance in Features Space

The evolution of the norm of the features vector for a given patient can correspond to the evolution of his/her state, under the assumption that the acquisition conditions are always the same. This can be reached using our imaging system because particular, patient dependent values are stored in the machine and always called by a new examination of this patient.

At first each feature was normalized between its lowest and highest value in order to decrease the importance of its absolute value (features can have exponents between 10^{-5} and 10^{+5}!). Using this normalized vector components the Euclidean distance from the origin is then computed. The observation of the evolution of this distance over the time can allow some conclusion like "the patient is going to have a new rejection".

STUDY DESIGN AND RESULTS

Study Population and Evaluation Steps

For this study we analyzed 165 echocardiographic images from 18 heart transplant recipients. The study includes 53 images at the time of cardiac rejection.

The ECG-triggered end diastolic images were on-line digitized in the image processing system. In each image a R.O.I. was interactively set in which the 51 texture parameters were computed over almost six windows. This features were then classified using both classifier at the first time of this study. We have soon given up the statistical classification because the results were the same but the computing time was remarkably longer. In all cases the multidimensional distance was computed.

In summary the classification system consists on the acquisition of the ecnocardiograms, followed by a parametrization of the acoustical properties of myocardial tissue using texture features and then either a classification using a neural network or the observation by the physician of the evolution of this properties over the time.

Results Obtained using the Different Decision Methods

Bayesian Classifier : The classification of the two learning sets, consisting in almost 100 features vectors extracted from echocardiographic images with clinical proven state (for rejection as well as for no rejection), was an ideal separation of both classes : the vectors of both learning set were affected to 100% to the class they have to belong. The learning set were assembled independently in a previous study step. None of the 165 images meant in this study belongs to the learning sets. As we already remark, classification results of the test sets were the same as with the neural classifier and are summarized in figure 4.

Neural Network : The back propagation multilayer neural network consists on three layers : the input layer having 51 neurons (getting the information contained in the features vectors), the single hidden layer 60 neurons and the output layer two neurons corresponding to the two possible decisions. A few thousands iterations were sufficient for the network to have learned both classes with a very low mean square error (about 10^{-6}). Here again the vectors of both learning set were correctly affected to 100%. The study results (the classification of the 165 echocardiograms) are presented in figure 4. As we can see the classifier had a specificity of 84% and a sensitivity of 96%.

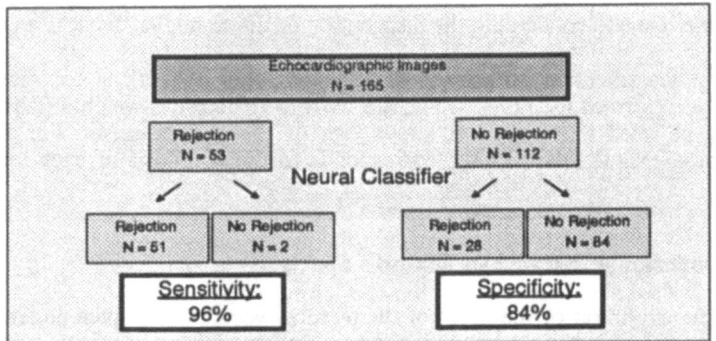

Figure 4. Study results.
This picture summarizes the classification results obtained with both classifier. Their performance is particularly satisfying because of the difficulty of diagnosing cardiac rejection.

Distance Evaluation : As a result of the distance evaluation we just present an example in figure 5, showing that maxima in the distance strongly correlate with cardiac rejection (even if the correlation is not perfect). This measure can be used to foresee a new cardiac rejection.

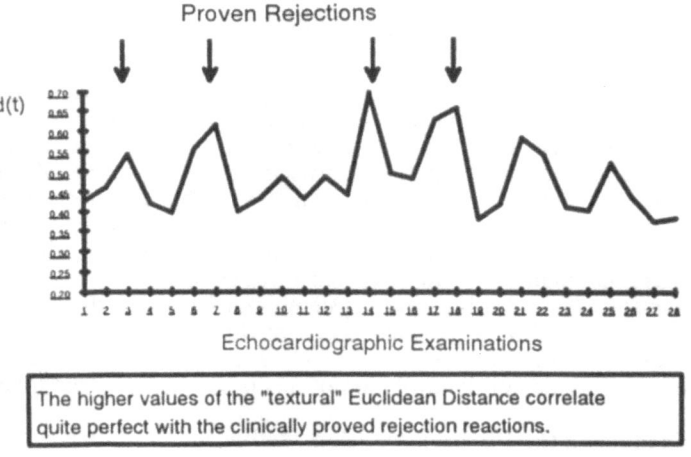

Figure 5. Evolution of the textural distance over the time for a patient of this study. As the arrow show, cardiac rejection and distance maxima are highly related.

CONCLUSION

In this study we have shown that changes in myocardial tissue in case of cardiac rejection can be described with texture features. This allows an objectivisation of the evaluation of echocardiographic images as far as the texture seen on it is then described with numerical values instead of visual impressions.

As far as statitical and neural classifiers give the same results, we prefer to use the second one because it works with a reduced computing time. In addition, for valid results the network does not need any assumptions concerning the distribution of the features vectors.

REFERENCES

1. A. Smolarz. Course Script CS19 at Université de Technologie de Compiègne (1987).
2. R. Haralick, Statistical and Structural Approaches to Texture, *Proceedings of the I.E.E.E.*, Vol. 67, n°5, 786:804 (1979).
3. R. Haralick, K. Shanmugan and Its'hack, Textural Features for Image Classification, *I.E.E.E. trans. on systems, man & cybernetics*, Vol. SMC-3, n°6 (1973).
4. M. Galloway, Texture Analysis Using Gray Level Run Lengths, *Computer Graphics and Image Processing*, n°4, 172:179 (1975).
5. E. Lieback. "Computer gestützte sonographische Gewebedifferenzierung", Springer Verlag Heidelberg (1992).
6. H. Ritter, T. Martinetz and K. Schulten. "Neuronale Netze: Eine Einführung in die Neuroinformatik selbstorganisierender Netzwerke", Addison-Wesley (1990).
7. E. Schöneburg, N. Hansen and A. Gawelczyk. "Neuronale Netzwerke. Einführung, Überblick und Anwendungsmöglichkeiten", Markt&Technik Verlag (1990).

ULTRASONIC RECONSTRUCTION OF
BRAIN PATHOLOGIC INCLUSIONS

E. L. Borodina[1], F. A. Kazulin[2], S. S. Sukhov[2],
A. I. Khil'ko[1], A. G. Zyganov[2]

[1]Institute of Applied Physics,
 46 Uljanov Str., 603600 Nizhny Novgorod, Russia
[2]Moskow, Ultramed, P.O.Box 117574

INTRODUCTION

The use of ultrasound for visualization in medical practice is based on its possibility to penetrate into the human body and to interact with tissue. Information on the body structure is coded in passed and scattered acoustical fields, and the aim of the visualization system is to interpret this information. These effects of reflection and refraction on media boundaries can be appreciable: in the frequency band $0.8 \ldots 15$ MHz used in ultrasonic diagnostics and therapy for tested inclusions the geometrical optics conditions are satisfied and reflection coefficients at right angles of incidence vary from -2 dB (bone/soft tissues or water) and -18 dB (crystalline lens/glassy body or liquid of the front chamber) to -50 dB (blood/cerebrum)[1].

The creation of visualization systems in ultrasonics in comparison with optical or roentgenological testing means a number of distinguishing features. The slowness of acoustical waves propagation through the body organs allows to construct image forming systems basing on echo-impulse methods. Difficulties in this way of image restoration are caused, firstly, by coherence of acoustic waves that produces significant interference effects. Secondly, up to now there is not the universal model of sound scattering on tissues. One of them, for example, represents a tissue as lots of randomly distributed point scatterers of definite density. In another model density and elasticity modula are concerned to be continuously varying within a volume. But all of them can not describe tissue characteristics at the back ultrasonic scattering. And, thirdly, it is especially significant for brain imaging that a multi-scattering on scull bones and a variation of a sound velocity in different parts of a cerebrum result in artifacts.

For visual presentation of measured data we employed two regimes[1,2]. In the simplest A-regime one-dimensional dependence of echo-signal amplitude versus time (A-echogram) was displayed. The most popular in medical practice format of image is B-regime providing two-dimensional (2-D) image corresponding the original distribution. This pattern can be realized by various ways but all of them use a lot of A-echograms

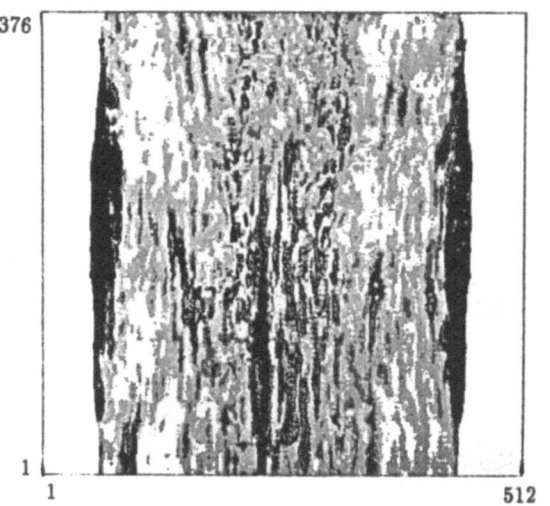

Figure 1. The initial image of human's brain

for modulation of a beam brightness on a display on which two signals of a fast sweep (along x−axis and y−axis) come simultaneously. The position of origin and the direction of line are determined by signal of origin and direction from the converter. 2-D pattern is acquired by translation of ultrasonic beam in (xy) plane. In C-regime a desired part of A-scan is fixed by means of gating corresponding to the position of the beam at fixed distance to transducer.

DATA ACQUISITION

Series of two A-echograms were obtained at frequency 1 MHz or 1.5 MHz using two acoustic sounders placed symmetrically at right and left sides of a patient's head. The transducers were manually translated around the head. The accuracy of localization and definition of geometrical dimensions by this way is 0.4 mm, visualization depth is about 200 mm. The resolution along a trace at 1 MHz is 3 mm (and 2 mm at 1.5 MHz). Received signals vary in a range 100 dB, the range of represented signal value is 40 dB.

After each step data were digitized on-line and stored on disk for subsequent image reconstruction. A single frame (or image) is usually stored as a matrix for which a pixel dimension corresponds the desired resolution (or some smaller). For example, for the depth of brain image restoration 200 mm discretization step is $d \sim 0.4$ mm. This resolution is enough for reconstruction of objects of character size $10^1 d \ldots 10^2 d$. It should be mentioned that data acquisition and keeping requires a bulky storage. Thus, the matrix 512×476 of 1−byte numbers takes up 0.244 MB of RAM. At the further stage of processing where operations with complex matrixes in the spectral region are supposed required storage is about 1.9 MB. Previous image processing (digitization and displaying) and a prior reconstruction were performed by *Kranioskop - 3*, a workstation for an encephaloscopy. Fig.1 shows the example of acquired B-image of patient's head *in vivo*.

DATA PROCESSING

The aim of the signal processing is an improvement of an image quality proceeding from definite criteria, such as decrease of a noise level or amplification of some fea-

462

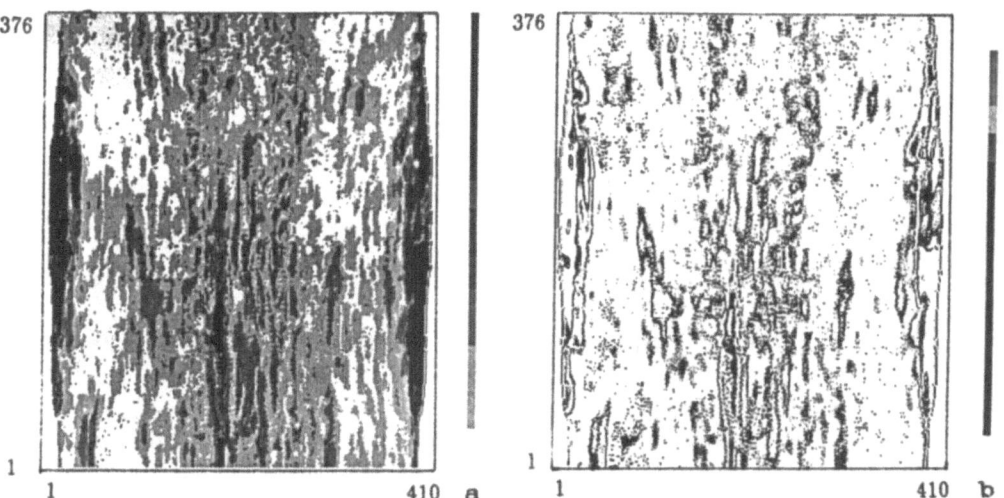

Figure 2. Images processed by smoothing (a) and outlining (b) filters.

tures, for example, object edges. The methods of smoothing and filtration are widely employed to solve the problem (earlier we studied possibilities of some other methods, hydrophysical methods among them[3]).

Images (and especially their fragments) can often be formed by indented brightness functions. When additional measurements are impossible a smoothing is usually carried out by the next ways. In step smoothing the range of arguments is divided into a several intervals and intensity value in each division is changed by the average in interval. Such image can be obtained at once by intensity value quantification into several levels.

The smoothed by moving summation image $\bar{I}(r)$ is obtained as a result of the convolution of initial image $I(r)$ and smoothing function $S(r)$, $r = (r_1, r_2 \ldots r_N)$ is quantified value of N−dimensional signal.

$$\bar{I}(r) = \sum_{r=-R}^{R} S(r) \times I(\rho - r)$$

The simplest and often used function is rectangular window

$$S(r) = \begin{cases} 1 & |r_i| \leq R_i, i = 1 \ldots N \\ 0 & |r_i| > R_i \end{cases}$$

Generally, the step smoothing can be interpreted as a filtration (as in smoothing by moving summation) and a further discretization.

The spatial smoothing by 9-element filter was carried out in the region of the coordinates by a convolution of image matrix elements and 8 weighted values of neighboring elements. This filter acts on each image element sequentially (Fig. 2a). (Filter dimension can be increased, but it requires the increase of calculation time.) If the strongly smoothed image is required, the filter elements are chosen very close and vice versa. 1-D filter can be employed also for time smoothing of a number of sequential frames. The average difference of filter elements determines the power of smoothing.

An outlining (i.e. object contours display) was done by insertion of negative values into filter matrix (Fig. 2b).

More complicated filtration algorithms can be realized in fourier frequency region by suppression or elimination of undesirable spectral components. Such operation in the region of spatial frequencies requires a simple multiplication of functions, whereas filtering in spatial region is fulfilled by the convolution procedure.

We employed usually used *Hann window*[1,4]

$$W(f) = \begin{cases} 0.5(1 + \cos(\pi|f|/f_0)) & |f| \leq f_0 \\ 0 & |f| > f_0 \end{cases}$$

and *Butterworth window* [Herman]

$$W(f) = (1 + (|f|/f_0)^{2p})^{-1}$$

where $W(f)$ is transmission characteristic, f_0, p are declaring values. At large frequency parameter f_0 the filtration yields contouring and emergence of HF noise. At low f_0 an image becomes smoothed.

Apparently, even for satisfactory spatial resolution of visualization system but for bad statistical characteristics of an image the smoothing is necessary, that in it's turn decreases the resolution.

Usually, initial signal contains noise. And the inverse filtration aiming to increase an image sharpness can decrease signal/noise relation. Apparently, an optimal correction level exists, when the sharpness is restored enough and a noise is not too significant. In this case the frequency characteristic $F_k(\xi_1, \xi_2)$ of an optimal reconstructing filter is determined by frequency characteristic $F(\xi_1, \xi_2)$ of corrected system

$$F_k(\xi_1, \xi_2) = \frac{F^*(\xi_1, \xi_2)}{|F(\xi_1, \xi_2)|^2 + \varepsilon^2}$$

where $*$ denotes conjugation, and ε^2 is equal a ratio of noise power fourier spectrum to power spectrum of ensemble of images.

The image sharpness can be improved also by means of the simple recoursive filter transforming elements of video-signal $I_{n,m}$ as follows

$$I'_{n,m} = I_{n,m} + \alpha \left(I_{n,m} - \frac{1}{(2N_1 + 1)(2M_1 + 1)} \sum_{\mu=-M_1}^{M_1} \sum_{\nu=-N1}^{N_1} I_{n+\nu, m+\mu} \right)$$

coefficient α and dimensions of region $(2N_1 + 1)(2M_1 + 1)$ are chosen on condition that the frequency characteristic of the above filter is approximated by correcting filter characteristic (we used 9-element filter $N_1 = M_1 = 1$ providing an enough accuracy).

Consider now an importance of magnitude and phase information on fourier image of the signal. A similarity between the signal and the result of its phase-only synthesis was obtained in crystalline structure fourier synthesis and in image processing by another investigators[5] on the base of coupling of an exact phase and different magnitude functions. In particular, if the signal is reconstructed only from phase information, the result contains many structure features of the original (see Fig.3).

Analogous experiments showed that neither magnitude-only synthesis (from spectrum magnitude and zero phase) nor combination of a magnitude and a phase averaged in ensemble produce the useful information.

Apparently, image reconstruction from phase distribution is preferable. From the point of view of image processing it was founded that even in the case of rough phase quantization (from 1 bit of phase information and constant magnitude) an acceptable image synthesis can be fulfilled [5]. The image restoration quality can be improved by combination of a phase and a magnitude averaged in ensemble of nearest values. Fig.4 shows the image processed by 17-element filter, the best result from those obtained by 5...33-element filters (in addition the image was equalized by the way described

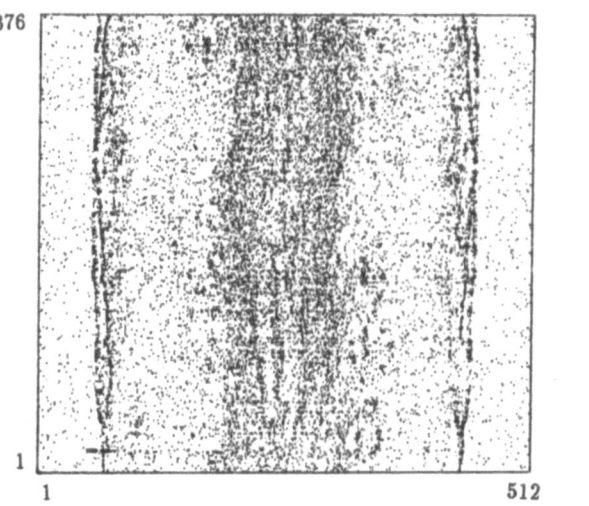

Figure 3. The result of composition of fourier image phase and constant magnitude.

in the next section). Thus, we obtained the defined images of a calloused body and a back horn of *ventriculus lateralis* (in the left cerebral hemisphere). And one can discern a thalamus among interference, it can be identified by its tapering upper outline and rounded lower outline.

In both cases (spectral phase and magnitude smoothing) we did not use the information on correlation properties of the image and its spectrum (approximate dimension of region containing unzero values (for example, components of spectrum magnitude) and a smoothness of the function in an area of this region).

Despite the operations in the region of spatial frequencies require a significant calculation time especially in the case of large data arrays, this processing is more rapid and effective than operations in the coordinate region. The latter processing is preferable when a numerical value of image element is connected in some way with values of neighbor elements.

DISPLAY AND TRANSFORMATION OF RESULTS

The ending purpose of data processing is the most effective display of image for selection of a diagnostic information. From this point of view a numerical representation is more preferable than analog one, because it allows to vary brightness gradations, select and transform different sub-levels. Fig.5 shows an example of transformation of full- gradation image to pick out levels of gray of the same order.

The often employed method of pattern quality increase is use of contrast colors for close brightness gradations, that yields the effect of contouring.

For better presentation of medium levels of gray we carried out additional processing. Fig.6 gives the initial histogram of the image shown at Fig. 3. White colors (above the right arrow) remained white, and black colors (below the left arrow) remained black, and medium gradations were amplified using smooth function.

Another way of transformation is intended for bright details separation. We left levels lower the threshold fixed and turned upper levels into white. Varying the threshold one can discern details of different intensity.

Notice once more, that a measured frame must be detailed, i.e. character objects

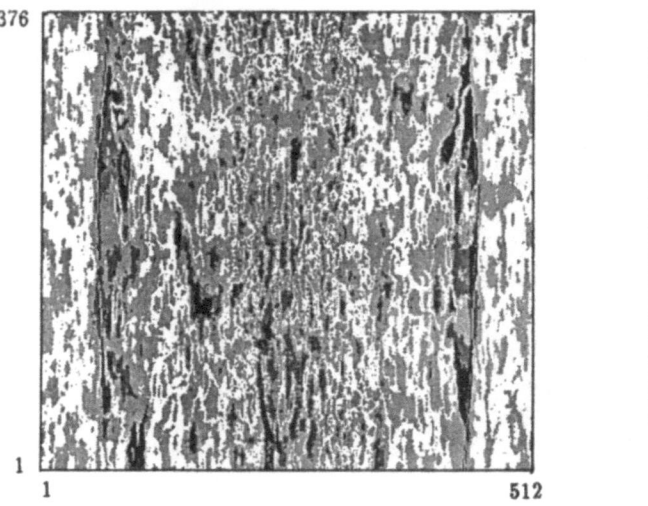

Figure 4. The result of composition of fourier image phase and magnitude averaged by 17-element mask.

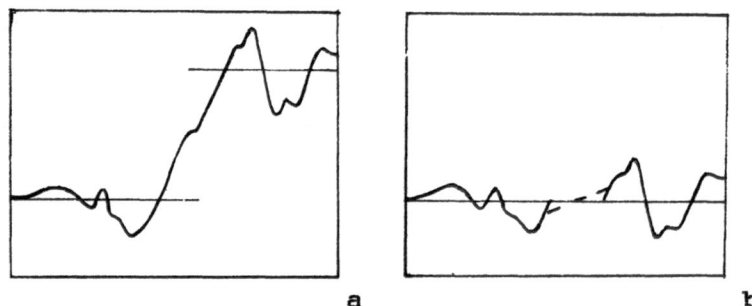

Figure 5. Result of equalization (b): the contrast of deviation from previous two levels (a) increases.

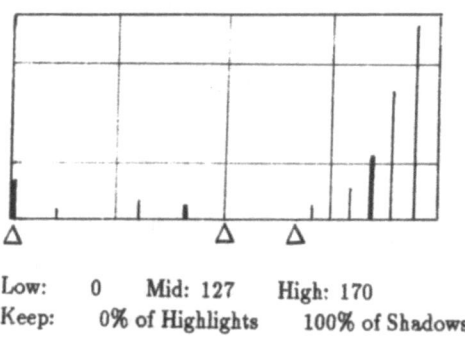

Low: 0   Mid: 127   High: 170
Keep: 0% of Highlights   100% of Shadows

Figure 6. Initial histogram of picture Fig.3 for 16 gradations of gray. Levels between two edge arrows were amplified.

exceed the element of resolution. In opposite case visible crossings of element edges reduce the diagnostic value of such images.

The considered simple methods allow to restore poorly defined images in the presence of low noise. Now the possibilities of reconstruction and restoration of objects of medical diagnostics by operations with amplitude part of spectrum (iterative methods among them) are investigated.

REFERENCES

1. J. Bamber and M. Tristam, Diagnostic ultrasound, in: "The Physics of Medical Imaging", S. Webb, ed., Adam Hilger, Bristol and Philadelphia (1988).
2. P. Greguss, "Ultrasonic Imaging", Focal Press Limited, London, Focal Press Inc., New York (1980).
3. E. L. Borodina, F. A. Kazulin, S. S. Sukhov, et al., in Proc. II Symp. RAS, Moscow (1993).
4. T. F. Budinger, G. T. Gullberg, and R. H. Huesman, Emission computed tomography, in: "Image Reconstruction from Projections", G. T. Herman, ed., Springer-Verlag, Berlin, Heidelberg (1979).
5. M. H. Hayes, Uniqueness of multidimensional signal restoration from magnitude and phase of its Fourier transform, in: "Image Recovery: Theory and Application", H. Stark, ed., Academic Press, New York (1987).

BLOOD FLOW ASSESSMENT USING WAVELET TRANSFORM BASED CROSS-CORRELATION TECHNIQUE

Xiao-Liang Xu, Hehong Zou, and James F. Greenleaf

Biodynamics Research Unit
Department of Physiology and Biophysics
Mayo Clinic and Foundation
Rochester, MN 55905

ABSTRACT

In this paper we present a wavelet transform based cross-correlation (WTCC) technique for assessment of blood flow velocity. Unlike the conventional windowed cross-correlation (WCC) technique, the proposed method has varying time-frequency resolutions by performing cross-correlation in the wavelet-domain. Because of this unique feature the WTCC has advantages over the WCC in performance. In particular, it has better performance for low signal-to-noise cases and is less sensitive to window size and system parameters in comparison with the WCC technique. Two simulation examples are used to show that the WTCC technique has a smaller failure rate and more accurate time–shift estimation in a single-cluster-of-scatterers case and provides a better estimation of the blood flow velocity profile in a simulated case of parabolic blood flow assessment.

INTRODUCTION

Time-domain blood flow estimation techniques have recently attracted more and more attention in the ultrasound research area.[1] One of these techniques is the windowed cross-correlation (WCC) method used for PW ultrasound, which has been extensively studied.[2,3] The WCC technique considers blood flow estimation as one of time delay estimation problems. In particular, it tracks the maxima of the windowed cross-correlation between consecutive ultrasonic pulse echoes; the location of each maximum indicates the movement (time-delay) of a cluster of scatterers at a specific location, from which the velocity of the cluster of scatterers is estimated.

For a single moving scatterer (or a single cluster of moving scatterers), two received consecutive echoes can be expressed as

$$\begin{cases} x_1(k) = s(k) + n_1(k) \\ \\ x_2(k) = s(k-d) + n_2(k) \end{cases} \quad \text{for } 1 \leq k \leq N, \tag{1}$$

where $x_1(k)$ and $x_2(k)$ denote the two sampled echoes with size N, $n_1(k)$ and $n_2(k)$ are the contaminating noise sequences which are assumed to be mutually independent zero-mean white Gaussian sequences, and $s(k)$ is the source signal (with a short duration) that is delayed by d from echo 1 to echo 2. The cross-correlation technique estimates the time delay d by maximizing the cross-correlation function of the two echoes

$$R(\tau) = \sum_{k=1}^{N-\tau} x_1(k)x_2(k+\tau) \tag{2}$$

over τ. That is,

$$\hat{d} = \max_{\tau} \; R(\tau) \tag{3}$$

When the pulse repetition rate, the medium sound speed and the Doppler angle (the angle between the beam axis and the scatterers moving direction) are given, the velocity of the moving scatterer can be estimated from $\hat{d}$.

For blood flow measurement, the measured echoes contain signals reflected from different clusters of scatterers moving in different velocities (e.g., the velocity profile is parabolic), the two received echoes can be expressed as

$$\begin{cases} x_1(k) = s(k) + n_1(k) \\ \\ x_2(k) = s(k-d(k)) + n_2(k) \end{cases} \quad \text{for } 1 \leq k \leq N, \tag{4}$$

where $d(k)$ implies that the time delay is a function of k and thus the location of scatterers. To estimate the velocity of a specific cluster of scatterers (e.g., in a range cell) in blood flow, the WCC technique is used. Specifically, to estimate the time delay $d(k)$ at $k = k_o$, the WCC technique first computes the windowed cross-correlation given by

$$R_w(k_o, \tau) = \sum_{k=k_o}^{k_o+N_o} x_1(k)x_2(k+\tau), \tag{5}$$

where N_o is the window size, and then provides an estimate of $d(k_o)$ by maximizing $R_w(k_o, \tau)$ over τ.

Note that Eq. (5) can be interpreted as the product of the inverse of windowed Fourier transform of x_1 and x_2. It is then clear that the performance of WCC depends on the window size employed in computing the cross-correlation. In other words, the WCC technique depends on the time-frequency-resolution of the window. It is well known that the WCC algorithm works well if the applied and actual time frequency resolutions are close, but it works poorly when they are not close.[4] Because the flow velocity is rarely of a uniform profile, the received signals have a nonuniform time-frequency resolution.

Therefore it is difficult to select an optimal window size for estimating $d(k)$ corresponding a wide range of k, i.e., different positions.

In this paper we employ the recently explored wavelet theory to overcome the aforementioned resolution related problems.[4] Basically, we use the wavelet transform to decompose the received echoes into subspaces spanned by the translates and dilates of a single function called the mother wavelet, and then apply the cross-correlation technique to the wavelet-domain signals. The wavelet transform based cross-correlation (WTCC) technique is less sensitive to window size and the other system parameters because the wavelet-domain signals cover a wide range of time-frequency resolutions.

METHOD

The basic idea of the wavelet transform is to decompose signals into a subspace spanned by the translates and dilates, $\left\{2^{-j/2}\varphi(2^{-j}t - k)\right\}_k^j$, of a single function called the mother wavelet, $\varphi(t)$. Such a wavelet can be generated through a so called scaling function $\phi(t)$ by

$$\varphi(t) = \sqrt{2}\sum_k d_k \phi(2t - k) \tag{6}$$

where $\phi(t)$ is defined as follows,

$$\phi(t) = \sqrt{2}\sum_k c_k \phi(2t - k) \tag{7}$$

In the above equations, $\{c_k\}$ and $\{d_k\}$ are the coefficients which form low pass and high pass filters $H(\omega)$ and $G(\omega)$ defined as,

$$\begin{aligned} H(\omega) &= \sum c_k e^{-j\omega k} \\ G(\omega) &= \sum d_k e^{-j\omega k} \end{aligned} \tag{8}$$

A particular example of a dyadic wavelet with linear phase was generated by Mallat.[5] The corresponding coefficients are

$$\{c_k\} = \{.0625, 0.25, .375, .25, .0625\} \times \sqrt{2} \tag{9}$$

and

$$\{d_k\} = \{-.0001, -.0164, -.1087, -.5926, 0, .5926, .1087, .0164, .0001\} \times \sqrt{2}. \tag{10}$$

With the above defined wavelets, the dyadic wavelet transform of a signal $f(t)$ can be defined as,

$$Wf(2^{-j}, n) = 2^{-j/2}\int f(t)\varphi(2^{-j}t - n)dt, \tag{11}$$

471

where $Wf(2^{-j}, n)$ denotes the wavelet transform at scale 2^{-j}. It has been shown that this transformation can be realized in a fast algorithm by the prototype filters $\{c_k\}$ and $\{d_k\}$ as,

$$Wf\left(2^{-(j+1)}, n\right) = \sum d_k Sf\left(2^{-j}, n - k\right), \tag{12}$$

where $Sf(2^{-j}, n)$ is the smoothed version of $f(t)$ at scale 2^{-j} defined as,

$$Sf\left(2^{-(j+1)}, n\right) = \sum c_k Sf\left(2^{-j}, n - k\right). \tag{13}$$

In the case of a discrete signal $f(t)$, the block diagram of this fast dyadic wavelet transform is shown in Fig. 1, in which $Sf(2^0, n)$ is assumed to be the signal $f(n)$ itself.[5]

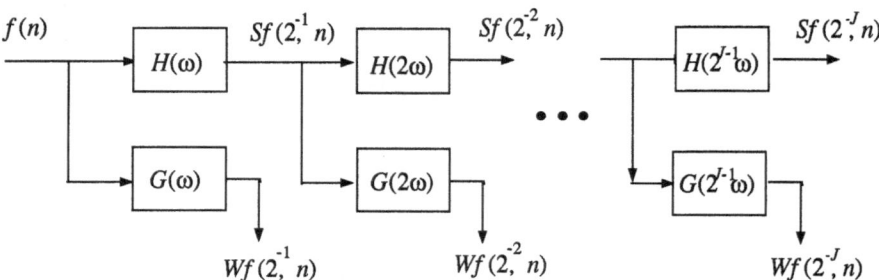

Figure 1 The block diagram of a fast dyadic wavelet transform algorithm.

It is well established that the above dyadic wavelet transform is a redundant signal representation like the windowed Fourier transform. Furthermore, the signal representation at each discrete point contains the information of the signal in its neighborhood related with the different scales (that is, the different window sizes). That is why the wavelet transform is often referred to as the time-scale signal representation. An advantage of this transform over the windowed Fourier transform is that the windows in obtaining the transformation have the multiple and varying sizes. Hence, in a particular time delay estimation problem, it will be of advantage to utilize this property for better estimation.

The cross-correlation technique based on the wavelet transform can now be presented with the following steps:

Step 1. Compute the wavelet transform for each observation in different scales:

$$\{Wx_1(2^{-j}, k); \ 1 \leq j \leq J\} \quad \text{and} \quad \{Wx_2(2^{-j}, k); \ 1 \leq j \leq J\}. \tag{14}$$

Step 2. Form an approximation of $x_1(k)$ and $x_2(k)$ in wavelet space:

$$\tilde{x}_i(k) = \sum_{j \in Z} Wx_i(2^{-j}, k), \qquad \text{for } i = 1, 2, \tag{15}$$

where Z is a collection of j's in the range of $1 \leq j \leq J$.

Step 3. Apply the WCC to $\tilde{x}_1(k)$ and $\tilde{x}_2(k)$ to estimate $d(k_o); \ 0 < k_o < N$.

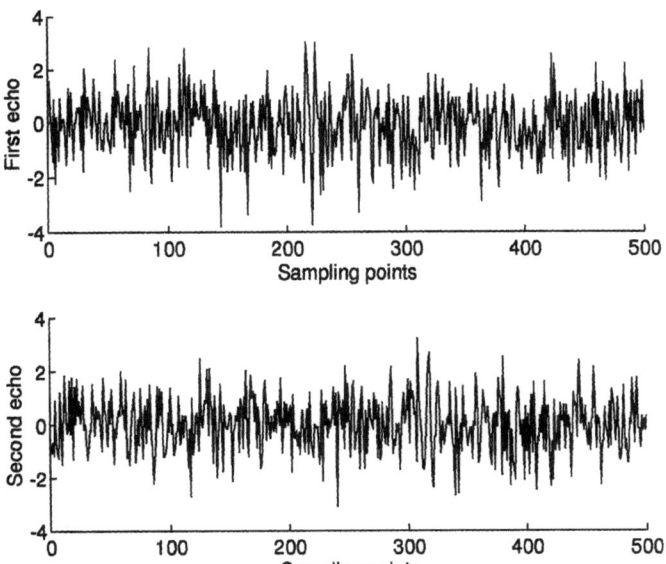

Figure 2 Two Consecutive echoes contaminated by mutually independent white Gaussian noise (SNR=0). The reflected signal of interest appears in the 200th – 230th sampling points in the first echo and appears in the 300th – 330th sampling points in the second echo.

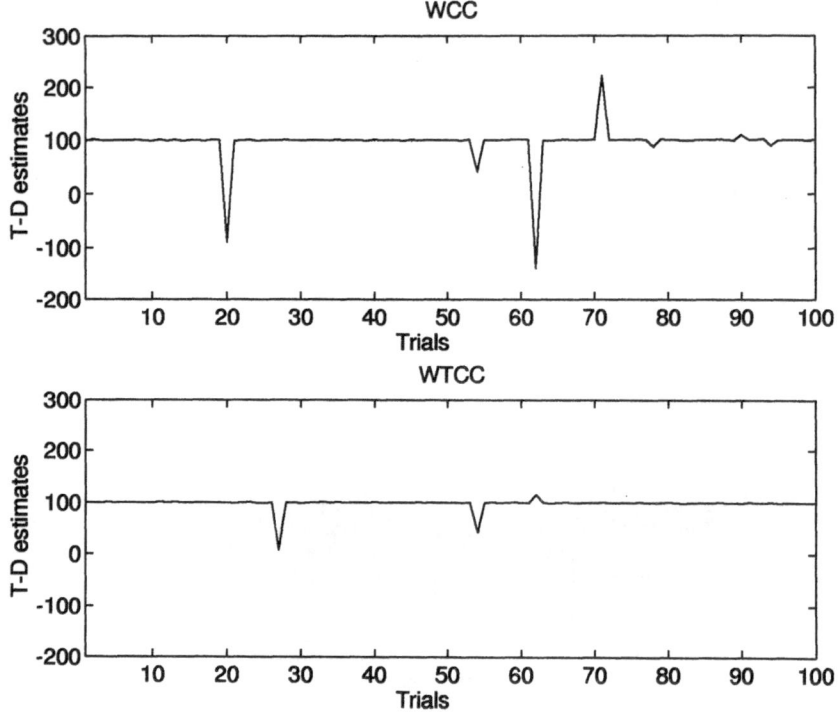

Figure 3 The time delay estimates from 100 independent trials using WCC and WTCC. (SNR = 0dB.)

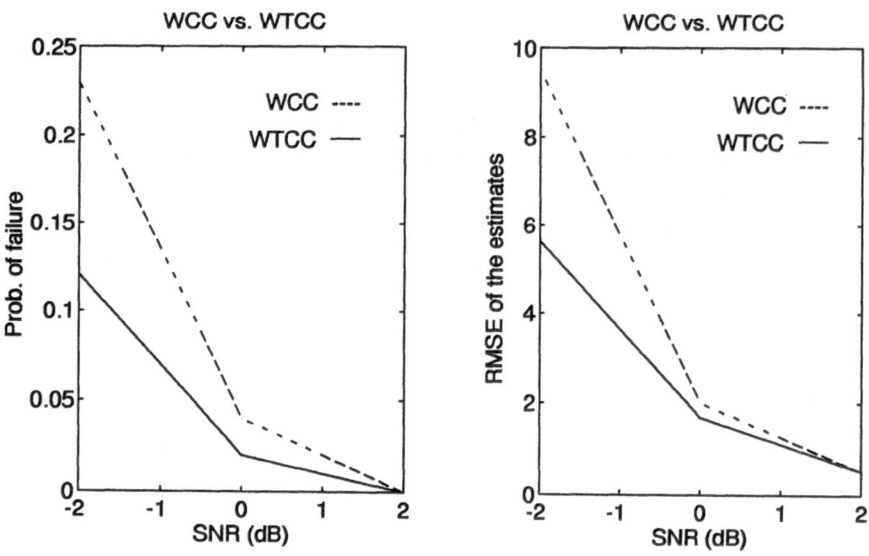

Figure 4 Failure rates vs. SNR, and the root mean squared errors (RMSE's) of estimates vs. SNR.

Notice that in step 2 Eq. (15) represents a projection of each original signal onto a subspace spanned by $\left\{2^{-j/2}\varphi\left(2^{-j}t - k\right)\right\}_{k}^{j \in Z}$. For certain j's, $\varphi\left(2^{-j}t - k\right)$ has a resolution very different from the original backscattering signals, such j's should not be included in Z.

The major difference of this algorithm from the WCC method is that the wavelet transform is used as for the matching between observations instead of the windowed Fourier transformation. The algorithm does not assume a known signal size because the wavelet transform has varying scales (resolution).

RESULTS

We have run a number of simulation examples to compare the performance of WCC and WTCC, and we show two of them in this paper. In both examples, Mallat's wavelet specified by Eq. (9) and Eq. (10) was used, and $Z = \{2, 3\}$, i.e., only the two wavelet transformed signals at scales 2^{-2} and 2^{-3} were used in Eq. (15).

In the first example, we assumed a case of a single cluster of moving scatterers. Figure 2 shows two simulated consecutive echoes that are contaminated by white zero-mean Gaussian noise (SNR = 0 dB). The reflected signal of interest appears in the 200th – 230th sampling points in the first echo and appears in the 300th – 330th sampling points in the second echo. Therefore the true time delay is the interval corresponding 100 sampling points. Figure 3 shows the time-delay estimates from 100 independent trials using WCC and WTCC. The window size for both methods was 30 points, which was the same as the true length of the signal of interest. For both methods, the obtained estimates are around 100 in most trails, but a few are far different from 100. We assume that the estimation fails if the difference between the estimate and the true delay is over 50%. Figure 4 shows the failure rate vs. the SNR and the mean squared error vs. SNR for the WCC and WTCC estimates. Clearly, Fig. 4 indicates the WTCC has a smaller failure rate and a smaller estimate error, especially when SNR is low.

In the second example, we simulated estimating the velocity profile of blood flow using the WTCC and WCC. In this example, the transducer was Gaussian focused, the focal length was $60mm$, the emitted pulse center frequency was $3.5MHz$, the pulse length was $1.0\mu s$, the pulse repetition rate was $2.5KHz$, and the vessel diameter placed in the focal plane was $6mm$, the velocity profile of blood flow was parabolic with maximum velocity $50cm/s$, the Doppler angle was 30°, the sampling rate was $35MHz$, and the sampling interval was $16\mu s$. Fig. 5 shows the two consecutive echoes (noiseless). We tested the WCC and WTCC on the noisy echoes (SNR = 4dB) using different window sizes. Fig. 6 shows the results from using the window size of $1.0\mu s$, indicating the WTCC has better estimation of the flow velocity profile. The window size of $1.0\mu s$ seems optimal for WCC in this example. When other window sizes were used, the performance of the WCC degraded rapidly, while the WTCC performed more consistently. Figure 7 shows estimates of the velocity profile obtained by using the window size of $1.6\mu s$, and Fig. 8 shows estimates of the velocity profile obtained by using the window size of $0.5\mu s$.

CONCLUSION

In this paper we have proposed applying the wavelet transform to the cross-correlation technique for blood flow assessment. The wavelet transform based cross-correlation technique has been shown in simulations to have better performance than the conventional cross-correlation technique. Furthermore, it is less sensitive to contaminating noise and the window size.

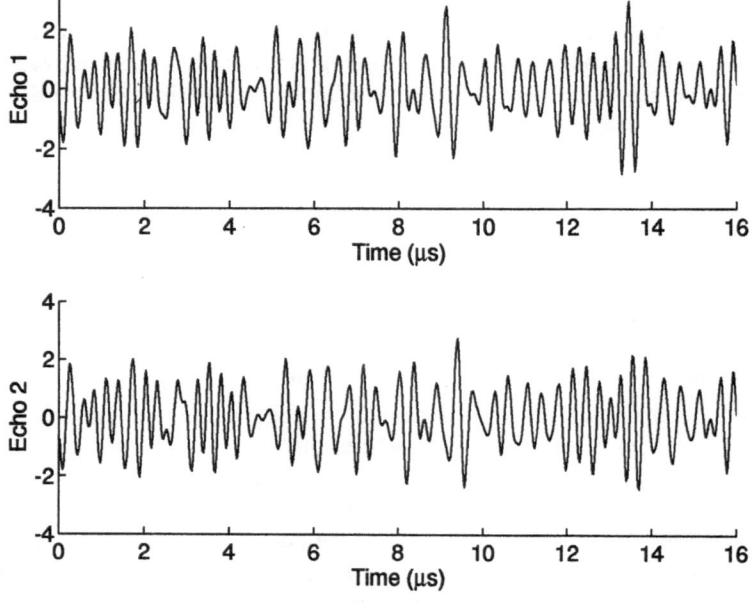

Figure 5 Two consecutive echoes (noiseless) for example 2 simulating blood flow assessment. (pulse: 3.5*MHz* center frequency and 1.0μs duration, vessel: *6mm* diameter and parabolic velocity profile with 50*cm/s* peak)

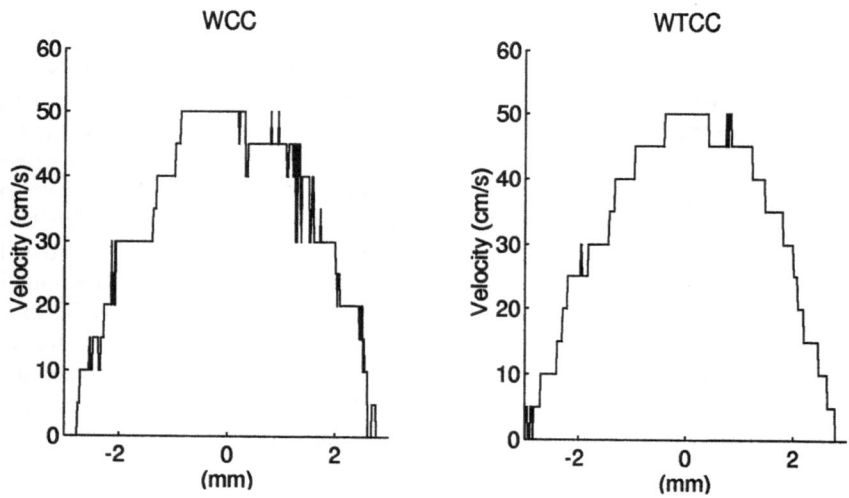

Figure 6 Estimation of the velocity profile (for signal of Fig. 5 with SNR = 4dB) using WCC and WTCC, the window size is 1.0μs, the horizontal axis represents relative positions of scatterers across the vessel with the diameter of 6mm (0 indicates the center of the vessel).

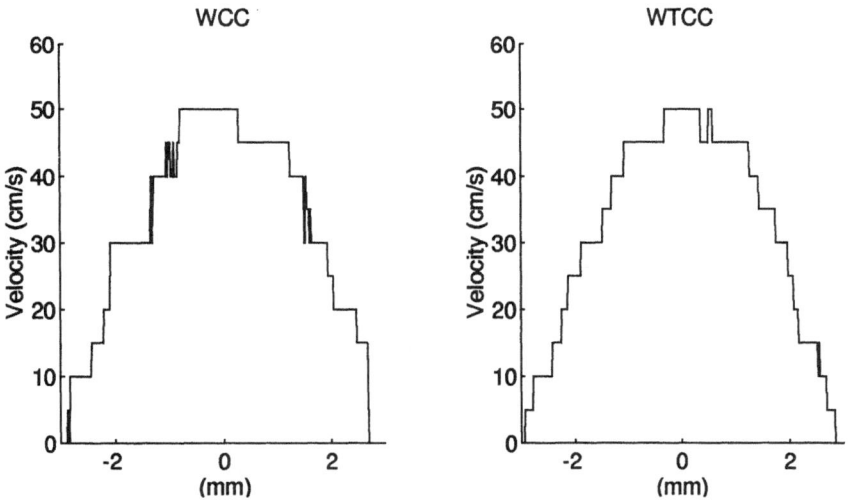

Figure 7 estimation of the velocity profile (for signal of Fig. 5 with SNR = 4dB) using WCC and WTCC, the window size is $1.6\mu s$, the horizontal axis represents relative positions of scatterers across the vessel with the diameter of 6mm (0 indicates the center of the vessel).

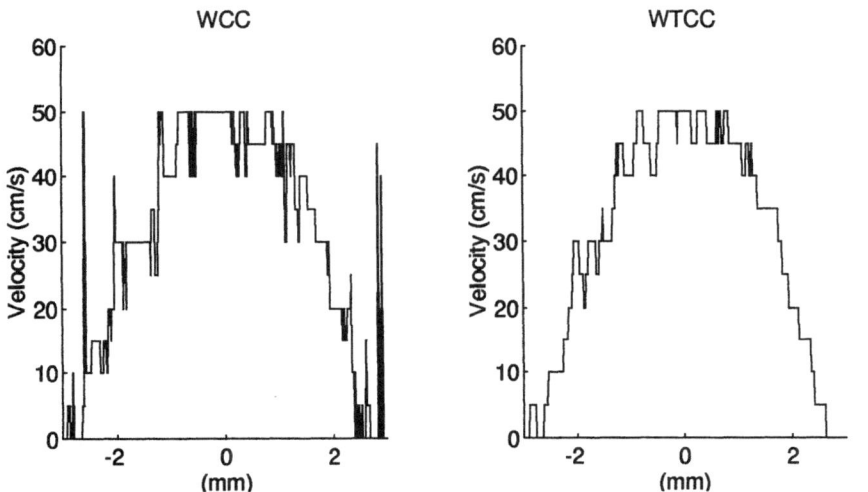

Figure 8 estimation of the velocity profile (for signal of Fig. 5 with SNR = 4dB) using WCC and WTCC, the window size is $0.5\mu s$, the horizontal axis represents relative positions of scatterers across the vessel with the diameter of 6mm (0 indicates the center of the vessel).

ACKNOWLEDGMENT

The work was supported in part by HL 41046 from NHLBI. The authors thank Elaine C. Quarve for her secretarial work.

REFERENCES

[1] I.A. Hein and W.D. O'Brien, Jr., Current time-domain methods for assessing the tissue motion by analysis from reflected ultrasound echoes — a review, *IEEE Trans. Ultrason. Ferroelec. Freq. Contr.* UFFC-40:84–102 (1993).

[2] S.G. Foster, P.M. Embree and W.D. O'Brien, Volumetric blood flow via time-domain correlation: error analysis and computer simulation, *IEEE Trans Ultrason. Ferroelec. Freq. Contr.* UFFC-37:164–175 (1990).

[3] P.M. Embree and W.D. O'Brien, Volumetric blood flow via time-domain correlation: experimental verification, *IEEE Trans Ultrason. Ferroelec. Freq. Contr.* UFFC-37:176-189 (1990).

[4] C.K. Chui, *An Introduction to Wavelets*, Academic Press, New York (1992).

[5] S. Mallat and W. Hwang, Singularity detection and processing with wavelets, *IEEE Trans. Inf. Theory*, 38(2):617–643 (March 1992).

AUTOREGRESSIVE RECURSIVE ALGORITHMS FOR MAXIMUM FREQUENCY ESTIMATE OF DOPPLER ULTRASONIC SIGNALS

L.Agostini[1], A.Fort[1], C.Manfredi[1], L.Masotti[1], F.Picchiarini[1], S.Rocchi[2]

[1] Dipartimento di Ingegneria Elettronica, Università degli Studi di Firenze, Via S.Marta 3, 50139, Firenze, Italia
[2] Dipartimento di Ingegneria Elettronica, Università degli Studi di Siena, Via Roma 77, Siena, Italia

INTRODUCTION

Doppler ultrasound technique is widely used in medical applications to extract the blood velocity distribution in human arteries from Doppler frequency shift. A real-time spectral analysis of Doppler shift signals provides the time evolution of the velocity distribution in the vessels. An interesting clinical application of Doppler ultrasound technique is the estimate of the maximum frequency, beyond which the Power Spectral Density (PSD) signal-to-noise ratio goes below 0 dB.

The maximum frequency estimate of noisy Doppler signals is a troublesome problem, since the spectrum falls off slowly due to transit time effects and stochastic uncertainty. The maximum frequency is supposed to be located at the falling edge of the spectrum. Consequently, the slope of the edge and its amplitude above the noise are fundamental to locate the maximum frequency.

The spectral analysis of non-stationary signals requires a signal processing tool that presents both good time and frequency resolution. The most commonly used method for spectral analysis is the Fast Fourier Transform (FFT).

However, methods based on FFT computation, such as the periodogram technique, are not suitable for estimating the maximum frequency due to the finite length of the observed frame which causes spurious peaks in the high frequency range. Moreover, the choice of the window function represents an inherent trade-off between time and frequency resolution in the spectrogram[1].

In the present application the frequency content of non stationary Doppler signals coming from fast varying flows is estimated; thus, short data segments have to be analyzed to fall into the assumed local stationariety of the process (5÷15 ms).

In this case the PSD can be estimated via the so called parametric approach, which has been shown to be a powerful tool for discrete spectral analysis since it implicitly extrapolates data outside the segment under consideration: in this way distortion effects due to windowing can be eliminated. In the parametric approach a model identification problem is faced: the underlying process is described by a suitable linear model whose order and parameters must be estimated.

In the present paper autoregressive (AR) models are used, which allow parameter estimation by solving a set of linear equations and generate well-defined peaks. A high frequency resolution is often associated to the latter characteristic; however it is a function of the Signal-to-Noise Ratio (SNR). A sequential algorithm is used to process data for updating the parameter estimate: this is useful when tracking a signal with a varying spectrum, as in our case. Sequential estimate techniques update the AR parameters each time a new sample is acquired; thus a plot of the PSD at any time instant can be obtained and its spectral time evolution can be followed. These methods are called "adaptive" since parameters are evalueted according to the time-varying signal characteristics.

For short data segments the correct AR model order determination is particularly troublesome because of the failure of traditional methods (Akaike Information Criterion AIC, Final prediction Error FPE etc.): a too low order estimation results in a smoothed spectral estimate, while too high an order introduces spurious details into the spectrum, mismatching the true signal characteristics. Thus, a correct order choice is of the utmost importance for a meaningful estimate of parameters. It must be stressed here that the choice of a fixed model order (as suggested in literature for this kind of Doppler signals), based on an a priori knowledge of data characteristics and on experimental results, must be considered with great care, since it is intimately bound to a correct signal sampling. In the present paper the AR recursive parametric modelling technique is proposed, combined with the Singular Value Decomposition (SVD) method. For each time window a matrix of data samples is constructed and its singular values are evaluated to recover the dimensions of signal and noise subspaces. A new adaptive method is proposed to select the model order by separating the singular values representative of signal subspace from the ones belonging to the noise subspace in order to bypass the problem of an a priori threshold definition in the SVD method.

The main drawbacks of the parametric approach are the dipendence of the spectral resolution on the SNR and the increased computational cost, but it must be stressed that in the present application the properties of noise rejection (typical of parametric methods) are of utmost importance, since the presence of spurious peaks in the high frequency range produces a false maximum frequency location.

Finally, an ad hoc definition of the maximum frequency is proposed and results obtained from experimental and simulated Doppler signals are presented.

SPECTRAL ANALYSIS BASED ON RECURSIVE AR MODELLING AND SVD

When short data segments have to be analyzed, the discrete signal spectral analysis can be performed by using the AR parametric approach for the PSD evaluation. It has been shown that this method is largely insensitive to noise and, if properly implemented, is numerically efficient at a low computational cost[2].

The linear difference equation that describes an AR model is:

$$x(k) = -\sum_{i=1}^{p} a_i x(k-i) + \epsilon(k)$$

(1)

where a_i, $i=1,...,p$, are the AR parameters, p is the model order, x(k) is the sequence which models the observed data and $\epsilon(k)$ is a white noise zero-mean sequence with variance ρ^2. The AR Power Spectral Density is:

$$P_{AR}(f) = \frac{\rho^2 \Delta t}{\left| 1 + \sum_{i=1}^{p} a_i e^{-j2\pi fi\Delta T} \right|^2}$$

(2)

where ΔT is the sampling period.

Among the AR adaptive algorithms, the Recursive Least Square (RLS) technique was preferred to the Least Mean Square (LMS) method, thanks to its better performance and higher speed of convergence.

In the present paper the RLS Fast Transversal Filter (FTF) technique in its stabilized version was used. In fast algorithms the computational complexity is reduced from $O(p^2)$ to $O(p)$ operations by adopting vectorial instead of matrix operations. The FTF algorithm is based on geometrical projections and eliminates unnecessary or redundant computations. The employed algorithm corrects instability problems related to the FTF provided that some computational redundance is reintroduced[3].

The RLS adaptive technique estimates the AR parameters by minimizing at each step the exponentially weighted squared error $\eta(t)$:

$$\eta(t) = \sum_{k=1}^{t} \lambda^{t-k} |e_t(k)|^2$$

(3)

where p is the model order and $e_t(k)$ is given by the following expression:

$$e_t(k) = x(k) + \sum_{i=1}^{p} a_i(t)x(k-i)$$

(4)

i.e. $e_t(k)$ is the prediction error on the k-th sample obtained using the estimated coefficients at time t, $a_i(t)$, $i=1,...,p$, and λ is the forgetting factor ($0<\lambda\leq1$). If $\lambda<1$, the factor λ^{t-k} weigths recent errors more than older ones. Samples taken more than $(1-\lambda)^{-1}$ instants ago influence eq.(1) about 36% less than those taken more recently; this allows the model to follow slowly-varying signal characteristics. If $\lambda=1$ all the samples are equally weighted, i.e. the memory is extended to the whole signal: this implies the loss of tracking properties.

Notice that eq.(4) represents a linear forward prediction equation that looks like eq.(1) where the noise input sequence $\epsilon(k)$ is substituted by the forward prediction error $e_t(k)$.

In practice, the number p of the model parameters is unknown; consequently, if the choosen model order is wrong, the AR model spectral estimate is not correct, and the frequencies of the underlying signal (i.e. the peaks of the signal spectrum) will be wrongly located.

The traditional methods for AR order assessment (AIC, FPE)[4,5] result inapplicable to short data records, since they only asymptotically converge to the true model order; hence, they most often produce an underestimation of the true order. This may be due both to the intrinsic lack of stability of the methods as well as to statistically fluctuations of the short data segments[6].

Since the length of each data segment cannot be arbitrarily increased, due to the nonstationarety of the Doppler signal, none of the traditional methods can be successfully applied in the present context. On the other hand, the practical rule that suggests an order selection between N/2 and N/3 (N=data sequence length) can be satisfactory, but it always implies a subjective judgement. In the literature a fixed order p=12 seems to be quite satisfactory for Doppler signals[7]; however experimental results show that this choice is not suitable for short data records, since the results rapidly deteriorate as the frame becomes shorter[8].

This problem can be solved by using the Singular Value Decomposition (SVD) of an appropriate data matrix, which allows the separation of the signal from noise contribution[9]. The application of this method in the analysis of a signal x(k) consists in finding a singular value decomposition of the following 2(N-L)xL data matrix A:

$$
A = \begin{bmatrix}
x(L) & x(L-1) & . & x(1) \\
x(L+1) & x(L) & . & x(2) \\
. & & & . \\
. & . & . & . \\
x(N-1) & x(N-2) & . & x(N-L) \\
x^*(N-L+1) & x^*(N-L+2) & . & x^*(N) \\
x^*(N-L) & x^*(N-L+1) & . & x^*(N-1) \\
. & . & . & . \\
. & . & . & . \\
x^*(2) & x^*(3) & . & x^*(L+1)
\end{bmatrix} \tag{5}
$$

where L is the maximum allowed order ($\leq$N/2) , N is the number of samples and the superscript * denotes conjugation.

Since the singular values and singular vectors of a matrix are relatively insensitive to perturbations in the matrix entries due to noise as well as to finite precision errors, the SVD technique is used to evaluate the numerical rank of a noise corrupted matrix. In fact, when L is greater than p and at a SNR >>1, L+1-p small singular values of the noisy matrix will generally be clustered together and separated in magnitude from the p stronger ones. However, in actual cases, large and small singular values do not neatly separate in the presence of strong noise (SNR $\cong$0 dB). The choice of a robust method for deciding which singular values correspond to the Signal Subspace (SS) and which to the Noise Subspace

(NS), is still one of the crucial points of the SVD method. The most common methods are based on the comparison with a static or dynamic threshold.

In the present paper a criterion tailored to the SVD method allowing an automatic separation between SS and NS on each data segment is proposed, making the overall procedure reliable and easily implementable.

The processing flow, named Dynamic Mean Evaluation (DME), is the following:

$$c=0;$$

$$\text{for} \quad k=1,2,..,\frac{L}{2}$$

$$\text{if} \quad \left|\frac{1}{k}\left(\sum_{i=1}^{k}\sigma_i\right)-\sigma_{k+1}\right|>\left|\frac{1}{k}\left(\sum_{i=1}^{k}\sigma_{L-i+1}\right)-\sigma_{k+1}\right|$$

$$\text{then:} \quad p=k;$$

$$c=1. \tag{6}$$

$$\text{if } c=0:$$

$$\text{for} \quad k=1,2,,..,\frac{L}{2}$$

$$\text{if} \quad \left|\frac{1}{k}\left(\sum_{i=1}^{k}\sigma_i\right)-\sigma_{L-k}\right|<\left|\frac{1}{k}\left(\sum_{i=1}^{k}\sigma_{L-i+1}\right)-\sigma_{L-k}\right|$$

$$\text{then:} \quad p=L-k.$$

$$\text{otherwise:} p=L$$

where σ_i is the i-th singular value of A and L is the maximum allowed order. In words, the procedure works as follows: for the singular values (s.v.) in decreasing order, $\sigma_1>\sigma_2>...>\sigma_L$, the model order p is selected through an iterative scheme comparing the (k+1)-th (k=1,...,L/2) s.v. distance both from the arithmetic mean of the first k s.v. and from that of the last L-k-1 s.v. (SS and NS respectively). If the first distance is larger than the second, the (k+1)-th s.v. belongs to the NS; consequently, the model order is p=k. If no value of k can be found satisfying the above condition, the procedure is repeated starting from the index of the smallest s.v. (i.e. the s.v. which most probably belongs to the NS), thus giving the SS and the model order p=L-k. Finally, if none of the above conditions is verified, the model order p is set equal to L, all the s.v. being of comparable dimension.

This method is reliable both in a well-defined situation, where the singular values of NS are clustered together and neatly separated from those of SS, as well as in the opposite case of uniformly decreasing singular values, when the two subspaces may not be correctly separated on a mere threshold basis, thus implying an erroneous model order selection. In fig.1 the AR order p computed by the SVD+DME technique for a Doppler signal is shown together with the spectrogram of the same signal. The time resolution considered is 10 ms. The proposed technique can be summerized in the following steps:

1. The length of the time window is set to N (40 or 80 points) and the maximum order L is chosen equal to 20.
2. The SVD of matrix A (eq.5) is performed.
3. The s.v. of A are separeted by applying the DME procedure: the dimension of the SS is the order p of the AR model.
4. The RLS algorithm is applied to the same time window. The forgetting factor is chosen according to the length N of the window ($1/1-\lambda$ is the number of samples that have the greater influence on eq.1). For each sample the AR parameters are updated, and the parameters estimated at the N-th sample are used to evaluate the PSD (eq.2).
5. A new nonoverlapped window is considered and the procedure is repeated.

A system for Doppler signal spectral analysis based on the RLS and SVD+DME technique has been implemented on a DSP TMS320C30 supported by the Comdisco SPW software. The complexity of the SVD algorithm doesn't allow a real time analysis: it requires about p^3 operations (multiplications and divisions) for each N sample time window.

EXPERIMENTAL RESULTS

The proposed technique was applied both to simulated and to experimental Doppler signals. Usually the performance of a Doppler signal analysis technique is evaluated through the quality of some spectral parameter estimate, which are meaningful for diagnostic purposes. Particular interest is addressed to the mean frequency, the variance of the Doppler PSD (directly related to the mean velocity and the turbulence of the blood flow) and to the maximum spectral frequency (related to the maximum blood velocity). The third parameter, even if extremely relevant for clinical applications, is rarely considered since many problems are encountered for its estimate. A significant assessment of the maximum frequency can be obtained only when the PSD estimate doesn't introduce relevant noisy peaks outside the signal band, and can be improved if the applied spectral estimator performs also a noise rejection action. Therefore the comparison among various spectral estimators is better performed through the examination of this critical parameter.

A unique definition for maximum spectral frequency doesn't exist; in the present work, the one given operatively by Hatle and Angelsen[10] is considered.

Given a PSD, two windows of the same length l_w, sliding on the frequency axis, are defined; a fixed frequency delay between the two windows Δf is assumed and the 'energy' content of each window (E_1, E_2) is calculated at each step:

$$E_1(f_i) = \int_{f_i}^{f_i+lw} P(f)w_{lw}(f)df$$

$$E_2(f_i) = \int_{f_i+\Delta f}^{f_i+\Delta f+lw} P(f)w_{lw}(f)df$$

(7)

where P(f) is the analyzed PSD, $w_{lw}(f)$ is the frequency window of length lw, and f_i represent the starting frequency of the first window.

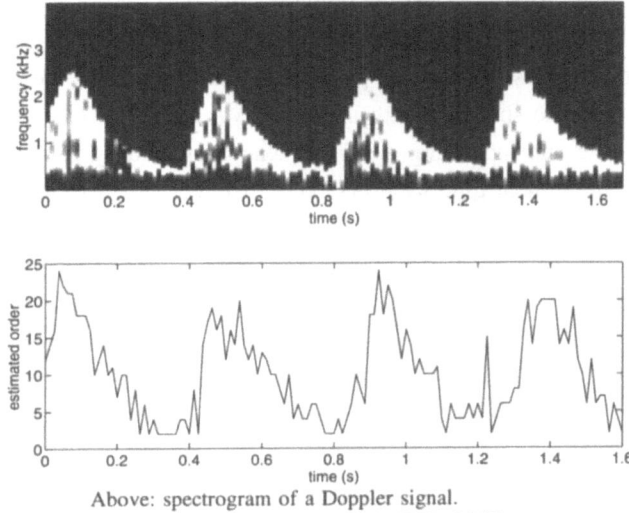

Above: spectrogram of a Doppler signal.
Below: AR order p estimated via SVD+DME.

Figure 1

The maximum frequency can be found when the energy (E_1) contained in the first window becomes well 'larger' than the one contained in the second (E_2), i.e. when the following equation becomes true:

$$\frac{E_1(f_i)}{E_2(f_i)} > \gamma \tag{8}$$

where γ is a conveniently selected threshold.

In the present paper a modified version of the method by Hatle and Angelsen[10] is proposed: the ratio of the energy content of the two windows is evaluated at each step and its value is weighted with the energy content of the first window. This allows to discard peaks in frequency regions where the spectral contribution is mainly due to noise. Therefore a frequency function $R(f_i)$ is evaluated according to the followimg expression:

$$R(f_i) = \frac{E_1^2(f_i)}{E_2(f_i)} \tag{9}$$

The maximum frequency is then given by the frequency of the maximum of $R(f_i)$, thus avoiding the problem of an appropriate threshold value selection.

The proposed recursive AR+SVD technique was tested on a set of simulated signals and then applied on a set of experimental signals obtained from continuous Doppler waveforms coming from fetal umbilical veins.

The simulated Doppler signals were produced as a sequence of adjacent stationary time windows; in each time window the signal is obtained by the superimposition of a set of sinusoids with random phases embedded in noise. Simulated signals are characterized by rectangular or asymmetric spectrum with increasing variance and fixed central frequency or with fixed variance and increasing mean frequency obtained by adding an appropriate set of sinusoids in subsequent time windows. The SNR was varied from 4 to 10 dB and a sampling frequency of 8 kHz is considered.

A first evaluation of the recursive AR+SVD performance was obtained by comparing it to AR spectral estimators with standard order selection criteria (AIC, FPE, etc).

The results of simulations agree with theoretical prediction: when the time window is sufficiently long and the SNR is high, results obtained with standard methods are equivalent to those obtained by applying the proposed method. The advantadge of the SVD+DME model order estimate becomes evident when the time window length or the SNR is reduced. In particular, with the considered SNR values, when the time window falls below 128 samples, the order estimate given by AIC becomes highly unstable. This often causes an overestimate of the order and the consequent appearence of noisy peaks, which severely affect the quality of the maximum frequency assessment, leading to an erroneous identification of the maximum frequency with the falling edge of a spurious spectral peak. On the other hand, an underestimate of the order produces a spectral broadening effect and a smoothing of the spectral falling edge with a consequent less neat definition of the maximum frequency.

The described technique was applied to Doppler signals recorded from the umbilical artery of pregnant women. The cardiac cycle of a healthy fetus is very irregular and has a beat rate which is almost twice that of an adult; for these highly non-stationary Doppler signals, time resolutions of 10 ms or 5 ms were considered. A comparison between the maximum frequency obtained from these signals via the proposed technique and the one obtained by means of a fixed order AR (order 12) is presented in figs.2-5, where the fixed order model parameters are evaluated with the modified covariance block algorithm[2]. For this application, involving both short time windows and a low SNR, the model order estimate provided by standard method is highly unreliable; this fact is confirmed also by other works[6].

The selected fixed order was derived as an average of the orders estimated by the SVD technique (when considering 10 ms time resolution), and it is in agreement with the value found by Sclindwein[7]. In figs 4-5 the better results obtainable by SVD are clearly shown: the adaptive order AR spectral estimator succeeds in avoiding the noisy spectral peaks, as it follows the time varying signal characteristics according to the selected window lingth. In fact, when the time window is shortened, the estimated order tends to become lower: the estimation of a smaller number of model parameters results in a reduced sensitivity to noise of the spectral estimator.

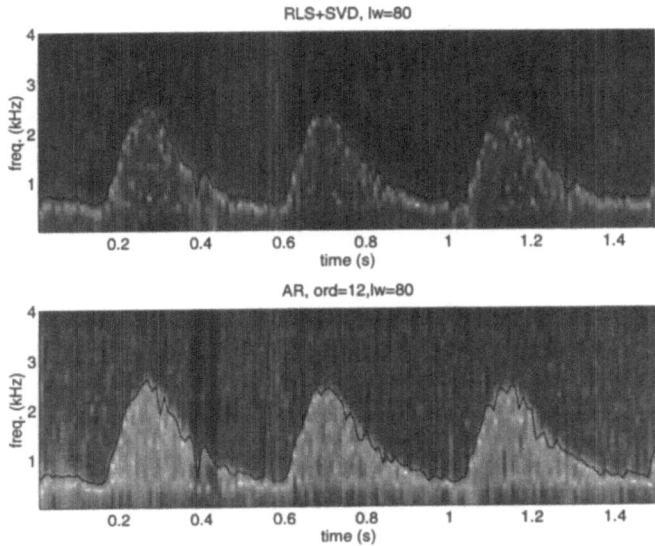

Figure 2 Max. freq. obtained by RLS+SVD (above) and by fixed order AR (p=12) (below). The estimated max. freq. is superimposed to the spectrograms. Time resolution is 10 ms.

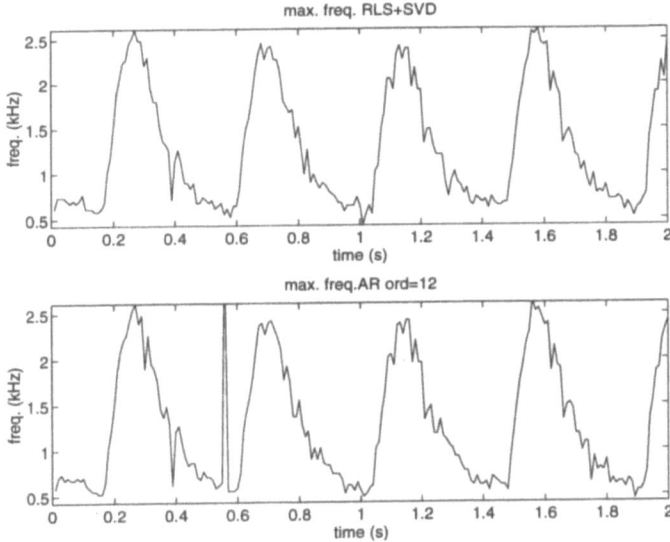

Figure 3 Maximum freq. estimated by RLS+SVD (above) and by fixed order AR (p=12) (below). The time resolution is 10 ms.

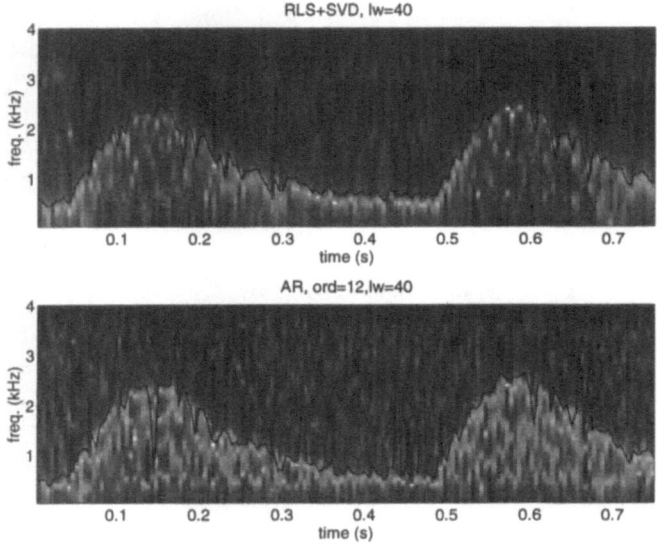

Figure 4 Max. freq. obtained by RLS+SVD (above) and by a fixed order AR (p=12) (below). The estimated max. freq. is superimposed to the spectrograms. The time resolution is 5 ms.

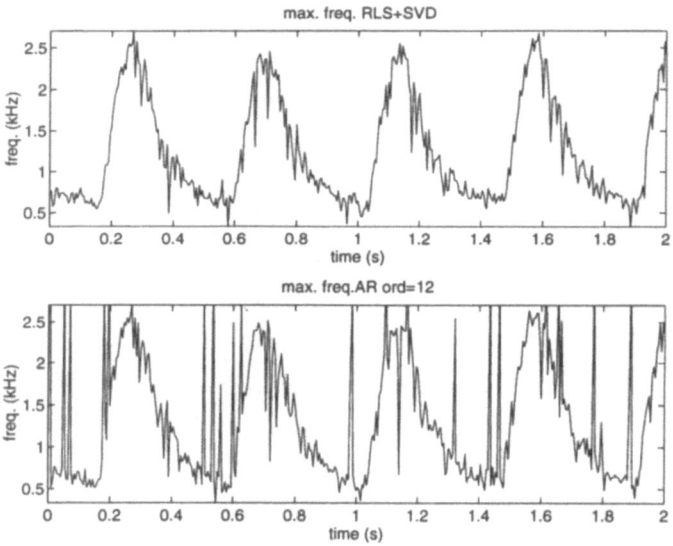

Figure 5 Maximum freq. estimated by RLS+SVD (above) and by fixed order AR (p=12) (below). The time resolution is 10 ms.

CONCLUSIONS

The analysis of the spectral characteristics of fast time-varying noisy Doppler signals via the traditional periodogram technique suffers from poor frequency resolution as the data frame becomes shorter, with obvious drawbacks as far as the applications are concerned. The AR parametric approach performs well in such critical situations, since it is characterized by good time and frequency resolution, but always presents the problem of the correct model order selection. Most commonly, this problem has been addressed via a trial procedure or classical criteria such as AIC, FPE and others, but no result of general validity has yet been obtained. Undoubtedly, the crucial point is a correct separation between the signal and the noise subspaces, which allows the signal PSD evaluation devoid of spurious peaks. The SVD method provides a solution to this problem, but still requires a threshold value, usually a priori fixed.

In the present paper, an adaptive model order selection technique is proposed, tied to the SVD method, which overcomes the above mentioned problem with a small amount of computation: the model order is adaptively chosen on the basis of the varying signal 4characteristics. The method, named Dynamic Mean Evaluation (DME), performs well also on short data sequences, whose length is prohibitive for the traditional techniques. Such advantages are confirmed by the reported experimental results.

A sequential algorithm (RLS) is used to update the AR parameter estimate, in order to correctly track the signal varying spectrum.

However, it must be stressed that problems may rise in the time regions where the signal is characterized by a narrow band centered at a very low frequency and where a low model order is correctly estimated by the SVD method. On the other hand in that region the signal is dramatically over-sampled: therefore an AR model which describes the present sample as a function of few past samples can be highly inaccurate. In fact, when taking into account only few samples noise and quantization error can completely hide the slow signal variation. For this reason, when analyzing Doppler signal a minimum order, 4 or 5, may be advisable.

REFERENCES

1. L.Cohen, Time-frequency distribution - a review, *Proc. IEEE 77-7.* 941:981 (1989).
2. S.L.Marple. "Digital Spectral Analysis with Applications," Prentice-Hall Inc., Englewood Cliffs, NJ (1987).
3. D.T.M.Slock and T.Kailath, Numerically stable fast transversal filter for recursive least square adaptive filtering, *IEEE Trans. on Signal Processing 39-1.* (1991)
4. H.Akaike, Power spectrum estimation through autoregression model fitting, *Ann. Inst. Stat. Math 21*, 407:419 (1969)
5. H.Akaike, A new look at the statistical model identification, *IEEE Trans. AC 19,* 716:723 (1974).
6. F.S.Schlindwein and D.H.Evans, Autoregressive spectral analysis as an alternative to fast fourier transform analysis of doppler ultrasound signals, *in:* "Diagnostic Vascular Ultrasound," K.H.Labs et al.Eds., E.Arnold, London, UK, 74:84 (1992).
7. F.S.Schlindwein and D.H.Evans, Selection of the order of autoregressive models for spectral

analysis of doppler ultrasound signals, *Ultras. in Med.and Biol. 16-1*, 81:91.(1990)

8. S.Dessi', A.Fort, C.Manfredi, S.Rocchi, Adaptive ar model order estimation for doppler signals spectral analysis, Proc. IFAC Sysid Symposium 4-6 July 94, Copenhagen DK (1994).

9. B.D.Rao and K.S.Arun, Model based processing of signals: a state-space approach. *Proc. IEEE 80-2*, 283:309 (1992)

10. L.Hatle and B.Angelsen. "Doppler Ultrasound in Cardiology", Lea & Febiger, Philadelphia (1982)

A POST-FILTERING TECHNIQUE FOR AUTOCORRELATION ESTIMATOR

IN REALTIME COLOR FLOW MAPPING SYSTEMS

Tai K. Song

Department of Information and Communication Engineering
Korea Advanced Institute of Science and Technology
P.O. Box 201, Cheongryang, Seoul, Korea

ABSTRACT

A new post-filtering technique to eliminate the clutter signals for the autocorrelation mean frequency estimation (ACE) method that is most commonly used in detecting the Doppler shift in Color Flow Mapping (CFM) systems is presented. In the general two dimensional Doppler imaging environment, the complex input Doppler signals contain the clutter components even after clutter filtering, which results in the estimation error or the loss of the ability to detect the low speed flow. A post-filter, which is a zero-phase FIR filter, is added after the autocorrelator to further remove the clutter components. The additional computational amount required for the post-filtering increases with the filter length until it becomes equal to the length of the autocorrelation function of the clutter filtered samples but does not change after then. This is because that only two terms of the filtered autocorrelation function are needed for the mean frequency and variance estimation. Since the post-filter can have any large number of taps, it can be designed to have arbitrary responses and find many applications. One example of applying this new technique to dramatically reduce the effect of the clutter components is provided, which is verified by computer simulation.

I. INTRODUCTION

During the last decade, Doppler ultrasound has experienced considerable growth both in the number and diversity of examinations performed. A number of different types of Doppler systems are available for the measurement of blood flow information which is of great clinical importance for understanding the function, control, and state of the vascular systems in human body [1-6]. Although each instrument is based on the Doppler principle for the acoustic measurement of blood velocity, the manner in which the Doppler signals are acquired, processed and displayed distinguishes one type of instrument from another. Among them, two dimensional (2-D) color flow mapping (CFM) is the most recent technique which provides the flow pattern through out the lumen visualized instantaneously as the hemodynamics change during the cardiac cycle and makes it easy to identify a small vessel that is too small to be discerned with real-time gray-scale imaging or to differentiate vascular and nonvascular structures more readily.

2-D CFM in real time requires very efficient and robust processing of Doppler signals because in comparison with conventional pulse wave (PW) Doppler systems the observation time is reduced to only a few repetitions (about 10 or less) to satisfy real time frame-rate requirements. Numerous techniques for determining the Doppler shift frequency have been suggested, many of which can be categorized as power spectrum centroid estimators where some of them rely on signal processing in the time domain and the others work on FFT or AR spectrum domain [7-16]. The major difficulties of these techniques are the accurate determination of the Doppler shift in a noisy environment and in the presence of low frequency signal components (Clutter signal) arisen from the movement of a transducer, tissue, and objects other than Red Blood Cells (RBC) such as heart muscles and vessel walls. The aliasing problem caused by the spectrum broadening of Doppler signals in the ultrasound Doppler process is another difficulty. Some time domain estimators such as the Auto-Correlation Estimator (ACE) and the cross correlation method have been proven to provide more accurate mean frequency and to be easy to implement than others [13-16]. However, these techniques are also susceptible to the estimation error introduced by the clutter components. This estimation error can be further reduced by a clutter filter applying to the received signal. The high order clutter filter (hence, large number of coefficients) is often desired to significantly suppress the clutter from the Doppler components. However, as will be shown in the next section, this clutter filter should be limited in order to achieve the good frequency estimation performance and keep the system noise down as well in most real applications. Consequently, the clutter components are usually insufficiently removed, especially when energy of the clutter component is much greater than that of the Doppler one.

We show in this paper that the filtered clutter components can further be removed by employing the post-filtering technique that is newly introduced in this paper. The post-filter is applied to the autocorrelated function of the filtered signal by the conventional Clutter filter. Therefore, the post-filter should have zero-phase at all frequencies for correct mean frequency estimation. It will be demonstrated that the new technique yields comparable results to the original ACE in weak clutter environments but shows superior results when strong clutter signals are present. Section II provides a brief review of the ACE regarding its input signal characteristics and the clutter filtering. In Section III, the post-filtering technique is described and examined for the applicability to CFM systems. A new algorithm for the Doppler shift estimation combining the ACE and the post-filtering technique is also introduced in this section. The performance of the new algorithm is tested by computer simulations in Section IV. The paper concludes in Section V with a summary and discussion of the advantages and disadvantages of the new post-filtering technique.

II. SIGNAL CHARACTERISTICS AND CLUTTER FILTERING IN ACE

A general block diagram of a CFM system is illustrated in Fig. 1. The input signal, $z_i(n)$, is composed of the noise contaminated Doppler and clutter components. When both the Doppler and clutter signals are pure sinusoidal waves with different frequencies, $0.35 \times$ PRF (Pulse Repetition Frequency) and $0.05 \times$ PRF, respectively, and no system noise exists, the spectrum of $z_i(n)$ would have sidelobes as shown in Fig. 2(a) due to the limited sample length N_i obtained in a given observation time. In this example, the Clutter-to-Doppler Ratio (CDR) of the input signal is assumed to be 25 dB, and hence the peak of the Doppler spectrum is less than surrounding sidelobe levels of the clutter spectrum. As a result, the spectrum of $z_i(n)$ appears to have no Doppler component at all. Fig. 2(a) also shows two clutter filtered spectrums in different ways illustrated in Figs. 2(b) and (c). Since the spectrum of the filtered signal in the way of Fig. 2(b) is that of $z_i(n)$ weighted by the filter response, it seems to be almost impossible to estimate correct Doppler frequency. To overcome this problem, a window must be applied after the clutter filtering as illustrated in Fig. 2(c), which is to throw away the transient responses and can be described as,

$$z(n) = \{z_i(n) * h_c(n)\} \cdot W_N(n), \tag{1}$$

where * denotes the discrete convolution and $W_N(n)$ stands for the window function defined as

$$W_N(n) = \begin{cases} 1 & if\ 0 \le n \le N-1 \\ 0 & otherwise \end{cases} \tag{2}$$

whose length is $N = N_i - N_c + 1$ where N_i and N_c are the length of $z_i(n)$ and $h_c(n)$, respectively. As a sequel, the spectrum of the Doppler component of $z(n)$ is clearly seen and hence the Doppler frequency can be estimated more accurately. The clutter filter used in this example is a 3-tap FIR filter with the impulse responses {-0.25, 0.5, and -0.25}.

While the window operation serves to further enhance the Doppler spectrum, it also broaden the spectrum of Doppler and clutter signals especially when the length of the clutter filter N_c is large. In the extreme case where the clutter filter has length $N_i - 1$, the length of $z(n)$ becomes only two (since $N = N_i - N_c + 1 = 2$) and its spectrum spreads over the entire frequency axis. This makes it almost impossible to estimate accurate mean frequencies and its variances in noisy environment. Moreover, it is evident that the estimation result becomes more sensitive to the system noise as the number of filtered output decreases [17].

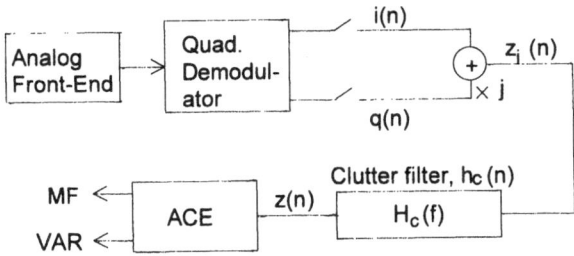

Fig. 1. Block diagram of a CFM system using ACE

It is now clear from the above discussions that the input signal of the mean frequency estimator (MFE) is the windowed output of a clutter filter whose length is often very small, usually not greater than 7, in a real situation. One might experienced difficulties in designing a clutter filter even with 7 taps which can provide sufficient clutter rejection while varying the cutoff frequencies. Consequently, the Doppler shift detection is still influenced by the clutter component.

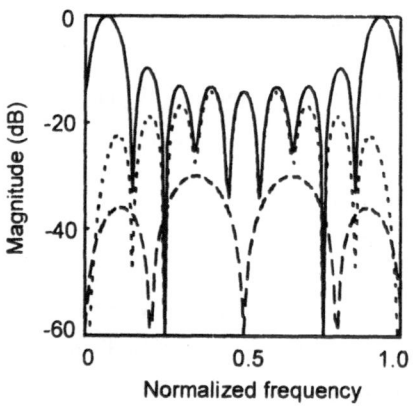

(a) Fourier spectrums of the real parts of input complex Doppler signal, $z_i(n)$, (solid line) and its filtered signal, $z(n)$, in two different ways (dotted line for scheme (b) and dashed line for scheme (c)).

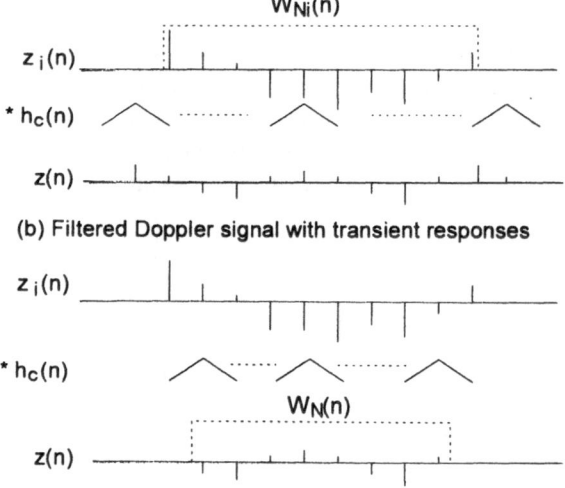

(b) Filtered Doppler signal with transient responses

(c) Filtered Doppler signal without transient responses

Fig. 2. Plots for illustrating the effect of clutter filtering.

III. POST-FILTERING TECHNIQUE FOR ACE

Since Eq. (1) implies that the filtered signal can be thought to have an infinite length, that is, with zero samples outside the window function, however, another filter, which is called a post-filter in this paper, can be employed on this signal to improve the performance of the mean frequency estimation. In this paper, we suggest a post-filtering technique, especially for the autocorrelation estimator, which can further reduce the filtered clutter power for more accurate mean frequency detection. The post-filter is placed after the autocorrelator, where the autocorrelation of $z(n)$, $R(n)$, is calculated for all n's, from $-(N-1)$ to $+(N-1)$. The $R(n)$ is filtered by the post-filter to give the filtered autocorrelation function,

$$R_p(n) = R(n) * h_p(n) \tag{3}$$

The mean frequency is then calculated from the filtered power spectrum, $S_p(f)$, as

$$\bar{f}_p = \frac{\int f \, S_p(f) \, df}{\int S_p(f) \, df}, \tag{4}$$

where

$$S_p(f) = S(f) \cdot H_p(f). \tag{5}$$

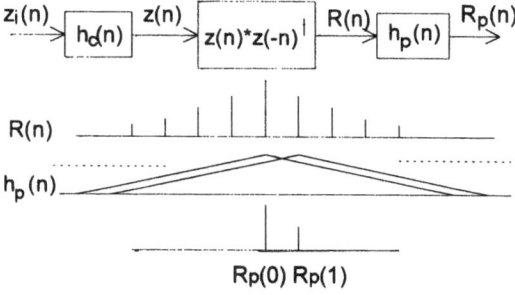

Fig. 3. Post-filtering scheme

The only requirement to get the correct mean frequency from the above equations is that $h_p(n)$ should be a zero-phase filter. If this requirement is met, one can easily verify that the mean frequency is calculated as

$$\bar{f}_p = \frac{1}{2\pi T} \tan^{-1} \frac{\text{Im}[R_p(1)]}{\text{Re}[R_p(1)]} \tag{6}$$

Fig. 3 shows a post-filtering scheme we suggested. The post-filter can also be placed in front of the autocorrelator. However, the post-filtering technique shown in Fig. 3 is much more efficient in computational complexity. In the proposed scheme, we need to calculate only two terms of $R_p(n)$, $R_p(0)$ and $R_p(1)$. Since $R(n)$ is zero for all n except for $n = -(N-1) \, to + (N-1)$, the computational amount for the post-filtering does not increase with the filter length when the filter length exceeds $2N-1$. In other words, an arbitrary high order post-filter can be utilized without increasing the computational complexity. On the contrary, when the post-filter is placed in front of the autocorrelator the computation amount increases with the filter length for both the filtering and taking autocorrelation.

IV. APPLICATIONS OF THE POST-FILTERING TECHNIQUE

The post-filter can find many applications for different purposes. One example of the post-filter is to reduce the influence of the clutter components with a sharp highpass filter. A single sideband filter can be utilized which may be used to detect the forward and reverse flows separately when they exist simultaneously. As another application, a bank of narrow bandpass filters can be used for spectral analysis because $R_p(0)$ after each bandpass filter is the power contained in its passband. We show in this paper only the usefulness of first application by computer simulation results. Our computer simulations are based on the synthesized Doppler and clutter signals using the procedures described in [14]. We assumed that both of the clutter and Doppler signals have Gaussian shaped Power Spectrum Density (PSD), which results in the signal-plus-noise distribution $P(f)$ formulated as

$$P(f) = -\ln[x(f)]\left\{\frac{10^{(CDR+SNR)/10}\,c(f)}{\displaystyle\sum_{f=-0.5PRF}^{0.5PRF} c(f)} + \frac{10^{(CDR+SNR)/10}\,d(f)}{\displaystyle\sum_{f=-0.5PRF}^{0.5PRF} d(f)} + \frac{1}{N}\right\} \qquad (7)$$

where N is the number of discrete frequencies where the spectral distribution, $x(f)$ is a set of random numbers uniformly distributed between zero and one to get a chi-squared PSD, and $c(f)$ and $d(f)$ are Gaussian PSD's with mean frequencies, f_c and f_d, and standard deviations, σ_c, σ_d. The complex signal, $z_i(n)$, is the Fourier inverse transform of $P(f)$ with random phases, between $-\pi$ and $+\pi$, added.

All the post-filters used in the simulations are even symmetric FIR filters with 63 taps using a Kaiser window with a minimum stopband attenuation of 60 dB, which can almost completely remove the clutter components lower than the filter cutoff. Such a sharp filter can introduce an estimation error which occurs when the Doppler signal is very weak or filtered out by the post-filter and CDR is very high. In this situation, the filtered spectrum contains high frequency components of clutter and system noise and MFE can wrongly detect as if there are high speed flows. We applied the common power thresholding technique to avoid this fault estimation. That is, when the signal-plus-noise power in the passband of the post-filter is less than the predetermined Power Masking Level (PML), the estimation frequency is set to be zero. To reject the wrong flow mapping caused when the clutter or Doppler signal with a mean frequency lower than the filter cutoff contributes to add the power in the passband of the post-filter so that the total power is greater than the PML, we introduce another power hresholding technique, Clutter Power Thresholding (CPT), named so because any spectral components lower than the filter cutoff is to be regarded as a clutter signal. Fig. 4 shows the autocorrelation estimator employing the post-filtering technique with CPT where CPTV is predetermined CPT value.

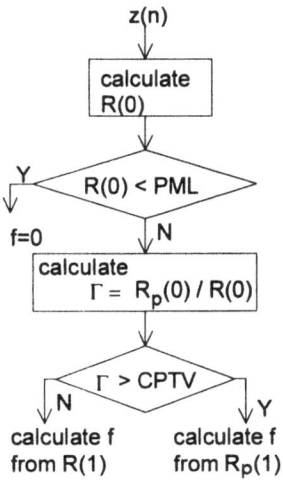

Fig. 4. Flow diagram of ACE using a post-filter for clutter rejection

First, if the power of $z(n)$, $R(0)$, is less than the power threshold level for noise rejection, the output estimation frequency is set to be zero. Otherwise, $R_p(0)$ is calculated to get the following quantity:

$$\Gamma = \frac{R_p(0)}{R(0)}. \tag{8}$$

If Γ is greater than CPTV, it is assumed that the Doppler signal from moving RBC exists with sufficient power in the passband of the post-filter and the mean frequency is calculated from $R_p(1)$. On the other hand, if Γ is less than CPTV, it can be thought that the Doppler signal is very weak or its mean frequency is less than the filter cutoff. In this case, it seems reasonable to output the mean frequency computed by the conventional ACE.

With the computer simulated signals, the performances of the conventional and post-filtered autocorrelation methods are tested and the results are illustrated in Figs. 6 - 9. We used 3, 5, and 7 tap clutter filters with different cutoff frequencies, some of them are plotted in Fig. 5 together with the post-filters used in the simulations.

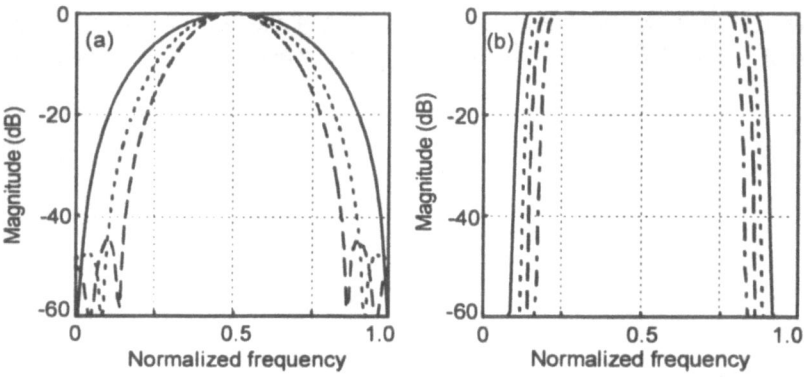

Fig. 5. Clutter and post filter responses. (a) A 3-tap (solid line) and two 7-tap (dotted line, dashed lines). (b) 63-tap post-filters with different cutoffs.

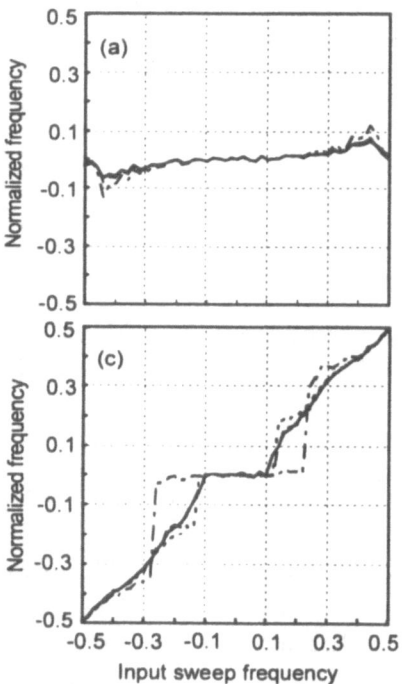

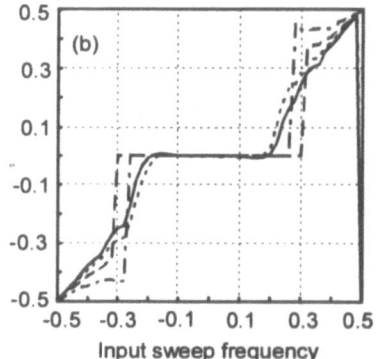

Fig. 6. Estimated mean frequencies by (a),(b) the conventional ACE and the (c) post-filtered ACE under the same signal condition: $f_c = 0$ $\sigma_c = 0.03$ CDR=40 dB $\sigma_d = 0.2$ $N_i = 10$ SNR = 18 dB (solid line), 12 dB (dotted line), 6 dB (dashed line), 0 dB (single dot and dashed line).

For the conventional ACE, we used all the clutter filters and chose the best results. From Fig. 6 through 8, panel (a) shows the estimation results of the conventional ACE using the 3-tap clutter filter and panel (b) shows the best result of the conventional ACE among all clutter filters. Panel (c) shows the estimation results for the post-filtered ACE. Fig. 6 shows the estimation results of two methods when the mean frequency of a Doppler signal (N_i = 10) is swept from -0.5PRF to 0.5PRF for different values of SNR while f_c, σ_c, σ_d, and CDR are fixed. When using the 3-tap clutter filter (panel (a)), the conventional ACE fails to work. We used a 7-tap clutter filter to get the results shown in panel (b) where the estimated mean frequency becomes biased higher as SNR decreases. Fig. 6(c)

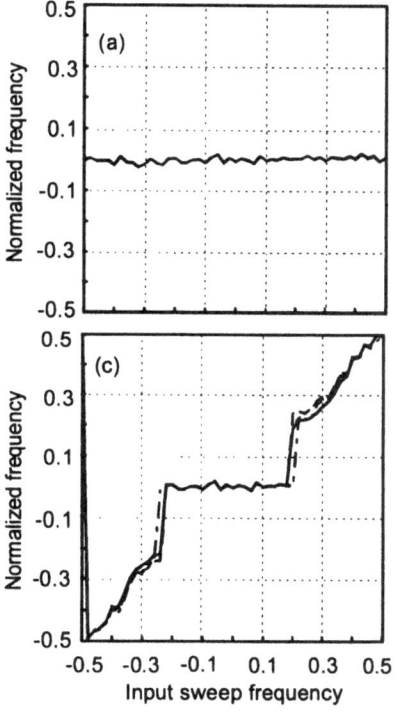

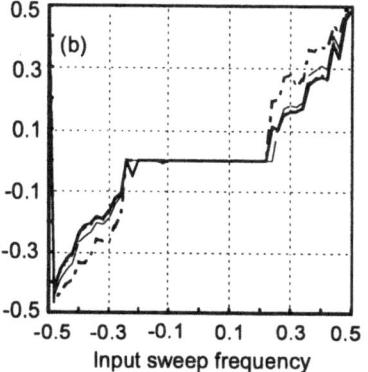

Fig. 7. Estimated mean frequencies by (a),(b) the conventional ACE and the (c) post-filtered ACE under the same signal condition: $f_c = 0$ $\sigma_c = 0.05$ CDR=40 dB $\sigma_d = 0.05$ $N_i = 8$ SNR = 18 dB (solid line), 12 dB (dotted line), 6 dB (dashed line), 0 dB (single dot and dashed line).

shows the superiority of the post-filtered ACE over the conventional ACE. In Fig. 7, the same plots are shown for a smaller number of PRF's (i.e., N_i) and larger Doppler and clutter variances. Fig. 8 shows the results for different values of σ_c. The power masking is not employed in the estimation process for panel (b). As can be seen from this plot, the power masking level should be properly chosen. Notice that even with an ideal power masking setup the ability to detect the low speed flow of the conventional ACE is worse than that of the post-filtered one. In Fig. 9, the number of PRFs is reduced to six and hence only the 3-tap clutter filter is used for the conventional ACE. The results show clearly that the post-filtering can improve the estimation performance in a strong clutter environment.

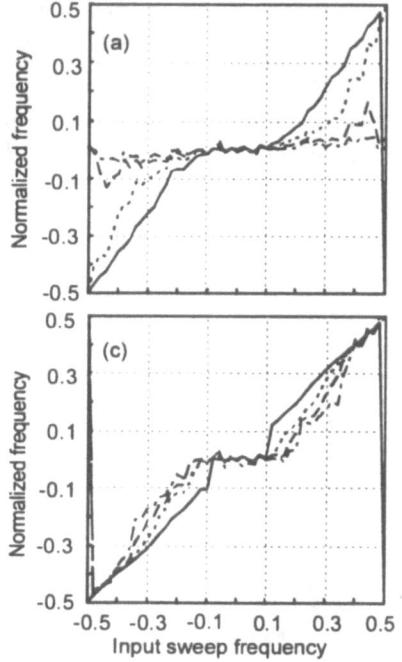

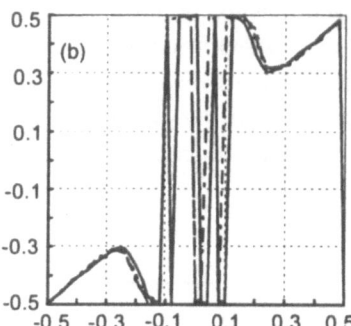

Fig. 8. Estimated mean frequencies by (a),(b) the conventional ACE and the (c) post-filtered ACE under the same signal condition: $f_c = 0.05$ $N_i = 10$ SNR = 18 dB σ_d = 0.03 CDR=40dB σ_c = 0.03 (solid line), 0.04 (dotted line) 0.05 (dashed line), 0.06 (single dot and dashed line).

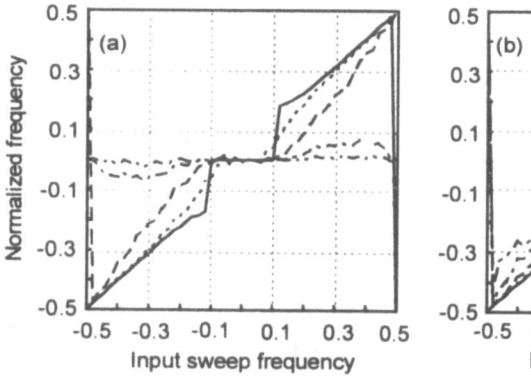

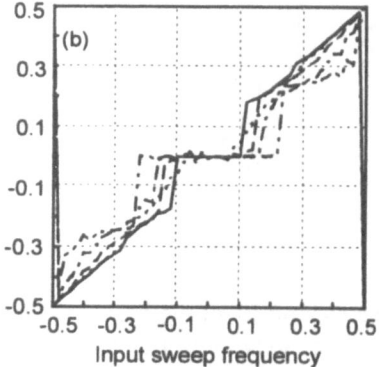

Fig. 9. Estimated mean frequencies by the (a) conventional ACE and the (c) post-filtered ACE under the same signal condition: $f_c = 0$ σ_c = 0.05, σ_d = 0.02, SNR = 18 dB, N_i = 6 CDR= 0 dB (solid line), 10 dB (dotted line), 20 dB (dashed line), 30 dB (single dot and dashed line), 40 dB (double dot and dashed line).

V. SUMMARY AND CONCLUSIONS

In the general realtime 2-D CFM systems, the number of input Doppler samples is limited by the finite observation time (a few pulse repetitions) to achieve the required frame rate. For the satisfactory noise averaging and the reasonable frequency domain spectrum shape in the Doppler mean frequency estimation, the length of the clutter filter should be restricted to a small number. In this situation, the clutter components cannot be sufficiently removed. This leads to the possible estimation error and could affect the detectability of low speed flows. The post-filtering technique suggested in this paper provides an effective way of reducing the influence of the clutter components on the estimation accuracy. It has been shown that the undesired low frequency components below a certain threshold frequency can be completely filtered without much increase in the computational complexity when applying the post-filter to the Autocorrelation estimator. Comparing with the conventional ACE, we found that the post-filtered ACE increases the total computation time no more than 20 % when they are programmed in the assembly code of a fixed-point DSP (ADSP2105, Analog Device). However, we believe that the advantages gained by such a post-filtering technique overcome the sacrifice in the computation amount. In all simulation results, the post-filtered ACE provides better results than the conventional ACE and can detect the flows with strong clutter components while the conventional one completely fails to work. In those simulations, CPTV was varied to get the best results. Notice that the proposed scheme would produce the same result as the conventional ACE when CPTV is not chosen properly. Since the clutter power remained in the stopband of the post-filter, $R(0) - R_p(0)$, can be easily calculated in the post-filtering scheme, it may be possible to develop an adaptive algorithm to adjust the CPTV automatically.

We think that the post-filtering technique developed in this paper should be more advantageous in real imaging. For instance, the Doppler signals obtained from the different imaging regions under different conditions could have different characteristics. Hence, it may be required to adjust the clutter filter case by case. This requirement can hardly be satisfied with the conventional signal processing techniques. However, it is possible and easy to design arbitrary post-filters to manipulate the input signals to achieve the best performance for each applications.

REFERENCES

[1] D. H. Evans, W. N. Mcdicken, R. Skidmore, and J. P. woodcock, Doppler Ultrasound: Physics, Instrument and Clinical Applications. Chichester: Wiley, 1989.

[2] D. L. Hykes, W. R. Hedrick, and D. E. Starchman, Ultrasound Physics and Instrumentation. St. Louis: Mosby Year Book, 1992.

[3] J. M. Reid, "Doppler ultrasound," IEEE Eng. Medicine and Bio. Magagine, pp. 14-17, 1987.

[4] A. R. G. Hokes, R. Reneman, and P. A. Peronneau, "A multigate pulsed Doppler System with serial data processing," IEEE Trans. Sonics Ultraso., vol. SU-28, no. 4, pp. 242-247, 1981.

[5] C. Kasai, K. Namekawa, A. Koyano, and R. Omoto, "Real-time Two-dimensional blood flow imaging using an autocorrelation technique," IEEE Trans. Sonics Ultraso., vol. SU-32, no. 3, pp. 458-463, 1985.

[6]   K. Namekawa, C. Kasai, M. Tsukamoto, and A. Koyano, "Realtime blood flow imaging system utilizing auto-correlation techniques," in Ultrasound '82, R. A. Lerski and P. Morley, Eds. New York: Pergamon, pp. 203-208, 1982.

[7]   M. Brandwstini, "Topoflow - A digital full range Doppler velocity meter," IEEE Trans. Sonics Ultraso., vol. SU-25, no. 5, pp. 287-292, 1978.

[8]   L. Gerzberg and J. D. Meindle, "Power-spectrum c entroid detection for Doppler systems application," Ultrason. Imaging, vol. 2, no. 3, pp. 232-261, 1980.

[9]   W. D. Barber, J. W. Everhard, and S. G. Karr, "A new time domain technique for velocity measurement using Doppler ultrasound," IEEE Trans. Biomed. Eng., vol. BME-32, pp. 213-229, 1985.

[10] A. P. G. Hokes, H. P. M. Peeters, C. J. Ruissen, and P. S. Reneman, "A Novel Frequency Estimation for sampled Doppler signals," IEEE Trans. Biome. Eng., vol. BME-31, no. 2, pp. 212-220, 1984.

[11] O. Bonnefous and P. Pesque, "Time domain formulation of pulse Doppler ultrasound and blood velocity estimation by cross correlation," Ultrason. Imaging, vol. 8, pp. 73-85, 1986.

[12] K. Kristofferson and B. A. J. Angelson, "A comparison between mean frequency estimation for multigated Doppler systems with serial signal processing," IEEE Trans. Biomed. Eng., vol. BME-32, no. 9, pp. 645-657, 1985.

[13] R. S. Jaffe, "Extended-range pulsed Doppler mean-frequency estimation based on mean-frequency prediction," Ph.D. dissertation, G556-3, Stanford Electronics Laboratories, Stanford University, Stanford, California, U.S.A., 1983.

[14] T. Loupas and W. N. McDicken, "Low-order complex AR models for mean and maximum frequency estimation in the context of Doppler color flow mapping," IEEE Trans. on UFFC, vol. 37, no. 6, pp. 590-601, 1990.

[15] Y. B. Ahn and S. B. Park, "Estimation of mean frequency and variance of ultrasonic Doppler signal by using second-order autoregressive model," IEEE Trans. on UFFC, vol. 38, no. 3, pp. 172-182, 1991.

[16] J. C. Willmetz, A. Nowick, and J. J. Meister, "Bias and variance in the estimate of the Doppler frequency induced by a wall motion filter," Ultrason. Imaging, vol. 11, pp. 233-244, 1989.

LATERAL VELOCITY PROFILE AND VOLUME FLOW MEASUREMENTS VIA 2–D SPECKLE TRACKING

L.N. Bohs[1], B.H. Friemel[1], B.A. McDermott[1,2] and G.E. Trahey[1]

[1]Duke University
Department of Biomedical Engineering
Durham, NC 27706

[2]Siemens Medical Systems, Inc.
Ultrasound Group
Issaquah, WA 98027

ABSTRACT

Current Doppler–based techniques for measuring velocity profiles and volume flow rates are plagued by inaccuracies due to errors in estimating the direction of flow with respect to the transducer axis. In addition, such techniques are incapable of quantifying lateral flow. We describe a system which tracks 2–D ultrasonic speckle patterns in order to quantify both the axial and lateral components of motion, thereby obviating the need for angle estimation. The method involves tracking the motion of small regions of an image from one acoustic interrogation to the next using a time–domain pattern matching algorithm. We report on the ability of this system to measure laminar flow velocity profiles and volume flow rates in a laterally flowing phantom. Results indicate that the technique provides accurate velocity and volume flow estimates over a wide range of flow rates.

INTRODUCTION

Current Doppler–based velocity estimators measure only the component of velocity in the direction of the transducer axis. For this reason, operators must estimate the direction of flow in order to obtain 2–D velocity measurements. Unfortunately, small errors in estimating this angle may result in substantial velocity measurement errors, particularly when the angle between the transducer axis and the blood vessel approaches 90° (Phillips, 1989).

Several researchers have evaluated a technique which overcomes this problem by tracking local speckle patterns from one ultrasonic acquisition to the next (Akiyama et al., 1986; Trahey et al., 1988; Chen et al., 1991). Recently, we have reported on a system which uses this speckle tracking method to directly measure two–dimensional motion in real time (Bohs et al., 1993). A brief description of the method follows, with reference to Figure 1.

First, along a fixed viewing angle, two sets of lines are acquired and digitized: the first is called the *kernel* set (darkly shaded sector in the center panel of Figure 1), and the second, larger set is called the *search* set (lighter shaded sector surrounding the kernel). Second, the axial length of the kernel and search regions are defined

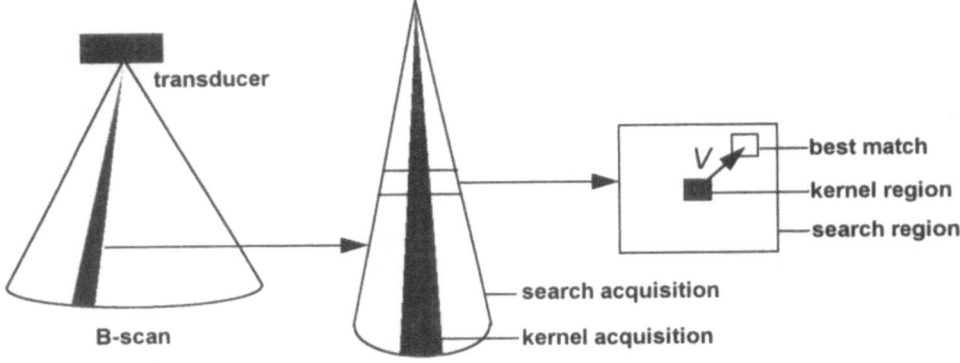

Figure 1. Geometry for two dimensional speckle tracking. At a fixed transmit angle, a small sector of *kernel* lines and then a larger sector of *search* lines are acquired. At each axial range, the best match between a kernel region and surrounding search region is determined, thereby defining the displacement vector for that kernel. Performing these operations over the lateral extent of the image yields a two–dimensional vector map of motion.

to produce the two–dimensional search geometry shown at the right in Figure 1. Third, an exhaustive search through possible matching regions is performed using the Sum–Absolute–Difference (SAD) algorithm (Bohs and Trahey, 1991)

$$\varepsilon_{\alpha,\beta} = \sum_{i=1}^{l} \sum_{j=1}^{k} |K_{i,j} - S_{i+\alpha,j+\beta}| \tag{1}$$

where K and S are samples within the kernel and search region respectively, $l \times k$ are the kernel region dimensions and (α, β) are the coordinates of a trial matching region within the search region. After performing the calculation over the entire search region, the (α, β) yielding the minimum $\varepsilon_{\alpha,\beta}$ indicate the location of the best match. Given the time between acquisitions, the local target velocity can then be determined. Repeating these operations for numerous kernel regions over the desired range results in a *vector line*. A vector velocity map is then obtained by computing vector lines over the lateral extent of the B–scan.

The technique has previously been shown successful for measuring the two–dimensional motion of a string target at various angles within the scan plane (Bohs et al., 1993). In this study we investigate the ability of the system to measure parabolic flow velocity profiles in a laterally flowing phantom, and to estimate volume flow rates based on these measurements.

METHODS

For this study, a test phantom consisting of a 1m long straight rigid entrance tube feeding a latex rubber vessel with a diameter of approximately 12mm was constructed. A solution of 40% glycerol and water was pumped continuously through this closed system by a computer controlled, calibrated flow unit (UHDC, Toronto). A concentration of 2g/l of cornstarch was added to the glycerol solution as a scattering agent. This test setup provided fully developed laminar flow with Reynolds numbers varying between approximately 150–800 over the range of flow rates utilized in these experiments.

The velocity measurement system for this study has been previously described in detail (Bohs et al., 1993). Briefly, the system consisted of a Siemens SI–1200 phased array scanner which provided analog envelope–detected signals. These signals were sampled to 8 bits at 27 MHz and transmitted to 14 SAD–tracking boards, which computed 2D velocity vectors in real time. The velocity vectors were color encoded and displayed in real time as well as stored for off–line analysis on a PC.

The tracking geometry utilized a 3×8 (i.e. number of lateral lines $\times$ number of axial samples) kernel region and an 11×32 pixel search region, yielding physical dimensions of 0.37mm $\times$ 0.25mm and 1.35mm $\times$ 1.02mm respectively at the 3cm range of the vessel.

Velocity profiles were measured at a transducer/vessel angle of 90 degrees across the tube diameter at five flow rates from 5 to 25 ml/s. At each flow rate, 100 independent velocity profile measurements were acquired and stored for analysis.

Two techniques were utilized for estimating volume flow rates. In the first, an estimate of the tube diameter was obtained by finding the first zero–velocity estimates in both directions from the peak velocity. Using the relationship between peak velocity and mean velocity under conditions of laminar flow, $\overline{V} = V_{pk}/2$, the first volume flow estimate was obtained from

$$Q_1 = \frac{V_{pk} \times \pi D^2}{8} \tag{2}$$

where V_{pk} was the peak measured velocity and D was the tube diameter.

The second volume flow estimate was obtained from the discrete velocity measurements by assuming circular symmetry about the velocity peak and averaging the measurements on either side of the peak according to

$$Q_2 = \sum_{i=0}^{N} \frac{\pi(r_i{}^2 - r_{i+1}{}^2)(V_i + V_{i+1} + V_{-i} + V_{-i-1})}{4} \tag{3}$$

where N was the index of the first zero–velocity estimate on the tube wall, V_i were individual velocity estimates and $i = 0$ was at the location of the peak velocity.

RESULTS

Figure 2 shows the velocity profiles measured at each of the five flow rates at 90 degrees. These profiles represent the mean of 100 estimates, after removal of false peaks as described previously (Bohs et al., 1993). In order to confirm that the velocity profiles were indeed parabolic, a least–squares parabolic fit was computed for each measured profile, as shown in Figure 2. The correlation coefficients between the best parabolic fits and the velocity measurements were 0.94, 0.96, 0.97, 0.97, and 0.93 at flow rates of 5, 10, 15, 20 and 25 ml/s, respectively.

Figure 3(a) shows the standard deviation of the velocity estimates for the measurements at 15 ml/s. Standard deviations for other flow rates were qualitatively similar. False peaks are large errors due to decorrelation and noise; such errors typically may be removed in real time via spatial filtering methods. False peaks represented 20–40% of the estimates as indicated for the case of 15ml/s flow in Figure 3(b).

Close examination of Figure 2 indicates that the tube diameter increased slightly with increasing velocity due to expansion of the tube under higher pressure. In addition, measurements at 20 and 25 ml/s are offset from those at lower flow rates due to adjustments in the phantom and transducer setup during the course of the experiments.

Figures 4(a) and 4(b) show the volume flow rates computed by Eqn. (2) and Eqn. (3) respectively versus the actual flow rates. The least–squares best fit lines are as indicated. For the peak velocity method (Figure 4(a)), the best fit line had a slope, y–intercept, and correlation coefficient of 1.47, -4.27 and 0.986 respectively. For the discrete integration method, (Figure 4(b)), the best fit line had a slope, y–intercept, and correlation coefficient of 1.17, -1.56 and 0.996.

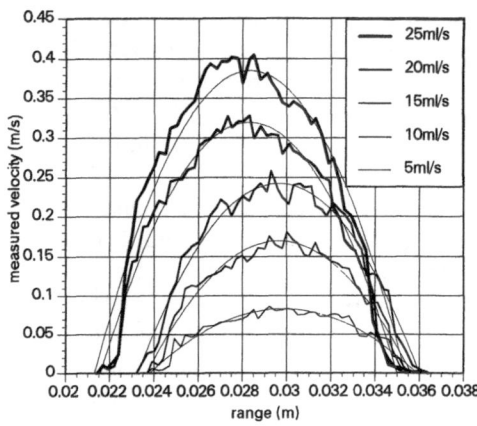

Figure 2. Measured velocity profiles versus axial range. Data represents the mean of 100 estimates at each flow rate, after removal of false peaks. Approximately 60 estimates were obtained across the vessel diameter in range. Thin smooth lines represent the best parabolic fit to each measured profile.

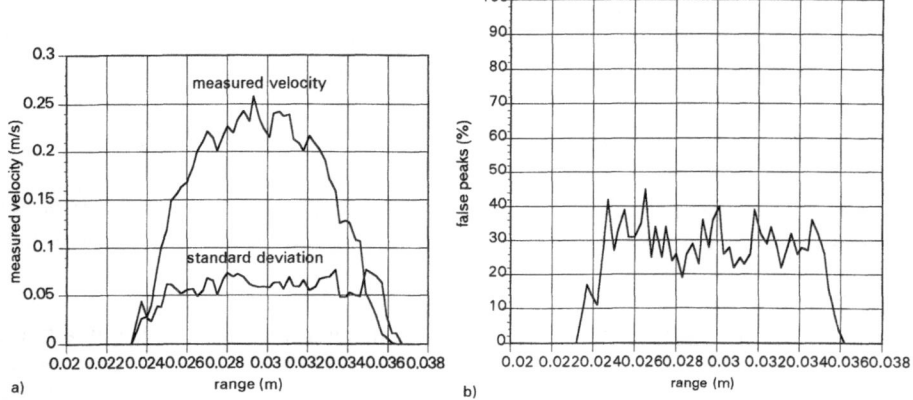

Figure 3. (a) Standard deviations of the measurements at 15 ml/s. (b) Percent false peaks, out of 100 total velocity estimates, removed from the measured data prior to plotting the data in Fig. 2 for the 15 ml/s velocity profile.

DISCUSSION

In this study, the two different methods of volume flow rate estimation produced somewhat different results. While the method based on peak velocity is simpler to compute, it assumes that the flow is fully developed, which limits its accuracy and clinical applicability. The discrete integration method provide better estimates of flow rates even in this idealized study; in more realistic cases where the flow profile is not parabolic, but remains circularly symmetric, this method is likely to prove far superior.

Several sources of error contributed to the measurements in this study. First, shadowing behind the top vessel wall and reverberations within the latex rubber vessel were observed, which would be expected to bias estimates downward. Second, slight angular and elevational displacements of the transducer relative to the exact longitudinal center–plane of the vessel were possible, since transducer placement was guided visually. Finally, errors in estimating speed of sound in the glycerol–water mixture, which was measured using time–of–flight trials yielding a value of 1716 m/s., were a large source of potential inaccuracies since both velocity computations and range determinations depend on this estimate. An over–estimate of the sound speed would result in velocity over–estimation, and vice–versa.

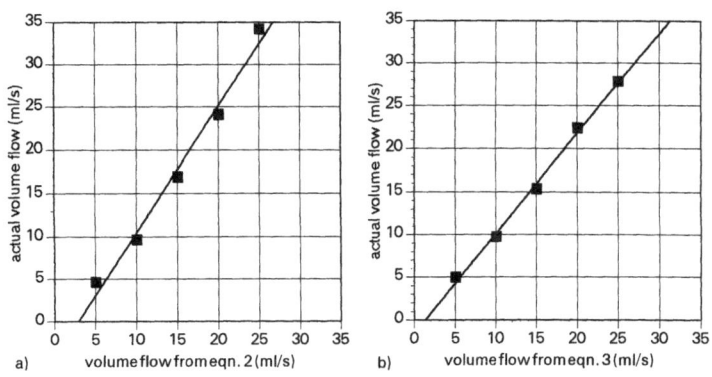

Figure 4. Volume flow rates computed from mean measured velocity profiles. (a) Flow rates based on peak velocity (Eqn. 2); (b) Flow rates based on discrete integration (Eqn. 3). Best linear fit lines are as shown.

It is important to emphasize that all measurements reported in this study were made at a transducer angle of 90 degrees, and no angle correction was required. Such an angle is convenient for measuring flow in vessels parallel to the body surface, such as the carotid artery and jugular vein. Measurements at this orientation are not possible with current Doppler instrumentation.

CONCLUSIONS

The performance of a system which tracks local speckle patterns to compute two dimensional velocities has been evaluated. The system provided accurate measurements of parabolic velocity profiles which were used to compute volume flow estimates. The results indicate that the technique obviates the need for angle estimation as in current velocity and flow rate estimators, and is capable of quantifying flow perpendicular to the transducer face.

ACKNOWLEDGEMENTS

This work was supported in part by NSF/ERC grant CDR–8622201, DHHS NCI research grant CA–43334 and The Whitaker Foundation.

REFERENCES

Akiyama, I., Hayama, A., Nakajima, M. Movement analysis of soft tissues by speckle patterns' fluctuation. *JSUM Proceedings*: 615–616; 1986.

Bohs, L.N. and G.E. Trahey, A novel method for angle independent ultrasonic imaging of blood flow and tissue motion; *IEEE Trans. BME* 38: 280–286, 1991.

Bohs, L.N., Friemel, B.H., McDermott, B.A., and G.E. Trahey, A real–time system for quantifying and displaying two–dimensional velocities using ultrasound, *Ultrasound in Med. & Biol.* 19: 751–761, 1993.

Chen, E.J, Hein, I.A., Fowlkes, J.B., Adler, R.S., Carson, P.L and W.D. O'Brien, Jr., A comparison of the motion tracking of 2–D ultrasonic B–mode images with a calibrated phantom, *Proc. 1991 Ultrason. Symp.*, pp. 1211–1214.

Philips, D.J., Beach K.W., Primozich J. and D.E. Strandness, Jr., Should results of ultrasound Doppler studies be reported in units of frequency or velocity?, *Ultrasound in Med. & Biol.* 1989; 15: 205–212.

Trahey, G.E., Hubbard, S.M. and O.T. von Ramm, Angle independent ultrasonic blood flow detection by frame–to–frame correlation of B–mode images. *Ultrasonics* 26: 271–276; 1988.

BENEFITS AND LIMITATIONS OF THE C-MODE DOPPLER PROCEDURE

U. Moser, P. M. Schumacher, M. Anliker

Institute of Biomedical Engineering, ETH and University Zurich, Gloriastr. 35, 8092 Zurich, Switzerland

Introduction

A new procedure called 'C-Mode-Doppler' has been developed for the quantitative measurement of 2-d velocity fields and the volume flow in large blood vessels. It combines conventional pulse Doppler and holographic techniques and makes use of a 2-d array transducer.

Fig. 1 shows an arrangement of the array transducer and a blood vessel. The measuring slice lies parallel to the transducer surface and is divided into small voxels. Within each voxel the velocity of the red blood cells can be measured using the Doppler effect. As the latter only assesses velocity components in direction to the transducer - which is perpendicular

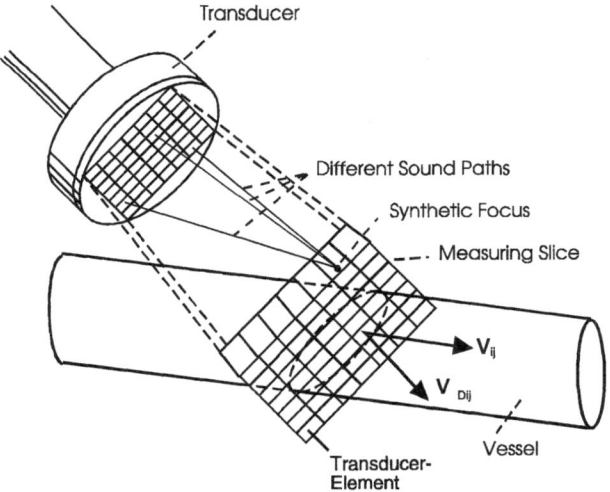

Figure 1. The C-Mode Doppler procedure uses a 2-d array transducer that allows to separate different voxels at a given distance in a chess board-like arrangement. This measuring slice is placed such that it covers the whole vessel cross section.

to the measuring slice - volume flow can be computed independently of the angle of incidence. This important feature of the C-Mode Doppler procedure is discussed in the paper by P.M. Schumacher in this volume of Acoustical Imaging.

The procedure

In order to reduce the measuring time for one frame of the 2-d velocity field, a parallel interrogation technique is used instead of the commonly used sequential one. For this purpose a large sound pulse is transmitted, which covers the whole measuring slice at once. An example of a wide sound beam is shown in fig. 2. It was generated by a small experimental array divided into 6 by 6 elements. In the transmit mode the 36 elements are tied together to 4 subgroups which are driven by 4 programmable transmitters. The left part of the figure shows a computer simulation and on the right side the corresponding measured sound field is displayed. The C-Mode Doppler procedure does not depend on a homogenous intensity distribution in the measuring slice: The velocity measurement is based on the Doppler frequency shift and therefore the signal power has relatively little influence on the estimated values. However, the intensity variations produce higher crosstalk between neighboring voxels and for this reason an adequate sound homogeneity is advantageous. The homogeneity can be improved if more transmitters are provided.

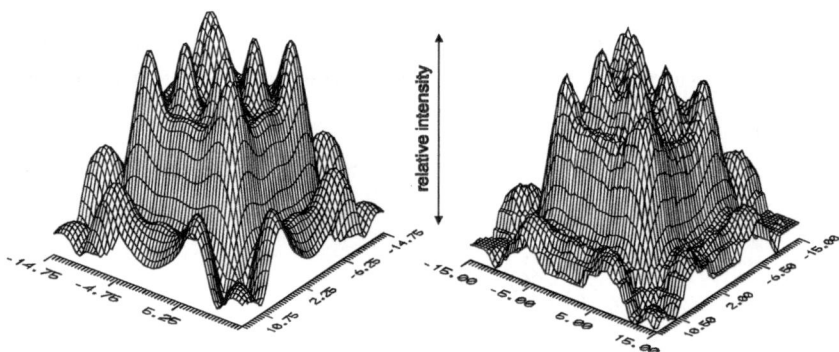

Figure 2: Wide sound beam generated by a 6 by 6 element 2-d array at a distance of 80 mm. Center frequency of the transducer: 4 MHz, pulse bandwidth: 1 MHz aperture of the array: 9x9 mm. Left picture: Computer simulation, right picture: Measurement in a water tank.

Separation of the echoes of the individual voxels is carried out at time of reception only. As indicated in fig. 1, signal paths from a specific voxel to the individual transducer elements have different lengths and therefore the echoes from that voxel arrive with different delays at the transducer elements. This fact can be used to separate the echoes from the individual voxels at reception by a synthetic focusing process.

Each structure in the measuring slice reflects a small amount of sound energy back to the transducer surface as a spherical wave (fig. 3). At a given time, lines of zero-phase of the back scattered waves are circles around a center point, located at the projection of the reflective structure on the array transducer surface. The echoes of all reflective structures produce a complex interference pattern, which is spatially sampled by the individual transducer elements.

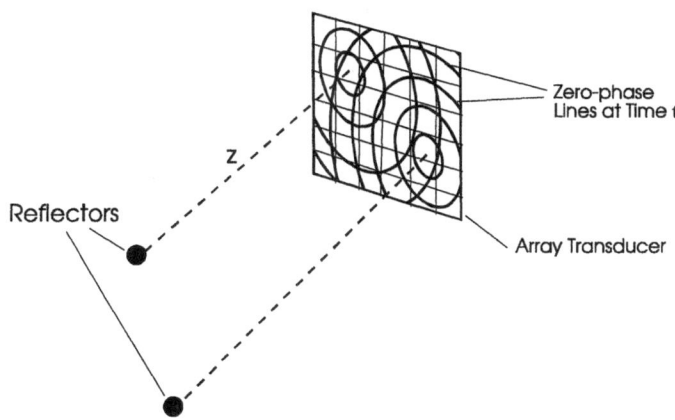

Figure 3. Interference pattern on the transducer surface. Each point scatterer reflects a spherical wave back to the transducer, where at any given instant a complex interference pattern develops.

To satisfy the spatial sampling theorem, at least 2 elements are required between 2 lines of zero-phase. The distance between lines of zero phase becomes smaller with increasing deflection of the synthetic receive pattern. Echoes from voxels located at the border of the measuring slice give raise to most dense lines of zero-phase on the opposite border of the array transducer. Therefore the separation of voxels which have large

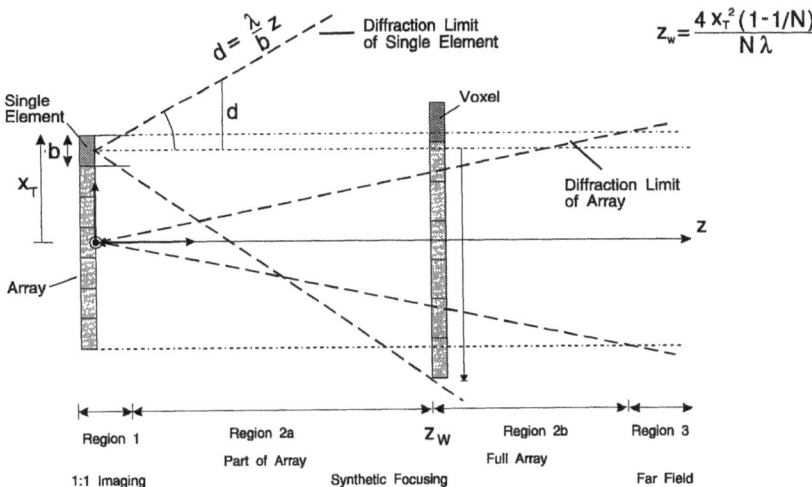

Figure 4. Characterization of 2-d array sound field with NxN elements. Region 1: Near field of single element. Region 2a: Part of array is used for synthetic focus. Region 2b: Full array used for synthetic focus. Region 3: Far field of array. For the experimental 4 MHz array with 6x6 elements and 9 mm aperture we find: Near field of single element: up to 3 mm, $z_w = 30$mm, far field: > 107 mm. .

off-axis distance and that are located near to the transducer requires a large number of array elements. If a loss of lateral resolution can be tolerated, it is possible to reduce this number by utilizing only part of the array for the synthetic focusing process.

Fig. 4 shows a coarse characterization of the sound field of a 2d array transducer. We have identified 3 regions. In the near field of a single transducer element only a 1:1 mapping of voxels and transducer elements is possible. In the following region 2 voxel separation is realized by synthetic focusing. Up to a distance z_w, i.e. in region 2a, elements exist that do not include the whole measuring slice in their field of view. Only part of the array should be used in these cases. This can be achieved by weighting the transducer elements with a gaussian function, which is centered around the voxel's projection point on the transducer surface. The width of this weighting function must be adapted to the distance between the transducer and the specified voxel. Fig. 5 gives an example of the receive pattern of an array with 6 by 6 elements and an aperture of 9 mm at a distance of 20 mm. The left picture shows strong grating lobes whilst on the right side due to the apodization grating lobes exhibit a lower level. However, the width of the main lobe is accordingly larger. In the region 2b the full array can be used and grating lobes generally lie outside the measuring slice. Finally region 3 depicts the far field region of the array where lateral resolution can not be improved by focusing.

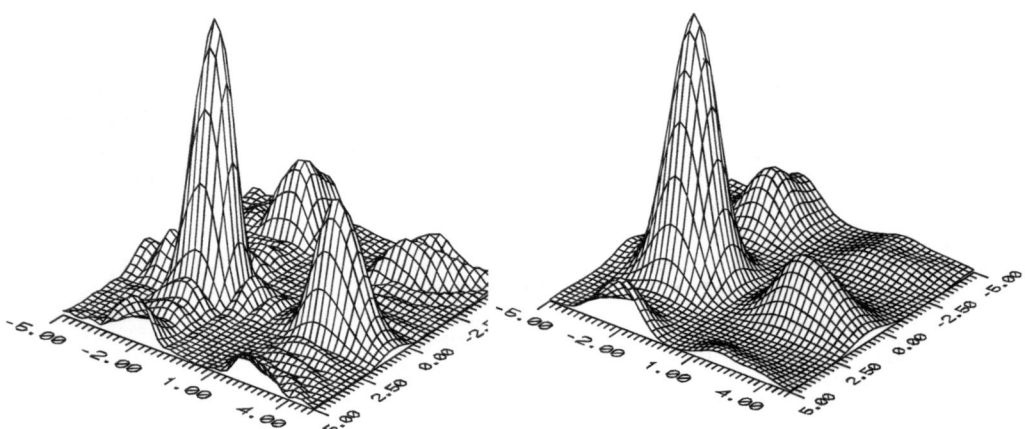

Figure 5: Characteristic of reception of experimental 4 MHz array with 6 by 6 elements and 9 mm aperture at z = 20 mm. Left side without apodization shows strong grating lobes. Right side with gaussian apodization with a -6 dB radius of 2.25 mm shows suppressed grating lobes.

Fig. 6 shows a simplified diagram of the electronic circuitry of our real-time C-Mode prototype. In the transmit mode, the array elements are connected to up to 8 subgroups which are driven by 8 programmable transmitters. Since no lateral deflection is needed for the transmitted sound pulse, a small number of transmitters is sufficient to achieve an adequate smoothing of the launched pulse. However, at time of reception, signals of each transducer element must be recorded in parallel and consequently each one needs its own receiving channel.

A receiving channel consists of a preamplifier and a synchronous demodulator with the necessary filters. A digitizer performs the analog-to-digital conversion with a resolution of 12 bits. A high-pass digital filter suppresses stationary echoes in order to reduce the dynamic range. Data is then temporarily stored in a fast RAM. A digital beam former separates the echoes from the different voxels. According to pre-computed delay times between each pair of voxels and transducer elements, the beam former selects the nearest time gate. In a next step the signal phase is adjusted such that the echoes originating from a specific voxel are in phase at all the channels. Superposition of these signals amplifies echoes from the selected voxel while all others are suppressed. This process is repeated for each voxel with the same data stored in RAM, but with the appropriate delay time corrections. The result is an "image" of the scatterer distribution in the measuring slice, obtained from a single transmitted pulse. To evaluate the Doppler shift within each voxel, this process is repeated for many pulse periods and the separated data must be rearranged in a circular buffer. A FFT-processor computes the Doppler spectrum (128 or 256 points) of the separated signals and a spectrum interpreter estimates the mean Doppler shift. Finally a graphics processor displays the velocity field in real-time.

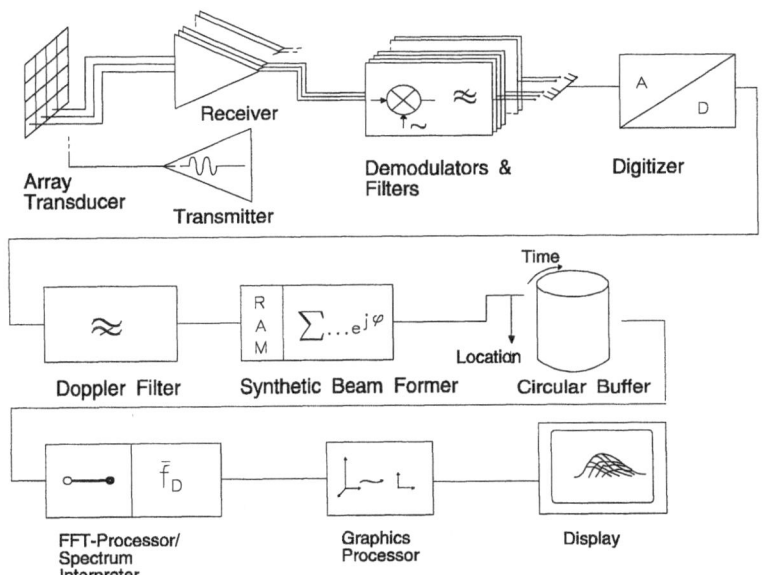

Figure 6. Bloc diagram of C-Mode Doppler system. See text.

Discussion

The accuracy of the C-Mode procedure is affected by the limited spatial resolution. Since a mean frequency detector is used, the mean velocity within each voxel is measured, so that the influence of the limited resolution on the estimated volume flow is minimized. However, an overestimation of the volume flow occurs due to partial volume effects at the vessel wall.

A disadvantage of the C-Mode Doppler procedure is the fact that many transducer elements are needed and that each of them must be equipped with the appropriate electronics, which results in a high hardware effort. Hence it is important to minimize the number of transducer elements, which can be achieved by adapting the transducer aperture to the size of the vessel and for short measuring distances by using only part of the array as mentioned above.

The C-Mode procedure has the advantage, that volume flow measurements do not depend on the angle of incidence. In addition it is relative insensitive to sound field variations since the frequency shift is evaluated and not the power of the Doppler signal. This also makes it insensitive to fluctuations of the reflection coefficient of the blood, which may occur in the presence of turbulence. The parallel interrogation allows for the measurement of a 2-d velocity field within short time intervals of 10 to 30 ms. In contrast to sequential scanning procedures, increasing the number of voxels or the spatial resolution does not affect the time resolution or the accuracy of the measured Doppler frequency. It primarily is a question of the technical effort, i.e. the size of the transducer and the number of its elements with the necessary electronic circuitry.

The real-time display of the 2-d velocity field allows the user for the correct placement of the measuring slice and gives additional hemodynamic information.

Conclusion

The C-Mode procedure inherently avoids most of the difficulties of the conventional ultrasonic volume flow techniques. It therefore can be considered as a reliable and robust procedure for volume flow measurements. Thanks to the fast progress of micro electronics, the relative high technical effort will be of decreasing importance in the future.

Literature

1. R. W. Gill: Measurement of Blood Flow by Ultrasound: Accuracy and Sources of Error, Ultrasound in Med. & Biol., Vol. 11, 4, pp. 625-641, Pergamon (1985).

2. U. Moser: Method and Configuration for Measuring a Two-Dimensional Reflective Structure, USA Pat. No. 5,117,692, 1992.

3. U. Moser, M. Anliker, P. Schumacher: Ultrasonic Synthetic Aperture Imaging Used to Measure 2-D Velocity Fields in Real-Time, IEEE International Symposium on Circuits and Systems, Singapore 1991.

4. P. M. Schumacher, A. Martinoli, A. Bührer, U. Moser: In-vitro Study of the Velocity Distribution and Volume Flow Using Quantitative C-Mode Doppler, Europ. J. Ultrasound, Su 1, Vol. 1, Euroson Abstracts 1993.

IN VITRO MEASUREMENTS OF TRUE VOLUME FLOW WITH C-MODE DOPPLER

Peter M. Schumacher, Urs Moser, Max Anliker

Institute of Biomedical Engineering and Medical Informatics, ETH and University Zürich

Abstract

'C-Mode Doppler' is a new real-time technique, that can measure the two dimensional velocity distribution and true volume flow in large blood vessels. An in-vitro study has been carried out for the evaluation of a real time prototype C-Mode Doppler system. The C-Mode system was equipped with a 2D array transducer with 63 elements arranged in a 7 by 9 matrix. The measured velocity profiles are in good agreement with the expected parabolic distribution. However, a slight overestimation of the vessel diameter has been observed due to crosstalk between neighbouring sample volumes and the limited resolution of the system. The evaluated independence of the angle of incidence has an error range of 2 % to 4 % depending on the distance between the array transducer and the measuring slice. The linearity was tested for flow towards and away from the probe at rates between 2 and 38 ml/s; the resulting correlation coefficient was 0.9986. The presence of disturbed 2D velocity profiles showed only minor influence on the evaluated flow rate, as documented with measurements up- and downstream of stenoses and with dynamically changing velocity profiles when using oscillatory flow.

Assessment of volume flow

Before considering the C-Mode measuring process we want to introduce the mathematical definition of flow through an arbitrary area A (Fig. 1).

$$\dot{Q}(t) = \int_A \mathbf{v}(t)\,\mathbf{n}\,dA = \int_A v_\perp(t)\,dA \qquad (1)$$

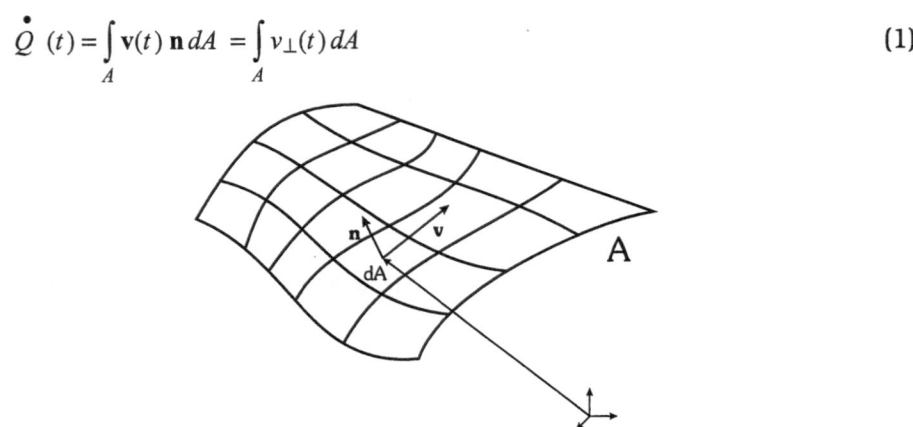

Figure 1 Flow through an arbitrary area A. **v** is the velocity vector of a particle and **n** is the normal vector on dA.

In contrary to conventional color flow mappers C-Mode Doppler uses a scan plane perpendicular to the beam axis of the transducer and therefore requires a two dimensional array transducer for the beam steering in the measuring slice. This measuring slice is subdivided into an array of sample volumes in each of which the Doppler shift is evaluated (Fig. 2). By summing up the products of the velocities v_{ij} (perpendicular on A_{ij}) with the known sample volume size A_{ij} the actual volume flow through the measuring slice is obtained:

$$\dot{Q}(t) = \sum_{ij} v_{ij}(t)\,A_{ij} \qquad (2)$$

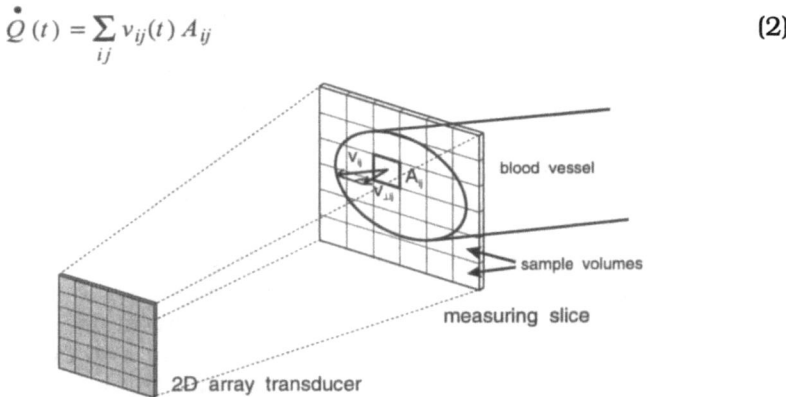

Figure 2 C-Mode Doppler volume flow measurement setup.

Equation (2) is the discrete form of the mathematical flow definition above, which means, that volume flow can be measured independently of the angle of incidence and in the presence of secondary flow motions as long as the blood vessel lies fully within the measuring slice. Because the *mean velocity* is detected in all sample volumes, it is possible to use a rather coarse arrangement of sample volumes and to get still an accurate flow estimation. If the knowledge of the exact 2D velocity distribution is of importance, then of course one might choose more sample volumes (and probably a matching 2D transducer). Thanks to the parallel illumination and echo recording of the sample volumes C-Mode Doppler yields a frequency resolution and stability that is comparable to that of a single gate Doppler system.

C-Mode Prototype System

The C-Mode prototype system is described in the paper "Benefits and Limitations of the C-Mode Doppler Procedure" by Moser et al. in this volume of Acoustical Imaging.

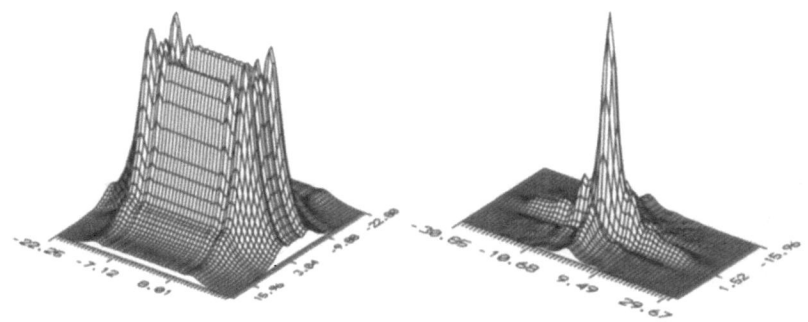

Figure 3 Large transmit pulse for parallel illumination of the measuring slice (left) and focused beam (right) for a measuring distance of 50 mm.

The 2D array transducer used for the measurements consists of 7 by 9 single elements with a total aperture of 16 by 32 millimetres. The transmit pulse shape is large for parallel illumination of the measuring slice (Fig. 3, left). The focused beam in 50 mm measuring distance has a -6 dB focus spot size of 2.2 by 3.4 mm (Fig. 3, right). This gives us the approximate resolution at that measuring distance.

To measure the accuracy of the C-Mode Doppler system a flow reference was needed with an error of less than one percent for steady flow conditions. Because no commercial system could provide that, we built our own pump system consisting of two cylinders with mechanically coupled pistons that are moved by a computer controlled stepper motor. The flow rate can be varied within 100 ml/s with the required precision. The test tube used, had an internal diameter of 10 mm and a sufficient inlet length in both directions to establish parabolic flow for the blood mimicking fluid as well as for blood itself.

Experiments

Angle of incidence   The first experiments documented here were carried out to prove the independence of the angle of incidence of the C-Mode Doppler measuring process. The results of the first two experiments are based on a blood mimicking fluid consisting of a suspension of fine lime powder in water.

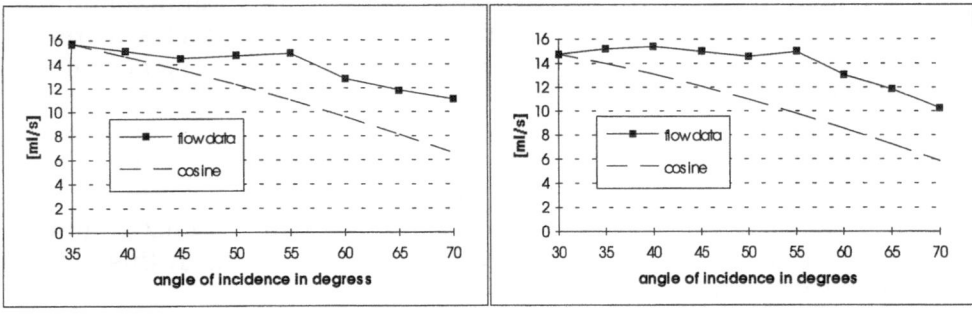

Figure 4 Measurements of the dependence of the angle of incidence at 40 mm (left) and 50 mm (right) measuring distance. At 60 degrees and above the tube was no longer completely inside the measuring slice.

The flow was set to 15 ml/s and the angle was varied between 30 and 70 degrees. In the left diagram of fig. 4 the measuring slice was at 40 and in the right at 50 mm distance from the transducer. The curves denoted 'cosine' are for reference, and show how the output of a conventional Doppler system would vary under the same conditions. The 'cosine' curve was normed so that it starts at the same value as the measured flow data.

Both plots show a reasonably constant value within 30 and 55 degrees; the standard deviation is 3 percent for 40 mm distance and 2 percent for 50 mm distance of the transducer of the respective flow value. At 60 degrees and above the tube was no longer completely inside the measuring slice, which accounts for the decreasing flow value. Smaller angles than 30 degrees could not be studied because of mechanical constraints of the large transducer.

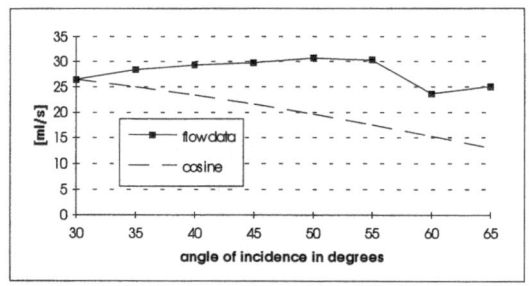

Figure 5 Same experiment as in fig. 5 but with blood at a measuring distance of 50 mm. Partial aliasing was observed at 30 and 35 degrees (see text).

As shown in fig. 5 the results can be reproduced with blood instead of the blood mimicking fluid. The flow rate was increased to 30 ml/s. This is the reason for the partial aliasing observed at 30 and 35 degrees when considering the used pulse repetition frequency of 6.8 kHz. The standard deviation of the valid measurements (40 - 55 degrees) is 2% of the flow rate.

Linearity The next experiment was done to estimate the linearity of the C-Mode Doppler measurements. The flow system supplied steady flow rates between 2 and 42 ml/s in steps of 2 ml/s. At about 40 ml/s aliasing was detected, these values were not used for the fit procedure. Apart from the measured flow data 2 additional curves can be seen in fig. 6. The reference shows the one to one mapping of the pumping power to the ordinate. The fit is a linear fit (least squares) of the flow data with a resulting correlation factor of 99.86 %. The slope of the fit line is not 45 degrees, it amounts to 46.8 degrees.

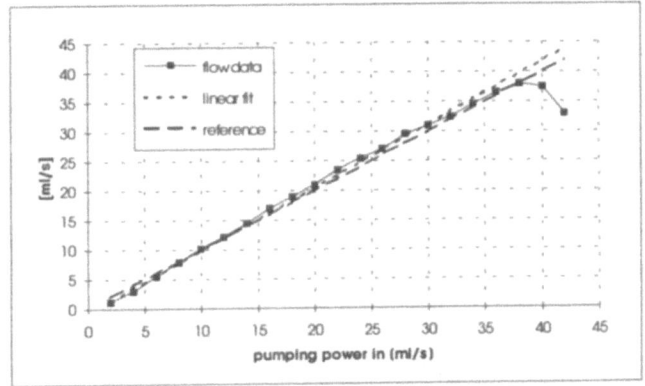

Figure 6 C-Mode Doppler linearity measurements. Note, that there is aliasing at 40 ml/s and above.

Figure 7 shows a sample of the 2D velocity profile measured in the 10 mm test tube (single, non averaged frame). The large plot was obtained by 2D interpolation of the original data shown on the right side. At an angle of incidence of 45 degrees a vessel diameter of about 14 mm in the long direction (projection) and of course 10 mm in the other direction should be measured.

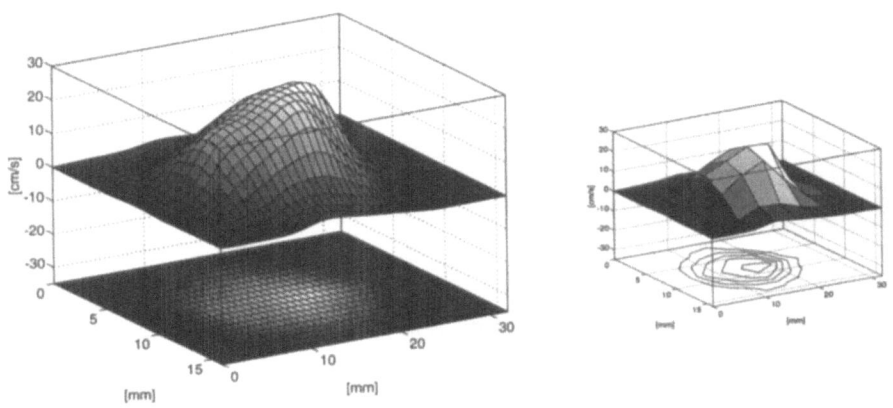

Figure 7 Sample velocity profile in a 10 mm tube in 50 mm measuring distance. The angle of incidence was 45 degrees and the flow rate was 20 ml/s. On the left side is a 2D interpolation of the original data shown on the right side.

However, if we read the scales, we get approximately 17 and 11 mm. This overestimation is due to some crosstalk between neighboring sample volumes and of course the limited resolution of the 7 by 9 array transducer as mentioned earlier. This accounts for the partial volume effects observed on the vessel wall.

Disturbed velocity profiles The next few measurements will document the performance of the C-Mode Doppler system in the presence of disturbed velocity distributions. First the influence of artificial stenoses will be shown with concentric and asymmetric obstructions, which are, due to construction constraints, 12 mm long. Thanks to the fact that oscillatory flow can be generated with our pump system, only one measuring position is needed in order to measure the flow up- and downstream of a stenosis. After that, the formation of a parabolic velocity distribution will be documented when using slow oscillatory flow motion.

The 2D velocity profiles shown in fig. 8 show single, non averaged frames, recorded during real-time processing. On the left plot the findings for a concentric 40 % stenosis are shown. The measuring slice was positioned 17 mm away from the centre of the stenosis. The left subplot shows the virtually undisturbed velocity distribution upstream of the stenosis. The right subplot shows the distribution downstream of the stenosis with a comparatively small jet in the middle and with some backward flow. The volume flow rate of 10 ml/s is correctly detected in both cases.

On the right plot of fig. 8 we have the results of an asymmetric stenosis. The measuring slice was positioned 14 mm away from the centre of the stenosis this time. The left subplot again shows the virtually undisturbed velocity distribution upstream of the stenosis. The velocity distribution downstream of the stenosis however shows a large amount of backward flow as would be expected. The flow of 8 ml/s is again correctly measured in both cases.

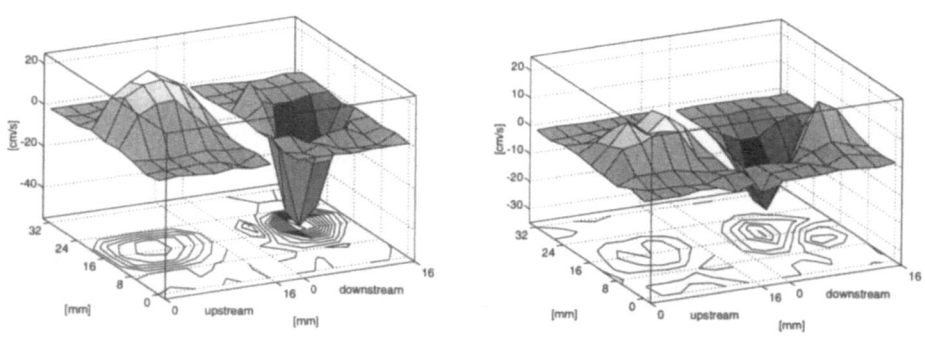

Figure 8 Velocity profiles of a symmetric (left) and asymmetric (right) stenosis with up- and downstream measurements (see text).

A difficulty during the measurements with stenoses were the high velocities that may appear in a jet downstream of a stenosis. For this reason we used comparatively small flow rates of 10 and 8 ml/s in the experiments. The problem of aliasing *must* be addressed when measuring stenotic vessels.

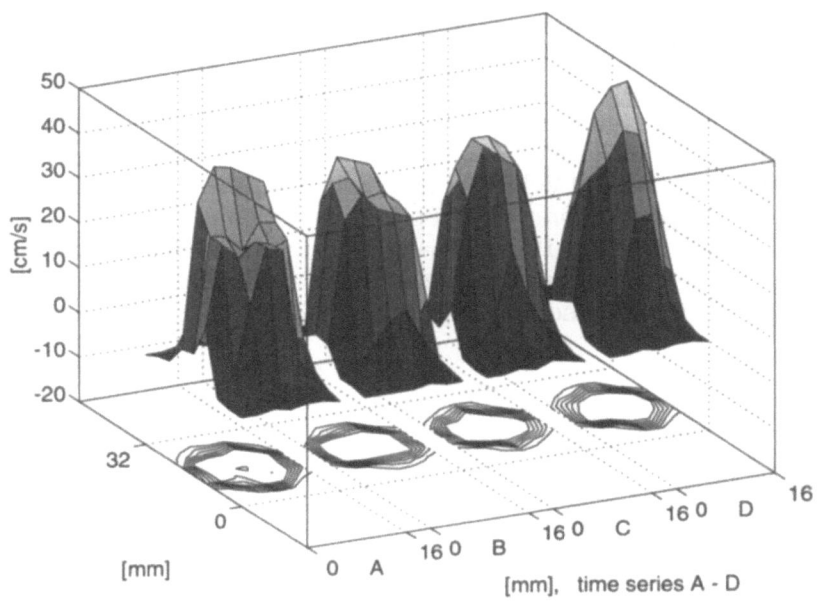

Figure 9 Formation of a parabolic flow profile during slow oscillatory flow motion.

The formation of parabolic flow in the 10 mm test tube is shown in fig. 9. The shown 2D velocity profiles are again single, non averaged frames and were taken at different times of the real-time recording. The first profile on the left side shows the distribution shortly after the reversal of the flow direction. Clearly visible is the dent in the centre of the velocity distribution. The next profile shows a rather flat distribution after 2 tenths of a second. Four tenths of a second later we already have a rounded shape of the distribution. A stable parabolic distribution can be observed approximately 2 seconds after the flow reversal. During the formation of the parabolic distribution the net flow through the tube remains of course constant. With the C-Mode Doppler the standard deviation of the measured flow rates is less than 4 % of the actual 25 ml/s in this case.

Conclusions

The common problems of ultrasonic volume flow measurements like the dependence of the angle of incidence, errors due to asymmetric and disturbed velocity distributions or secondary flow motions are inherently addressed by the C-Mode measuring procedure. The results in this paper, based on a prototype system, confirm the postulated properties of the C-Mode Doppler procedure within a comparatively small error margin.

References

[1] R. W. Cribbs, S. Arnon, "High Resolution Synthetic Aperture Imaging", Ultrasound in Medicine, Vol. 3B, 1977

[2] S. Shibata, T. Koda, J. Yamaga, "C-Mode ultrasonic imaging by an electronically scanned coaxial spherical receiving array", in Ultrasonics, March 1978.

[3] U. Moser, "Verfahren und Messanordnung zum Messen einer zweidimensionalen reflektierenden Struktur", Internationale Patentanmeldung PCT/EP89/01246, 1989

[4] U. Moser, M. Anliker, P. Schumacher, A.Vieli, P. Pinter, "'C-Mode Doppler': a quantitative Ultrasonic Real-Time Procedure for the Measurement of 2-D Velocity Fields and Volume Flow in large Blood Vessels", Eurodop Congress Proceedings, 1991

[5] U. Moser, M. Anliker, P. Schumacher, "Ultrasonic Synthetic Aperture Imaging used to measure 2D Velocity Fields in Real Time", IEEE Int. Symp. on Circuits and Systems, Vol 1, 1991

DEVELOPMENTS IN A PHOTOACOUSTIC PROBE FOR POTENTIAL USE IN INTRA-ARTERIAL IMAGING AND THERAPY

RJ Dewhurst[1], QX Chen[1], GA Davies[2], A Kuhn[1], KF Pang[1], PA Payne[1] and Q Shan[1]

[1]DIAS, UMIST, PO Box 88, Manchester, M60 1QD, UK
[2]Tissue Med Ltd., Astley Lane Industrial Estate, Astley Lane, Leeds, LS26 8XT, UK

INTRODUCTION

Interest in medical diagnostic ultrasound has progressively increased over the last two decades. The advent of sophisticated technology has enhanced the reliability and the user-friendliness of ultrasound systems, together with significant improvement in the quality of data available to clinicians. Much of the development work has been based on external scanning techniques; however, more recently, systems have been developed for invasive ultrasound investigations inside the lumen of blood vessels. Cardiovascular disease, the morbidity and mortality of which have reached epidemic proportions in the western world, remains a major challenge for the health care professionals, scientists and economists alike.

Various devices are available for treating arterial diseases (stenosed segments of arteries leading to a reduction or cessation of flow). As an alternative to coronary artery bypass grafting (CABG) performed surgically, such devices, from angioplasty balloons to instruments for cutting, grinding or ablating atheromatous plaque, may be more effective under certain specific vessel conditions. For example, a different device may be more appropriate in dealing with hard calcified plaque as opposed to soft plaque.

In current medical practice, balloon angioplasty, mechanical atherectomy and laser angioplasty techniques are effective for the removal of stenotic lesions and short occlusions within arteries. For the treatment of longer occlusions, some major difficulties remain, some of which are associated with inadequate device guidance[1]. As a consequence, recent developments in medical ultrasound have included intravascular ultrasonic imaging using miniature transducer catheters. Although the concept of intraluminal diagnostic ultrasound methods[2] has been reported since the early 1950s, it is only recently that strong research interest has been shown and significant research effort made. This is evident from several recent publications available on this subject[3-7]. Combining the ultrasonic imaging catheter with the therapeutic laser catheter has resulted in a more recent and useful catheter design and commercial products with this design have become available[8-9]. Otherwise, the medical procedure involves using an ultrasonic imaging probe to identify the stenosis, followed by withdrawal of this device and insertion of a laser catheter. Guidance up to the diseased region normally involves angiography. This present paper points to a new form of combined catheter design, concentrating on the development of ultrasonic imaging in a forward-looking configuration. It offers the advantages of simplified probe design arising particularly from the use of polymer based piezoelectric materials, plus the possibility of improved diagnosis of the form of tissue under interrogation.

First, we note that a fibre optic catheter for laser beam delivery has been used to deliver laser beams for photo-ablative removal of atherosclerotic plaque[10]. Multimode silica fibres are employed having diameters of some 600 μm. In other experiments, similar size fibres have been used to transport high peak power laser pulses for the generation of ultrasound[11,12]. The ultrasound has been in the form of transients on metallic surfaces[11] or vascular tissue[12,13]. Ultrasonic detection has sometimes used polyvinylidene fluoride (PVDF) transducers[14,15], or other forms of piezoelectric materials.

When ultrasonic waves are generated by laser irradiation, the ultrasonic detection device need only be a good acoustic receiver. This makes ultrasonic transducer design, with high receive sensitivity, much easier; and piezoelectric polymers are ideal materials for this purpose[14,15]. Compared with conventional piezoelectric ceramic materials, piezoelectric polymers are chemically more stable, mechanically more robust and acoustically better matched to water and blood environments. Due to their low dielectric permittivity, piezoelectric polymers have a large piezoelectric g constant, making them good acoustic receivers. Another advantage of polymer piezoelectric materials is their high internal dielectric and mechanical losses, making any probe that uses them intrinsically wide band.

Some features describing the behaviour of the laser probe are presented in this paper.

EXPERIMENTAL PROCEDURE

Figure 1 presents an outline of the experimental arrangement used to examine the behaviour of the probe. A conventional Q-switched Nd:YAG laser was used to provide laser pulses with a 20 ns duration at a wavelength of 1.06 μm or 0.53 μm. Neutral density filters were used to attenuate the pulse energy before focusing the beam into a 600 μm core silica fibre. As a consequence, $\approx$ 3 mJ was delivered from the probe head at a maximum repetition rate of 25 Hz. Laser peak power was $\approx$ 150 kW. Such power levels, delivered down a fibre, have already been shown to be sufficient for the ultrasound generation in solids and liquids[11].

The probe used in these experiments consisted of a conventional polymer-based ultrasound transducer with an optical fibre passing through its centre. The active area of the transducer was annular in shape, with an inner diameter of about 0.6 mm and an outer diameter of 3.2 mm. The outer sheath increased the overall diameter to 4.5 mm. It was constructed with a piezoelectric polymer element, PVDF, having a thickness of 28 μm. The performance of an earlier version of the probe with an overall diameter of 10.0 mm has been described elsewhere[16].

Not shown in Figure 1 are electrical leads taken from the transducer head and carried coaxially along the length of the optical fibre. After about 2 m, these leads were led out through a manifold. Using miniaturised connectors and cables, signals from the transducer were displayed on a fast digitising oscilloscope. No intermediate amplifier was used between the transducer and oscilloscope in measurements reported here. Waveforms were recorded on a fast digital oscilloscope, using an external trigger pulse derived from a beam splitter and photodiode placed close to the output port of the laser system.

In the present experiments, the probe was used to examine various forms of porcine tissue, such as samples from arterial wall. On irradiation, photoacoustic generation took place within the tissue, Figure 2. With the beam directed onto the intima, thermoelastic heating in the tissue occurred on timescales similar to those of laser pulse durations. Because irradiation was used with a wavelength of 1.06 μm or 0.53 μm, where the optical attenuation coefficient is expected[17] to be between 5-30 cm^{-1}, some laser pulse energy was expected to penetrate through the tissue, typically 1-2 mm thick, and travel to an aluminium base plate. Hence at these laser wavelengths, photoacoustic generation was also expected at this aluminium surface.

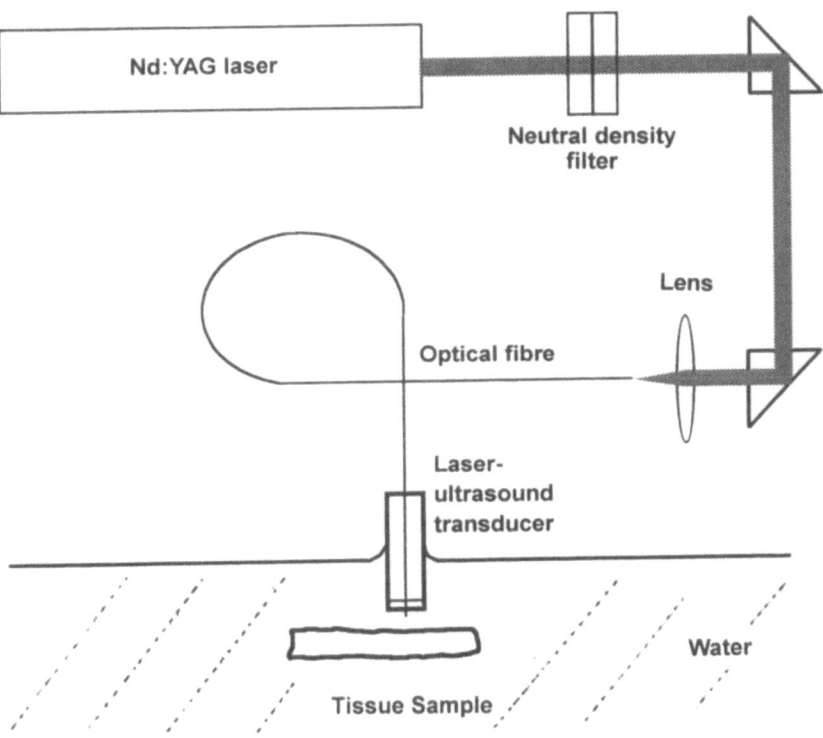

Figure 1. Experimental arrangement for characterisation of the integrated transducer.

RESULTS

Waveforms were recorded by the probe in-vitro using porcine artery just two hours old. The arteries were cut longitudinally to form a rectangular sample. These samples were then mounted in a sample holder so that they provided a horizontal and stable surface. Photoacoustic examination took place within a physiological saline solution, in a scheme depicted in Figure 2. Several characteristic signal features are shown in Figure 3 which show measurements taken at both laser wavelengths. Many features are similar, but higher ultrasonic amplitudes from porcine artery are achieved at a laser wavelength of 0.53 μm due to an expected decrease in the optical penetration depth.

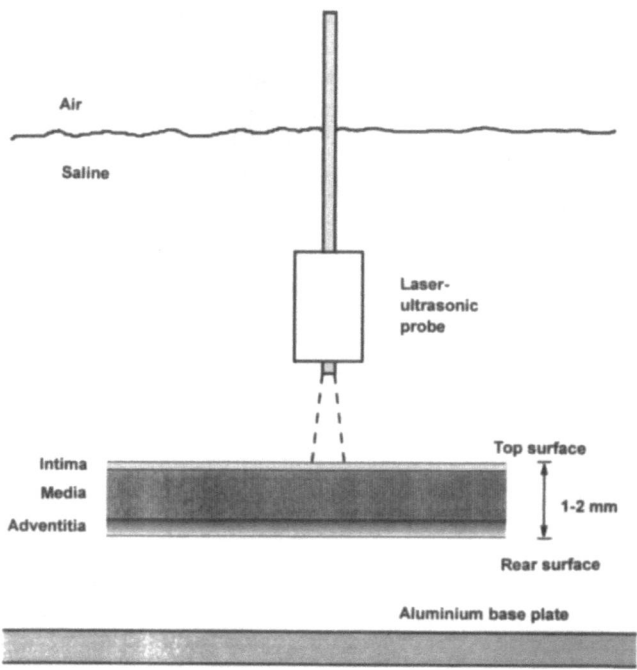

Figure 2. Location of the arterial wall with respect to the transducer probe and a rear aluminium base plate. Photoacoustic generation takes place primarily in the tissue, but some laser irradiation may penetrate through the tissue to cause photoacoustic generation on the aluminium base plate.

Time delays following firing of the laser pulses were used to help identify the signals seen in the waveforms, Figure 3. Signal A arose from photoacoustic generation in the porcine artery. It represented a pulse with its rising edge generated on the top surface of the sample and travelling backwards towards the probe. Signal D was of very low amplitude. It corresponded with the expected arrival time of an ultrasonic signal generated at the top sample surface, travelling forward through the sample to the adventitia where it was reflected at the boundary of connective tissue with saline. However, most of the energy associated with this forward going pulse proceeded to the aluminium base plate, where it underwent specular reflection and eventually reached the transducer, signal E. The pulse associated with the firing of the laser was unlikely to be an ultrasonic signal, but was instead a voltage generated by partial light absorption in the PVDF transducer. This light leaked from the circumference near the probe tip where the fibre cladding had been removed. Such a signal can be generated by the pyroelectric effect, and served as a useful time-marker. Corresponding heating of the transducer tip also caused additional local thermal expansion, launching an ultrasonic wave into the water medium. Signal H corresponded to such a pulse which had propagated to the aluminium base plate and then returned to the probe. Finally, signal C, only present from the infra-red irradiation source, arose from laser light that had penetrated the tissue to directly reach the aluminium plate. Here, local absorption caused another photoacoustic source, from which ultrasound travelled back to the probe head. Such signal features have been observed in a number of experiments where a base plate was used, and aligned to provide specular reflection.

Initial photoacoustic pulses emanating from different tissue types were similar in shape when using 1.06 μm irradiation. Figure 4 shows some of the signals, type A, obtained from pulmonary artery, aorta, heart muscle, femoral artery and lymph gland

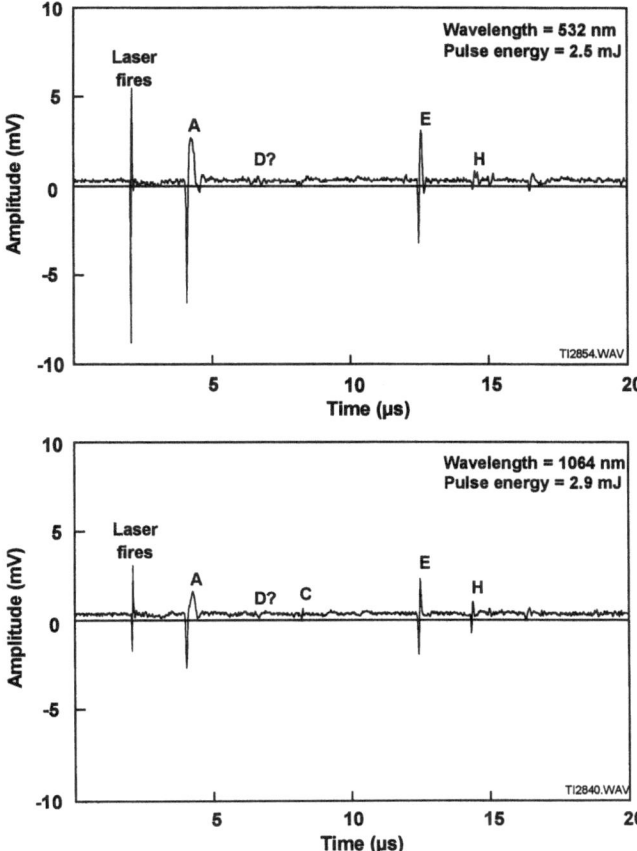

Figure 3. Waveforms recorded from laser irradiation of porcine artery wall, at laser wavelengths of 0.532μm and 1.06μm. Signal A is a photoacoustic signal arriving directly from the tissue. Signal D is associated with ultrasound reflection from the adventitia/saline boundary. Signal C arises from photoacoustic generation at the aluminium base plate. Signal E and H ultrasonic reflections from the aluminium base plate, with signal E arising from the photoacoustic signal generated at the arterial wall closest to the probe, and signal H inadvertently arising from ultrasonic generation in the probe itself.

obtained from porcine samples. In these experiments, a different laser system was used, and was an inferior system in the sense that it emitted an rf signal when Q-switching. The phenomena generated pick-up noise on the probe leads, as shown in Figure 4.

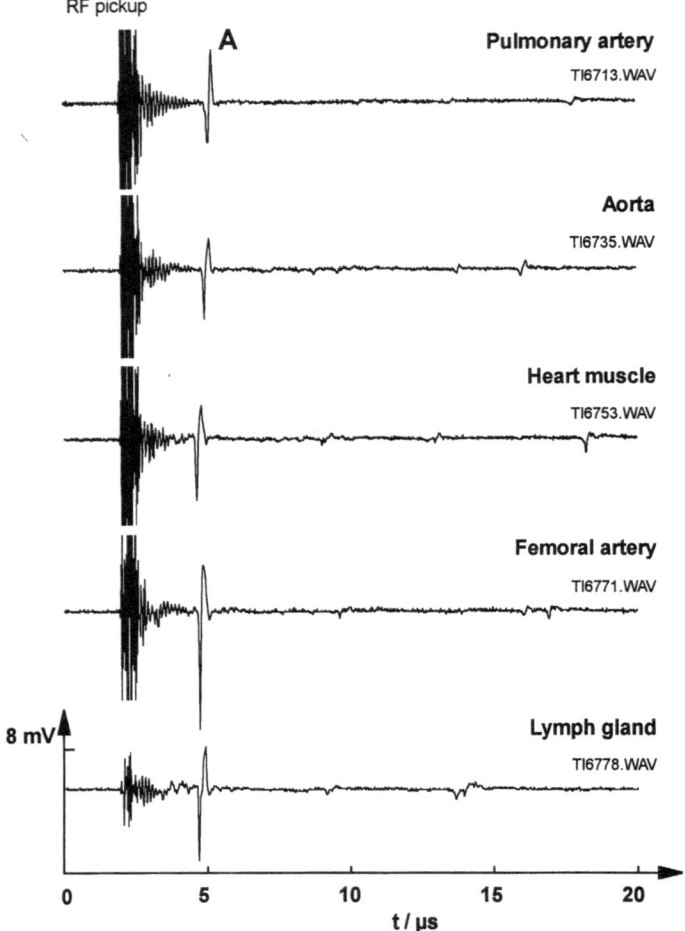

Figure 4. Series of photoacoustic signals recorded from different tissue types, namely pulmonary artery, aorta, heart muscle, femoral artery and lymph gland. Laser pulses of 5 mJ at 1.06 μm were used to irradiate the tissue samples.

Nevertheless, the photoacoustic signal, A, was still resolvable. Variations in amplitude and shape were noted in the bipolar pulse. However, just as pronounced were variations in signal from one location to the next on the same sample. We conclude that 1.06 μm irradiation, having significant optical penetration depth in these tissue types, could not be used to distinguish between these tissue types.

Returning to the examination of porcine artery wall, Figure 5 shows the effect of several laser pulses irradiating the same tissue location. The top trace displays all the signal features referred to earlier. Subsequent traces show that on multiple pulse irradiation, photoacoustic generation of signals arising from the tissue decreased, namely signals A, D and E. We believe this is the first time that such a phenomena has been reported. This figure also shows that the other signals were largely unaffected. For example, signal C arose from through-transmitted radiation reaching the aluminium base plate. Its amplitude was not significantly affected when the photoacoustic signal from tissue decreased on a shot-to-shot basis. Nor was signal H affected significantly, arising from ultrasound generated

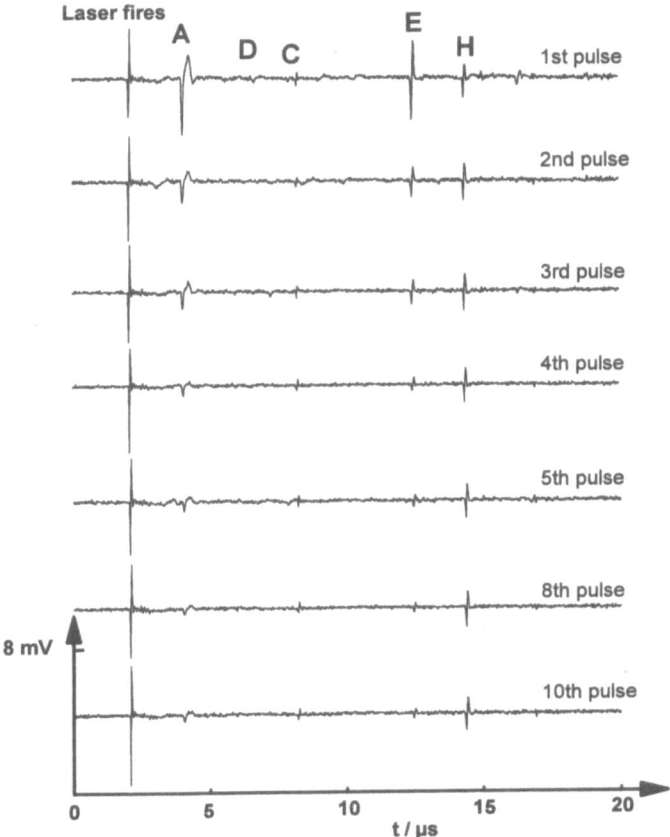

Figure 5. Series of waveforms showing photoacoustic signal decay for consecutive laser pulses incident on the same tissue location. Laser wavelength of 1.06 μm with a pulse energy of 2.9 mJ irradiated a porcine arterial wall just 2 hours old.

at the probe tip travelling through the artery wall and reflecting at the base plate. These waveforms help to confirm that signals A, D and E originate from photoacoustic interaction with the tissue sample.

Measurements showed that decreases in signal amplitude took place even when the laser was operated at the rate of one pulse every thirty seconds. Waiting several minutes before renewed irradiation did not recover the photoacoustic signal amplitude. A similar effect was also seen using 0.53 μm irradiation. Decreases in signal amplitude are not presently understood. With an energy fluence of less than 1 Jcm^{-2}, irradiation of the sample was well below expected ablation thresholds published in the literature[18]. These suggest that fluences should exceed 5 Jcm^{-2} for ablation to take place. Further investigations are required to find the cause of our decreasing signal since we estimate that the average temperature rise expected due to uniform heating would be less than 1°C. We noted that there was no visible damage seen on the sample after laser irradiation.

To confirm that these observations were due to physical changes within the tissue sample, the sample investigated was replaced by a neutral density filter serving as the sample, where uniform, volumetric absorption of laser light was known to take place. It was considered to be a phantom of homogeneous composition. Figure 6(a) shows a waveform prediction from a simulation model we have developed, and Fig 6(b) shows a typical experimental waveform. Experimental waveforms were reproducible on a shot-to-shot basis, without any amplitude degradation. Waveforms exhibit a series of pulses. The first signal, predominantly positive going, arose from a pulse emanating from the top surface of the neutral density filter and travelling directly to the transducer head. The next, a negative going signal, emanated from photoacoustic interaction with the rear wall of the filter. After that time, there were a number of ultrasonic pulse reverberations within the filter, with partial transmission through the front surface leading to a series of pulses arriving at the transducer head. There is good agreement between predicted and experimental waveforms. Results therefore show that features measured by the probe can be properly described in cases of photoacoustic interaction with well-defined targets.

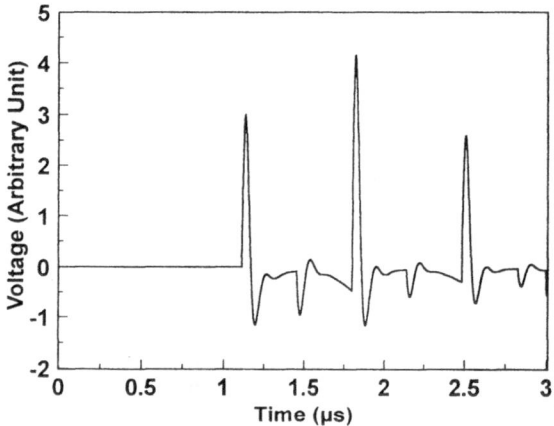

Figure 6a. Waveform prediction for laser irradiation of a neutral density filter in water. Parameters used in the model were laser wavelength = 1.06 μm, laser pulse energy = 9 mJ, pulse duration = 8 ns, glass thickness = 1.9 mm and glass absorption coefficient = 12.59 cm^{-1}.

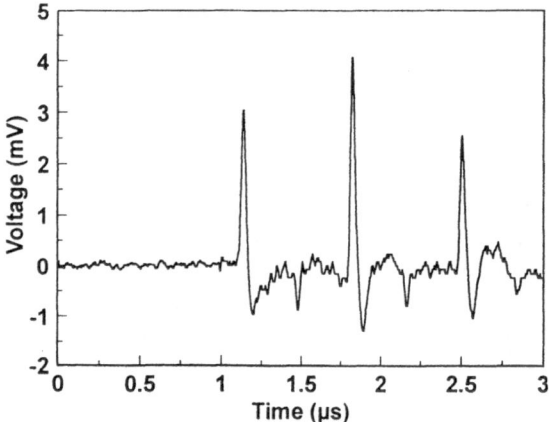

Figure 6b. Corresponding experimental waveform to that shown in Fig 6a.

530

CONCLUSION AND DISCUSSION

A photoacoustic probe has been described which may be used to examine animal or human tissue. Delivery of laser pulses to the sample was through a silica fibre, of the type which may also be used to transport laser beams for photo-ablative removal of atherosclerotic plaque. Presently, characteristic features of the probe are being examined, where return echoes to the probe tip have been detected by a polyvinylidenefluoride transducer mounted on the front face of the probe. Overall diameter of the probe is presently 4.5 mm, with the active area of the sensor being 3.2 mm in diameter. Future work will include a further reduction in probe diameter.

Measurements have been presented which demonstrate features of the probe when assessing various targets. Conducting experiments in-vitro, a decrease in photoacoustic signal amplitude has been observed on multiple laser shots, this being a phenomena present at both 0.532 μm and 1.06 μm wavelength irradiation. The energy fluence is below those published for ablation threshold, so that further investigation is required to explain the decreasing signal in repetitive laser shots. A plausible explanation may be based on experimental observations already noted by Oraevsky et al[19]. They discuss the possibility of absorption centres in arterial tissue at 0.532 and 1.06 μm. At the end of a short duration laser pulse, the temperature of the absorbing centre might have surpassed the temperature required for thermal denaturation or microvaporization, while the surrounding tissue remains well below this temperature. Our own observations indicate that some remaining blood in the tissue may act as the source of absorbing centres.

As expected, experiments have shown that the probe can produce a strong photoacoustic interaction with tissue. Waveforms have been interpreted to identify various sources of return signal. For tissue examination, laser wavelengths of 1.06 μm and 0.53 μm are not optimum, and it is anticipated that future work will include tissue examination at other laser wavelengths.

ACKNOWLEDGEMENTS

The authors wish to acknowledge with thanks financial support for this work, which has come from the UK Science and Engineering Research Council, Contract No GR/G54733, and studentship support from both the Science & Engineering Research Council and the National Heart Research Fund, Leeds, UK.

REFERENCES

1. R A White, G E Kopchok, M R Tabbara, D M Cavaye and F C Cormier, Intravascular ultrasound guided holmium: YAG laser recanalization of occluded arteries, *Lasers in Surgery and Medicine*, 12, 239-245, (1992).
2. N Bom, H ten Hoff, C T Lancee, W J Gussenhoven and J G Bosch, Early and recent intraluminal ultrasound devices, *International Journal of Cardiac Imaging*, 4, 79-88, (1989).
3. G E Kopchok, R A White, C Guthrie, Y Hsiang, D Rosenbaum, M Tabbaba and G H White, Intravascular ultrasound: A new potential modality for angioplasty guidance, *Angiology*, 785-792, (1990).
4. J M Tobis, J Mallery, D Mahon, K Lehmann, P Zalesky, J Griffith, J Gessert, M Moriuchi, M McRae, M L Dwyer, N Greep and W L Henery, Intravascular ultrasound imaging of human coronary arteries in vivo, *Circulation*, 83, 923-926, (1991).
5. S E Nissen, J C Gurley and A N DeMaria, Assessment of vascular disease by intravascular ultrasound, *Cardiology*, 77, 398-410, (1990).
6. D M Cavaye and R A White, *Intravascular Ultrasound Imaging*, Raven Press, New York, (1993).

7. P G Yock, P J Fitzgerald, K Sudhir, D T Linker, W White and A Ports, Intravascular ultrasound imaging for guidance of atherectomy and other plaque removal techniques, *International Journal of Cardiac Imaging*, **6**, 179-189, (1991).

8. K W Gregory, H T Aretz, M A Martinelli, E G LeDet, G F Hatch, R E Gregg, T Sedlacek and W C Haase, Intraluminal laser atherectomy, with ultrasound and electromagnetic guidance, *SPIE, Vol. 1425, Diagnostic and Therapeutic Cardiovascular Interventions*, 217-225, (1991).

9. R J Crowley, M A Hamm, S H Joshi, C D Lennox and G T Roberts, Ultrasound guided therapeutic catheters: Recent developments and clinical results, *International Journal of Cardiac Imaging*, **6**, 145-156, (1991).

10. R A White and W S Grundfest, *Lasers in Cardiovascular Disease*, Year Book Medical Publishers Inc. Chicago, (1989).

11. R J Dewhurst, A G Nurse and S B Palmer, A high power optical fibre delivery system for the laser generation of ultrasound, *Ultrasonics*, **26**, 307-310, (1988).

12. H Crazzolara, W von Muench, C Rose, U Thiemann, K K Hasse, M Ritter and K R Karsch, Analysis of the acoustic response of vascular tissue irradiated by an ultraviolet laser pulse, *J. Appl. Phys.*, **70**, 1847-1849, (1991).

13. F W Cross, R K Al-Dhahir and P E Dyer, Ablative and acoustic response of pulsed UV laser-irradiated vascular tissue in a liquid environment, *J. Appl. Phys.*, **64**, 2194-2201, (1988).

14. Q X Chen, P A Payne, F Moss, R E Banks and S Smith, Piezoelectric copolymers and their applications, *Proc. 2nd International Conference on Electrical, Optical and Acoustic Properties of Polymers, EOA II*, Plastics and Rubber Institute, London, UK, 3/1-6, (1990).

15. P A Payne and Q X Chen, The design and construction of high frequency broad band ultrasound transducers based on plastic film. In: *Reliability in Non-Destructive Testing, NDT-88*, Eds: C Brook & P D Hanstead, Pergamon Press, Oxford, UK, 319-330, (1989).

16. Q X Chen, R J Dewhurst, P A Payne and B Wood, An integrated laser optical fibre and polymer film ultrasound transducer for potential use in intra arterial imaging and therapy, *Ultrasonics*, Accepted for publication (1994).

17. N P Furzikov, Different lasers for angioplasty: thermo-optical comparison, *IEEE J Quant Elect*, 23, (10), 1751-1755, (1987).

18. P E Dyer and R K Al-Dahir, Transient photoacoustic studies of laser tissue ablation, *SPIE, Laser Tissue Interaction*, **1202**, 46-60, (1990).

19. A A Oraevsky, R D Esenaliev, V S Letokhov, Temporal characteristics and mechanism of atherosclerotic tissue ablation by nanosecond and picosecond laser pulses, *Lasers in the Life Sciences*, **5**, 75-93, (1992).

MAXIMUM AND MEAN BLOOD VOLUME FLOW MEASURED VIA TIME DOMAIN CORRELATION AND ULTRASONIC FLOWMETRY: A COMPARATIVE STUDY

Flemming Forsberg,[1] Ji-Bin Liu,[1] Sherry L. Guthrie,[2] and
Barry B. Goldberg[1]

[1] Department of Radiology
 Thomas Jefferson University
 Philadelphia, PA 19107
[2] Philips Ultrasound
 Santa Ana, CA 92704

INTRODUCTION

Measuring volumetric blood flow is of significant clinical importance. The methods currently available are either invasive (e.g. electromagnetic flowmetry) or non invasive Doppler ultrasound. However, significant errors are associated with the latter (Gill, 1985). Recently, time domain correlation has been developed as an alternative method for measuring blood flow velocities (Bonnefous and Pesque, 1986; Foster et al., 1990). This technique also allows for an accurate estimate of the one-dimensional velocity profile across a vessel (Foster et al., 1990). Several algorithms for obtaining volumetric flow estimates from the velocity profiles have been suggested (Embree and O'Brien, 1990). In this study the specific algorithm implemented under the name CVI-Q^{TM} on a commercial scanner (P700; Philips Ultrasound, Santa Ana, CA) was utilized.

The objective was to compare CVI-Q to an established, invasive volumetric flow measurement technique: ultrasonic flowmetry. The instantaneous maximum volume flow and the mean flow over 3 to 6 cardiac cycles were measured in vitro and in vivo.

MEASUREMENT TECHNIQUES

Ultrasonic Flowmetry

This invasive technique for volumetric blood flow measurement utilizes a calibrated perivascular flow probe placed around the vessel under examination. In these experiments a T101 flowmeter (Transonic Systems, Ithaca, NY) with 2-8 mm diameter flow probes (absolute accuracy: ±10 %) was used. Each probe contains two transducers at an angle of 45°. A steel bracket underneath the vessel acts as an ideal reflector. By alternating reception and transmission flow signals are obtained upstream and downstream. The difference in transit time allows the volume flow rate to be calculated.

Time Domain Correlation - CVI-Q

In this algorithm the volumetric blood flow is estimated by integrating the velocity profile over a diameter; assuming the velocity field to be circular symmetric. Specifically, the effective vessel diameter is found automatically as the distance between the points where the velocity profile equals zero. Each velocity in the velocity profile is assumed to exist in semi-circular symmetry through the vessel. The product of each semi-annulus area and its associated velocity is an estimate of the volume flow within this area, while the sum of all these products yields the total blood volume flow through the vessel.

The P700 scanner depicts the velocity profiles in a color M-mode display. The y-axis shows depth with time along the x-axis. The flow velocities, measured via time domain correlation, are shown color coded as in conventional color flow imaging. The operator selects the 3 to 6 cardiac cycles over which the average volume flow is calculated; displayed in the lower right corner (*VOL). The currently selected velocity profile is depicted on the left. An example of the M-mode display is given in Figure 1.

CVI-Q assumes (1): a circular vessel, (2): an axisymmetric velocity distribution, (3): the beam intercepting the center of the vessel and (4): a small sample volume compared to the vessel size. The dominant error is associated with estimating the correct beam/vessel angle. This error increases rapidly at angles less than 25° or greater than 80° (Picot and Embree, 1994). Hence, scan angles from 30° to 70° are recommended. Results from a multicenter trial indicate, that an overall accuracy of ±15 % is achievable for an experienced operator (Guthrie et al., 1993).

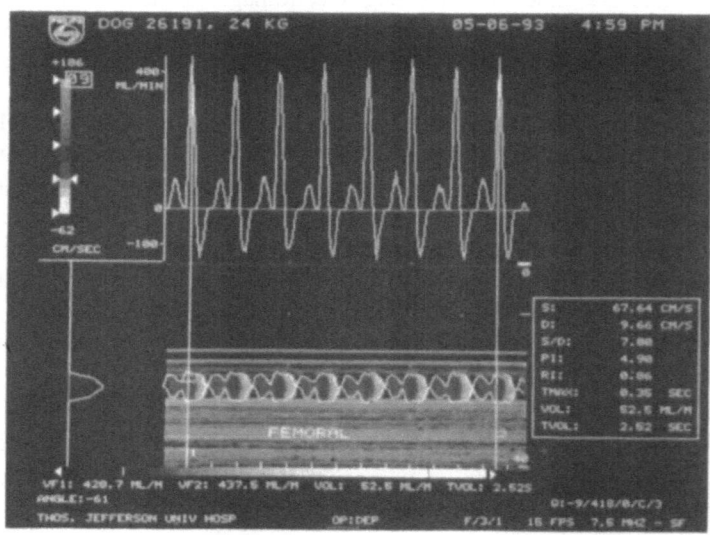

Figure 1. Example of a CVI-Q M-mode trace recorded in vivo in the femoral artery of a dog.

IN VITRO MEASUREMENTS

A pulsatile flowpump (model PFS-A-3-1-G; Quest Image Inc., London, Canada) was utilized. This pump is capable of simulating physiological waveforms with high accuracy; error < 1 % (Holdsworth et al., 1991). In these experiments carotid (i.e., low resistance) and femoral (i.e., triphasic) type waveforms were studied.

A blood mimicking fluid, containing crystaline cellulose particles with an average diameter of 20 μm (type 20 sigmacell, Sigma Chemical Co., St. Louis, MO), was pumped through an 8 mm diameter latex tube at flow rates from 50 ml/min to 550 ml/min. Mean volume flow estimates were measured simultaneously using the T101 flowmeter and CVI-Q. The latter were obtained with a 7.5 MHz linear array (model LA7530). After optimization of the M-mode trace the transducer was fixed using a clamp armature. The transducer was not manipulated for the duration of the experiment. At each pump setting five measurements were acquired.

The measurements performed on the carotid and femoral waveforms are presented in Figures 2 and 3 with the results of the statistical analysis summarized in Table 1. Both CVI-Q and the flowmeter achieved regression line slopes close to unity and correlation coefficients above 0.996. Both methods resulted in standard deviations below 4% and significantly smaller than the 95% confidence intervals. Moreover, t-tests on the linear relationship between CVI-Q and the flowpump, respectively, between the flowmeter and the pump were highly statistically significant (p < 0.001).

The absolute error of the volume flow estimates are less than ±5% for most volume flows; cf. Figure 4. However, for lower flow rates the error increases rapidly to over 15%. It should be noted, that in the carotid data set it is the T101 flow estimates which have the largest error, while the opposite is the case in the femoral data.

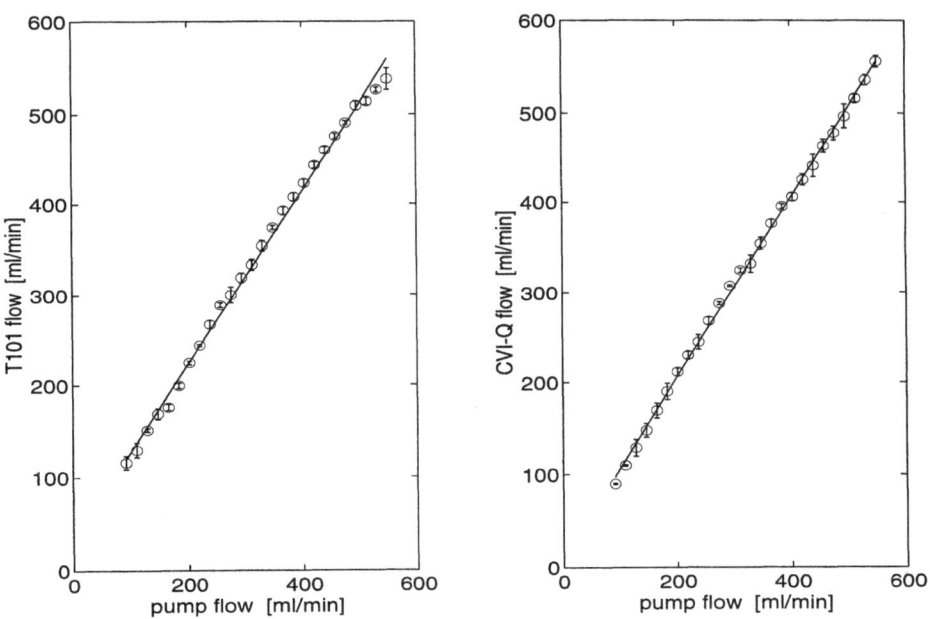

Figure 2. Carotid waveform; mean ±1 standard deviation and the best linear regression fit.

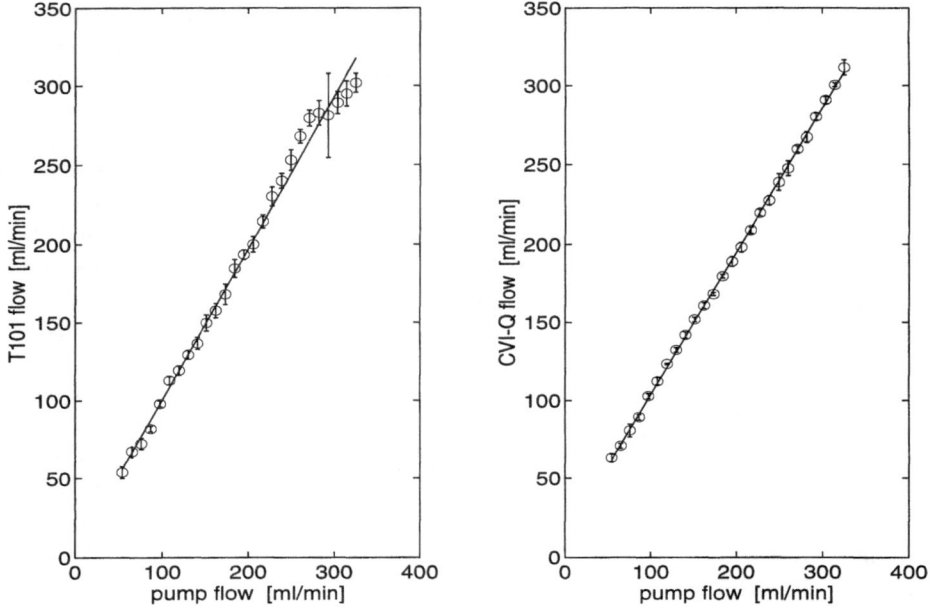

Figure 3. Femoral waveform; mean ±1 standard deviation and the best linear regression fit.

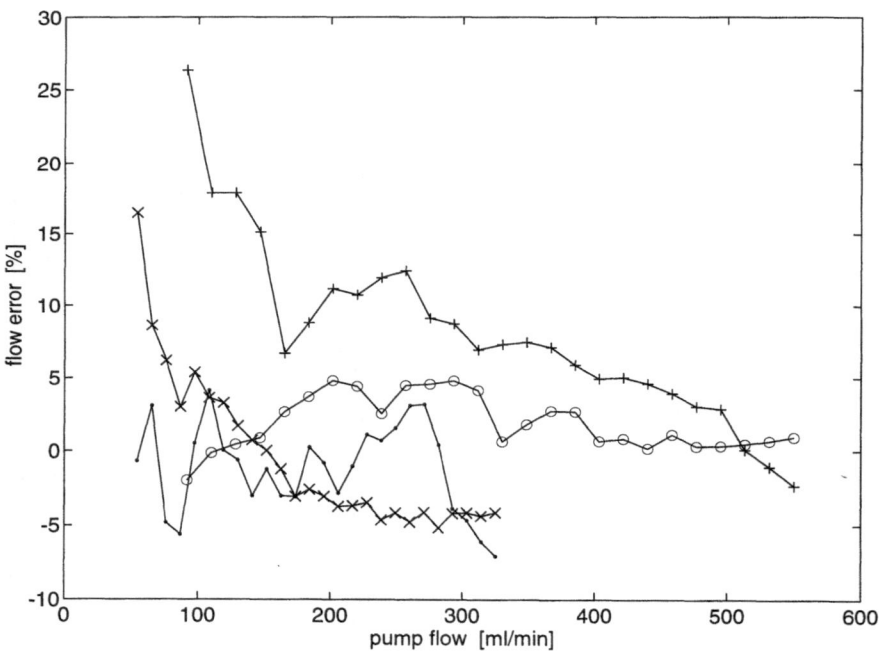

Figure 4. Error of flow estimates in percentage of actual flow rates. '+': cca T101; 'o': cca CVI-Q; '·': femoral T101; 'x': femoral CVI-Q.

Table 1. Linear regression analysis on in vitro data. All lines were statistically significant (p < 0.001).

waveform	variable vs pump	regression line	r	std dev ml/min (%)	95 % conf interval ml/min
carotid	T101	y=0.96x+31.99	0.9980	8.68 (2.56)	9.92
carotid	CVI-Q	y=0.999x+5.72	0.9995	4.63 (1.42)	1.8.59
femoral	T101	y=0.97x+3.74	0.9963	7.00 (3.75)	3.45
femoral	CVI-Q	y=0.91x+12.1	0.9998	1.61 (0.87)	15.00

IN VIVO MEASUREMENTS

Five mongrel dogs (mean weight 22.4 kg) had their common carotid and femoral arteries exposed under general anesthesia. If necessary branches were ligated to produce a straight vessel segment large enough to accommodate the T101 flowprobe and a 7.5 MHz linear array (for CVI-Q measurements) at the same time. A wedged standoff pad was utilized with the linear array to bring the vessels out of the transducer's near field and to secure an appropriate angle of insonation.

Data from the T101 flowmeter and CVI-Q were acquired simultaneously. The instantaneous maximum volume flow and the mean flow over 3 to 6 cardiac cycles were measured. To vary the flow rates each vessel was partially occluded halfway through the experiments; simulating a 90 % stenosis. A total of 186 mean flow datapoints were recorded, while 140 maximum volume flow estimates were obtained.

The protocol for these studies was approved by the University Animal Care and Use Committee. Moreover, the animal facilities used are fully accredited, and all experiments took place under the supervision of a veterinarian and a veterinary technologist.

The volume flow estimates acquired from the carotid artery are presented in Figures 5 and 6; showing mean and maximum flow rates, respectively. The best least squares, linear fit has been included in both cases. The results of the linear regression analysis are given in Table 2. Correlation coefficients (denoted r) larger than 0.86 and standard deviations from 20.3% to 27.6% were obtained. The validity of the linear model was examined with a t-test, and in both cases found to be statistically significant (p<0.001).

The results from the femoral artery are presented in Figures 7 and 8 with the linear regression analysis summarized in Table 3. The measurements exhibit more variation than those obtained in the carotid artery. This is confirmed by slightly lower correlation coefficients (r equal to 0.73 and 0.85), and a higher standard deviation (27.3% and 28.7%). However, as is the case for the carotid data, the standard deviations are still significantly lower than the 95% confidence intervals found (cf. Tables 2 and 3). Both regression lines were statistically significant (p<0.001).

To investigate the agreement, if any, between the two techniques all data points were studied as a function of time using a paired t-test. The null hypothesis being that the mean difference between simultaneously acquired data points is zero. In the carotid artery the analysis showed agreement between CVI-Q and ultrasonic flowmetry, while statistically significant differences were found in the femoral data set (cf. Table 4).

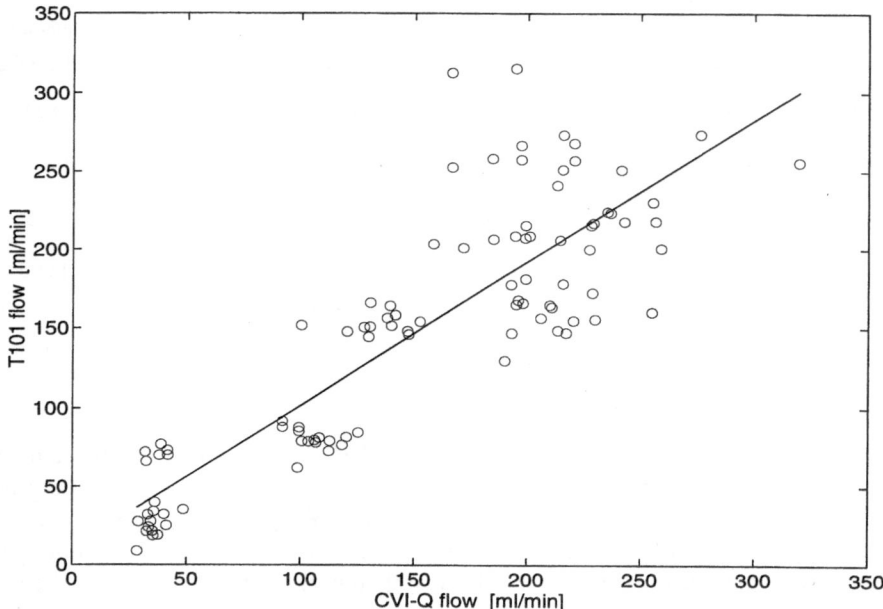

Figure 5. Carotid mean volume flow estimates and linear regression fit.

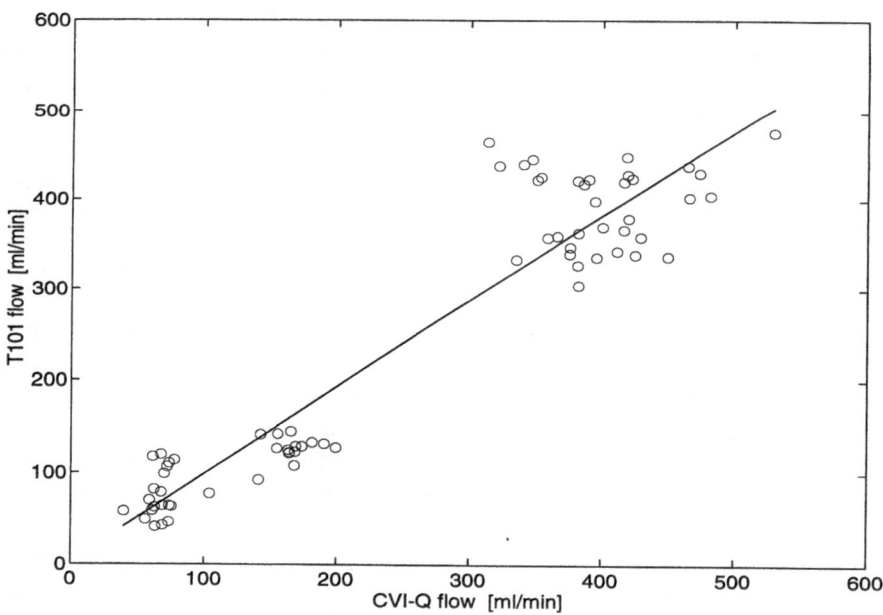

Figure 6. Carotid maximum volume flow estimates and linear regression fit.

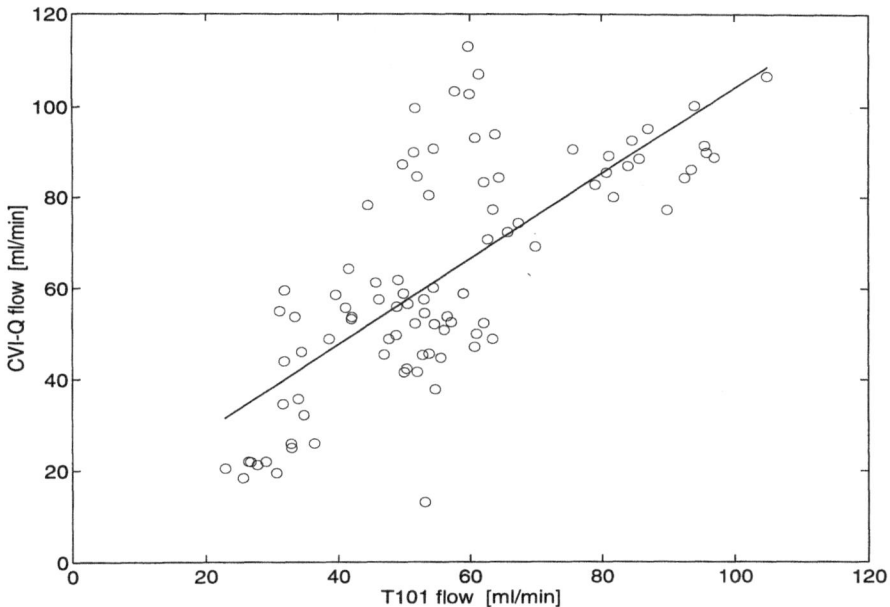

Figure 7. Femoral mean volume flow estimates and linear regression fit.

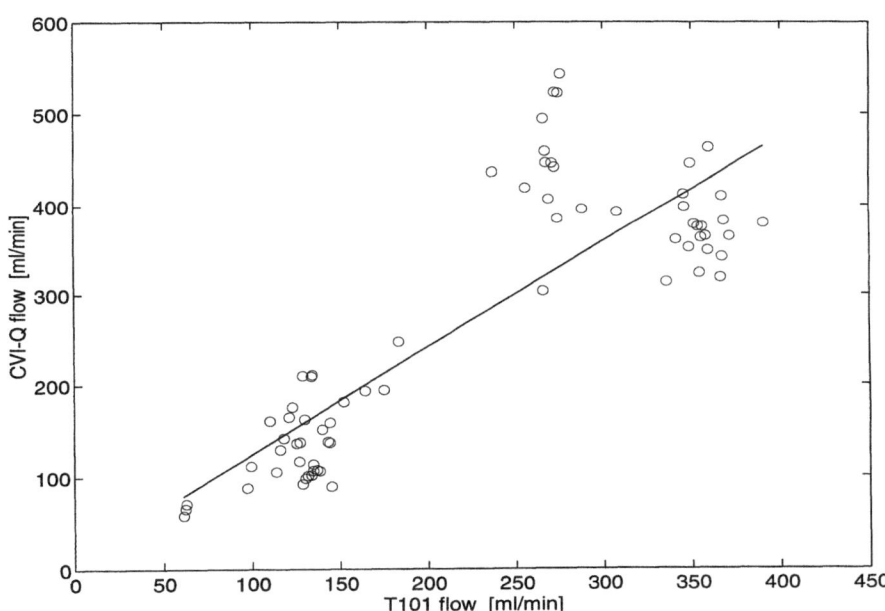

Figure 8. Femoral maximum flow estimates and linear regression fit.

Table 2 Linear regression analysis on carotid in vivo data. Both lines were statistically significant (p < 0.001).

flow type	regression line CVI-Q vs T101	r	std dev ml/min	std dev %	95% conf. interval ml/min
mean	y=0.91x+10.92	0.86	40.41	27.59	81.25
max	y=0.94x+4.48	0.95	49.78	20.31	100.47

Table 3. Linear regression analysis on femoral in vivo data. Both lines were statistically significant (p < 0.001).

flow type	regression line T101 vs. CVI-Q	r	std dev ml/min	std dev %	95% conf. interval ml/min
mean	y=0.94x+9.92	0.73	17.07	27.32	34.34
max	y=1.17x+7.19	0.85	77.40	28.72	156.21

Table 4. Agreement between CVI-Q and ultrasonic flowmetry evaluated via a paired t-test.

vessel	flow type	p-value	agreement
carotid	mean	0.46	yes
carotid	maximum	0.11	yes
femoral	mean	3.3×10^{-4}	no
femoral	maximum	7.1×10^{-6}	no

DISCUSSION

An extensive computer simulation study of the uncertainty in the CVI-Q volume flow estimate was recently undertaken (Picot and Embree, 1994). A 2° error in the assessment of the angle of incidence lead to uncertainties of ±10 % in the final flow estimate. Hence, the need for an experienced sonographer when using CVI-Q. To substantiate the theoretical analysis in vitro experiments were also conducted on a constant flow system. Errors of ± 5% were measured for flow rates above 150 ml/min. Our results, from the first in vitro validation study using pulsatile flow, are quite similar.

Apart from the operator dependence of CVI-Q another important source of error was identified in our in vitro studies. Positioning the T101 flow probe correctly was a critical factor in obtaining accurate flow estimates. The ultrasonic flowmetry results could be altered by up to 40 % by minor adjustments in the probe position; while this ambiguity could be resolved in the in vitro experiments, this is clearly not the case in vivo.

Individual, in vivo flow values obtained with CVI-Q and the flowmeter were clearly correlated (r from 0.73 to 0.95). The standard deviations ranged from 20 to 28% with 95% confidence intervals of 41% to 58% of the respective mean values (cf. Tables 2 and 3). The two techniques were in agreement in the carotid measurements. However, no such agreement existed in the femoral artery (cf. Table 4). The differences occurred mainly in one dog. A large part of this discrepancy appear to be an offset value, which might have been caused by the problems associated with positioning the T101 flowprobe. This phenomenon might also explain some of the variation found in the linear regression analysis. Other factors which will influence the measurements are the increased pulsatility and the reverse flow in the femoral artery compared to the carotid artery.

Other investigators have performed in vivo studies on the use of duplex Doppler ultrasound for volume flow measurements, and compared the results to invasive, electromagnetic flowmetry (Avasthi et al., 1984; Dauzat and Laynargues, 1989). In duplex Doppler a single mean velocity measurement is multiplied by the vessel diameter (i.e., area) to yield the volume flow; assuming a circular vessel. Avasthi and colleagues measured renal blood flow in three dogs (1984). From the 35 measurements correlation coefficients above 0.78 was calculated. The linearity was not as good as obtained in this study with CVI-Q; regression line slopes of 0.43, 0.86 and 1.12. A more extensive data set was acquired by Dauzat and Laynargues (1989); 304 simultaneous measurements in the portal vein of dogs. The mean difference was small 11ml/min, but the confidence interval was very large -267 to +239ml/min. This lead to the conclusion that duplex Doppler ultrasound was not the ideal method for portal vein blood flow measurement.

A technique similar to CVI-Q, only using standard color flow images instead, has recently been introduced (Fillinger and Schwartz, 1993). A reasonable degree of accuracy was achieved under laminar flow conditions; mean error 11 % compared to timed collection. Because of the improved axial resolution in time domain correlation measurements CVI-Q should produce better volume flow estimates; further studies are required to demonstrate this.

In summary the results of this study compare favorably with the literature, but the idealized experimental conditions should be noted. In particular the fact that all the in vivo measurements were acquired without any intervening tissue. The impact of transcutaneous measurements on the CVI-Q volume flow estimates is currently being evaluated in ongoing studies in humans (Deane et al., 1994).

CONCLUSION

Both techniques were validated in vitro. The absolute errors were below ±5 % under these ideal circumstances; except for slow flow situations where larger errors were encountered. Accurate flowmeter results depended critically on the positioning of the flow probe.

The respective manufacturers report accuracies of 10 - 15%. Hence, in vivo correlation coefficients ranging from 0.73 to 0.95 with regression line slopes close to unity demonstrate good correlation between the two techniques.

These studies indicate that non invasive CVI-Q measurements of blood volume flow in the carotid and femoral arteries are linear and accurate compared to invasive ultrasonic flowmetry.

ACKNOWLEDGMENTS

The support for this project from Philips Ultrasound, in particular Paul Embree and Dennis Paul, is gratefully acknowledged.

REFERENCES

Avasthi, P.S., Greene, E.R., Voyles, W.F., and Eldridge, M.W., 1984, A comparison of echo-Doppler and electromagnetic renal blood flow measurements, *J. Ultrasound Med.* 3:213-218.

Bonnefous, O, and Pesque, P., 1986, Time domain formulation of pulse-Doppler ultrasound and blood velocity estimation by cross correlation, *Ultrasonic Imaging* 8:73-85.

Dauzat, M., and Laynargues, G.P., 1989, Portal vein blood flow measurements using pulsed Doppler and electromagnetic flowmetry in dogs: a comparative study, *Gastroenterology* 96:913-919.

Deane, C.R., Besarab, A., Forsberg, F., and Goldberg, B.B., 1994, Hemodialysis access viability - a study using sequential ultrasound volume flow measurements. Submitted to *RSNA*; (abstract).

Embree, P.M., and O'Brien Jr., W.D., 1990, Volumetric blood flow via time-domain correlation: experimental verification, *IEEE Trans. Ultrason. Ferroelec. Freq. Contr.*, 37:176-189.

Fillinger, M.F., and Schwartz, R.A., 1993, Volumetric blood flow measurement with color Doppler ultrasonography: the importance of visual clues, *J. Ultrasound Med.*, 12:123-130.

Foster, S.G., Embree, P.M., and O'Brien Jr., W,D, 1990, Flow velocity profile via time-domain correlation: error analysis and computer simulation, *IEEE Trans. Ultrason. Ferroelec. Freq. Contr.*, 37:164-175.

Gill, R.W., 1985, Measurement of blood flow by ultrasound: accuracy and sources of error, *Ultrasound Med. Biol.*, 11:625-641.

Guthrie, S.L., Paul, D.E., Travers, D., and Gleason, L.L., 1993, Color velocity imaging quantification, the effect of interoperator variability and operator repeatability, presented at the *Soc. Vasc. Technol. Meeting*, Washington DC, June.

Holdsworth, D.W., Rickey, D.W., Drangova, M., Miller, D.J.M., and Fenster, A., 1991, Computer-controlled positive-displacement pump for physiological flow simulation, *Med. & Biol. Eng. & Comput.*, 29:565-570.

Picot, P., and Embree, P., 1994, Quantitative volume flow estimation using velocity profiles, *IEEE Trans. Ultrason. Ferroelec. Freq. Contr.*, 41, May. In press.

THE ATTENUATION EFFECT IN
PULSED DOPPLER FLOWMETERS

Nicholas Thomas [1] and Sidney Leeman [2]

[1] Department of Radiography,
[2] Department of Medical Engineering and Physics,
King's College, London

INTRODUCTION

Pulsed Doppler flowmeters are considered to estimate the velocity of blood flow by measuring the frequency shift of echoes, due to the Doppler effect [1]. Typical frequency shifts of the echoes are of the order of kHz. However, the time and frequency domain properties of ultrasound pulses travelling in tissue also undergo changes due to propagating and scattering effects. For example, the attenuation coefficient for a soft tissue such as liver [2] is $\alpha(f) = 0.4 f^{1.14}$. If the emitted pulse is assumed to be Gaussian, with a centre frequency of 5 MHz and a 3 dB bandwidth of 0.8 MHz, the shift in mean frequency with depth for such a medium is calculated to be $0.11\ \mathrm{MHz\,cm^{-1}}$. Hence, if pulsed Doppler flowmeters assess velocity by estimating a shift in frequency of the echoes, then attenuation should have a dramatic influence on the measurement outcome. Such a severe effect is not seen [1,3,4] and, although several investigators have considered the influence of attenuation, no clear reason for this discrepancy has been proposed and there is disagreement as to the size of the effect.

This paper examines the problem by constructing and analysing a time domain description of the demodulation process in conventional pulsed Doppler systems for the simplified case of a single constant velocity target. This description allows such systems to be compared with other methods of velocity estimation which are insensitive to attenuation. An electronic pulse injection system is used to test and confirm the theory.

THE DOPPLER EFFECT IN MULTIPLE ECHOES

Demodulation in pulsed Doppler flowmeters entails processing a number of sequential pulse echo data sequences. As a first step in the analysis, the nature of the rf-echo produced by a stationary target and one moving with constant velocity in a stationary medium will be considered. To simplify the analysis, only motion of the target away from the transducer will be examined. For a transducer insonating a stationary target, the (ideal) received signal for the sequence of detected echoes, is written as

$$E_s(t) = \sum_{n=0}^{n=N} g(t - Pn - \tau) \tag{1}$$

where $g(t)$ is a time domain function describing a single emitted pulse, P is the time separation of the emitted pulses ("pulse repetition period"), $\tau = 2R/c$, with R the range of the target at time $t = 0$, and c is the ultrasound velocity in the intervening stationary medium. For simplicity, it is assumed that the nature of the target is such, that the echo is returned with no phase change.

On the other hand, for a moving target, receding from the transducer with velocity v, the signal for each single echo $e(t)$ is perceived to have a shift in wave frequency, such that,

$$e(t) = g(\phi(t - \tau_m)) \tag{2}$$

where,

$$\phi = \frac{1 - \beta}{1 + \beta} \qquad \tau_m = \frac{\tau}{1 - \beta} \qquad \beta = \frac{v}{c} \tag{3}$$

Here, the Doppler effect (the change in **frequency** perceived by an observer (transducer) due to relative motion between it and the scatterer) has been expressed as an associated transformation in the time domain (the change in **time scale** perceived by an observer (transducer) due to relative motion between it and the scatterer) [5]. Hence, there is a perceived expansion in the wave period and, conversely, the carrier frequency of the received echo will decrease.

The signal describing the series of echoes from a constant velocity interface can be written as,

$$E(t) = \sum_{n=0}^{n=N} g(\phi(t - Pn(1 + \varepsilon) - \tau_m)) \tag{4}$$

Eq. 4 shows that the echoes are not separated by the same pulse repetition period as in Eq. 1. Because of the progressively increasing distance of the moving reflector, each successive echo is delayed by an additional amount εP relative to that for a stationary reflector, where

$$\varepsilon = \frac{2\beta}{1 - \beta} \tag{5}$$

and is referred to as the time delay. Although the echoes are separated by a fixed "echo separation period", $ESP = \varepsilon P(1 + \varepsilon)$, it is no longer equal to the emitted pulse temporal separation P. Note that a measurement of ESP would, if possible, provide an estimate of the velocity (actually: v/c), provided that the (emitted) pulse repetition period is known. While such a method may be directly possible when there is only a single moving target, the general case -- with many overlapping and interfering echoes -- rules out such a direct approach. But, in fact, it will be noted below that this is the essential principle by which pulsed Doppler systems measure target velocity -- rather than by estimating the actual frequency shift exhibited by the echo.

CONVENTIONAL DEMODULATION

The term conventional demodulation is used in this paper to refer to a technique based on that used in coherent continuous wave demodulation systems [6].

Continuous wave devices

In these systems, the received echo signal $A \sin(\phi \omega_o(t - \tau_m))$ is multiplied by a continuous reference signal at the emitted frequency, viz.: $\sin(\omega_o t)$. This produces a superposition of two signals, one with a frequency equal to the sum of the reference and echo, the other with a frequency equal to the difference,

$$D_c(t) = \frac{A}{2} \{\cos((\phi + 1)\omega_o t - \phi\omega_o \tau_m) + \cos((\phi - 1)\omega_o t - \phi\omega_o \tau_m)\} \tag{6}$$

Since $\phi + 1 = 2/(1 + \beta)$, $\phi - 1 = -2\beta/(1 + \beta)$, and $v \ll c$, appropriate low pass filtering removes the signal of frequency $(\phi + 1)\omega_o$ and passes that at frequency $(\phi - 1)\omega_o$. The latter is the shift in wave frequency of the (continuous wave) echo due to the Doppler effect.

Pulsed Doppler devices

In a conventional pulsed Doppler device [7] the echo is multiplied by a delayed reference burst signal $r(t)$, of radial frequency ω_o,

$$r(t) = \sum_{n=0}^{n=N} \sin \omega_o(t - Pn)[H(t - Pn - t_1) - H(t - Pn - t_2)]$$

(7)

where,

$$H(t) = 1 \quad t \geq 0$$

$$= 0 \quad t < 0$$

(8)

and $t_2 - t_1$ determines the length of reference gate which is mixed with the returning signal. In the present analysis it is assumed that the length of the reference gate is long enough for the envelope of a single component of the received pulse echo sequence to go to zero within the gate. For an emitted wave of narrow bandwidth it is usually convenient to describe the pulse, $g(t)$ using,

$$g(t) = a(t)\sin(\omega_o t)$$

(9)

where $a(t)$ is referred to as the "pulse shape". Demodulation is usually carried out using in-phase and quadrature gated signals to give directional information. For the present, in-phase demodulation only will be considered. Multiplication of the echo signal and reference also produces a superposition of two signals, one with a frequency equal to the sum of the reference and echo, the other with a frequency equal to the difference of the reference and echo,

$$D_2(t) = \sum_{n=0}^{n=N} a(\phi(t - \Delta_n))\sin(\omega_o\phi(t - \Delta_n))\sin(\omega_o(t - Pn)) \times \ldots$$

$$[H(t - Pn - t_1) - H(t - Pn - t_2)]$$

$$= \frac{1}{2}\sum_{n=0}^{n=N} a(\phi(t - \Delta_n))[\cos(\omega_o(t(\phi - 1) - \phi\Delta_n + Pn)) - \ldots$$

$$\cos(\omega_o(t(\phi + 1) - \phi\Delta_n - Pn))] \times \ldots$$

$$[H(t - Pn - t_1) - H(t - Pn - t_2)]$$

(10)

where,

$$\Delta_n = Pn(1 + \varepsilon) + \tau_m$$

(11)

Hence, low pass filtering gives,

$$D_3(t) = \frac{1}{2}\sum_{n=0}^{n=N} a(\phi(t - \Delta_n))\cos(\omega_o(t(\phi - 1) - \phi\Delta_n + Pn))$$

(12)

where the gating term has been dropped because of the assumption that the envelope goes to zero within the gate. The low pass filter output is sampled once within each pulse-echo sequence and this procedure can be represented mathematically by the sampling function,

$$s(t) = \sum_{n=0}^{n=N} \delta(t - Pn - \gamma_s)$$

(13)

where γ_s is the time after each pulse emission at which the signal is sampled. Hence the n th sampled output is

$$D_3(Pn) = a(-\phi\varepsilon Pn + \sigma_e)\cos(\omega_o(-\phi\varepsilon Pn + \sigma_o)))$$

(14)

where,

$$\sigma_e = \phi(\gamma_s - \tau_m) \qquad \sigma_o = \phi(\gamma_s - \tau_m) - \gamma_s$$

(15)

$D_3(Pn)$ is held until the demodulation and sample-and-hold processes begin again in the next period. Thus the "ideal" sample-and-hold output is given by,

$$D_4(t) = \sum_{n=0}^{n=N} a(-\phi\varepsilon Pn + \sigma_e)\cos(\omega_o(-\phi\varepsilon Pn + \sigma_o)) \times \ldots$$

$$[H(t - Pn - \gamma_s) - H(t - P(n + 1) - \gamma_s)]$$

(16)

The signal is bandpass filtered to remove the effects of the holding term and it can be shown [8] that this returns the signal,

$$D_5(t) = a(-\phi\varepsilon t + \sigma_e)\cos(\omega_o(-\phi\varepsilon t + \sigma_o)) \qquad (17)$$

This has the form of a time-scaled version of the envelope of the emitted pulse, multiplied by a similarly time-scaled sinusoidal factor.

In conventional pulsed Doppler systems the low pass filtered signal is usually integrated over the range gate and the final value held using the sample and hold circuit controlled by a logic unit. The effect this has on the demodulated signal can be examined by considering a pulse shape, $a(t)$, of the form $\exp(-b^2t^2)$. The nth integrated value, I_n, is,

$$I_n = \frac{1}{4}\int_{t_1}^{t_2} \exp(-b^2\phi^2(t-\Delta_n)^2)\{\exp(+i\omega_o[t(\phi-1)-\phi\Delta_n+Pn])+\ldots$$

$$\exp(-i\omega_o[t(\phi-1)-\phi\Delta_n+Pn])\} \qquad (18)$$

which becomes,

$$I_n = \frac{\sqrt{\pi}}{2b\phi}\exp(-[\omega_o(\phi-1)/2b\phi]^2)\cos(\omega_o[\varepsilon Pn+\tau_m]) \qquad (19)$$

I_n is sampled and held until the demodulation and sample-and-hold processes begin again in the next period, so that the sample-and-hold output over a number of pulse repetition periods becomes,

$$D_6(t) = \frac{\sqrt{\pi}}{2b\phi}\exp(-\omega_o(\phi-1)/2b\phi]^2)\sum_{n=0}^{n=N}\cos(\omega_o[\varepsilon Pn+\tau_m])\times\ldots$$

$$[H(t-Pn-t_2)-H(t-P(n+1)-t_2)] \qquad (20)$$

Bandpass filtering removes the holding term to produce the demodulated signal,

$$D_7(t) = \frac{\sqrt{\pi}}{2b\phi}\exp(-[\omega_o(\phi-1)/2b\phi]^2)\cos(\omega_o[\varepsilon t+\tau_m]) \qquad (21)$$

a sinusoidal signal whose frequency is dependent on ε only. The importance of this result, with regard to the effect of attenuation on pulsed Doppler flowmeters, will be considered in the following sections.

EVIDENCE FOR THE ROLES OF THE TWO "DOPPLER" EFFECTS

Since ϕ may be measured, in principle, from a single echo, whereas ε requires measurement with at least two echoes, it is interesting to observe the effect of each individually. Indeed, it is possible to envisage situations where each is measured separately [9]. Removing the wave frequency shift is equivalent to setting $\phi = 1$. In this case, the mean frequency of the demodulated signal becomes $\varepsilon\overline{\omega}_o$, and the target velocity is still recoverable. However, removing the time delay effect is equivalent to setting $\varepsilon = 0$: in this case, the mean frequency of the demodulated signal is zero and the target velocity is no longer recoverable from the demodulated signal. Thus, although the two effects are present in the demodulated signal, two points are clear. First: the time delay is essential, in pulsed Doppler systems, for the measurement of target velocity. Second: the role of ϕ highlights the effect of frequency dependent processes on pulsed Doppler flowmeters. Equivalently, if the system assesses target velocity from a measurement of ε alone, it is insensitive to frequency-dependent artefacts (unless the modified form of the pulse shape enters into the assessment). Time domain correlation techniques would fall into the latter category. Indeed, an often claimed advantage of these systems over conventional pulsed Doppler techniques is that they measure the time delay only and, hence, are insensitive to frequency-dependent attenuation effects [10]. However, the above analysis shows that conventional demodulation schemes are also dependent on the time delay factor in order to measure target velocity. Also, as Eq. 21 indicates, processing techniques in conventional pulsed Doppler flowmeters can be envisaged which allow velocity to be assessed via ε only: in such cases, the velocity estimate would not be expected to be significantly influenced by attenuation artefacts.

In order to experimentally validate the effect of changes in ϕ and ε on the output from pulsed Doppler flowmeters, a system for changing these parameters independently is required. This can be substantially achieved in practice by using an electronic phantom, as in Fig. 1a, for example. A signal generator (1), triggered by the transmit signal of the pulsed wave device under test, produces a tone burst of a predetermined frequency and number of cycles. The triggered tone burst is modulated with a continuous low frequency wave from a second signal generator, (3), and is injected back into the receiver input of the system under test. It is readily shown that there is both a change in frequency of the tone burst as well as a cyclic change in tone burst amplitude. It can be shown that the latter change is equivalent to a change in ε, while the former is, more obviously, equivalent to a change in ϕ. Fig. 1b shows the mean frequency of the demodulated signal of a pulsed Doppler device as the (electronic phantom's) modulating frequency is changed. The tested device uses a direct sampling method [11] for demodulation -- producing a similar result to Eq. 17, with the same influence of ϕ and ε as described above. As shown in Fig. 1b, the measured mean frequency of the outputted demodulated signal is equal (within the constraints of aliasing, which occurs at 2.4kHz) to the phantom's modulating frequency. From this result it may be - erroneously - inferred that pulse Doppler systems measure the shift in the actual wave frequency. In fact, the pulse Doppler system output is merely the response to the cyclic variation of the tone bursts' *amplitude*. This observation may be tested directly in the following way. The signal generator (1) can change the signal frequency of the individual tone bursts without changing their repetition frequency (which is controlled by the trigger of the pulsed Doppler system). Fig. 1c shows the mean frequency of the demodulated signal (o), for a fixed modulating frequency (500 Hz), as the tone burst frequency is changed in steps of 50 kHz. If the system were to truly measure the shift in tone burst signal frequency, large fluctuations in the demodulated frequency should occur, with aliasing taken into account (--). However, the measured mean frequency of the outputted demodulated signal remains constant, at approximately 500 Hz -- confirming that it is by reason of the change in ε that the pulsed Doppler system measures velocity.

THE EFFECT OF ATTENUATION

From Eq. 2, Eq. 17 and Eq. 21, the mean frequency of the rf-echo, demodulated signal derived by sampling and demodulated signal derived by integrating and sampling are respectively,

$$\overline{\omega}_R = \phi\overline{\omega}_o$$

$$\overline{\omega}_S = \phi\varepsilon\overline{\omega}_o$$

$$\overline{\omega}_I = \varepsilon\omega_o \tag{22}$$

where $\overline{\omega}_o$ is the mean frequency of $g(t)$. For the demodulation scheme involving integration, $\overline{\omega}_o = \omega_o$ since $a(t)$ in Eq. 9 was assumed to be Gaussian. The change in perceived (mean) frequency of a single rf-echo is given by

$$\Delta\overline{\omega} = (\phi - 1)\overline{\omega}_o$$

$$\sim -2\frac{v}{c}\overline{\omega}_o \tag{23}$$

Hence, knowing $\overline{\omega}_o$ and c and measuring $\Delta\overline{\omega}$ allows the velocity of the target to be recovered. On the other hand, the mean frequency of the demodulated signals of Eq. 9

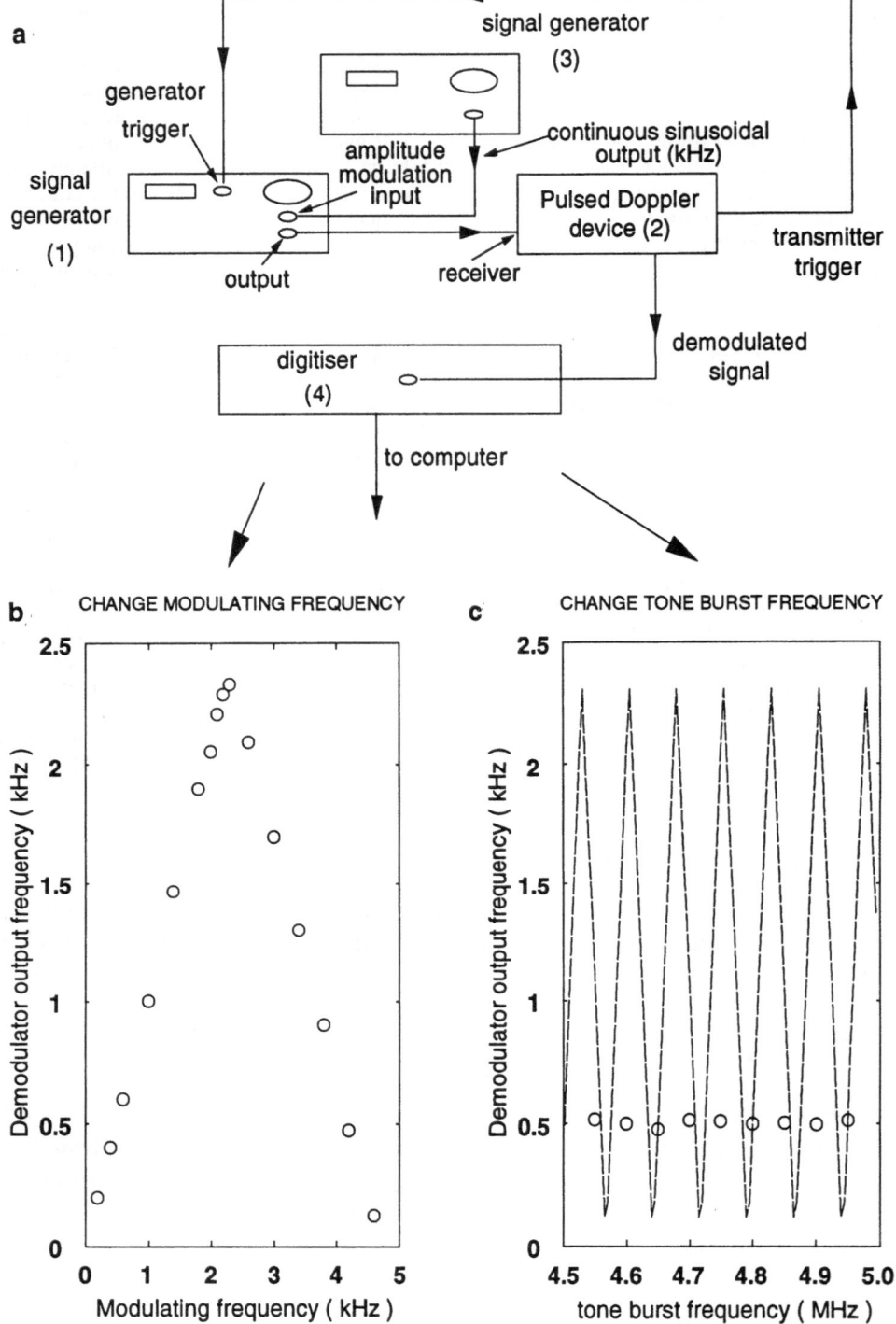

Fig. 1 - a) electronic pulse injection system b) mean frequency of demodulated signal as **modulating frequency** *is changed* (o) *c) mean frequency of demodulated signal as* **tone burst** *frequency is changed* (o) *and the theoretical mean frequency of demodulator output of pulsed Doppler device considering it to measure shift in wave frequency (--).*

and Eq. 21 can be approximated as,

$$\bar{\omega}' \sim 2\frac{v}{c}\bar{\omega}_o \qquad (24)$$

from which the velocity of the target can also be recovered. Note that, in Eq. 23, it is a mean frequency *shift* that has to be calculated in order to estimate the velocity, while in Eq. 24 the target speed is recovered from a mean frequency itself (directly).

The ϕ-factor is associated with (wave) frequency-dependent changes in the echo signals. Consider now the possibility of frequency-dependent artefacts which modify the ϕ-factor, causing it to take on an effective value of $q\phi$. Under these circumstances, the mean frequencies of the respective signals of Eq. 22 become

$$\bar{\omega}_{RA} = q\phi\bar{\omega}_o$$

$$\bar{\omega}_{SA} = q\phi\varepsilon\bar{\omega}_o$$

$$\bar{\omega}_{IA} = \varepsilon\omega_o \qquad (25)$$

Note that $\bar{\omega}_I = \bar{\omega}_{IA}$, since the frequency of the demodulated signal obtained by integrating and sampling is independent of ϕ. For the other signals, $q = 1$ represents the absence of frequency dependent artefacts, $q < 1$ a downshift in mean frequency of the received echo (attenuation artefact) and $q > 1$ an upshift in mean frequency of the received echo (scattering and nonlinearity artefacts). If the velocity of the target is to be recovered from a single echo signal, by using Eq. 23, then the ratio of the artefact-affected estimated velocity, v_A, to the (frequency-dependent) artefact-free estimated velocity, v, is given by

$$\frac{v_A}{v} = 1+\left(\frac{\phi}{\phi-1}\right)(q-1) \qquad (26)$$

If the appropriate velocities are estimated in the more usual manner, viz. from the demodulated signal, via Eq. 24, then the ratio of the artefactual to the correct value is given by

$$\frac{v_A}{v} = q \qquad (27)$$

for the demodulated signal derived by sampling alone and is unity for the demodulated signal derived by integrating and sampling. Hence, if $q = 0.9$, the ratio of the velocities estimated from the rf echo signal is $v_A/v = 1-0.1(c-v)/2v$, whereas the ratio of the velocities estimated from the demodulated echo signals is either 0.9 or 1. Since $c \gg v$, the error due to attenuation in the velocity estimated from the demodulated signal is zero or much less than the error in velocity estimated from the rf echo.

In order to experimentally measure the effect of attenuation, the electronic phantom of Fig. 1a, was modified. The low frequency modulating signal (signal generator (3)) and the trigger from the pulsed Doppler device were disconnected from the other signal generator (1). The outputted pulses from the signal generator (1) were triggered internally in the generator and thus their repetition period was not synchronised with the sampling time of the pulsed Doppler device. Tone bursts with a fixed pulse repetition period were fed directly into the pulsed Doppler system and the signal frequency of the individual bursts was varied from 4.4 to 5.8 MHz in 40 kHz steps. Fig. 2a shows the frequency that would be produced if the pulsed Doppler device measured the actual shift in frequency of the tone bursts. The large fluctuations are seen to produce large oscillations in the theoretical frequency of the demodulator output. Fig. 2b shows the measured frequency of the demodulator output (o) of the pulsed Doppler device for a series of tone bursts with a repetition period of $\sim 212\,\mu s$ in the same frequency range. The large oscillations in the measured frequency of the demodulated signal are not seen. Instead, there is a gradual

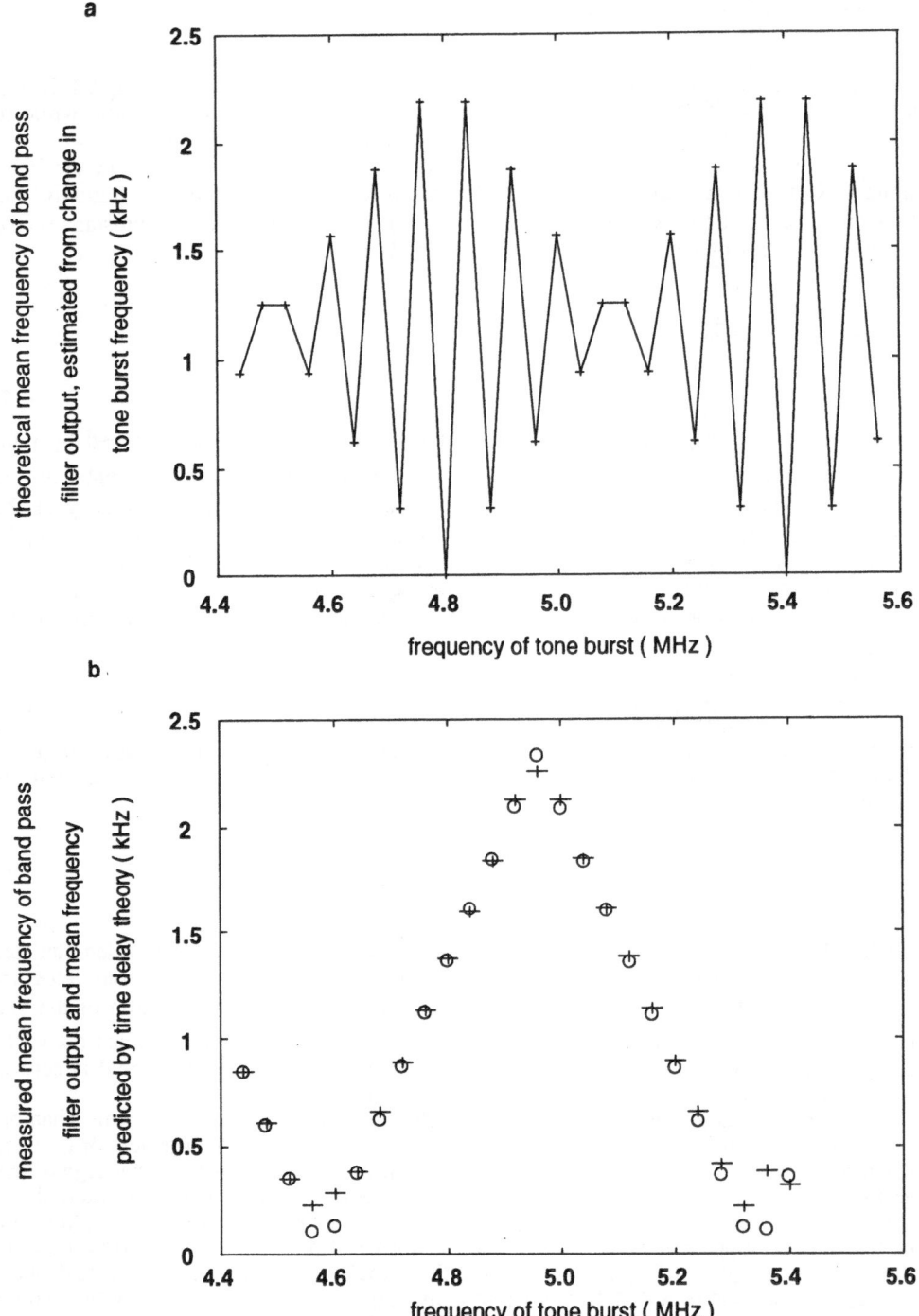

Fig. 2 - a) Theoretical mean frequency of demodulator output of pulsed Doppler device considering it to measure shift in wave frequency of series of tone burst inputs b) measured mean frequency of demodulator output of pulsed Doppler device (o) for a series of tone burst inputs and mean frequency predicted by time delay theory (+).

increase in the measured frequency for input frequencies of 4.6 to 4.88 MHz. Above 4.88 MHz aliasing occurs and the measured frequency decreases.

Fig. 2b can be understood by using the time delay theory presented above. The repetition period of the tone bursts from signal generator (1) is not the same as the sampling period of the pulsed Doppler device (213.3 µs). Hence, each successive tone burst shifts relative to the sampling point and a demodulated tone burst is produced. The repetition period of the tone bursts produces a demodulated signal with a frequency above the aliasing frequency of the pulsed Doppler device. Hence, a series of long input tone bursts (~ 100 cycles) had to be used to obtain a demodulated signal of reasonable duration. The frequency of the demodulated signal was predicted using Eq. 7 - Eq. 17, with aliasing taken into account. Using the repetition period displayed by the signal generator (212 µs) the estimated results were shifted laterally from the experimental findings (o). The theoretical estimate (+) in Fig. 2b was obtained with a repetition period of 212.023 µs.

CONCLUSION

For conventional pulsed Doppler systems the demodulated signal is seen to be affected by two effects associated with the pulse echoes from a constant velocity target. The first effect is the 'classical' Doppler effect, viz. a wave frequency shift of the individual echoes; the second effect is a time delay of successive echoes, which manifests itself as a shift in the pulse repetition frequency of the received echo signals. It is primarily by virtue of the second effect that pulsed wave Doppler systems measure the velocity of the target. Since frequency-modifying propagation and scattering artefacts (including attenuation) cannot influence the second of the two above effects, they have either no effect on the mean frequency of the demodulated signal or one that is very much reduced in comparison with that expected on the basis of the classical Doppler effect. The magnitude of the attenuation artefact has been shown to depend on the precise nature of the demodulation procedure. This may explain why various investigators have disagreed as to the effect of frequency dependent attenuation. Since the ϕ-factor is associated with (wave) frequency-dependent changes in the echo signals, the presented theory can also be used to explain the effect of other propagation and scattering artefacts such as frequency dependent scattering, non-linear propagation, interference and diffraction [9].

ACKNOWLEDGEMENTS

The Wellcome Trust is thanked for an equipment grant which made this research possible. The MRC is also thanked for funding this research.

REFERENCES

1. Azimi M. And Kak A. C., 1985 - An Analytical Study of Doppler Ultrasound Systems. *Ultrasonic Imaging*, **7**, pp. 1-48.

2. Lin T. and Ophir J., 1987 - Frequency-Dependent Ultrasonic Differentiation of Normal and Diffusely Diseased Liver, *J. Acoust. Soc. Am.*, vol. **64**, pp. 423-457.

3. Holland S. K., Orphanoudakis S. C. and Jaffe C. C., 1984 - Frequency- dependent attenuation effects in pulsed Doppler ultrasound: experimental results, *IEEE Trans. Biomed. Eng.*, vol. **BME-31**, pp. 626-631.

4. Thomas N. and Leeman S., 1991 - Blood Velocity Estimation in Medical Imaging - Artefacts and Errors, in: *"Ultrasonics International '91"* (Butterworth, Guildford), pp. 95-98.

5. Leeman S., Roberts V.C. and Willson K., 1986 - Quantitative Doppler with ultrasound pulses. In: *"Physics in Medical Ultrasound"*, J.A. Evans, ed., IPSM, London, pp. 134-140.

6. Atkinson P. and Woodcock J. P., 1982 - Doppler Ultrasound and its use in Clinical Measurement. (Academic Press : London).

7. Advanced Technology Laboratories, February 20th 1984 - Service Manual for 459

Series Pulsed Doppler / Flow Analyzer, Adaptive Doppler, P/N 106-23155-02, Rev B, **4**, pp. 14.

8. Oppenheim A. V., Willsky A. S. with Young I. T., 1983 - *Sampling*, Chapter **8**. In : *Signals and Systems*, Prentice-Hall.

9. Thomas N. and Leeman S., 1991 - Mean Frequency Via Zero Crossings, *IEEE Ultrasonics Symposium Proceedings*, pp. 1297-1300.

10. Foster S. G., Embree P. M. and O'Brien W. D., 1990 - Flow Velocity Profile via Time-Domain Correlation : Error Analysis and Computer Simulation. *IEEE Trans. Ultrason. Ferroelec. Freq. Contr.*, vol. **37**, N° **2**, pp. 164-175.

11. Thomas N. and Leeman S., 1993 - The Double Doppler Effect, *International Conference on Acoustical Sensing and Imaging*, Publ. No. **369**, (IEE : London), pp. 164-168.

CORRELATION OF HISTOLOGY AND ACOUSTIC PARAMETERS OF LIVER TISSUE BY SCANNING ACOUSTIC MICROSCOPY

A.F.W. van der Steen[1], J.M. Thijssen, J.A.W.M. van der Laak, P.C.M. de Wilde, G.P.J. Ebben

University Hospital Nijmegen
P.O.Box 9101
6500 HB Nijmegen
The Netherlands

INTRODUCTION

In former studies clinical trials have been performed to test the possibilities to differentiate between different types of eye melanomas i.e. spindle cell and mixed/epitheloid cell types melanomas using ultrasound methods (Romijn et al. 1991, Thijssen et al. 1991). The latter type is much more malignant than the first type. In these studies a sensitivity and a specificity of differentiation of about 90 % was reached. However, when correlations between the acoustical parameters and the histology of the tumours were investigated, hardly any correlation was found. This means that the underlying acoustical models were not adequate. The present project was started to investigate in detail the acoustical properties of biological tissues. The measurements in this project were performed in vitro at frequencies between 10 and 55 MHz. These frequencies are higher than the frequencies used in the clinical study (7.5 MHz), but still low enough to relate the results to those of the clinical study.

A new method has been developed to correlate acoustical spectral parameters and histology at a microscopic scale (van der Steen et al. 1994). The local acoustical parameters of a section of 250 μm thickness were correlated to quantitative histological parameters of an adjacent section of 10 μm thickness. These methods were tested on White New Zealander Rabbit liver.

---

[1]   A.F.W. van der Steen is now working for the Interuniversity Cardiology Institute of the Netherlands at the Laboratory of Experimental Echocardiography of the Thorax Centre, Room Ee2302, P.O. Box 1738, 3000 DR Rotterdam, The Netherlands

METHODS

Tissue processing

Liver samples, obtained from White New Zealander rabbits (n=10), were fixed in 4% buffered formalin. This has no major influence on the acoustical properties of the

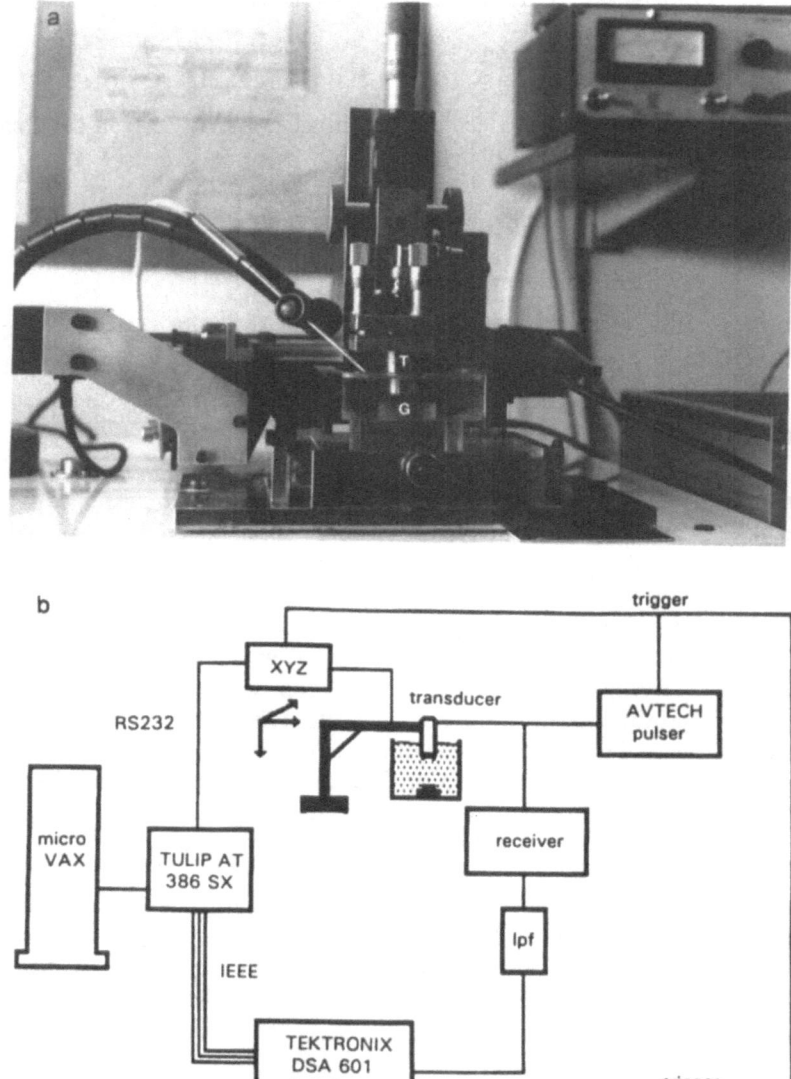

Fig.1. The instrumental set up that is used for the acoustic measurements: a) picture of scanning table with transducer (T) and glass plate (G) b) scheme of measurement and processing devices of C-SAM

tissue (van der Steen et al. 1991, van der Steen et al. 1992). The liver was cut into blocks of approximately 2 x 1.5 x 1 cm. In these blocks, parallel markers were placed. These markers could be identified both acoustically and optically. The liver was sectioned using a freeze sled microtome (Leitz Weltzlar GmbH). Sections were cut alternating in thickness between 10 μm and 250 μm. The 10 μm sections were used for optical examination and the 250 μm sections for acoustical examination. The optical sections were stained by a trichrome staining technique according to Goldner, resulting in sections with red parenchymal tissue, green collagen and dark red nuclei.

Acoustic

Equipment. The custom designed cross sectional scanning acoustical microscope (C-SAM) that was used in this study has been extensively described in another paper and it is displayed in Fig. 1 (van der Steen et al. 1994a). It consists of a pulser (Avtech Inc.)-transducer (PVDF, Fulmer Inc.)-receiver-system that has a -6 dB bandwidth of 10 up to 55 MHz. The transducer is mounted in an XYZ-stepper motor translation system (Märzhauser GmbH) which has an accuracy of 1 μm in all three directions. RF-signals

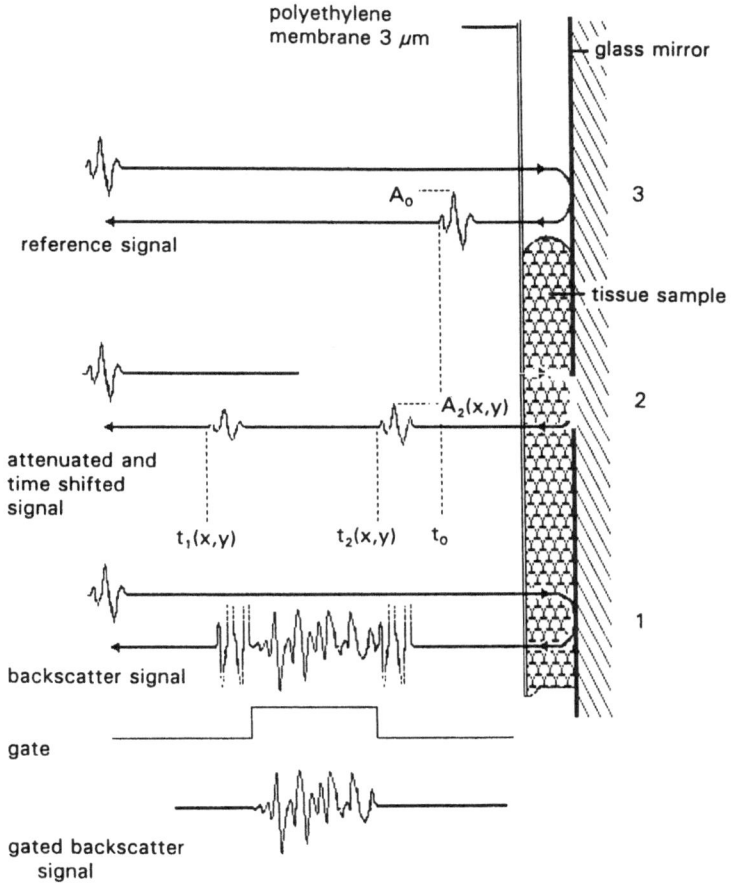

Fig. 2. Scheme of the reflection-/ and backscatter measurements for the estimation of the acoustic parameters

are digitized in 8 bit using a digital oscilloscope (DSA 601, Tektronix Inc.) at a sampling rate of 250 MHz. The data are stored in an IBM compatible 386 PC that also controls the XY-movements of the translation system. Data are processed off-line on a Microvax 3200 (Digital Inc.)

Acquisition. The acoustical sections (250 μm) were placed on a glass block in a watertank that was filled with a physiological saline solution. It was covered by a polyethylene membrane of 3 μm thickness which is practically transparent for ultrasound in the used frequency range. The surface of the block was parallel to the XY-plane of the translation system. The transducer position was vertically adjusted until the section was in the focal zone. Measurements were performed at room temperature (20 $\pm$ 2 °C). The position of the markers was assessed prior to the measurements. Three C-scans (100 x 100, $\triangle$x=$\triangle$y=25 μm) were performed around the position corresponding to the centre between the two markers. The first C-scan was made while using a transmission pulse of -500 V, the second with a transmission pulse of -100 V, and the third one, after removal of the tissue section also with a transmission pulse of -100 V (Fig. 2).

Data processing. The following acoustical parameters were obtained:
Velocity: The local ultrasound velocity was calculated from the time of flight measurements of the glass plate reflection with and without the tissue interposed, the membrane reflection when the tissue was interposed and the known sound velocity in the physiological saline solution. The local thickness was calculated from the local velocity and the glass plate and membrane reflection when the tissue is interposed.
Attenuation: The attenuation was obtained using the substitution method: The spectra of the glass plate reflections with and without tissue are obtained. From these spectra and the local thickness the logarithmic attenuation spectrum was calculated (dB/cm).
From this spectrum, the slope of the linear regression fit and the attenuation at the central frequency were used as parameters.
Backscattering: The backscattering parameters were calculated from the RF-signal of the segment between the membrane reflection and the glass plate reflection. The spectra of these signals were corrected for the system properties and the ultrasound attenuation. The backscattering at the central frequency and the slope of linear regreesion fit to the spectrum were taken as backscatter parameters.

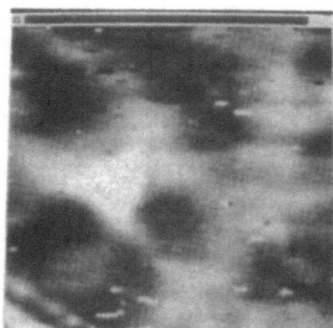 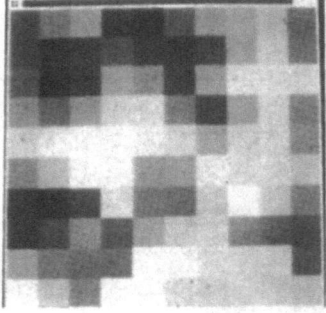

Fig. 3. The attenuation slope in a liver section (2.5 x 2.5 mm.): left: a 100 x 100 points image; right: the image after averaging over 10 by 10 points (c.f. van der Steen et al. 1994a)

Imaging of parameters. After the calculation of the parameters, parameter images were reconstructed (100 x 100 points, 2.5 x 2.5 mm). For the correlation study the parameters were averaged over adjacent subimages of 10 x 10 points. Since the thickness of the section was about 250 μm and the points were acquired 25 μm apart this procedure resulted in an image of 100 cubic voxels of 250 μm sides. An example for the attenuation slope is shown in Fig. 3

Optical

Instrumental setup. The optical images were acquired by a CCD RGB videocamera (Sony 3CCD CA-325AP) that was mounted on top of a light microscope (Axioscop, Zeiss). On the table of this microscope two transducers for positioning feedback were connected (Sony MSH 707, $\triangle x = \triangle y = 1$ μm). The images were acquired and processed on an image processing system VIDAS[plus] (Kontron Inc.,F.R.G.).

Data acquisition. The optical section was placed under the microscope. The markers in the section were found by visual inspection and exactly localized by means of the positioning feedback device. From the localization of the markers in the optical and acoustical section the exact location and orientation of the acoustical voxels was calculated and projected on the light microscopic image. Each of the fields (250 x 250 μm) corresponding to such a voxel was acquired by the system.

Segmentation. The optical section after staining presented red parenchymal tissue, green collagen and dark red nuclei. An RGB-segmentation was performed on the acquired images. From each of the fields the percentage collagen content, the relative area lumen and interstitial spaces and the number of nuclei were segmented automatically by the developed software.

Correlation

The Pearson correlation coefficient was calculated for the relation of all acoustic parameters to the optical parameters. In addition, the partial correlation coefficients of the optical parameters were calculated to eliminate the influence of one optical parameter to the correlation of another optical parameter to the acoustical parameters. These correlations were calculated for each liver and then averaged over ten slices of livers.

RESULTS

An extensive discussion of the results can be found in another paper (van der Steen et al. 1994b). High correlations were found between the attenuation parameters and collagen content and area lumen. After partialling out the area lumen there still was a high correlation between the attenuation and the collagen content. After partialling out the collagen content, the correlation between area lumina and attenuation had totally disappeared. This means that the correlation between lumina and attenuation can be fully explained from the correlation between collagen and attenuation and collagen and lumina.

No significant correlation was found between the number of nuclei or the interstitial spaces and any acoustical parameter. There was no statistical significant correlation found between the backscatter parameters and the ultrasound velocity with any investigated histological feature.

DISCUSSION

A powerful tool has been developed to investigate the correlation of light microscopic and acoustical properties of biological tissues at a microscopic scale. The methods are illustrated by a liver study. The methods will be applied in near future to other tissues, with special emphasis on human intraocular tumours, human skin melanomas that were subcutaneously implanted in nude mice, Hamster Green Melanomas and human eye melanomas from enucleated eyes.

ACKNOWLEDGEMENT

This project has been financially supported by a grant from the Dutch Cancer Foundation (KWF-NUKC 89-03).

REFERENCES

Romijn, R.L., Thijssen, J.M., Oosterveld, B.J., Verbeek A.M., 1991, Ultrasonic differentiation of intraocular melanomas: parameters and estimation methods. Ultrasonic Imaging 13, 27-55

van der Steen, A.F.W., Cuypers M.H.M., Thijssen, J.M., de Wilde, P.C.M., 1991, Influence of histochemical preparation on acoustic parameters of liver tissue: a 5 MHz study. Ultrasound in Med. and Biol. 17, 879-891

van der Steen, A.F.W., Thijssen, J.M., Ebben, G.P.J, de Wilde, P.C.M., 1992, Effects of tissue processing techniques in acoustical (1.2 GHz) and light microscopy. Histochemistry 97, 195-199

van der Steen, A.F.W. Thijssen, J.M. van der Laak, J.A.W.M., Ebben, G.P.J., de Wilde, P.C.M., 1994a, A new method for correlation of acoustical spectroscopic microscopy (30 MHz) and light microscopy. J. Microscopy (in press)

van der Steen, A.F.W., Thijssen, J.M., Ebben, G.P.J., van der Laak J.A.W.M., de Wilde, P.C.M., 1994b, Correlation of histology and acoustical parameters of liver tissue on a microscopic scale. submitted Ultrasound in Med. and Biol. 20, 177-186

Thijssen, J.M., Verbeek, A.M., Romijn, R.L., de Wolff-Rouendaal, D, Oosterhuis J.A., 1991, Echographic differentiation of histological types of intraocular melanoma. Ultrasound in Med. and Biol. 17, 127-138

IMAGING OF NORMAL AND ATHEROSCLEROTIC ARTERIES BY ACOUSTIC MICROSCOPY

P. Anthony N. Chandraratna, Jacqueline Gallet, Parakrama Chandrasoma, Joie P. Jones, and Satish Choudhary

From the Division of Cardiology, Los Angeles County + University of Southern California Medical Center, USC School of Medicine, Los Angeles, CA, and Department of Radiological Sciences, U.C. Irvine

INTRODUCTION

Atherosclerotic vascular disease is a major cause of morbidity and mortality in this country. A technique that accurately diagnoses atherosclerotic lesions and is capable of detecting regression of such lesions, will be of considerable value to both the clinician and researcher. Intravascular ultrasound which uses high frequency transducers has been shown to be a useful technique in the diagnosis of atherosclerosis.

Acoustic microscopy uses very high frequency ultrasound to obtain high resolution images. At a frequency of a 1000 MHz, the wave length of ultrasound is similar to that of light and the resolution approaches one micron. The scanning acoustic microscope has a resolving power that is several hundred times that of ultrasonic diagnostic systems. This instrument permits the visualization of tissues on a microscopic scale, similar to that provided by light microscopy. However, unlike light microscopy, this technique does not require staining of the specimen. Acoustic microscopy is at present mainly used in industry for non-destructive testing. Biomedical applications of acoustic microscopy were first described by Lemons and Quate (1-4). The use of acoustic microscopy for evaluation of myocardial pathology was first reported by our group (5,6). The purpose of this study was to evaluate the role of very high frequency ultrasound (acoustic microscopy) in the diagnosis of atherosclerotic lesions in vitro.

METHODS

Fifteen blood vessels (12 coronary arteries and 3 specimens of abdominal aorta) obtained at the time of autopsy were sectioned at 5 μm and unstained, deparaffinized sections

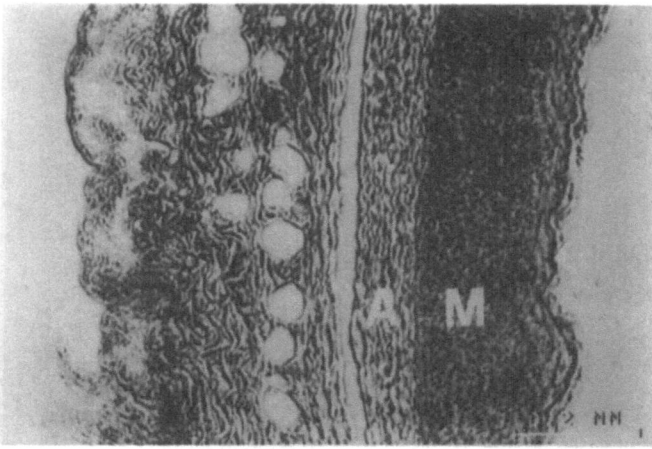

Figure 1. The acoustic microscope image of a normal coronary artery is illustrated in this figure. Intima is thin and an echogenic internal elastic lamina (arrow) is seen. The media (M) and adventitia (A) are clearly seen.

were used for acoustic microscopy. Six vessels were normal and the rest contained atherosclerotic plaque. After acoustic microscopy was performed, the sections were stained with hematoxylin and eosin, and light microscopy was done. The acoustic and light microscopic images were interpreted by two separate observers.

Olympus UH3 Scanning Acoustic microscope

A radio frequency (RF) signal stimulates a piezoelectric crystal, which emits ultrasound waves which are focused by a sapphire crystal. The ultrasound wave traverses the specimen of myocardial tissue on the slide and is then reflected from the slide and on returning to the piezoelectric crystal, a RF signal is emitted and collected by a receiver. The acoustic signal's intensity is converted to brightness for display as a point of light on a

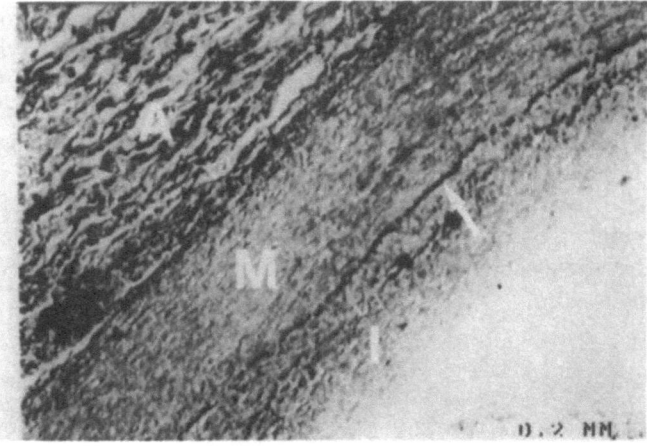

Figure 2. This figure depicts a fatty plaque which shows expansion of the space between the endothelium and the internal elastic lamina (arrow) with hypoechoic material representing fat (I = intima).

cathode ray tube. The specimen is scanned in the horizontal plane in a raster fashion, allowing point by point analysis of the elastic properties of a cross-section of tissue. A 600 or a 1000 MHz transducer was used for this study. The acoustic image is displayed on a screen.

Generation of Images

The source of contrast for the ultrasound images is differing attenuation between various components of myocardial tissue. Structures with the greatest attenuation appear black, while those components of tissue with the least attenuation appear white. When the section contains significant amounts of structural protein, particularly collagen, both attenuation and impedance play a role in the generation of the image. The instrument is also

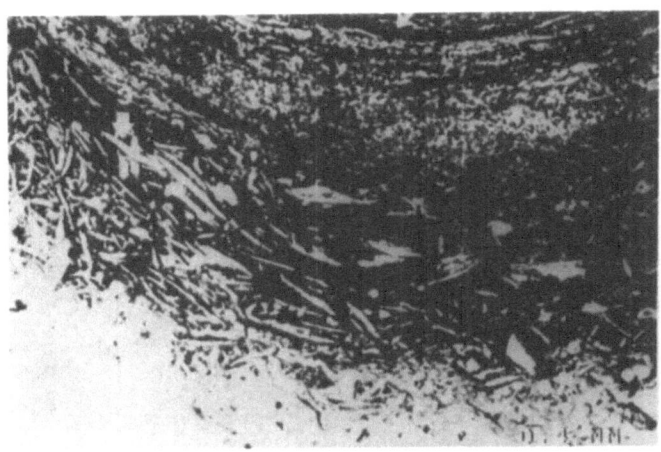

Figure 3. An ulcerated plaque in the abdominal aorta with numerous cholesterol clefts is depicted in this figure.

sensitive to changes in viscosity. No staining was required to generate the images. The acoustic microscopic images and the light microscopic images were evaluated by two observers who were blinded to the results obtained by the alternative technique.

RESULTS

Acoustic microscopy of normal coronary arteries revealed a thin intima, an echogenic internal elastic lamina, media, adventitia and perivascular adipocytes (Figure 1). Expansion of the space between the internal elastic lamina and endothelium was identified in atherosclerotic vessels. Fatty plaque (Figure 2) produced hypoechoic zones interspersed with collagen fibers (fibrous component). Large numbers of cholesterol crystals were seen in some fatty plaques (Figure 3). Fibrous plaques consisted mainly of collagen fibers, which appeared

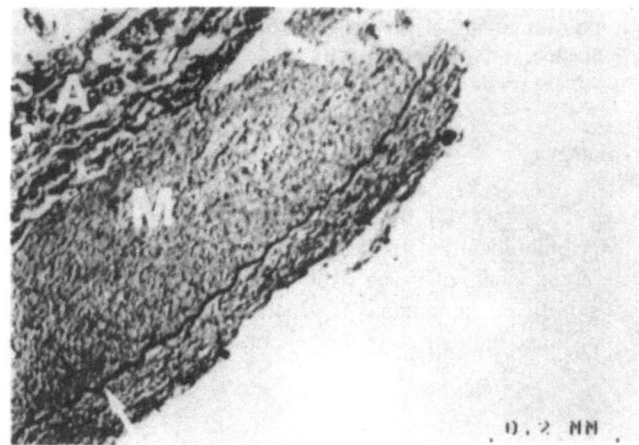

Figure 4. A fibrous plaque is shown in this figure. The thickened intima shows dark linear echoes representing fibrous tissue.

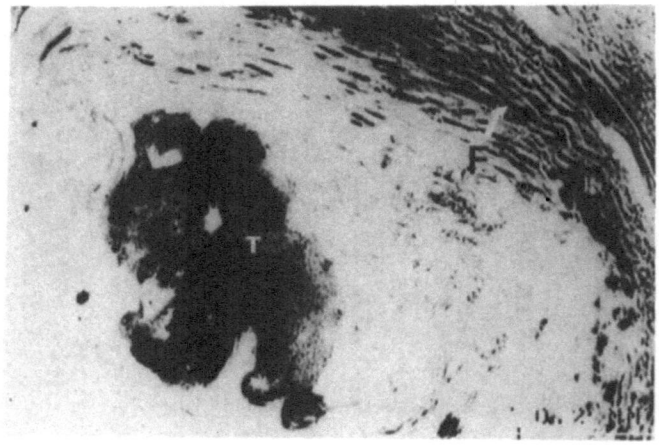

Figure 5. A partially organized thrombus (T) overlying a plaque is illustrated in this figure. Note the discrete dark echoes within the thrombus. Collagen fibers (F) are seen as dark linear strands.

as dark linear echoes on the acoustic image (Figure 4). Calcific plaques were very echogenic. The acoustic microscope findings were confirmed by light microscopy in all cases.

Intraluminal thrombus could be clearly differentiated from fatty plaque, a distinction that cannot be easily made by other forms of ultrasound imaging, including Intravascular ultrasound. The thrombus consisted of clumps of dark discrete echoes which probably represented red cells and/or clumps of platelets (Figure 5). Vascular spaces and hemorrhage into the wall of the artery could be identified by acoustic microscopy (Figure 6).

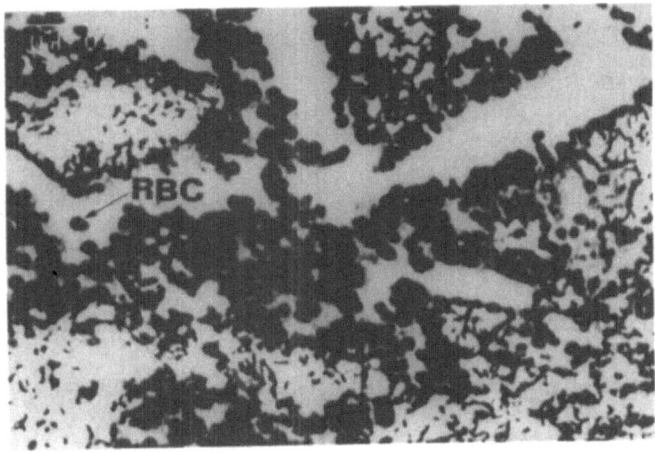

Figure 6. Vascular channels containing red cells (discrete round echoes) and hemorrhage into the wall of the aorta are illustrated in this figure.

DISCUSSION

Atherosclerotic coronary artery disease predisposes to myocardial infarction, which is a major cause of death in Western countries. A variety of invasive and non-invasive techniques have been used to diagnose coronary artery disease. Coronary arteriography provides images of the coronary arterial tree, and this technique is useful in the detection of coronary artery stenoses. More recently, Intravascular ultrasound imaging using a 20 or 30 MHz transducer mounted at the tip of a cardiac catheter, has been used to image both the arterial lumen and atherosclerotic plaque in the vessel wall. Non-invasive techniques such as stress thallium scintigraphy, stress echocardiography and treadmill exercise testing, may be employed to diagnose myocardial ischemia, which is usually caused by coronary atherosclerosis. However, none of these techniques permit the study of histological detail of atherosclerotic plaques.

Normal vascular structure and atherosclerotic plaques were clearly identified by acoustic microscopy. Fatty, fibrous and calcified plaques could be identified by this technique. Pathological phenomena, such as infiltration with cholesterol crystals, intraplaque hemorrhage and Intravascular thrombus were detected. This technique is capable of identifying multiple abnormalities in atherosclerotic vessels, which are not easily evaluated by currently available techniques. Since these images were obtained without staining, further development of the technique for "in vivo" imaging is warranted.

REFERENCES

1. Lemons RA, Quate CF. Acoustic microscope-scanning version. Appl Phys Lett 1974;24:163-165.

2. Johnston RN, Atalar A, Heiserman J, Jipson V, Quate CI. Acoustic microscopy:

Resolution of subcellular detail. Proc Natl Acad Sci U.S.A. 1979;76:3325-3329.

3.   Hildebrand JA, Rugar D, Johnston RN, Quate CI. Acoustic microscopy of living cells. Proc Natl Acad Sci U.S.A. 1981;78:1656-1660.

4.   Neild TO, Attal J, Saurel JM. Images of arterioles in unfixed tissue obtained by acoustic microscopy. J Microsc 1985;139:19-25.

5.   Chandraratna PAN, Choudhary S, Jones JP, Chandrasoma P, Kapoor A, Gallet J. Visualization of myocardial cellular architecture using acoustic microscopy. Am Heart J, 1992;124:1358-1364.

6.   Chandraratna PAN, Choudhary S, Jones JP. Visualization of isolated myocardial cells by acoustic microscopy {Abstract}. Circulation 1990;82:III-69.

WHEN A-PRIORI INFORMATION CANNOT RESOLVE
TOMOGRAM AMBIGUITIES

Steve Isakson, A. Meyyappan and G. Wade

University of California at Santa Barbara
Santa Barbara, California 93106

ABSTRACT

The goal of scanning tomographic acoustic microscopy (STAM) is to image unambiguously at least a portion of a 3-dimensional object. In STAM, a source is always on one side of the object at different locations for different projections and a scanning sensor on the other. STAM is a limited angle tomographic system and the scanning does not necessarily pass over an edge of the object. The authors show that under these conditions it is mathematically impossible to compute an unambiguous map of the absolute transmissivity of the layers of interest in the object and that a priori information can often be used to resolve the ambiguities. However in actual practice the variations in the transmissivity may be too small to be discerned.

We demonstrate the existence of these ambiguities with tomographic simulations and show that in some cases we can remove the ambiguities by invoking the a priori constraint that transmission coefficients cannot be greater than unity or by scanning past an edge of the object.

A similar ambiguity exists for phase information within an object. While multi-frequency scanning can reduce the effects, it does, of course, come at increased computational cost.

INTRODUCTION

While tomography has been around for a long time, scanning tomographic acoustical microscopy (STAM) is a relatively recent development. STAM by nature is a limited angle tomographic system and is subject to the ambiguities common in these types of tomographic systems. One such ambiguity causes the absolute amplitudes of transmissivity of any given cross section to be indeterminate. The origins of this and other ambiguities are associated with the impracticality of insonifying an object over a large range of incident angles. As a consequence ambiguities may well develop in producing cross sectional images from the data.

This paper first examines the features of STAM that cause the ambiguities and show that these ambiguities are associated with the limited angle nature of the system. We will then examine the kinds of prior information that will eliminate the ambiguity.

LIMITED SCAN AREAS AND LIMITED SCAN ANGLES

A typical STAM configuration is shown in figure 1. In this example, varying the insonification angle creates the multiple projections. Other methods, such as rotation of the sample, produce the same end result.

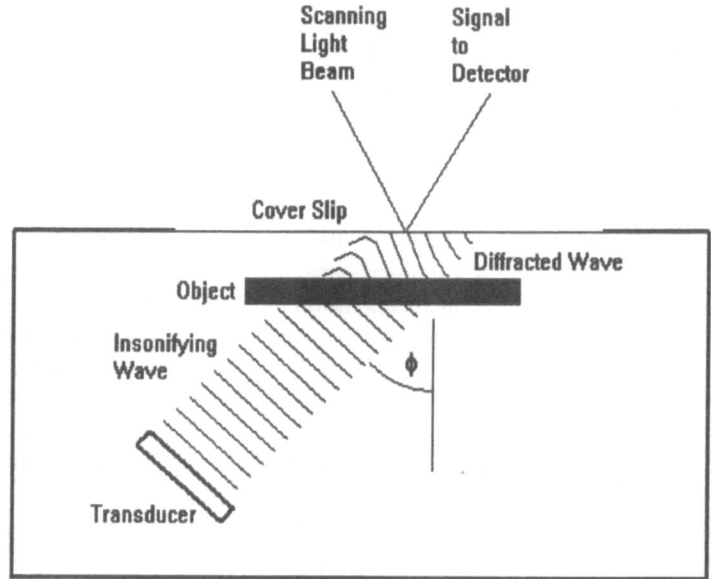

Figure 1. Diagram of a typical STAM setup. ϕ is the insonification angle and can be varied to achieve multiple projections

To achieve high resolution, short wavelengths are used. An operating frequency of 100 MHz gives a wavelength of 15 microns in water. At this wavelength a reasonable scan area measures 5 millimeters square; however, samples are frequently much larger. A silicon chip can be several centimeters on a side -- or larger if mounted. A biological sample is arbitrarily large. While it is theoretically possible to scan the entire sample area, there are practical limits. The size of the transducer providing a uniform insonification field is one limit. The deflection of the laser beam another.

Beyond this, there is a computational limitation. To prevent aliasing the sampling frequency must exceed the maximum spatial frequency on the cover plate by at least a factor of two. As a consequence the number of samples increases with the area sampled. The computational complexity increases even faster. At some point the area becomes too large to scan and image in a reasonable time.

An additional problem is imposed on the insonification angle ϕ. If the acoustical impedance of the sample is different from the surrounding water, the reflected acoustic energy increases as ϕ increases. As ϕ becomes large the signal to noise ratio decreases.

Additionally, in order to keep the distance between the sample and the cover plate at a minimum to reduce propagation losses, the sample is generally much thinner than either the x or y dimensions. Therefore at high ϕ, the distance through the sample increases, which causes higher propagation losses. Finally, if the impedance of the sample is smaller than water, at some angle total reflection occurs and no signal is received. For all of these reasons, high angle insonification of samples in current configurations of STAM provides little information content due to poor signal to noise ratio.

THE CONSEQUENTIAL AMBIGUITY

The above STAM limitations result in an unavoidable ambiguity in the resulting image. An example, illustrated in figure 2, is used to demonstrate this ambiguity. In this example we will assume that the sample consists of only two intensity modifying layers and uniform material between them. The numbers in the figure represent the transmissivities of the individual regions of each layer. We also assume we know a priori where these layers are and the layers are significantly larger than the scan area. The above simplifications help illustrate the conclusions with no loss of generality. The goal of this example is to determine the intensity modifying profile of each of these layers.

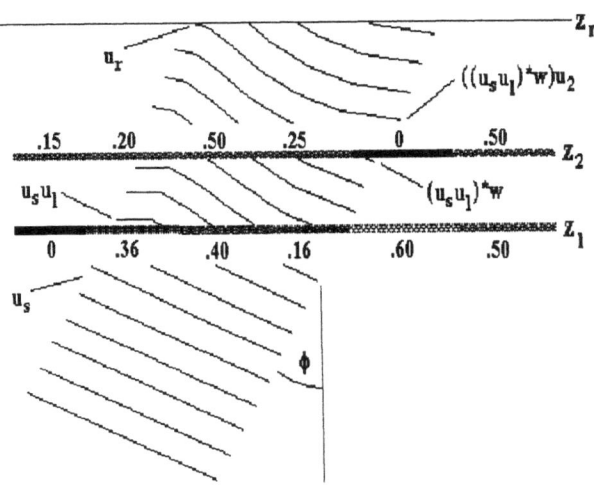

Figure 2. Example to demonstrate the origins of the ambiguity.

In order to achieve these results, the sample is insonified from several different angles. Each different projection is represented in the following equation by a different value of ϕ. For each angle the received signal is:

$$u_r(x,y,\phi) = [((u_s(x,y,\phi)u_1(x,y)) * w(x,y,z_1,z_2))u_2(x,y)] * w(x,y,z_2,z_r)$$

where

$$w(x,y,z_1,z_2) = \frac{|z_2 - z_1|}{2\pi} \frac{1 - jkr}{r^3} e^{jkr},$$

z_1, z_2, and z_r are the z positions of layer 1, layer 2, and the receiver, respectfully, and * denotes convolution. Taking each term of the right hand side of the above equation in order we first see the insonifying plane wave (u_s) modified by the first layer (u_1). The convolution with function w then represents the propagation of the signal from layer one to layer two. The second layer (u_2) then modifies the wave front followed by the propagation of the wave to the receiver (again represented by the convolution with w). Therefore, the above equation represents the propagation of the insonifying wave through the sample and onto the receiving cover slip. These equations are developed in [2].

The above situation has an inherent mathematical ambiguity. If the attenuations of one layer is increased by some factor α and the other layer decreased by the same amount, the received signal will be identical. This is illustrated in the following equation:

$$u_r(x,y,\phi) = [((u_s(x,y,\phi)(\alpha u_1(x,y))) * w(x,y,z_1,z_2))(\tfrac{1}{\alpha}u_2(x,y))] * w(x,y,z_2,z_r)$$
$$= [((u_s(x,y,\phi)u_1(x,y)) * w(x,y,z_1,z_2))u_2(x,y)] * w(x,y,z_2,z_r)$$

It is seen that the received signal is the same in both cases for any proportionality factor α and is independent of the insonification angle ϕ that causes the signal to pass through both layers. Therefore, while the relative attenuations for a single cross sectional area can be reconstructed, the image is ambiguous as far as the absolute attenuations from cross section to cross section.

Figure 3 illustrates this point. Figures 3(a) and 3(b) will both produce the same received signal for any angle ϕ that is available to STAM. Therefore, STAM cannot differentiate the received field for the two figures for any set of projections.

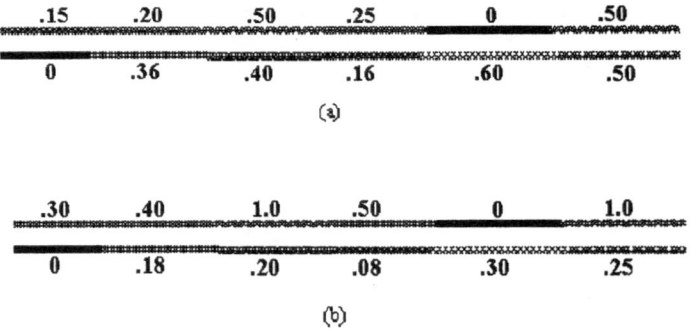

Figure 3. Example of two different objects that produce the exact same received signals for all projections that pass through both planes. In this case, $\alpha = 0.5$.

Finally, there is no loss of generality in this example. If more layers are included, the equation expands, but the only requirement would be the product of the proportionality factors for all the layers be 1. For instance, if the object was modeled as three layers and the layers modified by the factors α, β, and γ, then $\alpha\beta\gamma$ must equal 1. Likewise, the sample could be represented by a continuous solid. That model would be represented by a similar integral equation with the same proportionality ambiguity. A different set of cross sections still results in similar equations. We had simply used these cross sections in this example because they fit the model for the object.

RESOLVING THE AMBIGUITY WITH A PRIORI INFORMATION

There are methods of introducing a priori information to resolve the ambiguities. One is simply to scan off the edge of the sample. Figure 4 is a simplified example of how this is effective. The lines represent the signal passing through the object for a single projection. Each projection will pick scan an area at the edge of the sample. The reconstruction algorithm knows the transmissivity of one portion of each layer -- the area off the end, presumably very close to one. This is sufficient information to set the absolute attenuation of each layer -- no ambiguity will exist.

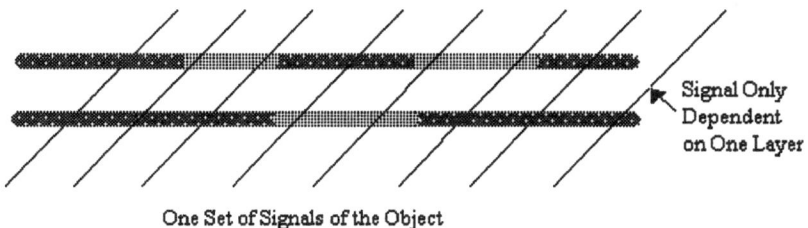

One Set of Signals of the Object

Figure 4. Obtaining a priori information by scanning off the edge of a sample.

Unfortunately, this is not a practical solution. Often the area of interest is in the middle of the sample, and it is impractical to reach the edge. Even if a portion of the edge is scanned, the errors in that small sample of edge information will propagate as larger errors in the center.

Prior information about the specific sample may resolve the ambiguities more effectively. For example, knowledge of uniformity from layer to layer in the sample is used in figure 5 in two different ways to determine absolute transmissivities in each layer. In the first instance, a uniform section through a complete cross section is used to resolve the ambiguities. The second case uses a combination of several partial cross sections to define the transmissivities. While this method works well when the a priori information is available, it usually requires manual intervention and is therefore not cost effective.

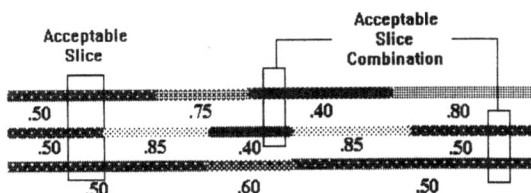

Figure 5. Using areas of uniformity in the differing layers to resolve ambiguities.

A very useful piece of a priori data was observed in our first example – the transmissivity cannot be greater then one. This is sufficient to completely resolve the ambiguities when the object has large areas of high transmission, such as a silicon chip. Figure 6 shows an object not susceptible to the ambiguity when this information is invoked.

Figure 6. An object with transmission areas of unity. The ambiguity can be completely resolved.

While the above method is very successful for samples containing the a priori information, a very distorted picture can develop for samples containing no such information. One such example is depicted in figure 7. This can be typical of a soft biological specimen. Figure 7a contains no areas of unity transmission, and figure 7b demonstrates one possible outcome of invoking this particular knowledge. Vertical slices of this image do not accurately represent the actual object. In this instance it is sometimes

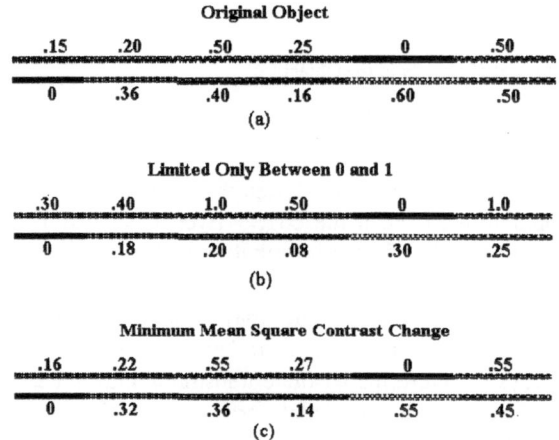

Figure 7. Showing the effects of limiting the transmission to a maximum of unity for an ill- conditioned sample (b). Also shows a compromise solution (c).

beneficial to not seek an exact answer, but to produce a useful one. Often in a sample of this nature there is relative continuity in the sample. Minimizing the mean squared contrast between layers often produces a more acceptable image, although not a correct solution. This is illustrated in figure 7c for the sample object. In a similar manner care must be used with this method. If the original object had high contrast between layers, the contrast would again be distorted and likely produce a less desirable result.

PHASE AMBIGUITIES

Other ambiguities can also develop. The phase delay information for layers illustrated in figure 8 creates an ambiguity very similar to the amplitude ambiguities discussed above. The main difference is that transmission ambiguity is multiplicative while phase ambiguity is additive. An additive increase in phase delay of 0.5π is shown for one layer and a decrease of 0.5π for the other. Since the received signal is the same for both figures at any given angle usable by STAM, the possible solutions cannot be differentiated by any number of projections. Similar a priori information can often be applied to resolve this problem as well.

An additional phase ambiguity exists, however. The receiver cannot differentiate between a received phase shift of π, 3π, 5π, etc. While prior knowledge of the object often resolves this issue, the proper use of multiple frequencies and multiple angles can also automate the process with little extra knowledge, although at a significant computational increase.

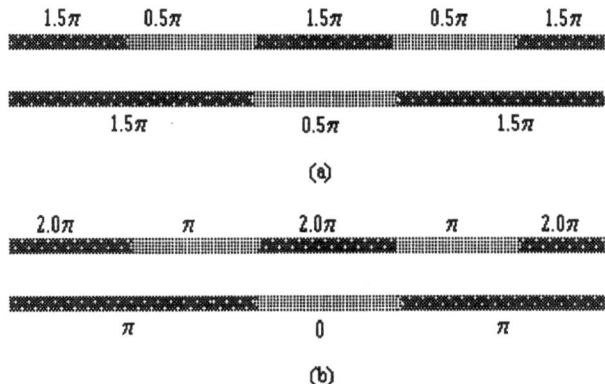

Figure 8. Illustration of phase ambiguity. (a) shows the original object. (b) shows the same object with different phase shifts on each layer, but the received signals are the same.

SUMMARY

In this paper we reviewed some unavoidable ambiguities that occur in typical STAM applications. We found that these ambiguities cannot be removed by any algorithm without some a priori knowledge about the object. In many cases, elementary knowledge can be invoked to remove the ambiguities. However, we find that these same methods, when applied to ill-conditioned samples result in highly distorted images.

The best images result when all the prior knowledge of the object is used to decide the appropriate method to remove the imaging ambiguities. Sometimes, however, the best approach results in only an approximation at increased computational cost.

ACKNOWLEDGMENT

This work is partially supported by NSF under grant number MSS-9020556.

REFERENCES

[1]   A. Meyyappan, <u>An Iterative Algorithm and Refined Data Acquisition for Scanning Tomographic Acoustic Microscopy</u>, Ph.D. Dissertation, University of California, Santa Barbara, 1989.

[2]   A. Meyyappan and G. Wade, "An Iterative Algorithm for Scanning Tomographic Acoustic Microscopy", Ultrasonic Imaging 13, pp334-346 (1991).

ACOUSTIC MICROSCOPY FOR SPHERICAL CAVITIES IN SOLIDS

Oleg Lobkis[1*], Tribikram Kundu[1], and Pavel Zinin[2*]

[1]Department of Civil Engineering and Engineering
Mechanics, University of Arizona, Tucson, AZ 85721

[2]Institute for Material Science and Structure Research
University of Bremen, 28334 Bremen, Germany

[*]on leave from Institute of Chemical Physics
Russian Academy of Science, Kosygin str. 4, 117334
Moscow, Russia

INTRODUCTION

The main advantage of Scanning Acoustic Microscopy (SAM) over conventional microscopes is the possibility of subsurface imaging of optically opaque materials. From the pioneer paper of Lemons and Quate[1], where the detection of subsurface defects in the integrated circuits have been demonstrated, the number of publications involving the images of different subsurface inclusions increases drastically[2]. Numerous experimental works exhibit the great capability of SAM to detect subsurface inhomogeneities located under a solid-liquid interface.

However, the images of subsurface structures are difficult to interpret and may not be directly used for the characterizing the internal inclusions. This is because the signal in SAM contains not only the information from the focal plane but also from the layers above and below the focal plane. For subsurface particle in a homogeneous material SAM output signal is affected by the wave reflected from the solid-liquid interface as well as the waves scattered from the particle.

The problem of the visualization of spherical particles in reflection SAM was considered elsewhere[3]. The purpose of the present paper is to consider the effect of the solid-liquid interface on the image of a the spherical inclusion located under this surface. Effects of acoustical properties of immersion liquid and solid on the image are also studied. Spherical particles are chosen for the reason that only for spherical particles it is possible to obtain the analytical solution for the scattering problem. Thus one can investigate the basic features of the imaging properties of spherical subsurface inclusions.

THEORY

Let the spherical cavity center O_s be placed at a distance z_s from the center of curvature of the spherical focused transducer along the transducer axis and at a horizontal distance r_s from the transducer axis. The radius of the cavity is a. The distance between solid-liquid interface and the center of cavity O_s is d (Fig.1).

To obtain the analytical solution of the output signal of microscope for this system we will make several assumptions. The duration of the RF signal of the microscope is chosen so long that the signal scattering by the cavity may be described

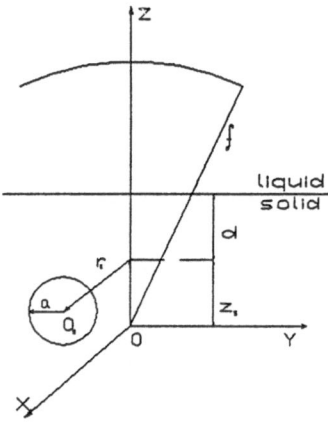

Fig.1. Geometry of system.

in terms of steady state solution of the diffraction problem. For example[4], the field of the radiator may be described as continuous wave solution if the signal duration Δt contains more than several oscillation of the carrier frequency ν, that is $\nu\Delta t > 1$.

On the other hand duration of the signal is adjusted sufficiently short so that the components reflected from liquid-solid interface, and longitudinal and transverse waves scattered by the cavity are separated in time. Hence the signal duration and depth of the cavity must be related in the form $(d - a) > (z_s + d)(c_1/c)(1-cos\alpha) + c_1\Delta t/2$, where c is the longitudinal wave speed in immersion liquid, c_1 is the longitudinal wave speed in solid, and α is the half aperture angle of the transducer.

At the same time satisfaction of both the conditions mean that the cavity is located deeper than the wavelength and hence not affected by the Rayleigh wave and the reflected wave at the interface is not disturbed by the cavity. So this signal can be represented in the standard form of $V(z)$ curve for a half space[5].

To determine the incident acoustical field in solid one has to represent incident field as the set of plane waves[5]. The longitudinal and transverse plane waves at arbitrary point in solid can be obtained by multiplying the incident waves by transmission coefficients from liquid to solid $L(\theta_i)$ and $T(\theta_i)$ for the longitudinal and the transverse waves[6].

To find the acoustical field scattered by the cavity the incident wave is expressed as a solution of the wave equation in spherical coordinates with origin at the cavity center. The velocity of a particle in spherical system is given in book[7]. The total scattered field be represented in the same form as the incident field that is in terms of spherical wave functions with unknown coefficients. These coefficients are determined from boundary conditions. When the obstacle is a spherical cavity (it can be considered as an inclusion of density equal to zero) in an otherwise solid medium, the boundary conditions at $r = a$ are $\sigma_{rr} = 0$; $\sigma_{r\theta} = 0$; $\sigma_{r\varphi} = 0$.

After the scattered waves from spherical cavity have been determined one needs to solve the problem of transmission of these waves through the plane interface from solid to liquid. Hence the scattered fields once again have to be represented as a set of plane waves. The spherical wave function for scattered wave is expressed in terms of plane waves[8]. Then the transmission of the scattered waves through the solid-liquid interface from solid to liquid is performed in the same manner as that one from liquid to solid[6].

After calculating the scattered field can get the output signal by integrating the field on transducer surface. Besides, it is convenient to write the output signal expressions corresponding to different types of waves which arrive at different times. The first signal that is received by the transducer is the wave reflected from liquid-solid interface. This signal has the form of $V(z)$ curve and we will denote it as V_R (R for reflection). The first wave from cavity that is received by the transducer is the signal that is transmitted as the longitudinal wave into the solid and scattered by the cavity again as the longitudinal wave. We will denote this wave as ll-wave and output signal from this wave will be denoted as V_{ll}. After ll-wave two types of waves reach the transducer simultaneously. These two waves are lt (transmitted into solid as the transverse wave and scattered by the cavity as longitudinal wave) and tl (this is reverse of lt-wave). The output signal corresponding to these two waves is $(V_{tl} + V_{lt})$. The final wave received by the transducer is tt - wave. This wave is transmitted into the solid as the transverse wave and scattered by the cavity as the transverse wave also. The voltage corresponding to this signal is V_{tt}.

Then the total output signal voltage V generated by the scattered waves is equal to

$$V = V_R + V_{11} + (V_{t1} + V_{1t}) + V_{tt} , \tag{1}$$

Where

$$V_{11} = \frac{2V_0}{1 - cos\alpha} \frac{\rho_s c_1}{\rho c} \sum_{n,m} (-1)^n (2-\delta_{0m}) \, \tau_n^{11} \, L_{nm}^2 , \tag{2}$$

$$V_{t1} + V_{1t} = \frac{2V_0}{1-cos\alpha} \frac{\rho_s}{\rho c} \sum_{n,m} (-1)^n (2-\delta_{0m})(c_1 \tau_n^{1t} + c_t \tau_n^{t1}) L_{nm} T_{nm}^{SV}, \tag{3}$$

$$V_{tt} = \frac{2V_0}{1-cos\alpha} \frac{\rho_s c_t}{\rho c} \sum_{n,m} (-1)^n (2-\delta_{0m}) \left[\tau_n^{tt} \left(T_{nm}^{SV} \right)^2 - \tau_n^{SH} \left(T_{nm}^{SH} \right)^2 \right]. \tag{4}$$

Here V_0 is the output signal of the transducer when a perfect reflector is placed at the transducer's focal plane, ρ and ρ_s are the densities of the liquid and the solid respectively, c and c_1 have been defined earlier and c_t is the transverse speed in the solid, δ_{0m} is the Kroneker delta symbol, and τ_n^{ij} are the scattering coefficient for spherical cavity[9,10]. The integral coefficients L_{nm}, T_{nm}^{SV}, and T_{nm}^{SH} depend on the cavity position and elastic properties of the solid and can be expressed as

$$L_{nm} = (-i)^m \int_0^\alpha L(\theta_1) \, exp(i\Phi_1) \, J_m(kr_s sin\theta_1) \, \overline{P}_n^m(cos\theta_1) \, sin\theta_1 \, d\theta_1, \tag{5}$$

$$T_{nm}^{SV} = (-i)^m \int_0^\alpha T(\theta_1) \, exp(i\Phi_t) \, J_m(kr_s sin\theta_1) \, \frac{\partial \overline{P}_n^m(cos\theta_t)}{\partial \theta_t} \, sin\theta_1 \, d\theta_1, \tag{6}$$

$$T_{nm}^{SH} = (-i)^m \int_0^\alpha T(\theta_1) \, exp(i\Phi_t) \, J_m(kr_s sin\theta_1) \, \frac{m\overline{P}_n^m(cos\theta_t)}{sin\theta_t} \, sin\theta_1 \, d\theta_1, \tag{7}$$

where J_m is the cylindrical Bessel function of the first kind, $\overline{P}_n^m(cos\theta)$ is the normalized associated Legendre polynomials, k is the wave number in immersion liquid, k_1 and k_t are the wave numbers of the longitudinal and the transverse wave in the solid, $\Phi_1 = -k(z_s + d)cos\theta_1 + k_1 dcos\theta_1$, $\Phi_t = -k(z_s + d)cos\theta_1 + k_t dcos\theta_t$, and the angles $\theta_{1,1,t}$ follow Snell's law, $ksin\theta_1 = k_1 sin\theta_1 = k_t sin\theta_t$.

RESULTS

For small cavity ($ka<1$) only the terms with $n=0,1,2$ can be retained in equations (2)-(4). In this case the output signal of the microscope is actually proportional to the intensity of the incident field at a point where the cavity is located. The $V(z_s)$ curve and the image curve, $V(r_s)$ for a small cavity in plexiglas are presented in fig.2. The V due to ll signal is larger than that due to lt and tt signals which almost coincide with horizontal axis. This is because the transverse wave speed in plexiglas is less than wave speed in water and the incident focused beam of the transverse wave is diverged in plexiglas.

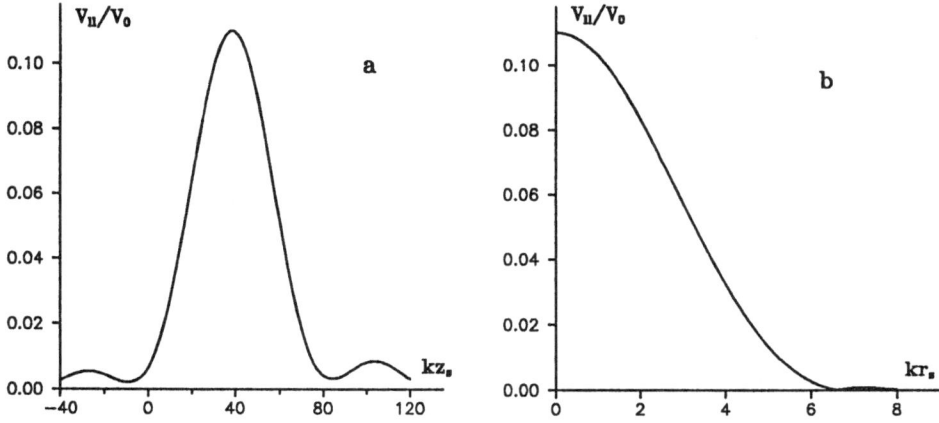

Fig. 2. (a) $V_{11}(z_s)$ curve, and (b) image $V_{11}(r_s)$ curve for a small cavity in plexiglas ($ka = 1$, $kd = 30$, $\alpha = \pi/6$).

The $V(z_s)$ curves for large cavities [$ka(1-cos\alpha) > 1$] in plexiglas and aluminum is presented in fig.3 a,b. One can see that for aluminum the $V(z)$ from tt wave is larger than those from other components scattered by the cavity. This is because in aluminum the transverse wave speed is about two times but the longitudinal wave speed four times the wave speed in water.

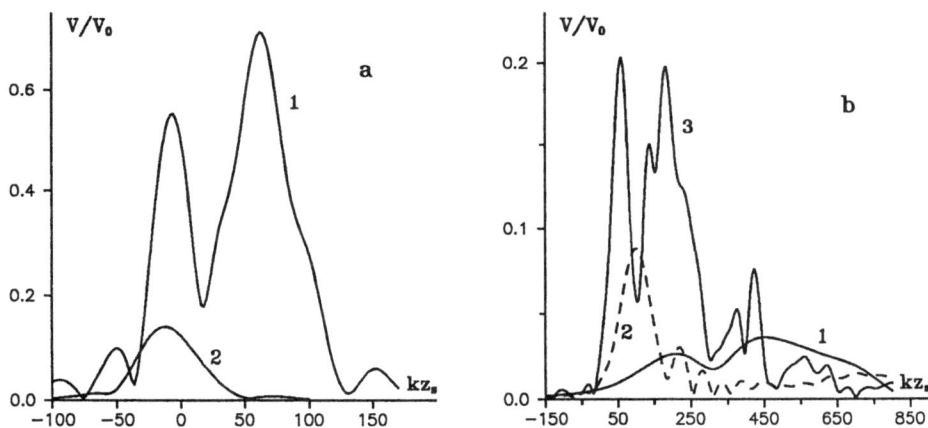

Fig.3. $V(z_s)$ curves for large cavities (a) in plexiglas, $ka=30$, $kd=50$, (b) in aluminum, $ka=50$, $kd=100$, 1-ll wave, 2-lt wave, 3-tt wave. $\alpha=\pi/6$.

The $V_{11}(z_s)$ and $V_{tt}(z_s)$ curves for a large cavity have two maxima. These two maxima arise when the focus of the incident beam coincides with the front surface of the cavity (position z_f) and with its center (position z_c) respectively. For a spherical object in the immersion liquid the distance between this maxima is equal to its radius[3], but for spherical cavity in a solid this distance is different because of the focused beam aberrations. From equations (2) and (4) one can show that for small aberrations z_f and z_c are equal to

$$z_f^{1,t} = -d + \frac{(d - a)\, cos\theta_0^{1,t}}{\sqrt{cos^2\theta_0^{1,t} - 1 + (c/c_{1,t})^2}}\,, \tag{8}$$

$$z_c^{1,t} = -d + \frac{d\, cos\theta_0^{1,t}}{\sqrt{cos^2\theta_0^{1,t} - 1 + (c/c_{1,t})^2}}\,, \tag{9}$$

where $cos\theta_0^{1,t} = (1 + cos\theta_m^{1,t})/2$, and $\theta_m^{1,t} = \alpha$ if $(c_{1,t}/c)sin\alpha < 1$ and $\theta_m^{1,t} = arcsin(c/c_{1,t})$ for opposite case. Hence one can determine the true radius of the cavity from z_f and z_c.

It should be noted that the $V_{1t}(z_s)$ curve has only one peak because in the case of focusing at the center of the cavity there is no transformation from longitudinal to transverse wave for normally incident wave on the cavity surface. When aberrations increase (due to larger difference in wave speeds in solid and liquid) the $V(z_s)$ curve becomes flatter making it difficult to determine the peak positions. One can see this situation for $V_{11}(z_s)$ curve for cavity in aluminum (fig.3b). But in this situation it is possible to find the size of the cavity and its position using the positions of the maxima of $V_{1t}(z_s)$ and $V_{tt}(z_s)$ curves and waves reflected from the solid-liquid interface that is $V_R(z_s)$ curve.

Typically $V(z_s)$ curves for large cavities have two maxima and it is interesting to investigate image of cavity ($V(r_s)$ curve) for these positions - z_f and z_c. These front focus (z_f) and center focus (z_c) images of the cavities in plexiglas and aluminum are shown in fig.4.

In papers[11,12] it was shown that for a spherical object in immersion liquid the size of central image r_c do not depend on the sphere radius and is equal to the half of the Airy spot radius that is $r_c = 0.3\lambda/sin\alpha$, where λ is the wavelength in immersion liquid. This size remains almost same for a cavity but the main maximum is some blurred and the amplitude of side lobes increases because of aberrations (fig.4 a,b). The front image of spherical objects in immersion liquid depends on the radius of the sphere and equal to[12] $r_f = a\, sin\alpha$. For a spherical cavity in solid this size can be estimated from expression $r_f^{1,t} = a \sin\theta_m^{1,t}$, where $\sin\theta_m^{1,t} = (c_{1,t}/c)\sin\alpha$ if $(c_{1,t}/c)\sin\alpha < 1$ and $\sin\theta_m^{1,t} = 1$ for opposite case. True cavity radius can be obtained from this equation also and results obtained from the earlier technique can be verified.

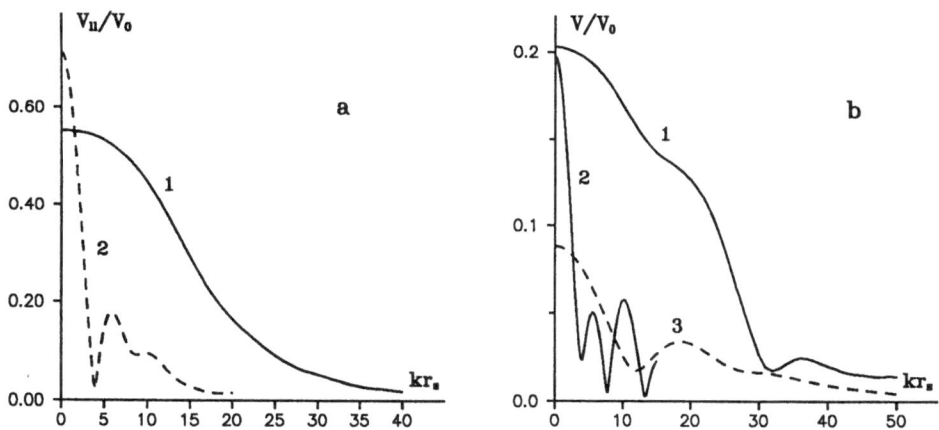

Fig.4. (a) $V_{11}(r_s)$ of a spherical cavity in plexiglas for focusing at the front surface of the cavity (curve 1) and its center (curve 2), $ka = 30$, $kd=50$, $\alpha = \pi/6$. (b) $V(r_s)$ of cavity in aluminum, curve 1 – $V_{tt}(r_s)$ with front focusing, 2 – $V_{tt}(r_s)$ for center focusing, 3 – $V_{1t}(r_s)$ curve. $ka = 50$, $kd = 100$, $\alpha = \pi/6$.

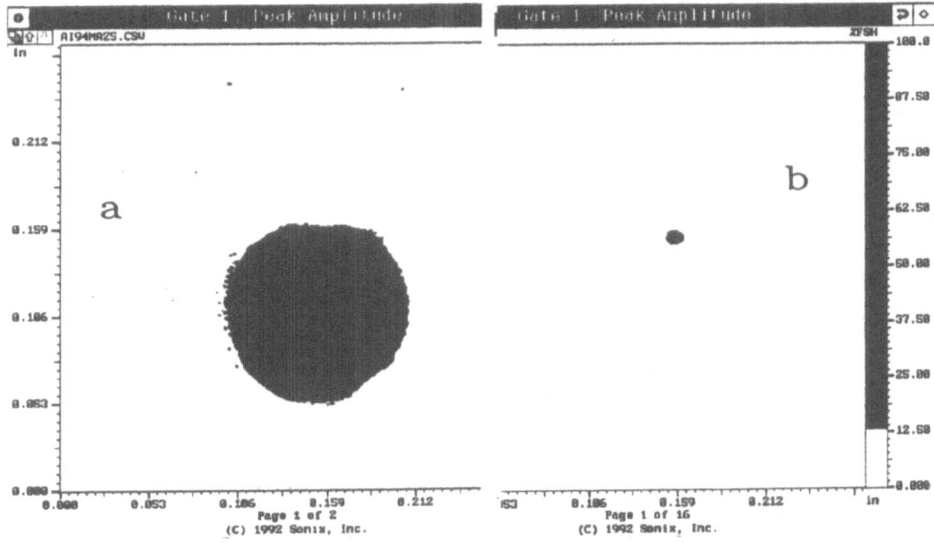

Fig.5. (a) Experimental front focus image of the hemispherical cavity in plexiglas, the cavity radius $a = 0.1588$ cm, the depth of localization $d = 0.3416$ cm. (b) Experimental center focus image of the hemispherical cavity in aluminum, the cavity radius $a = 0.1588$ cm, the depth of localization $d = 0.2248$ cm (spherical focusing transducer, frequency 25 MHz,

wavelength in water $\lambda = 60~\mu$m, focal distance $f = 0.953$ cm, diameter $D = 1.016$ cm, half aperture angle $\alpha = 32^{\circ}$).

ACKNOWLEDGEMENT

Financial support of the CAST program from National Research Council , Washington D.C. is gratefully acknowledged. Partial financial support of NATO Linkage grant HTECH. LG 931353 and NSF grant MSS-9310528 also helped this investigation. The third author was a Humboldt scholar during the investigation and support him from Humboldt Foundation is acknowledged.

REFERENCES

1. R.A.Lemons, and C.F.Quate, Integrated circuits as viewed with an acoustic microscope, Appl.Phys.Lett. 25:251 (1974).
2. A. Briggs."Acoustic Microscopy", Clarendon Press, Oxford (1992).
3. O.I.Lobkis, and P.V.Zinin, Imaging of spherical objects in acoustic microscopy, Sov.Phys.Acoust. 37:343 (1991).
4. J.Krautkremer, and H.Krautkremer. "Ultrasonic Testing of Materials", Springer Verlag, N.Y. (1983).
5. A. Atalar, An angular spectrum approach to contrast in reflection acoustic microscopy, J.Appl.Phys. 49:5130 (1978).
6. L.M. Brekhovskih. "Waves in Layered Media", Academic Press, N.Y. (1960).
7. P.M. Morse, and Y. Feshbach. "Methods of Theoretical Physics", McGraw-Hill, N.Y., Vol.II (1953).
8. A. Erdelyi, Zur theorie der kugelwellen, Physica, 2:107 (1937).
9. C.F.Ying, and R.Truell, Scattering of a plane longitudinal wave by a spherical obstacle in an isotropically elastic solid, J. Acoust. Soc. Amer. 27:1086 (1956).
10. N.G. Einspruch, E.J. Witterholt, and R.Truell, Scattering of a plane transverse wave by a spherical obstacle in an elastic medium, J.Acoust.Soc.Amer. 31:806 (1960).
11. A. Atalar, A backscattering formula for acoustic transducers, J.Appl.Phys. 51:3093 (1980).
12. P.V. Zinin, O.V. Kolosov, O.I. Lobkis, and K.I. Maslov, Imaging of spherical objects in a reflection acoustic microscope, Acoust.Phys. 39:343 (1993).

SCANNING NEAR-FIELD ACOUSTIC MICROSCOPY (SNAM) WITH NANOMETER RESOLUTION IN THE KILOHERTZ FREQUENCY RANGE

A. Kulik, C. Wüthrich and G. Gremaud

Ecole Polytechnique Fédérale de Lausanne
Institut de Génie Atomique
CH-1015 Lausanne, Switzerland

INTRODUCTION

The spatial resolution of Scanning Acoustic Microscopes is determined by the wavelength of the excitation. The maximum working frequency in ambient temperature (and the shortest wavelength) is limited to ca. 2 GHz, due to the extensive attenuation in the coupling fluid (water) which increases as the square of the frequency.

One of the possible ways of increasing the spatial resolution of an Acoustic Microscope is near-field detection. An acoustic wave is sent through the specimen. The subwavelength detector is scanned in close proximity to the sample surface. In this case the spatial resolution is limited by the detector size, rather than the wavelength.

The concept of near-field detection is not new [1-8] and the major improvements were obtained after the development of Scanning Tunneling Microscopy (STM) and Atomic Force Microscopy (AFM) [6,7]. This class of instruments uses a very precise piezoelectric scanning system and detects the sample's topography with excellent spatial resolution.

Several ideas were realized based on STM instruments. Ultrasonic waves can modulate the gap distance between the STM tip and the sample surface. The resulting signal can be detected directly, providing that the current amplifier used has a sufficiently wide bandwidth [9]. One can also use the nonlinear dependence of the tunneling current as a function of the gap distance and use nonlinearity to mix the ultrasonic frequency with the reference signal [10-12]. The resulting signal of differential frequency is then easily detected by a standard low bandwidth amplifier.

Atomic Force Microscopes use a very soft spring (cantilever) with a tip touching the sample surface. Displacements of the cantilever are detected usually by an optical system. The sample is scanned and its topography is imaged. AFM and similar instruments can detect easily superimposed low frequency displacements of the sample surface (so called: force modulation experiments). Similar experiments were scaled up [13] in order to check relations between the radius of the tip, spatial resolution and penetration depth. When the wavelength used has the frequency much higher than the resonant frequency of the cantilever, wide bandwidth detectors [14] must be used, or ultrasonic waves can be detected by a nonlinear dependence between tip-sample forces and displacements (mechanical diode) [15]. The high frequency experiments have the advantage of using the inertia of the vibrating cantilever, increasing the sensitivity of the method to small variations of the elasticity on the sample surface.

There is no clear relationship between the applied frequency and the obtained spatial resolution.

EXPERIMENTAL

We performed low frequency experiments based on a modified Atomic Force Microscope (Park Scientific SPC400). Additional vertical oscillations of the sample were generated using the z-axis piezoelectric scanner provided in every AFM. The frequency was chosen below the principal resonant frequency of the system and far from other weak resonances (f=7804 Hz). The amplitude of the acoustic vibrations of the sample was detected on the surface via the Si cantilever. The vibration frequency was separated from the total displacements detected optically by means of the lock-in amplifier. The amplitude of the received signal was recorded simultaneously with the topographic image, forming the additional image of the local elasticity of the surface.

The test sample was prepared using a 250 nm AsGa grating. The surface was covered by a thin layer of polyimide under which the grating was completely submerged. The polyimide was etched (RIE) to obtain a flat surface containing both stiff (AsGa) and soft (polyimide) materials. We found as well areas of the sample completely covered by polyimide and they were used to test the subsurface sensitivity of our setup.

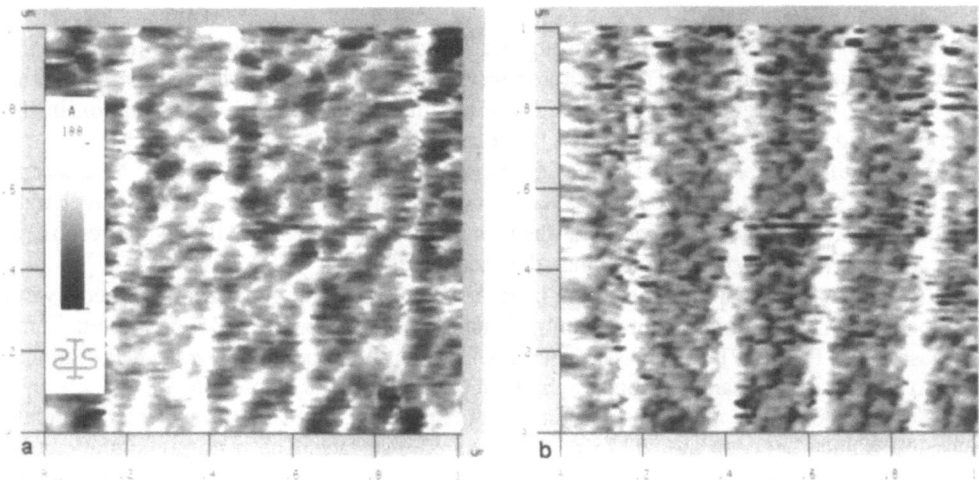

Fig. 1. Topographic (a) and acoustic (b) images of the AsGa grating in polyimide.
Frequency: 7804 Hz, vibration amplitude: 0.7 nm. Scanning field: 1 μm x 1 μm.

RESULTS AND DISCUSSION

Figure 1 shows a pair of simultaneously recorded images (topographic and acoustic) of the region where the AsGa grating emerged from the polyimide. In the topographic image (Fig. 1a) the grating is barely visible. The acoustic image (Fig. 1b) shows the grating fully resolved and develops small details in the polyimide. The estimated spatial resolution is of the order of 30 nm. On a different area (Fig. 2) the grating was covered by the very thin layer of the polyimide and invisible in the topographic image (Fig. 2a) whereas the grating is easily visible in the topographic image (Fig. 2b). The region where the grating is covered by a thicker layer of polyimide (thickness estimated to be 100 nm) shows that the subsurface grating (Fig. 3b) can be detected acoustically. In this experiment the static force (and resulting surface deformation) had to be increased, causing decrease of the spatial resolution and the recorded signal. Despite the poorer Signal to Noise (S/N) ratio, the deeply hidden grating was easily visible. The region where the grating was above the surface in the topographic image (Fig. 4a) reveals in the acoustic image (Fig. 4b) several additional details of the AsGa surface as well as inside the polymer.

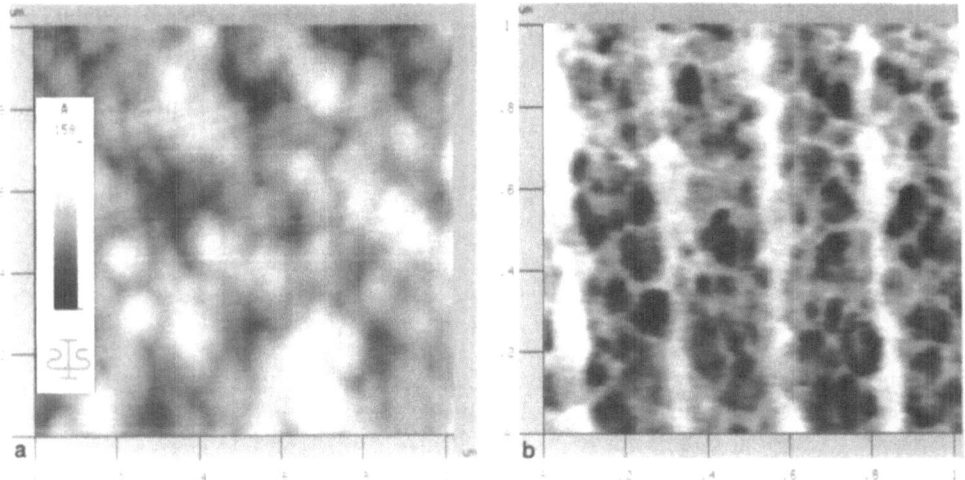

Fig. 2. Topographic (a) and acoustic (b) images of the AsGa grating under the very thin layer of polyimide.

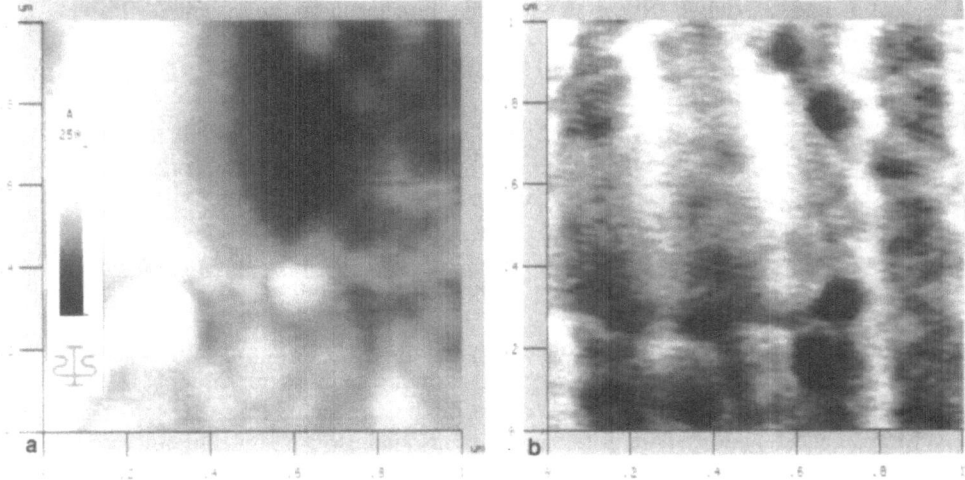

Fig. 3. Topographic (a) and acoustic (b) images of the AsGa grating under the ca. 100 nm thick polyimide layer. The grating in acoustic image (b) is barely visible.

CONCLUSIONS

Extension of the Scanning Probe Microscopes (STM, AFM etc.) to Acoustics (SNAM) proves that acoustic imaging may be performed with unprecedented nanometer spatial resolution. The main advantage of the SNAM over the classical STM/AFM can be appreciated in its ability to image subsurface features, even if it is mainly sensitive to the thin layer close to the surface. SNAM's unique contrast mapping of the acoustic properties will widen the field of applications of the Scanning Probe Microscopes.

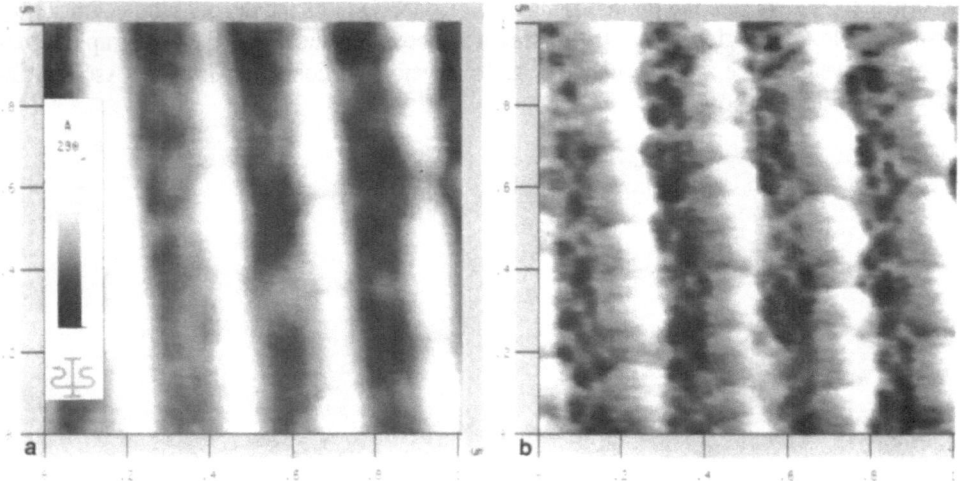

Fig. 4. Topographic (a) and acoustic (b) images of the AsGa grating higher than the polyimide layer. Please note the many additional details visible in the acoustic image as compared to the topographic image.

REFERENCES

1. E. A. Ash and G. Nichols, Superresolution aperture scanning microscope, *Nature*, 237:510 (1972).
2. W. Dürr, D. A. Sinclair and E. A. Ash, High resolution acoustic probe, *Electronic Letters*, 21:805 (1980).
3. J. K. Zieniuk and A. Latuszek, Non-conventional pin scanning ultrasonic microscopy, *in:* "Acoustical Imaging vol 17", H. Shimizu et al, Plenum Press, New York (1989)
4. B. T. Khuri-Yakub et al, Near field acoustic microscopy, *in:* "Scanning Microscopy Instrumentation", SPIE, San Diego (1992).
5. K. Takata et al, Tunneling acoustic microscope, *Appl. Phys. Lett.* 55 (17):1718 (1989)
6. K. Uozumi and K. Yamamuro, A possible novel scanning ultrasonic tip microscope, *Jpn. J. Appl. Phys.* 28(7):1297 (1989).
7. P. Günther et al, Scanning near-field acoustic microscopy, *Appl. Phys. B*, 48:89 (1989)
8. A. Kulik, J. Attal and G. Gremaud, Nearfield scanning acoustic microscopy, *in:* "Acoustical Imaging vol 20", You Wei and Benli Gu, Plenum Press, New York (1993).
9. A. Moreau and J. B. Ketterson, Detection of ultrasound using a tunneling microscope, *J. Appl. Phys.* 72(3):861 (1992).
10. W. Rohrbeck et al, Detection of Surface Acoustic Waves by Scanning Tunelling Microscopy, *Appl. Phys. A*, 52:344 (1991).
11. E. Chilla et al, Probing of surface acoustic wave fields by a novel scanning tunneling microscopy technique: Effects of topography, *Appl. Phys. Lett.* 61(26):3107 (1992).
12. E. Chilla et al, Amplitude and Phase Variation of Surface Acoustic Wave Field in the Nanometer Level Measured with a Scanning Tunneling Microscope, *in:* "Ultrasonic International 93 Conference Proceedings", (1993).
13. B. Cretin and F. Stahl, Scanning microdeformation microscopy, *Appl. Phys. Lett.* 62(8):829 (1993) and this conference.
14. U. Rabe and W. Arnold, Acoustic Microscopy by Atomic Force Microscopy to be published in *Appl. Phys. Lett.*
15. O. Kolosov and K. Yamanaka, Nonlinear Detection of Ultrasonic Vibrations in an Atomic Force Microscope, *Jpn. J. Appl. Phys,* 32 (22) (1993).

ACOUSTIC MICROSCOPY BY ATOMIC FORCE MICROSCOPY

U. Rabe and W. Arnold

Fraunhofer-Institute for Nondestructive Testing
University, Bldg. 37
D-66123 Saarbrücken, Germany

INTRODUCTION

High-Resolution acoustic imaging is a powerful tool for materials investigation[1,2]. By using such techniques, elastic properties and defects in materials can be determined with a resolution given by the wavelength. Various optical schemes have been developed in order to detect the displacements of surface-, longitudinal, and shear acoustic waves[3], either for local analysis or combined with a scanning technique such as Scanning Laser Acoustic Microscopy[4]. But due to Abbe's principle a lateral resolution not better than about a wavelength is obtained in techniques where focused beams are used. This limit can be surmounted by using near-field techniques. Recently, it has been successfully demonstrated that the high lateral resolution of near-field techniques can be exploited for the detection of acoustic waves by using a Scanning Tunneling Microscope[5-9], and even images have been obtained[9,10]. Vibrating the sample at frequencies below the resonance frequency of the cantilever is used for elasticity mapping with AFM[11], and is a standard technique in non-contact Scanning Force Microscopy[12].

In Atomic Force Microscopy (AFM)[12,13] the speed of signal pick-up is generally limited by the lowest mechanical resonance frequency of the free cantilever, which usually is 10 to 100 kHz. To record signals at higher frequencies, mixing techniques have been applied, for example in the Electrostatic Force Microscope[14], the Scanning Maxwell Stress Microscope[15] and the Ultrasonic Force Microscope[16,17]. In a recent paper[18] we showed that ultrasonic vibrations with spectral components of up to 20 MHz can be measured on the surface of a conventional AFM cantilever, provided a fast enough detector - for example a knife-edge detector - is installed. In this paper, we report on further experiments combining Atomic Force Microscopy and Acoustic Microscopy (AFAM).

EXPERIMENTAL APPARATUS

The AFM used in our experiments (Nanoscope III, manufactured by Digital Instruments, Santa Barbara, CA, USA) is supplied with a beam deflection sensor. The optical beam from a laser diode (670 nm wavelength) is focused on the top surface of the cantilever, reflected and directed onto a segmented photodiode by a mirror (Fig. 1). By this photodiode the lateral and vertical deflections of the cantilever can be measured with high sensitivity and linearity. The frequency response of the photodiode extends from DC to several kHz, and therefore ultrasonic vibrations at MHz frequencies are not recorded. We therefore added an optical beamsplitter which couples half of the light intensity reflected from the cantilever to an external knife-edge-detector (KN). The KN consists of a focusing lens, a fast photodiode (Si-PIN diode, risetime 1 ns) and a razor blade which blocks half of the light beam. Knife-edge detection is a well-known technique for optical detection of ultrasound and can be viewed as a special form of optical interferometry[19]. The output of the additional fast photodiode is amplified by 60 dB using an amplifier of bandwidth 1-500 MHz and then connected to one input channel of a digital oscilloscope.

For generation of the ultrasonic waves, an ultrasonic transducer is inserted between the AFM scanner and the sample and bonded to it by a coupling gel. The piezoelectric transducer of 10 MHz center frequency emits ultrasonic pulses of longitudinal polarization which travel through the sample and cause out-of-plane displacements of the sample surface. These vibrations also propagate into the AFM cantilever and can be monitored by the knife-edge detector. The electrical pulses exciting the transducer are either short spikes of 400 V amplitude with 3 ns risetime and approximately 15 ns decay time, or tone-bursts with a defined center frequency and an adjustable number of oscillations, typically 10 to 50 cycles.

As a reference, we also measured the absolute vibration amplitude of the sample surface using a Michelson heterodyne-interferometer (MI) designed for detection of ultrasonic displacements[3]. The output signal of the MI is connected to the second input channel of the digital oscilloscope. In this way, it becomes possible to compare the delay and the form of the cantilever vibrations measured by the knife-edge detector with those of the surface of the sample adjacent to the tip. In order to use the MI, a sample of sufficient optical reflectivity had to be used. We chose a laser mirror, which was a 9.5 mm thick glass (BK7) cylinder covered with aluminum and a hard protective layer.

Fig. 2 displays the signals detected after the transducer was excited by bursts of 5 MHz center frequency and 7.5 μs duration. In this experiment the cantilever was of triangular shape, made of gold coated Si_3N_4. It was 115 μm long, 0.6 μm thick and had a spring constant of approximately 0.6 N/m. As can be seen from the A-scan (i.e. the ultrasonic amplitude versus the time-of-flight in the sample) (Fig. 2a) measured by the MI, the sample surface vibrates nearly sinusoidally for several microseconds. The vibration amplitude is 0.13 nm. The cantilever follows the surface vibration, however, slightly amplitude modulated (Fig. 2b). We were able to excite and detect the harmonic cantilever vibrations up to 20 MHz by sweeping the carrier-frequency. This upper limit is probably only given by the detector sensitivity.

TRANSMISSION OF ULTRASOUND TO AFM CANTILEVERS

The transmission of ultrasound to an AFM cantilever can be described as follows. When the tip is in contact with the surface, it is attracted by adhesion forces, and repulsive forces arise from the deformation of the tip and the sample. The cantilever with the tip can be regarded as a point mass m suspended between two surfaces, the cantilever mount and the sample surface as shown in Fig. 3. The position of the point mass, which for simplicity is thought to be the tip position, is called z(t). z_0 is the tip position in the absence of forces when the cantilever is not deflected. Close to the surface, the tip is in a new equilibrium position z_e where the sum of the cantilever force $F_c = k \cdot (z_0 - z_e)$ and the tip-sample interaction force F_s vanishes:

$$k \cdot (z_0 - z_e) + F_s (z_e) = 0 \qquad (1)$$

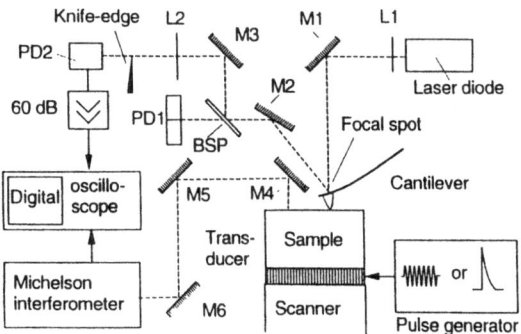

Figure 1. In a conventional beam-deflection position-detector of an AFM, the beam of a laser diode is focused by a lens (L1) onto the cantilever and directed onto a segmented photodiode (PD1). In our experiment half of the light intensity is coupled out by the beamsplitter (BSP) and focused by L2 onto the fast photodiode PD2. The knife-edge blocks half of the laser beam. M1-M3 are mirrors. Furthermore, by a Michelson interferometer the amplitude of the surface displacement can be measured. With the help of mirrors M4-M6, the He-Ne laser beam of the MI is adjusted such that it is back reflected from the sample surface.

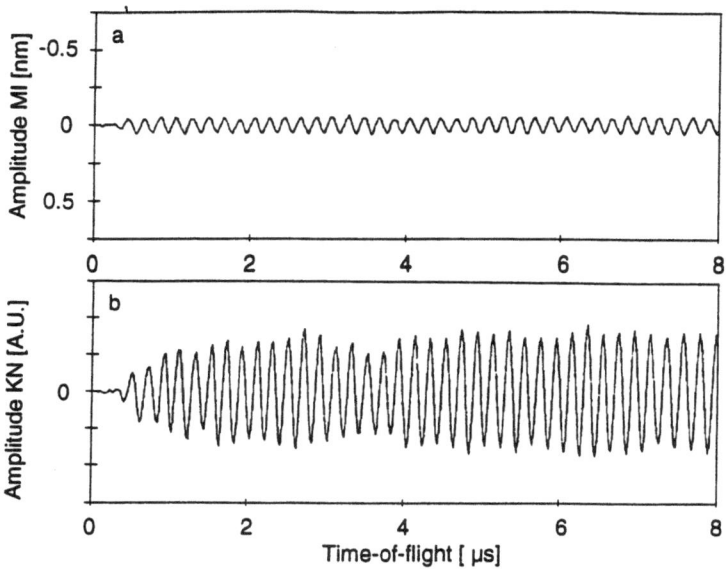

Figure 2. a) Signal of the Michelson interferometer showing the vibrations of the sample surface next to the tip. The surface vibration amplitude is 0.13 nm. Positive amplitudes correspond to an elevation of the surface; b) cantilever vibration as measured by the signal from the knife-edge detector.

where k is the spring constant of the cantilever. A movement a(t) of the surface caused by the ultrasonic displacement will induce a displacement $\Delta z(t)$ of the cantilever so that:

$$z(t) = z_e + \Delta z(t) \tag{2}$$

The equation of motion for the point mass m is then:

$$m\ddot{z} = k \cdot (z_0 - z_e - \Delta z) + F_s(z_e + \Delta z - a(t)) \tag{3}$$

If the surface vibration amplitude is small, a first order expansion of F_s at z_e is sufficient:

$$F_s(z) = F_s(z_e) - k^* \cdot (\Delta z - a(t)) \tag{4}$$

where

$$k^* = - \left. \frac{\partial F_s}{\partial z} \right|_{z = z_e} , \tag{5}$$

is the negative derivative of F_s at the equilibrium position of the tip. With the equilibrium condition (1) a simple equation for a forced vibration is obtained:

$$m\,\Delta z + (k + k^*)\,\Delta z = k^*\,a(t) \tag{6}$$

The left side of equation (6) shows, that the resonance frequency of the cantilever in contact with the surface $\omega_0' = \sqrt{\dfrac{k + k^*}{m}}$ is different from the resonance frequency of the free cantilever $\omega_0 = \sqrt{k/m}$. When the tip is in contact with the sample surface, the derivative $\partial F_s/\partial z$ is negative and k^* is positive. At larger tip-sample distances k^* becomes negative. Tip positions with negative k^* and $|k^*| \geq |k|$ are unstable, and if this happens the tip jumps towards the sample surface. The decrease of the resonance frequency of the cantilever due to tip-sample attraction is well-known and used in dynamic force microscopy where the tip does not touch the surface. When the tip is in contact with the surface, however, the stronger adhesion forces can shift ω_0' to the MHz region.

The linear approximation of $F_s(z)$ is only valid if the ultrasonic vibration amplitude is sufficiently small, so that the tip-sample interaction force-curve can be linearized. At larger amplitudes as a consequence of the nonlinearity of F_s there is also a shift in the equilibrium position z_e. If the vibration of the sample surface is harmonic, $a(t) = a_0 \cos\omega_{us} t$, where $\omega_{us} = 2\pi/T$ is the angular frequency, the shifted equilibrium position $\bar{z}_e$ can be calculated from the time average of F_s[16,20]:

Fig. 4a shows the low frequency cantilever deflection in our experiments as it is recorded by the Nanoscope when the surface deflection amplitude is several nm. In this case, the transducer was excited by a voltage spike. The signal recorded by the fast KN is displayed in Fig. 4b. Note that for experimental reasons the output signal of the photodiode of KN is negative, thus both in 4a and 4b the excursion of the cantilever is away from the surface of the sample. The absolute calibration of the graphs can be obtained from the force-calibration curve[20]. The amplitude of the cantilever response is much larger than the amplitude of the surface vibration which was a few nm. This behaviour is quite remarkable and is probably caused by the much lower effective elastic constant of the cantilever as compared to a bulk solid made of the same material as the cantilever. The observed

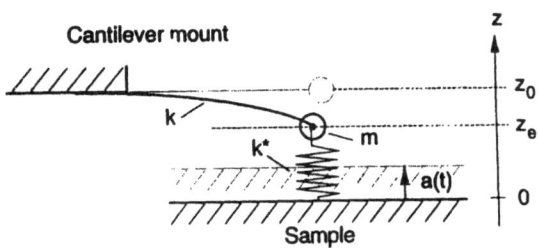

Figure 3. Point-mass model for an AFM cantilever vibrating under the influence of a tip-sample force F_s. z_0 is the rest position of the tip in the absence of forces. In the equilibrium position z_e, F_s - represented by its "spring constant" k^* - is balanced by the cantilever force F_c (spring constant k). All distances are measured from the sample surface in the absence of vibrations, a(t).

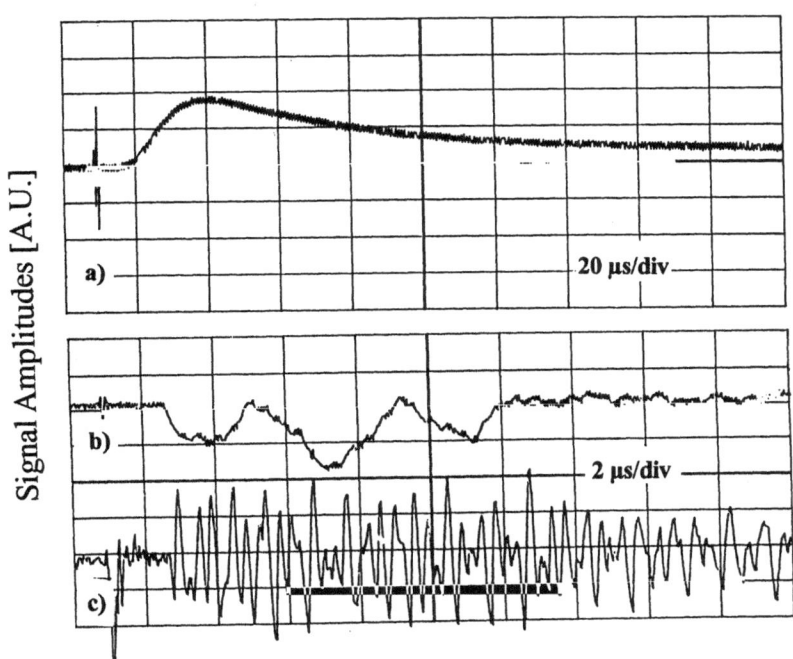

Figure 4. a) Cantilever deflection as recorded by the Nanoscope when the ultrasonic surface deflection is induced by exciting the transducer with a spike; b) the same cantilever deflection recorded by the additional fast KN. Here, the photodiode was connected directly to the digital oscilloscope of bandwidth DC to 80 MHz. The frequency components above 1 MHz become only visible after further amplification; (c) without the DC-components displayed in a). Note that the direction of excursion is the same for a) and b).

deflection amplitude changes with the surface vibration amplitude and with the tip equilibrium position z_e. Furthermore, non-linear phenomena such as mixing of the different frequencies become noticeable at large ultrasonic amplitude.

The simple point mass model can only describe *the coupling* of the cantilever to the vibrating surface *but not the propagation* of ultrasound in the cantilever. For a more precise description, the cantilever must be treated as a flexible bar clamped at one end and coupled at the other end via the tip to the vibrating surface[20]. A first understanding of our experimental data can be achieved by considering the flexure vibrations of the cantilever (Fig. 5). For a homogeneous beam of uniform cross section, the equation of motion for free flexural vibrations is

$$EI\frac{\partial^4 y}{\partial x^4} + \rho A\frac{\partial^2 y}{\partial t^2} = 0 \tag{7}$$

where E is the modulus of elasticity, ρ is the mass density, $A = a \cdot b$ is the cross section, and I the area moment of inertia[21]. Here, a and b are the width and the thickness of the cantilever, respectively. The area moment is $I = ab^3/12$ in the case of a beam with rectangular cross-section. A solution of the type $y = y_0 e^{i\kappa x - i\omega t}$ inserted into equation (7) yields a dispersion relation:

$$EI\kappa^4 - \rho A\omega^2 = 0. \tag{8}$$

This gives the phase velocity $v_{ph} = \omega/\kappa$ and the group velocity $v_{gr} = d\omega/d\kappa$ in a bar of infinite length:

$$v_{ph} = \sqrt[4]{\frac{EI}{\rho A}}\sqrt{\omega} \quad \text{and} \quad v_{gr} = 2\,v_{ph}. \tag{9}$$

If the cantilever beam is of finite length L, boundary conditions have to be fulfilled by the solutions of equation (7). Therefore, a set of modes with discrete frequencies is obtained[20]:

The behaviour of the cantilever as an acoustic waveguide can be corroborated by further experimental evidence. An ultrasonic pulse needs a finite time to travel from the cantilever tip to the end. By moving the focal spot of the laser diode beam along the cantilever, we measured this time delay between the beginning of the cantilever vibration detected near the tip and near the fixed end of the cantilever[20]. With a distance between the two measurement points of about 200 µm, this results in a group velocity of about 1 mm/µs, which agrees well with the group velocity $v_g = 2v_{ph}$ expected for a frequency of 10.4 MHz.

IMAGING

For imaging the experimental procedure is as follows: The amplitude of the first ultrasonic signal is selected by a gate and is fed into a boxcar integrator, yielding a voltage proportional to the amplitude of the ultrasonic pulse. The integrated signal is stored in the second input channel of the microscope and is displayed as a function of the position of the tip. The feedback loop maintains a constant low interaction force during the scan, also keeping the working point of the external knife-edge detector constant. A special vibration isolation is not necessary because mechanical vibrations are well below the 3 dB bandwidth of the knife-edge detector (1-28 MHz).

Fig. 6b shows an image of the ultrasonic amplitude distribution obtained on an alumina sample reinforced with Si-fibres. In contrast to the topographical image of the same area

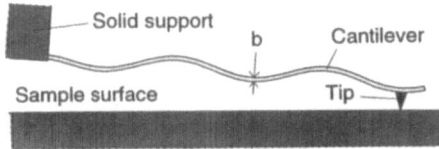

Figure 5. Schematic drawing of cantilever exhibiting flexural vibrations. b is the cantilever thickness.

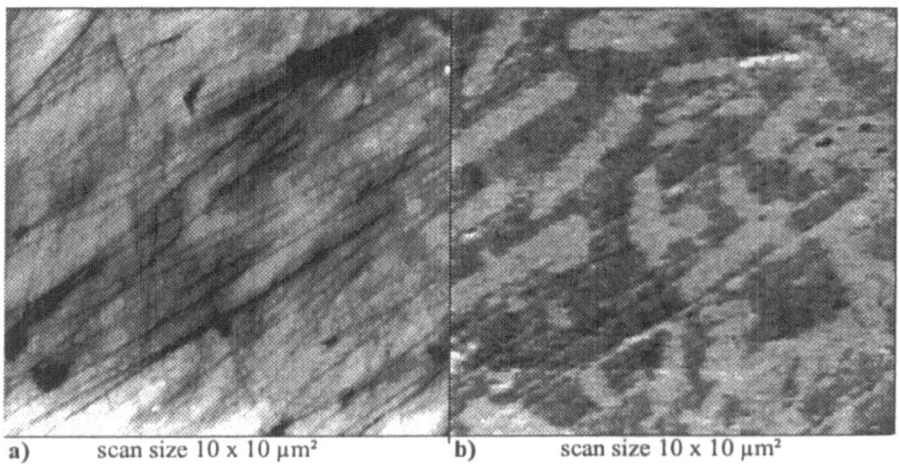

a)   scan size 10 x 10 µm²   b)   scan size 10 x 10 µm²

Figure 6. Images taken of a Al_2O_3-ceramic containing SiC-fibres: a) topography, gray scale covers 250 nm from dark (low) to light (high): and b) ultrasonic amplitude.

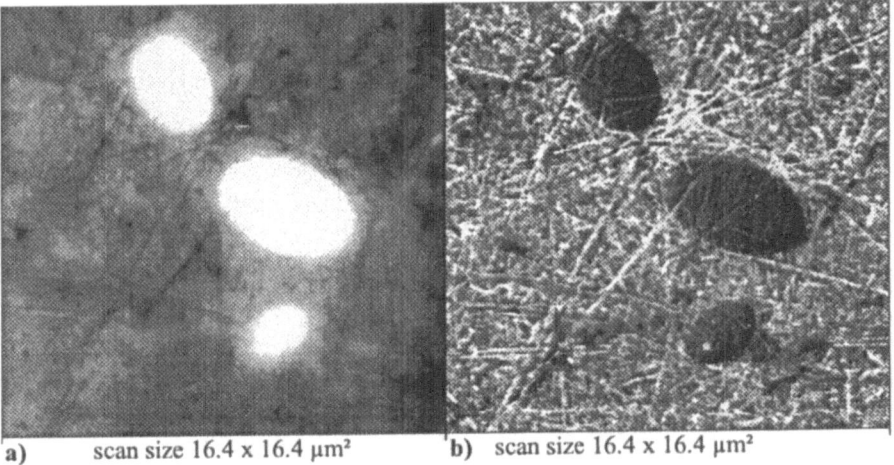

a)   scan size 16.4 x 16.4 µm²   b)   scan size 16.4 x 16.4 µm²

Figure 7. Image of an aluminum sample: a) topography and b) AFAM image. Note that a wealth of additional features become visible in b) which are not seen in a).

(Fig. 6a) the SiC-fibres become clearly visible. The scan size is $10 \times 10 \ \mu m^2$. The scan rate was 1 Hz and 512 points per line were sampled. The repetition frequency was adjusted to be at least $1/\Delta t$ where Δt is the time interval between two image points. The gray scale covers 250 nm of corrugation. Fig. 7b shows an AFAM image of an aluminum sample reinforced with SiC-particles and Fig. 7a displays the corresponding topographical image. In both cases, the AFAM images contain a wealth of additional features not seen in the topographical images. However, some of these additional features become visible in Lateral Force Microscopye images. At present, our efforts are concentrated in elucidating the contrast observed in AFAM, in particular in as much local elasticity variations and adhesion properties contribute to the contrast.

CONCLUSION

We have demonstrated experimentally that a conventional AFM cantilever with a lowest free resonance frequency of several kHz can be excited to ultrasonic vibrations in the MHz-frequency range and that imaging is possible at these frequencies. This can be understood by regarding the cantilever as an acoustic waveguide with a distributed mass. We gave a first interpretation of the experimental results in terms of flexural vibrations of the cantilever and its eigenfrequencies. Furthermore, images taken with an AFAM instrument reveal additional information not present in topographical images.

REFERENCES

1.	see for example collected papers in these Proceedings.
2.	G.A.D. Briggs, "Acoustic Microscopy", Oxford University Press, Oxford, 1992.
3.	J.-P. Monchalin, IEEE Trans. UFFC-33 (1986) 485.
4.	L.W. Kessler, Proc. 18th Symp. Acoustical Imaging (1991). Eds. H. Lee and G. Wade, Plenum Press, New York and London (1992).
5.	H. Heil, J. Wesner, and W. Grill, J. Appl. Phys. 64, 1939 (1988)
6.	W. Rohrbeck, E. Chilla, H.J. Fröhlich, and J. Riedel, Appl. Phys. A52, 344 (1991).
7.	K.J. Strozewski, S.E. McBride, and G.C. Wetsel, Ultramicroscopy, 42-44, 388 (1992).
8.	A. Moreau and J.B. Ketterson, J. Appl. Phys. 72, 861 (1992).
9.	E. Chilla, W. Rohrbeck, H.J. Fröhlich, R. Koch, and K.H. Rieder, Appl. Phys. Lett. 61, 3107 (1992).
10.	K. Takata, T. Hasegawa, S. Hosaka, S. Hosoki, and T. Komoda, Appl. Phys. Lett. 55, 1718 (1989)
11.	P. Maivald, H.J. Butt, S.A.C. Gould, C.B. Prater, B. Drake, J.A. Gurley, V.B. Elings, and P.K. Hansma, Nanotechnology 2, 103 (1991).
12.	D. Sarid, "Scanning Force Microscopy", Oxford University Press, Oxford, 1991.
13.	G. Binnig, C.F. Quate, and Ch. Gerber, Phys. Rev. Lett. 56 (1986) 930.
14.	R.A. Said, G.E. Bridges, and D.J. Thomson, Appl. Phys. Lett. 64 (1994) 1442.
15.	H. Yokoyama, M.J. Jefferey, and T. Inoue, Jpn. J. Appl. Phys. 32 (1993) L1845.
16.	O. Kolosov and K. Yamanaka, Jpn. J .Appl. Phys. 32 (1993) 22.
17.	K. Yamanaka, H. Ogisio, and O. Kolosov, Appl. Phys. Lett. 64 (1994) 178.
18.	U. Rabe and W. Arnold, Appl. Phys. Lett. 64 (1994) 1493.
19.	R.L. Whitman and A. Korpel, Appl. Opt. 8 (1969) 1567.
20.	U. Rabe and W. Arnold, to appear in "Annalen der Physik", 1994.
21.	M.C. Junger and D. Feit, "Sound, Structures and their Interaction", MIT Press, Cambridge, USA 1972.
22.	C.M. Harris and C.E. Crede, Eds., "Shock and Vibration Handbook", McGraw Hill, New York, USA 1976.

IMAGING DEEP MICRO DEFECTS IN STRUCTURAL CERAMICS BY PRECISION C-SCAN TECHNIQUE

Jianzhong Shen[1], Futang Jiang[2], Jingjun Deng[1] and Lugun Wang[1]

[1]Institute of Acoustics, Academia Sinica, Beijing 100080
[2]Beijing Research Institute of Materials and Technique
Beijing 100076, P.R. China

INTRODUCTION

New advanced materials such as structural ceramics exhibit excellent performance, but mostly they are very sensitive to tiny defects. The critical size of defect caused failure of structural ceramics is typically of 10 to 100 micron. And it is needed to detect such micro defects in large depth inside the ceramic components. The precision ultrasonic C-Scan is one of the new nondestructive techniques which could be used to detect micro defects in solids[1].

For both surface and deep detection, a precision C-Scan system was set up in our lab on the design and making by ourself[2]. Basically it consists of a micro dot focus ultrasonic probe, a narrow pulse transducer driver, a low noise receiver with gated high gain amplifier, an 8 bit A/D converter, a 2-D precision mechanical scanner and its controller, a PC AT286 personal computer with a high resolution display. It could work in a wide range of frequencies from 20 MHz to 150 MHz. The system is simple but effective. It could find tiny slits with width of $30\mu m$ as deep as 10mm on the bottom of a specimen of silicon nitrogen in depth 10mm at frequency of 100MHz. And two grooves separated great than $70\mu m$ could be distinguished in this deep.

THE FACTORS AFFECTING DEFINITION

There are mainly four factors affecting the definition of an image formed by the scanning imaging system. The global resolution of the image system are determined by
1. The resolution of the ultrasonic wave beam. The focus system is the essential part. It is needed to treat the probe with the couplant and the sample as a whole focusing

system. Proper working frequency is needed. The details are discussed below.

2. The resolution of the mechanical scanner. It is obviously that the step of mechanical scanning, i.e., the resolution of scanning, should affect the global resolution. The precision, the stability and the reliability of motion are most important.

3. The resolution of data collection. This is dependent on the quality of the electrical circuits and the precision of the A/D converter. A low noise electrical receiver with high gain amplifier is prefer. And a high precision A/D converter as well. The signal-to-noise ratio(SNR) of the circuits and the precision of the A/D converter affects the definition of an image significantly, but they are not changed much with deep detection in general.

4. The resolution of the displaying screen. It is no doubt by use of a high resolution displayer to improve the definition of an image. These aspects, of which are depth depending, are discussed following and some improvements are made for deep detection.

THE WORKING FREQUENCY

To achieve the detecting sensitivity of 10 to 100 micron for deep detection, say in depth of 10mm, the working frequency need proper chosen.

The resolution limit is usually considered as the order of half of the wave length. Thus the higher frequency is desirable for good definition of the testing results.

Besides, generally speaking, it is benefit to get stronger back scattered waves from tiny scatters when higher frequency is used to improve the SNR, hence the sensitivity.

The structural ceramics exhibit extremely high velocity of ultrasonic waves, e.g., the velocity of P wave in Si_3N_4 ceramic might reach 10700m/s. It leads that the wavelength would be large enough than the size of the most small defects to be detected. Even at 100MHz, the wavelength is equal to 107 microns, which is still large than the size of the defects to be detected. Thus, using higher frequency, say more than several tens MHz is necessary.

On the other hand, at RF frequency, the attenuation is proportional to the square of frequency. In water the absorb coefficient even is as high as 2.2 dB/mm at 100MHz[3]. The strong attenuation leads the detectivity to be decreased seriously with depth at high frequency. Besides, nonlinear absorbance at high frequency would cause the deformation of the probing signals.

As a conclusion, the range of frequencies from 30MHz to 120MHz would be proper to match the requirements in different cases of detecting structural ceramics. By the way the noise caused by the background scattering is negligible and the absorption is not so severe in this range of frequency.

THE FOCUS SYSTEM

The focal properties are depend on the whole focus system. It is ideal to have a spot of focus small enough in transverse and very long in longitudinal in the detecting area. And in order to get good longitudinal resolution, the system of ultrasonic probe combining with electric circuits should be with large relative bandwidth. Some of these requirement are conflict somehow.

The focus probe which is both as an emitter and receiver is most important part of the system. The probe we used consists of a Y-cut $LiNbO_3$ piezoelectric crystal transducer, a buffer rod and a spherical acoustical lens. The sensitivity of this kind probe is high but encounter some drawbacks. Some modifications are made[4].

The Focal Distance F

According to the Snell law, great deviation of the refracted wave in the sample occurs because of its extremely high velocity, which results in two effects: great refocus and more severe defocus of the ultrasonic beam. This decreases the detectivity with depth rapidly. A very long focal distance Fo in water is required because of the refocus effect. The F, the one in the sample is

$$F=(F_o-h_w)\frac{V_w}{V_s} \tag{1}$$

where V_w, V_s is the velocity of ultrasonic wave in the couplant(water), and sample respectively. The h_w is the distance between lens and sample. As V_w is about 1480m/s, which is about 1/7 of V_s, the velocity of the longitudinal wave of Si_3N_4. Then F is far less than F_O.

The Focal Length L and Focal Spot Size D

The focal length L_w and focal spot size D_w in water for a spherical acoustical lens according to the classical theory are

$$L_w=4\lambda\frac{F_o^2}{d^2} \tag{2}$$

and

$$D_w=\lambda\frac{F_o}{d} \tag{3}$$

respectively, where d is the diameter of the aperture of lens. From formula (1) one can estimate that the focal spot size D in the sample will remain the same as D_w and the focus length L is deduced to L_wV_w/V_s, provides the h_w is short.

As the large focal distance F_o is needed for deep inspection, the D will be larger and then the definition be worse in deep inspection. To solve this problem, the practical way is to enlarge the aperture d of the lens at the cost of some mismatch with the electrical circuits.

The Length of Buffer Rod

The ordinary consideration on the buffer length is that the lens should be located in the far field beyond the so called Fresnel length F_1 to get uniform illumination according to the theory of continuous waves. Because large d and small λ should be used for high resolution, The F_1 would be large.

However, the transducer works in state of pulse in fact. The pulse field exhibits only direct wave and edge wave but no complex near field structure as in the case of continue wave. Fig.6 in Ref.5 shows this situation. The direct wave does be monotonous change during propagating. Thus the lens could get uniform illumination anywhere by the direct wave pulse.

Hence the length of buffer rod might be chosen according to another principle that there are no, or very weak if any, the disturbing signals in the time interval where the echoes from defects might appear[4]. The length chosen could be shorter than F_l. As the interfering echoes causing by multi-reflection in the long buffer are coherent and hardly to be eliminated, on this principle could raise the SNR significantly.

THE PRECISION MECHANICAL SCANNER

The precision of scanning is important. To improve the stability of motion, it is changed to provide quasi-continue motion in one direction in our system. The speed of motion is adjustable, the fast one for surveying, the normal one for detail detection. Two raster meters were used to provide 2 micron absolute precision locating.

In order to save the time of CPU of PC286, two CPU are used. The precision mechanical scanner was managed by a microcomputer MCS8051 which is controlled by the main controller PC AT286. The step of scanning could be chosen each integer value through keyboard between $1\mu m$ to $999\mu m$ for precision measurement or fast surveying.

TESTING RESULTS

For testing, some artificial defects of grooves and bottom holes were made by ablation of the focus impact of laser pulses on silicon nitrogen ceramic samples, of which the thickness are 3mm to 10mm. The width of grooves is between 30 to 100 micron and the diameter of bottom holes 30 to 200 micron. AS the depth of detection increased, the definition of the image will become worse. The definition could be improved by use of proper working frequency and suitable ultrasonic beam size as above mentioned.

Fig.1 is an image of a grove with width round $30\mu m$ in 10mm deep. The mechanical scanning step was $10\mu m$. The width of the image measured is much more large than the real one. It may be explained mainly because of the ultrasonic beam spread. But it could be improved somehow by choosing proper displaying threshold.

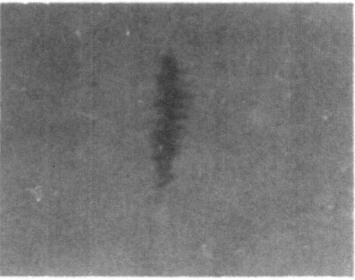

Fig.1. An image of a grove of $30\mu m$ width, in 10mm depth

Fig.2a and Fig.2b show two images of a group of parallel grooves also in 10mm deep. Fig.2a is the acoustical image and Fig.2b the optical one. There are 7 grooves which are separated in various distance by 360, 160, 50, 60, 35, 6μm respectively. The width of the grooves if round 100μm. In Fig. 2a, four lines are displayed clearly. The 3rd, 4th grooves and the 5th, 6th 7th ones are as two groups. One might distinguish the two groups which is separated about 60μm. The focus distance is round 80mm in water and frequency 100MHz are used. In Fig.2a, the dot in the right down corner is caused by the inner flaw of the sample.

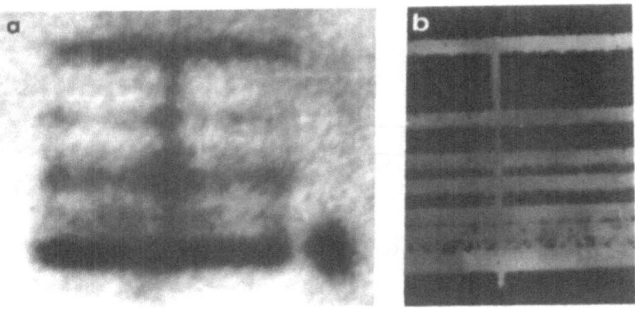

Fig.2 Image of A Lattice, in 10mm depth, a) Acoustical, b) Optical

Fig.3 shows the boundary effect we so called. There are four bottom holes being located in the bottom of a 5mm thickness sample. These holes are separated from 1mm distance and with diameter of 160, 180, 90, 80μm from left to right respectively. And they are near the edge of the sample about 1mm. One can see that these holes are partially immersed in the background founding by the edge of the sample. This situation could be illustrated by Fig.4. As the lens with a large aperture for deep detection, only a part of the acoustic waves enters the sample when it near the boundary of the sample. Boundary effect might hide the defects in its affecting area. However one could distinguish the bottom holes by noticing the little change of the boundary background line as Fig.3 shown. It might be displayed more clearly if some post image processing are carried out.

Fig.3 The boundary effect. Bottom holes detected, sample thickness 5mm, distance from the boundary 1mm

CONCLUSION

A PC controlled precision C-Scan System was tested. It showed that it has potential to image tiny defects effectively. In our case, the grooves with width narrow as 30μm in 10mm deep could be imaged. And the resolution between two grooves could be better than 80 micron or more at a depth of 10mm, with the frequency at 100MHz. All the bottom holes with diameters more than 80μm can be detected at 5mm deep. The threshold of displaying is important and needs the proper chosen depth for obtaining a good image definition. The boundary effect is noticed, which may be important in the inspection when the defect is near the boundary of components.

Fig.4. Illustration of the boundary effect

ACKNOWLEDGMENT

This work is partially supported by the National Sciences Foundation of China

REFERENCES

1. U.Netzelmann, H.Stolz, and W.Arnold, Defect sizing in ceramic materials by high-frequency ultrasound techniques, in "Abstracts of 20th International Symposium on Acoustical Imaging", Southeast University, Nanjing(1992).
2. Shen Jianzhong, Jiang Futang, Deng Jingjun and Wang Lugen, A precision C-scan system testing fine materials, Seventh Asian-Pacific Conference on Nondestructive Testing, Shanghai. 349-351(1993).
3. L.W.Kessler, Acoustic microscopy commentary:SLAM and SAM, IEEE Trans. on Sonic and Ultrasonics, SU-32-2:136(1985).
4. Deng Jingjun, Zhang Suoyu, and Shen Jianzhong, Research of RF Ultrasonic Focused Probe and Pulsed Driver, Seventh Asian-Pacific Conference on Nondestructive Testing, Shanghai. 131-134(1993).
5. Ying Chongfu, Photoelastic visualization and theoretical analyses of scattering of ultrasonic pulse in solid, in "Physical Acoustics", Vol.XIX, W.Mason and R.N. Thurston, ed., Academic Press, Inc., New York(1990).

DIGITAL ACOUSTIC MICROSCOPY OF CELL-SUBSTRATE ATTACHMENT: EXPERIMENTS AND MODEL PREDICTIONS

A. Cambiaso, T. Tommasi, M.T. Parodi, and M. Grattarola

Bioelectronics Laboratory
Biophysical and Electronic Engineering Department, University of Genoa
Via Opera Pia 11A, 16145 Genova, Italia

INTRODUCTION

Cell adhesion is an important component of numerous biological responses including growth, differentiation, and motility. In many cultured cell lines, immunofluorescence studies have verified the co-localization of the fibronectin receptors talin and vinculin at specific cell-substrate attachment sites, which have initially been detected by interference reflection microscopy (IRM) and called "focal contacts" (Izzard and Lochner, 1976).
More recently, scanning reflection acoustic microscopy (SAM) has been successfully used for detection of focal contacts (Hildebrand, 1985). By operating the acoustic microscope at gigahertz frequencies, one can easily detect single focal contacts (typical dimensions: 1 μm x several μm) . As compared with IRM, SAM offers the unique opportunity of obtaining information about elasto-mechanical properties of cell adhesion areas by imaging such areas through the thickness of a whole cell, with minimal perturbations of the cell structures. From this point of view, it is worth noting that the very recent application of scanning force microscopy (SFM) (Binning et al., 1986) to the study of living cells has shown that SFM complements rather than replaces SAM.
In the following, experiments will be described, which aimed to utilize SAM to compare images of focal contacts in the absence/presence of detachment-inducing chemical perturbations of the culture medium (Wolf, Gingell, 1983). A model simulating these phenomena is also proposed, which is based on changes in the so-called "V(z) curve" (Briggs, 1992) as a function of variations in the impedance of cell adhesion regions.

MATERIALS AND METHODS

3T3(ATCC)cells were cultured on a glass slide in D-MEM (culture medium),which was supplemented with 10% foetal calf serum(FCS). After a 24-36 hour incubation period in 5% CO_2 at 37 ºC, the culture slide was transferred to an especially designed chamber for

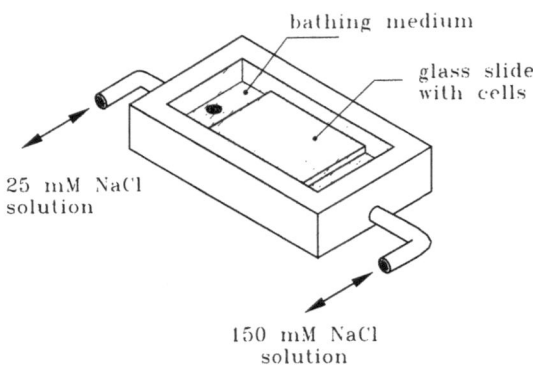

Figure 1. Chamber used for living cell measurements.

acoustical microscopy observations (see Figure 1). The chamber was filled with a physiological solution (150 mM NaCl), which, during the time-course observations, was gradually replaced with a 25 mM Dextrane solution and then restored:. A decrease in ionic strength is known to cause a decrease in cell adhesion (Wolf, Gingell, 1983). Observations were performed at room temperature.

Acoustic images were acquired with an ELSAM (Leica,Germany) acoustic microscope which operated in the 1-1.5 GHz range. The coupling with the lens was obtained by direct immersion in the bathing medium. A 486 PC, equipped with a DT2851 (Data Translation,USA) video frame grabber, was employed for image digitization and processing. The computer simulation programs were developed in Fortran language on an HP360 Unix Workstation (Hewlett Packard).

MODELLING AND SIMULATIONS

Simulations of V(z) microscope responses were performed, to evaluate possible changes in the V(z) curves due to the presence or the absence of adhesion sites. Usually, quantitative V(z) analyses of cell specimens are made using different techniques (Briggs, 1992; Kundu et al.,1991) for the purpose of characterizing the cell layer (thickness, impedance, density, acoustic velocity, attenuation, etc.). But the relation between V(z) changes and adhesion sites has not been extensively investigated. To the best of our knowledge, a quantitative approach was developed only in (Hildebrand, 1985), based on image contrast and not on the V(z) curve.

An easy way of evaluating V(z) changes lies in considering the periodicity of V(z) peaks, as it is related to surface acoustic waves (SAWs). In (Litniewski, *1989)*, a theory based on ray tracing proves that a liquid layer on a substrate does not change the velocity of SAWs, v_{saw}, hence, the periodicity of V(z). It has also been proved that the presence of a liquid layer determines a shift in the whole V(z) curve, and that this shift can be utilized for cell investigations (Litniewski and Bereiter Hahn, 1992). Starting from field theory, we have proved that a thin liquid layer induces changes in v_{saw} (Bianco et al., 1992). But the changes in V(z) periodicity are too small to be utilized for cell characterization, as cell impedance (even when adhesion sites are present) is too close to water impedance. On the other hand, a significant change in the V(z) curve occurs in the distance, Δz, between the main and second peaks. The change in Δz can be explained by means of ray theory: the main peak is due to the characteristic lens response, whereas the other peaks and valleys result from the interference between the axial and SAW rays. The presence of a liquid layer induces shifts in the peaks due to SAWs, whereas the position of the main peak should not change significantly.

Preliminary simulations were carried out, which confirm the above analysis. As a model, a planar multilayer was assumed, consisting of a substrate (glass (crown), longitudinal velocity v_l= 5660 m/s, shear velocity v_s = 3420 m/s, density ρ = 2240 Kg/m^3), a thin liquid layer 0.15 μm thick, a cell layer, and water as a coupling fluid between cell and lens. The cell was modelled as a medium with elastic parameters slightly different from those of water (e.g., v_l= 1540 m/s, ρ =1020 Kg/m^3). The thin liquid layer was assumed as a water medium (no cell adhesion) or as a liquid layer representing an adhesion site. Results confirm that variations in V(z) periodicity are very small (less than 0.5%) for cells and adhesion sites. On the other hand, the more the adhesion site impedance increases, the more the distance Δz decreases. For a spherical lens of focal length 80 μm and with a half-aperture angle of 40 deg, and by operating at a frequency of 1 GHz, the following results where obtained: Δz = 5.62 μm when no adhesion site was present (only a water layer of 0.15 μm was between cell and substrate); when the attachment-site impedance was changed from 1.6· 10^6 Kg/m^{2}s to 2.2 · 10^6 Kg/m^{2}s, Δz ranged from 5.45 μm to 5.02 μm. This V(z) feature can be utilized to obtain a quantitative evaluation of adhesion sites. In particular, a comparison of an actual V(z) curve with a simulated one can lead to a characterization of adhesion site impedance. It should be stressed that real curves are formed by a region larger than an adhesion site. As a consequence, measures correspond to average values of the elastic properties of the region involved. On the other hand, by analyzing the V(z) curve for depth values up to about 6 μm, the region investigated can be estimated to be inside a circle of radius about 3 μm (assuming to use the same lens as described above). This is a limited area which can be regarded as a homogeneous attachment site.

EXPERIMENTAL RESULTS

In this section, two examples of results (experiments (I) and (II)) are given, which refer to two sequences of acoustic images, shown in the Figures 2 and 3 respectively.

Experiment (I)
All the images of the sequence used to describe this experiment were taken at a 1.5 GHz frequency. The displayed field is about 300 μm large. Figure 2.a shows an acoustic image of a 3T3 cell in isotonic condition, at the beginning of the experiment: in the image, dark streaks indicate the attachment sites (see arrows). The image in Figure 2.b was acquired 15 minutes after replacing 50% of the physiological solution with a Dextrane solution to reduce the ionic strength. It is easy to observe that the number and the size of the focal contact sites are reduced. The physiological solution was then diluted again to obtain a 75% Dextrane solution. Figure 2.c shows a cell 5 minutes after the second dilution: the adhesion sites appear further reduced. The image in Figure 2.d was taken 10 minutes after the complete replacement of the physiological solution with a Dextrane solution, a few minutes before the complete cell detachment; the focal-contact sites completely disappeared from the image.

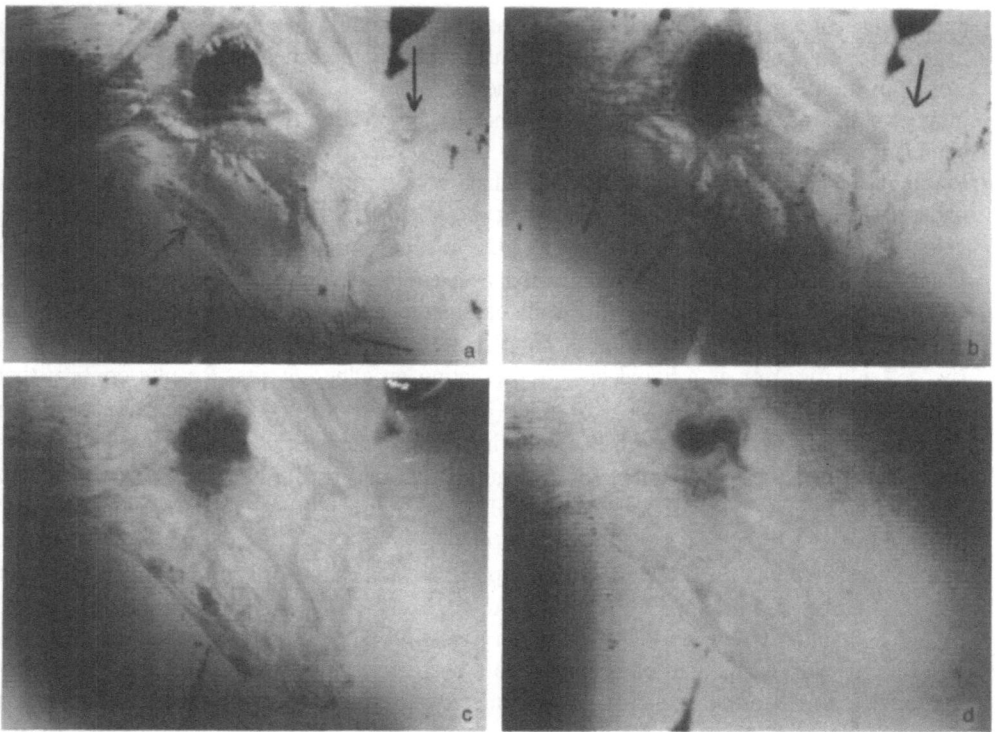

Figure 2. Experiment (I). a: physiological condition. b: 15 min. after 50% salt replacement. c: 5 min. after 75% salt replacement. d: 10 min. after complete salt replacement.

It can also be noticed that, in the last condition, the contrast of the whole cell in the image is very reduced, as compared with the starting condition. This effect was observed in all performed experiments, it suggests a strong dependence of elastic cell properties on the ionic strength of the culture medium.

Experiment (II)

The images of the sequence used to describe this experiment were taken at a 1.1 GHz frequency and the displayed field is about 200 μm large. Figure 3.a shows an acoustic image of two 3T3 cells (called A and B) in isotonic condition, at the beginning of the experiment. The arrows indicate the regions where the focal contact sites are more evident. The image in Figure 3.b was taken 15 minutes after the replacing 75% of the physiological solution with Dextrane. The size and the number of the adhesion sites are reduced; the interference fringes that appear on cell B (see arrow), seem to indicate a swelling of the cell, perhaps due to the detachment process. Figure 3.c shows the two cells after 25 minutes: no adhesion sites are observable in the image, and image contrast is in general very low, as in experiment (I). At this point (after the complete detachment of cell A), the starting condition (100% of physiological solution) was restored. Figure 3.d shows the situation 20 minutes after this operation: the other cell was detected again with a high contrast, and new adhesion sites were formed.

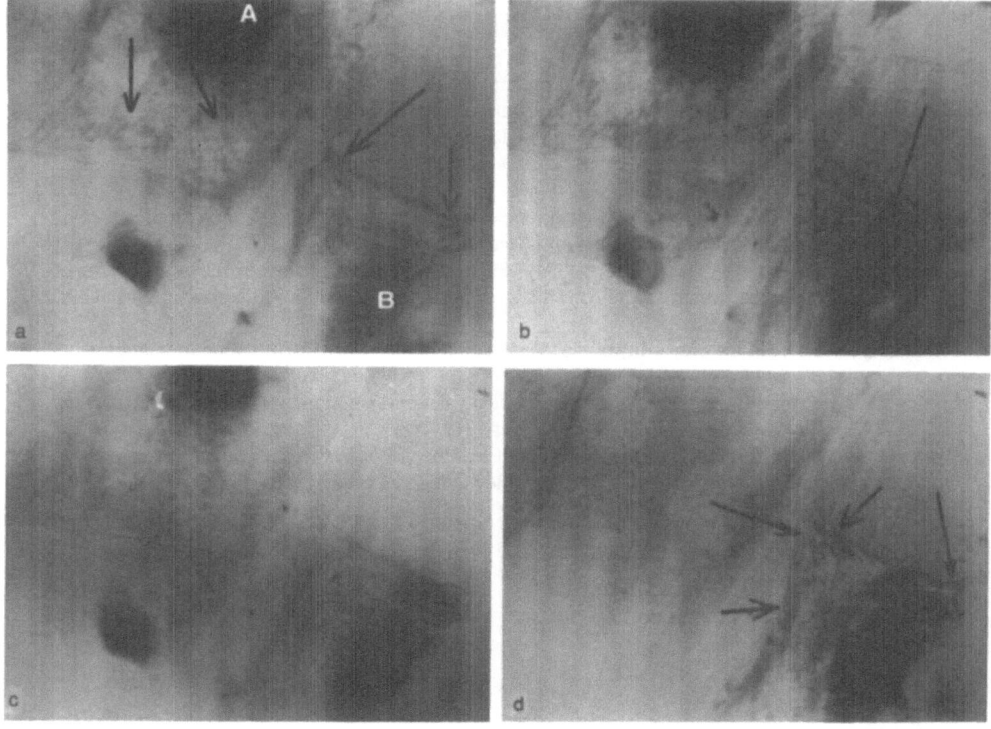

Figure 3. Experiment (II). a: physiological condition. b: 15 min. after 75% salt replacement. c: 25 min. after 75% salt replacement. d: 20 min. after restoring the initial conditions.

CONCLUSIONS AND PROSPECTS

The experiments performed indicate that acoustic microscopy is a valuable technique for time-course characterization of living cells. Computer simulations confirm that focal adhesion sites can induce significant changes in the image contrast.

Other methods, able to decrease cell adhesion, should be employed in the future to better monitor the cell detachment mechanism. For example, an experimental protocol based on the addition of trypsine at low concentrations to the bathing medium, can be compared with the described protocol which is based on decreasing the ionic strength. Immunofluorescence techniques, employing specific monoclonal antibodies (e.g., a-Talin) and an epifluorescence microscope coupled with an acoustic one, will be used to verify the correct interpretation of acoustic images.

V(z) measurements, correlated with simulations, can provide a more quantitative estimates of density changes at adhesion sites.

Acknowledgment

Work supported by the Italian Ministry for the University and Research (MURST).

References

Bianco, B., Cambiaso, A., Paradiso, R., Tommasi, T., 1992, Experimental and theoretical surface acoustic wave analysis of thin-film lipid multilayers, *Appl.Phys.Lett* 61:402.

Binnig, G., Quate, C.F., and Gerber C., 1986, Atomic force microscopy, *Phys. Rew. Lett.*930:56.

Briggs, A.,1992, "Acoustic microscopy", Oxford Claredon Press.

Hildebrand, J.A., 1985, Observation of cell-substrate attachment with the acoustic microscope, *IEEE Trans. Sonics Ultras.* 332:SU-32.

Izzard, C.S., Lochner, L.R., 1980, Cell-to-substrate contacts in living fibroblasts:an interference-reflexion study with an evaluation of the technique, *J. Cell Sci.* 21:129.

Kundu,T., Bereiter-Hahn,J., Hillmann,K., 1991, Measuring elastic properties of cells by evaluation of scanning acoustic microscopy V(z) values using simplex algorithm, Biophysical Journal 59: 1194.

Litnienski, J., 1989, On the possibility of the visualization of the velocity distribution in biological samples using SAM, IEEE Trans.UFFC 36:134.

Litniewski, J., Bereiter Hahn, J., 1992, Acoustic velocity determination in cytoplasm by V(z) shift, in: "Acoustical Imaging Vol. 19", Plenum Press, New York.

Wolf, H., Gingell, G., 1983, Conformational response of the glycocalyx to ionic strength and interaction with modified glass surface: study of live red cells by interferometry, *J. Cell Sci.* 101:63.

CONTRAST FORMATION FOR SIMPLE SHAPED OBJECTS IN ACOUSTIC MICROSCOPY

Pavel Zinin,[1,3] Wieland Weise,[1] Oleg Lobkis,[2,3] and Siegfried Boseck[1]

[1]Institute for Material Science and Structure Research
University of Bremen
Bremen, 28334, Germany
[2]Department of Civil Engineering
University of Arizona
Tucson, AZ 85721, USA
[3]on leave from Institute of Chemical Physics
Russian Academy of Sciences
Moscow, 117334, Russia

INTRODUCTION

Reconstruction of 3-dimensional objects in Scanning Acoustic Microscopy (SAM) is now of great interest. From the mathematical point of view, the simplest kind of 3-dimensional objects are spherical particles. Investigations of the image formation of spheres have revealed different phenomena, which are of current scientific concern. First theoretical treatments[1] showed that the image formation of spheres in transmission configuration of SAM is easily to understand, and is close to the imaging in optical microscopy. In transmission acoustic microscopy, the dimension of the image is only slightly dependent on the scanning plane position. The strongest image contrast arises when the focal plane coincides with the center of the sphere. The diameter of the image in this position is nearly equal to the particle size. In contrary, in reflection acoustic microscopy, image formation of spherical particles has revealed extraordinary properties[2,3]. The purpose of this report is to present an interpretation of the images of spheres at different focal plane positions, using Fourier spectrum approach in reflection SAM.

THEORY

Following the approach outlined in paper[4], the lens of the microscope is modeled by a focusing transducer with the pupil function of the lens $\mathcal{P}(\Theta)$, and with radius of curvature f being equal to the focal length of the lens, and an (half) aperture angle α, which is equal to the half opening angle of the lens subtended with respect to the focal point. The center of a sphere with radius a is placed at a distance R from the acoustic transducer axis and at a distance Z from the focal plane. The images arise when the object is scanned in the x-y plane with respect to the lens. An expression for the output signal of the reflection acoustic microscope for spherical particles in Fraunhofer approximation ($kZ^2/f \ll 1$, $ka^2/f \ll 1$, with k: wave number) was obtained by Lobkis and Zinin[4] and corrected by Zinin et al.[5]:

$$V(R, Z) = \frac{2V_0}{1 - \cos \alpha} \sum_{n=0}^{\infty} \sum_{m=0}^{n} (-1)^{n+m} (2 - \delta_{0m}) A_n L_{nm}^2(R, Z) \tag{1}$$

$$L_{nm}(R, Z) = \int_0^{\alpha} \mathcal{P}(\Theta) \exp(-ikZ \cos \Theta) J_m(kR \sin \Theta) \overline{P_n}(\cos \Theta) \sin \Theta d\Theta \tag{2}$$

where $V_o = 2\pi f^2(1 - \cos \alpha)(v_0/k) \exp(i(2kf - \pi/2))$ is the signal from a rigid plane surface, placed in the focal plane, with v_0 the oscillating velocity of particles on the transducer surface in the ultrasound field. A_n are the constants derived from the boundary conditions, δ_{om} is the Kronecker delta symbol, $J_m(x)$ are the cylindrical Bessel functions, $\overline{P_n}(\cos \Theta) = P_n(\cos \Theta)N_{nm}$ are the normalized associated Legendre polynomials, $P_n(\cos \Theta)$ the associated Legendre polynomials, $N_{nm} = \sqrt{(2n + 1)(n - m)!/(n + m)!/2}$ are the normalizing coefficients. Further on we will refer to the solution (1) and (2) as the exact solution. In case of a rectangular pupil function,

$$\mathcal{P}(\Theta) = \begin{cases} 1 & \Theta \leq \alpha \\ 0 & \Theta > \alpha \end{cases} \tag{3}$$

expressions (1), (2) may be simplified by expanding the integral in a series.

$$V(R, Z) = \frac{8V_0}{1 - \cos \alpha} \sum_{n=0}^{\infty} \sum_{m=0}^{n} (-1)^{n+m} A_n (2 - \delta_{0m})$$

$$\left[\sum_{l=0}^{\infty} (-i)^l j_{l+m}(kR_s) \quad \overline{P_{l+m}}(\cos \Theta_s) \overline{G_{nl+m}} \right]^2 \tag{4}$$

Where $R_s = \sqrt{R^2 + Z^2}$ is the distance between the focal point and the center of the sphere. $\cos \Theta_s = Z/R_s$, and $j_{l+m}(kR_s)$ is the spherical Bessel function. For $n \neq l$ the coefficients $\overline{G_{nl+m}}$ have the form:

$$\overline{G_{nl}} = \frac{(1 - q^2)\left[\left[\frac{\partial}{\partial q}\overline{P_n}(q)\right]\overline{P_l}(q) - \overline{P_n}(q)\left[\frac{\partial}{\partial q}\overline{P_l}(q)\right]\right]}{(n - l)(n + l + 1)} \tag{5}$$

Here $q = \cos \alpha$. For $n = l$, integral $\overline{G_{nn}}$ can be calculated using interactive equation:

$$\overline{G_{nn}} = \overline{G_{nn}^{m-1}} - \sqrt{\frac{(1 - q^2)}{(n + m)(n - m + 1)}} \overline{P_n}(q)\overline{P_n^{m-1}}(q) \tag{6}$$

where $\overline{G_{nn}^0}$ has the form:

$$\overline{G_{nn}^0} = 1/2 \left[1 - qP_n^2(q)\right] + \sum_{k=1}^{n-1} P_k(q)\left[P_{k+1}(q) - qP_k(q)\right], \tag{7}$$

In (7) it is assumed to put the sum equal to zero if, the upper limits is smaller than the lower one. So for $n = 0$ and $n = 1$ we have: $\overline{G_{00}^0} = (1 - q)/2$, $\overline{G_{11}^0} = (1 - q^3)/2$.

RESULTS AND DISCUSSION

The model given by eq. (1) - (4) describes the contrast formation in the reflection acoustic microscope for spherical particles. From now on we will restrict our consideration to the case of rigid spheres. The image of spherical particles is formed by scanning in the xy-plane at a fixed Z position. If $Z = 0$, the focal plane passes through the center of the sphere. If $Z = -a$, the focal plane touches the front surface of the sphere. Due to the axial symmetry, the image of a sphere can be characterized by a curve which only depends on the distance R between the sphere's center and the axis of the lens: a so-called $V(R)$ curve. The theoretical treatments[4] have shown that the maximal contrast in the images appears when the front surface of a sphere ($Z = -a$, top maximum) or its center ($Z = 0$, center maximum) coincide with the focal plane. At focal plane position laid in the middle between center and top of the sphere, the signal has a minimum. This is also the position at which the focus of the transducer coincides with the focal point of a perfect reflecting spherical mirror. So in order to characterize the imaging of spherical particles in the reflection acoustic microscope, it is prompt to consider the image formation in this peculiar focal plane positions.

Center Images

Strong contrast is obtained by focussing on the top and to the center of sphere. If the sphere's center coincides with the focal point, all reflected rays seem to originate from the center of the sphere. Now the virtual focus of the lens shall be moved a distance $R_s \ll a$ from the sphere's center. In paraxial approximation, all reflected rays have a virtual focus at the point O_T, which is located on the opposite side of the sphere's center, in respect of the virtual focus of the lens [7]. The distance between the sphere's center and point O_T is equal to R_s. This means that the reflected field is the same as the field of a focused transducer which faces the lens and which focus coincides with the point O_T. Using this model and Fourier spectrum approach [6] it is possible to show that the output signal of the reflection acoustic microscope has the form:

$$V(R, Z) = S_o \int_0^1 G^2(\rho) \exp\left(-i\frac{1}{2}w\rho^2\right) J_0(v\rho)\rho d\rho \qquad (8)$$

Where we introduce the notations: $w = 2kZ \sin^2\alpha$, $v = 2kR \sin\alpha$, $S_o = 2\pi f^2 \sin^2\alpha$. For a lens with circular pupil function ($G(\rho) = 1$ if $\rho \leq 1$, $G(\rho) = 0$, if $\rho > 1$), the output signal $V(R, Z)$ near the sphere's center coincides with the expression of the field distribution in the vicinity of the focal point of the focusing transducer contracted twice in R_s direction. For a lens with circular pupil function and for scanning in the focal plane $Z = 0$, expression (8) may be rewritten:

$$V(R) = S_o \frac{J_1(2kR \sin\alpha)}{2kR \sin\alpha} \qquad (9)$$

Expression (9) is the well known formula for the focal plane field distribution of the focused transducer (Airy pattern). The radius of the first minimum is equal to $0.3\lambda/\sin\alpha$, which is twice smaller than the radius of the Airy disk. The same conclusion was derived by Atalar[2]. In Figure. 1 the central image of a rigid particle and the Fourier optics approximation are presented. It can be seen that the Fourier optics description approximates the exact solution quantitatively well. The experimental results[3] are in good coincidence with these conclusions.

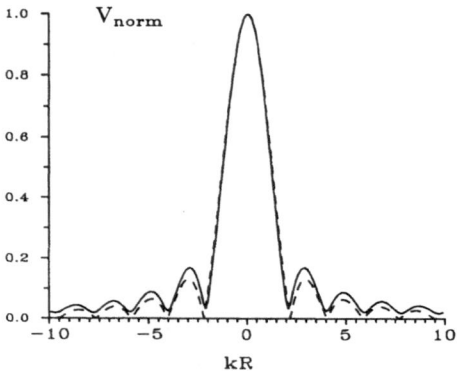

Figure 1. Comparison between central image of large sphere ($ka = 100$) and the modulus of the jinc function $|2J_1(2kR\sin\alpha)/(2kR\sin\alpha)|$. Half aperture angle of the lens $\alpha = 60°$. Solid line - image of sphere, dashed line - modulus of jinc function.

Now consider the $V(Z)$ curves formation near the center. If we put $R = 0$ in equation (8), we obtain the expression which coincides with the well known formula of Atalar [6]. For a lens with a circular pupil function, the $V(Z)$ curve when the focal point crosses the center coincides with the $V(Z)$ curve of a perfect reflecting half space:

$$|V(0, Z)| = \left[\frac{\sin(0.5kZ\sin^2\alpha)}{0.5kZ\sin^2\alpha}\right] \tag{10}$$

Middle Images

As the scanning plane moves towards the center from the top of sphere, the contrast and the dimension of the image decrease. The lowest level of signal is reached when z is equal to $a/2$. In this position the focal point of the lens coincides with the paraxial focus of the perfect reflecting mirror represented by the sphere. Let us consider the image formation at the middle position in terms of paraxial approximation. The SAM images was prepared when the focal point scans an object in the x-y plane at fixed Z-coordinate positions. But consider the movement of focal point on the half radius spherical surface: For small R/a, the image taken at this Z position can be approximated by one taken along the spherical surface with radius $a/2$. The reason is that for small R/a we can neglect deviations in Z between these two types of images, because they are of second order in R/a. For large spheres, the image taken along this arc can be easily understood. If we move the lens in such a way, that the focal point coincides with a half-radius surface, the reflection field can be presented as a tube of parallel rays. The direction of propagation of this rays coincides with a line, connecting the center of sphere and the focal point. Using this model and the first order method of stationary phase for double integrals, the output signal of SAM can be written:

$$V \approx \frac{S_o}{ka\sin^2\alpha}\,\mathcal{P}^2(\Theta,\varphi) \tag{11}$$

Here $\Theta = \arcsin(2R/a)$, φ is the azimuthal angle, $S_o = 2\pi f^2\sin^2\alpha$. So the output voltage is determined by only one point of the pupil, given by the direction of the returning plane wave. The signal is by a factor of approximately $ka(\sin^2\alpha)$ weaker than the signal obtained by focussing to the center or the top of the sphere. Hence if we scan with distance $a/2$ between focal point and the sphere's center, we can directly determine the square of the 2-dimensional pupil function.

608

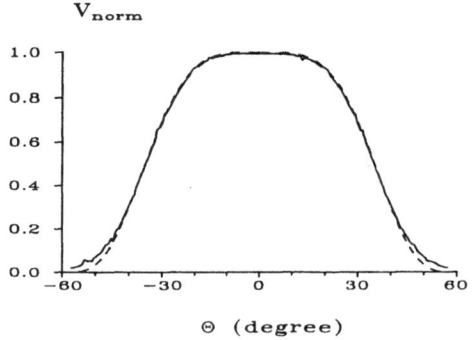

Figure 2. Images taken along the arc at half radius of sphere with ($ka = 100$) and (square of) pupil function. Half aperture angle of the lens $\alpha = 60°$. Solid line - image, dashed line - pupil function.

Figure 2. shows a calculated image $V(\Theta)$, taken along the arc at $a/2$ for a sphere with $ka = 100$ with a smooth pupil function which is proposed by Sasaki et al.[8], also having a half aperture angle $\alpha = 60°$:

$$
\mathcal{P}(\Theta) = \begin{cases} \dfrac{1}{2}\left[1 + \cos\left(\pi\dfrac{\cos\Theta - 1}{\cos\alpha - 1}\right)\right] & \Theta \leq \alpha \\ 0 & \Theta > \alpha \end{cases}
\tag{12}
$$

All curves are plotted as a function against angle Θ, where Θ corresponds to $R = a/2\sin\Theta$. The image is nearly equal to the pupil function.

Top Images

Next we want to consider the images we obtain by scanning with constant Z the plane that coincides with the top of the sphere. Remaining within the ray picture we can say that the pressure field under the focal point will be restricted to a cone. Now for $R << a\tan\alpha$ the width of the insonified spot on the sphere's surface will extend only over a small solid angle, if α is sufficiently small. So this spot can be approximated by an inclined plane

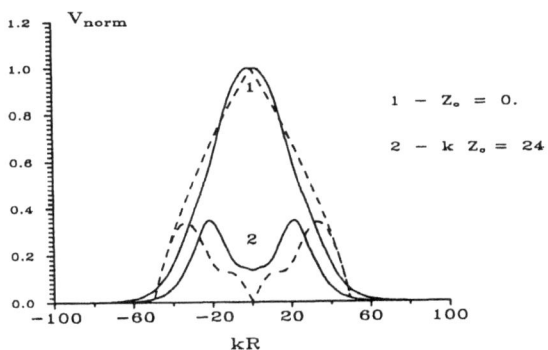

Figure 3. Subtop images of spheres ($ka = 100$) for different Z_o positions. Half aperture angle of the lens $\alpha = 30°$. a) Exact solution. b) Fourier optics approach.

surface. Following the method of proposed by Tsukahara et al.[9] we can approximate the exact solution up to a constant by the formula:

$$V(R, Z_o) = \frac{\sin\left\{0.5(ka)\ \left[1 - \left(\frac{R}{a}\right)^2\right]\ \left[1 - \frac{Z_o}{a} - \sqrt{1 - \left(\frac{R}{a}\right)^2}\right]\left(\sin\alpha - \left(\frac{R}{a}\right)\right)^2\right\}}{0.5(ka)\sin^2\alpha\ \left[1 - \left(\frac{R}{a}\right)^2\right]\ \left[1 - \frac{Z_o}{a} - \sqrt{1 - \left(\frac{R}{a}\right)^2}\right]} \tag{13}$$

Here Z_o is the distance between the scanning plane and the top of the sphere. Z_o is positive if the scanning plane is under the top of the sphere. The image of the top of a sphere ($Z_o = -a$) can be represented as a light spot with a radius equal to $a\sin\alpha$. Indeed the signal of the top image (13) is equal to zero in that position. This conclusion is in agreement with experimental results of paper[3].

During defocusing (increasing Z_o in (13)) the contrast will decrease but the dimension of the spot will be the same (see Figure 3.). With increasing the defocusing Z_o, the subsurface image becomes a ring. The contrast maximum of the ring occurs at a radius that is at most as big as the radius, where the scanning plane crosses the sphere's surface. With increasing value of defocusing the coincidence between the exact solution and the approxi-

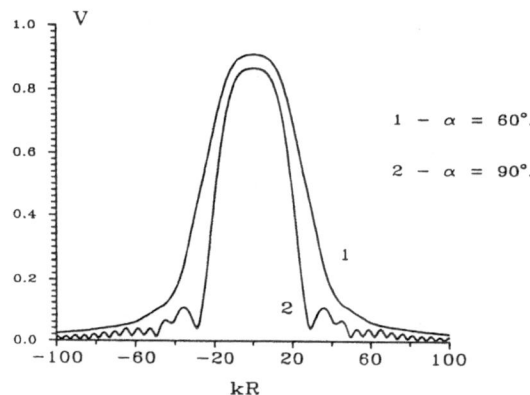

Figure 4. Top images of spheres ($ka = 100$) for different aperture angle. Exact solution.

mate solution reduces (see Figure 3.), but it can be seen that even far from the top, formula (13) qualitatively well describes the behavior of the images.

Translating the lens along the z axes ($R = 0$) we obtain the expression for the $V(Z)$ curve of a rigid half space, since we modeled the top of a large sphere by a plane.

$$V(0, Z_o) = \frac{\sin\left(0.5Z_o\sin^2\alpha\right)}{0.5Z_o\sin^2\alpha} \tag{14}$$

The $V(Z)$ curves obtained by Fourier optics approach, are in good agreement with the exact solution in the range of the main $V(Z)$ curve maximum.

For large ka and large aperture angle α, the dimension of the radius of the inner light spot of the top image will be smaller than $a\sin\alpha$ (see Figure 4.).

This obviously occurs when the sinus in (13) reaches zero,

$$\left\{0.5(ka) \quad \left[1 - \left(\frac{R}{a}\right)^2\right] \quad \left[1 - \frac{Z_o}{a} - \sqrt{1 - \left(\frac{R}{a}\right)^2}\right]\left(\sin\alpha - \left(\frac{R}{a}\right)\right)^2\right\} = \pi \tag{15}$$

inside the interval $R \ll a \sin\alpha$. Calculations of formula (13) show that the deviation from the law: $a \sin\alpha$ appears when $ka > 2000$. However, for big α this deviation appears for rather small ka. So for big α, eq.(13) qualitatively well describes the features of the images.

ACKNOWLEDGEMENT

This work was supported by the Alexander von Humboldt Foundation and is a part of the doctoral dissertation of Wieland Weise.

REFERENCES

1. P.V. Zinin, W. Weise, and S. Boseck, Theory of elastic sphere imaging in a different types acoustic microscopes, *in*: "Acoustoelectronics'93," Proc. Int. Conf., Varna, (1993).
2. A. Atalar, Backscattering formula for acoustic transducers, *J. Appl. Phys.* 51:3093 (1980).
3. O.V. Kolosov, O.I. Lobkis, K.I. Maslov, and P.V. Zinin, The effect of the focal plane position on the images of spherical objects in the reflection acoustic microscope, *Acoust. Lett.* 16:84 (1992).
4. O.I. Lobkis and P.V. Zinin, Acoustic microscopy of spherical objects. Theoretical approach, *Acoust. Lett.* 14:168 (1991).
5. P.V. Zinin, O.V. Kolosov, O.I. Lobkis, and K.I. Maslov, Visualization of spherical objects by the reflection acoustic microscope. *Phys. Acoust.* 39:343 (1993).
6. A. Atalar, An angular-spectrum approach to contrast in reflection acoustic microscopy, *J. Appl. Phys.* 49:5130 (1978).
7. O.V. Kolosov, K. Yamanaka, O.I. Lobkis, and P.V. Zinin, Direct spatial evaluation of point-spread-function of the axially symmetric and asymmetric acoustic imaging focusing system using spherical ball reflection, *in* "Ultrasonics International'93," Proc. Int. Conf., Vienna, (1993).
8. Y. Sasaki, T. Endo, T. Yamagishi, and M. Sakai, Thickness measurement of a thin-film layer on an anisotropic substrate by phase-sensitive acoustic microscope, *IEEE Trans. UFFC* 39:638 (1992).
9. Y. Tsukahara, N. Nakaso, and K. Ohira, Angular spectral approach to reflection of focused beams with oblique incidence in spherical-planar-pair lenses, *IEEE Trans. UFFC.* 38:468 (1991).

IMAGING GEOLOGICAL MATERIALS WITH
THE ACOUSTIC MICROSCOPE

Nicholas E. Pingitore Jr.[1,2], Cindy L. Gillespie[1,2], Kate C. Miller[1],
Leon DuPlessis[2], and Lawrence E. Murr[2]

[1] Department of Geological Sciences
[2] Materials Research Institute
The University of Texas at El Paso
El Paso, TX 79968-0555

INTRODUCTION

Acoustic microscopy, a technique best known in materials science for non-destructive testing, presents exciting and intriguing opportunities for imaging geological materials and for micro-characterization of their elastic properties. In the last dozen years, acoustic microscopy has won broad acceptance in materials science and high-technology manufacturing, particularly for non-destructive testing of electronic devices. Several vendors market commercial instruments designed and used chiefly to detect delaminations in layered electronic components. Due to the scarcity of commercial or purpose-built instruments in universities, very few studies of geological materials have been made by acoustic microscopy. Notable exceptions include examination of mineral specimens by Bonner (1978) and Morales *et al.* (1991), of rocks by Dansburg and Yuhas (1978) and Rodriguez-Rey (1990), of fossils by Kustov *et al.* (1987) and Scott and Hemsley (1991), and of a variety of geologic materials by Pingitore *et al.* (1993). The objective of our research program, then, is to explore the possible applications of acoustic microscopy in the geosciences, appraise their merit, and pursue the most promising lines of investigation.

The Acoustic Microscope

Acoustic microscopy exploits a confocal transducer source and receiver shaped into a hemispherical lens and coupled to the solid specimen through a fluid, normally water. Pulsed acoustic waves in the MHz to GHz range are focused on the surface of the solid, and the travel time and strength of the reflected wave is recorded back at the lens. Focus is achieved by vertical motion of the lens; source pulsing provides a time window for reception of the reflected signal. By rastering the lens over the surface and storing and processing the accumulated data, a pixel-by-pixel image is created. For the images we present herein, the process normally took 2 to 10 minutes. In theory, and in practice, acoustical images of surfaces can rival light microscopy in resolution. For example, at 2 GHz the acoustic wavelength is less than 0.75 μm, approximately that of red light. This gain in resolution with frequency is accompanied, unfortunately, by a deterioration in subsurface imaging. A simple description of acoustic microscopy is found in Briggs and Hoppe (1991) and a more advanced treatment is available in Briggs (1992).

Advantages of Acoustic Microscopy in the Geosciences

Every microscopy (e.g., light, SEM, TEM, STM/AFM, etc.) presents a complex set of advantages and disadvantages to the user, and thus different microscopies generally are viewed as complementary. Four salient advantages of acoustic microscopy with potential relevance to problems in the geosciences are:

Surface Topographic Imaging. With the lens focused on the average surface, a detailed map or image of this topography is obtained. Cross sections of surface "elevations" are routinely obtained.

Subsurface Imaging. By focusing the lens into the specimen, reflections of subsurface features (voids, inclusions, layers) are recorded. An image of a particular depth level may be obtained, or a tomographic image created from scans at multiple depths.

Impedance Contrast Imaging. A flat and polished surface yields no topographic contrast. However, the strength of the reflected signal varies with the acoustic impedance of the materials present at the surface (e.g., grains of different minerals). Thus a map of impedances can be created, or impedances can be measured on a sub-mm scale.

Surface (Rayleigh) Wave Contrast. The hemispherical acoustic lens can generate waves at angles sufficient to induce surface waves in the sample. Coupling of the sample surface with the fluid excites waves in the water, which can interfere with the reflected signal wave. The interference highlights cracks and provides contrast between grains of a mineral with different crystallographic orientations.

In our research, we are exploiting these advantages in attempts to solve specific problems in such diverse areas of geoscience as paleontology, mineralogy, igneous and sedimentary petrology, hydrology, and geophysics. Examples of some of these efforts are illustrated and described in this paper.

TECHNIQUES

Instrumentation

Our research was conducted using a Hitachi AT 6000C scanning acoustic tomograph equipped with 10 and 25 MHz acoustic sources. Images were collected and generated with the proprietary Hitachi software and no further processing of the images was done.

One advantage offered by the configuration of the Hitachi AT 6000C instrument, as well as several others, is its ability to image a large area, up to approximately one square foot (0.1 m²) in dimension. Thus slabbed oil-well drill core, as well as hand specimens and thin sections, can be examined with no special arrangements, and the sample tray could be modified to accept a piece of drill core, say, 6 feet (2 m) long. By scanning large areas in a large-pixel--low-resolution mode (a high-resolution image would take quite some time to obtain) the instrument acts as a macroscope, providing excellent images of porosity and grain distributions.

Sample preparation

The geological samples examined comprised a wide variety of materials including individual mineral specimens, rocks, natural glasses, and fossils. These had been prepared by several different traditional techniques. Oil-field drill cores were slabbed horizontally or vertically, and then ground smooth and polished. Smaller rock and mineral specimens similarly were cut, ground, and polished.

Thin sections, fabricated for examination with a standard petrographic (transmitted polarized light) microscope, were produced by gluing the cut and ground surface of the specimen block to a glass slide. The block was cut again, parallel and close to the glass slide, and then ground and polished to a thickness of $30\,\mu$m. To prevent damage, porous, friable materials were vacuum-impregnated with a plastic resin prior to fabrication of the thin section. Degree of polish varied from shiny on most of the cores to a mirror finish ($0.05\,\mu$m alumina powder) on some thin sections prepared for electron microprobe analysis.

Specimens for analysis of surface features, e.g., fossils, were examined without any preparation.

RESULTS

Current projects

Table I presents a number of specific projects which we have begun to explore or will explore in the near future. These by no means are intended to be exhaustive, but rather represent our initial ideas of productive lines of investigation, based, in part, on materials and knowledge readily at hand. We further expect that imaging a wide range of materials, even where no immediate purpose is evident, will yield serendipitous results.

It will be noted that many of these projects involve oil-well cores. This reflects both the petroleum industry background of the geology members of our research team and our belief that acoustic microscopy offers special benefits for this area. Petrophysics -- the investigation of various physical properties of bulk rock -- is fundamental to the petroleum industry at all stages of operations from initial exploration, through field development, to enhanced oil recovery. The acoustic microscope is well suited to the all-important study of porosity, including its geometry, orientation, and distribution. Further, this instrument presents the opportunity for the traditional petrographer (who examines thin slices of rock with a polarized-light microscope) to image samples with the same probe, acoustic waves, used on much larger scales to create sonic logs of oil wells and seismic profiles of sedimentary basins.

Examples of images

All of the images which follow were acquired using a 25 MHz acoustic lens, with the sample immersed in distilled water at room temperature.

Table 1. Initial projects in acoustic imaging of geological materials.

Project	Materials	Principle Exploited
In situ micro-impedance mapping and measurement	oil-well cores	impedance contrast
Mapping location and geometry of porosity in three dimensions	corals, sandstones, limestones	subsurface imaging
Mapping porosity in 2-D to detect anisotropy	oil-well cores, rock slabs	topography
"Point counting" grains by impedance recognition and automatic calculation of relative areas	oil-well cores, thin sections, slabs	impedance contrast
Detection and mapping of deformation which has changed elastic properties of a rock	metamorphics, tectonically strained sedimentary rocks	impedance contrast
Micro-crack detection and mapping of permeability to reveal fluid-flow paths in sedimentary, igneous, and metamorphic rocks	oil-well cores	surface wave interference
Detection of fluid inclusions in opaque minerals	pyrite, other sulfides and metallics	subsurface imaging
Enhanced imaging of fossil impressions	graptolites, insects, fish	topography
Enhancement of distribution pattern of cements and voids	oil-well cores	topography, impedance
Imaging of etched fission tracks - difficult to see in optical microscope, but crack enhancement may help	suitable fission-track minerals, e.g., apatite	surface wave interference
Grain orientation in replacement fabrics	dolomite: oil-well cores, slabs, thin sections	surface wave interference

Fossil fish. Figure 1 is the acoustic image of a fossil coelacanth fish, *Diplurus newarki*, which is preserved as a delicate impression in a black shale. The specimen was exposed by carefully tapping the fissile shale with a hammer, thereby splitting the rock to reveal the fossil. The dark color and shallow depth of the impression make it difficult to see details of the structures visually or with an optical stereographic microscope. The acoustic image clearly delineates the discrete radiating elements of the caudal fin (tail) of one individual superposed on the bony structure of the head of a second individual.

Acoustic imaging of impression-type fossils is rapid and non-destructive, involving no sample preparation, other than placing the specimen in distilled water. It is an alternative to optical macrophotography and photomicroscopy, where providing appropriate illumination could be time-consuming or impossible with such specimens.

Fossil coral. Figure 2 presents the acoustic image of the calcium carbonate skeleton of the colonial Caribbean reef coral, *Montastrea cavernosa*. The specimen had been prepared as a standard thin section for examination by transmitted polarized-light microscopy, and the surface therefore is both flat and polished. Individual corallites (complex cylindrical voids in which individual polyps resided, and which appear circular in the reduced dimension of the thin section) are filled with an epoxy resin, added to harden the material during cutting, grinding, and polishing. Of particular interest in

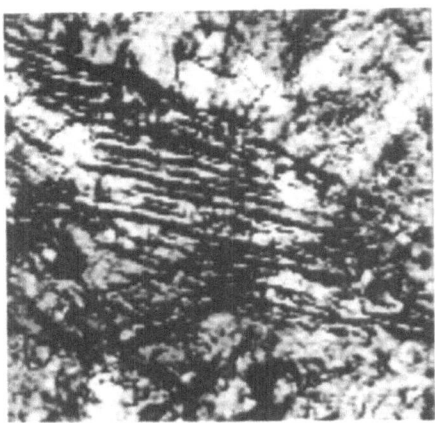

Figure 1. Fossil impression of fish tail in black shale. Field of view 23 mm.

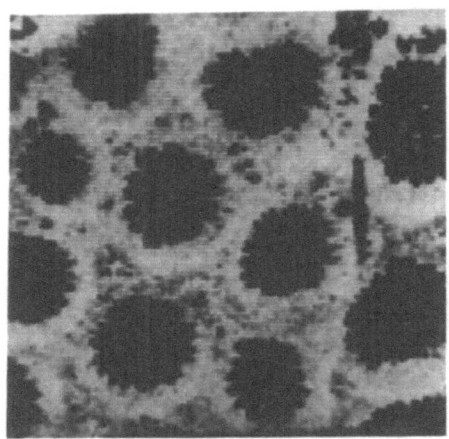

Figure 2. Fossil coral composed of calcium carbonate. Field of view 20 mm.

this specimen is the contrast seen within the skeleton between white and darker patches. The brighter, more extensive regions are the original coral skeleton, composed of CaCO₃ in the form of the mineral *aragonite* (orthorhombic). At an ultrastructural level, this is a composite biomaterial of needles of aragonite and organic binder, organized at several scales. The darker areas are a void-filling natural cement composed of a different polymorph of $CaCO_3$, the mineral *calcite* (rhombohedral). The calcite cement formed during post-mortem exposure of the coral to fresh water, subsequent to tectonic uplift of the entire reef from the shallow marine environment.

The acoustic contrast between aragonite skeleton and calcite cement likely is a function of two factors. First, the minerals aragonite and calcite have different inherent acoustic impedances. Second, the aragonite is a complex, organized biocomposite which presumably differs significantly in its elastic properties from pure, bulk crystalline aragonite. Currently we are attempting *in situ* measurement of the impedances of these materials to permit quantification of such impedance mappings.

Oil-well core. Figure 3 is the acoustic image of part of a horizontally slabbed drill core, of 4.25-inch (11 cm) diameter. The drill core is from the pay zone, a carbonate

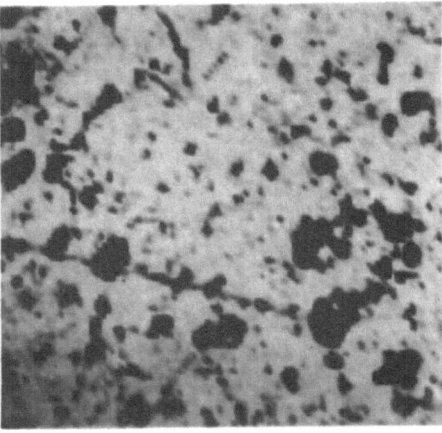

Figure 3. Oil-well core in carbonate mudstone. Field of view 25 mm.

mudstone with macropores technically known as a *loferite*, of a producing well in the Permian Basin in southeastern New Mexico. In the acoustic photograph, the porosity is well highlighted, and even more strikingly exhibited in the false-color image unfortunately but understandably not reproduced here. In the hand specimen, or under a low-power optical stereo-zoom microscope, the porosity distribution is more difficult to visualize. The acoustic image does not reveal any anisotropy in the distribution of porosity in the horizontal plane.

Numerous cracks are evident in the acoustic image, but these can be seen optically only when a directional light source is tilted to just the right angle. These cracks are probably crucial features of the permeability of this rock, determining the ability of hydrocarbons to be removed from the unit.

Pisolites. Figure 4 presents images of pisolites, accretionary grains of $CaCO_3$ found typically in travertine, cave deposits, and soils. The pisolite develops by intermittent precipitation of $CaCO_3$ around an asymmetric nucleus, usually a carbonate grain. Separation of the layering in a number of places is visible in the surface slide.

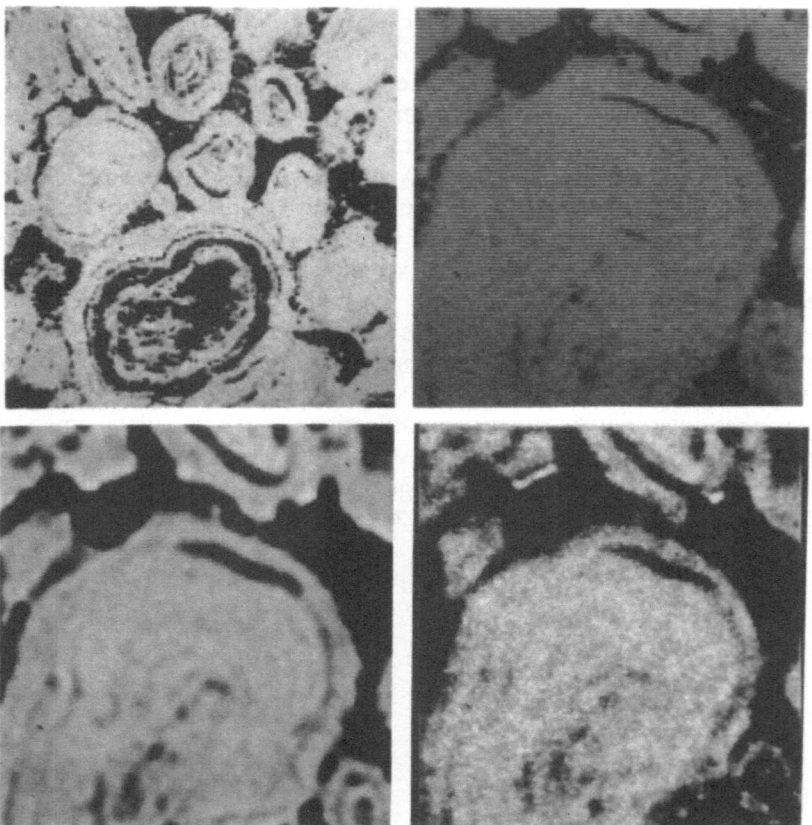

Figure 4. Surface and subsurface views of pisolites, concentrically layered accretionary grains of calcium carbonate. **Upper left.** Surface view of polished rock specimen composed of cemented pisolites. Field of view 32 mm. **Upper right.** Close-up of individual pisolite, surface view. Field of view 12 mm. **Lower left.** Subsurface view of same pisolite, depth of 1 mm. Field of view 12 mm. **Lower right.** Subsurface view of same pisolite, depth of 2 mm. Field of view 12 mm.

Subsurface scans reveal the 3-dimensional external morphology of the pisolite, as well as the geometry of the surrounding pores. The acoustic scans at greater depths indicate that the focal plane passed increasingly beyond the center of the largest pisolite and thus its diameter diminished in successive images. Concomitantly, the pore space increased in volume, and its geometry changed.

Grain-pore relationships are of fundamental importance in the recovery of subsurface hydrocarbons. Because of its unique ability to image in three dimensions, the acoustic microscope presents an opportunity to break free of the two-dimensional views of classical optical microscopy. Because of recent evidence (Yang and Kundu, this volume) that images of internal pores may be biased inherently and therefore incorrect, we are in the process of comparing acoustic views such as Figure 4 with optical views of serial sections produced by grinding the original specimen. These destructive optical images will provide a "ground truth" for the sub-surface acoustic images. Surface-breaking pores such as seen in Figure 4 may not be as problematical as confined interior pores.

CONCLUSIONS

Acoustic microscopy presents significant opportunities for research in a broad range of subdisciplines of the geosciences, especially those related to the petroleum industry. In particular, images created by impedance contrast, subsurface focus, and Rayleigh contrast provide visual information unavailable with other conventional microscopies. Our current challenge is to overcome the expected technical difficulties and perfect the application of acoustic microscopy to geomaterials and thereby fulfill the promise of this technique.

ACKNOWLEDGEMENT

C.L.G. was supported by NSF Award RII-8922191 to N.E.P.

REFERENCES

Bonner, B.P., 1978, Detecting internal defects in single crystal olivine with the acoustic microscope, *Eos, Trans., Amer. Geophysical Union.* 43:1696 (abs.).

Briggs, A., 1992, "Acoustic Microscopy," Oxford Univ. Press, Oxford.

Briggs, G.A., and Hoppe, M., 1991, Acoustic microscopy, *in:* "Images of Materials," D.B. Williams, A.R. Pelton, and R. Gronsky, eds., Oxford Univ. Press, Oxford.

Dansburg, J.S., and Yuhas, D.E., 1978, Acoustic microscope images of rock samples, *Geophysical Res. Letters.* 5:885.

Kustov, A.I., Kulakov, M.A., Morozov, A.I., and Erlanger, O.A., 1987, On the potential use of the scanning acoustic microscope in paleontology, *Paleont. Zhur.* (Scripta Technica trans.). 2:117.

Morales, J.G., Rodriguez, R., Durand, J., Ferdj-Allah, H., Hadjoub, Z., Attal, J., and Doghmane, A., 1991, Characterization and identification of berlinite crystals by acoustic microscopy, *J. Materials Res.* 6:2484.

Pingitore Jr., N.E., Gillespie, C.L., Miller, K.C., Niou, C-S., DuPlessis, L.P., and Murr, L.E., 1993, Applications of acoustic microscopy in the geosciences, *Eos, Trans., Amer. Geophysical Union.* Oct. 26:626 (abs.).

Rodriguez-Rey, A., Briggs, G.A., Field, and M. Montoto, 1990, Acoustic microscopy of rocks, *Jour. Microscopy.* 160:21.

Scott, A.C., and Hemsley, A.R., 1991, A comparison of new microscopical techniques for the study of fossil spore wall ultrastructure, *Rev. Palaeobotany Palynology.* 67:133.2

NONDESTRUCTIVE IMAGING OF STRESS DISTRIBUTION IN METAL
USING NONLINEAR ELASTO-ACOUSTICS

Masaru Kato, Takuso Sato, Keisuke Kameyama
and Hideyuki Ninoyu

Interdisciplinary Graduate School of Science and Engineering,
Tokyo Institute of Technology,
4259 Nagatsuta, Midori-ku, Yokohama 227, Japan

INTRODUCTION

Nondestructive detection of stress characteristics in metal is desired in many fields. For example, the residual stress in metal sometimes induces destruction of mechanical parts and structures. In order to predict the destruction, measurement of the distribution of the value of static stress in metal is desired. The nonlinear relation between the stress in metal and ultrasonic sound velocity can be used as a powerful means for the purpose[1].

In this paper, first, the nonlinear dependence of ultrasound velocity on the stress in aluminum alloy (A5052) is observed and the basic principle of the imaging method of the stress distribution in metal by using such a nonlinear dependence is shown. On the basis of the above principle, the stress perturbation is given to the observing region and the corresponding velocity change of ultrasound is detected as the phase change. Then, the procedure of stress distribution reconstruction by using the detected phase change and an inverse operation is formulated in a matrix form. Furthermore, the proposed method is confirmed by a simulation. Finally, the measuring system to detect the phase change is constructed and it is applied for the estimation of the stress distribution in aluminum alloy.

THE PRINCIPLE OF ESTIMATION OF STRESS DISTRIBUTION

The schematic construction of measuring system of the dependence of ultrasound velocity on the stress in metal is shown in Fig. 1. We use aluminum alloy as the metal sample whose shape is restricted as shown in Fig. 1 to show only the basic idea of our method. The observing region is limited close to the surface of the metal where 50MHz ultrasonic probing wave is passed through. Now, if uniform stress σ is loaded normal to the direction of the propagation of probing wave by compressing from the upper and lower parts, the ultrasound velocity is changed. The ultrasound velocity can be obtained by observing the sound velocity change due to the loaded stress. It is measured as the phase difference between the reference probing wave (ref) and the wave passed through the metal sample (RF). As the observed phase difference includes the change due to the strain in the direction of the propagation of the probing wave, the strain of the metal sample in this direction is measured at the same time and used for the compensation to get real sound velocity.

An experimental result is shown in Fig. 2(a). It actually shows nonlinear dependence. Using this result, the derivative of sound velocity with respect to stress is obtained as shown in Fig. 2(b). It shows one-to-one correspondence. Hence, if a stress perturbation Δσ is given as shown in Fig. 2(a), the sound velocity change is different according to the value of compressive stress in metal. Moreover, the sound velocity change due to a given amount of stress perturbation has one-to-one correspondence with the stress in metal by the characteristics of Fig. 2(b). Using this one-to-one correspondence, we can estimate the stress value in metal by acquiring the data of phase difference due to the stress perturbation. This is the basic idea of estimation of the stress in metal.

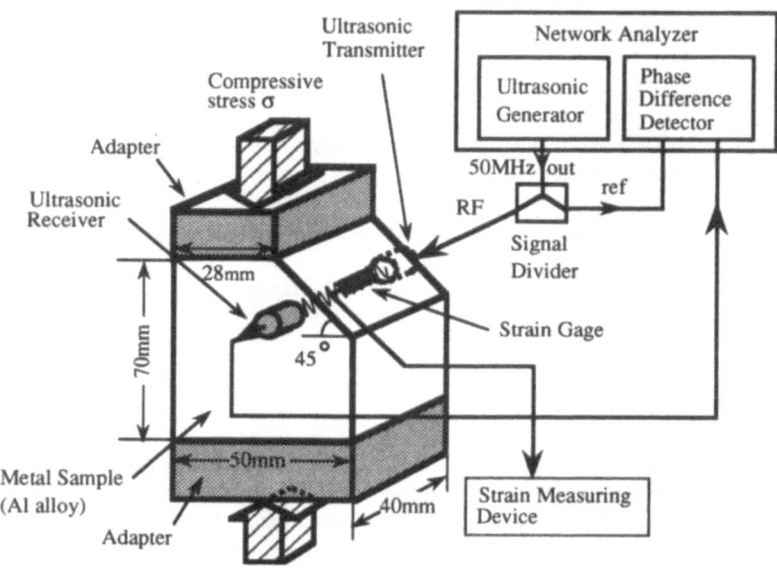

Fig. 1 Schematic construction of measuring system of the dependence of ultrasound velocity on the stress in metal. The ultrasonic wave is passed at the depth of 10mm from the slanted surface.

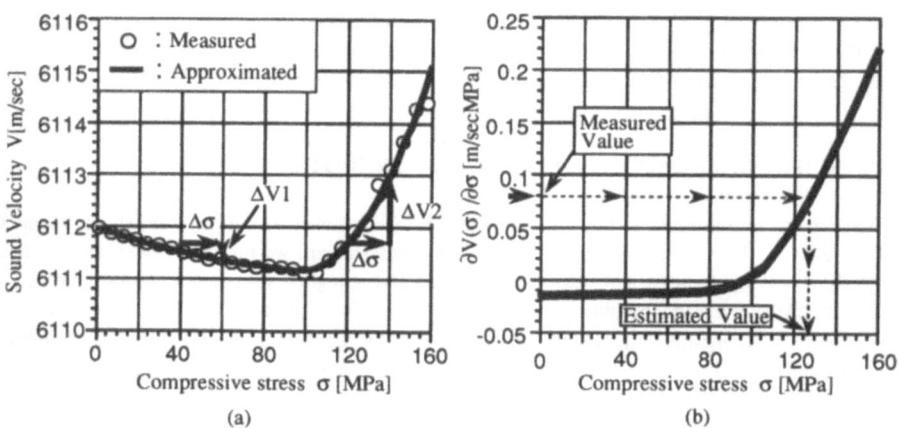

Fig. 2 (a)Nonlinear dependence of ultrasound velocity on the stress in metal. (b)The derivative of sound velocity with respect to stress.

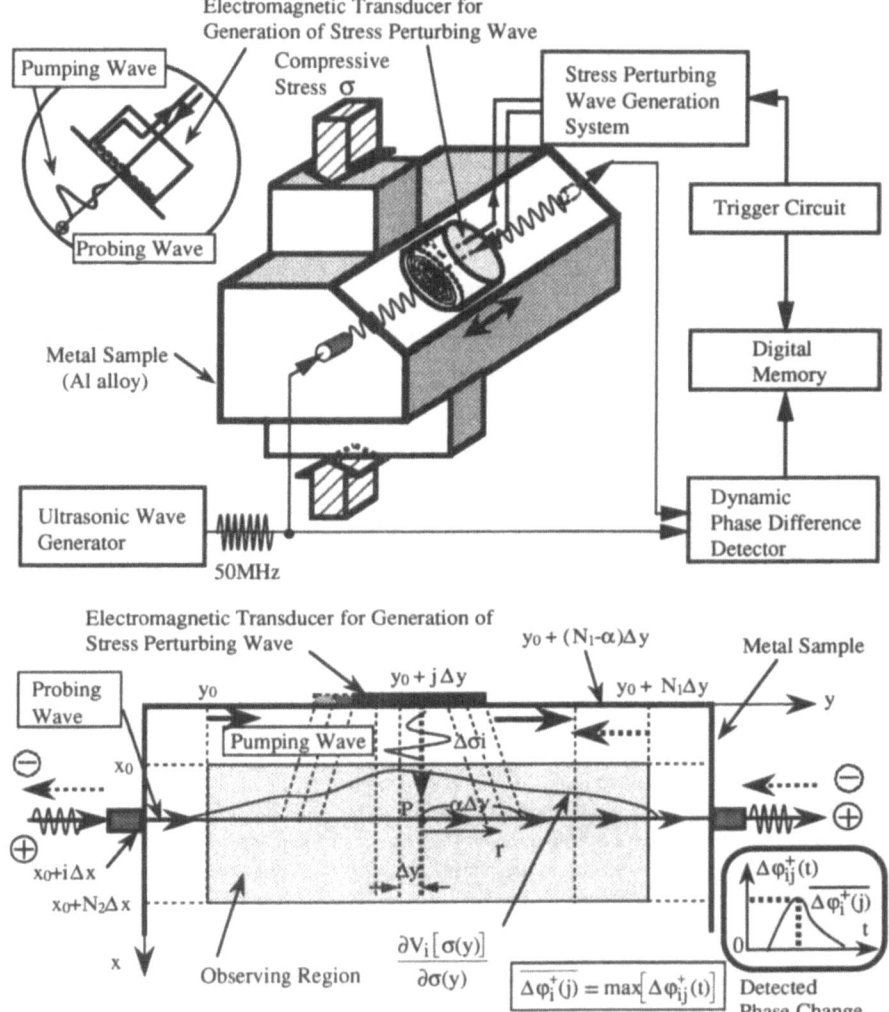

Fig. 3 Schematic construction of stress distribution measuring system.

Next, we consider the principle of imaging of stress distribution in the observing region under the surface of metal. A schematic construction of stress distribution measuring system is shown in Fig. 3. Here, the stress in the metal sample is not uniform but distributed along the path of the ultrasonic probing wave and we estimate this stress. On the basis of the above idea, the phase change of the probing wave is observed. For the purpose of measuring of the phase change, the probing wave is passed through the observing region close to the surface, the stress perturbing pumping wave is generated by the electromagnetic transducer mounted on the slanted surface of the metal sample. In the observing region where the probing and pumping waves are crossed, the sound velocity of the probing wave is changed by the stress perturbation and we can measure the sound velocity change as the waveform of phase change. When the transducer of pumping wave is scanned along the probing beam, a set of waveforms of phase changes are obtained. They are related to the stress through the relation shown in Fig. 2(a)(b). But, as the crossing region of two waves is not pin point, we can not derive the stress distribution directly from the obtained data. Actually, the phase change of the probing wave due to the pumping wave can be expressed using the derivative of sound velocity with respect to the stress, $\partial V/\partial\sigma$, and the spatio-temporal distribution of the stress perturbing pumping wave, $\Delta\sigma$.

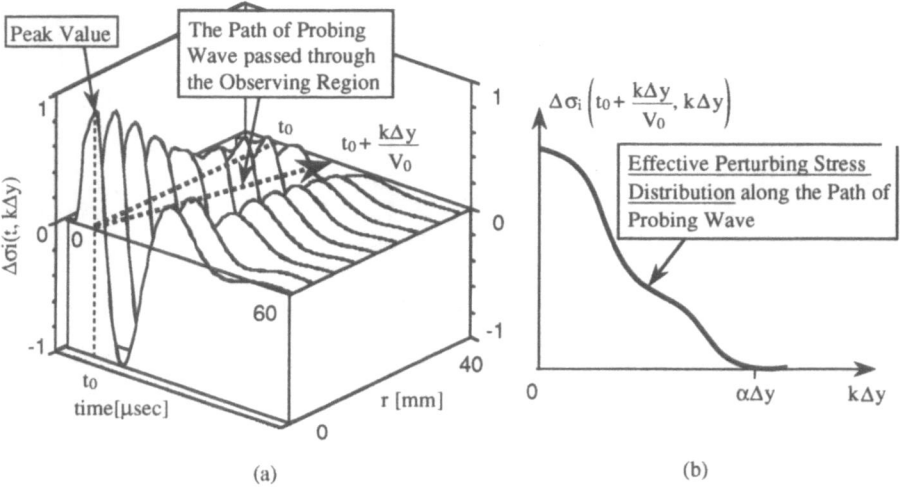

Fig. 4 (a)Spatio-temporal distribution of the pumping wave at the depth of x=x₀+iΔx from the surface. (b)The effective perturbing stress distribution.

When the center of the transducer for generation of the pumping wave is mounted at the position y = $y_0+j\Delta y$ (j = 0,···, N₁) on the slanted surface, the phase change will have the waveform $\Delta\varphi_{ij}^+(t)$ as is shown in Fig. 3. That is, the phase change caused by the stress perturbation shows spread curve which has a peak value $\overline{\Delta\varphi_i^+(j)}$ and can be evaluated as the accumulation of the effect of the stress along the propagation path of the probing wave through the metal. To get reliable data, we decided to use only the first peak value of the phase change, which is related to the first peak value of spatio-temporal distribution of the pumping wave. Actually, it is obtained by tracing the spatio-temporal distribution of the pumping wave starting from t₀ which is the time when the first peak value of the pumping wave arrives at the point P on the probing beam as is shown in Fig. 4(a). This is the effective perturbing stress distribution $\Delta\sigma_i$ as is shown in Fig. 4(b), and it generates the peak phase change $\overline{\Delta\varphi_i}$. If the pumping wave transducer is scanned in +y direction along the probing beam, the peak value of the phase change can be derived as follows

$$\overline{\Delta\varphi_i^+(j)} = -\frac{\omega}{v_0^2}\sum_{k=0}^{\alpha}\frac{\partial V_i\left[\sigma(y_0 + j\Delta y + k\Delta y)\right]}{\partial\sigma(y_0 + j\Delta y + k\Delta y)}\Delta\sigma_i\left(t_0 + \frac{k\Delta y}{V_0}, k\Delta y\right)\Delta y, \quad (j = 0,\cdots, N_1 - \alpha) \tag{1}$$

where ω is angular frequency, V₀ is ultrasound velocity without stress, Δy is scanning pitch, 2αΔy shows the spread of the effective perturbing stress distribution (see Fig. 4(b)), ∂Vi[σ(y)] / ∂σ(y) is the distribution of the derivative of sound velocity with respect to stress at the depth of x = x₀+iΔx (i = 0, ···, N₂) from the surface and Δσi is effective perturbing stress distribution. In the same way, by scanning the pumping wave transducer in -y direction, the peak value of phase change can be obtained as

$$\overline{\Delta\varphi_i(j)} = -\frac{\omega}{v_0^2}\sum_{k=0}^{\alpha}\frac{\partial V_i\left[\sigma(y_0 + j\Delta y - k\Delta y)\right]}{\partial\sigma(y_0 + j\Delta y - k\Delta y)}\Delta\sigma_i\left(t_0 + \frac{k\Delta y}{V_0}, k\Delta y\right)\Delta y, \quad (j = N_1 - \alpha + 1,\cdots, N_1) . \tag{2}$$

The equations (1) and (2) can be expressed in a matrix form as Eq. (3) or as Eq. (4). Here, the effective perturbing stress distribution matrix Δσ is known. Hence, the distribution of the derivative of sound velocity with respect to the stress in metal, ∂Vi[σ(y)] / ∂σ(y), is derived as Eq. (5) by means of an inverse matrix operation.

Finally, if we use the one-to-one relation between the derivative of sound velocity with respect to stress and the stress in metal which is given by Eq. (6), the stress distribution in metal is obtained as Eq. (7).

This is the process of estimation of stress distribution in metal by taking into account the spatio-temporal distribution of stress perturbation.

$$\begin{bmatrix} \overline{\Delta\varphi_i^+(0)} \\ \overline{\Delta\varphi_i^+(1)} \\ \vdots \\ \vdots \\ \overline{\Delta\varphi_i^+(N_1-\alpha)} \\ \overline{\Delta\varphi_i(N_1-\alpha+1)} \\ \vdots \\ \vdots \\ \overline{\Delta\varphi_i(N_1)} \end{bmatrix} = -\frac{\omega\Delta y}{V_0^2} \begin{bmatrix} \Delta\sigma_i(t_0,0) & \cdots & \Delta\sigma_i\left(t_0+\frac{\alpha\Delta y}{V_0},\alpha\Delta y\right) & 0 & \cdots & \cdots & 0 \\ 0 & \Delta\sigma_i(t_0,0) & \cdots & \Delta\sigma_i\left(t_0+\frac{\alpha\Delta y}{V_0},\alpha\Delta y\right) & 0 & \cdots & 0 \\ \vdots & & & & & & \vdots \\ & & & & & 0 & \\ 0 & \cdots & \cdots & 0 & \Delta\sigma_i(t_0,0) & \cdots & \Delta\sigma_i\left(t_0+\frac{\alpha\Delta y}{V_0},\alpha\Delta y\right) \\ 0 & \cdots & 0 & \Delta\sigma_i\left(t_0+\frac{\alpha\Delta y}{V_0},\alpha\Delta y\right) & \cdots & \Delta\sigma_i(t_0,0) & 0 & \cdots & 0 \\ \vdots & & 0 & & & & \\ \vdots & & & & & & 0 \\ 0 & \cdots & 0 & 0 & \cdots & 0 & \Delta\sigma_i\left(t_0+\frac{\alpha\Delta y}{V_0},\alpha\Delta y\right) & \cdots & \Delta\sigma_i(t_0,0) \end{bmatrix} \begin{bmatrix} \frac{\partial V_i[\,\sigma(y_0)]}{\partial\sigma(y_0)} \\ \frac{\partial V_i[\,\sigma(y_1)]}{\partial\sigma(y_1)} \\ \vdots \\ \frac{\partial V_i[\,\sigma(y_0+(N_1-\alpha)\Delta y)]}{\partial\sigma(y_0+(N_1-\alpha)\Delta y)} \\ \frac{\partial V_i[\,\sigma(y_0+(N_1-\alpha+1)\Delta y)]}{\partial\sigma(y_0+(N_1-\alpha+1)\Delta y)} \\ \vdots \\ \frac{\partial V_i[\,\sigma(y_0+N_1\Delta y)]}{\partial\sigma(y_0+N_1\Delta y)} \end{bmatrix} \quad (3)$$

$$\left[\Delta\varphi_{ij}\right] = -\frac{\omega\Delta y}{V_0^2}\left[\Delta\sigma_{ij}\right]\left[\frac{\partial V_{ij}}{\partial\sigma}\right] \quad (4) \qquad\qquad \left[\frac{\partial V_{ij}}{\partial\sigma}\right] = -\frac{V_0^2}{\omega\Delta y}\left[\Delta\sigma_{ij}\right]^{-1}\left[\Delta\varphi_{ij}\right] \quad (5)$$

$$\frac{\partial V_{ij}}{\partial\sigma} = f(\sigma_{ij}) \quad (6) \qquad\qquad\qquad \sigma_{ij} = f^{-1}\left(\frac{\partial V_{ij}}{\partial\sigma}\right) \quad (7)$$

SIMULATION OF IMAGE RECONSTRUCTION

In order to confirm the process of stress distribution reconstruction, some computer simulations have been carried out for the same configuration as Fig. 3 under the following conditions. The 50MHz ultrasonic probing wave is passed at the depth of 10mm from the slanted surface of the metal sample whose length is 80mm as is shown in Fig. 5(a). The pumping wave transducer which gives the stress perturbation with a maximum value of about 0.3MPa at the observing region is scanned from y=10mm to y=70mm with the pitch of 1mm. When the stress distribution is given by compressing the sample as shown in Fig. 5(a), the phase change is calculated by Eq. (3) and the obtained values of the set of peak phase changes are shown in Fig. 5(b). The image of stress distribution reconstructed by using the inverse matrix operation and the relation of Fig. 2(b) is shown in Fig. 5(c). It shows fairly good agreement with the given stress distribution.

EXPERIMENTAL RESULT

The measuring system of the configuration as Fig. 3 was constructed for estimation of the stress distribution. 50MHz ultrasonic probing wave is passed at the depth of 10mm from the slanted surface of Al alloy metal sample whose length is 80mm as is shown in Fig. 6(a). The observed effective perturbing stress distribution is shown in Fig. 6(c). Electromagnetic transducer for generation of the pumping wave is 20mm in diameter and it gives the stress perturbation with a maximum value of about 0.38MPa at the observing region. This transducer is scanned from y=20mm to y=60mm with the pitch of 2.5mm. Compressive stress of 110MPa is loaded only around the center of the sample to give stress distribution. The result of stress distribution estimation by the proposed method is shown in Fig. 6(b). We can see fairly good agreement between the given and observed ones. This estimated distribution, however, does not have sharp edges, may be due to the fact that the actual distribution at 10mm under the surface does not coincide with the one at the surface.

CONCLUSION

First, the nonlinear dependence of ultrasound velocity on the stress in Al alloy was confirmed. Then, the stress perturbation was introduced and the procedure of stress distribution reconstruction by means of an inverse matrix operation was shown. The validity of the method was confirmed by computer simulations and, finally, actual measuring system for estimation of the stress distribution in metal was constructed. The estimated result showed fairly good agreement with the expected one. The system for observation under more general conditions is under study.

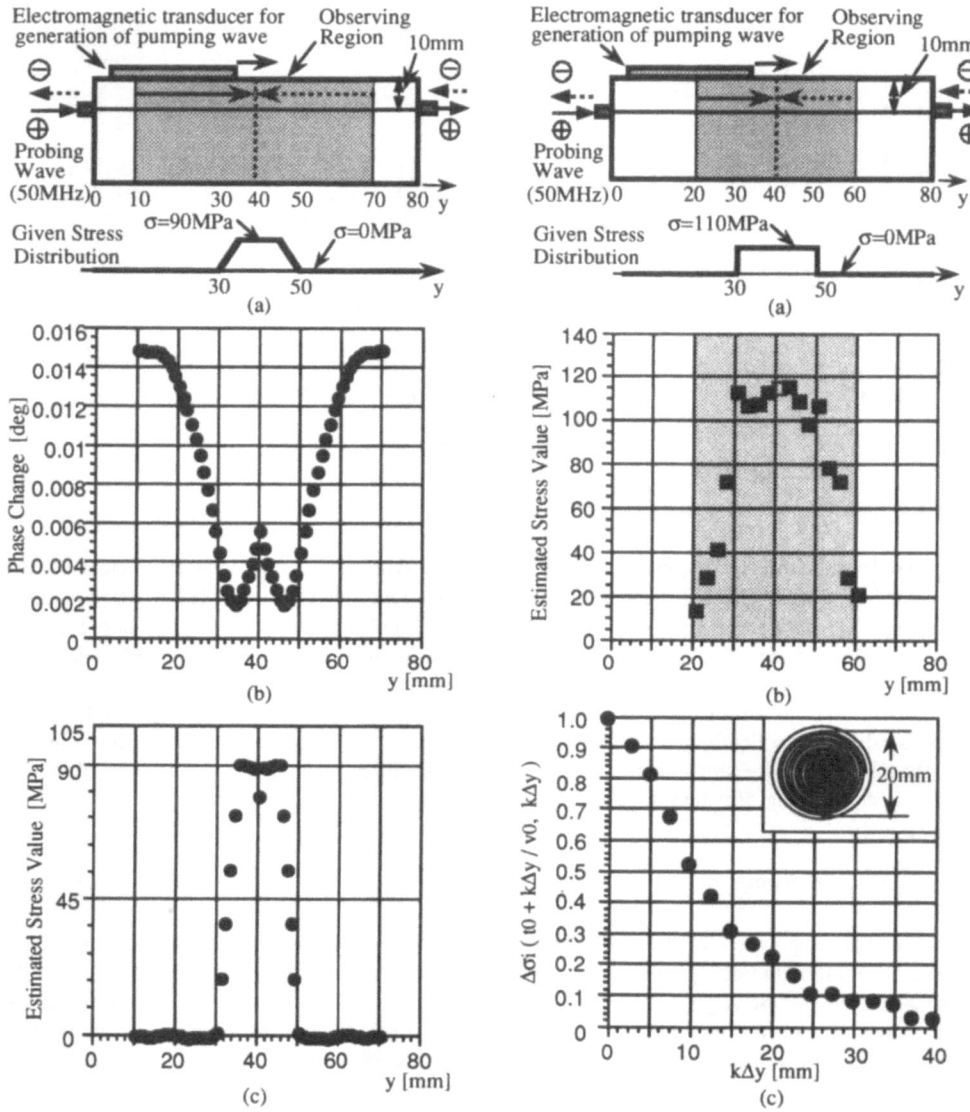

Fig. 5 Computer simulation. (a)Simulation conditions. (b)Calculated phase change. (c)Reconstructed stress distribution.

Fig. 6 Experimental result. (a)Experimental conditions. (b)Estimated stress distribution. (c)The effective perturbing stress distribution obtained by using a disk type transducer (20mm in diameter).

REFERENCE

1. T. Sato, W. Ma, H. Ninoyu, K.Y. Jhang and Y. Kosugi, Estimation of the stress state inside metals using stress perturbing waves and probe waves, NDT&E International 26(3):119-126 (1993).

REAL-TIME FLAW CLASSIFICATION IN WELDS
WITH THREE DIMENSIONAL ULTRASONIC IMAGES

C.G.Windsor[1] and L. Capineri[2]

[1] AEA Industrial Technology, Harwell Laboratory
   Didcot, Oxfordshire OX11 0RA, UK
[2] Dipartimento di Ingegneria Elettronica, Universita' di Firenze
   Via S. Marta 3, 50139 Firenze, Italy

INTRODUCTION

Defect characterisation problem in large welded components, such as pressurised vessels or pipes, is of practical importance for the evaluation of the structural integrity and lifetime. Generally ultrasonic imaging techniques are used for non destructive controls of welds in mechanical structures. The detection of meaningful defects is still largely made by operators with visual inspection of the ultrasonic image series. Even the most experienced operators suffer from fatigue and loss of concentration and the final response is affected by subjectivity of the analysis and lack of consistency.

This work investigates classification methods able to assist automatic defect characterisation with potentials of fast response and for reducing the subjectivity of the classification process. Here we present the results obtained on laboratory defect samples with classifiers based on the adaptive receptive field (ARF) method as a conventional classifier and a multi-layer-perceptron (MLP) neural network. The two classifiers are included as new modules in the software of a ZIPSCAN instrument for testing an on line demonstrator.

Finally we discuss the advantages of computer aided system in which morphological information of the weld geometry and the a priori knowledge of the position of certain defect types are merged together for increasing the classification reliability.

In a previous work of Burch and Bealing[1] several types of defects were identified in welds but the most important factor is to discriminate between the dangerous crack and the more innocuous volumetric defects. The different types can be grouped into four classes: (i)porosity and (ii)slag for the volumetric defects, (iii) smooth cracks and (iv)rough cracks for the crack like defects.

DATA COLLECTION FOR 3-D ULTRASONIC IMAGES

Investigations on the reflected signals from various type of defects revealed that additional information of the defect type may be obtained with the analysis of the backscattered signals from the entire volume of the defect and over different inclinations.

It's found necessary to use three or four dimensional images, where the third dimension is the depth of the defect beneath the plane of the receiving aperture, and the fourth dimension is the angle of the incident ultrasound beam. A set of three-dimensional (3D) images were formed by stacking a set of parallel B-scan acquired at different position along the weld direction Y (see Figure 1). The fourth dimension is provided by the acquisition with different probe angles usually with 20° difference. A data set with of four dimensional image of real defects was collected by Burch et al.[1]. In that work the defects of known type were introduced artificially into metal test pieces. The data took the form of three-dimensional ultrasonic reflected intensity as a function of position X, Y and depth Z in the material. After the data acquisition a preprocessing of raw data is essential to obtain a compact version of the input ultrasonic image suitable for subsequent classification by a neural network.

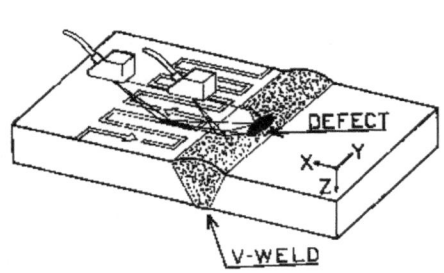

Figure 1 Ultrasonic scanning system for three/four dimensional acquisition: the 3D image is formed by compounding the B-scans.

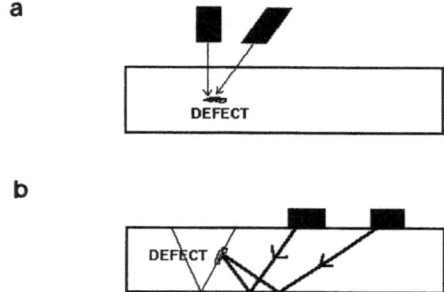

Figure 2 Inspection geometry: (a) with single immersion probe on planar defects, (b) with 2 angle contact probes for V welds.

The first step for setting up a classifier is the collection of a real dataset that is used for training and testing. In this work are considered two types of steel blocks containing artificial buried defects:

1) the first type is a series of 66 planar defects at a given depth from the top surface. The direct pulse-echo reflected signals were acquired at 0° and 20° from the vertical. The measurements were carried out with a single immersion probe above the defect, coupled by water to the metal specimen (see Figure 2a) and calibrated with a side drilled hole.

2) the second series is formed by 38 defects in V-welds. The forward and the backward propagation of ultrasonic pulses is always via the bottom wall reflection. The movement of the probes along the X-Y directions allows the inspection of the whole defect volume (see Figure 1). In this case we used shear wave contact probes with central frequency 5 MHz with two different inclinations: the first probe angle is the same of the V-weld in order to have specular reflections from defects along the weld surface and the second angle with an offset 20-30° from the main one (see Figure 2b). To accommodate different V-weld inclinations we devised the following angle pairs for the acquisition: 60°-45°,60°-70°,70°-45°.

PREPROCESSOR OF ULTRASONIC DATA

From the above considerations a preprocessor program was developed to create a set images in a standard format. The main task of the preprocessor is the conversion of the original images into standard size images centered on the defect volume with the most appropriate resolution for classification. The centering operation is based on the definition of the centre of gravity of the original image. Once this characteristic point is found, the original image is converted to a new standard size image with its center onto the centre of gravity.

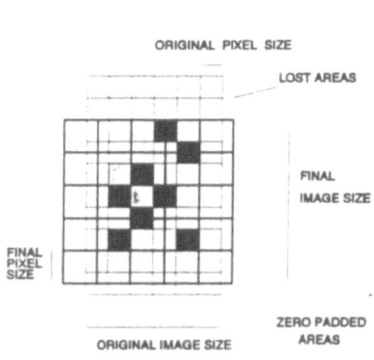

Figure 3 Averaging and centering task of the original ultrasonic image (raw data) for the two dimensional case.

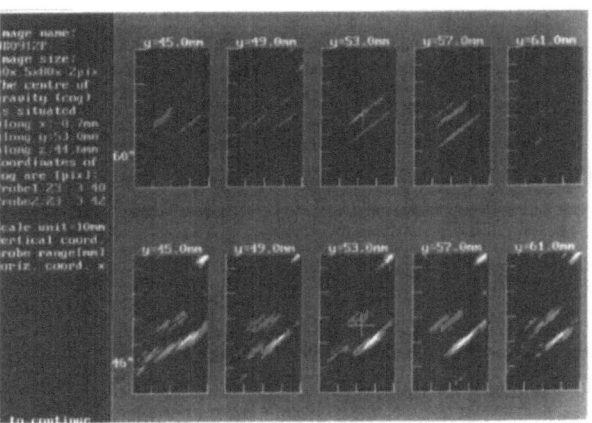

Figure 4 Graphic display output of the preprocessor. The series of 5 B-scan images at top and bottom containing a porosity are acquired at 60° and 46° respectively.

The following process places a square scratch array of standard size on the image with the central pixel over the centre of gravity of the original image.

In Figure 3 there are illustrated, for a two dimensional image, the three processes of centering, placing and averaging carried out by the preprocessor. We can observe that areas of the original image not contained by the new array are truncated and wasted. Areas of the new image that are not filled by the original image are padded with zeros.

The optimum voxel size for averaging should be large as possible while preserving the essential features of the image.

The last operation of the preprocessor is the amplitude normalization of the ultrasonic images made with rectified radiofrequency signals. It is assumed that only the image intensity distribution retains the essential features and the phase information is neglected. Images are then formed with detected ultrasonic signals digitized with 8 bit resolution. The peak intensity of each image was normalized to a constant value.

The graphic display output of the preprocessor on a high resolution color monitor is shown in Figure 4 (colors are here converted with grey scale). We developed this type of integrated presentation for displaying 3D information and for comparing the defect responses at different angles of ultrasound. In the example the investigated defect is a porosity artificially introduced in a V-weld of a 25 mm thick steel block. The upper and the lower series of 5 images correspond respectively to 60° and 46° ultrasound beam angle. The white cross on central images indicates the position of the calculated centre of gravity.

RECEPTIVE FIELD METHODS FOR DIRECT CLASSIFICATION OF REAL IMAGES

At an early stage of this work Burch et al.[2],Windsor et al.[3,4] tackled the problem of defect characterisation using conventional and neural network classifiers based on feature extraction. However the classification in the feature space revealed two main drawbacks for solving our problem: 1) *the advantages of a fast classification offered by the MLP neural network are minimised because the computational demand for the feature extraction is too long (tens of minute on a VAX workstation)*, 2) *the choice of the best set of feature is based on the past experience but remains subjective and may change according to the particular problem.*

A possible solution comes by direct classification methods based on the concept of the <u>receptive field</u>. The method of receptive fields was already used by Fukushima in his neocognitron neural net model and applied to character recognition[6].

In this work we used two different types of classifiers: an adaption of MLP for the receptive field as a neural method and an adaptive receptive field as a conventional method. Outlines of the operation principles of the above methods are reported below[5].

The Receptive Field Multi-Layer-Perceptron Method.

We used a variation of the basic form of a MLP neural network in which the inputs are given by the pixels of the receptive field (see Figure 5). The network was trained in the supervised learning mode with 4 outputs (one for each class), one hidden layer with a variable number of units between 3 and 6.

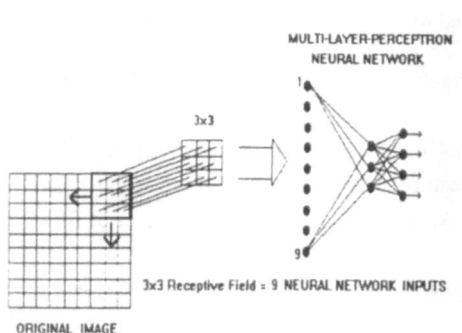

Figure 5 The receptive field adaption of multi-layer-perceptron neural network: example for two dimensional case.

Figure 6 Display output of the Adaptive Receptive Field classifier. The image represents a rough crack. The four characteristic fields are shown at the bottom.

Here the advantage of using a receptive field is the reduction of the adjustable weights of the network. Previous tests shown that large input images led to large difference between the number of adjustable weights and the number of distinguishable pixels in the image causing the problem of overfitting in the learning phase.

In this work we investigated two different methods for the selection of a portion of the image: the first stops on the portion of input image with highest central pixel of the scanning window, the second verifies also where the sum of outer pixels in the receptive field is maximum. Though the two methods gave similar results the first one has the advantage to be faster.

The Adaptive Receptive Field Method.

In this method four receptive fields, one for each class, are calculated using an iterative procedure. The classifier was trained and tested with the database of 66 ultrasonic images containing various types of planar defects. An initially random receptive field was swept across the image containing one defect type. The receptive field stops with the centre on the position where is found the minimum distance, calculated in a pixel by pixel fashion, between the current receptive field and the underlying image. The receptive field is updated by adding a percentage of the selected part of the image to the current field. This process is repeated iteratively by scanning over the training set for each class until a stable field is obtained. The same process is repeated for each defect type.

During testing in turn each of the four characteristic fields is scanned over the image. The class whose field gives the best fit is chosen. The receptive field method is non sensitive to spatial variance because the center of the scanning window is chosen on the local minimum distance criterion so that it preserves the relative amplitude variations of adjacent pixels.

Figure 6 shows an example of the graphic display set up for the adaptive receptive field classifier. The preprocessed images with a coarse voxel size (2x2x2 mm) represents a rough crack. On the bottom are shown the four receptive field for both angles (3x3x3x2 pixel). They corresponds respectively from left to right to porosity, slag, smooth crack and rough crack. The white square rectangle superimposed to the images, indicates the position where the best match is found for the current receptive field. On the left side are reported numerical information about the image under test and classification results.

Table 1. Performances with database of images with amplitude calibration

Method	Type	Rec. Field size [pixel]	Speed [s] 386SX/16MHz	% Success Rate
Adaptive Rec.Field	Conventional	3x3x3x2	10	93.9
MLP + Rec. Field	Neural Network	3x3x3x2 6 Hidden Units	4	93.9

RESULTS

The performances of the two type of classifiers for the characterisation of ultrasonic images from data base with 66 planar defect samples are summarized in Table 1.

Other tests are performed on a set of 38 images of various type of defects buried in V-welds with the classifiers trained on the previous data set. The performance of the two classifiers are shown in Table 2. In this case the contact probe amplitude calibration was not taken into account during acquisition.

Table 2. Performances with 38 defect samples in V-welds without amplitude calibration

Method	Type	Rec. Field size [pixel]	Speed [s] 386SX/16MHz	% Success Rate
Adaptive Rec.Field	Conventional	3x3x3x2	10	73
MLP + Rec. Field	Neural Network	3x3x3x2 6 Hidden Units	4	48

These results show that the good success rate obtained the data base of planar defects is not replicated by the set of images acquired in V-welds. The analysis of images relative to misclassified cases pointed out the inconsistency of the average intensity between images at both angles. We identified the influence of several factors on the classification. In different experiments we noticed a lack of coupling of contact probes due to the roughness of the weld surface. Afterwards the amplitude calibration of the two angle contact probes revealed a different sensitivity in order of 2-3 dB. This minor amplitude correction was sufficient to recover some of the misclassified defects. However, in real applications we must considered the different attenuation of the two angle probes due to the different ultrasonic path length and the angle dependent attenuation for anisotropic material as well. Additional difficulties are due to the different sizes and resolutions between the planar and the inclined defect series.

The classification time of the neural method is in the order of four seconds on a processor 386SX/16 MHz without math coprocessor nor source code optimization and it is at least twice faster than the adaptive receptive field method. Both methods are fast enough for an on-line performance. In fact the data acquisition time is in the order of several minutes for a three dimensional scan.

MERGING WELD GEOMETRY AND ULTRASONIC 3D IMAGES INFORMATION FOR DEFECT CHARACTERISATION

A tricky part of the classification of V-weld is the discrimination of cracks from ultrasonic signals coming from geometrical weld reflectors such as the V-weld root. In order to improve the reliability of the defect characterisation method, we developed software that shows to the user the weld sketch in two dimensions superimposed to a B-scan image in real coordinates. The aim of this type of presentation is to use morphological information about the weld geometry and the position of certain type of defects inside the weld. Especially in doubtful classifications the user recognizes easily the class of a defect by visual inspection: for example a lack of fusion has the highest probability to stay on the weld separation line instead of the bulk of the weld. This method has also the additional advantage to explain the significance of bright spots due to geometrical weld reflectors.

An example of the display output is illustrated in Figure 7 representing one of the B-scans relative to the porosity defect in Figure 4. In Figure 7 is clearly individuated the top surface of the weld and the weld root by the high intensity pixels of the image in real coordinates while the region of porosities is in the middle of the weld. Likely the same bright points when shown with conventional B-scan images as in Figure 4, can be misinterpreted as defects by a poorly trained user.

We envisage further investigation of the possibility of data fusion of a priori morphological information of the defect with a computer aided system containing a geometrical description of the weld.

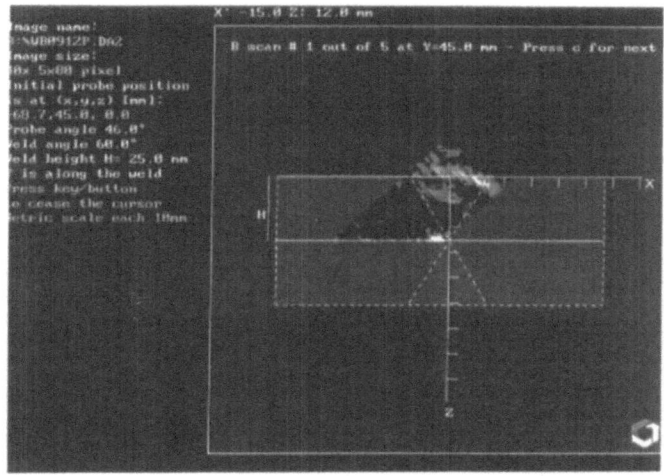

Figure 7 Graphic display of a B-scan superimposed to a sketch of the V-weld in real coordinates.

ACKNOWLEDGMENT

The authors wish to thank the British Council for supporting this research work.

REFERENCES

1   Burch S F and Bealing N K, A physical approach to the automated ultrasonic characterization of buried weld defects in ferritic steel, NDT International, 19, 145-153 (1986)

2   Burch S F, Lomas A R and Ramsey A T, Practical Automated Ultrasonic Characterisation of Welding Defects, British Journal of NDT, Vol 32, No 7, July pp. 347-350, (1990)

3   Baker A R and Windsor C G, The classification of defects from ultrasonic data using neural networks: the Hopfield method, NDT International 22, 97-105 (1989)

4   Windsor C G, Neural Networks from Models to Applications, ed. Personnaz and Dreyfus, IDSET, Paris, 592-601 (1989)

5   Windsor C.G., Anselme F., Capineri L., Mason J.P., The classification of weld defects from ultrasonic images: a neural network approach, The British Journal of Non-destructive Testing, Vol 35, No.1, pp.15-22, (1993)

6   Fukushima K and Miyake S, Neocognitron: A new algorithm for pattern recognition tolerant deformations and shifts in position. Pattern Recognition, 15, pp. 455-469, (1982)

ACOUSTIC EMISSION DATA FROM PULL-OUT TESTS OF REINFORCED CONCRETE ANALYSED WITH RESPECT TO PASSIVE US-TOMOGRAPHY

Christian U. Grosse[1], Hans W. Reinhardt[1], György L. Balázs[2]

[1]Forschungs- und Materialprüfungsanstalt Baden Württemberg
Pfaffenwaldring 4
D-70569 Stuttgart, Germany
[2]Department of Reinforced Concrete Structures
Technical University of Budapest
H-1521 Budapest, Hungary

ABSTRACT

A series of pull-out tests with different load histories have been conducted in order to study the cyclic bond behaviour of a reinforcing bar in concrete. In parallel, the acoustic emissions (AE) were recorded with an array of 8 piezo-transducers in that way that the broadband waveforms were stored. A number of 50 up to 300 events were recorded during the different experiments. The analysis of our data shows an interesting pattern (in time and location) of the AE events. To interpret the fracture behaviour we compared the waveforms and the localizations of the events with our models. A couple of typical AE waveforms were found, which can be interpreted as the signals of different fracture processes. The similarity of the events of one cluster was checked by calculating the coherence of the signals. We found that the coherence was high in the frequency range below the noise level (< 300 kHz).

Our interest was focused to the passive US-tomography, that means the investigation not only of the 3D-time delays, but of the locations of the events at the same time. As a first step the problem of the localization of acoustic emissions in concrete under the assumption of homogeniety and isotropy was solved. In a second step we used the standard deviations of the traveltimes to evaluate significant lateral or horizontal inhomogenieties. At present we are working on a comprehensive 3D-tomography algorithm to improve the accuracy of the localization data.

INTRODUCTION

It has been stated (Grosse and Reinhardt, 1992a), that the localization of the sound sources is an essential basis for the analysis of AE data. Only in this way an analysis of damage progress, a study of the fracture mechanisms, and a discrimination of background noise is reliably practicable. Although several authors have already worked successfully on the 3D-localization of acoustic emissions (Ohtsu, 1991; Berthelot, 1987; Labuz, 1988), the appropriate localization algorithms either were not available or not transferable to our problem. That is why we developed an appropriate software in co-operation with the Institute of Geophysics at the University of Karlsruhe (Dr. Lani Onescu). This program was tested during pull-out tests performed within a DFG-project for the research of the bonding behaviour of stell in concrete. The high data quality enabled us to apply a number of comprehensive methods of analysis. Before a short description of the program, we want to decribe the test setup and some fundamental results of the mechanical and ultrasonic measurements.

PULL-OUT TESTS OF REINFORCED CONCRETE

During the DFG-Project we were able to conduct some US-measurements and to test the **Hypo**[AE] software (see below) for the first time. Altogether, 14 pull-out tests on test cubes with an embedded steel bar were carried out, applying monotonic, as well as cyclic and long term loads. The test setup is shown in Figure 1a. The concrete cube with a side length of 10 cm was reinforced with a steel bar of 1.6 cm diameter. The bonding length was chosen as 2 cm according to the double rib spacing of the steel bar (Figure 1b). In addition to the commonly registrated mechanical parameters like tensile force (resp. bond stress) and slip (registered by an LVDT), the occuring acoustic emissions were recorded by 8 transducers. The monitoring system used in these tests has already been described previously in detail (Grosse and Reinhardt, 1992a).The AE-signals were recorded by 8 piezoelectric transducers (broadband and resonant, resp.) and amplified by 8 preamplifiers. They were converted and stored digitally by an 8-channel transient recorder with a sampling rate of 1 MHz and an amplitude resolution of 12 bit.

Analysis of the steel-concrete interaction

In each test we could record 50 to 300 events. The duration of the tests varied between 5 minutes and 3 days, according to the load history. First, the data was evaluated statistically with regard to the amplitude distribution. In this method, which is standard in acoustic emission technique, the maximum amplitudes of the AE signals are determined automatically and plotted per time unit.

The resulting histograms give a first idea of the energy transformed into acoustic signals during the test and, hence, of the fracture processes occuring inside the test specimen. Because of the automatic evaluation, however, the deficiency of this method is that the transducer location, the different wave modes, and background signals were not considered. It turned out that the histograms of the 8 transducers generally look rather different, according to the distance between the source of the signals and the actual transducer. This represents another argument for a localization of the AE-sources. Concerning the evaluation of the pull-out tests with regard to bond stress, slip, and the statistical analysis of the AE data, we may refer to earlier papers (Balazs, Grosse, Koch and Reinhardt, 1993).

LOCALIZATION OF ACOUSTIC EMISSIONS BY HYPO[AE]

The program **Hypo**[AE], for the most part, is based on an idea developed by Ludwig Geiger (of the math.-phys. class of Emil Wichert in Goettingen 1910). Also at that time, as usual in seismology, the epicenters of earthquakes have been determined by evaluating the arrival times of the "first and second precursor", this means of compression and shear wave. Carrying out the localization only with the arrival times of N station, Geiger solved this problem. His reflection:

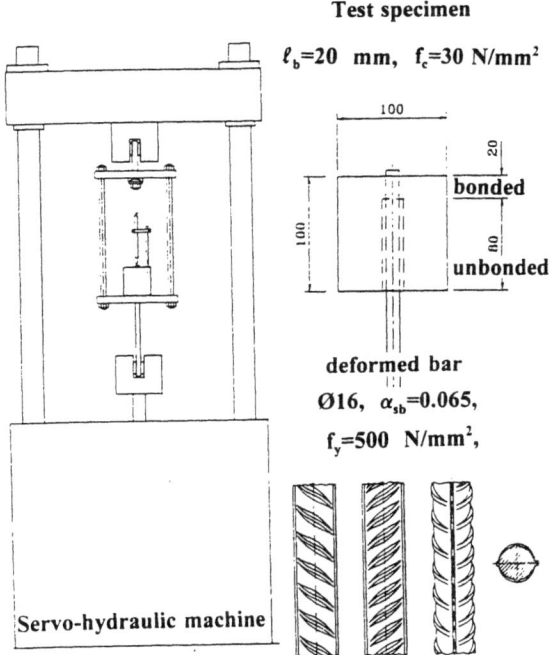

Test specimen

ℓ_b=20 mm, f_c=30 N/mm^2

100

20

bonded

100

80

unbonded

deformed bar

Ø16, α_{sb}=0.065,

f_y=500 N/mm^2,

Servo-hydraulic machine

Figure 1. Test setup and concrete specimen
         a) servo-hydraulic testing machine
         b) cubic specimen with partly bonded reinforcement
         c) details of the rib pattern and the reinforcing bar (BSt500)

"If it were possible to give a method of center determination that is merely based on the arrival times and the travel time functions of the first precursor, the center of every earthquake could be determined with high precision." The idea invented by Geiger and the proceeding described in the 'Bericht der Koeniglichen Gesellschaft der Wissenschaften zu Goettingen' (Geiger, 1910) has been extended later to the localization of the hypocenter, this means the real location of the earthquake in the depth. In our days, it is still in use, being adapted to the modern numerical possibilities of evaluation (e. g. Buland, 1976). It describes the solution in general by the data of N stations. For the determination of the hypocenter at least 4 stations are necessary; in this case, the problem is determined unequivocally and yields a certain position as center and a certain time as time of the event. Having more than 4 stations, the problem is overdetermined. Thus it it possible to realize a calculus of observations using the method of least error squares.

Adaption to the acoustic emission technique

We had to adapt this algorithm to our requirements. Unlike in geophysics, in material testing with acoustic emission methods it is possible to fix the stations - in our case the transducers - also below the sound source. While in geophysics often a "plain-surface-projection" is performed, this is impossible here, because source and receiver may be arranged in any position. Our algorithm enables the specimen geometry, the number of transducers, and their arrangement to be chosen optionally. Besides, in addition to (or instead of) the compression wave arrival times, also the arrival times of the shear waves can be used for the evaluation. When there are more than 4 seismograms available, the focus calculation is performed numerically by an iteration algorithm. The calculation itself can be influenced in different ways, for example:

- different weighting of each station
- different convergence criteria
- causality conditions
- storing and using of station corrections

As the latter possibility is a useful help for improving the results, we want to describe the particulars. Usually, the program **Hypo^AE** localizes an AE signal of which the arrival times of N stations have been evaluated. After the numerical computation of the focus, the difference between the theoretically determined arrival time and the one that was read out from the seismogram is calculated for each single station. This error, the so-called O-C value (*Origin minus Calculated*), represents the determined deviation from the iteratively calculated focus for the event and the station in question.

Thus, the option "station corrections", which is an automatic routine, can be used to consider the systematic errors of a station (transducer). These errors affect all events in the same way. However, therefore it is necessary to record several events. The more events can be evaluated, the better the systematic error can be determined. The program **Hypo^AE** uses a data file containing the arrival times of all AE data. For all events, it carries out a localization and stores the average of all O-C values into another file. After that, it repeats the localization of all foci, but this time considering the average station corrections determined during the first iteration.

Input and output of data is accomplished using ASCII textfiles that can be edited by every commercial editor. The program's conception permits a maximum control over the numerical calculation. This requires the possibility to survey the single iteration steps on the screen. For each iteration step, the following information is shown:

1. The first column indicates the number of the iteration step.
2. The second column shows the time delay of the focus time in msec compared to the one of the last iteration. The iteration starts with the arrival time values that are

farthest in the past. The sum of all these time delays ("dt") yield in the end the theoretical travel time of the wave between focus and nearest transducer.

3. The 3rd, 4th, and 5th column represent the appropriate changes in the x, y, and z-coordinates (in m) from one iteration to the next one. Here the iteration starts in the geometric center of the specimen, as specified before.

4. In the sixth column, the standard deviation ("SD-value") of the iteration step is given.

5. Finally, the seventh column shows the number of station informations used in this step of calculation.

After the calculation being completed, the result is shown on the screen (Table 1) and can be printed. The specification of the specimen geometry that was mentioned before, is also accomplished by an ASCII textfile. In this context, the input of the wave velocity is of eminent importance. Usually, we determine it directly in a number of ultrasonic transmissions. As shown before (Grosse and Reinhardt, 1992b), however, it is also possible to determine the compression wave velocity with suffient accuray by the impact-echo method, or at the surface, using two transducers. In addtition, the program **Hypo**[AE] gives the user a helpful possibilty to control the velocity measurement. After the input of the "model-velocity", the localization for all events of the test is performed. Now the calculated travel path as a function of the measured travel time or each transducer can be plotted in a diagram. The gradient of the resulting straight line gives a new velocity that fits better to the real velocity. Certainly, it's possible to perform the calculation several times as an iteration. In the tests described below, however, the error reached a value of less than 1 % already after the first iteration.

Three-dimensional localization of acoustic-emission-sources

Additionally, a series of other results have been attained by the localization of the AE data using the **Hypo**[AE] software. Before we enter into the particulars, it is important to give a description of the localization accuracy. In these tests on rather small test cubes, it was limited especially by the sampling rate of the A/D-converter.

Table 1. Result of an iteration by program **Hypo**[AE]. The first two lines represent the results of the localization (time and focus X,Y,Z), the third gives the calculated errors resp. the number of iterations and the following shows the results for each single station. The O-C value gives the standard deviation of this particular station for the certain event. The "INFO"-column represents the importance of the different stations for the iteration (1.00 is highest).

EVENT-ID	ORIGIN-TIME	EV X	EV-Y	EV-Z	MD	SD	NL/NM
027	-35.33µs	.052m	.047m	.055m	.0	4.51	8/ 0
ERRORS :	2.11µs	.011m	.009m	.013m	IT: 13		

STATION	R	AZM	EMG	ARRIVAL-TIME	WAVE	O-C	WT	INFO
1 S3	.052	248	78	-23.00	P	-2.61	1.00	.86
2 S2	.058	160	73	-20.00	P	-1.36	1.00	.78
3 S4	.059	294	120	-14.00	P	4.23	1.00	.80
4 S5	.061	65	73	-16.00	P	1.63	1.00	1.00
5 S1	.064	203	119	-17.00	P	-.18	1.00	.52
6 S8	.064	203	149	-16.00	P	.79	1.00	.57
7 S6	.065	114	116	-13.00	P	3.68	1.00	.56
8 S7	.071	48	141	-21.00	P	-6.17	1.00	.70

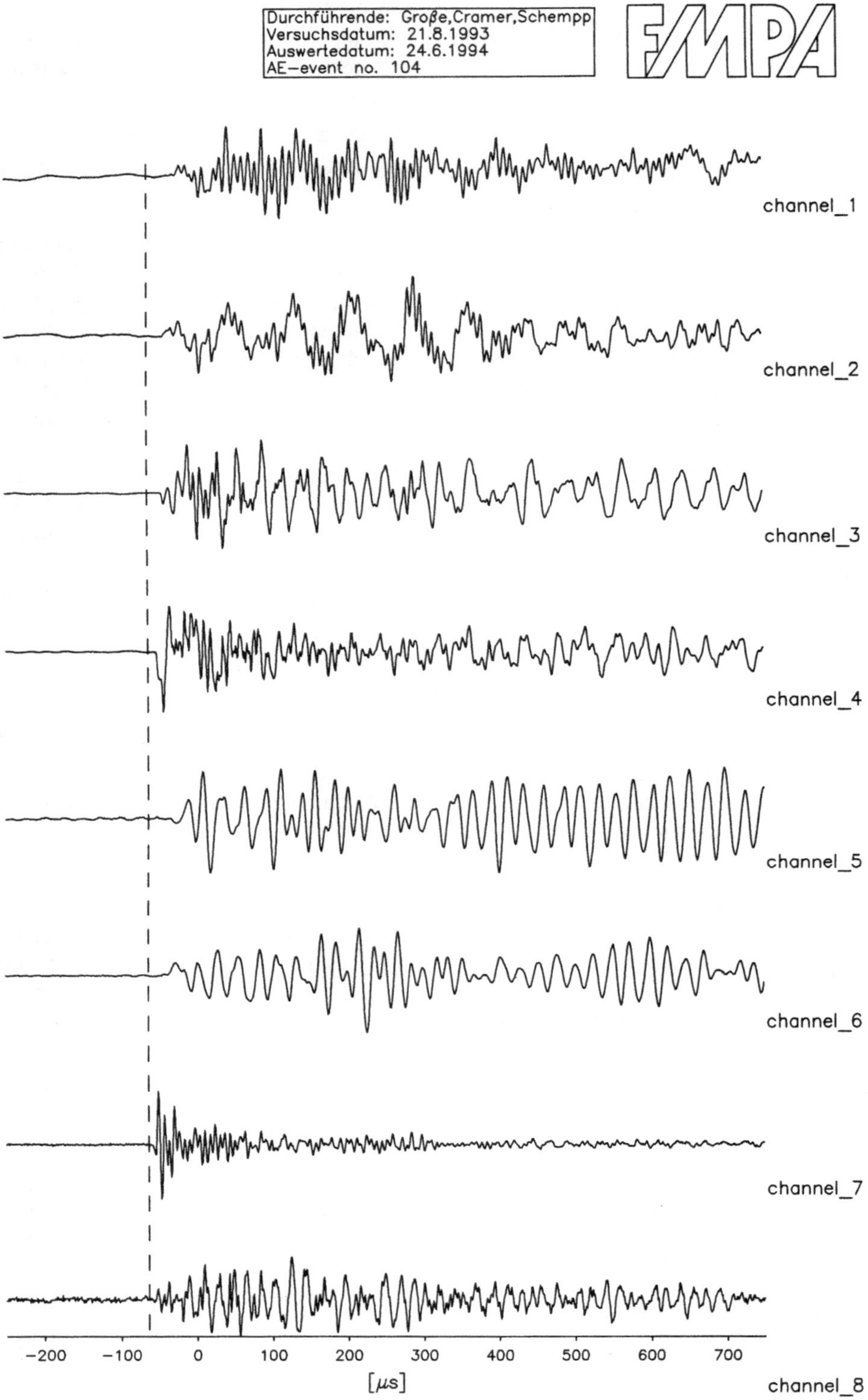

channel_1

channel_2

channel_3

channel_4

channel_5

channel_6

channel_7

-200   -100   0   100   200   300   400   500   600   700

$[\mu s]$

channel_8

Figure 2. Seismograms of acoustic emissions during the pull-out tests; channel 1 to 8. Unfiltered and broadband seismograms of event No. 104.

Figure 2 shows the seismograms of a typical acoustic emission event during a pull-out test. For the evaluation, the arrival times (in μsec) of the compression waves of all channels have been read off. Presuming a good signal-to-noise ratio, the accuracy in reading off the arrival times is limited particularly by the sampling rate (= horizontal resolution) of 1 MHz of each channel of the transient recorder. Hence follows a bandwidth of 500 kHz and, with an average wave velocity of 4 km/s in concrete, an accuracy of 8 mm. The statistical inaccuracy due to the iterative computation of the hypocenter as shown before is far below this value.

Figure 3a shows the localization of a typical pull-out test. First, Figure 3a indicates the three-dimensional location. In the cartesian coordinate system, the bond length is located at the bottom side of the cube. Hence, the steel bar, located in the middle of this picture, was pulled out upwards. The large points indicate the coordinates of the 3-D localization in the cube, while the small points indicate the projections on the planes. As expected, most of the sound sources are located near the bar or in the area somewhat above, which is the area of highest load amplitude in the cube. Looking especially at the x-z-plane, one realizes chains of points starting from this cluster and running to the right and the left, respectively, reaching the upper border of the picture. These points are obviously marking the area of cracks which are leading, in this case, to the failure of the specimen. The cleavage faces correspond very well with these localizations.

Analysing the x/y projection, which is separately shown in Figure 3b, further interesting details become obvious. First, the area without sound sources in the very middle of the graphic is striking. As we expect no sources inside the steel bar, the events localized in the inner shell zone of the bar surely originate directly from the surface of the bar. However, they are still within the limits of error. Besides, a certain eccentricity of the data with crucial point on the upper left-hand side and the lower right-hand side of the bar are observed. Being compared with Figure 1c, it becomes evident, that this is the image of the ribs on the steel bar, also having the shape of an ellipsoid. The ribs cause a considerably higher acoustic activity, because also the load level is higher in these areas.

INVESTIGATION OF HETEROGENEITIES WITH PROGRAMM HYPO^{AE}

Another simple means of analysis provided by this program shall be presented in conclusion. To begin with, we investigated the influence of the bar area to the accuracy of the localization. Therefore, we looked at the statistically determined errors of the stations that were lying exactly in the shadow area of the bar. And in fact, just these transducers showed significantly delayed arrival times - a residual - compared to the calculated times (O-C values). This means, an AE signal arriving later at the transducer, is lying in the shadow zone of the bar. Consequently, we carried out the localization with **Hypo**^{AE} one more time, giving a lower weight to these stations during the iteration. This procedure gave much smaller errors in most of the localizations.

Of course, this knowledge can also be applied in another way. Figure 4 shows once more a graphic of an x/y projection. On the margins, the transducer positions are drawn (transducers 1, 3, 5). The localizations showing a distinct residual ($\geq$ 5 μsec) after the iteration have been accentuated by filled symbols according to the transducers in question. The steel bar causes a clear "residual-tail". With a sufficient number of transducers and a good shadowing of the specimen, thus a rough picture of heterogeneities inside the specimen can be given. In contrast to conventional tomographic methods, additionally the source of the sound signal has to be inverted. This renders the inversion of the heterogeneity even more difficult. In the course of a research work, we are actually working on this topic at the FMPA. The presented method describes a possible way to find significant inhomogenieties roughly but very fast. In most cases it is not necessary to increase the accuracy by an exact US-tomography. Especially for our data the localization accuracy was improved, what led to some further results.

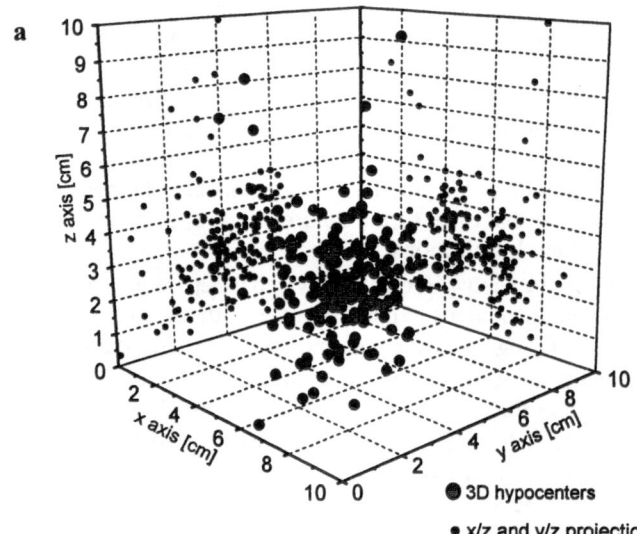

a

3D hypocenters

x/z and y/z projection

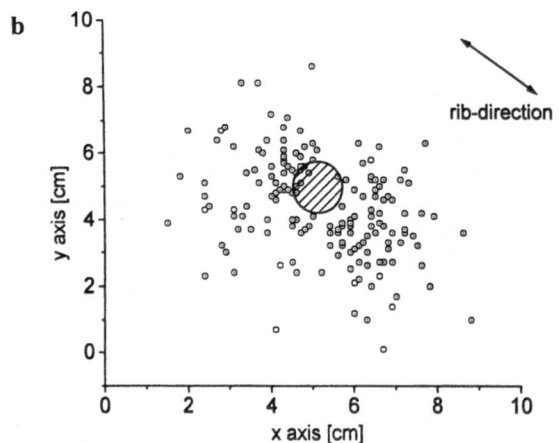

b

rib-direction

Figure 3. a) 3-D Locations of AE-Events.
b) x/y Projections and Steel Bar.

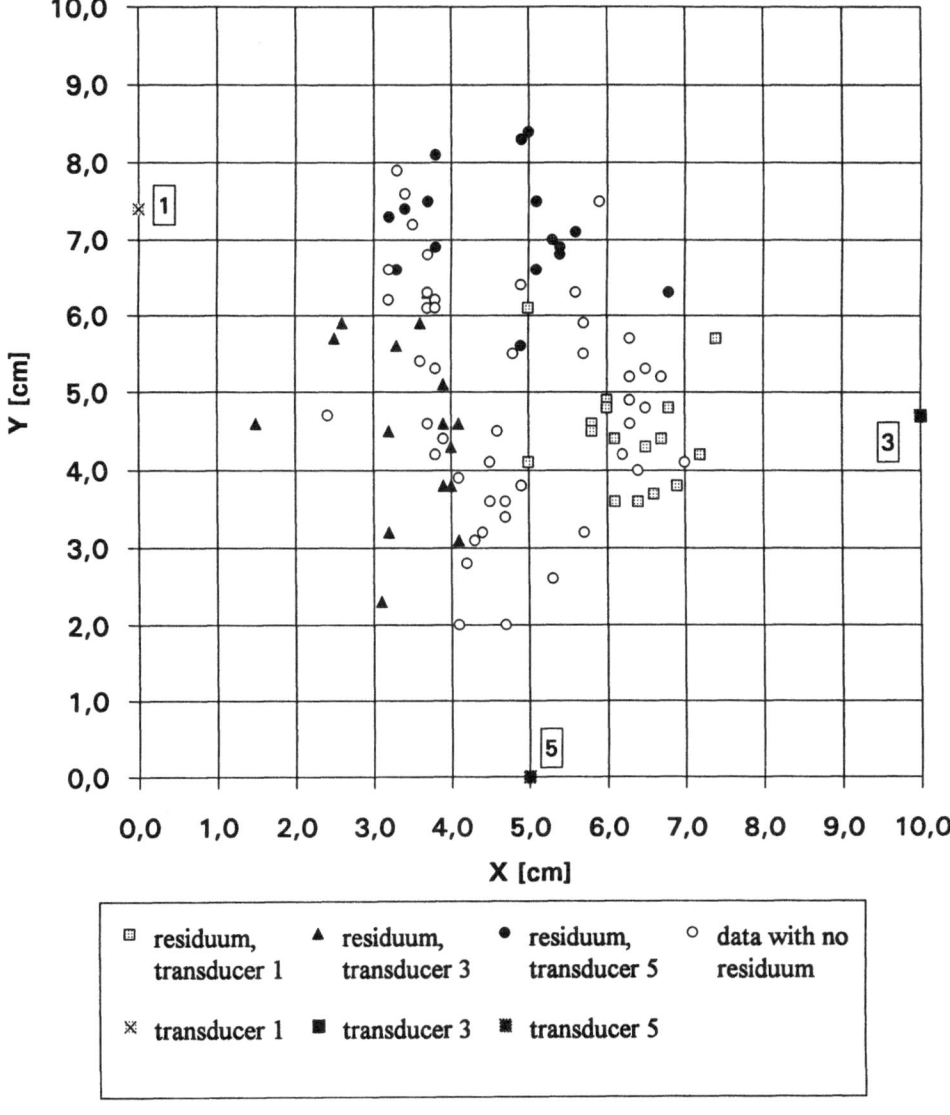

Figure 4. Travel-time-residuals (≥ 5 μsec) in the x/y plane; shadow-zone of the slab.

CLASSIFICATION OF AE-SOURCES WITH THE MAGNITUDE SQUARED COHERENCE (MSC)

With the test setup described above altogether 2500 complete AE events have been observed during 14 pull-out tests. Comparing the time signals (especially those recorded by one and the same transducer), the following was observed. During one experiment there are groups of signals with a similar or almost the same waveform. Other signals show significant differences. To explain the term 'similarity', in Figure 5 another AE-event is plotted. Compared to the event shown in figure 2, it is clear to see that the signals of all channels matches very close even in the coda of the P-wave (that means the oscillations following the first puls of the compression wave).

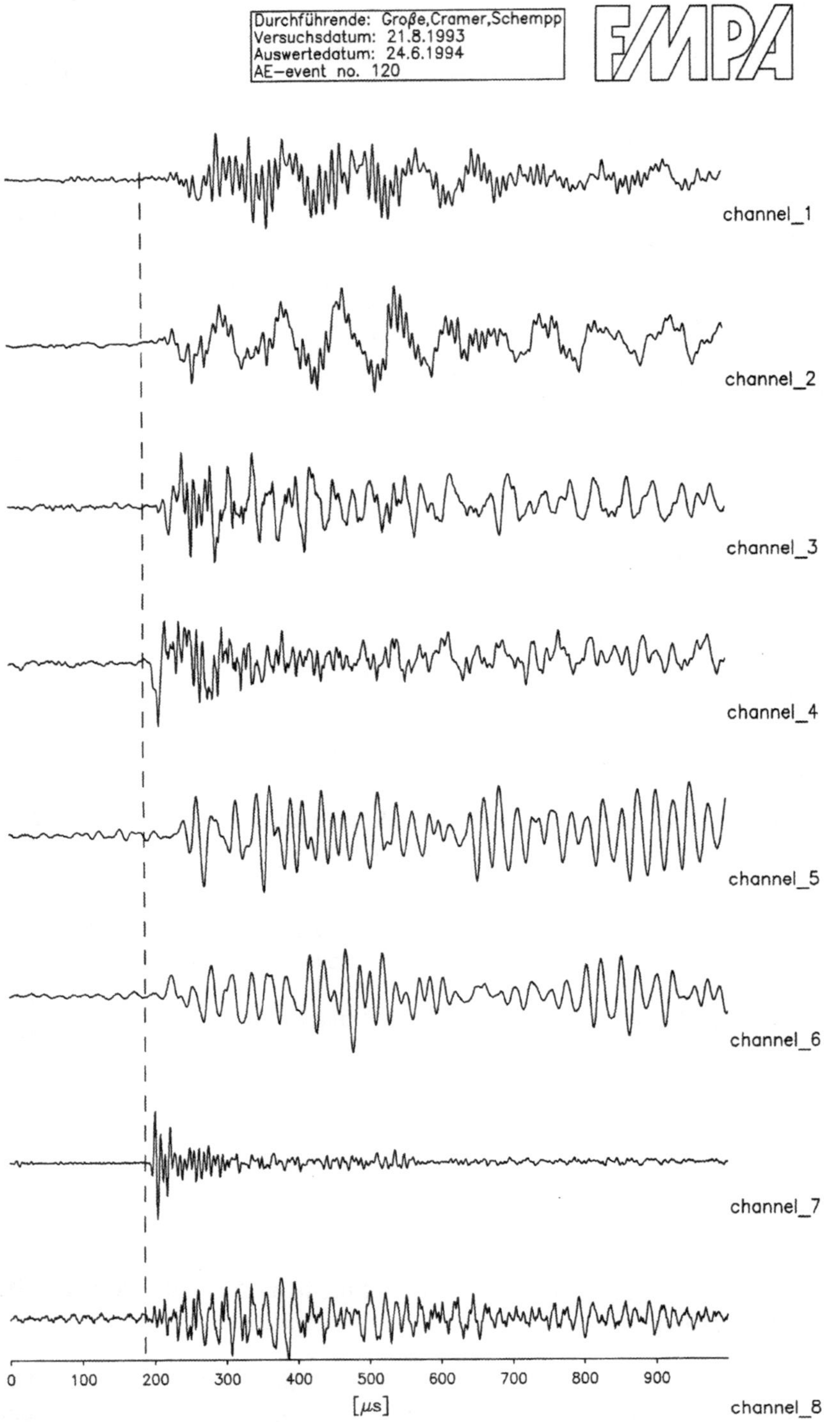

channel_1

channel_2

channel_3

channel_4

channel_5

channel_6

channel_7

| | | | | | | | | | |
|0|100|200|300|400|500|600|700|800|900|

$[\mu s]$

channel_8

Figure 5. Seismograms of acoustic emissions during the pull-out tests; channel 1 to 8. Unfiltered and broadband seismograms of event No. 120.

644

To quantify the above observations, the similarity of two signals X and Y can easily be determined using the mathematical tool called magnitude squared coherence (MSC) (Carter and Ferrie, 1979). The analysis of the signals is based on the discrete Fourier transform (DFT), converting the signals from the time to the frequency domain. It should be stated that besides also the autospectra G_{XX} and G_{YY} and the cross spectral density G_{XY} of the signals has to be calculated. Then the coherence spectrum C_{XY} is defined by the squared cross spectrum divided by the product of the two autospectra

$$C_{XY}(v) = \frac{|G_{XY}(v)|^2}{G_{XX}(v) \cdot G_{YY}(v)} \qquad (1)$$

where v indicates the frequency-domain. A complete derivation of the formulas can be found in Balazs, Grosse, Koch and Reinhardt (1993).

We calculate the *mean coherence* $\overline{C}_{XY}$ of the coherence functions of all comparable channels of two signals. As an example the functions of the mean coherence for two data pairs are given in Figure 6a and 6b. Figure 6a was calculated using the two similar events of Figure 2 and Figure 5. In the frequency band below the noise level (300 kHz), a high coherence is found. A perfect coherence is only obtainted by two identical signals and would result in a coherence of the value one over the hole frequency band. In comparison the coherence function of two not similar events is shown in Figure 6b.

Now it is possible to find a figure that represents the overall coherence. Using the mean coherence functions in figure 6a and 6b, we calculate the *coherence sum* C as the area below the curves using a numeric integration in the range of 0 to 300 kHz. This results ina coherence sum of 229 respectively 72. In this case a perfect coherence would result in a value of 300.

With a transformation of the signals in the frequency domain and the calculation of the coherence functions it is, therefore, possible to find a quantitative relationship between the waveforms of similar signals. On the basis of an exact localization as shown in the chapters above, this method is intended to be used for systematic classification of the AE signals with the automatic deduction of the coherence sum to recognize similarities and differencies in signal pattern very fast.

CONCLUSIONS

The program **Hypo**[AE] enables a three-dimensional localization of acoustic emission signals. It offers a number of possibilities to control the calculation of the hypocenters. The algorithm has successfully been tested during an investigation of the bonding behaviour of steel in concrete. It could be shown on the one hand, that the accumulation of sound sources describes the eccentric ribs of the steel bar, and on the other hand, that the shadow zones of the steel bar cause a residual (time delay). The latter shall be applied within the frame of a passive travel time tomography. This could provide a possibility to detect heterogeneities and could also increase the accuracy of the localization. Ths use of transducers of the same type and an optimized arrangement of them is another possibility for a further improvement.

The similarity of different acoustic emission signals has been proved applying their coherence function. We wanted to verify the assumption, that (apart from the influence of the specimen medium and of the different transducer characteristics) similar source mechanisms result in similar acoustic emission signals. The transfer functions of transducer and medium being well-known, the shape of the AE signal is determined, above all, by the fracture mechanism.

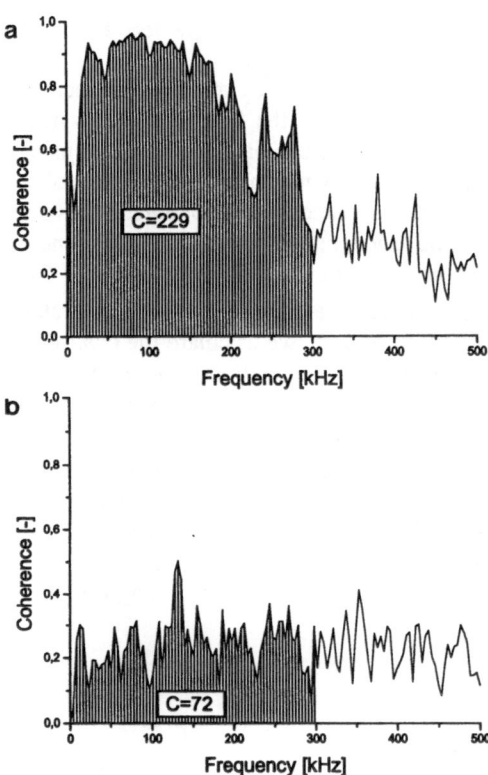

Figure 6. Mean coherence functions of two signals
   a) with good similarity (coherence sum= 229) and
   b) with bad similarity (coherence sum = 72)

This is expressed in the frequency spectrum. The similarity of two signals can be quantified with the use of their Fourier transforms (FFT or FHT, resp.). Therefore, the square of the cross-correlation of both signals is divided by the product of the according autospectra. For better comparability a single value can be extracted from the coherence function by a numerical integration. The <u>coherence sum</u> allows to recognize similarities and differencies in signal pattern very fast.

Another method have been used to increase the information content of the measurement. Although it have not been described before, we want to mention that we have now the possibility to investigate the localizations spatially with the use of a <u>visualization program</u> called **AcroSpin**©. The usual graphic programs (Figure 3) shows the spatial structure of the sources as a 2-dimensional projection. The program, however, provides a pseudo-three- dimensional visualization of the data - similar to an animation. It enables the rotation of the graphic with any speed in realtime, zooming, and the drawing and cancelling of the different layers of the graphic. Further information can be received from the authors.

ACKNOWLEDGEMENTS

The friendly support and engaged help of Dr. Lani Oncescu regarding the programm **Hypo**^{AE} is gratefully acknowledged. The authors are also grateful to Dr. Rainer Koch, Wolfgang Staudenmeier, Wolfgang Albert, Bernd Weiler, Till Cramer and Stefan Schempp for their help during the experiments. The experimental research connected with this topic was partly supported by the German Research Society (Deutsche Forschungsgemeinschaft).

REFERENCES

Balazs, G., Grosse, C. U., Koch, R., Reinhardt, H.W., 1993, Acoustic emission monitoring on steel-concrete interaction, *Otto Graf Journal*, vol. 4, pp. 56-90

Berthelot, J.M., Robert, J.L., 1987, Modeling concrete damage by acoustic emission, *J. of ac. emission*, vol. 6, no. 1, pp. 43-60

Buland, R., 1976, The mechanics of locating earthquakes, *Bull. of the seis. soc. of am.*, vol.66, no.1, pp.173-187

Cater, G.C., Ferrie, J.F., 1979, A coherence spectral estimation program, *in "Programs for digital signal processing"*, Digital Signal Processing Committee, C.J. Weinstein et al., ed., IEEE Press, pp. 2.3-1 - 2.3-18

Geiger, L., 1910, Herdbestimmung bei Erdbeben aus den Ankunftszeiten, *Nachrichten von der Königlichen Gesellschaft der Wissenschaften zu Göttingen*, vol. 4, pp. 331-349

Grosse, C., Reinhardt, H.W., 1994, Lokalisierung von Schallemissions-Signalen in Stahlbeton bei Pull-out-tests, 10. Schallemissionskolloquium in Jena, DGZf, in press.

Grosse, C., Reinhardt, H.W., 1992a, Fortschritte bei der Anwendung der Schallemissionsanalyse zur Untersuchung von Betonbauwerken, 9. Schallemissionskolloquium in Oybin der DGZfP

Grosse, C., Reinhardt, H.W., 1992b, The resonance method - application of a new nondestructive technique which enables thickness measurements at remote concrete parts, *Otto Graf Journal*, vol. 3, pp. 75-94

Grosse, C., 1991, Detection of cracks in reinforced concrete - an introduction to the problem with some measurements, *Otto Graf Journal*, vol. 2, pp. 72-90

Labuz, J.F., Chang, H.S., Dowding, C.H., Shah, S.P., 1988, Parametric study of acoustic emission location using only four sensors, *Rock mechanics and rock engineering*, vol. 21, pp. 139-148

Ohtsu, M., Shigeishi, M., Iwase,H., Koyanagi, W., 1991, Determination of crack location, type and orientation in concrete structures by acoustic emission, *Mag. of Concrete Research*, vol. 43, No. 155, June, pp. 127-134

ACOUSTIC EMISSION IMAGES OF CRACKS IN SOLIDS

Konstantin A. Chishko

Institute for Low Temperature Physics & Engineering
National Ukrainian Academy of Science
47 Lenin Ave. Kharkov 310164 Ukraine

INTRODUCTION

Cracks that are created and propagate in plastically deformable crystals are sources of strong acoustic emission which carry information about the dynamic properties of the fracture process. Although in the interpretation of experimental results it is often assumed that cracks act as sources of acoustic waves, the theoretical description of the effect is far from complete. The elastic field of the crack which opens up in the bulk of an isotropic medium were calculated rigorously in[1] . The dynamics of an elastic semiinfinite space with a surface crack was studied in[2]. The approach developed in[1,2] is suitable for fields due to a deep crack, but it is difficult to apply it in a study of motion of the edges of the crack. Moreover, the result obtained in Refs.1 and 2 are not concerned directly with radiation fields as defined in field theory[3].

On the other hand, it is possible to describe nucleation and propagation of cracks using a model of two-dimensional dislocation pile-up[4,5]. However, the dynamic behavior of such a pile-up needs to be describe more rigorously since pile-ups corresponding to cracks consist of dislocations which cannot undergo slip in the plane of the defect[4]. It is quite natural to use dislocation models in the theory of fracture since the evolution of dislocation ensembles is the true physical origin of crack formation[6]. Finally reducing the problem of crack formation and its time development to the dynamics of a dislocation pile-up, we can use a well-developed quantitative theory[7] to study acoustic emission which occurs in this process.

Our aim is to calculate acoustic emission fields using the dislocation model both a brittle and a viscous crack created on in infinite isotropic medium. A crack of variable width $2L(t)$ is regarded as a two-dimensional pile-up of a rectilinear edge dislocations continuously distributed with a density $\rho_{ik}(\mathbf{r}, t)$. As radiation fields we mean the displacement velocities field $\mathbf{v}(\mathbf{r}, t)$ and the stress fields $\sigma_{ik}(\mathbf{r}, t)$ in the wave zone at distances r from the observation point satisfying $r \gg 2L$.

Acoustical Imaging, Vol. 21, Edited by
J.P. Jones, Plenum Press, New York, 1995

FORMULATION OF THE PROBLEM. DISLOCATION MODEL OF A BRITTLE CRACK

The field describing the velocities of elastic medium displacements containing mobile dislocations satisfies the following dynamic equation of the elasticity theory[5,7]

$$\rho\frac{\partial^2 v_i}{\partial t^2} - \lambda_{iklm}\frac{\partial}{\partial x_k}\frac{\partial}{\partial x_l}v_m = \frac{\partial}{\partial t}f_i(\mathbf{r}, t) \tag{1}$$

Here, $\mathbf{v}(\mathbf{r}, t) = (\partial/\partial t)\mathbf{u}(\mathbf{r}, t)$; $\mathbf{u}(\mathbf{r}, t)$ is the displacement field of a medium with a density ρ; λ_{iklm} is the tensor of elastic moduli. Bulk forces are related to the dislocation current density tensor $j_{ik}(\mathbf{r}, t)$ (see Refs. 5 and 7)

$$\frac{\partial}{\partial t}f_i(\mathbf{r}, t) = \lambda_{ijnp}(\partial/\partial x_j)j_{np}(\mathbf{r}, t) \tag{2}$$

We shall consider only an isotropic medium with Hooke's tensor

$$\lambda_{iklm} = \rho(c_l^2 - 2c_t^2)\delta_{ik}\delta_{lm} - \rho c_t^2(\delta_{im}\delta_{kl} + \delta_{il}\delta_{km}) \tag{3}$$

where c_l and c_t are the velocities of longitudinal and transverse acoustic waves.

We shall study cracks with rectilinear edges parallel to the OZ axis lying in the $y = 0$ plane (Fig.1). To solve the problem we have formulated, it is necessary to describe the dynamics of a crack as the evolution of a pile-up of continuously distributed edge dislocation[4,5], whose extra planes lie in the plane of the crack $y = 0$ and Burgers vectors are infinitesimally small. It is assumed that the law of motion

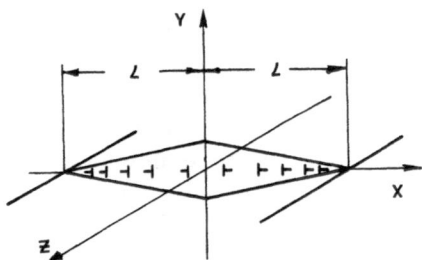

Figure 1. Configuration of a brittle crack.

of the crack end points $x = \pm L(t)$ is a specified time function. The only nonzero component of the dislocation current density tensor is then j_{yy} and the only nonzero component of the dislocation density tensor is ρ_{zy} (Refs. 5 and 7) (both quantities are functions of the two-dimensional radius vector $\mathbf{R} = (x, y)$ lying in a plane perpendicular to dislocation lines forming the dislocation pile-up). These two components are related by an continuity equation[7]

$$\frac{\partial}{\partial t}\rho_{zy} + \frac{\partial}{\partial x}j_{yy} = 0 \tag{4}$$

In the calculation of acoustic field we need not only the tensor j_{ik}, but also the related integral value, i.e. the total dislocation dipole moment of the system[5,7], whose time derivative can be obtained from Eq.(4)

$$\frac{\partial}{\partial t}D_{yy}(t) = \int j_{yy}(\mathbf{R}, t)dxdy = -\int_{-\infty}^{\infty} dy \int_{-L}^{L} dx \int_{-L}^{x} d\xi \frac{\partial}{\partial t}\rho_{zy}(\xi, y, t) \qquad (5)$$

The dislocation density $\rho_{ik}(\mathbf{R}, t)$ in a moving pile-up should be determined from the solution of the corresponding dynamic problem, but such a solution is not required in the case under study. Since a thin crack is equivalent to a dislocation pile-up, it follows that displacements around the crack are identical with the geometric displacements in the medium containing a dislocation pile-up with a density $\rho_{ik}(\mathbf{R}, t)$ which generates the same stress fields on its open surfaces. From the dynamic point of view, these are two objects have nothing in common: dislocations in a pile-up modeling a crack cannot slip in the plane $y = 0$, whereas a crack opens up at a high rate. A distribution of displacements at the edges of defect is reached in a time $\tau \propto L/c_R$ (c_R is the velocity of Rayleigh waves) and we may assume in the case of interest when the velocity of the crack ends satisfies $V_c = dL/dt \ll c_R$ so that at every time moment ρ_{ik} is equal to its static distribution[5]

$$\rho_{zy}(\mathbf{R}, t) = \frac{2ax\delta(y)\theta(t)}{[L^2(t) - x^2]^{1/2}}; \quad a = (1 - \sigma)p_0/\mu \qquad (6)$$

where $\mu = \rho c_t^2$ is the shear modulus; σ is the Heaviside step function. It is assumed that $L(0) = L_{cr} = 2\alpha\mu/[\pi(1 - \sigma)p_0]$ is the half-length of a critical crack[5] (α is the crystal surface extension coefficient). We should replace Eq.(6) at distances $d \propto 10a_0$ (a_0 is the lattice constant) near the ends of a freely growing crack by $\rho(x) = (L^2(t) - x^2)^{1/2}$ (see Ref. 5), but we are not using here this result since it would be necessary to retain a small correction $d/R \ll 1$ in the integral quantity defined by Eq.(5). Equation (6) describes a thin crack growing according to a law which states that the ratio of its length $L(t)$ to its width $h(t) = \int_{-\infty}^{\infty} dy \int_0^{L(t)} \rho(x, t)dx$ is a small constant quantity at all times, i.e., $h(t)/L(t) = 2a = const(t) \ll 1$. The pile-up defined by Eq.(6) whose total Burgers vector is equal to zero consists of two identical pile-ups (superdislocations) of opposite signs with the Burgers vectors $B_+ = -B_- = h(t) = 2aL(t)$, having the same magnitudes.

Substituting Eq. (6) in Eq. (5), we obtain $\partial D_{yy}/\partial t = \pi a(\partial/\partial t)L^2(t)$. It follows that a dipole consisting of two superdislocations which move apart at velocities $V_c \ll c_R$ representing the source of fields in a medium containing the crack. This result holds for long cracks satisfying $L(t) \gg L_{cr}$. Using the well-known results for dislocation pile-ups[7], it is easy to show that the time derivative of the dipole moment for short cracks, whose dimensions are of the order of the critical length L_{cr}, is given by[5] $\partial D_{yy}/\partial t \propto [\partial L(t)/\partial t]L_{cr}$.

It is clear that our model describes correctly all the main features of a crack regarded as source of sound waves in a medium. Our model assumptions are confirmed by the observation of the emission of acoustic waves from dislocation pile-ups[8]. These experimental results indicate that the "nidus" of emission is located at the rapidly moving head of a pile-up.

EMISSION OF ACOUSTIC WAVES BY A LINEAR BRITTLE CRACK IN AN INFINITE MEDIUM

A general expression for the acoustic field emitted by a system of rectilinear parallel dislocations that move arbitrarily in an infinite isotropic medium was derived in Ref.9. However, these results are not applicable to nonconservative motion of dislocations considered here and need to be generalized within the framework of the elasticity theory using the concept of a Green function $G_{ik}^{(0)}(\mathbf{r}, \mathbf{t})$ for an infinite medium[10]. The general solution of the system (1) is given by

$$v_i(\mathbf{R}, t) = \int_{-\infty}^{t} dt' \int d^2 R' \tilde{G}_{ij}^{(0)}(\mathbf{R}, \mathbf{R}'|t - t') \frac{\partial}{\partial t'} f_j(\mathbf{R}', t') \tag{7}$$

where $\tilde{G}_{ik}^{(0)}(\mathbf{R}, t)$ is the Green function of the dynamic two-dimensional elasticity theory for an infinite isotropic medium[9]. Using the method of Ref. 9, we obtain the following expression[11] for the spectral components $\mathbf{v}^\omega(\mathbf{R})$ of the velocity field in the wave zone $\omega R/c \ll 1$:

$$v_i^\omega(\mathbf{R}) = -\frac{c_t^2}{(2\pi i \omega R)^{1/2}} \sum_{\lambda=l,t} c_\lambda^{-5/2} \Psi_{ik}^{(\lambda)}(\mathbf{N}) J_k^{(\lambda)\omega}(\mathbf{N}) \exp[-i\omega \frac{R}{c_\lambda}] \tag{8}$$

Here,

$$\mathbf{J}^{(\lambda)\omega}(\mathbf{N}) = i\omega \mathbf{s}(\mathbf{N}) \int d^2 R' j_{yy}^\omega(\mathbf{R}') \exp[-i\frac{\omega}{c_\lambda}\mathbf{R}'\mathbf{N}] \tag{9}$$

$\mathbf{N} = \mathbf{R}/R$; $\mathbf{s} = \{(1-2\gamma^2)N_x, N_y, 0\}$; $\gamma = c_t/c_l$ and $\Psi_{ik}^{(\lambda)}(\mathbf{N})$ are tensors which determine the angular distribution of the emitted acoustic waves

$$\Psi_{ik}^{(l)}(\mathbf{N}) = N_i N_k, \quad \Psi_{ik}^{(t)}(\mathbf{N}) = \delta_{ik} - N_i N_k \tag{10}$$

In writing down Eq.(8), we assumed that the field is detected far from the source[3] : $|\mathbf{R} - \mathbf{R}'| \simeq R - \mathbf{R}'\mathbf{N}$, and $R' \propto L$. The spectral components $\sigma_{ik}^\omega(\mathbf{R})$ of the stress field in the wave zone can obtained from Eq.(8) using Hooke's law

$$\sigma_{ik}^\omega(\mathbf{R}) = \frac{\rho c_t^4}{(2\pi i \omega R)^{1/2}} \sum_{\lambda=l,t} c_\lambda^{-7/2} \Lambda_{ikm}^{(\lambda)}(\mathbf{N}) J_m^{(\lambda)\omega}(\mathbf{N}) \exp[-i\omega \frac{R}{c_\lambda}] \tag{11}$$

where $(\beta = \gamma^{-1})$

$$\Lambda_{ikm}^{(l)}(\mathbf{N}) = [2(\delta_{ik} - N_i N_k) - \beta^2 \delta_{ik}] N_m \tag{12}$$

$$\Lambda_{ikm}^{(t)}(\mathbf{N}) = 2N_i N_k N_m - \delta_{im} N_k - \delta_{km} N_i \tag{13}$$

It should be noted that the vector $\mathbf{J}^{(\lambda)\omega}$ and the tensors $\Lambda_{ikm}^{(\lambda)}$ differ from similar expressions obtained from the acoustic fields emitted by a system of conservative moved (slipping) dislocations[9].

The following result holds in the dipole approximation (the zeroth approximation in the parameter $\omega R'/c \ll 1$):

$$\mathbf{J}^{(\lambda)\omega}(\mathbf{N}) = \pi a \mathbf{s}(\mathbf{N})[-i\omega L^2(0) + (i\omega)^2 (L^2)^\omega] \tag{14}$$

where the first term is due to nucleation of a crack of critical length $L(0) = L_{cr}$ at $t = t_0$; the contribution of this term is analogous to the radiation during annihilation of dislocations[9]. The second term in Eq.(14) refers to the "bremsstrahlung" due to time dependent evolution of a growing crack. It should be noted that the current density defined by Eq.(14) is proportional to the square of the crack length. We shall now find the inverse Fourier transforms with respect to time, of Eqs.(8) and (11) and Eq.(14) to determine the spatial and time dependencies of the acoustic fields emitted by a crack in an infinite medium

$$v_i(\mathbf{R}, t) = \frac{ac_l^2}{2(2R)^{1/2}} \sum_{\lambda=l,t} c_\lambda^{-5/2} s_k(\mathbf{N}) \Psi_{ik}^{(\lambda)}(\mathbf{N}) \mathbf{F}(t - \frac{R}{c_\lambda}) \qquad (15)$$

where

$$\mathbf{F}(\xi) = L(0) V_c(0) \frac{\theta(\xi)}{\sqrt{\xi}} - \int_0^\infty \frac{d\tau}{\sqrt{\tau}} \theta(\tau - \xi) \frac{\partial^2}{\partial \xi^2} L^2(\xi - \tau) \qquad (16)$$

It follows that

$$\sigma_{ik}(\mathbf{R}, t) = -\frac{a\rho c_l^2 c_t^2}{2(2R)^{1/2}} \sum_{\lambda=l,t} c_\lambda^{-7/2} s_m(\mathbf{N}) \Lambda_{ikm}^{(\lambda)}(\mathbf{N}) \mathbf{F}(t - \frac{R}{c_\lambda}) \qquad (17)$$

The general structures of of Eqs.(15) and (17) are similar to the structure of analogous expressions describing acoustic emission of annihilated dislocations [9] . However, the meaning of the functions which determine the spatial and time evolution of acoustic pulses is quite different. The "initial velocity" $\tilde{u}_0 = L(0) V_c(0)$ and the "acceleration" $\tilde{W}(t) = (\partial^2/\partial t^2) L^2(t)$ represent, during subsequent expansion of the crack, are not the true velocity and acceleration of defect ends but they refer to the time derivatives of the square of the crack width. By this means the amplitude of radiation field is proportional to the second derivative with respect to time of the free volume per unit length opened up by the crack (in full agreement with results of Ref.12) i.e.

$$v \propto \frac{\partial^2}{\partial t^2}(aL^2) \qquad (18)$$

These results are obtained because our model of a crack assumes that it is a pile-up of continuously distributed dislocations equivalent to a pair superdislocations of opposite sign with time dependent Burgers vectors equal to $aL(t)$ and separated from one another by a distance $2L(t)$. The acoustic fields emitted by an infinitely long plane crack include cylindrical shear and compression waves, as expected from the translation symmetry of the system in the z - direction. The component v_z of the velocity field and the components of the stress field σ_{iz} are equal to zero.

Finally, we shall estimate the amplitude of acoustic pulses emitted by a crack of length $L \propto 10^{-4} cm$ created in a time τ_0 ($R \propto 1cm$, $V_c \propto 10^{-2} c_t$):

$$v \propto c_t (L/R)^{1/2} (L/c_t \tau_0)^{3/2} \propto c_t (L/R)^{1/2} (V_c/c_t)^{3/2} \propto 10^{-5} c_t \qquad (19)$$

As usual, the stress is given by $\sigma \propto \rho c v \propto 10^{-6} \mu$. Not surprisingly, we find that the acoustic emission due to formation of crack is quite strong. The amplitude of the "initial peak" due to formation of a critical crack at the beginning of fracture is comparable with the amplitude of acoustic emission during annihilation of dislocations[9]. Using the result that the Burgers vector of a single superdislocation in a crack of critical length L_{cr} is given by $B_+ = -B_- = (2a/3)[dL_{cr}]^{1/2} \propto 10b$ (b is the Burgers vector of a lattice dislocation), we find that the amplitude of acoustic emission from a crack is a factor 2-3 greater (the intensity is greater by ten times) than the amplitude corresponding to annihilation of superdislocations.

DISLOCATION MODEL OF A VISCOUS CRACK

The acoustic emission by a crack, whose expansion is interpreted as a non-diffusive climb in a planar pile-up of continuously distributed edge dislocations with infinitesimal small Burgers vectors, is calculated in previously sections. A model of this kind [11], which is quite satisfactory in the case of brittle fracture, makes it possible to treat the acoustic emission of a defect as a superposition of waves from sources (dislocations) distributed with a variable density (6) in the plane of the crack opening $y = 0$.

In the case of plastic materials in which cracks appear as a result of advanced strain hardening a more realistic mechanism treats the motion of the crack edges as a consequence of annihilation of lattice dislocation currents, characterized by a Burgers vector such that at any time moment in the y-plane the distribution of the dislocation density (6) is maintained in this plane. This in turn makes it possible to treat the acoustic emission from a crack as the annihilation radiation due to self-consistent dislocation currents. We shall adopt this formulation and solve the problem of the emission by a growing crack on the assumption that all the sound radiation is due to dislocations which "bombard" free surfaces of crack and its edges expands in accordance with the known law of motion $x = \pm L(t)$.

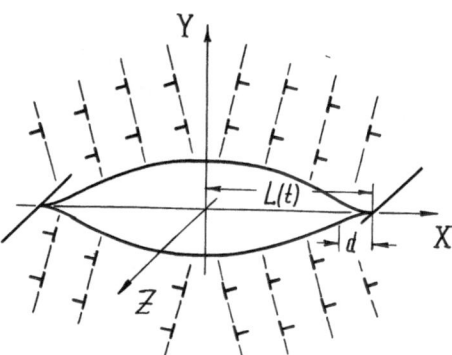

Figure 2. Configuration of a viscous crack

We shall consider a crack (Fig.2) which opens up in the y-plane in such a way that its edges diverge along y-axis. The crack is infinite along z-axis and its edges remain parallel to this axis. The main assumption of actual model is that the distribution of dislocations described by Eq.(6) is at each moment determined by a self-consistent dislocation currents which become annihilated at the crack surfaces (Fig.2). In our case the displacements of these surfaces along y-axis can be maintained if there are currents of edge dislocations with the nonzero components b_y of the Burgers vector and the nonzero components V_y of the dislocation velocity in the dislocation current. In this case the only nonzero component of the dislocation current density tensor j_{ik} is j_{xy} , which is responsible for the growth of a crack in accordance with the assumed mechanism. It follows from the continuity equation [5] that

$$\frac{\partial}{\partial t}\rho_{zy} - \frac{\partial}{\partial y}j_{xy} = 0 \tag{20}$$

The relationship (20) is valid everywhere where the density of mobile dislocations differs from zero. The distribution (6) applies to a planar pile-up lying in the y-plane.

Therefore, using Eq.(6), we can obtain from Eq.(20) the dislocation current gradient at the free surface of the opening crack:

$$\left[\frac{\partial}{\partial y}\ \dot{j}_{xy} \right]_{y=0} = -\frac{4ax\delta(y)}{[L^2(t) - x^2]^{3/2}} L(t) \frac{\partial}{\partial t} L(t) \tag{21}$$

The expression (21) yields a current of dislocations which should become annihilated at surfaces that become free as a result of cracking, so as to maintain the assumed motion of the ends of the crack described by $L = L(t)$.

We shall now consider the emission by a crack in an infinite isotropic medium. We shall regard it as the emission due to the annihilation of dislocations in the y-plane within the interval $|x| \leq L(t)$. The force f_i in the right-hand side of Eq.(1) has only one nonzero component:

$$\frac{\partial}{\partial t} f_x = \rho c_t^2 \left[\frac{\partial}{\partial y}\ \dot{j}_{xy} \right]_{y=0} \tag{22}$$

which corresponds to perturbation of the medium as a result of annihilation of dislocations at the crack free surfaces.

ACOUSTIC EMISSION OF A VISCOUS CRACK

We cannot use directly the expressions for the fields of the annihilation radiation emitted by edge dislocations [9], written in the dipole approximation, and we have to calculate the quadrupole terms for the radiation fields emitted by a system of rectilinear edge dislocations parallel to the OZ axis. We then obtain [13] from (22) and (7):

$$v_i(\mathbf{R}, t) = \frac{\cos\varphi}{2\pi\rho(2R)^{1/2}}$$

$$\sum_{\lambda=l,t} c_\lambda^{-5/2} \Psi_{ix}^{(\lambda)}(\mathbf{N}) \int_0^\infty \frac{d\tau}{\sqrt{\tau}} \frac{\partial^2}{\partial t^2} Q_{xx}(t - \tau - R/c_\lambda) \tag{23}$$

where R and φ are the polar coordinates of the point of observation (polar angle φ measured from positive direction of x-axis) and $\Psi^{(\lambda)}(\mathbf{N})$ are functions (10) describing the angular distribution of radiation. The symbol $Q_{xx}(t)$ denotes the component of the quadrupole moment tensor:

$$\frac{\partial}{\partial t} Q_{xx}(t) = \int_{-L(t)}^{L(t)} x \frac{\partial}{\partial t} f_x(x, t) dx =$$

$$= -2\rho c_t^2 a\theta(t) [\frac{\partial}{\partial t} L^2(t)] \int_0^{L(t)} \frac{x^2 dx}{[L^2(t) - x^2]^{3/2}} \tag{24}$$

The integral (24) diverges at the upper limit because Eq.(1) is invalid in the end regions of a crack of dimensions $d \propto L$. In these regions the cracks have "beak" where the edges merge smoothly [5], and dislocation density ρ_{zy} in an equivalent pile-up approaches zero proportionally to $\propto (L^2 - x^2)^{1/2}$. The size of the end region d is independent of the size of the crack L. Bearing this point in mind, we can estimate the integral of Eq.(11) from

$$\frac{\partial}{\partial t} Q_{xx}(t) = -4a\rho c_t^2 \left(\frac{2L}{d} \right)^{1/2} L(t) \frac{\partial}{\partial t} L(t) \tag{25}$$

The stress field of created by a growing crack can be obtained from Eq.(23) by the Hooke's law

$$\sigma_{ik}(\mathbf{R}, t) = -c_t \frac{\cos \varphi}{2\pi (2R)^{1/2}}$$

$$\sum_{\lambda=l,t} c_\lambda^{-7/2} \Lambda_{ikx}^{(\lambda)}(\mathbf{N}) \int_0^\infty \frac{d\tau}{\sqrt{\tau}} \frac{\partial^2}{\partial t^2} Q_{xx}(t - \tau - R/c_\lambda) \tag{26}$$

It therefore follows that the radiation of interest to us consists of cylindrical compression and shear waves propagating at velocities c_l and c_t, respectively. The amplitudes of the fields given by Eqs.(23) and (26) are governed, as is easily seen from Eq.(25), by the velocities $dL(t)/dt$, and by the accelerations $d^2 L(t)/dt^2$ of the ends of the crack. Essentially this means that radiation sources are the ends of a crack; this result is fully understandable since the dislocation currents are maximal near the ends of a growing crack (at distances $\propto L - d$ from the origin of the coordinate, where $d \ll L$).

As pointed out already, the radiation investigated by us is a quadrupole nature. In contrast to quadrupole bremsstrahlung of the prismatic dislocation loops, the amplitude of which is proportional to the time derivative of the acceleration of the source [14] (in full analogy with the quadrupole emission of electromagnetic waves by a system of charges [3]), we are dealing here with the transient radiation that is generated during annihilation of dislocations at the surfaces of growing crack. In the dipole approximation the annihilation radiation emitted by dislocations is proportional to their velocity at the moment of annihilation or emergence of the surface [15,16] . In fact, it follows from Eqs.(23), (25) and (26) that the amplitudes of the displacement velocities of the points in the medium (and of the stresses) are proportional to

$$\frac{\partial^2}{\partial t^2} Q_{xx} \propto \left(\frac{2L}{d} \right)^{1/2} \left\{ \frac{\partial^2}{\partial t^2}(aL^2) + \frac{1}{2L} \frac{\partial L}{\partial t} \frac{\partial}{\partial t}(aL^2) \right\} \tag{27}$$

i.e. in addition to the expected dependence in the radiation fields of the brittle crack, there is a term due to the presence of the factor $(L/d)^{1/2} \gg 1$ in the rate of change of the quadrupole moment of Eq.(23). This factor may generally be quite large because of the condition $L \gg d$, so that the quadrupole annihilation radiation may be sufficiently intense, in spite of the fact that it is characterized by a higher order of smallness compared with the dipole radiation in terms of the parameter $(1/c)(dL/dt) \ll 1$.

The spectral components of radiation fields can be derived [14] from Eqs.(23) and (26) during inverse Fourier transformation:

$$v_i^\omega(\mathbf{R}) = -\frac{\cos \varphi}{(2\pi i\omega R)^{1/2}} \left[\frac{(i\omega)^2}{\rho} Q_{xx}^\omega \right] \sum_{\lambda=l,t} c_\lambda^{-5/2} \Psi_{ix}^{(\lambda)}(\mathbf{N}) \exp\left[-i\frac{\omega R}{c_\lambda}\right] \tag{28}$$

$$\sigma_{ik}^\omega(\mathbf{R}) = \frac{c_t^2 \cos \varphi}{(2\pi i\omega R)^{1/2}} [(i\omega)^2 Q_{xx}] \sum_{\lambda=l,t} c_\lambda^{-7/2} \Lambda_{ikx}^{(\lambda)}(\mathbf{N}) \exp\left[-i\frac{\omega R}{c_\lambda}\right] \tag{29}$$

Here Q_{xx}^ω is the spectral amplitude of Q_{xx}.

The acoustic emission by a brittle crack was considered above as the radiation emitted by a planar dislocation pile-up of Eq.(6) expanding in a non-diffusive climb along the x-axes, i.e. with nonzero component j_{yy} of the dislocation current tensor j_{ik}. In the case being in question the radiation is related to the component j_{xy} of this tensor for dislocations dropped on the free surfaces of crack. In general, when both these factors governing the crack growth are active, the continuity equations (4) and (20) should be rewritten by the form

$$\frac{\partial}{\partial t}\rho_{zy} - \frac{\partial}{\partial y}j_{xy} + \frac{\partial}{\partial x}j_{yy} = 0 \qquad (30)$$

If the components j_{yy} of the dislocation current predominates, the crack emits sound in the same way as predicted above for brittle crack. However, if the crack surfaces open up mainly because of the current component j_{xy} of the external dislocations, which are "dumped" into the crack, then the radiation is described by Eqs.(23) and (26). If both these components of the dislocation current are of the same order of magnitude, the situation becomes more complex: the total emission of sound represents a superposition of the quadrupole transient radiation described by Eqs.(23) and (26) and of the dipole bremsstrahlung [5] . The contribution of each of the mechanisms to the total acoustic signal can be determined if Eq.(30) is supplement by the relationship between the components j_{xy} and j_{yy} of the dislocation current, which then makes it possible to express these components in terms of the function ρ_{zy}. Such a relationship can be obtained by solving the equation of motion of the crack, which is a very difficult task. However, in a real experimental situation the expected nature of the radiation can be estimated as follows: in the case of brittle crystals, in which the glide of dislocations meets with may obstacles, the radiation should be emitted in accordance with the mechanism described in [11] . In plastic crystals, containing a large number of easy gliding dislocations, the mechanisms investigated in present section predominates provided the dislocation currents incident on the surfaces of the crack are not limited by some volume effects in the parts of a crystal adjoining directly the growing crack.

It is important to stress once again that the problem solved in the present work (and also that in Refs.11,14) is limited to the treatment in terms of an equivalent dislocation pile-up, the evolution of which requires - because of the continuity equation (30) - the presence of dislocation currents j_{xy} and j_{yy} in the $y = 0$ plane. This approach makes it possible to reduce the complex task and the dynamic theory of elasticity in a doubly connected region [3,4] to a calculation of the acoustic radiation emitted by dislocations the currents of which j_{xy} and j_{yy} are given functions of the coordinates and time. However, we must remember that the dislocation currents in Eq.(30) do not describe the motion of real dislocations, but simply determine the balance of the Burgers vector (i.e., the total geometric displacement) at the surfaces of a crack which expands in accordance with Eq.(6). Therefore, although we do use the term "annihilation dislocations", we must remember that we are speaking here only of the method of description which gives rise to a mechanism analogous to the generation of annihilation radiation.

RADIATION PATTERN AND ACOUSTIC IMAGES OF CRACKS

The crack propagation may be detected through the registration of the acoustic emission generated during the fracture process in crystals. For the identification of growing cracks it is necessary to know the acoustic image of the expected fracture mechanism. Now we shall discussed the most essential features of these images in two cases of brittle and viscous cracks.

Actual registration of the sound pulses radiated by a crack is rather simple problem which can be experimentally solved, for example, with the use of rapid oscillography. As evident from comparison of Eqs. (18) and (27), the form of acoustic signal is essentially different for the brittle and viscous cracks. However, in real situations the detail reconstruction of pulse form is very difficult task because of the limited time resolution of experimental design. In our opinion the unambiguously identification of the propagated crack can be based on the measurement of acoustic radiation pattern of compressive and shear waves emitted by a crack. The angle distribution of sound irradiation constitutes the peculiar features of the fracture mechanism and contains a wealth of information on speed or acceleration as well as on the spatial orientation of a crack. Furthermore, the above mentioned types of cracks demonstrate obviously different radiation pattern.

Radiation fields generated both brittle and viscous cracks (Eqs.(15), (17) and (23), (26) respectively) are described by the expression of the same structure. They are the superposition of cylindric compressive and shear waves. Hence, for stress field, for example, we can write:

$$\sigma_{ik}(\mathbf{R}, t) \propto \frac{1}{\sqrt{R}} \sum_{\lambda=l,t} \Phi_{ik}^{(\lambda)}(\varphi) \mathbf{F}(t - R/c_\lambda) \tag{31}$$

where $\mathbf{F}(t)$ determines the space-time distribution of sound irradiation and components of tensor $\Phi_k^{(\lambda)}(\varphi)$ describe the angle distribution of stresses. For pressure waves radiated from a brittle crack (see Eq.(17)) we can find[17]:

$$\Phi_{ik}^{(l)B}(\varphi) = (1 - 2\gamma^2 \cos^2 \varphi) \begin{pmatrix} 1 - 2\gamma^2 \sin^2 \varphi & \gamma^2 \sin 2\varphi & 0 \\ -\gamma^2 \sin 2\varphi & 1 + 2\gamma^2 \cos^2 \varphi & 0 \\ 0 & 0 & 1 + 2\gamma^2 \end{pmatrix} \tag{32}$$

and for shear waves we have

$$\Phi_{ik}^{(t)B}(\varphi) = \begin{pmatrix} -\gamma^2 \sin^2 2\varphi & \frac{1}{2}\gamma^2 \sin 4\varphi & 0 \\ \frac{1}{2}\gamma^2 \sin 4\varphi & \gamma^2 \sin^2 2\varphi & 0 \\ 0 & 0 & 0 \end{pmatrix} \tag{33}$$

In the case of viscous crack (see Eq.(26)) the result is

$$\Phi_{ik}^{(l)V}(\varphi) = \cos^2 \varphi \begin{pmatrix} 1 - 2\gamma^2 + 2\cos^2 \varphi & \frac{1}{2} \sin 2\varphi & 0 \\ \frac{1}{2} \sin 2\varphi & 1 - 2\gamma^2 + 2\sin^2 \varphi & 0 \\ 0 & 0 & 1 - 2\gamma^2 \end{pmatrix} \tag{34}$$

$$\Phi_{ik}^{(t)V}(\varphi) = \sin \varphi \begin{pmatrix} \sin 2\varphi & \cos 2\varphi & 0 \\ -\cos 2\varphi & -\sin 2\varphi & 0 \\ 0 & 0 & 0 \end{pmatrix} \tag{35}$$

for pressure and shear waves respectively. As an example the radiation pattern of the longitudinal part one of the stress field component σ_{yy} are shown in Fig.3 (brittle crack) and Fig.4 (viscous crack). The distinction between these pattern is evident.

If the currents of dislocations annihilating on the both surfaces of crack are equal, the radiation from the viscous crack is substantially quadrupole one. It remains so even in the case of inequality of mentioned currents provided that the glide planes of dislocations are normal to the surfaces of defect. On an inclined dislocation drop to a crack a dipole component in the radiation field appears. Its amplitude depends on the angle of incidence α and has maximum value when $\alpha = \pi/4$. Unlike above mentioned situation the acoustic emission of brittle crack is always dipole radiation.

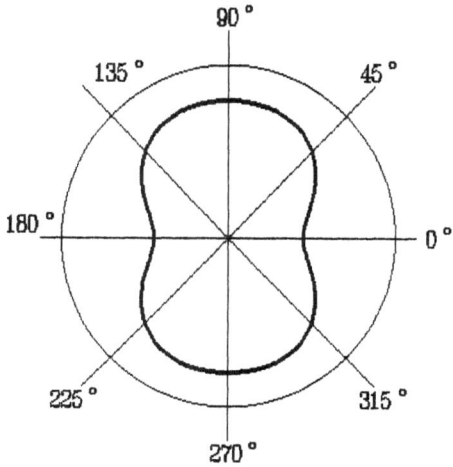

Figure 3. Radiation pattern for the longitudinal part of the stress field generated by a brittle crack.

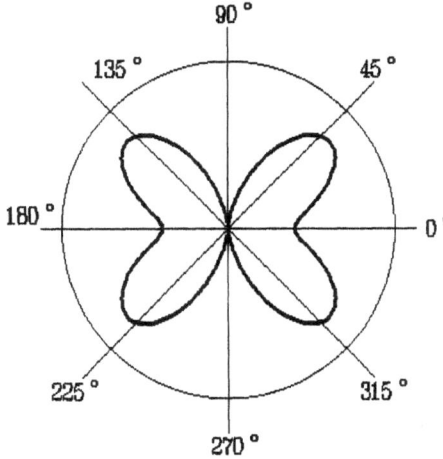

Figure 4. Radiation pattern for the longitudinal part of the stress field generated by a viscous crack.

CONCLUSION

The dynamic behavior and acoustic emission of a real crack are defined, of course, by the joint action of two mechanisms of brittle and viscous expansions. The recording of emission signal with due to account of present theory makes, in principle, possible to ascertain what mechanisms are predominate in each particular case. This example illustrates the possibility of the acoustic emission method for investigation of plastic behavior and fracture of solids. Generally, the theoretical distribution of the acoustic emission of a crack must be based on the rigorous solution of the dynamic problem for crack evolution in presence of external forces and outer lattice dislocations. This solution is still absent and thus the acoustic emission theory for a crack is far from its completion.

REFERENCES

1. V.V.Krilov, On sound radiation of developing crack, *Sov.Phys.Acoust.* 29:468 (1983).
2. V.V.Krilov, E.P.Ponomarev, Acoustic emission accompanying generation of surface microcracks, *Sov.Phys.Acoust.* 31:122 (1985).
3. L.D.Landau and E.M.Lifshitz. "The Classical Theory of Fields", 3-rd ed., Pergamon Press, Oxford (1970).
4. J.P.Hirth and J.Lothe, "Theory of Dislocations", McGraw-Hill, New York (1968).
5. L.D.Landau and E.M.Lifshitz, "Theory of Elasticity", 3-rd ed., Pergamon Press, Oxford (1970).
6. J.Friedel, "Dislocations", Pergamon Press, Oxford (1964).
7. A.M.Kosevich, "Dislocations in the Theory of Elasticity" [in Russian], Naukova Dumka, Kiev (1978).
8. V.S.Boiko, R.I.Garber, L.F.Krivenko, and S.S.Krivulya, Sound radiation of twinning dislocations, *Sov.Phys. Solid State* 11:3041 (1970).
9. V.D.Natsik, K.A.Chishko, Sound radiation during annihilation of dislocations, *Sov. Phys. Solid State* 14:2678 (1973).
10. W.Nowacki (ed.), "Theory of Popular Elasticity", Springer Verlag, Berlin (1974).
11. K.A.Chishko, Emission of sound due to formation of a crack in an infinite elastic medium and at the surface of an elastic semiinfinite medium, *Sov. Phys. Solid State* 31:476 (1989).
12. J.D.Achenbach, K.-I.Hiroshima, K.Ohno, Acoustic emission due to nucleation of a microcrack in the proximity of a macrocrack, *J. Sound and Vibr.* 89:523 (1983).
13. K.A.Chishko, Acoustic emission during annihilation of prismatic dislocation loops and kinks in line dislocations *Ukrainian Phys.Journ.* 19:1264 (1974).
14. K.A.Chishko, Dislocation mechanism of the emission of sound during growth of a crack in a crystal, *Sov. Phys. SolidState* 34:462 (1992).
15. V.D.Natsik, K.A.Chishko, Emission of sound from dislocations moving near the surface of a crystal, *Sov. Phys. Solid State* 20:264 (1978).
16. V.D.Natsik, K.A.Chishko, Acoustic emission caused by dislocation reaching the surface of a crystal, *Sov. Phys. Acoust.* 28:225 (1982).
17. T.N.Antsygina, K.A.Chishko, Angular pattern of sonic emission from thin cracks *Sov. Phys. Solid State* 35:1797 (1993).

PHASE-INSENSITIVE DETECTION OF BACKSCATTERED ULTRASOUND FOR VOID CONTENT DETERMINATION IN FIBER REINFORCED COMPOSITES

D. Grolemund and C.S. Tsai

University of California, Department of Electrical and Computer Engineering, and Institute for Surface and Interface Science
Irvine, CA USA 92717

INTRODUCTION

Fiber reinforced composites use one of many polymeric resins to bind the fibers into the proper configurations and to provide the compressional elastic properties missing in the fibers. Various problems during the resin cure cycle can produce a distribution of small voids ranging in size from 20 - 300 μm. This porosity can degrade the compression related strength, and induce structural and safety problems. The parameter of central engineering interest in these porous laminates is the void content, C, defined as the percentage of the total volume occupied by the voids. Void content can be related empirically to various mechanical strengths, therefore, providing a potential platform for nondestructive measurement of mechanical properties. The PMR-15 resin system was chosen for this study because of its increasing use in high-temperature, primary structures and its propensity to create porosity even during nominally correct cure cycles.

Considerable attention has been devoted in the past few years to the problem of acoustically estimating the void content [1-7]. Most techniques have related the acoustical field measured with large-aperture piezoelectric transducers to C either empirically or theoretically. These phase-sensitive receivers, such as PZT, are subject to errors in the resultant amplitude. Studies have shown that the spatial and temporal distortion of the wavefront by an inhomogeneous medium results in measurement artifacts by phase cancellation [8-9]. One solution for spatially inhomogeneous fields has been to employ phase-insensitive receivers such as the semiconductor CdS [10-11]. There are several practical difficulties with this approach because experimental implementations can be cumbersome. In addition, CdS has lower sensitivity than most piezoelectrics and this can cause severe S/N problems.

A second solution to the wavefront distortion errors implemented by Johnston and Miller [8-9] was the use of small aperture, phase-sensitive receivers to simulate a larger aperture, phase-insensitive transducer. In this method, small piezoelectric transducers were used to independently measure the wavefront at a number of regularly spaced locations and then to combine the signals in a phase-insensitive manner. The present work used the phase-insensitive summation techniques of Johnston and Miller to acquire a more accurate empirical correlation between ultrasound backscattered from porous fiber-reinforced laminates and their void content. The phase-insensitive detection results were compared to the measurements made with a single 13.5 mm dia. PVDF transducer that possesses the same receiving area as the pseudo-array.

EXPERIMENTAL METHODS

In order to explore the validity of the phase-insensitive summation method, a set of well-characterized porous laminates was required. For this study, we produced eleven laminates of varying void content. The manufacture and characterization of the composite samples used in this study has been described elsewhere [2]. Sixteen-ply satin harness fabric impregnated with PMR-15 resin were autoclave cured. The PMR-15 resin is a condensation reaction system that produces copious amounts of methanol and water as a by-product of the cross-linking mechanism. Variations in the magnitude and time of application of the autoclave pressure were used to purposely trap these volatiles in the form of distributed voids for study. Although this process is an inefficient way to produce porous laminates, it provides a void distribution that very closely mimics porosity seen in production PMR-15 composites. The characteristics of the void distribution were destructively measured by acid digestion and optical microscopy of cut and polished sections. The results of the destructive tests are shown in Table 1 where C_a and C_o represent the void content measurements acquired by acid digestion and optical microscopy, respectively.

Table 1. Void content from acid digestion (C_a) and optical image analysis (C_o).

Sample	C_a (%)	C_o (%)	Sample	C_a (%)	C_o (%)
A	0.3	0.4	G	3.7	3.4
B	0.5	0.8	H	4.3	4.6
C	1.2	0.9	I	5.4	5.5
D	1.8	1.6	J	7.2	7.8
E	2.4	2.1	K	8.1	9.5
F	2.8	3.2			

Custom transducers were manufactured from PVDF film so that direct comparison of phase-insensitive and phase-sensitive detection methods could be made. Element sizes were chosen such that the combined area of the small-aperture psuedo-array and the single large aperture transducer were identical. Two circular, plane transducers were designed and manufactured, one with a 1.5 mm diameter, and one with a 13.5 mm diameter. The diameter of 1.5 mm was chosen as an empirical trade-off between sensitivity (receiving area) and the desire to reduce the size to achieve a fairly constant pressure field across the face of the transducer, i.e., reduce its phase-sensitivity. PVDF film of 110 μm thickness with Cu-Ni (Constantine) electrodes (Atochem North America) was used to fabricate the hydrophone and the large aperture receiver.

Standard photolithographic techniques were used to etch the metallization into a circular shape and along with a short bonding electrode. The film was then bonded to an optically polished, backing consisting of a loaded polyethylene resin by spinning low viscosity epoxy onto the backing with a photoresist spinner, and applying pressure during room temperature cure.

The single 1.5 mm dia. PVDF transducer was used to simulate an array placed nearly perpendicular to the incident beam axis in the backscatter mode. The physical arrangement of the transducers prevented acquisition of the data directly on axis. Figure 1 shows the experimental schematic. We note however that this configuration is not strictly in the backscatter mode because of physical restrictions of the transducers. Movement of the PVDF receiver was controlled by a mechanical, multi-axis scanner with a positional accuracy of better than 0.1 mm. The transmitting transducer was a 12.7 mm dia. focused PZT with a center frequency of 2.25 MHz and a focal length of 100 mm. The received signal was fed through an low-noise, selectable gain amplifier to an 8-bit PC based A/D system (Sonix STR-8100). The digitization rate was set at 100 MHz.

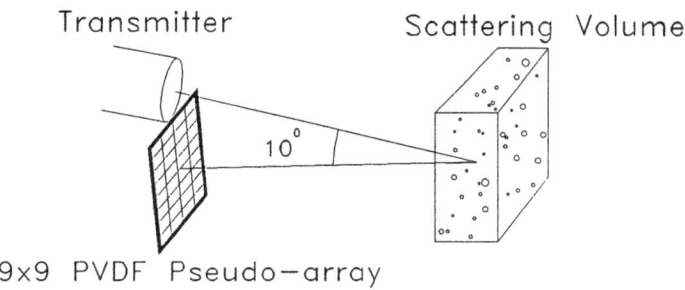

Transmitter   Scattering Volume

10^0

9x9 PVDF Pseudo−array

Figure 1. Schematic of backscatter measurements.

The pseudo-array in Fig. 1 consisted of a 9 x 9 rectangular arrangement with diameters of adjacent elements touching. The phase-insensitive summation was calculated as a zeroth-order spatial moment [9],

$$m_0 = \sum_{i=1}^{9} \sum_{j=1}^{9} \mid V_{ij}(f) \mid^2 \tag{1}$$

where $\mid V_{ij}(f) \mid$ is the voltage magnitude of the spectrum of the ij th pseudo-array element. The reference signal in the experiment consisted of the field reflected from a planar stainless steel plate placed at the same location and orientation as the samples. Complete data sets were acquired at six spatially distinct locations on each panel, and the measured spectra, $V_{ij}(f)$, normalized by the response from a steel plate. Johnston and Miller refer to these normalized spectra as the backscatter transfer function which we denote as $T_B(f)$.

The ease with which PVDF film can be worked allows custom transducers to be manufactured to suit the required application. To directly compare the phase-insensitive summation to a phase-sensitive transducer, a 13.5 mm dia. transducer was manufactured from the same PVDF film. The area ratio between the pseudo-array (9 x 9 grouping) and this single 13.5 mm dia. transducer is unity. Therefore, given the same incident field, any variation in resultant amplitude can be attributed to the different summation method employed, i.e., summation according to Eq. (1) for the pseudo-array, and summation across the face of the 13.5 mm transducer.

RESULTS

First, we examine the backscatter transfer function, T_B (f), which is simply the ratio of the spectra from the composite to that of the steel plate. Figure 2 shows the typical variation of T_B (f) as a function of void content and frequency for four different void contents.

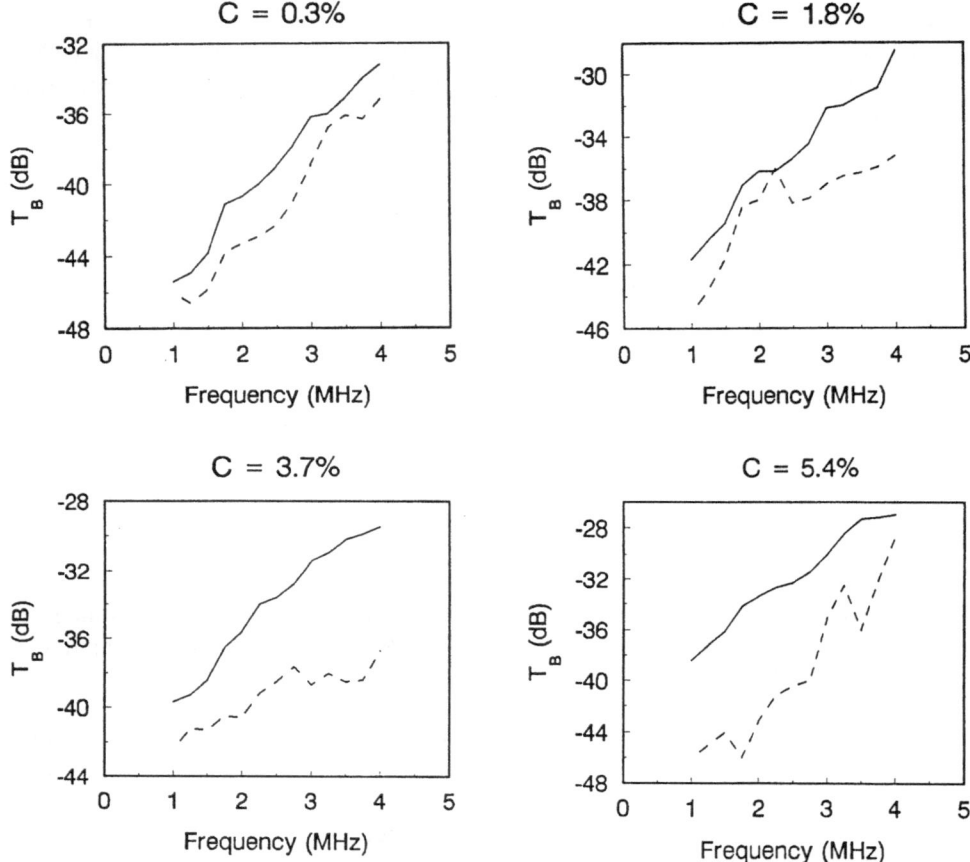

Figure 2. Backscatter transfer function as a function of void content. Data taken from single site on each panel. Solid line (----) phase-insensitive, dotted line (- - -) phase-sensitive.

Previous work have shown the echo return from porous composites to be a linearly increasing function of void content [2]. Figure 3 depicts the ensemble or spatial mean of phase-sensitive and phase-insensitive measurements from samples A-K. Each data point represents the average amplitude, referenced to the steel plate echo, from six randomly selected sites on each panel. The mean was calculated through the integrated backscatter [9]. Several important details are evident in this data. First, the scatter of the data about the linear fit is less for the phase-insensitive data than for the phase-sensitive as witnessed by the χ^2 values of 0.12 and 0.62, respectively. Secondly, the overall level of the phase-insensitive curve has been shifted upwards due to reduction in phase cancellation. In addition, the slope of the regression lines increases for phase-insensitive detection, namely 2.3 for the phase-insensitive and 1.8 for the phase-sensitive.

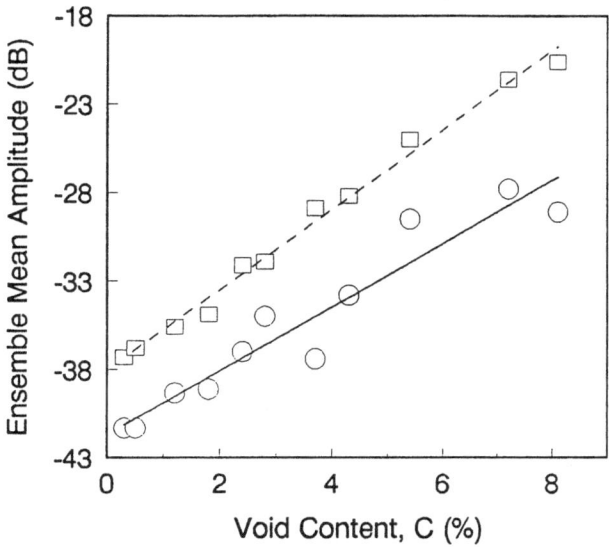

Figure 3. Spatial mean amplitude as a function of void content. Data points are the mean of six sites per sample. Phase-sensitive ($\circ$), phase-insensitive ($\square$).

CONCLUSIONS

Sources of error in porous composite void content measurements are not completely understood. Techniques work adequately in controlled experiments but inadequately when examining a wide range of production composites produced under less well controlled circumstances. There is some evidence that the deviations may be due to departure of the void size distribution from the ideal assumptions used in most models. Johnston and Miller [9] have shown that phase cancellation effects by fields from inhomogeneous media have a direct impact on data measured by piezoelectric transducers.

We have demonstrated that the phase-insensitive summation technique reduces the scatter in the void content estimate. With condensation reaction resins such as PMR-15, the prevention of porosity is difficult. Consequently, engineers now conservatively specify allowable voids for different stress regions of the airframe taking into account the error in ultrasonic void content measurement. A more accurate void detection scheme could potentially save a significant number of components from rejection by allowing more confidence in the determination of void content and relaxing the margins.

ACKNOWLEDGEMENTS

The authors would like to thank D. Cram and M. Jacobs of Rohr Industries for supplying the PMR-15 samples and the destructive testing.

REFERENCES

[1] D. Grolemund and C.S. Tsai, Statistical moments of Rayleigh scattered fields in porous fiber reinforced composites, *Proc. IEEE Ultrasonics Symp.*, 92CH3118-7: 815-818, (1992).

[2] D. Grolemund and C.S. Tsai, Statistical moments of backscattered ultrasound in porous fiber reinforced composites, I. Experiments, to be submitted to *IEEE Trans. UFFC*.

[3] D. Grolemund and C.S. Tsai, Statistical moments of backscattered ultrasound in porous fiber reinforced composites, I. Monte Carlo Simulations, to be submitted to *IEEE Trans. UFFC*.

[4] R. A. Roberts, Analysis of ultrasonic backscatter for porosity characterization in graphite epoxy, *Rev. Prog. Quan. NDE,* D.O. Thompson and D.E. Chimenti, eds., Plenum, New York, 8B:1559-1566, (1987).

[5] E.D. Blodgett, L.J. Thomas III, and J.G. Miller, Effects of porosity on polar backscatter from fiber reinforced composites, *Rev. Prog. Quan. NDE*, D.O. Thompson and D.E. Chimenti, eds., Plenum, New York, 5B:1267-1274, (1986).

[6] D.E. Yuhas, C.L. Vorres, and R.A. Roberts, Variations in backscatter attributed to porosity, *Rev. Prog. Quan. NDE*, D.O. Thompson and D.E. Chimenti, eds., Plenum Press, New York, 5B:1275-1284, (1986).

[7] R.A. Roberts, Porosity characterization in fiber-reinforced composites by use of backscatter, *Rev. Prog. Quan. NDE*, D.O. Thompson and D.E. Chimenti, eds., Plenum, New York, 6B:1147-1156, (1987).

[8] P.H. Johnston and J.G. Miller, A comparison of backscatter measured by phase-sensitive and phase-insensitive detection, *Proc. IEEE Ultra. Symp.*, 85CH2209-5: 927-831, (1985).

[9] P.H. Johnston and J.G. Miller, Phase-insensitive detection for measurement of backscattered ultrasound, *IEEE Trans UFFC*, 33: 713-721, (1986).

[10] L.J. Busse and J.G. Miller, Response characteristics of a finite aperture, phase-insensitive ultrasound receiver based on the acoustoelectric effect, *J. Acoust. Soc.Am.*, 70:1370-1376, (1981).

[11] L.J. Busse and J.G. Miller, Detection of spatially nonuniform ultrasonic radiation with phase-sensitive (piezoelectric) and phase-insensitive (acoustoelectric) receivers, *J. Acoust. Soc. Am.*, 70: 1377-1386, (1981).

ULTRASONIC NON-DESTRUCTIVE EVALUATION: HIGH RESOLUTION TECHNIQUES FOR CERAMIC MATERIALS

E.Biagi, A.Fort, L.Masotti, L.Ponziani

Dipartimento di Ingegneria Elettronica
via di S. Marta 3
50139 Firenze, Italia

INTRODUCTION

The aim of this research was to obtain high resolution ultrasonic images for flaw detection in ceramic materials. Non destructive tests of both green and completely sintered materials were taken into account, for manufacturing process optimization and final material quality assessment.

Ultrasonic waves gather thorough information on the material microstructure since they interact with the inhomogeneities in the bulk material, grain boundaries, pores and inclusions. For these reasons ultrasonic non destructive techniques are widely used in the microstructural characterization of ceramic materials.

In the present work the experimental determination of the forward-scattered wave characteristics, attenuation and propagation velocities were correlated to the density of scatterers in green materials.

Specimens of fine grained Zirconia stabilized with Yttria, characterized by a large variation of the fractional volume of pores (from 0% to 50%) were investigated. Ultrasonic velocities and attenuation versus the porosity degree reached in different phases of the manufacturing process were found. A physical understanding of the experimental results was achieved by comparing them with the theoretical predictions provided by the conventional elastic theory and a self consistent scattering theory introduced in the 1980s. The results pointed out the capability of the ultrasonic parameters in assessing the quality of the manufacturing process in terms of conformity and repeatability.

For quality control of the sintered material, high resolution ultrasonic images were obtained for inner defects detection. An immersion pulse echo technique was employed by using a focussed wide band transducer with a central frequency of 50 MHz a 50 mm focal length and a 0.2 mm spot size.

Porosity maps were obtained, allowing the detection of porosity spatial gradient representing the first phase of cracks generation in the sintered material.

Two techniques were developed for different defect types. The first one, used for large defects visualization, is based on the spectral analysis of the echo coming from a flaw; the second, aiming at the detection of small defects (less than a few wavelengths) is based on the processing of specimen back wall echoes. In particular, when the defect mean size is large enough to generate a detectable backscattered echo, an improvement of the image quality is obtained by a compound of the information related to the amplitude and to the spectral content of the echo signal. In this way, more information can be gained about the defect orientation and surface roughness. For this purpose the attenuation and the centroid of the spectral distribution of the echo signals were computed.

On the other hand, when the amplitude of the defect backscattered signal has the same magnitude of the noise, the presence of a flaw is detected through the observation of its filtering effect on the forward propagated acoustic field. If the defect has shape, thickness or roughness such as to induce a frequency selective attenuation, the flaw presence is highlighted by analyzing the spectral distribution of the specimen back wall echo in terms of frequency centroid shift. In order to collect more information the back wall echo amplitude was evaluated as well. The obtained images show that this new investigation method improves the system resolution.

The proposed defect detection techniques were applied to ceramic specimens of fine grained Zirconia stabilized with Yttria, with bulk flaws and superficial calibrated defects. It is important to point out that the proposed techniques allow the observation of flaws that weren't detectable by X-ray microfocus investigation.

ULTRASONIC EVALUATION OF GREEN ZIRCONIA

The assessment of porosity degree in green ceramic materials is of the utmost importance since through the knowledge of this parameter, at various manufacturing steps, a thorough understanding of the sintering process and material phases transformation can be gained.

A porous medium can be treated with respect to ultrasonic wave propagation as a random two phase medium in which the second phase is void. That is, as a random spatial distribution of scatterers with various dimension and shape. By micrograph analysis it was shown that the used elastic wavelength was much larger than the mean pore size, so that the scattering media behaves like a random distribution of Rayleigh scatterers in a uniform ceramic matrix. In this case the ultrasonic velocities v_l (longitudinal) and v_t (transversal) can be expressed as a function of fractional volume of pores δ and of the elastic characteristic of the ceramic medium according to the following equations[1]

$$v_t = v_{t1} \sqrt{\frac{2a}{-b - \sqrt{b^2 - 8a}} \frac{1}{1 - \delta}}$$

where:

$$a = (1 - 2\delta)\left[1 - \frac{3}{4}\left(\frac{K_1}{k_1}\right)^2\right]$$

(1)

$$b = \left[(6\delta - 5) + \frac{3}{4}(3 - \delta)\left(\frac{K_1}{k_1}\right)^2\right]$$

$$v_l = v_t \sqrt{\frac{4}{3} \frac{(1 - p) + \frac{3}{4}p\left(\frac{K_1}{k_1}\right)^2}{1 - p\delta + \frac{3}{4}p\delta\left(\frac{K_1}{k_1}\right)^2}}$$

(2)

Where:

$$p = \frac{-b - \sqrt{b^2 - 8a}}{2a}$$

668

where K_1 and k_1 are the shear and longitudinal wave number in the ceramic matrix. This equation holds also for large values of pore fractional volume ($\delta < 0.3$).

It can be shown that the relationship between longitudinal velocity and porosity δ, is almost linear in the whole validity range.

The longitudinal attenuation coefficient can be expressed as a function of δ, by the following equation [2]:

$$\alpha = \frac{\delta}{2} k_1^4 a^3 C \tag{3}$$

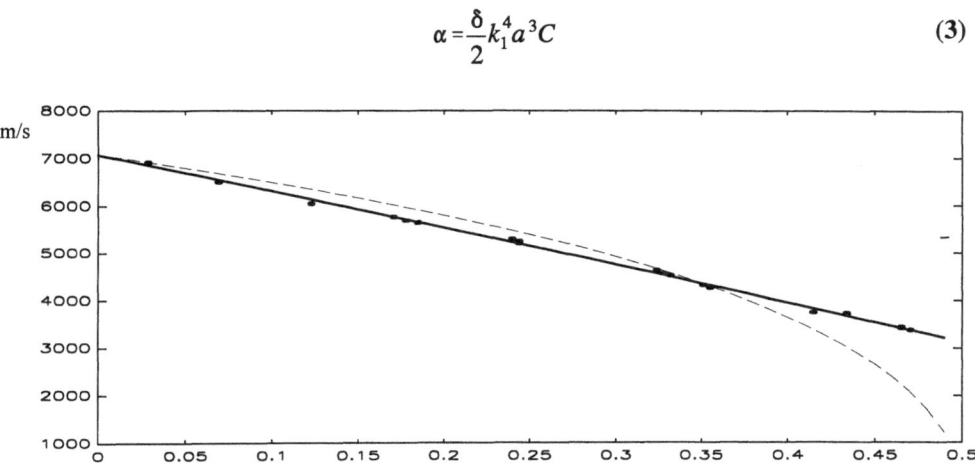

Figure 1. Longitudinal velocity. Experimenta data "o"; experimental curve "———"; Theoretical prediction "---".

where a is the mean pore radius, and C is a constant depending on the void shape (supposed spherical) and on the matrix elastic characteristic.

This expression can be considered valid only for $\delta < 0.1$. Moreover the attenuation depends on the mean pore dimension, thus mean void size must be estimated to obtain absolute value of porosity from attenuation measurements. For these reasons, better results can be obtained for porosity evaluation of green ceramic materials by measuring the ultrasonic velocity. Measurements were performed by using longitudinal and shear transducers, with central frequency ranging from 20 MHz to 50 MHz in a pulse echo scheme. Both dry coupling and immersion transducers were used. The ultrasonic signal was digitized and processed to evaluate time of flight (cross-correlation and phase detection techniques)[3], attenuation coefficient (spectral deconvolution) and spectral centroid frequency (spectral analysis).

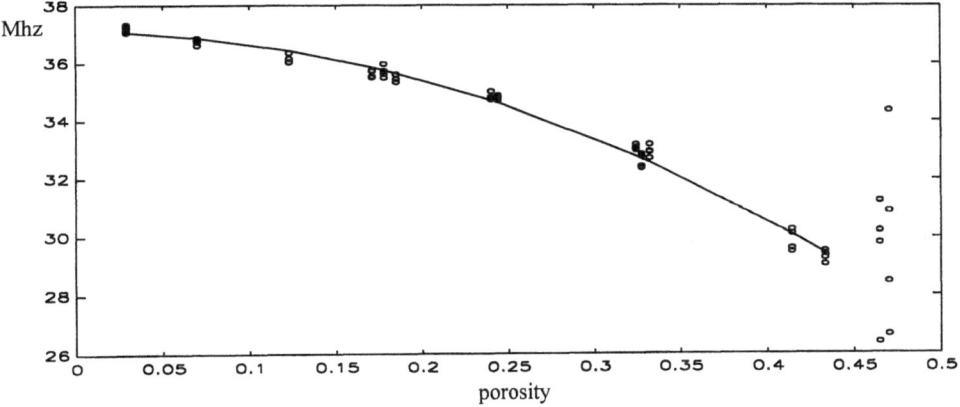

Figure 2. Centroid shift versus fractional volume of pores. Experimental data "o"; interpolating curve "———".

The theoretical prediction for ceramic Zirconia stabilized with Yttria were confirmed by experimental measurements on a set of samples at different sintering stages with porosity in the range $0<\delta<0.5$. In fig.1 the experimental results for longitudinal wave are compared with the theoretical curve coming from eqq.1-2.

The comparison between the measured data and theoretical predictions can be carried out only in the limited range of pore volume fraction, up to 0.3, above which the hypothesis assumed for calculations are satisfied. Outside this range simple relationships between ceramic density and ultrasonic velocities were found which well fit the experimental data:

$$v_l = 7044(1 - 1.11\delta - .09\delta^2)$$

$$v_t = 3676(1 - 0.51\delta - 1.04\delta^2) \tag{4}$$

In fig 2 the centroid shift data versus fractional volume of pores is reported. The interpolating curve is evaluated on the data are obtained with a 37.4 MHz central frequency transducer. The curve has the following expression,:

$$f_c = 37.40(1 - 0.15\delta - 0.64\delta^2) \tag{5}$$

where f_c is the spectral centroid frequency expressed in MHz.

By means of experimental relationships[4], porosity maps were produced for Zirconia specimens with unknown density. In fig.3 a density map is shown (lateral resolution is 1 mm in both direction). For the time of flight estimation, the threshold was set to 1.7 the estimated r.m.s. value of noise. This kind of image

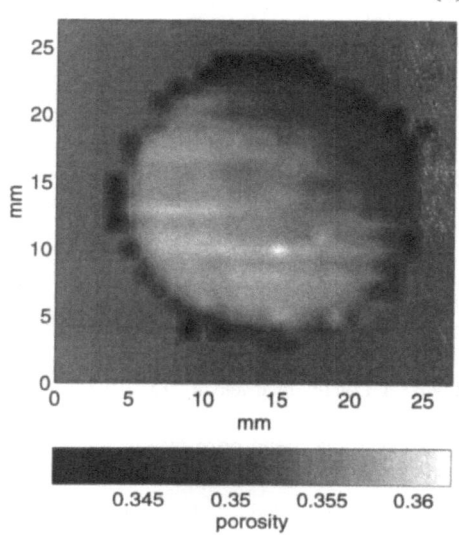

Figure 3. Porosity map.

obtained at different manufacturing stages helps to determine how density variations and voids evolve during the sintering phases.

ULTRASONIC DEFECT DETECTION IN SINTERED CERAMICS: HIGH RESOLUTION IMAGES

Ultrasonic high resolution images of sintered ceramic materials were obtained using immersion pulse echo techniques based on a wide band focused 50 MHz transducer, with 0.25 mm focal spot and 50 mm focal length. A three-axis mechanical scanning system with a 4 µm resolution was used to move the ultrasonic probe over the specimen. A bidimentional scanning was performed and a set of ultrasonic echoes was digitally acquired for subsequent processing.

For defects characterized by a large dimension (more than a few wavelengths) able to generate a signal above the noise, portions of the echoes which correspond to the defect depth were processed. Thus a partial reconstruction of the defect shape was obtained through the assessment of time of flight, signal energy (maximum amplitude) and spectral centroid frequency of the echoes coming from the defect. This technique cannot be applied for the detection of small defects (such as cracks), with a non-planar geometry or any subtle flaws that do not appear as a large obstacle to the propagating ultrasonic beam. This problem was overcome by observing the filtering effect that the defect induces on the

670

forward propagated wave. Hence an indirect detection method is proposed based on the processing of the echo reflected by the sample back wall.

A wave propagating in a material is scattered by any discontinuities encountered: the scattering section of the defect α depends on frequency, dimension and shape of the scattering element[5]. For instance, in Rayleigh scattering regime, a dependence on the fourth power of frequency and sixth power of scatter size of α is expected for a spherical defect. For this reason the echo coming from the specimen backwall contains information about any interacting structure. As mentioned above, the effect of a defect on the propagating ultrasonic wave depends both on its shape and the elastic wavelength. In particular, when the defect size is close to wavelength, the higher the frequency, the larger the energy scattered by the flaw. Thus, when analyzing the spectral energy distribution of the backwall echo, a reduction of the spectral components in the frequency range corresponding to larger scattering effect will be noted. Hence the presence of a flaw produces a decreasing of the backwall echo energy (maximum amplitude), since some energy is scattered away, and a shift of the centroid frequency, since the scattering is frequency dependent.

Ultrasonic images of ceramic samples were obtained by presenting the backwall echo energy and the spectral centroid frequency, allowing the observation of defects undetectable with traditional echographic techniques.

The energy has proved to be a less robust parameter. In fact, the energy reduction is very small and it fluctuates due to back wall irregularities, so that images based on the backwall echo energy are used only when compared with centroid shift based images.

EXPERIMENTAL RESULTS

The traditional echographic images were produced when a signal which overcame a selected threshold occurs. The threshold value was chosen depending on the subsequent signal processing to be performed. For the estimation of the signal energy (maximum amplitude) the threshold was set at twice the r.m.s value of the noise. For spectral analysis

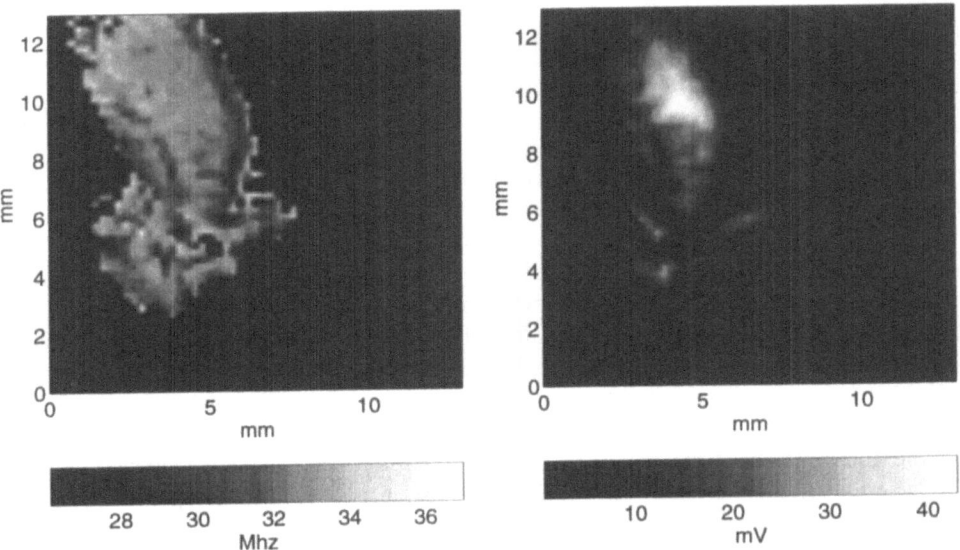

Figure 4. Stratigraphic images of a defect. Centroid image on the left and maximum amplitude image on the rigth.

performed to obtain the spectral centroid frequency the threshold was higher, namely three times the r.m.s. noise value.

In the images hereafter presented, regions where a defect echo cannot be found are represented by the black color.

In fig. 4 two traditional stratigraphic images of a Zirconia cylinder are showed, based on the defect echo maximum amplitude and spectral centroid frequency respectively. For this specimen strong echoes coming from a region close to the sample edge were detected at a depth of 2.7 mm (from the upper surface). This indicates a defect with a 9 x 4 mm area, with the central zone almost planar and parallel to the sample's surface reflecting a strong echo and a margin with a different orientation producing weak signals. The centroid map, confirms this analysis, and indicates a smooth defect surface: the centroid values are uniform and there is no spectral shift. On the analyzed sample an X-ray micrographic analysis couldn't detect any anomaly.

In fig. 5 images obtained by applying the indirect detection technique are shown, based on the energy variation and spectral centroid shift of the backwall echo signal. The defect

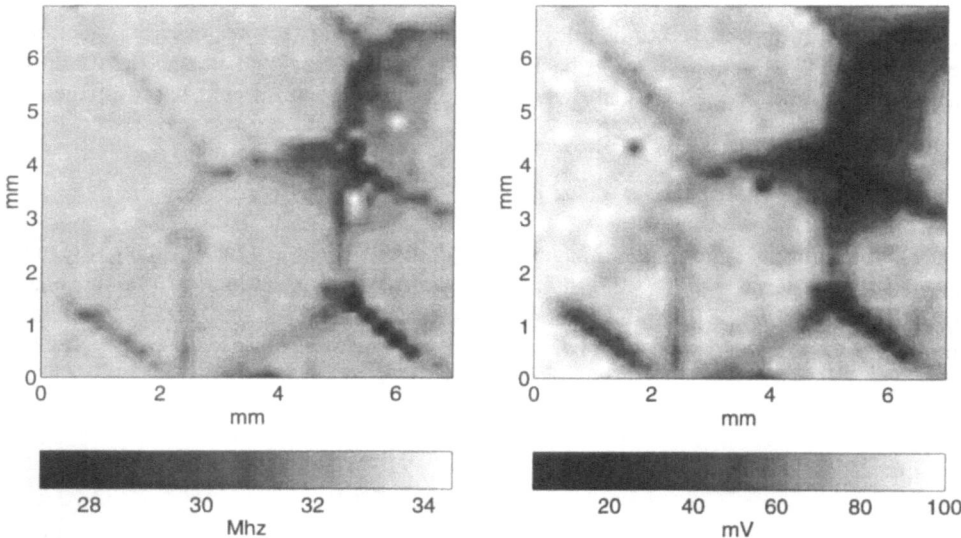

Figure 5. Defect detection by using the indirect method. Centroid image on the left and maximum amplitude image on the rigth.

presented in this figure has a branched structure, developing from a black center (corresponding to a backwall signal loss), which is the characteristic crack shape typical of ceramics. The structure of the defect is visible only through the indirect technique; in fact the traditional stratigraphic image shows only the central portion of the crack. Moreover it is possible to observe an increased image quality when comparing the centroid image with the energy based one.

The porosity map of the tested sample shows that the crack nucleus is located in the region where a strong density gradient is detected (see in fig.3 the different porosity region in upper right portion of the specimen). In the porosity map the defect structure isn't visible since a coarser scanning map was used.

To test the performance of the proposed technique, images of a Zirconia slip-casting sample with calibrated defects were produced and compared with the one obtained by a 50 MHz acoustic micrography (fig 6 left side). In figure 6 this comparison is shown: ultrasonic

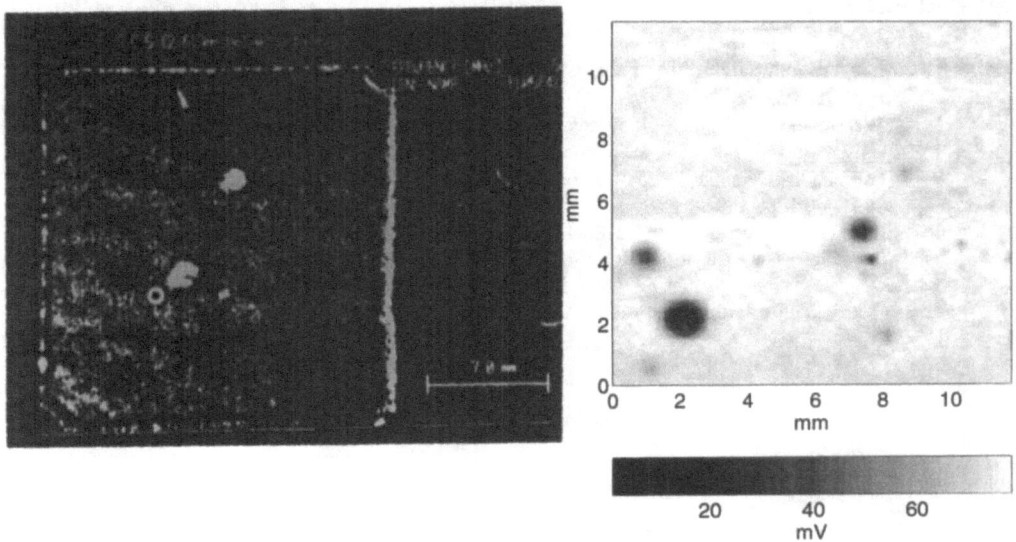

Figure 6. Left: defect image by a 50 MHz Acoustic micrography. Right: centroid image of the defect by using the proposed indirect method.

image, reported on the rigth side, was obtained by the proposed indirect technique.

The calibrated defect were two superficial holes with cylindrical shape and flat bottom with diameters of 1.134 mm and 0.61 mm, and with depths of 0.6 mm and 0.3 mm, respectively. The ultrasonic images were obtained by putting the surface with the defect at the opposite side of the transducer. In figure 6 a structure located at 22 mm below the surface is indicated with a circle.

CONCLUSIONS

Experimental results highlighted that ultrasonic velocity is a suitable parameter to evaluate porosity for defect detection and manufacturing process optimization. The time of flight measurement accuracy is limited by two error causes: an error < 2% due to the cross-correlation algorithm and an error <1% in the sample thickness measurement. In the accuracy evaluation the error due to the digitizing of the signal, 4 ns, is neglected.
Therefore, the resulting accuracy on porosity degree is equal to 3%.

The porosity maps obtained for partially sintered specimens pointed out in some cases the presence of strong porosity gradients, probably due to a lack of uniformity in the forging system.

Stratigraphic images were obtained in order to give qualitative information about shape dimension and location of the defects that reflect a detectable echo signal. Defects with 300 μm mean diameter are observed by this technique (50 MHz central frequency wideband transducer).

The defect detection based on the indirect method gives a qualitative description about the defect size and shape, obviously no information is achieved about its depth. Anyway, to assess the integrity of the product the primary interest is to detect the defect, while the defect location has no discriminant importance.

REFERENCES

1. C.M.Sayers, R.L.Smith, " The Propagation of Ultrasound in Porous Media", Ultrasonics, Sept. 1982, pp. 201-205.
2. P.C.Watermann, R.Truell,"Multiple Scattering Waves" Jour. Math. Phys, 2, (1961), pp. 512-537.
3. E.Biagi, L.Masotti, S.Rocchi, P.Fassina,"Green Ceramic Material Non Destructive Testing: Porosity Estimation" Ceramics Today-Tomorrow Ceramics, Ed. P.Vincenzini, Elsevier Science Publishers 1991, pp.1885-1892.
4. E.R.Generazio, D.G.Roth, G.Y.Baaklini,"Acoustic Imaging of Subtle Porosity Variation in Ceramic",Mat. Evaluation, Vol n°46, Sept. 1988, pp.1338-1343.
5. C.F.Ying, R.Truell,"Scattering of a Plane Longitudinal Wave By a Spherical Obstacle in an Isotropically Elastic Solid", Jour. Appl. Phys., 27, (1956), pp. 1086-1097.

HIGH RESOLUTION ULTRASONIC IMAGES FOR DEFECT DETECTION IN WOOD

E.Biagi, G.Gatteschi, L.Masotti, A.Zanini

Dipartimento di Ingegneria Elettronica
via S. Marta 3
50139 Firenze, Italy

ABSTRACT

The object of the study in this paper was to obtain, for the first time, stratigraphic images for defect detection in wood. Amplitude images characterized by a high signal to noise ratio and frequency images with high axial resolution were obtained.

INTRODUCTION

Wood is a very heterogeneous medium for the acoustic wave propagation. The microstructural details such as fibers, vessels, and parenchyma interact differently with the ultrasonic wave, depending on the ratio between their mean size and the ultrasonic wavelength. The macroscopic result of this interaction is a strong scattering of the ultrasonic energy that leads to a frequency selective attenuation. In the preliminary phase of the research the acoustic characterization of the material is performed and the behaviour of the ultrasonic beam, in terms of spectral distribution centroid shift versus propagation depth, is pointed out. Moreover, the different spectral content of the echo-signal coming from the defects in respect to that backscattered from the morphological constituent elements is put in evidence.

Two different signal processing algorithms are applied on the digitalized radiofrequency signal. The first one employs a time domain signal processing to simulate a superheterodyne beating; the second one works in the frequency domain.

In the superheterodyne system, the radiofrequency signal beats with a sinusoidal frequency sweep in order to shift the radiofrequency band to a fixed band pass filter, whose bandwidth is controlled by the axial resolution. The frequency sweep versus time is controlled taking into account the centroid shift versus the investigated depth induced by the wood. The algorithm allows the investigation of each depth with the best signal to noise ratio. The obtained images detect the presence of defects and also give qualitative information about the internal structure.

The processing in the frequency domain is designed to point out the defects from the structural elements. The employed digital filter neglects the signal components which present the typical scattering spectral distribution produced by the structural elements. The radiofrequency signal is divided into appropriate intervals, to which a correction filter has been applied to each. To avoid errors and false indications, and also to improve the image quality in terms of axial resolution, another effect must be considered. When the ultrasonic pulse is transmitted, the echoes generated by closely spaced reflectors, along the beam axis, return to the transducer overlapping each other and giving therefore a nearly continuous signal (the so-called spectrum scalloping). A corrective algorithm is applied firstly to avoid an uncorrected evaluation of the power spectral distribution and then to increase the axial resolution. The final images, obtained on artificially induced defects, show an improvement of about 30% in the axial resolution.

EXPERIMENTAL MEASUREMENT SETUP

A monoelement focalized ultrasonic transducer is employed with a bandwidth of 2 MHz at -6dB and a central frequency of 2.25 MHz. A pulse-echo immersion technique is used. A x-y-z mechanical movement system is employed to move the transducer over the sample. All the system is controlled by a PC. The radiofrequency echo signal is acquired by a digital oscilloscope with a 400 MS/s sample rate. The data are transferred and stored in a PC. The signal processing is carried out in MATLAB by using a workstation SUN SPARC STATION 10.

HARDWARE IMPROVEMENT

As mentioned above, the interaction of the wood constituent elements with the ultrasonic wave induces a strong scattering of the ultrasonic energy inside the material. In the frequency domain this effect corresponds to a selective filtering action on the ultrasonic wave. The ultrasonic signal intensity A_Z at a depth Z, inside the medium, can be described as [1]:

$$A_Z = A_0 \exp[-\int_{Z_0}^{Z} \alpha(z,f) \ dz] \tag{1}$$

where:
- A_o is the amplitude at a reference depth Z_0;
- $\alpha(z,f)$ is the overall attenuation coefficient.

The medium induces on the signal which propagates an exponential decreasing amplitude which in turn is dependent on time (depth Z) and frequency. This exponential decrease leads to a signal to noise ratio reduction when the investigated depth inside the material increases. An immediate consequence is the loss of power resolution associated to deeper investigated zones. So is necessary to apply a time varying amplification on the ultrasonic signal: TGC (Time Gain Control)[2]. A versatile hardware implementation of the TGC is realized. The TGC board can be externally programmed to take into account the different absorption material characteristics, the material thickness and also the transducers geometric features.

The medium frequency selective attenuation is taken into consideration by an appropriate signal processing.

SIGNAL PROCESSING

On the spectrum of the ultrasonic propagated signal, a shift towards the lower frequencies occurs which leads to a lowering of the centroid position. This situation is noted only as far as a depth of approximately 10 mm. For depths in excess of this value the centroid remains constant at 1.5 MHz. This effect can be explained by considering the frequency dependent global attenuation coefficient [1]:

$$\alpha(z,f) = \alpha_A(z,f) + \alpha_S(z,f) \qquad (2)$$

The first term is the attenuation due to absorption which is related to the bulk material characteristics, such as the density. The second term is related to the material scattering action and can give information about the microstructural organization of the medium under test. The scattering coefficient has been classified by the ratio of sound wavelength λ to the mean grain diameter D. When this ratio is high (Rayleigh regime) the scattering attenuation coefficient $\alpha_S(z,f)$ depends on f^4.

The spectrum of the signal backscattered from the microstructural elements, which are of small dimension with respect to signal wavelength, presents a shift towards the higher frequencies as it was demontrated in a previous work [2]. This spectral variation gives a widening of the band and a shift of the centroid.

Two different signal processing algorithms, devoted to increase the signal to noise ratio and to improve the images accuracy and resolution, have been designed. The first one employs a time domain signal processing; the second one works in the frequency domain.

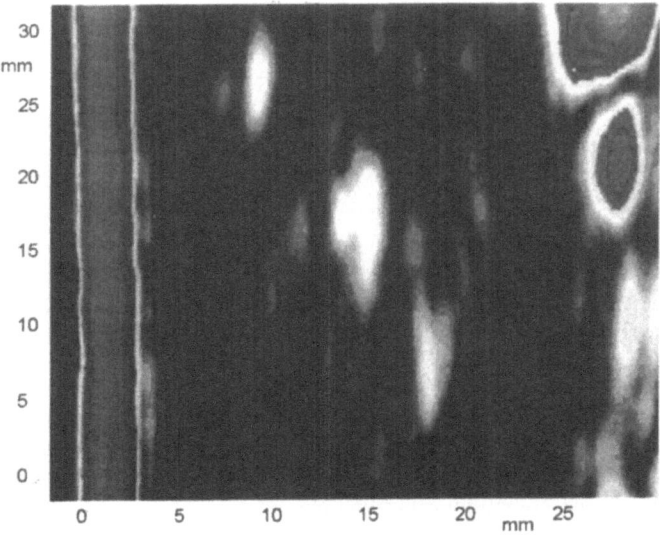

Figure 1. Super-heterodyne image of three artificially induced defects (dia 1.5 mm).

To maximize the signal-noise ratio (S/N) it is necessary to effect a pass band filtering of the signal in order to reduce the equivalent noise band. A narrow band improves the S/N but reduces the system axial resolution. Filters with a bandwidth from -6dB to -10dB of the useful signal have been found the best compromise.

Taking into account the frequency centroid shift versus depth induced by the frequency-selective attenuation of the medium, the central frequency of the band pass filter must be varied with the depth. The proposed solution is to move the band of the signal as the depth varies in such a way as to keep a filter fixed band. This is possible by beating the signal with a variable frequency to compensate for the centroid shift with the investigated depth. A super-heterodyne beating is simulated. Figure 1 reports the image of three artificially induced defects (1.5 mm dia.). The figure is the black and white reproduction of the original image where the signal amplitude modulates the cromatic code.

The wood thickness is along the horizontal axis; the transducer scan path is reported along the vertical axis. The first wood wall is positioned on the left side of the image (continuous strip); the sample back wall appears on the right side.

With the utilized wavelength, a specular reflection occurs for interface dimensions above 1 mm, which coincides with the minimum dimension of the detectable defect. However, the structural elements' backscattered echoes have a spectral distribution which in part coincides with those of specular reflection. With this algorithm it is impossible to completely ignore the structural element and highlight the defects.

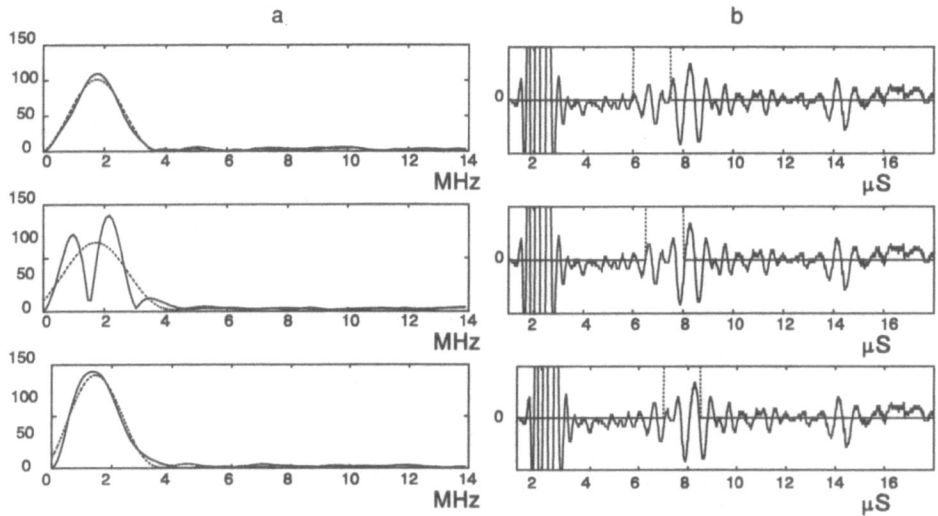

Figure 2. a) Spectral distribution: —— is referred to the selected time window, --- is the corrected spectrum. b) Radiofrequency echo signal; --- shows the selected time window.

As mentioned above, the spectrum of the backscattered signal is characterized by high frequency spectral contents with respect to the spectrum reflected from the defect. For this reason, the other proposed solution is to use software able to elaborate a selective frequency analysis to ignore the signal contributions characterized by spectral modifications typical of the backscattered signal. These modifications consist in a relative widening of the spectrum and in a centroid shift toward the higher frequencies. In fact the signal intensity A_S backscattered at a depth Z and collected by the transducer assumes the following expression:

$$A_S = A_0 \ C_S(Z,f) \exp[-2 \int_{Z_0}^{Z} \alpha(z,f) \ dz] \qquad (3)$$

where the term $C_S(Z,f)$ takes into account the backscattering effect depending on the fourth power of the frequency.

An analysis of the successive temporal windows of the signal is needed to obtain the necessary spatial correlation. A Hamming window with 1.5 μs temporal length and a window overlapping of 0.5 μs is employed[3]. The applied algorithm taking into consideration the so-called "spectrum scalloping effect" is able to correctly evaluate the echo spectral distribution and to increase the system axial resolution. When the ultrasonic pulse is transmitted, the echoes generated by closely spaced reflectors, along the beam axis, return to the transducer overlapping themselves. This gives a nearly continuous signal which can induce errors and false indications. When only two reflectors are illuminated by the ultrasonic beam the power spectrum $|X(w)|^2$ assumes the following simplified expression[4]:

$$|X(\omega)|^2 = |S(\omega)|^2 \cdot [r_1^2 + r_2^2 + 2r_1r_2 \cdot \cos(\omega \cdot \frac{2l}{v})] \qquad (4)$$

where:
- l is the distance between the two reflectors;
- $|S(w)|^2$ is the power spectrum of an isolated reflector with reflection coefficient equal to 1;
- r_1 is the reflection coefficient of the first reflector;
- r_2 is the reflection coefficient of the second reflector;

In the frequency domain an oscillation with a period equal to 2l/v is superimposed on the single reflector power spectrum $|S(w)|^2$.

A significant image, based on the evaluation of spectral distribution parameters, is obtained if an appropriate correction is performed for each time window to avoid the errors induced by the scalloping effect.

In fig. 2 the radio-frequency echo signal of a single sight line is reported on the right side, the dotted line selects the chosen time window. On the left side of the figure the continuous line is the spectral distribution of the selected time window, and the dotted line is the spectrum with the applied correction. No significant correction takes place if no oscillation induced by the scalloping effect is found. The filter action is evident when a partial overlapping of two echo signals inside the selected time window occurs. In figure 3 the centroid amplitude versus the windows' number of the same sight line of figure 2B is reported.

In order to increase the system axial resolution the algorithm detects if two or more reflections occur in the same time window and if their distance is such that they separately appear in surrounding windows. If this happens the image contribution of the current window is neglected. Figure 4 shows the

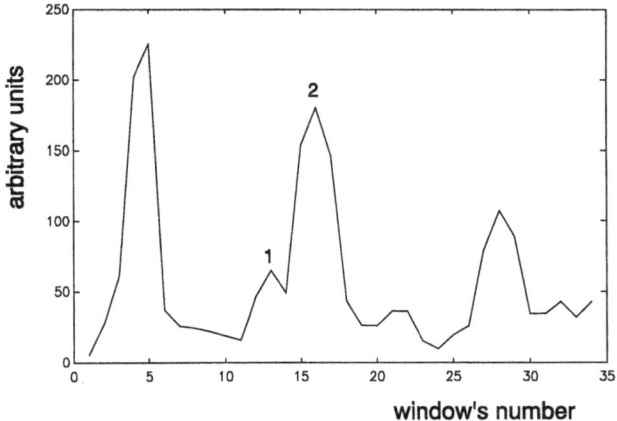

Figure 3. Centroid amplitude of the echo signal of fig.2b, versus the windows' number, after the scalloping correction.

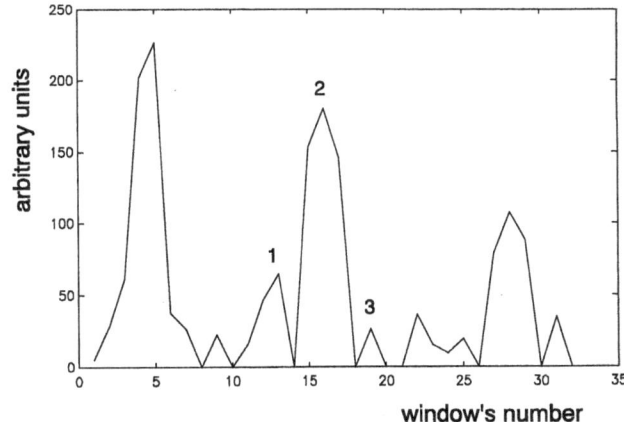

Figure 4. Centroid amplitude. Improvement of the axial resolution of figure 3 obtained by neglecting the windows with scalloping.

result of the application of this algorithm on the same signal of figure 3. It is possible to separate the reflections coming from the three interfaces spaced less than the axial resolution of the employed ultrasonic transducer. An improvement of about 30% in the axial

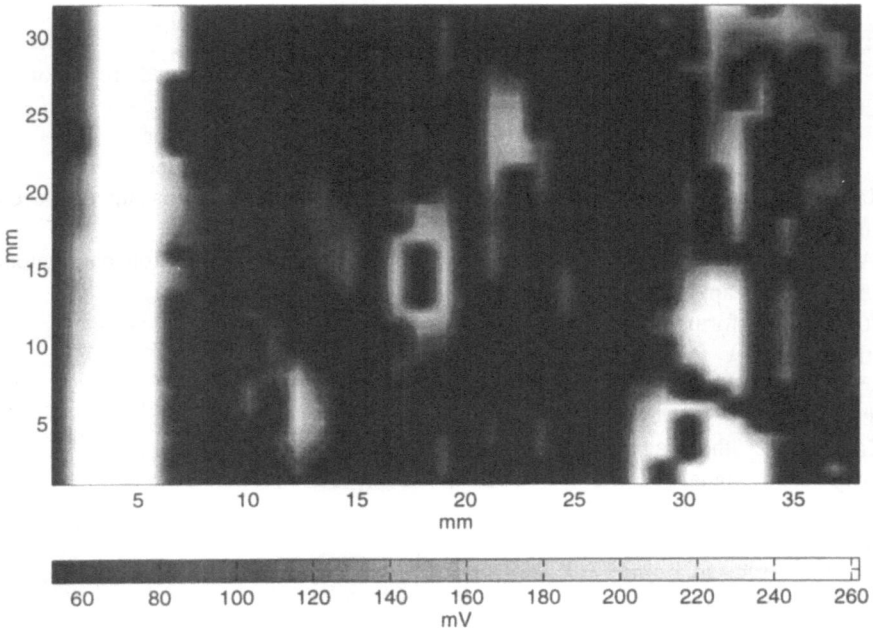

Figure 5. Frequency image of three artificially induced defects (dia 1.5 mm).

resolution is obtained.

By applying the above mentioned frequency selective analysis to ignore the signal contribution due to the structural scattering and employing the scalloping correction, the image reported in figure 5 is obtained. The three artificially induced defects as reported in figure 1 are shown. The image gray code is modulated by the amplitude of the centroid of the signal spectral distribution. The wood thickness is reported along the horizontal axis. The ultrasonic transducer scan path is reported along the vertical axis. Observing the image, it can be affirmed that the applied software correction is able to avoid the structural interferences and also to solve the dimension of the middle defect. In fact the first and the second wall of the defect are clearly detected.

CONCLUSION

The realization of a particular T.G.C. has allowed to investigate each depth zone with the same power resolution.

The signal processing based on superheterodine beating simulation has led to a noise reduction. The obtained images detect the presence of defects and also give qualitative information about the internal structural elements. To highlight the defects a digital filter able to neglect the signal component induced by the structural element has been designed. The application of the scalloping correction filter has produced an increase in the axial resolution.

REFERENCES

1. J. Saniie, T. Wang, N.M. Bilgutay, " Analysis of Homomorphic Processing for Ultrasonic Grain Signal Characterization ", IEEE Transaction on Ultrasonics, Ferroelettrics and Frequency Control, Vol. 36, N° 3, Maggio 1989.
2. E.Biagi, M.Cerofolini, G.Gatteschi, A.Lorenzi, L.Masotti, A.Zanini,"Tomografia ad Ultrasuoni per la Caratterizzazione Defettologica del Legno", Alta Frequenza, Marzo 1994.
3. A.V. Oppenheim, R.V. Schafer, " Digital Signal Processing ", ed. Prentice-Hall, (1975).
4. E.Biagi, G.Castellini, L.Masotti, S.Rocchi, "Ultrasonic Spettroscopy of Composite Materials", Ultrasonics International'89, Butterworth Scientific Ltd., Madrid, Giugno 89.

RECONSTRUCTION OF SOUND VELOCITY PROFILES IN THE ATMOSPHERE FROM IMPULSE MEASUREMENTS

Detlef Englich and Volker Mellert

Dept. of Physics
Carl von Ossietzky Universität
D–26111 Oldenburg
Germany

INTRODUCTION

Wind, temperature and moisture profiles and their fluctuations describe the micro-meteorological conditions in the atmospheric boundary layer (ABL). Generally, these profiles are experimentally determined by local measurements with meteorological masts. Sound velocity is mainly affected by wind and temperature. By measuring sound propagation it is possible to determine the wind and temperature profile of the stratified atmosphere. Sound velocity is measured with a short acoustic impulse travelling over a distance of several hundred meters. The impulse response is compared with a synthetic response gained by an advanced ray tracing model taking into account phase and amplitude history of the "rays". The sound rays carry the information of wind and temperature along their propagation path. With this spatially integrated information the real profiles are retrieved by adjusting profile parameters from micro-meteorological theories. Since the atmosphere is not stationary, the parameters are only valid for ensemble averages.

SYNTHETIC IMPULSE RESPONSE

One simplification for calculating sound propagation is the ray tracing technique, which is justified as long as the acoustic wave number is large compared to the gradients of the field of refraction index. This situation is often the case outdoors except in the direct vicinity of the ground. The used ray tracing method is applied for a layered, moving medium, i.e. the refraction index depends on the height (or depth) of the medium and not on horizontal location[1,2]. In the simulations logarithmic wind and temperature profiles proved in micrometeorology are used determining the refraction index field[3]. Especially wind fluctuations caused by turbulence, diurnal or weather dependent variations lead to variations of the index of refraction creating changes in the amplitude– and phase structure of measured acoustic sound signals for a fixed source-receiver–geometry. Interferences appear if the direct ray and the rays, which are ground reflected once or more times, arrive at the receiver. Because of the reflections the complex and frequency dependent impedance of the ground influences the impulse response. The ray tracing programme calculates very fast all eigen rays. Every eigen ray has its own history containing absolute travel time, path length, intensity level at receiver, angle of incidence, number of caustics, number of reflections and reflection angle. It is possible to calculate a "ray tracing impulse response" combining the information of all rays.

The expected measured sound signal is obtained, if the used emitted sound signal is convolved with the "ray tracing impulse response" with regard to the ground impedance and the air absorption. The acoustic impedance of the ground is not known, but well established theories exist to model it by a set of one to four soil parameters[4,5]. The air absorption is known from theory and depends mainly on the moisture of the air[6]. For calculating the air absorption only a mean moisture value is used instead of a profile.

For our measurements we use a spark source radiating a powerful broadband impulse, which is highly reproducible in phase and amplitude (figure 1). It was developed especially for outdoor sound propagation measurements. For this application it is necessary to use pulses of short duration, because the temporary changes of the mean values and of the deviations, for instance of the wind, are not recognizable with any other signal[7,8]. Mean values and higher statistical moments do not become stationary. It is assumed that for each realisation frozen turbulence hypothesis holds, at least during the duration of the sounding impulse.

The spark source signal convolved with the "ray tracing impulse response" results a "synthetic impulse response" consisting of a pulse train. It is justified to call the convolved signal an *impulse response* because of the used broadband and very short pulse. Every pulse carries the integrated information about the index of refraction along its ray path. Every path is a "prick" through the ABL like a cut in tomographic methods. The part of the ABL which could be investigated is defined by the source-receiver–geometry with the ground as a bottom boundary and the maximal height of the direct eigen ray as the top. It is called transfer channel[9].

It is necessary to use functional correlations established in the boundary layer meteorology in order to determine wind and temperature conditions from measured impulse responses. The used correlations are wind and temperature profiles which are described by several parameters like friction velocity u_*, temperature scale T_* and roughness length z_0. A systematic investigation of sound propagation and ray tracing with such profiles is given by Huisman[10]. It is possible to bring a "synthetic impulse response" in line with a measured one by varying such parameters[11,12]. Also this technique allows to draw some conclusions from the fluctuations in the measured impulse response into the fluctuations of the index of refraction field[13].

MEASUREMENT

In a May afternoon the profiles of figure 2 were ascertained with data monitored by a meteorological mast. The sky was cloudy, perhaps overcast, so that no strong temperature gradient was built up near the ground. The stratification was neutral.

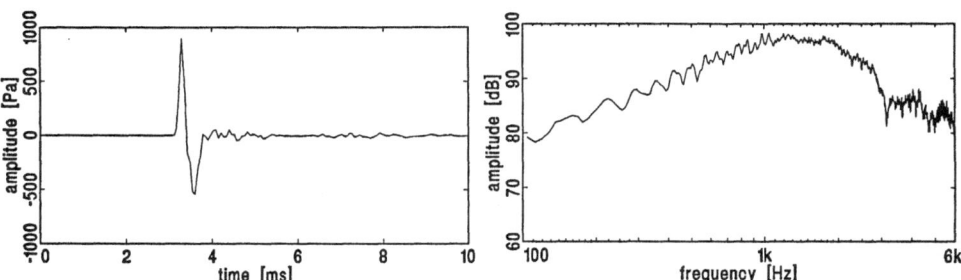

Figure 1. Left picture shows the time signal of the spark source monitored with a bandwidth of 6.4 kHz at a distance of 2 m and corrected for 1 m. Note the amplitude scale. The corresponding spectrum is depicted on the right side. It has a maximum nearly at 1.5 kHz which mirrors the shortness of the impulse of less than 1 ms. For low/high frequencies increases/decreases the spectrum with 6 dB per octave.

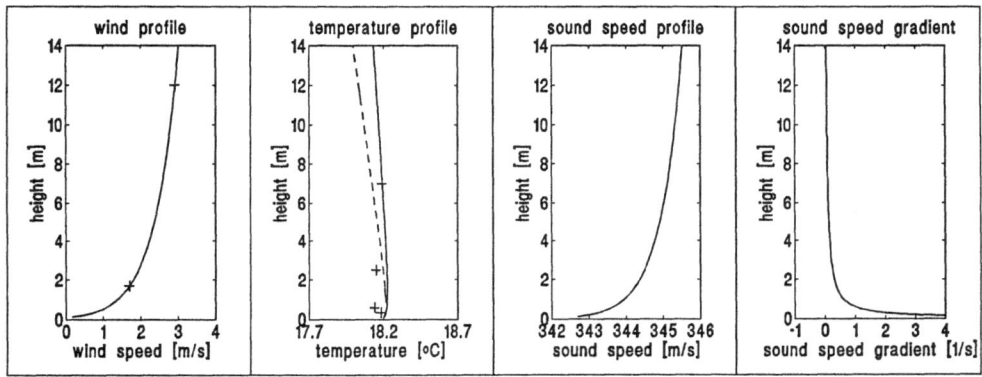

Figure 2. These profiles are calculated with monitored wind and temperature data of a meteorological mast (figure 3) using the Monin–Obukhov similarity theory known in micrometeorology. Crosses are sensor data. Solid lines are adjusted curves for wind and temperature, broken line shows the potential temperature. The adjusted parameters are friction velocity $u_* = 0.245$ m/s and temperature scale $T_* = 0.006°C$ corresponding to an Obukhov length $L = 745$ m, i.e. it is a temperature profile typical of a neutral stratification. Hence, the sound speed profile and the sound speed gradient profile are mainly influenced by the wind profile. A strong gradient appears especially near ground.

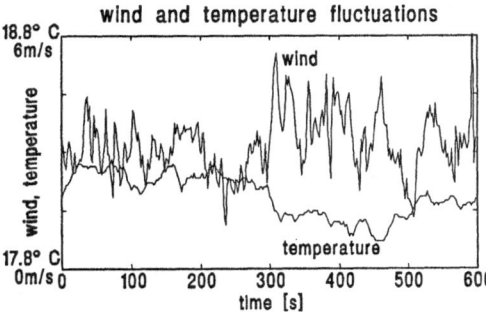

Figure 3. Two time series of wind and temperature data were monitored by sensors on a meteorological mast. The wind sensor was fixed at 12 m height, the temperature sensor at 7 m height. The interval of two successive data points is 2 s. A change of the mean values is visible at 300 s. For the interval from 0 s to 300 s the mean value of the wind is 2.9 m/s with a standard deviation of 0.6 m/s and the mean value of the temperature is 18.2°C with a standard deviation < 0.1°C.

685

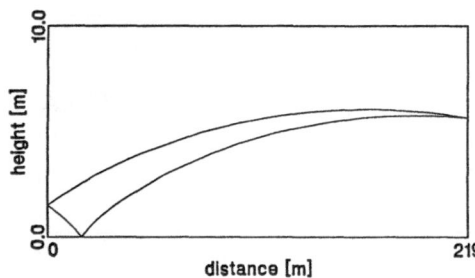

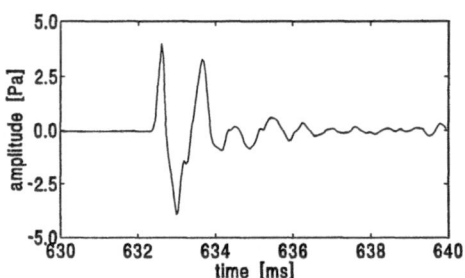

Figure 4. Using the profiles of figure 2 this ray tracing simulation was made. The source height is 1.5 m, the receiver height is 5.7 m and the distance is 219 m. Only two eigen rays eject from the microphone: the direct ray and a once reflected ray. At half the distance both rays reach a height above 5 m and move in a layer from 5 m to 6 m. of this region only.

Figure 5. For the geometry of figure 4 a measured pulse is shown. The time axis is the absolute travel time monitored by a radiofrequency trigger signal. The first impulse stems from the direct ray, the second one from the reflected ray. Latter impulse shows phase changes because of the reflection at the ground.

The fluctuations of wind and temperature over a time gap of 10 minutes are shown in figure 3. The order of magnitude of the temperature fluctuations is less than 0.1°C, but the order of magnitude of the wind fluctuations is 1 m/s to 2 m/s like the mean wind itself. At 300 s a small decrease of the mean temperature and an increase of the mean wind and especially its fluctuations is visible. Because of this weather change the following results are based on data measured before 300 s.

Using these profiles a ray tracing was calculated depicted in figure 4. Only two rays arrive at the receiver, the direct and a once reflected ray. The time lag between both rays is 0.8 ms. The direct ray moves in a layer from 1.5 m height to 6.1 m height, but it moves above a height of 5.5 m for more than half the distance. In comparison the reflected ray moves through the whole layer from the ground to the receiver height, but it is also above 5 m height at half the distance.

In figure 5 one of a series of measured signals is shown. The relative time lag between direct and reflected pulse is of the same order as expected from theory, but the amplitude of the reflected impulse is higher indicating deviations from the theoretical profile near ground. The absolute travel time was monitored by a radiofrequency trigger signal. The impulse of figure 1 was emitted 65 times with a repetition time of 20 s. The mean travel time is 632.7 ms with a standard deviation of 1.1 ms. Assuming that these values are caused mainly by the wind above 5 m height it is possible to estimate a mean wind at 2.4 m/s and with a deviation at $\sigma_u = 0.5 \ldots 0.6$ m/s. The latter value corresponds well with the deviation of $\sigma_u = 0.6$ m/s determined by the wind sensor data at 12 m height. From boundary layer meteorology is known that in a neutral stratification the wind standard deviation σ_u in wind direction and the friction velocity u_* are proportional: $\sigma_u = Au_*$ with $A \approx 2.47$. In our case this proportionality leads to $u_* = 0.202 \ldots 0.243$ m/s, which is a little smaller than $u_* = 0.245$ m/s determined by the meteorological mast data. The fluctuations of the temperature are negligible. According to the temperature deviation of $< 0.1°C$ the changes of the sound speed are < 0.06 m/s.

There is another method to determine the friction velocity u_*. It is based on the increase of the travel time difference between direct and once reflected ray[11]. The time

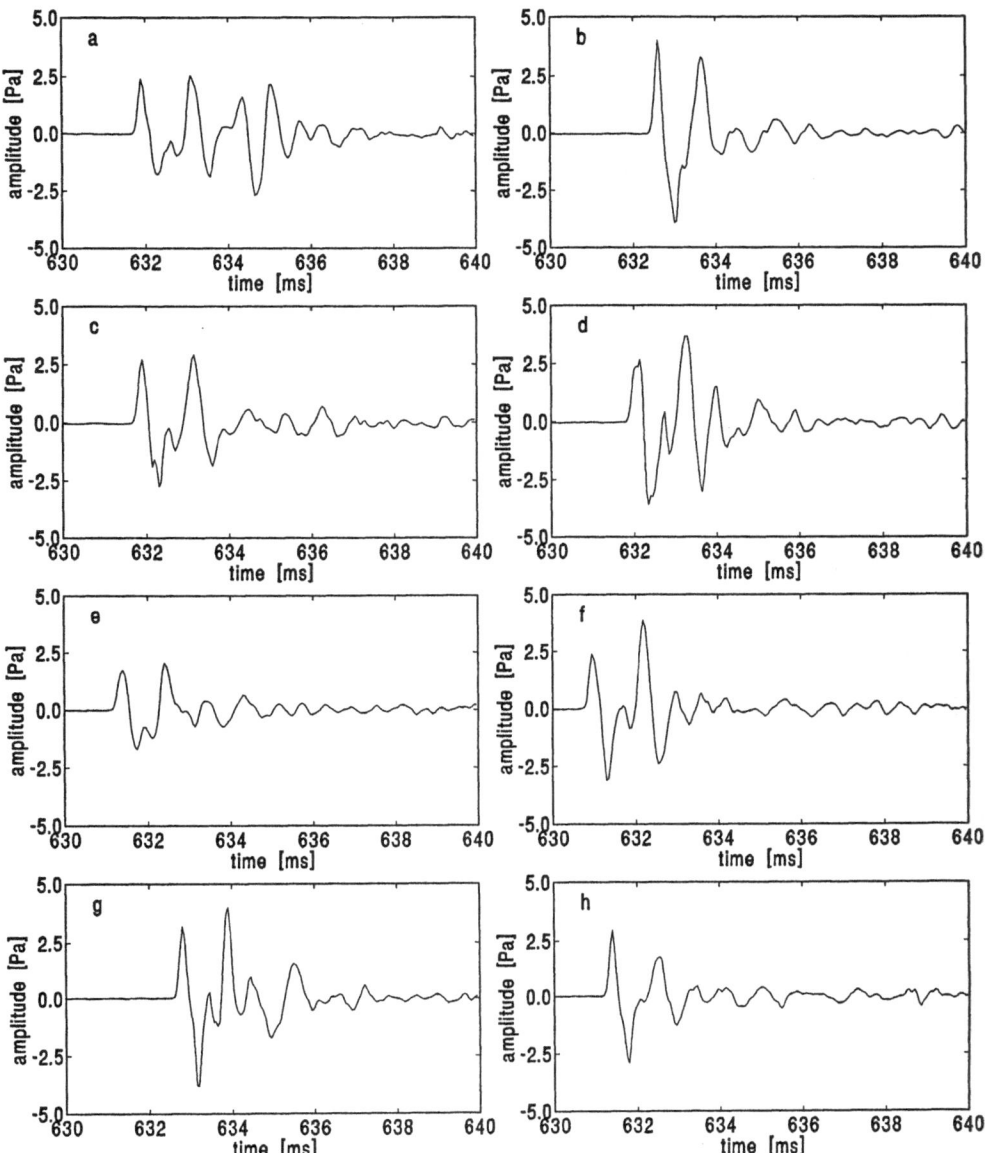

Figure 6. Successively monitored impulse responses measured with the geometry of figure 4 are depicted. The time lag was 20 s, the time axis shows the absolute traval time. In figure b–h two pulses are visible, which interact slightly. These three impulse responses agree well with the ray tracing using the profile of figure 2. The additional third pulse in figure a is explainable with a strong sound speed gradient at a height of 5 m. The second pulse in figure a shows a phase shift of nearly $-\pi/2$ because of the ground reflection. The third pulse has an additional phase shift which indicates that it touches a caustic.

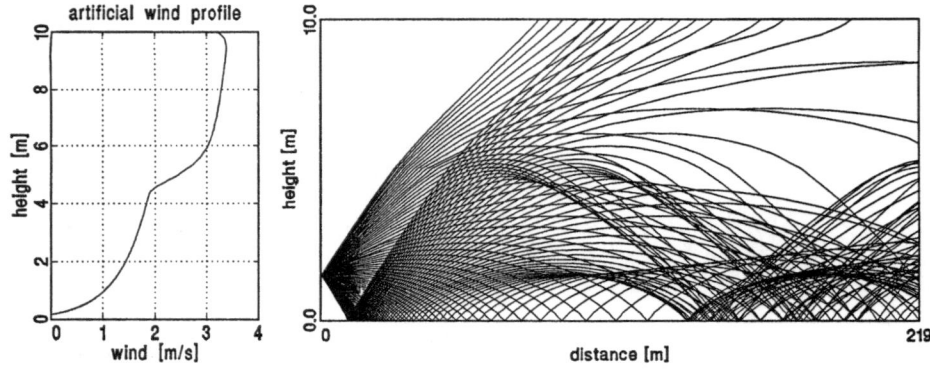

Figure 7. Left side shows an artifical wind profile used to explain the appearance of the third pulse in figure 6.a. Using this wind profile a ray tracing simulation was made. The ray group in the right picture shows strongly bent rays at a distance of nearly 80 m and 5m height, which are reflected at about 120 m and reach the receiver, too.

difference is 0.9 ms with a standard deviation of 0.15 ms calculated by the magnitude of the autocorrelation function. This is slightly longer than 0.8 ms of the ray tracing simulation. The 0.9 ms lead to a friction velocity $u_* = 0.201$ m/s, which is 20% smaller than the $u_* = 0.245$ m/s of the meteorological mast data. Considering the time difference deviation of 0.15 ms u_* is $0.158 \ldots 0.244$ m/s. All the results agree well and show the possibility to determine by sound measuremments micrometeorological parameters.

The measured impulse response in figure 6.a contains three pulses. The third one is not explainable with the mean profiles of figure 2. This fact points to an increased sound speed gradient, because the travel time of the direct ray is not affected. Some tests were made to simulate a situation. For that purpose the wind profile was changed in the limits of its standard deviation and a good result was reached with an artifical wind profile with a 'smooth' step at a height of 5 m from nearly 2 m/s to about 3 m/s (figure 7). In a ray tracing simulation with this artifical profile the 'smooth' step leads to another emitted ray reflected at a distance near 120 m. The ray group figure shows that these additional ray(s) produce a focusing effect at its (their) reflection distance. The additional ray graze a caustic leading to another phase shift visible in the third pulse of figure 6.a. Such a step in the wind profile could be caused by an eddy with a diameter of about 5 m that is a possible situation.

CONCLUSION

By using the very fast ray tracing technique and by taking the individual ray history into account it is possible to calculate impulse responses for certain profiles and profile variations. With a set of analytical descriptions of wind and temperature profiles it is possible to determine profile parameters and their variations by acoustic impulse measurements. An extension of the method including the amplitude structure to determine structure functions and fluxes is possible.

REFERENCES

1. R.J. Thompson, Ray theory for an inhomogeneous moving medium, *J. Acoust. Soc. Amer.* 51:1675 (1972).

2. R.J. Thompson, Ray acoustic intensity in a moving medium (I+II), *J. Acoust. Soc. Amer.* 55:729 (1974).

3. R.B. Stull, "An Introduction to Boundary Layer Meteorology", Kluwer Academic Publishers, New York (1988).

4. M.E. Delany and E.N. Bazley, Acoustical properties of porous materials, *Applied Acoustics* 3:105 (1970).

5. K. Attenborough, Acoustical impedance models for outdoor ground surfaces, *J. Sound Vib.* 99(4):521 (1985).

6. ISO 9613, Acoustics – Attenuation of Sound during Propagation Outdoors, Part 1: Calculation of Absorption of Sound by the Atmosphere.

7. D. Englich and V. Mellert, Impulsschallquelle für tiefe Frequenzen (in German), *Fortschritte der Akustik — DAGA Dresden*, is being printed (1994).

8. U. Radek, H. Klug and V. Mellert, Impulsive sound source of high intensity for outdoor sound propagation measurements, *Proc. 13th Int. Congr. on Acoustic (ICA), Belgrade* 2:23 (1989).

9. D. Englich, V. Mellert, U. Radek and R. Schmidetzki, Measurement of meteorological parameters of the atmospheric boundary layer by tomographic sounding, *Proc. Inst. Acoust., Keele* 13:365-372 (1991).

10. W.H.T. Huisman, "Sound propagation over vegetation–covered ground", Dissertation, Katholieke Universiteit Nijmegen, Netherlands, (1990).

11. H. Klug, Sound speed profiles determined from outdoor sound propagation measurements, *J. Acoust. Soc. Amer.* 90:475 (1991).

12. D. Englich, R. Schmidetzki, U. Radek, Meteorologische Einflüsse auf die Schallausbreitung über große Entfernungen, Teil 1 und 2 (in German), *Fortschritte der Akustik — DAGA Bochum* 733 (1991).

13. D. Englich and V. Mellert, Strahlenakustische Untersuchung von in der atmosphärischen Grenzschicht gemessenen Impulsen, *Fortschritte der Akustik — DAGA Frankfurt* 291 (1993).

WAVEFORM TOMOGRAPHY FOR TWO PARAMETERS IN ELASTIC MEDIA APPLIED TO CROSS-WELL GEOPHYSICS

Feng Yin and Jerry M. Harris

Department of Geophysics
Mitchell Building
Stanford University
Stanford, CA 94305-2215

ABSTRACT

In this paper, we developed one waveform tomography method for P velocity and density inversion in elastic media. Starting from the elastic wave equation, we derived one P wave equation which includes the scattered terms of P-wave to P-wave conversion and S-wave to P-wave conversion. By discretizing the scattering integral equation corresponding to this equation directly, we can obtain one equation corresponding to each point on the waveform. The result of this formulation is a very large system of algebraic equations that is solved using ART and SIRT. The results of computational simulation applied to cross-hole geometry show that our method is valid when the velocity and density perturbation is not very large. Next we will extend this method to heterogeneous media and apply it to real field data.

INTRODUCTION

Recently, the wave equation tomography methods based on acoustical wave equation have been applied to the real cross-hole data (Harris and Wang,1993). Therefore, these methods become more attractive in the inversion field. But, there are many complicated wave events contained in seismogram data, e.g., S wave, mode converted wave, Rayleigh waves, etc. , which can not be described by an acoustical equation. However, each of the different events provides useful information about the subsurface compressional and shear wave velocities and all these events can be modeled well by the elastic wave equation. Therefore, we should develop the inversion methods based on elastic wave equation to obtain more physical parameters of the media. In elastic wave equation inversion studies, one method is to find the parameters (P- and S-wave velocities and density) that minimize the square error between the wavefield computed using this model and the observed wavefield (Tarantola, 1984; Mora, 1986). But in this method, the seismogram must be calculated by solving the 2D elastic wave equation numerically in each iteration, therefore, the computation of this method is very costly. In order to simplify the inverse problem, some researchers considered a one dimensional inverse problems (Norton and Testard, 1988) and obtained a better inversion solution. However, the media of the earth usually are two dimensional, and in this case, this method is unvalid. Additionally, some only used the SH wave in their inversion method (Hooshyer and Weglein, 1986). In this case, the equation for inversion can be reduced to a scalar equation, but the SH wave is independent of P wave and SV wave, and therefore, the SH wave can only invert shear modulas and density. In order to obtain Lamé parameters and density in two dimensional elastic media effectively, we should develop the inversion methods based on the P wave equation.

In this paper, starting from the 2D isotropic inhomogeneous elastic wave equation, the scattering theory in weak inhomogenous media is studied. We also establish a scalar P wave equation which includes the scattered terms of P-wave to P-wave conversion and S-wave to P-wave conversion. From this equation, a scattering integral equation in the frequency domain is derived. Under the Born and geometrical optical approximation, we show how the time-domain scattered fields can be related to density and velocity perturbations through a generalized Radon transform. Therefore, time-domain P-wave scattered fields can be used as projection data to relate each point in the scattered waveform to the elastic parameters. The result of this formulation is a very large system of algebraic equations that is solved using ART and SIRT. The algorithm is tested on numerically simulated data generated for the cross-hole geometry of sources and receivers. The test results illustrate the ease of implementation and robustness of the method.

THEORY OF P WAVE SCATTERING IN WEAK INHOMOGENEOUS MEDIA

From 2D isotropic inhomogeneous elastic wave equation, we have

$$\frac{\partial^2 \mathbf{u}}{\partial t^2} = c_p^2 \nabla(\nabla \cdot \mathbf{u}) - c_s^2 \nabla \times (\nabla \times \mathbf{u}) + \frac{1}{\rho}[(\nabla \lambda)\nabla \cdot \mathbf{u} + (\nabla \mu) \cdot (\nabla \mathbf{u}) + (\nabla \mathbf{u}) \cdot (\nabla \mu)] \quad (1)$$

where $\mathbf{u} = \mathbf{u}(\mathbf{x}, t)$, it is displacement field, c_p, c_s are velocity of P wave and S wave, μ, λ, ρ are shear modula, Lame parameter and density of elastic media.

Assuming that the background is homogeneous and its elastic parameters are $c_{p0}^2, c_{s0}^2, \lambda_0, \mu_0, \rho_0$, and corresponding perturbation parameters due to heterogeneities are $\delta c_p^2, \delta c_s^2, \delta \lambda, \delta \mu,$ and $\delta \rho,$ we have

$$c_p^2 = c_{p0}^2 + \delta c_p^2, \qquad c_s^2 = c_{s0}^2 + \delta c_s^2,$$

$$\lambda = \lambda_0 + \delta \lambda, \qquad \mu = \mu_0 + \delta \mu, \qquad (2)$$

$$\text{and } \rho = \rho_0 + \delta \rho.$$

Putting equation (2) into equation (1), we have

$$\frac{\partial^2 \mathbf{u}}{\partial t^2} = c_{p0}^2 \nabla(\nabla \cdot \mathbf{u}) - c_{s0}^2 \nabla \times (\nabla \times \mathbf{u}) + \mathbf{B}, \qquad (3)$$

where

$$\mathbf{B} = \delta c_p^2 \nabla(\nabla \cdot \mathbf{u}) - \delta c_s^2 \nabla \times (\nabla \times \mathbf{u}) + \frac{1}{\rho_0}[(\nabla \delta \lambda)(\nabla \mathbf{u}) + (\nabla \delta \mu) \cdot (\nabla \mathbf{u}) + (\nabla \mathbf{u}) \cdot (\nabla \delta \mu)] . \quad (4)$$

By defining

$$\Phi = \nabla \cdot \mathbf{u}(\mathbf{x}, t) \qquad (5)$$

and taking divergence of both sides of equation (3), we have

$$\frac{\partial^2 \Phi}{\partial t^2} - c_{p0}^2 \nabla^2 \Phi = \nabla \cdot \mathbf{B} \tag{6}$$

where

$$\nabla \cdot \mathbf{B} = \nabla \cdot [(2\delta c_p^2 + \frac{\delta \rho}{\rho_0} c_{p0}^2)] \nabla \Phi - (\delta c_p^2 + \frac{\delta \rho}{\rho_0} c_{p0}^2) \nabla^2 \Phi$$

$$- \nabla \cdot [(2\delta c_s^2 + c_{s0}^2 \frac{\delta \rho}{\rho_0}) \nabla \times (\nabla \times \tilde{\mathbf{u}})] \tag{7}$$

in which the second derivative of $\delta\lambda, \delta\mu$ and $\delta\rho$ with respect to space position are omitted due to the weak inhomogeneous media assumption.

By taking the Fourier transform over time t on both sides of equation (7), we derive the equation in frequency domain,

$$\nabla^2 \tilde{\Phi} + k_{p0}^2 \tilde{\Phi} = -\frac{1}{c_{p0}^2} \nabla \cdot \tilde{\mathbf{B}}, \tag{8}$$

where $\tilde{\Phi}$ and $\tilde{\mathbf{B}}$ are the Fourier transform of Φ and $\mathbf{B}$ with respect to time t.

By defining

$$b_1(\mathbf{x}) = 2\delta c_p^2 / c_{p0}^2 + \delta\rho / \rho_0 , \tag{9}$$

$$b_2(x) = \delta c_p^2 / c_{p0}^2 + \delta\rho / \rho_0 , \tag{10}$$

$$b_3(x) = 2\delta c_s^2 / c_{p0}^2 + c_{s0}^2 / c_{p0}^2 \frac{\delta\rho}{\rho_0}, \tag{11}$$

therefore, equation (8) can be recast as

$$\nabla^2 \tilde{\Phi} + k_{p0}^2 \tilde{\Phi} = -\nabla \cdot [b_1(\mathbf{x}) \nabla \tilde{\Phi}] + b_2(x) \nabla^2 \tilde{\Phi} + \nabla \cdot [b_3(x) \nabla \times (\nabla \times \mathbf{u})] \tag{12}$$

In the right hand side of above equation, there are three terms, the first two terms belong to the scattering sources of P to P wave, the third term comes from the S wave to P wave conversion due to the inhomogeneous perturbation in elastic media.

Let

$$\tilde{\Phi} = \tilde{\Phi}_{in} + \tilde{\Phi}_{sc} , \tag{13}$$

where $\tilde{\Phi}_{in}$ is the incident field in the background media, which satisfies

$$\nabla^2 \tilde{\Phi}_{in} + k_{p0}^2 \tilde{\Phi}_{in} = s(\omega)\delta(\mathbf{x} - \mathbf{x}_s) . \tag{14}$$

In this equation $s(\omega)$ is source function, $\mathbf{x}_s$ is the position of the source, and the scattering field $\tilde{\Phi}_{sc}$ can be expressed as

$$\tilde{\Phi}_{sc}(\mathbf{x}, k_{p0}) = -S(\omega) \cdot \int_\Omega d\mathbf{x}' \{\nabla \cdot (b_1(\mathbf{x}') \nabla \tilde{\Phi}) - b_2(\mathbf{x}') \nabla^2 \tilde{\Phi}$$

$$- \nabla \cdot [b_3(\mathbf{x}') \nabla \times (\nabla \times \mathbf{u})]\} G(\mathbf{x}, \mathbf{x}', k_{p0}) \tag{15}$$

where $G(\mathbf{x}, \mathbf{x}', k_{p0})$ is the Green function, which satisfies

$$\nabla^2 G_0 + k_{p0}^2 G_0 = \delta(\mathbf{x} - \mathbf{x}') \tag{16}$$

From equation (15), we know that the scattering fields are produced by three perturbation parameters. Next we will develop one waveform tomography method to invert for parameters $b_1(\mathbf{x})$ and $b_2(x)$. For $b_3(x)$ inversion, the S wave scattering problem should be consider, it will be studied in the future.

WAVEFORM TOMOGRAPHY METHOD FOR P VELOCITY AND DENSITY

In our imaging method, the P wave point source is used to illuminate the object region. Then, applying the Born approximation to equation (15), the total field $\tilde{\Phi}$ and $\mathbf{u}$ in the integral equation (15) can be replaced by incident field $\tilde{\Phi}_{in}$ and $\mathbf{u}_{in}$, where $\mathbf{u}_{in}$ is displacement field of incident P wave. Therefore, $\nabla \times \mathbf{u} = 0$ in equation (15), and the third term is omitted. Notice that $\tilde{\Phi}_{in}$ satisfies equation (14),. Therefore, the second term in equation (15) can be simplified as (Chen and Wei, 1988)

$$-\int dx' b_2(x') \nabla^2 \tilde{\Phi}_{in} G(\mathbf{x}, \mathbf{x}', k_{p0}^2) = k_{p0}^2 \int dx' b_2(x') \tilde{\Phi}_{in} G(\mathbf{x}, \mathbf{x}', k_{p0}^2) . \tag{17}$$

Under the Born approximation, equation (15) can be expressed as

$$\tilde{\Phi}_{sc}(\mathbf{x}, k_{p0}) = -S(\omega) \cdot \int_\Omega dx' \{ \nabla \cdot (b_1(\mathbf{x}') \nabla \tilde{\Phi}_{in}) + k_{p0}^2 b_2(\mathbf{x}') \tilde{\Phi}_{in} \} G(\mathbf{x}, \mathbf{x}', k_{p0}^2) \tag{18}$$

Applying the geometrical optical approximation, we have

$$\tilde{\Phi}_{sc}(\mathbf{x}, k_{p0}) = -S(\omega) \cdot \int_\Omega dx' \{ \nabla \cdot (b_1(\mathbf{x}') \nabla A(\mathbf{x}', \mathbf{x}_s) e^{-i\omega T(\mathbf{x}', \mathbf{x}_s)})$$

$$+ k_{p0}^2 b_2(\mathbf{x}') A(\mathbf{x}', \mathbf{x}_s) e^{-i\omega T(\mathbf{x}', \mathbf{x}_s)} \} \cdot A(\mathbf{x}, \mathbf{x}') e^{-i\omega T(\mathbf{x}, \mathbf{x}')} \cdot \frac{c_0}{i\omega} , \tag{19}$$

where the amplitude A and travel time T satisfy the transport equation and eikonal equation, respectively.

By using Green's theorem in the plane and assuming that b_1 and $b_2 = 0$ on the boundary of the imaging region, we have

$$\tilde{\Phi}_{sc}(\mathbf{x}, k_{p0}) = i\omega S(\omega) \int_\Omega dx' a(\mathbf{x}, \mathbf{x}', \mathbf{x}_s)(b_1(\mathbf{x}') \cdot \cos(\theta) + b_2(\mathbf{x}')) e^{-i\omega \tau(\mathbf{x}, \mathbf{x}', \mathbf{x}_s)} \tag{20}$$

where

$$a(\mathbf{x}, \mathbf{x}', \mathbf{x}_s) = A(\mathbf{x}', \mathbf{x}_s) A(\mathbf{x}, \mathbf{x}') / c_0, \tag{21}$$

$$\tau(\mathbf{x}, \mathbf{x}', \mathbf{x}_s) = T(\mathbf{x}', \mathbf{x}_s) + T(\mathbf{x}, \mathbf{x}') . \tag{22}$$

and θ is the angle between $\nabla T(\mathbf{x}', \mathbf{x}_s)$ and $\nabla T(\mathbf{x}, \mathbf{x}')$.

Taking the inverse Fourier transform of both sides of equation (22), we obtain the scattering field in time domain as

$$\Phi_{sc}(\mathbf{x}, t) = W(t) * \int_\Omega dx' a(\mathbf{x}, \mathbf{x}', \mathbf{x}_s)(b_1(\mathbf{x}') \cdot \cos(\theta) + b_2(\mathbf{x}')) \delta(t - \tau(\mathbf{x}, \mathbf{x}', \mathbf{x}_s))$$

$$= W(t) * \int_{I(\tau)} a(\mathbf{x}, \mathbf{x}', \mathbf{x}_s)(b_1(\mathbf{x})\cos(\theta) + b_2(\mathbf{x}))ds \qquad (23)$$

where $W(t) = S'(t)$, convolution is denoted by *. The above equation is called a generalized Radon transform. $I(\tau)$ is an isochronic plane with a fixed time τ, from which the perturbation parameters can be related to the waveform. For each point on the waveform, we can derive one equation for inversion. For each pair of source and receiver $(\mathbf{x}, \mathbf{x}_s)$, all the isochronic lines can cover the object region. when the spatial position of the source or receiver is moved, the direction of set of isochronic lines belonging to the same pair of source and receiver $(\mathbf{x}, \mathbf{x}_s)$ is changed. Therefore, the projection in different direction can be derived, and we can use the waveform as a projection to invert for the elastic parameters.

Considering k^{th} isochronic plane $I(t_k)$, equation (23) can be discetized as

$$d_k = W_k * \sum_m a(l_{km}) \cdot (b_2(l_{km})\cos(\theta_{km}) + b_1(l_{km}))\Delta s \quad (1 \le k \le K, 1 \le m \le M) \quad (24)$$

where $d_k = \Phi_{sc}(\mathbf{x}, \mathbf{x}_s, t_k)$, $W_k = W(t_k)$, $K = L \times S \times R$, L, S and R are the total number of sampling points on the waveform, source and receiver, l_{km} is the length of kth isochronic line from the first point to m^{th}point , M is the total numbers of the integral step along m^{th} isochronic line, $a(l_{km})$, $b_1(l_{km})$ and $b_2(l_{km})$ are the values of $a(\mathbf{x}, \mathbf{x}', \mathbf{x}_s)$, $b_1(\mathbf{x}')$ and $b_2(\mathbf{x}')$ at l_{km}, θ_{km} is the value of θ at point l_{km}, and Δs is the integral step. The object region is divided into $I \times J$ pixels. When the coordinate of l_{km} satisfies, $i\Delta x \le x \le (i+1)\Delta x$, $i\Delta z \le z \le (i+1)\Delta z$, where Δx and Δz are the width of pixel in x and z direction, $1 \le i \le I$, $1 \le j \le J$.

By defining

$$\rho_{nkm} = \begin{cases} 1 & when \quad n = (j-1) \times J + i \\ \\ 0 & when \quad n \neq (j-1) \times J + i \end{cases} \qquad (25)$$

we have

$$b_1(l_{km}) = \sum_n \rho_{nkm} \cdot b_{1n} \qquad (1 \le n \le I \times J) \qquad (26)$$

$$b_2(l_{km}) = \sum_n \rho_{nkm} \cdot b_{2n} \qquad (1 \le n \le I \times J), \qquad (27)$$

where b_{1n} and b_{2n} are the values of $b_1(\mathbf{x}')$ and $b_2(\mathbf{x}')$ at n^{th} pixel. Puting equations (25)-(27) into equation (24), we have

$$d_k = W_k * \sum_n (c_{1kn} \cdot b_{1n} + c_{2kn} \cdot b_{2n}) \qquad (28)$$

where

$$c_{1kn} = \sum_m a(l_{km}) \cdot \rho_{nkm} \cdot \Delta s \qquad (29)$$

$$c_{2kn} = \sum_m a(l_{km}) \cdot \rho_{nkm} \cdot \cos(\theta_{km}) \cdot \Delta s \qquad (30)$$

Let

$$g_{kn} = \begin{cases} c_{1kn} & when \quad 1 \le n \le I \times J \\ \\ c_{2kn} & when \quad I \times J \le n \le 2 \times I \times J \end{cases} \qquad (31)$$

$$f_{kn} = \begin{cases} b_{1n} & \text{when} \quad 1 \leq n \leq I \times J \\ \\ b_{2n} & \text{when} \quad I \times J \leq n \leq 2 \times I \times J \end{cases} \tag{32}$$

Then equation (28) can be written as

$$d_k = W_k * \sum_{n}^{2IJ} g_{kn} \cdot f_n \tag{33}$$

Let

$$h_{kn} = \sum W_{k-l} g_{ln} \qquad (1 \leq l \leq L) \tag{34}$$

then equation (33) can be written as

$$d_k = \sum_{n}^{2IJ} h_{kn} \cdot f_n \tag{35}$$

From equation (35), we can invert for P velocity and density using the waveform as the projection. In order to solve equation (35) effectively, the algorithms ART and SIRT are used to solve equation (35) to get the fast execution speed of the inversion. Then we have the following ART and SIRT iterative form:

$$f_n^{(q+1)} = f_n^{(q)} + \alpha \frac{h_{kn}}{\sum_n h_{kn}^2}(d_k - \sum_n h_{kn} \cdot f_n) \tag{36}$$

$$f_n^{(q+1)} = f_n^{(q)} + \beta \sum_k \frac{h_{kn}}{\sum_n h_{kn}^2}(d_k - \sum_n h_{kn} \cdot f_n)/z_n \tag{37}$$

where α, β are the weight factors, z_n is the non-zero numbers of h_{kn} $(1 \leq k \leq L \cdot S \cdot R)$.

By the above methods, the waveform can be used as the projection to invert for the parameters $b_1(x)$ and $b_2(x)$, and the P velocity and density can be obtained from them.

COMPUTATION SIMULATION

Now we apply the above method to the cross-well imaging system. At first, three sources are located on the surface to illuminate the imaging region, and for each source, 20 receivers located in the left well at x=-50m and 20 receivers located in right well at x=50m are used to receive the signal, respectively, Then 10 sources located in left well are used to illuminate the imaging region, and 20 receivers located in right well are used to receive the signal. The depth of the well is 200m. The source function we used is a ricker wavelet, the center frequency is 200 Hz. The imaging region is divided into 20 by 20 pixels, where he width of each pixel is 3 meters. Figure 1 and figure 2 are the P velocity and density perturbation models for the reconstruction test. The parameters of the background are: $c_{p0} = 2500$ m/s, $\rho_0 = 2.3$ g/cm^3 .The perturbation of P velocity and density are 10% with respect to the background values. Figure 3 and figure 4 are the inverted results of P reconstruction and density by SIRT. Figure 5 and figure 6 are the reconstruction results of P velocity and density by ART. From figure 4 and figure 6, we can see the vertical resolution is not high in this imaging geometry. This is because there are not enough projections in the vertical direction. From the reconstruction results, we can see that the tomography results derived by SIRT are better than that by ART.

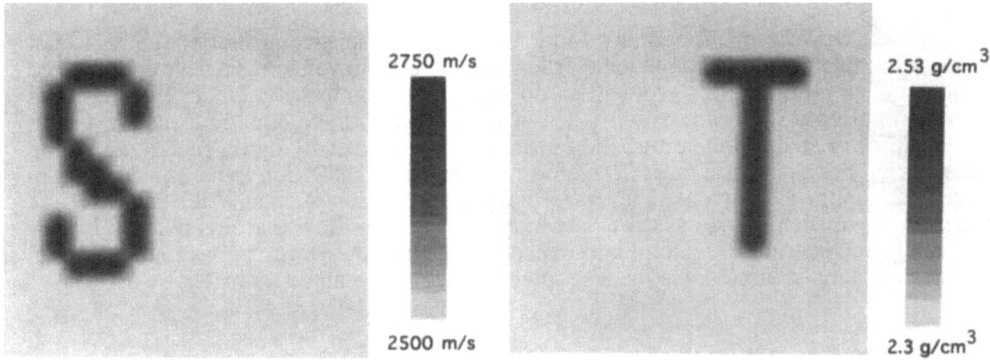

Figure 1. The synthetic P wave velocity model   Figure 2. The synthetic density model

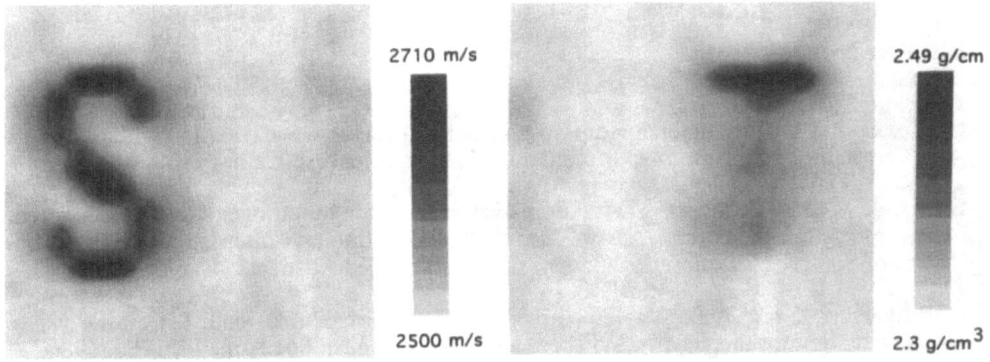

Figure 3. P velocity reconstructed result
by SIRT method

Figure 4. Density reconstructed result
by SIRT

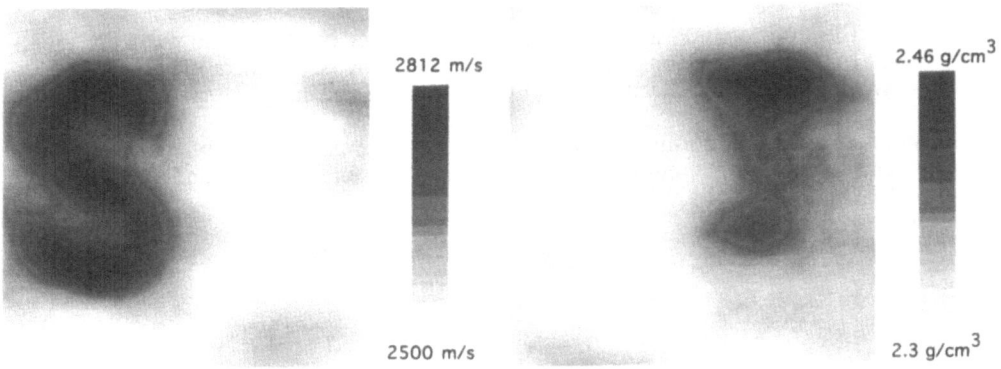

Figure 5. P velocity reconstructed result
by ART method

Figure 6. Density reconstructed result
by ART

CONCLUSIONS

A new waveform tomography for P velocity and density in elastic media is put forward in this paper. Although the elastic wave equation inversion is a very complicated mathematics problem; this problem can be simplified immensely. Under the Born and geometrical optics approximation when starting from P wave equation, the waveform data can be related to the perturbation of P velocity and density by a generalized Radon transform. For each point on the waveform, discretizing this Radon transform can lead to one equation for these two parameters. Then back projection methods ART and SIRT are used to solve such a huge system. The numerical results show that this method is very robust to reconstruct P velocity and density. Although the Born approximation in real data application is limited, the main structure in the media can be estimated by use of it.

ACKNOWLEDGMENTS

This work is supported by the Seismic Tomography Project of Stanford University and a research consortium of the oil and gas industry

REFERENCES

Chen, J.C. and Y. Wei, Inversion of two-dimensional elastic wave equation with two-variables within Born approximation, Apllied Sciencese(China),1988

Harris, J. M. and G. Wang, Diffraction tomography for inhomogeneities in a layered background medium, Sixty-Third Annual Meeting and International Exposition of Society of Exploration Geophysicists, Expanded Abstracts, 49-52, 1993.

Hooshyar, M.A. and A.B. Weglein, Inversion of the two-dimensional SH elastic wave equation for the density and shear modulus, J.Acoust. Soc.Am., 78, 1280, 1986.

Norton, S.J. and L.R. Testard, Reconstruction of one-dimensional inhomogeneneities in elastic modulus and density using acoustical dimensional resonances, J.Acousti. Soc. Am., 79(4), 932-942,1988.

Miller, D., M. Oristaglio, and G. Beylkin, A new slant on seismic imaging: Migration and integral geometry, Geophysics, 57,7, 943-946, 1984.

Mora, P., Elastic wavefield inversion and adjoin operator for the elastic wave equation, in: Mathematical Geophysics, p.117-137, Edited by N. J. Valaar, G. Nolet, M.J.R. Wortel & S.A.P.L. Cloetingh, 1988 by P. Reidel Publishing Company.

Tarantola,A., 1984, The seismic reflection inverse problem, Inverse Problems of Acoustic and Elastic waves, edited by F. Santosa, Y.H.Dao, W.Symes and Ch. Holland, SIAM, Philadelplia.

ULTRASONIC WAVE INVESTIGATION FOR SOLID IMPURITIES DETECTION IN OPAQUE PHARMACOLOGICAL PRODUCTS.[1]

L.Affortunati, E.Biagi, M.Calzolai, L.Masotti, A.Zanini

Dipartimento di Ingegneria Elettronica
via S. Marta, 3
50139 Firenze, Italy

ABSTRACT

This paper reports the result of a study for the application of ultrasonic investigation techniques to detect optically transparent impurities in opaque fluid. Two different detection methods are proposed for the control of vials containing pharmaceutical products and for the application in an industrial environment.

INTRODUCTION

Usually, in pharmacological vials light wave detection techniques are employed to detect solid impurities introduced during the manufacturing process. If the fluid contained in the vials is optically opaque different kind of investigation wave must be used. The aim of the research is to develope an investigation method by ultrasuond and to test its capability to solve this problem.

The work can be divided into two phases. The feasibility study, the first phase, is devoted to obtaining an acoustical characterization of the investigated medium and to compare the results obtained by different ultrasonic techniques in terms of resolution, detectability and accuracy. In the second phase an automatic defect detection system is realized. The system is designed in such a way as to fit the industrial manufacturing requirements such as high noise immunity and fast defect detection processing. And finally it must be able to overcome the vials' heavy geometric tolerance. The mechanical tolerance of the vial movement system must also be taken into account. By adopting an echo immersion technique and by using a particular signal processing the influence of the geometrical tolerance is avoided.

---

1 Work supported by Brevetti CEA S.p.A. via del Commercio 28, Sovizzo, Vicenza, Italia. Undergrant 1257

The solid impurities to be detected are deposited on the bottom of the vial and therefore they cannot be properly detected in static condition due to the vial bottom tolerance. The vials are then spun on their own longitudinal axis, and stopping the action rapidly lifts the impurities out of the bottom so that their detection is possible. The ultrasonic transducer is positioned so as to have an angled beam with respect to the longitudinal vial axis. This allows the reduction of the dead investigation zone caused by the signal reverberation inside the first vial glass wall.

The automatic system is composed of a mechanical movement module, an electronic part and a PC for image processing. All of the system is controlled by a Digital Signal Processor (DSP) also used for a fast elaboration of the digitalized echo signal to reduce the computational time. A broadband ultrasonic transducer with 10 MHz central frequency is employed in pulsing mode. Pulser and receiver have been realized for this particular research.

The system resolution and detectability are tested by employing defects with different shape and density. The defects are classified with the equivalent radius in terms of ultrasonic wavelength. Different volumes inside the vial are investigated by changing the relative angle and depth of the transducer with respect to the longitudinal vial axis. The statistical analysis, after an appropriate data compression using a DSP, has allowed to establish, for each kind of defect, its most probable position inside the vial for a given rotating velocity. Moreover, it has been possible to extract the defect motion law in regime condition and also during the transient condition referred to the rotation movement. That is, the defect path in time is qualitatively reconstructed by processing the acquired echo signals coming from the whole inner analyzed volume. The analysis of the data has permitted the realization of an industrial prototype[1] by minimizing the necessary analysis time and by determining the best position of the transducer to reach high detection probability. The last version of the realized equipment is able to control up to 7200 ampoules/h with a 98% detectability for glass contaminant, 83% for rubber and 90% for Teflon.

FEASIBILITY STUDY

A laboratory experimental system is used to conduct the feasibility phase, which is necessary to establish the most appropriate investigation technique and the principle measurement conditions and parameters.

The impurities detection, by using a transmission technique, is strongly affected by the geometrical vials tolerances which cause false readings in the impurities detection. Moreover the alignment of the two transducer under industrial conditions is very difficult to realize. To reduce this effects a pulse echo technique is proposed. In addition, an immersion technique reduces the tolerances due to coupling problems coming from a dry contact among the vial and transducer.

The system detectability is related to the transducer central frequency. Working with high frequencies leads to the increasing of the detectability[2]. On the other hand, it is necessary to limit the medium attenuation in such a way as to not diminish the signal to noise ratio. The transducer central frequency, 10 MHz, has been chosen on the basis of the acoustical characterization of the medium shown in figure 1. The centroid position of the spectral distribution of the signal coming from a specular reflector placed at different distances inside the medium is reported. Two different central frequency transducers, 15 MHz and 10 MHz, are employed for this test.

The ultrasonic signal intensity A_z at a generic depth z, inside the medium, can be expressed as[3]:

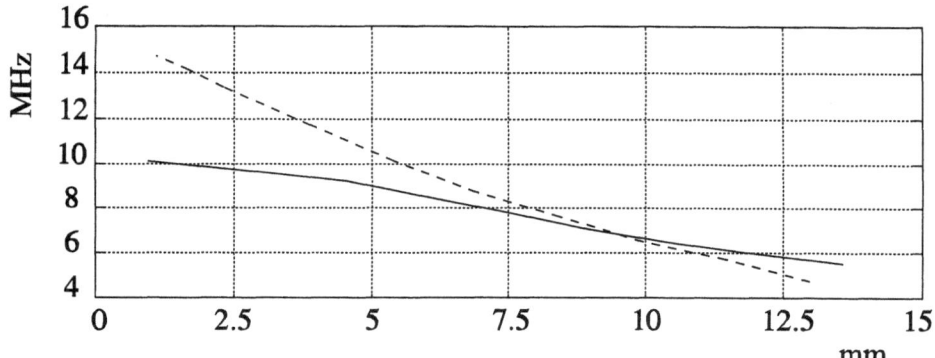

Figure 1. Centroid position versus the specular reflector depth. —— 10 MHz transducer, ---- 15 MHz transducer.

$$A_z = A_0 \, \exp[-\int_{Z_0}^{Z} \alpha(z,f) \, dz] \qquad (1)$$

where A_0 is the amplitude at a reference depth Z_0 and $\alpha(f)$ is the overall attenuation coefficient.

The medium induces on the propagating signal an exponential decreasing amplitude which in turn is dependent on time (distance z) and frequency. The attenuation coefficient $\alpha(f)$ can be expressed:

$$\alpha(z,f) = \alpha_A(z,f) + \alpha_S(z,f) \qquad (2)$$

The first term is the attenuation due to absorption which is related to the bulk medium characteristics. The $\alpha_S(f)$ term is related to the material scattering action and can give information about the medium selective attenuation. The scattering coefficient is classified by the ratio of sound wavelength to the mean dimension of the constituent medium elements.

The investigated medium is a homogeneous compound with particle suspension. From the acoustical point of view, the compound can be assimilated, in the considered frequency range, to a random distribution of Rayleigh scatterers in an uniform medium[3].

Referring to the 15 MHz transducer, a lowering of the spectral position (fig.1), according to Rayleigh regime (α_S depends on the fourth power of frequency), occurs as far as a depth of approximately 7 mm. After a path of 7 mm inside the medium, the high frequency signal contents are backscattered and the attenuation due to absorpion $\alpha_A(f)$ prevails. In fact the centroid value of 15 MHz signal follows the same trend of that 10 MHz. Depending on the medium backscattering coefficient, higher frequencies cannot be employed for defect detection.

The investigated volume portion is proportional to the active transducer diameter whose maximum value is imposed by geometrical dimension of the vial following the refraction Snell law. The position of the transducers in relation to the vial is limited by goemetry of the vial. The best solution to minimize the investigation time is to illuminate the vial through the bottom but its shape and tolerances make the defect detection impossible. In the proposed measurement configuration the transducer beam axis is perpendicular to the vial longitudinal axis. The transducer altitude position is defined as the beam axis position with reference to the vial bottom.

A wave propagating in a material is scattered by any discontinuities encountered: the scattering section of the defect depends on frequency, dimension and shape of the scattering element. The medium selective attenuation prohibits the use of recognition techniques based on the defect scattering. The proposed methods works under geometrical acoustic hypothesis. As a consequence, the theoretical minimum dimensions of the detectable defect is the ultrasonic wavelength λ. The detection of the defect depends on the dimension of its surface illuminated by the acoustic wave. Hence the method detectability is tested by using calibrated wire positioned in different directions inside the vials to simulate different defect surface back projections.

MEASUREMENT SYSTEM

The feasibility phase has demonstrated the high sensibility of the proposed investigation method to detect solid impurities inside the vial. To reach an industrial implementation of the method it is necessary to adapt the system to manufacturing process requirements. For this job an automatic measurement system able to simulate the overall operative conditions, has been realized (fig. 2).

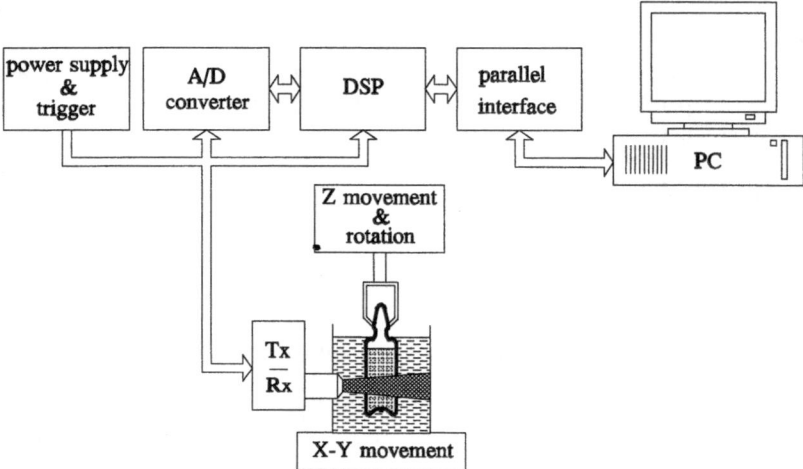

Figure 2. Measurement system.

The system, completely controlled by Digital Signal Processor board, is composed of:
-   a mechanical module which positions the vial on the X-Y-Z axis with respect to the transducer and also rotates the vial for lifting the impurities out of the its bottom;
-   an electronic pulser-receiver front-end hardware with high gain (60 dB) and low noise output: $V_{rms\ noise}$ = 30 mV with an output dynamic range of 700mV on 75 Ω output impedance. High noise immunity is obtained by assembling the front end directly on the transducer, which together make up the sensor, and by transmitting the envelope of the echo signal to the central unit.
-   a central unit composed by a 20 MHz data acquisition board, a Digital Signal Processor unit (T.I. 320C25) and a parallel interface board to PC.
-   a personal computer for data storing and processing.

The DSP generates the timing sequence of each measurement system constituent block, selects the appropriate acquisition time interval based on the first glass wall recognition, and finally performs a data compression. In figure 3 the envelope of a single

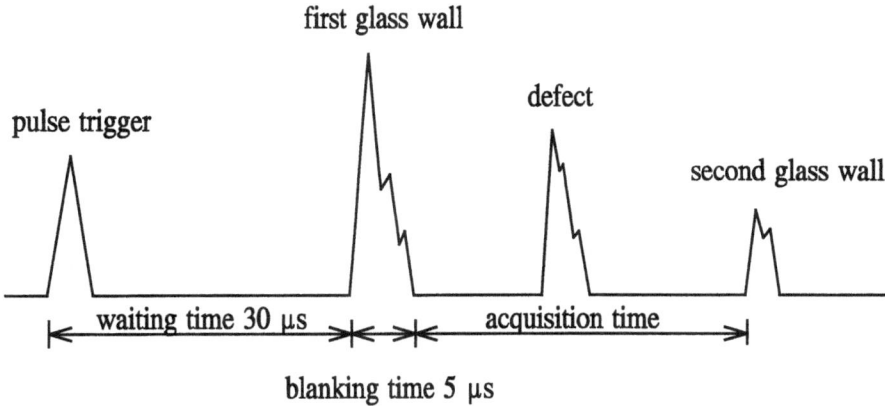

Figure 3. DSP timing sequence on the echo signal envelope.

sight line is reported with the time intervals selected by the DSP. After the waiting time related to the water path, a threshold detection is applied to individuate the first vial glass wall. The useful time interval for data acquisition begins after a programmable blanking interval. The mechanical tolerances of the vial movement system are avoided by adopting the first glass wall identification as the trigger for data acquisition. By utilizing the blanking interval, it's possible to overcome the geometric vial tolerances. All the data gathered within the acquisition interval are acquired.

Due to the quantity of data, it's necessary to compress them. The DSP data compression consists in processing the data to extract the temporal position, the time duration and the amplitude of the echo signal reflected by defects.
Only this information is stored in the 128 kbyte DSP memory and transferred to the PC for the subsequent processing. This data compression procedure permits the processing of 26.10^3 consecutive sight line. With a 3.3 kHz pulse repetition rate it is possible to investigate the overall vial volume by moving the vial during the acquisition or to check the same volume portion inside the vial for 8 seconds.

RESULTS

The acquired data are processed to supply a synthetic view of the defect motion law. In the hereafter presented images, the investigation depth inside the vial is reported versus the number of consecutive sight lines.

In figure 4, 640 consecutive sight lines acquired during the vial rotation are reported. The fact that 640 sight lines with a 3.3 kHz pulse repetition rate correspond to four vial rotation indicates that the periodicity of the figure reflects that the defect rotates together with the glass wall. The end of first glass wall echo is visible in the bottom side of figure.

For defects with different density (e.g. glass, rubber, teflon) it is absolutely necessary to identify the altitude of the defect during rotation to reach an high defect detection frequency. For figure 4, the transducer-vial position details are reported in the caption.
The results for the defect detection carried out on still vial are reported in figure 5. The same volume (sight zone) is investigated for 2.5 seconds, corresponding to 8300 sight lines, beginning at the moment of blocked rotation. After approximately one second, the defect enters in the sight zone and is detected in positions which progressively near the vial axis. Repeating the measurement procedure at different vial positions with respect to the transducer the movement of the defect can be reconstructed during its entire trajectory. This

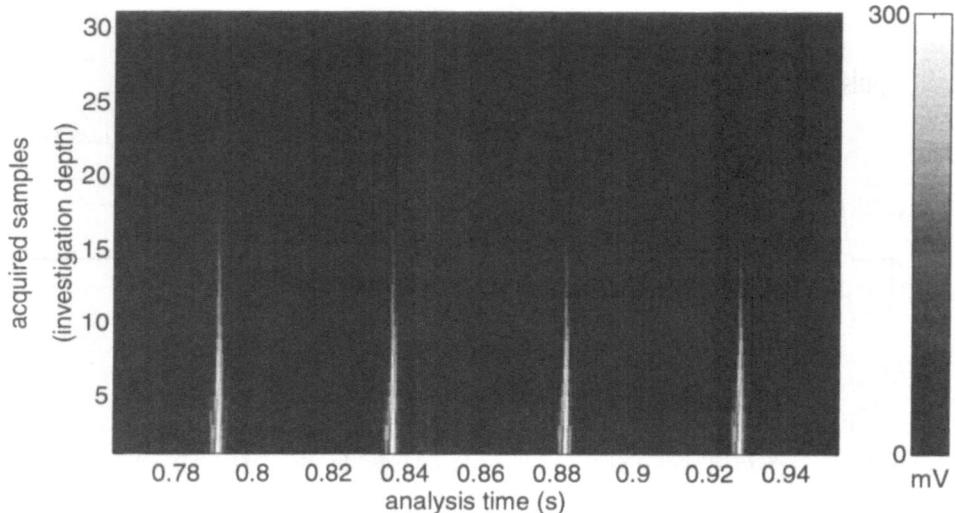

Figure 4. 640 consecutive sight lines acquired during vial rotation.

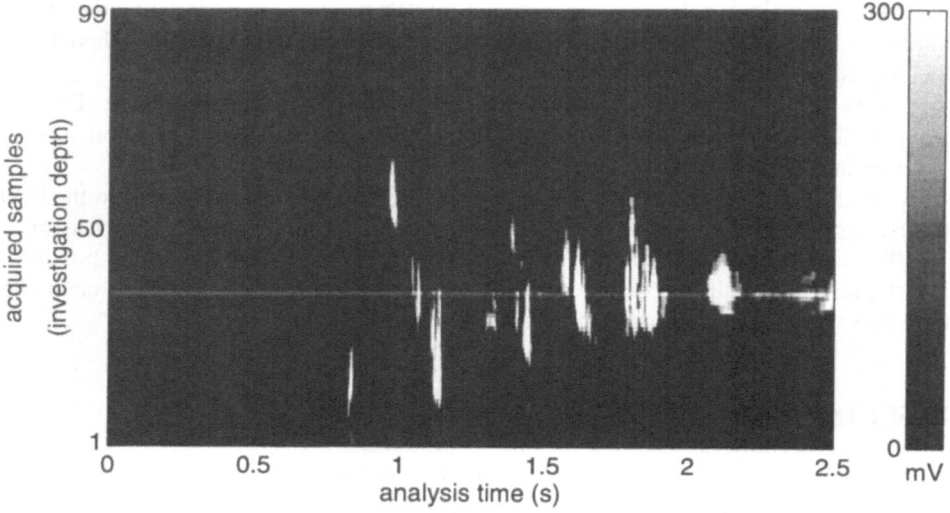

Figure 5. 8300 consecutive sight lines acquired with still vial. The continuos gray line is the middle vial axis.

reconstruction indicates that the defect follows an inverted helix pattern.

It is possible to employ two different investigation methods based on the knowledge of the defect motion law. The defect detection can be performed during vial rotation and after blocking the rotation. The first investigation method is characterized by a high defect detectability if the glass wall tolerances do not mask the defect presence. The principle feature of this method is the short time interval necessary to detect the defect. Keeping in mind that the defect rotates together with the vial, the minimum analysis time, 100 ms, is determined by the maximum obtainable rotation speed, 1500 rpm.

In the second method, as shown in figure 5, the maximum defect detection frequency occurs after some delay time from the blockage of rotation. The evaluation of the appropriate delay time is the most important parameter for this detection method. Different measurement setup parameters such as vial rotation speed, defect density and transducer altitude with respect to the vial, affect the delay time. The maximum vial rotation speed is imposed by the industrial limitation: 1500 rpm.

The experimental work carried out for each kind of defect has led to identify the suitable measurement setup parameters (delay time, transducer altitude) to reduce the analysis time in the range: 180 ms - 280 ms.

By using the appropriate measurement setup, defects equivalent radii varying in the range [250 μm - 700 μm], which in ultrasonic wavelength correspond to 1.8 λ - 3.4 λ are considered.

The following detection frequency are obtained:

glass defects	98%
iron defects	50%
teflon defects	90%
rubber defects	83%.

For iron defects a 98% detection frequency is reached by performing the control during vial rotation.

In industrial application the obtained results are not possible due to resolution of the transducer vial mechanical positioning system. By evaluating the effect of positioning errors a decreasing of the detection probability was proven to be no more than 4%. An altitude transducer error of 0.5 mm and an error of axiality of 0.5 mm are considered. These results are industrially acceptable.

CONCLUSIONS

A comparison between the two proposed defect detection methods permits the assertion that the first method is characterized by a greater defect detection probability, by a smaller analysis time and by a fewer number of critical positioning parameters.
However, the detection capability is strongly constrained by the vial geometrical tolerances.

The main feature of the second method is its immunity to the geometrical tolerances of the vial.

The industrial system implementation has required two separate control positions to obtain high frequency detection for all kinds of defect. One position for vials in movement and the other for still vials. Each position is composed of three different sensors which are placed at different altitudes with respect to the vial.

AKNOWLEDGMENTS

The authors wish to express thanks to Brevetti C.E.A. and to Dott. P. Pacini and to Mr. A. Signorelli for the many stimulating discussion.

REFERENCES

1. Patent n°FI9381 del 21.04.93, "Sperlatura ad US", Brevetti C.E.A. S.p.A..
2. G. Kino, "Acoustic Waves Devices, Imaging, & Analog Signal Processing", Englewood Cliffs, N.J.: Prentice-Hall, Inc., 1987.
3. J. Saniie, T. Wang, N.M. Bilgutay, " Analysis of Homomorphic Processing for Ultrasonic Grain Signal Characterization ", IEEE Transaction on Ultrasonics, Ferroelettrics and Frequency Control, Vol. 36, N° 3, May 1989.

HIGH RESOLUTION MULTISENSOR ULTRASONIC SYSTEM FOR ROBOT NAVIGATION

V. Gabbani[a], S. Rocchi[b], V.Vignoli[a]

[a]Dipartimento di Ingegneria Elettronica - Università degli Studi di Firenze
 Via Santa Marta, 3 - 50139 Firenze, Italy
[b]Facoltà di Ingegneria- Università degli Studi di Siena
 Via Roma 77- Siena, Italy

ABSTRACT

In this paper an ultrasonic multisensor acquisition and processing system is presented, that can handle up to 32 air-ultrasound transducers in the frequency range of 40-200 kHz. The system was developed for an obstacle avoidance applications in the robotics field. Both the processing and the sampling capability of the apparatus (it is based on the Digital Signal Processor TMS320C25 and can sample up to five parallel channels) allow to perform tasks of different difficulty degree, laying between the target-ranging and the pattern recognition.

INTRODUCTION

A multisensor ultrasonic system with up to 32 air-ultrasound transducers in the frequency range of 40-200 kHz was developed for applications in the robotics field. The system is composed of an acquisition and processing board (DSPDRD) and, for each sensor, a front-end electronics module (SONAR). The DSPBRD is based on the Digital Signal Processor (DSP) TMS 320C25 and is provided with both a VME and a RS232 interface: it can therefore run as a slave board in a VME rack or as a stand alone board controlled by a personal computer. The DSPBRD can control and acquire data from up to five transducers at the same time.

Each SONAR module is a 10 cm x 7 cm board containing the transducer driving and receiving electronics.

This system was developed for providing a map of the obstacles present in an area of a 3 m radius around a robot moving in a structured environment. To this purpose, twenty

50 kHz transducers were used, each operating in pulse-echo mode. Only the first echo was processed in this application.

The system has higher capabilities than those exploited for the above application. In terms of hardware resources, as mentioned above, the use of up to five parallel acquisition channels (multi-aural perception) is possible. This allows to improve the ultrasonic system lateral resolution by compounding the echoes from different transducers while preserving the phase coherence. In addition, the presence on the DSPBRD of the TMS320C25 allows the on-line implementation of several preprocessing algorithms such as, e.g., matched filtering, inverse filtering and correlation technique to improve the time of flight detection; cepstrum domain filtering for multiple echo suppression; neural network processing to solve specific object-recognition problems.

In particular in this work three adjacent transducers were used at the same time, the middle-one as a transmitter and all as receivers. This technique, also known as tri-aural perception[1], is used to enhance the spatial resolution for detecting objects with particular shapes and displacements.

The aim was to resolve a problem of particular interest in the "docking" phase of a robot navigation, that is the recognition of both the displacement and the inclination of a plane reflector with respect to a transducer plane array.

A neural network approach had proved to be a powerful processing tool to achieve the goal. In addition, the neural network processing allowed to overcome the inevitable errors (due e.g. to the transducer radiation-lobe width, the finite transducer dimension and the plane reflector inclination) made in the evaluation of the time of flight of the echo signal.

HARDWARE-DESIGN

The system presented is composed of an air-ultrasonic transducer array (up to 32 elements with center frequency in the range 40-200 kHz), a front-end electronics module (SONAR) for each transducer, and an acquisition and processing board (DSPBRD).

In fig. 1a-b is reported a picture of the system mounted on a mobile robot designed and carried out by the 'Telerobot s.n.c' company (Genova, Italy) within the 'Robotica targeted project' (subproject URMAD) of the Italian National Research Council. An example of connection for the whole system is sketched in fig. 2.

The DSPBRD board, whose block-scheme is shown in fig. 3, is equipped with[2]:
-    a DSP TMS 320C25, that is used to control the overall system and to implement specific signal preprocessing algorithms;
-    five acquisition channels (8 bits dynamic range, 1 Msample/s) that allow the using of transducers with center frequency up to 200 kHz;
-    a slave VME BUS interface;
-    a serial RS232 link (it allow stand-alone running under control of a PC).
-    two equivalent interfaces to the SONAR modules.

The front-end electronics SONAR modules are composed of two logic blocks (fig. 4): the programmable selection unit and the transmit/receiving unit. The former, based on Programmable Logic Devices (PLDs), enables the transducer to transmit or receive according to an arbitrary temporary and spatial schedule, and is the interface between the DSPBRD and the transmit receive unit. The latter allows to drive transducers with center frequency up to 300 kHz, using sinusoidal burst with minimum length of about 200 μs, and has a receiving stage provided with two variable-gain voltage amplifiers characterized by a sensitivity of 200 μV_{pp} (peak to peak) and a maximum gain of 76 dB.

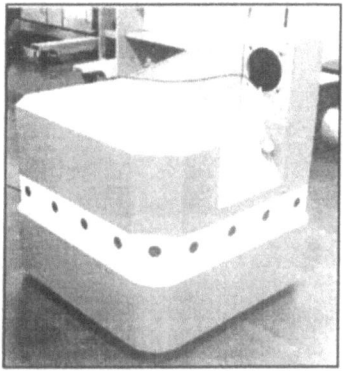

Figure 1a. The URMAD robot (without arm).

Figure 1b. The URMAD robot: view of the internal structure.

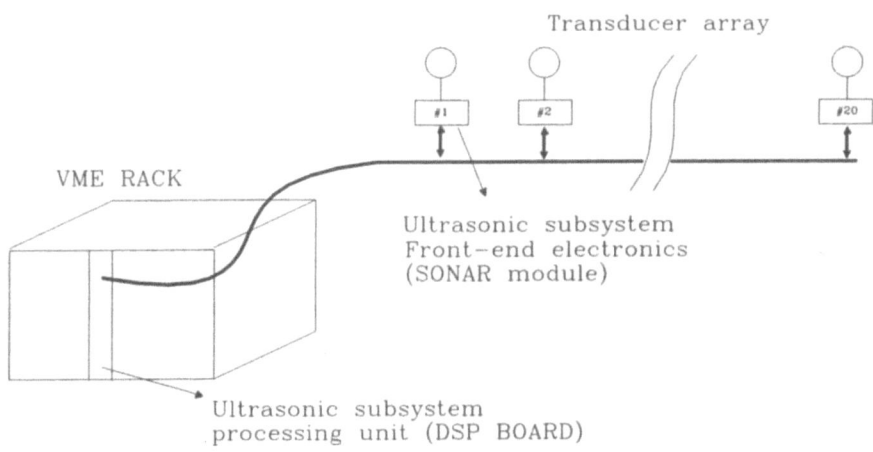

Figure 2. Sketch of the system interconnections.

VME backplane

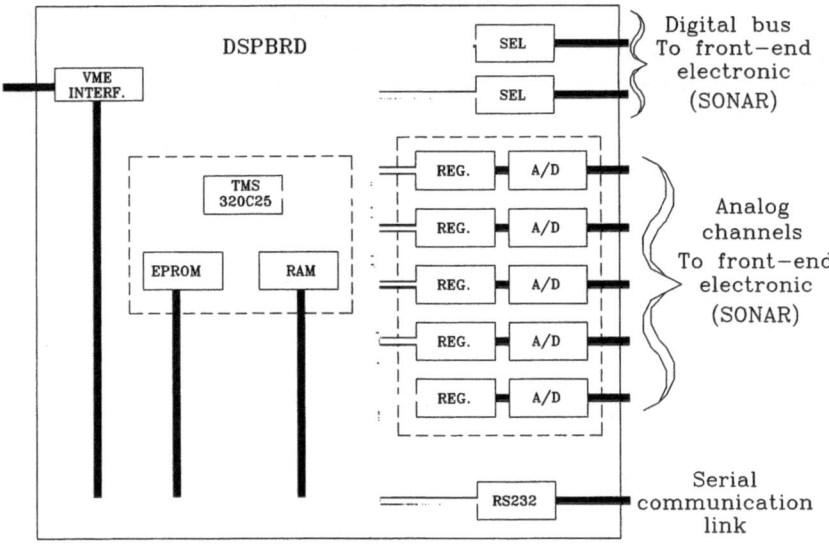

Figure 3. Block scheme of the DSPBRD module.

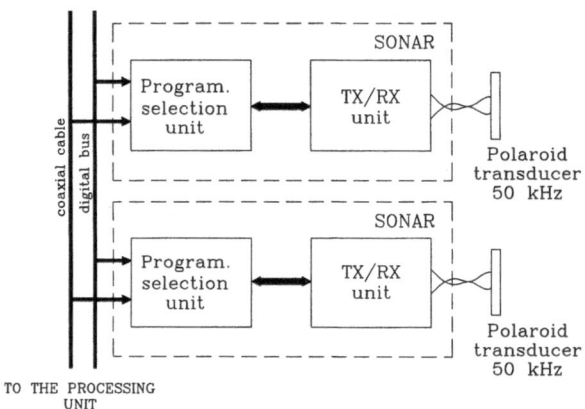

Figure 4. Block scheme of the SONAR module.

The connection between the DSPBRD and the SONAR modules is obtained with a mixed analog and digital bus (radio frequency signals, address and control signals). Among the digital signals, the SONAR module issues a 1-bit sampling of the r.f. signal, obtained with a tunable threshold, that can be used for the Time-Of-Flight (TOF) measurement.

SIMULATIONS AND EXPERIMENTAL DATA

In this section the problem of evaluating distance and inclination of a plane reflector with respect to a three-transducer plane array, that is of particular interest in the "docking" phase of a robot navigation, is considered. The problem was tackled both performing computer simulation and dealing with experimental data acquired using the system described in the previous section. During the simulations and the experimental phase, the central transducers was used as a transmitter, while all the transducers were used as receivers.

A neural network processing approach was used to obtain the parameters A (inclination) and R (distance) of the plane reflector (fig. 5), using as inputs the TOFs detected from the signals received by each transducer of the array. In the simulation phase a Back Propagation Network was selected, in the arrangement shown in fig. 6.

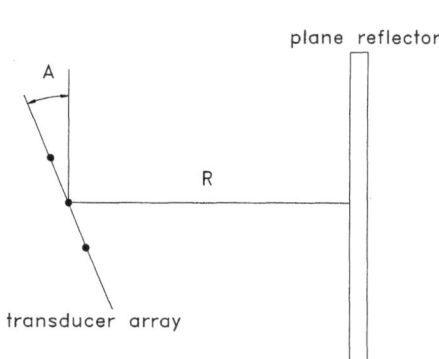

Figure 5. The plane reflector problem: parameter definition.

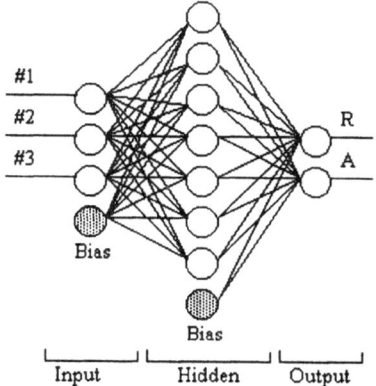

Figure 6. The 'back propagation' neural network used.

The RMS (Root Mean Square) of the learning process for this network, generally does not converge if a "good" initial condition is not used for the weight-matrix. A "good" initial condition can be obtained by performing the learning process in two steps:
in the first step the weight-matrices for two one-output neural networks (to obtain separately the parameters A and R), but with the same number of neurons in input and in the hidden layer with respect to the net of fig. 6, are evaluated. In the second step the two weight-matrices from the previous step are used to obtain the initial conditions for the global network.

The optimum learning-set was found to be composed of 16 examples (4 displacement values R, 4 angles A and their combinations).

In Table 1 the simulation results concerning both the 3-8-1 (one-output) and the 3-8-2 networks are summarized.

The experimental data processing are in good agreement with these theoretical results, as summarized in Table 2.

The experimental data were obtained using a plane 50 kHz transducer array whose elements were 15 cm spaced.

The target was a wooden plane reflector moving with step of 10 cm in the distance of 0.36-1.36 m from the array, and with step of 4° in the interval from -12° to 12°.

Table 1. Simulation summary.

	RMS Error
3-8-2 network: 1 step learn	0.340
3-8-1 network: 1st step learn (A and R)	< 0.015
3-8-2 network: 2nd step learn	< 0.015

Table 2. Experimental data summary.

	RMS Error
3-8-1 network: 1st step learn (A)	< 0.07
3-8-1 network: 1st step learn (R)	< 0.015
3-8-2 network: 2nd step learn.	< 0.07

CONCLUSIONS

A complete data acquisition and processing system was designed and carried out for application in the robotics navigation field.

Both the use of up to five independent acquisition channels and the processing capability of the DSP TMS320C25 allow the solution of the typical problems involved in ultrasound navigation.

The neural approach has proved to be a powerful processing tool, used together with synthetic-aperture-like techniques, to recognize the displacement and the inclination of a plane reflector. In particular, the neural network processing allows to overcome the inevitable errors (due to the transducer radiation-lobe width, the finite transducer dimension and the plane reflector inclination) performed in the evaluation of the time of flight.

ACKNOWLEDGEMENTS

This research was supported by the Italian National Research Council - Targeted Project "Robotica", subproject "URMAD".

REFERENCES

1. H. Peremans, K. Audenaert, and J. M. Van Campenhout, A high resolution sensor based on tri-aural perception, *IEEE Trans. Robotics Automat.* 9(1):36 (1993).
2. V. Gabbani, G. Neri, S. Rocchi, and V. Vignoli, Interfaccia slave VME realizzata con logiche programmabili, *REPORT 931201 of the Electr. Eng. Dept. of the University of Florence*, (1993).

SOUND FIELDS OF ULTRASONIC CIRCUMFERENTIAL ARRAYS - CASE OF AN ARRAY FOR STEAM GENERATOR HEAT EXCHANGER TUBES

Bernd Köhler

Fraunhofer-Institute for NDT Saarbrücken, Department EADQ Dresden
Krügerstraße 22
D-01326 Dresden

MOTIVATION FOR THE DEVELOPMENT OF AN ULTRASONIC PHASED ARRAY

Corrosion damage on steam generator (SG) heat exchanger tubes of nuclear power plants has repeatedly been the cause of limitations in availability and for steam generator replacements. This applies not only to the Russian nuclear power plants of the VVER type but is in fact a world-wide problem. If corrosion damage in the tube wall permits activity penetration into the secondary loop, the defective SG heat exchanger tubes must be plugged by welding. This causes a loss of heating surface and a reduction in the output and availability, and finally also a shortening of the life time.

There are also problems of safe operation because considerable amount of corrosion damage does not penetrate completely.

The existing quality of the tubes of the chosen Russian SG type means that eddy current techniques (ECT) can only be conditionally used to attain an appropriate level of confidence. Because of ovalities in the of U-bend area of the tube application of ECT is often impossible. Furthermore, the ECT used at present has weaknesses in the verification of small crack-type flaws and in the interpretation of indications. So the objective is to develop a method of ultrasonic testing (UT) which enables damage phenomena on SG heat exchanger tubes to be detected and assessed. The method is based on the use of miniaturized ultrasonic array probes of piezoelectric composite materials whose characteristic impedance can be well matched to the load medium (water) and guarantees therefore high signal efficiency with good time resolution.

PROBE DESIGN - NEED FOR MODELING

Heat exchanger tubes of steam generators of the VVER-440 type nuclear power plants are made of AISI 321 stainless steel. With a wall thickness of 1.4mm only they are the components of the thinnest walls between the primary and secondary loops. Due to chloride induced local corrosion[1], a wide spectrum of damages may occur depending on temperature, local pollutant concentrations and stress states. The main damage areas are the brackets of the heat exchanger tubes particularly adjacent to hot collectors. Due to thermohydraulic processes in their crevices pollutant enrichments of up to 10^5 may occur[2].

From studies of corrosion phenomena[1], the below conclusions are important for Nondestructive Testing[3]:

- mainly pitting corrosion and pitting induced stress crack corrosion occurs,
- pits and cracks are growing from the outer tube jacket towards the inner one, and
- crack faces are mainly directed towards the tube axis (longitudinal cracks).

To detect these defects, as a first step a miniaturized circular array probe with small strip-like elements arranged on its circumference parallel to the probe axis is developed. These elements are formed simply by dividing the electrode on one surface and not by cutting the piezoelectric material. Appropriate electronics (ARGUS[4]) allows to scan the tube circumference by exciting a successive subgroup of elements and to focus and steer the sound beam by delayed triggering the elements.

By certain tube test requirements the choice of several array parameters is considerably restricted. The normally used array design principles as they are usually applied to medical arrays, even lead us to contradictory conclusions. On one hand, to achieve an adequate depth resolution for the separation of the entry and rear wall echoes of the tube at a given wall thickness $(d_a - d_i) = 1.4$mm, a minimum ultrasonic frequency of 10 MHz is required. This corresponds to a wavelength in water of $\lambda_{H_2O} = 0.15$mm. Accordingly an element centre-to-centre spacing $x \leq \lambda/2 = 0.075$mm is necessary if grating lobes are to be avoided. The technical limitation of the overall number of elements to $N \leq 80$, on the other hand, demands an element centre to centre spacing of at least 0.27mm with an array probe diameter of $d_s = 7$mm. Similar contradictions arise at other points[3], and consequently the design must incorporate unconventional solutions and compromises.

A possible starting point is to allow $x > \lambda$ with the grating lobes supressed using very short acoustic pulses and limitating the steering angle. Thereby, a sufficiently small near field length is ensured by triggering only a small number of array elements (sub-group). All these formulations are to be checked which is most effectively achieved by model calculations.

In these calculations the novel situation, viz. the piezoelectric material radiates directly into the fluid and the element forming is by electrode shaping only*, has to be taken into consideration carefully. These considerations are important for other cases, too.

SOUND FIELD CALCULATIONS

The layered problem

Let us discuss the problem under consideration (Fig. 1) as a general layered problem. Supposing isotropic and homogeneous single layers with displacement field $\vec{u}(\vec{r}, t)$, velocity field $\vec{v}(\vec{r}, t) = \partial \vec{u}(\vec{r}, t)/\partial t$, stress field $\bar{\bar{\sigma}}(\vec{r}, t)$, body force density $\vec{f}$, mass density ρ, and Lamé elastic constants λ, μ, the equations of motion:

$$\rho \ddot{u}_i - \sigma_{ij,j} = \rho f_i$$
$$\sigma_{ij} = \lambda u_{k,k} \delta_{ij} + \mu(u_{i,j} + u_{j,i}),$$

(1)

together with initial and boundary conditions, have to be solved in each layer. It is important that the initial conditions $(\vec{u}_0 = \vec{u}(\vec{r}, t = 0), \vec{v}_0 = \vec{v}(\vec{r}, t = 0), \vec{r} \in V$) and appropriate boundary

---

* i.e. there are no cuts inbetween the elements

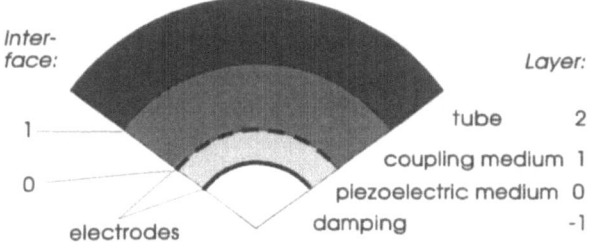

Inter-
face:

Layer:

tube   2

1

coupling medium 1

0

piezoelectric medium 0

electrodes

damping   -1

Figure 1. Scheme of the physical arrangement

conditions (e.g. given surface displacements $(\vec{u}(\vec{r},t),\ \vec{r} \in \partial V^{+}$) or surface tractions $(\vec{n}\vec{\vec{\sigma}}(\vec{r},t),\ \vec{r} \in \partial V)$) leads to unique solutions[5]. So, a possible way of solving the layered problem is to divide it into a sequence of half space and interface problems and proceeding as follows:

a) Supposing a known sound field incident from layer i to the interface between layer i and i+1 (denoted as interface number i - see Fig. 1) the traction distribution $\vec{t}(\vec{r}_i) = \vec{n}\vec{\vec{\sigma}}(\vec{r}_i)$ in this interface is calculated under the assumption that both layers are infinitely thick, i.e. are half spaces.

b) The half space i is removed. Instead of, the stress distribution is generated by surface force densities equal to the tractions $\vec{t}(\vec{r}_i)$. This step is possible because of the uniqueness theorem[5] of the solution.

c) The sound field in the half space i+1 (with parameters λ_{i+1}, μ_{i+1}) is determined for the given surface force distribution at surface i.

d) This sound field is taken as an incident field for the next interface problem between media having elastic properties λ_{i+1}, μ_{i+1} and λ_{i+2}, μ_{i+2} , respectively, proceeding with step b above, i.e. determining the traction distribution $\vec{t} = \vec{n}\vec{\vec{\sigma}}(\vec{r}_{i+1})$.

An alternative to d is to calculate the traction distribution in the interface i+1 directly from that in layer i

d′) A half space solution for a unit point surface force in one of the orthogonal directions m=1..3 $\vec{f}^m = e^{i\omega t}\vec{e}_m\delta(\vec{r} - \vec{r}_i)$ is determined. For this solution the corresponding traction per source unit in the next interface i+1 $:\vec{\vec{t}}_{\vec{r}_i}^{(m)}(\vec{r}_{i+1})$ is calculated. Then the traction distribution in this interface for a given traction distribution $t^{(m)}(\vec{r}_i)$ at surface i is obtained by summation over all directions m and integration over surface i

$$\vec{t}(\vec{r}_{i+1}) = \sum_{m} \int_{S_i} \vec{\vec{t}}_{\vec{r}_i}^{(m)}(\vec{r}_{i+1}) t^m(r_i)d^2\vec{r}_i .$$

In this work we made use of this last possibility. The half space solutions have been obtained in farfield approximation by applying the reciprocal identity[5]. Besides this, the sound transmission through the interfaces has been calculated in plane wave approximation. Both are good approximations if the distance between the layers are large as compared to the wavelength. Of course the described procedure neglects the occurrence of backward travelling waves and hence also multiple reflection.

The idea of using the reciprocity identity for calculating the sound transmission in farfield approximation is by Wüstenberg[6]. But in accordance with step b above, it must be applied to the corresponding half space problems subjected to surface forces and not to the interface problems. Fortunately, the case Wüstenberg applied his formulae to is characterized by a large mismatch of the acoustic impedances. So the deviation from the correct values are not too serious. In other cases, involving layers of similar acoustic impedances, the deviations are important.

$^{+}$ ∂V means the boundary surface of the volume V

We may consider our circular array problem as a three layer one: the piezoelectric layer "0", the coupling layer "1" and the tube·layer "2". For the special case with the shear modulus vanishing in "layer 1", some simplifications occur. Especially only normal forces are transmitted at corresponding interfaces, and $\vec{f}$ can be written as $\vec{f} = f\vec{n}$.

The source problem

The discussion above is not applicable to the first interface between the piezoelectric and the coupling medium. There is no incident wave given. Considering a fluid first layer, in previous work solutions for three configurations have been described theoretically and have been proved by experiment as well. This is the solution for a given surface deformation, given surface traction and for "free field" conditions[7]. Given surface traction is a good model in the case where a soft medium generates an elastic wave in the half space, and free field conditions are a limiting case for an interface with equal acoustic properties on both sides. These two evidently are not applicable to our present case. The variation in excitation (by phase and amplitude) of the array is given by electrode forming and electric excitation, only[+]. So we cannot suppose that a single region of the composite piezoelectric material is vibrating independently of its neighbourhood. Therefore, the third opportunity - prescribtion of the surface deformation - is also not applicable.

The piezoelectric action of a disc (at least in an one dimensional approximation[8]) can be described by force densities acting on its boundaries. So, as a good model for the exciting interface between the piezoelectric material and the load, we studied the problem of a given (non-constant) normal force density acting on it.

The point source farfield solution for this problem can be obtained again by applying the reciprocal identity[5] (not to the free half space but to the locally plane interface problem). Supposing a solution for the sound pressure in the form

$$p_\omega(\vec{r},t) = \frac{p_0}{r} e^{i\omega t} e^{-ikr} R(\vartheta) \qquad \vartheta = \angle(\vec{n},\vec{r}-\vec{r}_0) \qquad r = |\vec{r}-\vec{r}_0|, \tag{1}$$

for $R(\vartheta)$ in the fluid medium (index 1)[*] we get:

$$R(\vartheta) = \frac{2\cos(\vartheta)}{N(\vartheta)} \tag{2}$$

with

$$N(\vartheta) = 1 + \rho\cos(\vartheta)\frac{\left[1 - 2\frac{c_{0,t}^2}{c_1^2}\sin^2(\vartheta)\right]^2}{\sqrt{\frac{c_1^2}{c_{0,l}^2} - \sin^2(\vartheta)}} + 4\frac{c_{0,t}^4}{c_1^4}\rho\cos(\vartheta)\sin^2(\vartheta)\sqrt{\frac{c_1^2}{c_{0,l}^2} - \sin^2(\vartheta)}$$

and $\rho = \rho_0/\rho_1$. Thereby the $c_{0,l}$ and $c_{0,t}$ are the longitudinal and shear sound velocities in the piezoelectric medium, respectively, c_1 is the sound velocity in the liquid and ρ the density ratio. The solution in the fluid depends on both, the elastic properties of the fluid and the exciting solid.

[+] There are no cuts in-between the array elements
[*] see also Kühnicke[9]; where this formula is deviated for an other case

All these considerations, especially result (2), remain valid also for the two-dimensional case. Instead of point sources we have to consider line sources. The only substitution refers to the dependence on the distance in (1) which changes from $1/r$ to $1/\sqrt{r}$.

In the upper part of Fig. 2 the calculated directivity function $R(\vartheta)$ for the material constants of PZT and water is given. Because of the strange behaviour of $R(\vartheta)$ shown (this is the directivity function in water, not in a solid) we have tested it experimentally.

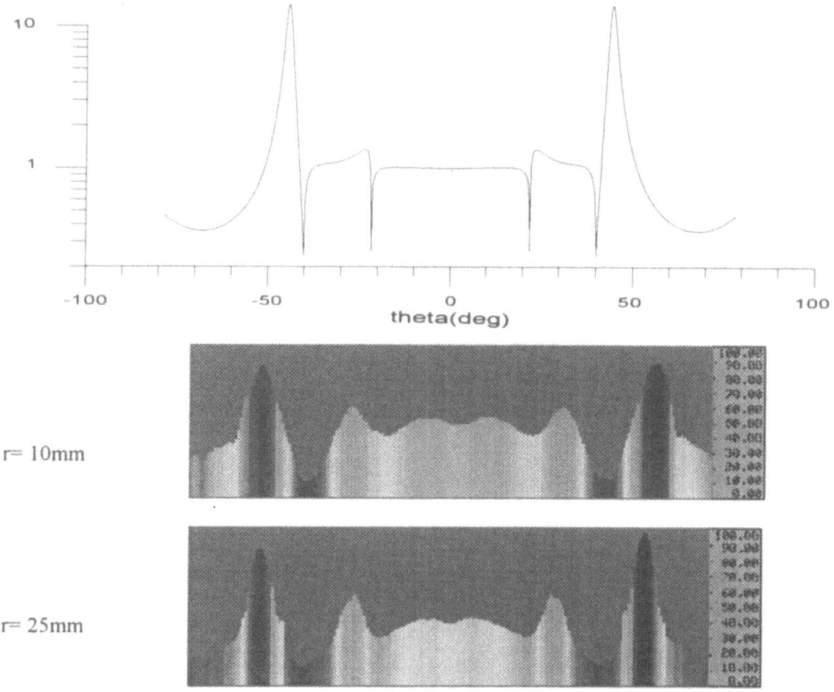

Figure 2. Calculated directivity function in water for a point force acting in a water-PZT interface (above) and correspondingly measured sound amplitudes at two distances (below). The abscissa is also valid for the measured curves.

Fig. 2 below shows the measured sound amplitudes for a thick PZT plate excited via a full electrode on the plate rear surface and a small (200 μm wide) strip on the front one for two distances. Because of the concentration of the electric field towards the electrode strip, indicated in the middle of Fig. 3, this arrangement is a good relization of a driving line force acting on an interface. The essential structures given by the calculated directivity functions are represented by the measurements correctly.

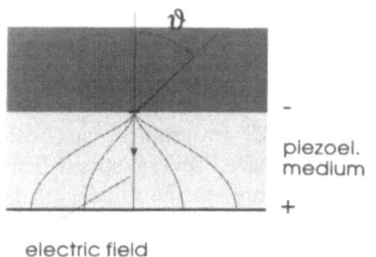

Figure 3. Configuration for the line directivity function measurement

717

The model calculations suppose isotropic material parameters even in the exciting piezoelectric effect can be considered as a small perturbation regarding the elastic properties this should approximately be valid also in the case of the PZT plate. But as composite material is mechanically anisotropic its directivity function has to be tested. The measured amplitude distribution (Fig. 4) does not show the characteristic structure according to (2). If we assume an isotropic line directivity function, it can be reproduced qualitatively. This needs further examination but it is supposed that it results form the anisotropy in sound velocity and especially damping of the composite material.

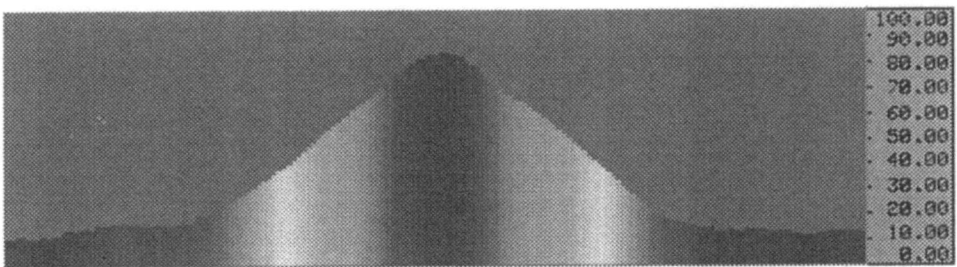

Figure 4. Measurement of relative sound pressure distribution of a composite plate, frequency 1.4 MHz, electroded by a strip, width b=1.5 λ, immersed in water.

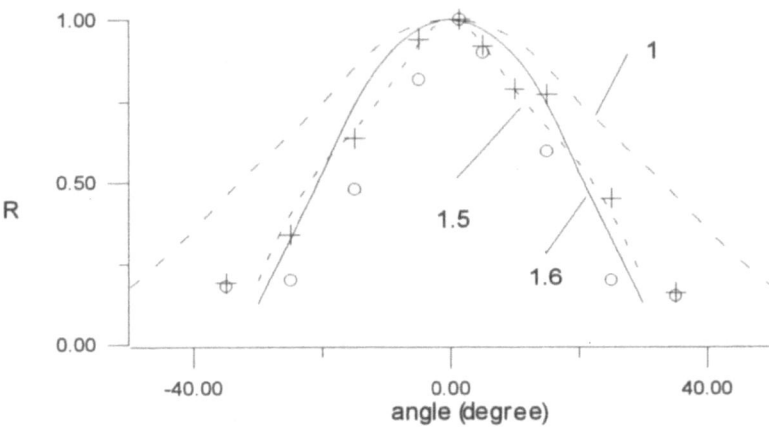

Figure 5. Comparison between measured directivity functions ("+": d/λ=1, "o": d/λ=1.5) and calculated ones (curves with b/λ ratio given as numbers). For the calculations an isotropic directivity function was assumed.

Model calculations for probe design

A program for sound field calculations of circular phased arrays basing on the discussed principles has been developed. To obtain a high calculation speed, this has been done as a 2D program, supposing the array strips being infinitely long. This program is able to account for the line directivity function in the exciting interface. The excitation phases of single elements can either be prescribed by the user or calculated in the program supposing a given geometric focus or direction of the sound beam (geometric focus at infinite distance). The quality of this 2D approximation had to be proven. So an available 3D program[10] has been used for comparison. As this program handles only fluids we compared several sound fields in water with both programs for the intended geometry of the array. In the middle plane always a good agreement was observed.

Fig. 6 exemplifies the sound field in water both in a plane perpendicular as well as parallel to the array axis. Indicated is also the position of the inner tube surface if the tube had been inserted. It is seen clearly that the sound field is strongly structured perpendicular to the axis and that it is nearly constant over the entire length of the array. This justifies the 2D approach, too.

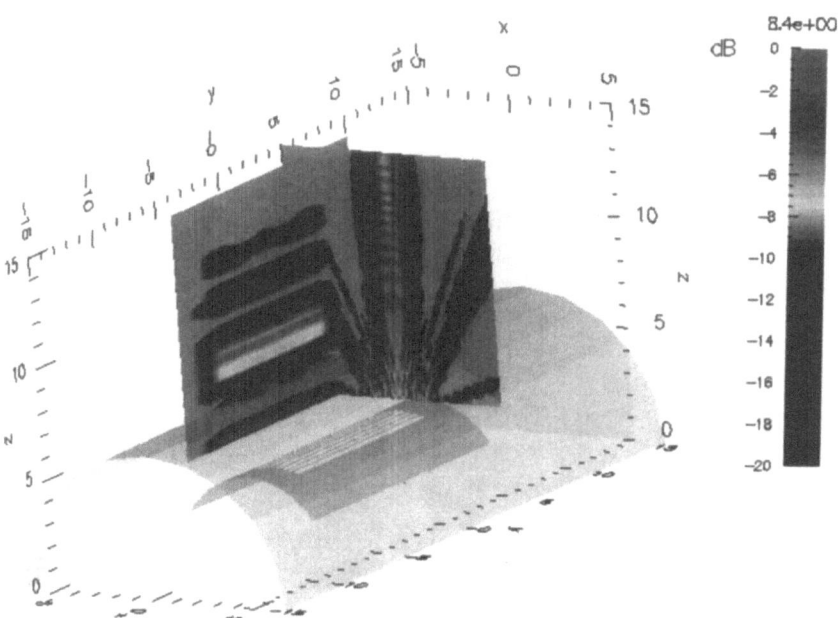

Figure 6. 3D sound field in water of an array with axis x=z=0. 8 are strips excited with such time delay that the geometric focus is at infinite distance on the z axis.

A large number of calculations has been done using the 2D program for the sound field in the water plus tube. Fig. 7 shows some of them, supposing an excited subgroup of 8 elements having an element centre to centre distance of double the wavelength in the adjacent water and the same element width. In steel the longitudinal component of the field is shown.

For equal phase excitation the subgroup acts like one large curved element so that the sound field in water and in the adjacent steel is defocused. Next, the phases have been chosen is such a way, that the geometric focus is at infinite distance. Remarkable is the strong fragmentation in water. In comparison, the sound field in the tube wall is relatively smooth. This can be explained with the large wavelength difference at the interface. It can also be seen, that the curvature of the tube wall has a focusing effect. The resulting 3 dB beam width of 1mm is completely adequate for the required flaw detection.

Focusing to the inner tube wall which seems to be a good idea at the first glance, proved inefficient. The grating loops in steel are enhanced - there are three nearly equal strong maxima - and defocusing occurs in the wall. The sound fields in the last two pictures of Fig. 7 have been calculated with phases of the single elements corresponding to a geometric focus in water at infinite distance and at angles of 5 and 10 degrees to the z-axis, respectively. They show that steering remains restricted to below 10° if the magnitude of grating lobes in the tube wall is to be limited.

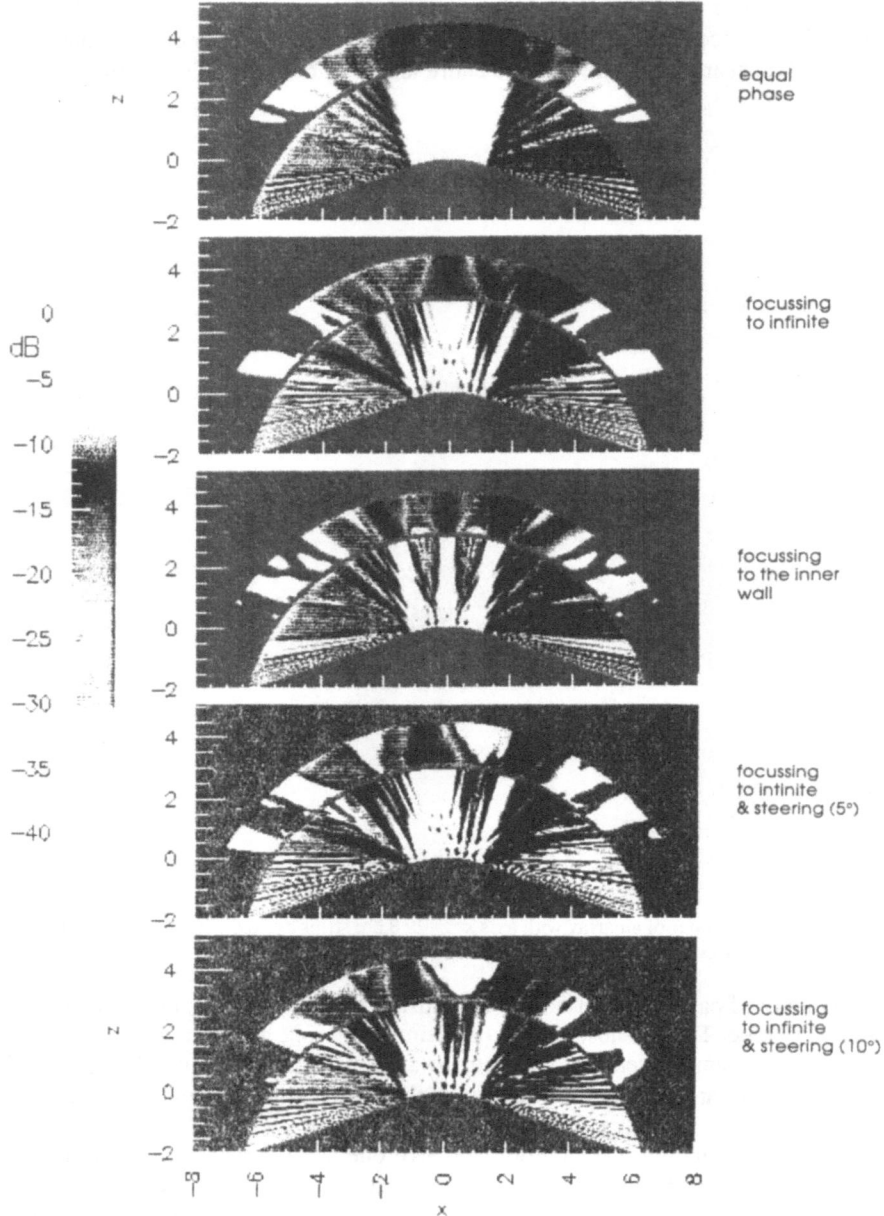

Figure 7. Sound field intensity in water (inner sections) and intensity of the longitudinal wave in the tube wall (outer sections) for different modes of excitation of the 8 elements; the element width and element centre to centre distance is 2λ.

Another parameter, which can be varied, is the number of elements excited. Usually (e.g. in medical applications) an even larger number compared with 8 are excited in a subgroup. We tested what happens if this number is reduced. Modeling results showed that the ratio between main lobe and grating lobes is improved! Under our conditions a number between 3 and 5 will be best. This conclusion is also true for focused and steered sound fields.

SUMMARY

The testing of heat exchanger tubes of nuclear power plants usually performed by ECT should be supplemented by an ultrasonic testing method using a circular phased array. The design of this array is subjected to various constraints and has to be optimized by model calculations. The circular array is based on piezoelectric composite material where the single elements simply are formed by strip like electrodes on the front surface. It has been shown that for such an element formation without cuts and isotropic elastic properties supposed, the directivity function of the exciting surface has a complicated behaviour, hitherto not reported in literature. It has been proved by experiment.

The directivity function of anisotropic piezoelectric composite material has been studied experimentally with the result that a supposed isotropic line directivity function forms a solid basis for numerical calculations.

A fast 2D program has been developed for the calculation of sound fields in the immersing fluid and in the tube. Comparison of numerical 2D and 3D programs for the sound fields in water shows that a 2D calculation is adequate for the present geometry.

Due to the restricted number of elements, the element centre to centre distance must be larger than the wavelength. Nevertheless grating lobes are small for $\leq 10°$ steering angles in water. An improvement in the sound fields can be obtained when lowering the number of excited elements from 8 to a number smaller than six.

Work on the relation between sound fields and pulse length is under way and will be reported elsewhere.

REFERENCES

1. I. Tscheike, P. Finke, K, Mummert, I. Bächer, Beschreibung und Bewertung ausgewählter Schadensbilder der lokalen Korrosion an Nadelrohren des WWER-440, Symp. Sicherheit von KKW Dampferzeugern, Ostrava, May 1989, Proceedings Vol. I, pp. 158-165.
1. P. Cohen, Chemical thermohydraulics of steam generating systems, Nucl. Techn. 55:105 (1981).
3. V. Liebig, B. Köhler, W. Gebhardt, M. Kröning, K. Mummert,
   Circumferential Arrays used for UT Endoscopy on Heat Exchanger Tubes
   12th Int'l Conference and Exhibits on NDE in the Nuclear and Pressure Vessel Industries, 11.-13. Oct. 1993, Philadelphia, PA.
4. W. Gebhardt, F.Walte, R. Hoffmann,
   ARGUS- ein neu entwickeltes Phased-Array-Gerät für den Nachweis und die Analyse von Werkstoffehlern, Proc. of the Annual Conf. of the German NDT Society (DGZfP) 6.-8. May 1991, Luzern, p. 498.
5. J.D. Achenbach. "Wave propagation in elastic solids", Elsevier Science Publishers. b.v., Amsterdam (1984).
6. H. Wüstenberg. "Untersuchungen zum Schallfeld von Winkelprüfköpfen für die Materialprüfung mit Ultraschall," thesis, Berlin (1972).
7. B. Delannoy, et al., The infinite baffles problem in acoustic radiation and its experimental verification, J. Appl. Phys. 50:5189 (1979).
8. B. Köhler, Impulse response of a piezoelectric layer, Acustica 73:144 (1991).
9. E. Kühnicke, A. Schamlott, Berechnung der Intensitätsverteilung von Winkelprüfköpfen und experimentelle Ergebnisse, in: Proc. Ultraschallmeßtechnik in Forschung und Praxis, Stollberg, 20-23 Nov. 1989, ISBN 3-86010-276-1.
10. B. Tylkovsky, private communication.

3-D UNDERWATER IMAGING SYSTEM

Pierre Alais, Nicolas Cesbron, Pascal Challande, François Ollivier.

Laboratoire de Mécanique Physique
Université Pierre et Marie Curie (Paris 6) - CNRS URA 879
2 Place de la Gare de Ceinture
78210 Saint-Cyr-l'Ecole, France

INTRODUCTION

Since the experiments of Sokolov in 1930[1] many versions of an underwater camera have been proposed. The Sokolov technique is using an analog 2-D phonosensitive retina and, while very interesting by its real time possibilities and relative simplicity, remains limited in resolution[2]. More recently, the proposed techniques are using linear or 2-D arrays. The natural physical solution is to use a 2-D matricial array[3,4] so that from a simple illumination of space, an adequate parallel treatment of the N^2 echographic signals delivered by the elementary transducers of the array may lead very fast to a 3-D information about the illuminated space. Difficulties come from the fact that even a low value of 64 for N imposes to build a 4096 element array, connections between this array and a specific integrated circuitry able to make the parallel treatment on 4096 parallel channels and a signal treatment fast enough to reconstruct an image without losing the benefit of quasi real-time. Nevertheless, due to progress in integration, this approach may be proposed[5,6].

Our work, issued from a Eureka Euromar project[7], presents a system which is not the fastest one but remains fast enough for many applications and leads to a relatively simple and cheap realization. The use of two orthogonal linear arrays for selecting an azimutal θ_x direction at reception and a site angle θ_y at emission is well known[8,9], but requires much care to preserve the intrinsic advantages of the concept. In this purpose, we have built special 64 element arrays permitting to avoid grating lobes artefacts with a .75° angle resolution over a 30° angle aperture and a dedicated electronics able to treat in real time the information delivered by the 64 receivers to that the system is able to give a 3-D information of space in the order of 1 second at a few meters ranges and a fast frontal imaging information at higher ranges (10 frames/second at 50 m).

I. GEOMETRY AND CHARACTERISTICS OF THE ANTENNAS

The cylindrical emitting and receiving arrays are identical and have 64 elements with a 4 mm spacing working at 500 kHz. Each cylindrical element is made of composite ceramics open on a sector of 40° with a curvature of 60 mm. This shape permits to obtain a beam pattern orthogonal to the array (Fig. 1) wide enough to cover the required 30° field while rejecting, below 35 dB, the grating lobes of the expected tomographic beam pattern associated with the orthogonal array (Fig. 2). This pattern presents a .75° resolution with low side lobes, except for grating lobes which have to be reduced by the effect of the orthogonal pattern of the other array. An advantage of the cylindrical shape is also to ensure a larger and more powerful emitting surface while the receiver impedances are lowered.

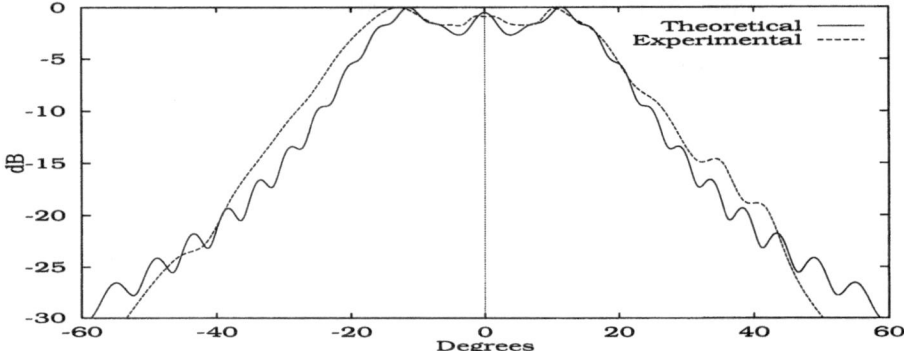

Figure 1. Orthogonal beam pattern

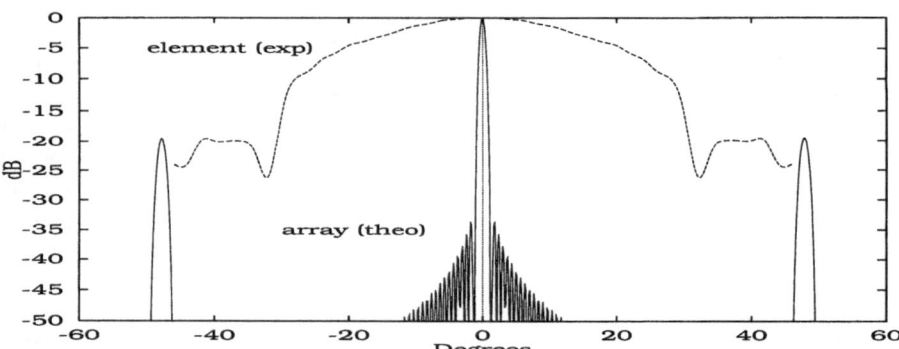

Figure 2. Tomographic beam pattern

II. THE RECEIVING PROCESS

The receiving system uses the horizontal array and is able to give, in quasi real time, a sectorial image open on 30° in azimuth angles obtained from a simple illumination either open in site angles (frontal imaging mode) or focused on a chosen site angle (3-D imaging mode). The sectorial image is built from 512 (R) x 64 (θ) pixels covering the total range x 30° azimutal aperture. Total range may be chosen as 3, 6, 12 up to 192 m, and the corresponding 500 kHz acoustical illumination is time modulated according to a gaussian shape and a duration equivalent to 4 pixels in range. The 64 signals obtained from the receiving array are amplified with a time varying gain and complex demodulated with sine and cosine 500 kHz reference signals. The 128 X and Y obtained signals are numerically converted on 8 bits and stored in a RAM memory according to 512 samples for the whole range.

Then, each pixel of the reconstructed image is computed with an algorithm operating simultaneously a phase compensation and time delays. The phase compensation is done with a classical Fresnel kernel adapted to the distance of the pixel from the antenna according to 32 focal zones stored in dead memories and covering the total range of 1.3 m to 200 m, so that the error in phase compensation remains as low as possible. The time delay is obtained by selecting the XY samples in range order. This operation is necessary only for short ranges when the time of flight differences are of the same magnitude as the illumination duration. Although this delay compensation is very rough, due to the poor sampling of the emitting signal (4 pixels only), it is very effective as it may be seen on Figs. 3 and 4, where theoretical results are compared with experimental ones obtained from a small emitting probe located at 2 m from the antenna, with .25° and - 9.75° steerings.

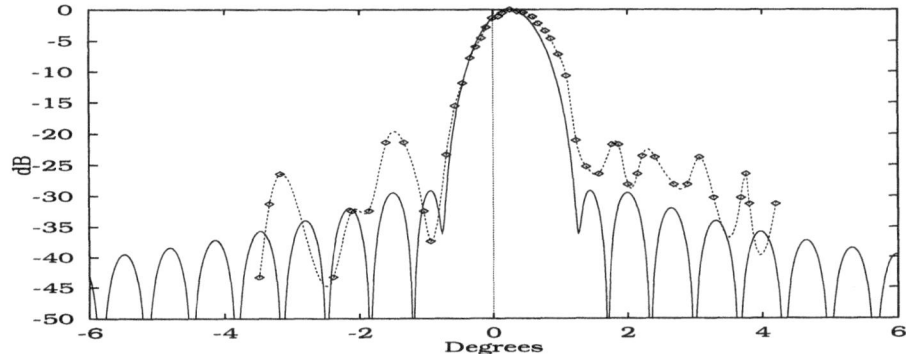

Figure 3. Focusing pattern for 0.25° steering

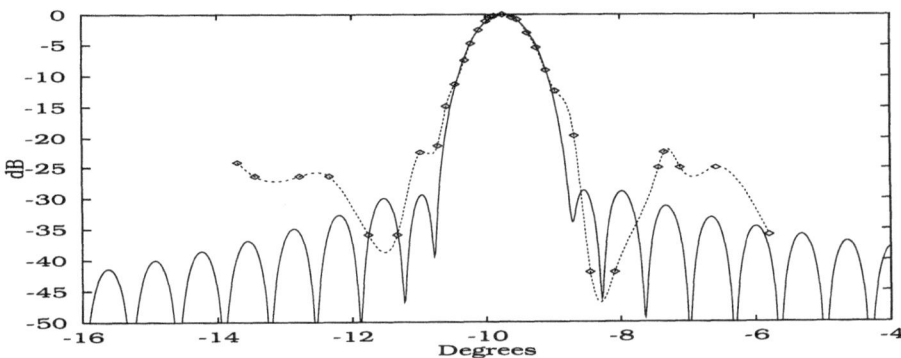

Figure 4. Focusing pattern for - 9.75° steering

Theoretical results are classical Fresnel results associated with a permanent harmonic illumination and could not be fitted in this manner by such a short illumination with a 2 m range and 10° steering if the time delay compensation was not effected.

By using dedicated electronics, the computation of each pixel, which operates the phase shift of 64 complex signals and sommation of the results, requires only 2 μs, so that obtaining a sectorial picture with 100 pixels in range, chosen among the 512 possible ones, requires only 12,8 ms added to the ultrasonic time of flight, which gives, for exemple, 30 frames/second for a 15 m range.

Then, this R-θ image is converted in real time in an X-Y video image using an interpolating transfer from an R-θ memory to an X-Y memory read at a classical video frame rate.

III. THE EMITTING PROCESS

The emitting system uses the vertical array and gives either a diverging illumination covering the 30° site angle aperture (frontal imaging mode), or an illumination focused at a selected range and site angle (3-D imaging mode). Obviously, an optimal vertical resolution can be obtained in elevation pictures only for the retained focal distance, and there will be only a limited field of view unlike the dynamic focusing obtained for the horizontal direction in the receiving process. This is important essentially at short ranges for which another degradation comes from the fact that, for sake of simplicity, we have chosen a simple phase compensation for the emitting focusing operation so that it may be seen, from figures 5 and 6, that the Y experimental resolution is not as good as expected at a 2 m range and especially for an important steering. The phase control is operated numerically using dead memories and permits to select 64 site angles and 32 focal zones identical to the receiving ones.

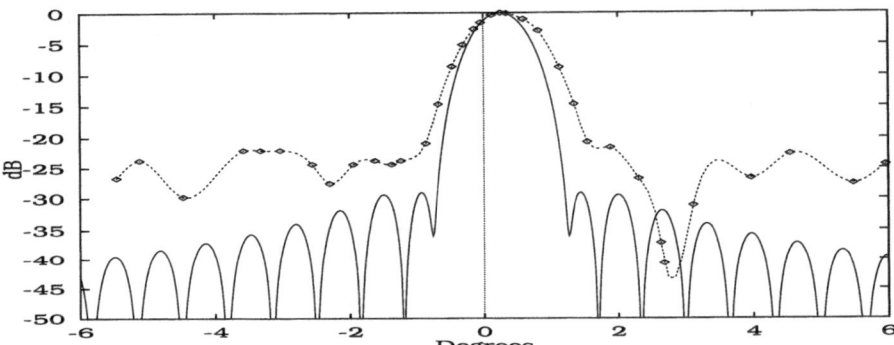

Figure 5. Transmit beam pattern for 0.25° steering

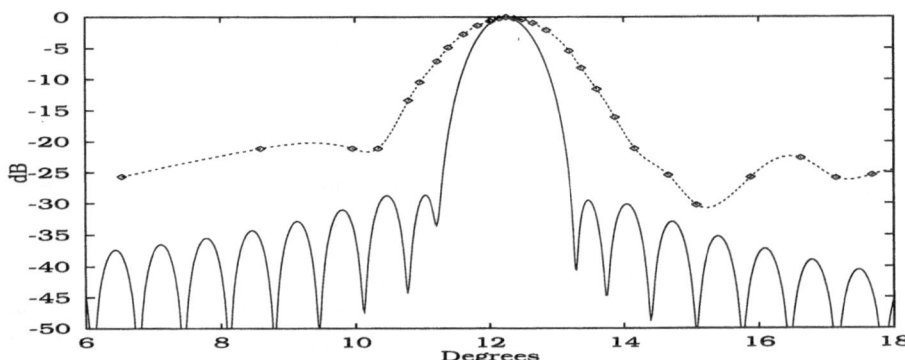

Figure 6. Transmit beam pattern for 12.25° steering

IV. IMAGES

The video sectorial image is displayed in real time for short ranges. Only for large ranges the refreshment time is, for the most important part, the acoustical time of flight (130 ms for 100 m range). At this moment, our system has not been packed in a waterproof casing so that our imaging experiments are reduced to an operation in a 3 m tank, and the quality of images is reduced due to the small number of wavelengths in the available dimensions of targets. In these conditions, good shadows and reverberation contrast are very difficult to obtain, but nevertheless we succeeded in getting interesting elevation pictures by retaining the maximal echographic 3-D information encountered for a θ_x - θ_y direction, inside a range zone [R1 - R2] which may be selected with the help of a range marker operating on the real time (θ_y evoluting) sectorial picture.

Our targets have been coated with sand to improve the acoustic diffusion. Figures 7 show the optical and acoustical view of a Siemens test-pattern, 60 cm in diameter, set at 2 m. Figures 8 are relative to a target representation of an element of an off-shore structure. It is interesting to check on the acoustical view the possibility of a 3-D perception due to the fact that, even at a short distance of 2 m, the depth of field remains large enough to appreciate the 3-D shape and the orientation of the object. Figures 9 and 10 are acoustical views obtained from targets made of electrical wire.

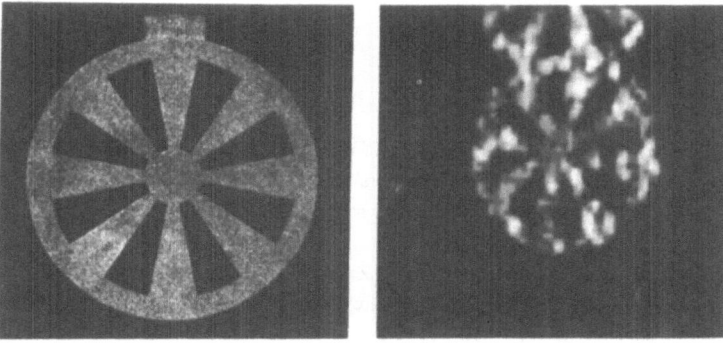

Figures 7. Optical and acoustical views of a Siemens test pattern at 2 m.

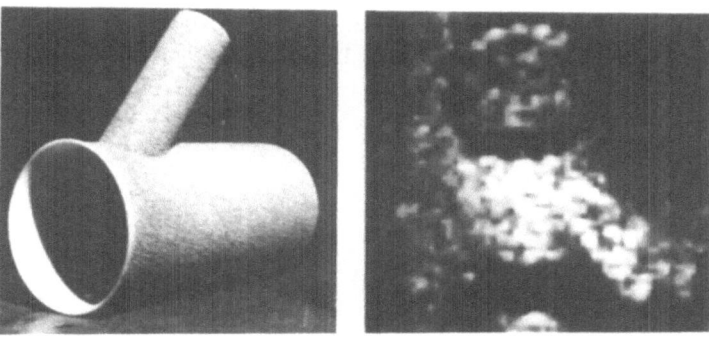

Figures 8. Optical and acoustical views of a tubular structure set at 2 m.

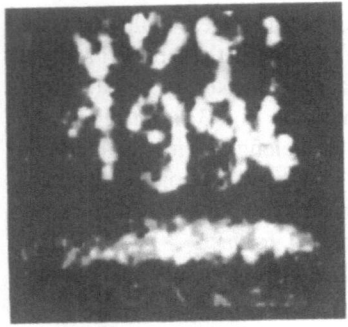

Figure 9. "1994" (electrical wire at 2 m)

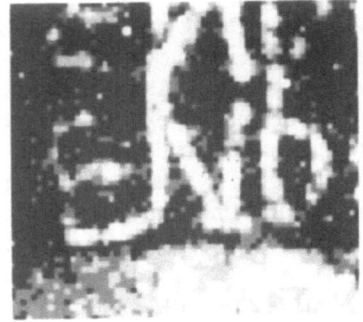

Figure 10. "END" (electrical wire at 2 m)

CONCLUSION

The great difficulty in conceiving a 3-D underwater acoustical imaging system is to keep the complexity and cost of the system versus acoustically permitted performances in practical and reasonable limits. Our choice leads to a system relatively simple and cheap, which keeps a true real time imaging in the frontal imaging mode useful for navigation uses and permits a relatively fast 3-D identification which could be useful for various off-shore applications. In the frame of Euromar, we should cooperate with various companies to develop an industrial version of this system.

REFERENCES

1. S. Sokolov, "Means for Indicating Flaws in Materials", US Patent 2, 164, 125 (June 27, 1939).
2. J.L. Dubois, "Large Aperture Acoustical Image Converters", Acoustical Holography, 2, 29-68, 1969.
3. H.R. Farrah, E. Marom and R. K. Mueller, "An Underwater Viewing System Using Sound Holography", Acoustical Holography, 2, 173-183, (1969).
4. E. Marom, R. K. Mueller, R.F. Koppelmann and G.Zilinskas, "Design and Preliminary Test of an Underwater Viewing System Using Sound Holography", Acoustical Holography, 3, 191-209 (1970).
5. Wide Acoustic Imaging and Classification System Maracous MAST Project "W.A.I.C.S." (1990).
6. R.K. Hansen, P.A. Andersen, "3-D Acoustic Camera for Underwater Viewing", Acoustical Imaging, 20, 723-728 (1994).
7. "High Resolution 3-D Imaging System", Euroka Euromar Project EU 406.
8. P. Alais, P. Challande and L. El Jaafari, "Development of an Underwater Frontal Imaging Sonar, Concept of a 3-D Imaging System", Acoustical Imaging, 18, 431-440 (1989).
9. P. Alais, "Techniques for Underwater Acoustic Imaging", Proceedings of 'Ultrasonics International 91', Le Touquet, France (1-4 July 1991).

ACOUSTICAL IMAGING OF UNDERWATER PLUMES

David R. Palmer

National Oceanic and Atmospheric Administration,
Atlantic Oceanographic and Meteorological Laboratory
4301 Rickenbacker Causeway, Miami, Florida 33149

INTRODUCTION

We review the research done to acoustically image naturally occurring underwater plumes. The review was undertaken to gain perspectives that might help in our efforts to image high-temperature, black-smoker, hydrothermal plumes in the deep ocean (Palmer, Rona and Mottl, 1986). Specifically, we wanted to identify common themes in hardware development, experimental design, and data analysis and interpretation.

Since our interest is in the use of acoustical techniques to study fluxes of material and heat across the ocean-seafloor interface, the review is limited to studies of those plumes that emanate from vents located on the seafloor. Consequently, we do not consider acoustical imaging of plumes that result from sediment re-suspension, turbidity currents, water-mass intrusions from rivers, or the dumping or discharge of dredge material, sewage sludge, or other potential marine pollutants. The literature on acoustical imaging of these types of plumes may be traced from the references cited in Palmer, Rona and Mottl (1986).

After describing the general structure of an underwater plume and discussing the advantages of acoustical imaging, we survey the individual experimental efforts. These include observations of offshore and deep-ocean hydrocarbon plumes, buoyant plumes from a submarine spring, and buoyant and neutrally-buoyant hydrothermal plumes. We then make some general statements about the characteristics of imaging sonars and their platforms, sonar calibration, scattering mechanisms, and the potential for Doppler measurements. Finally, we discuss the importance of *in situ* observations.

Structure of an underwater plume

Figure 1 illustrates the structure of an underwater plume. At the vent on the seafloor the flow can be jet-like with the pressure drop across the orifice, as well as buoyancy, providing the kinetic energy. Alternatively, the plume can originate from a submarine seep, where the plume material oozes from the seafloor. The buoyant force is then the only source of kinetic energy. In its initial stage of development the flow is primarily vertical with perhaps some bending due to a prevailing current. The plume is called a "buoyant plume" in this stage. Because of density stratification, a height of neutral buoyancy may exist. The plume overshoots and then spreads horizontally as one or more neutrally buoyant, effluent layers. In this stage of development, it is referred to as a "neutrally buoyant plume."

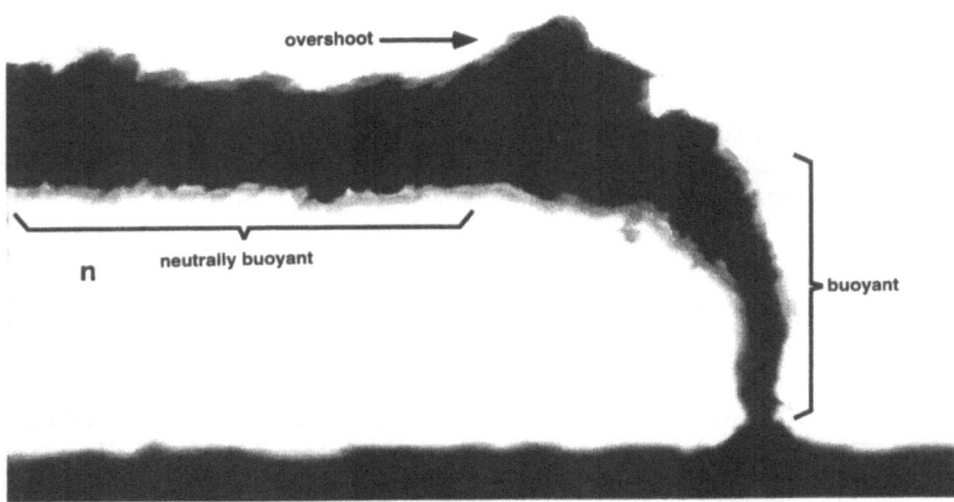

Figure 1. Structure of an underwater plume.

If a height of neutral buoyancy does not exist, the plume continues to rise until it meets the sea surface. In shallow water this encounter may result in a surface expression detectable from a ship or aircraft. In deep water the rising plume loses its organized structure long before it reaches the surface because of turbulent entrainment and the influence of currents. Understandably, surface expressions of deep-water plumes have never been observed.

The plume material itself may consist of a clear or particle-laden fluid having a different temperature, density and dissolved mineral content than that of the seawater. It may consist simply of gas bubbles or of solid matter having acoustic properties different from those of seawater. The composition and physical properties of the plume material determine the scattering mechanisms. These mechanisms can be classified as being either particulate matter or fluid inhomogeneities.

Advantages of Acoustical Imaging

Defining a plume's boundary is the most common application of acoustical imaging. Acoustical imaging is clearly the best way of obtaining an overall picture of the plume. In many cases, it is the only way. One needs to know the shape of a plume to guide chemical and physical sampling and to determine the relationship between the plume and its ocean setting.

The spatial and temporal variations of a plume can be characterized. One can determine spatial variations with scales of meters to hundreds of meters. Using existing platforms, one can determine temporal variations with scales from tens of seconds to an hour. With a bottom-mounted sonar, a system that does not now exist, one could measure temporal variations having scales up to one year.

Acoustical imaging has the potential for determining ocean-seafloor fluxes. This advantage will be discussed in greater detail when we consider the potential for Doppler measurements.

Acoustical imaging may be a useful tool in the search for plumes. Shallow-water hydrocarbon plumes have been located acoustically. It is possible that other types of plumes, including deep-ocean ones, could also be located. Information can be provided about the

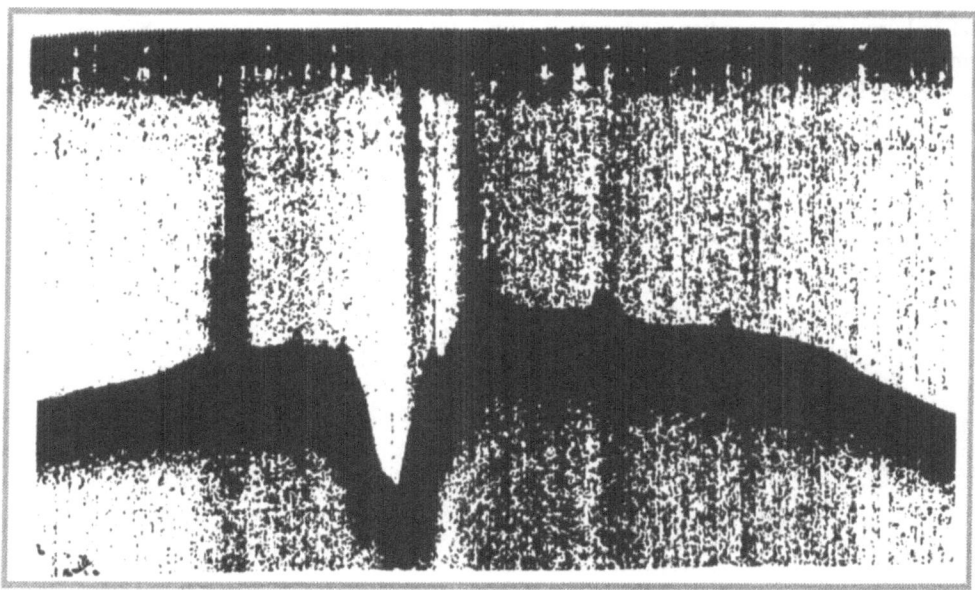

Figure 2. Sonar image of offshore hydrocarbon plumes [adapted from Sackett, (1977) with permission].

plume in real time. This advantage is the basis of the use of imaging to support the collection of *in situ* data. Finally, imaging provides the opportunity to non-invasively observe plumes and their ocean settings.

EXPERIMENTAL STUDIES

The literature in this area is not extensive. Two efforts in shallow water have been reported; the imaging of offshore hydrocarbon plumes and an effort undertaken in the Canadian Archipelago to image the plume from a submarine vent. All the other studies were in deep water.

Offshore Hydrocarbon Plumes

Hydrocarbon plumes have been imaged many times in shallow water using standard echo sounders and side-scan profilers mounted on ships (Bryant and Roemer, 1983; Sackett, 1977). These buoyant plumes are globally distributed on continental shelves and margins. They consist of gas bubbles and oil droplets rising from submarine seeps. They result in seepage into the marine environment of as much as 6×10^5 metric tons of hydrocarbons per year (Wilson et al., 1974).

Figure 2 is a sonar image obtained with a 12 kHz echo sounder of at least three shallow-water, hydrocarbon plumes (Sackett, 1977). The plumes in this image extended all the way to the sea surface, suggesting the existence of a surface expression.

Buoyant Submarine Spring Plume

Images have been obtained of a plume from a submarine spring in Cambridge Fjord, Baffin Island in the Canadian Archipelago (Hay, 1984; Colbourne and Hay, 1987; 1990). A commercial 192 kHz portable echo sounder mounted on a launch recorded the images. The plume consisted of turbulent, brackish water from a nearby lake. It rose from a submarine vent located on the ocean floor at a depth of 47 m.

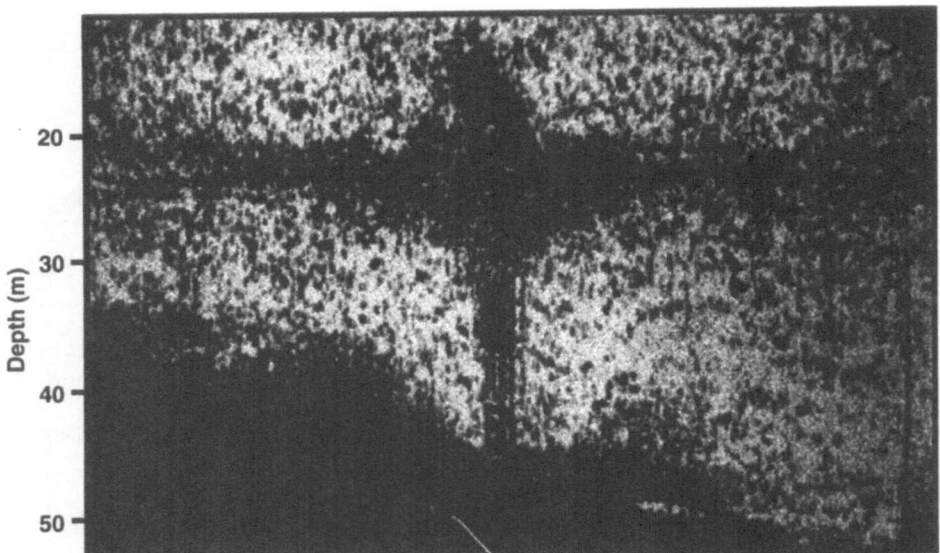

Figure 3. Submarine spring plume, Baffin Island, Canadian Archipelago [adapted from Colbourne and Hay, (1990) with permission].

Figure 3 is an image of the plume (Colbourne and Hay, 1990). It was constructed from backscattered data collected as the launch moved slowly over the submarine vent. The image shows both the buoyant and neutrally-buoyant portions of the plume, as well as the overshoot above the level of neutral buoyancy.

The spatial characteristics of the plume were determined from the ocean floor to the plume's cap at about 9 m depth and the vertical velocities were estimated by tracking discrete scattering structures while the launch was moored over the spring. This work represents the first published account of the monitoring of a naturally occurring, buoyant plume using acoustics.

Deep-Ocean Hydrocarbon Plumes

Merewether et al., (1985), reported observations of plumes in the Guaymas Basin of the Gulf of California made with the up-looking 23.5 kHz sonar and the 110 kHz side-scan sonar mounted on the Scripps Marine Physical Laboratory's instrument platform DEEP TOW. These plumes extended hundreds of meters above the seafloor. The water depth varied from 1600 m to 2000 m in the region. They were interpreted as light hydrocarbons escaping from natural seeps on the seabed. While shallow-water hydrocarbon plumes are routinely observed, this was the first observation of hydrocarbon plumes in deep water.

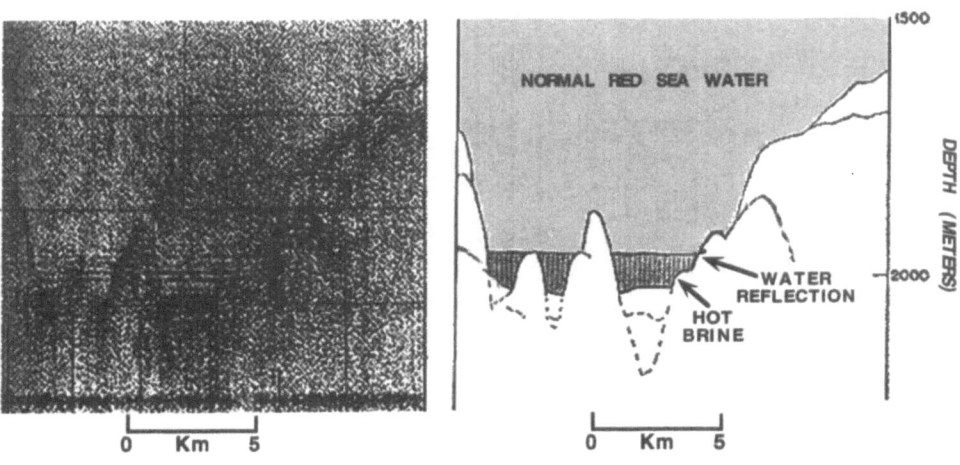

Figure 4. Red Sea brine pools [adapted from Hunt et al., (1967) with permission].

Red Sea Brines

The Red Sea brine pools were discovered from an examination of the records obtained from standard echo sounders mounted on research ships (Degens and Ross, 1969). The pools consist of hot, high-salinity (44° C, σ_T = 89) hydrothermal solutions contained in seafloor basins. They are separated from the overlying normal seawater (22° C, σ_T = 28) by a stable, narrow, horizontal interface. It is from this interface that acoustic signals are reflected.

The pools are a manifestation of the type of hydrothermal system that exists at seafloor spreading centers. The seafloor is rifting due to the slow separation of tectonic plates. Sea water seeps into the ocean crust through fissures in the seafloor created by the rifting. It is heated by proximity to magma and takes up minerals in solution. The hot, mineral-rich, reducing solution rises and is discharged at vent sites where it mixes with cold, oxidizing seawater. Precipitates are formed by the resulting chemical reactions.

The rifting is in its early stage in the Red Sea and the hydrothermal circulation is not vigorous. In later stages there is an upwelling of the underlying crustal material resulting in mid-ocean ridges. The circulation is much more vigorous during these later stages.

Shown in Fig. 4 is an acoustical image of Red Sea brine pools that was obtained with a standard, ship-mounted echo sounder. The line drawing shows the location of the pools in deep basins on the seafloor. The acoustic reflections from the brine interfaces are probably the result of scattering from suspended particulate matter precipitated from the hydrothermal solutions and not from impedance contrasts or turbulent temperature fluctuations.

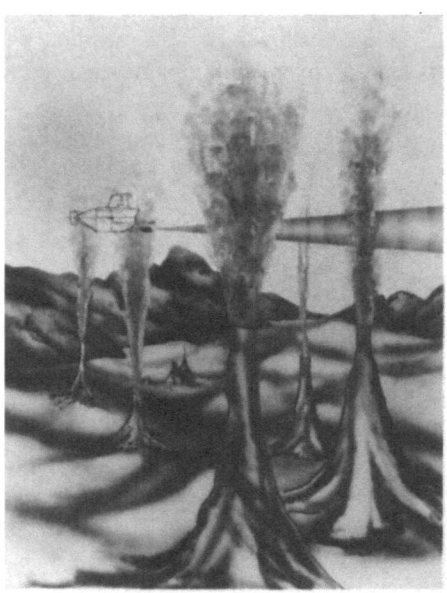

Figure 5. Artist's conception of a sonar mounted on a manned submersible being used to image a "black smoker" hydrothermal plume.

Buoyant Hydrothermal Plume - 11° N, East Pacific Rise

We have used sonars mounted on submersibles to image buoyant, black smoker hydro-thermal plumes in deep water at 11° N and 21° N on the East Pacific Rise in the Pacific Ocean. Black smoker plumes are formed when hydrothermal solutions are discharged from chimney-like vents at seafloor spreading centers. The "black smoke" is really a cloud of metallic-sulfide precipitates. The temperatures can be as high as 350° C and the flow rates can reach 5 m/sec; comparable to the flow from a fire hose.

The general goal of our effort is to develop a plume imaging sonar system that can provide images of plumes as a function of height above the ocean floor beyond the limits of photographic or video observations. These images will be used to guide physical and chemical sampling of the plume and to provide a better understanding of plume dynamics. Figure 5 is an artist's conception of a sonar mounted on a manned submersible being used to image a plume.

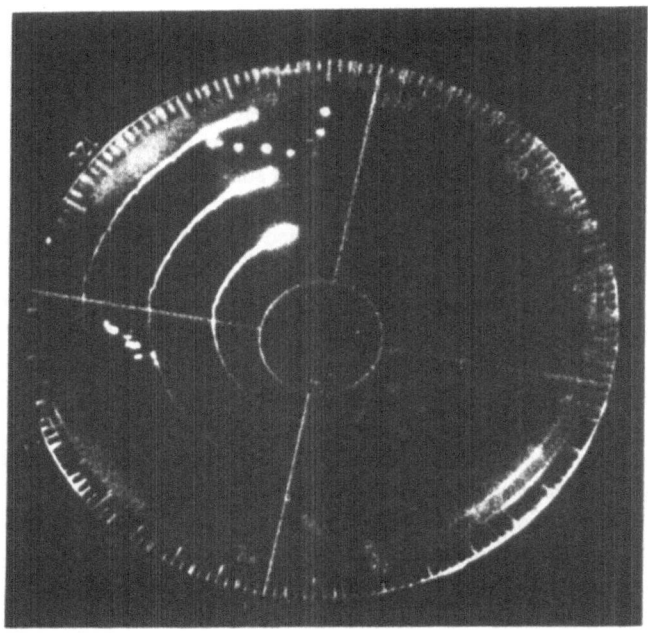

Figure 6. Photograph of a sonar display of a black smoker complex at 11° N on the East Pacific Rise (Palmer, Rona and Mottl, 1986).

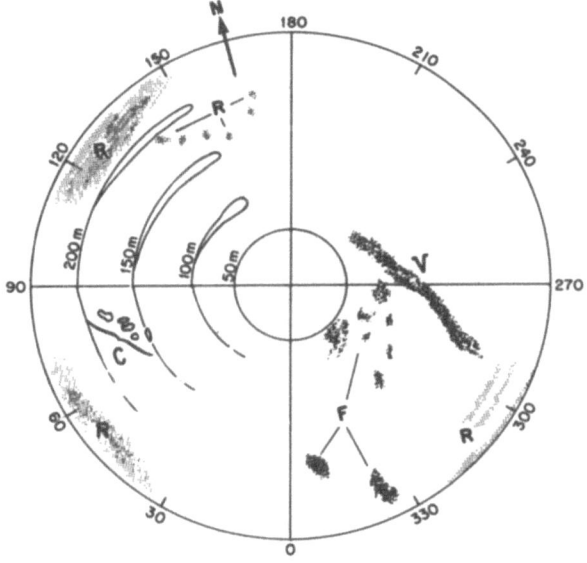

Figure 7. Line diagram corresponding to Figure 6 indicating important features. The plume complex is labeled "C" in the diagram.

The images collected at 11° N were recorded using the CTFM navigational sonar mounded on DSRV ALVIN at a water depth of about 2500 m (Palmer, Rona, and Mottl, 1986). The vent field consisted of six to eight black smoker vents. Figure 6 is a photograph of the sonar display and Fig. 7 is a corresponding line drawing indicating the plumes and other features. Although the sonar was not specifically designed for this application, it did record images of plumes. This study represents the first example of acoustical imaging of buoyant, hydrothermal plumes and the first use of a submersible as a platform for a plume imaging sonar.

Palmer, Rona and Mottl, (1986), also obtained estimates of the minimum detectable concentration of precipitates as a function of the range between the sonar and the plume. Comparison with measured concentrations indicated that black smoker plumes can be detected acoustically at ranges up to about 200 m using commercially available sonars.

Buoyant Hydrothermal Plume - 21° N, East Pacific Rise

For acoustical imaging studies at 21° N on the East Pacific Rise, we modified a 330 kHz, commercially available, imaging sonar in the inventory of the U.S. Navy Submarine Development Group One and mounted it on the U.S. Navy DSV TURTLE (Rona et al., 1991). The plumes rose from vents at a depth of 2600 m at the location where black smoker hydrothermal plumes were first discovered (RISE Project Group, 1980).

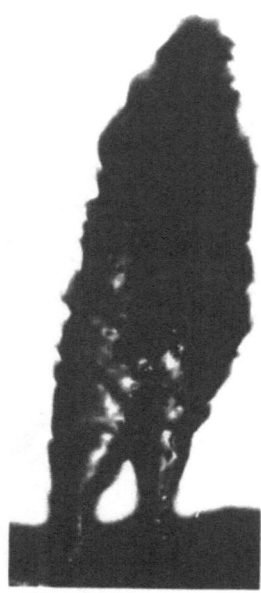

Figure 8. Three-dimensional view of two "black smoker" plumes constructed from backscattered intensity data obtained at 21° N on the East Pacific Rise (generously provided by C. Jones).

Figure 8 is a three-dimensional view of two black smoker plumes reconstructed from the backscattered intensity data collected during one of the eleven dives that took place during the sea trial. The two plumes originated from vents located about 3.5 m apart. The submersible was resting on the bottom about 7 m from the nearest vent. The sonar range was set at 50 m. The plumes bent to the right in response to the observed current flow.

This work demonstrated that acoustical imaging can define the three-dimensional boundary of a black smoker plume, can characterize its structure on spatial scales of meters to tens of meters and on temporal scales of minutes to hours, and can provide initial detection at ranges of the order of 100 m.

Neutrally Buoyant Hydrothermal Plume

Acoustical images have been reported of the neutrally buoyant portion of a hydrothermal plume (Thomson, Gordon, Dymond, 1989; Thomson, Gordon, and Dolling, 1991; Thomson et al., 1992). The plume emanating from vents on the Endeavour Ridge in the Northeast Pacific. The water depth was approximately 2200 m. The observations were made using a 150 kHz acoustic Doppler current profiler (ADCP). The ADCP was part of an instrumentation package lowered from a ship to a number of pre-selected depths. At each depth level, vertical profiles of backscattered intensity were obtained. They were compared with profiles of conductivity, temperature, and transmissometer data obtained simultaneously.

The observations indicated a puzzling absence of backscatter from the plume. Palmer and Rona (1990) argued that the observations could be explained if a significant attenuation mechanism was not taken into account in the calibration of the sonar. In particular, failure to account for the attenuation of the acoustic beam as it propagates through the plume because of scattering of acoustic energy out of the beam by the particulates in the plume; a multiple scattering effect, could explain the observations. Thomson, Gordon, and Gast (1990) discounted this explanation arguing that particulates could not contribute to the scattering because of their small size and concentration. The issue remains unresolved, however, since increased particulate loading was present in the plume, as determined by transmissometer readings, and should have resulted in some increase in backscattering. Moreover, no scattering mechanism can be ruled out using an argument based solely on particle size and concentration. Sonar characteristics, propagation loss, and noise level must also be considered. This issue illustrates how difficult the identification of the dominant scattering mechanism can be without physical sampling of the plume.

Buoyant Hydrothermal Plume - Guaymas Basin

Armishev and Berezutskii (1988) have reported the acoustic detection of hydrothermal plumes in the Guaymas Basin using a towed instrumentation platform. The observations were made in a region of hydrothermal activity previously investigated by Lonsdale and Becker (1985). The acoustic equipment consisted of a 78 kHz side-looking sonar and an 8 kHz acoustic profiler. The 78 kHz system, but not the 8 kHz one, clearly imaging ascending plumes. This study also suggested that hydrothermal plumes are such good sonar targets that they may create acoustic shadows.

SONAR DESIGN AND PERFORMANCE

Sonar Characteristics

No specialized custom-made sonars were used in any of the reported studies. All sonars were either commercially available or standard equipment on a platform. Modifications were made to only two of the sonars (Thomson, Gordon, Dymond, 1989; Rona et al., 1991) and those modifications were minor. All the sonars used were monostatic and active. There is no application for passive sonars in this field. Plumes radiate very little sound because the Mach numbers are so low.

All the sonars provided amplitude verses range data, either by transmitting a pulse or a CTFM signal. Isometric (bearing) information was then obtained either by moving the platform, as with a side-scan sonar, or by moving the transducer using stepping motors. Doppler data were collected by only one sonar (Thomson, Gordon, Dymond, 1989). These data describe the ambient currents rather than the flow characteristics of the plume.

The Potential for Doppler Measurements

Doppler data clearly have potential value. The most important characteristics of a naturally occurring plume are the material and heat fluxes across the ocean-seafloor boundary. Doppler can provide estimates of those fluxes. Traditionally, they have been determined in one of two ways. By a measurement of the time it takes a discrete structure in the plume to rise a given distance, together with an *in situ* measurement of the density of the physical quantity being transported. The second way uses a flow meter to estimate the discharge rate. Both techniques are difficult and many times not successful.

Doppler measurements could certainly be used to estimate discharge rates. There is a possibility Doppler could also be used to locate plumes. However, it is unlikely that Doppler could provide estimates of the velocity field within the plume because of a lack of spatial and temporal resolution.

Calibration Issues

We distinguish between the correction that is applied to the signal for spreading and absorption losses, the time varying gain (TVG), and the actual calibration of the system, which allows for the determination of target strengths.

It is routine for sonar processors to correct for spreading and absorption losses by applying TVG to the received signal. The purpose of doing this is to reduce the dynamic range needed by the recording system. The TVG function is determined from the sound speed and absorption coefficient for the ambient sea water, or, equivalently, from conductivity-temperature-depth data. In modern systems it is applied digitally and can easily be extracted from the data.

The calibration of the sonar is quite different. It consists of knowing two relationships. First, the relationship between the magnitude of the rms signal used to drive the transducer and the magnitude of the rms pressure field it radiates. Second, the magnitude of the received rms pressure field and the magnitude of the rms signal that is sent to the recording system. Both relationships are dependent on the acoustic frequency and the direction of propagation of the pressure field.

In general high-resolution imaging sonars are not well calibrated. For example, there are no procedures or facilities for calibrating deep-submergence, imaging sonars (J. E. Blue, private communication, 1992). *In situ* calibration may be possible using a standard target such as a tungsten-carbide sphere (Blue, 1984). The technique has not yet been attempted, however.

Without a calibration, researchers have resorted to the use of a sonar equation to describe performance and determine the nature of the scatterers. This approach is limited because sonar equations are most useful for comparative studies; comparing the performance of one sonar with another or one possible scattering mechanism with another.

SCATTERING MECHANISMS

Scattering mechanisms fall into two categories; particulate matter and fluid inhomogeneities. The first category includes solid particles, bubbles, and oil droplets as well as biota that might congregate within the plume or near its boundary. In the second category are steep gradients or discontinuities in the temperature, salinity, and velocity fields. A plume typically has more than one scattering mechanism. Determining the dominant one has been a ubiquitous problem in this field of research.

Scattering from Particulate Matter

The scattering is always incoherent. The relative change in particle positions between sonar pings is always large compared to one half of the wavelength. This simplifies the development of analytical and numerical models of the scattering process.

Although multiple scattering is usually ignored, it should not be. It occurs when the acoustic field incident on a particle has been altered because of the presence of other particles. Two characteristics of multiple scattering are echo lengthening and absorption. Echo lengthening occurs because the acoustic field spends additional time in the plume being scattered by more than one target. Absorption is indicated when the backscattered, integrated energy does not increase linearly with particle density.

It probably is enough in this research area to compensate for shadowing. Shadowing occurs when the sonar pulse is attenuated as it propagates through the plume. It is primarily a consequence of second-order multiple scattering and accounts for most of the absorption. Although there is a good empirical technique for correcting for shadowing based on transport theory (see, for example, Eberhard, McNice and Troxel, 1987), it has not yet found wide application in underwater acoustics.

Scattering from Fluid Inhomogeneities

Neutrally buoyant plumes are characterized by sharp, stable, impedance gradients that can backscatter sound. The scattering process can be studied using established procedures [see, for example, Weston (1958)].

Buoyant plumes are always turbulent. Scattering occurs from the fluctuations in the index of refraction. Two formalisms can be used to study the process. First, acoustical scattering from turbulence; a formalism developed by Tatarskii (1971) and others. Second, the dynamic description of a turbulent plume in a stratified fluid (see, for example, Turner, 1973 and Rodi, 1982). Tatarskii's formalism cannot be directly applied because a knowledge of the spatial correlations of the index of refraction is lacking. In the case of black-smoker hydrothermal plumes, the required instrumentation does not exist. The situation is not improved by assuming the turbulence is in the Kolmogorov inertial subrange since the needed rms fluctuation is also not known.

An alternative approach is to use the second formalism to build a model of the plume and then compare the properties of the model with the characteristics of the backscattered intensity. Hay, (1984), has done the most with this approach. Using it he was able to argue that the recorded images of the Cambridge Fjord plume where the result of turbulent scattering.

DESIGN AND CONDUCT OF EXPERIMENTS

Sonar Platforms

Three types of sonar platforms were used in these studies: surface ships, submersibles, and instrumentation platforms. Ships were used for imaging the Red Sea brines and all the shallow water plumes. My group used submersibles as platforms for imaging hydrothermal plumes. Towed instrumentation platforms have been used in deep water to image hydrocarbon and hydrothermal plumes. A tethered platform provided vertical profiles through a neutrally-buoyant, hydrothermal plume. A bottom-mounted platform has potential for monitoring plumes on long times scales up to one year. So far one has not been used.

The most important platform requirement is that the sonar be operated near the seafloor, close to the plume. To provide images of value, the sonar must have a footprint at the location of the plume small compared to the plume's dimensions. In addition, high-resolution sonars operate at frequencies above 50-100 kHz and have an operational range of the order of a hundred meters.

For a shallow-water plume this positioning requirement does not restrict the type of platform. For deep-water imaging of buoyant plumes, however, it does rule out a surface ship. In deep water, any ship-mounted sonar would have a footprint on the seafloor large compared to the dimensions of a buoyant plume.

The requirement does not necessarily restrict the choice of platform for deep-ocean, neutrally buoyant plumes, however. The Red Sea brine pools have been imaged with ship-mounted echo sounders having frequencies from about 10 to 20 kHz. The "image" is the reflection from the brine interface; a stable, reflecting surface having a horizontal extent of the order of 1 km^2; large compared to the sonars' footprint, even if the sonar is mounted on a ship.

The positioning requirement imposes constraints on deep-ocean imaging beyond choice of platform. The transducer must be pressure compensated to withstand the tremendous pressures and the system certified to pose no health risks to the crew. Meeting these requirements is usually not difficult but does introduce a degree of schedule risk. There are also navigational considerations. A navigational system is needed not only to locate a plume but also to determine the relative position of the sonar. To obtain quality images, the platform must be stable. For example, notwithstanding Fig. 5, we have not been able to think of a way of obtaining three-dimension images of black smoker hydrothermal plumes from a hovering submersible. All our three-dimensional images were obtained while the submersible was sitting on the seafloor. Finally, a suitable platform be available. This is the most important constraint since instrumentation platforms and submersibles are valuable assets and in great demand.

The Importance of *In Situ* Observations

It is difficult to imagine a situation where acoustical imaging could replace *in situ* observations. Hydrographic data are needed for determining the optimal TVG function and for determining the local sound speed, for constructing plume models, and for understanding the influence of currents on the plume shape. One also needs *in situ* measurements of the physical and chemical properties of the plume. In almost every experiment there has been a question of what scattered the sound. The only way to resolve this question is with *in situ* observations.

There is the point of view that acoustical imaging can be used as a remote sensing tool to determine the type and concentration of plume constituents. That is, knowledge about plumes can be obtained from solving the inverse problem. Research in this area is not sufficiently developed for this point of view to be valid. It is instructive to compare this research area with the use of acoustics in fisheries research. In fisheries research the inverse problem consists of estimating the size of fish stocks using echo sounders. Experiments are relatively easy to conduct and there is a degree of repeatability and control. The scattering mechanisms are understood. One can place the scatterers in a tank and conduct laboratory-style experiments to measure target strengths and other sound-scattering characteristics. Nonetheless, the solution of the inverse problem has been difficult and the results are controversial (MacLennan and Forbes, 1984).

In the area of acoustical imaging of plumes the situation is different. Experiments are difficult and tax researchers and their equipment. They are not conducted under controlled conditions and seldom are repeatable. The nature of the scatterers in the plume is not usually known. One cannot take a plume and place it in a laboratory for study. Until acoustical imaging of plumes becomes simple and routine and until the acoustical characteristics of underwater plumes are better understood, the development of a formalism for solving the inverse problem represents misplaced effort.

There are important and useful applications of plume imaging that do not involve solving the inverse problem. One can learn a great deal about plumes and the geophysical processes that create them by conducting imaging experiments in conjunction with a program of *in situ* chemical and physical measurements.

VIEW OF THE FUTURE

This survey suggests a number of directions for future research. In the area of sonar design and performance we have:

- There is no indication that a sonar will be designed and build specifically for imaging naturally occurring plumes in the forseeable future. Commercial sonars, perhaps with modifications, will continue to be used.
- There is a clear need for Doppler sonars. They can provide important, unique data.
- It will be very difficult to solve the inverse problem. Fortunately, most of the advantages of acoustical imaging do no depend on its solution.
- Multiple scattering should be compensated for in the received signal. This is not being done.

In the area of the design and conduct of experiments we have:

- The direction research in deep-ocean imaging takes will be dictated by platform availability. It will determine what sonars are used, what plumes are studied, and what groups are able to participate in the research.
- The use of bottom-mounted sonar platforms can provide data that will open up a new field of research; long-term characterization of plume dynamics.
- Acoustical imaging will never replace *in situ* observations. In fact, it is not extreme to adopt the point of view that the primary purpose of acoustic imaging is to support the collection of *in situ* data. Successful experiments will include a quality *in situ* observational component.

Several exciting research opportunities are suggested by this list. We are optimistic they will be realized.

ACKNOWLEDGMENTS

I have benefitted from the many discussions I have had with my colleagues at NOAA and the Naval Research Laboratory concerning the topic of this paper. I am very grateful to James Paisley for help in preparing the figures.

REFERENCES

Armishev, S.V. and Berezutskii, A.V., 1980, Sonar detection and location of hydrothermal plumes, *Akust. Zh.*, 34:942 (in Russian); English Translation: *Sov. Phys. Acoust.*, 34:541.

Blue, J.E., 1984, Physical calibration, *Rapp. P.-v. Réun. Cons. int. Explor. Mer*, 184:19.

Bryant, W.R., and Roemer, L.B., 1983, Structure of the Continental Shelf and slope of the northern Gulf of Mexico and its geohazards and engineering constraints, *in*: "CRC Handbook of Geophysical Exploration at Sea," R.A.Geyer, ed., CRC Press, Boca Raton.

Colbourne, E. and Hay, A.E., 1987, Remote acoustic mapping of a submarine spring plume, *in:* "Progress in Underwater Acoustics," H.M. Merklinger, ed., Plenum, New York.

Colbourne, E.B. and Hay, A.E., 1990, An acoustic remote sensing and submersible study of an Arctic submarine spring plume," *J. Geophys. Res.*, 95:13,219.

Degens, E.T. and Ross, D.A., eds., "Brines and Recent Heavy Metal Deposits in the Red Sea," Springer-Verlag, New York, (1969).

Hay, A.E., 1984, Remote acoustic imaging of the plume from a submarine spring in an Arctic fjord, *Science*, 225:1154.

Hunt, J.M., Hays, E.E., Degens, E.T. and Ross, D.A., 1967, Red Sea: Detailed survey of hot brine areas, *Science*, 156:514.

Lonsdale, P. and Becker, K., 1985, Hydrothermal plumes, hot springs, and conductive heat flow in the southern trough of Guaymas Basin, *Earth Planet. Sci. Lett.*, 73:211.

MacLennan, D.N. and Forbes, S.T., 1984, Fisheries acoustics: a review of general principles, *Rapp. P.-v. Réun. Cons. int. Explor.* Mer, 184:7.

Merewether, R., Olsson, M.S. and Lonsdale, P., 1985, Acoustically detected hydrocarbon plumes rising from 2-km depths in Guaymas Basin, Gulf of California, *J. Geophys. Res.* 90:3075.

Palmer, D.R., Rona, P.A., and Mottl, M.J., 1986, Acoustic imaging of high-temperature hydrothermal plumes at seafloor spreading centers, *J. Acoust. Soc. Am.*. 80:888.

Palmer, D.R. and Rona, P.A., 1990, Comment on 'Acoustic Doppler Current profiler observations of a mid-ocean ridge hydrothermal plume' by R.E. Thomson et at., *J. Geophys. Res.* 95:5409.

RISE Project Group, 1980, East Pacific Rise: Hot springs and geophysical experiments, *Science*, 207:1421.

Rodi, W., ed., "Turbulent Buoyant Jets and Plumes," Pergamon Press, Oxford, 1982.

Rona, P.A., Palmer, D.R., Jones, C., Chayes, D.A., Czarnecki, M., Carey, E.W. and Guerrero, J.C., 1991, Acoustic Imaging of Hydrothermal Plumes. East Pacific Rise. 21° N, 109° W., *Geophysical Research Letters*, 18:2233.

Sackett, W.M., 1977, Use of hydrocarbon sniffing in offshore exploration, *J. Geochemical Exploration*, 7:243.

Tatarskii, V. I., "The effects of a turbulent atmosphere on wave propagation," Israel Program for Scientific Translation, Jerusalem, 1971. (Available from the National Technical Information Service, Springfield, VA).

Thomson, R.E., Gordon, R.L. and Dymond, J., 1989, Acoustic Doppler current profiler observations of a mid-ocean ridge hydrothermal plume, *J. Geophys. Res.*, 94:4709.

Thomson, R.E., Gordon, R.L. and Gast, J.A., 1990, Reply to Comment on 'Acoustic Doppler current Profiler observations of a mid-ocean ridge hydrothermal plume' by D.R. Palmer and P.A. Rona, *J. Geophysical Res.* 95:5413.

Thomson, R.E., Gordon, R.L. and Dolling, A.G., 1991, An intense acoustic scattering layer at the top of a mid-ocean ridge hydrothermal plume, *J. Geophys. Res.* 96:4839.

Thomson, R.E., Burd, B.J., Dolling, A.G., Gordon, R.L., and Jamieson, G.S., 1992, The deep scattering layer associated with the Endeavour Ridge hydrothermal plume, *Deep-Sea Res.* 39:55.

Turner, J.S., "Buoyancy Effects in Fluids." Cambridge University Press, New York, 1973.

Weston, D.E., Observations on a scattering layer at thermocline, *Deep-Sea Res.* 5:44.

Wilson, R.D., Monaghan, P.H., Osanik, A., Price, L.C. and Rogers, M.A., 1974, Natural Marine Oil Seepage, *Science*, 184:857.

MOTION COMPENSATION IN SYNTHETIC APERTURE SONAR IMAGING

John M. Silkaitis, Bretton L. Douglas, and Hua Lee

Department of Electrical and Computer Engineering
University of California, Santa Barbara
Santa Barbara, California 93106–9560

INTRODUCTION

Ideally, the receiving array in any synthetic aperture imaging system will follow a linear trajectory during aperture formation. However, medium turbulence and currents will necessarily deflect the platform from the ideal trajectory. Uncompensated error between the ideal trajectory and the true trajectory of a fraction of an acoustic wavelength will cause a loss in phase coherence over the synthetic aperture and degrade the ensuing image. Hence, deviations from the ideal trajectory, and the corresponding phase errors, should be estimated and compensated to realize the improved resolution possible from synthetic aperture correlations.[1]

In synthetic aperture radar imaging systems, there are several methods for estimating and compensating platform motion. Inertial navigation systems can be used to measure the radar platform's pitch, yaw, roll, and velocity vector. Deviations from the nominal trajectory can then be estimated and the errors can then be compensated in software.[2] Since the target field is generally illuminated several thousand times, far-field approximations and data redundancy can be used to develop autofocus techniques to estimate platform motion parameters. Compensation is then applied in software.[3, 4]

It has long been known that estimating and compensating platform motion errors are the dominant limiting factors in the performance of synthetic aperture sonar systems. [5, 6] Acuostic wave propagation in the ocean is roughly 2 x 10^5 slower than microwave propagation in air. Slow wave propagation limits the forward speed of the sonar platform which makes the sonar platform more susceptible to motion errors during aperture formation. The target field can no longer be illuminated several thousand times and far-field effects no longer dominate over near-field effects. Hence, the algorithms used in synthetic aperture radar imaging for estimating and compensating motion errors are not as effective in synthetic aperture sonar imaging. Since accelerometer imprecision gows in proportion to the time squared, inertial navigation systems for sonar imaging must be much more accurate than in radar, and thus more cost prohibitive.[7]

This paper presents an autofocus algorithm, using data from subsequent physical aperture images, for estimating and compensating platform motion errors during synthetic aperture formation. This autofocus technique is similar in principle to techniques used in synthetic aperture radar, however, the algorithms differ in the data formats.

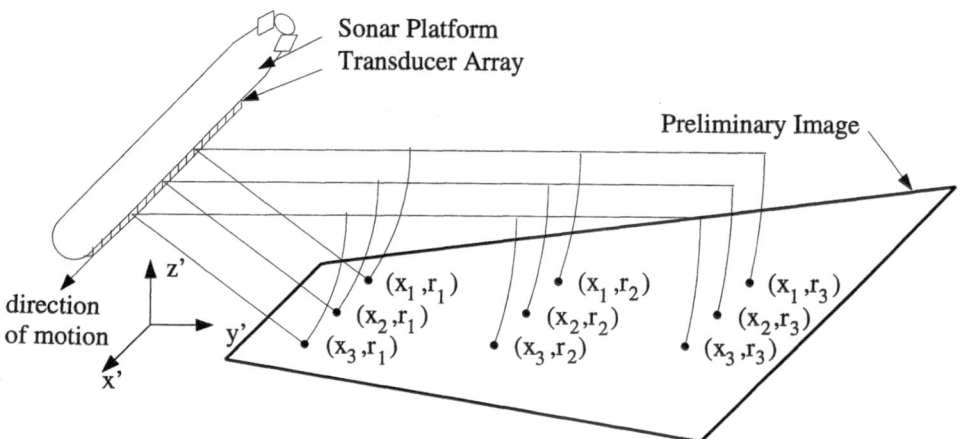

Figure 1 — Relationship between ocean floor and sonar platform.

MOTION PARAMETERIZATION

Model

Figure 1 depicts the ocean floor in the reference frame of the sonar array. In each physical aperture image, the three-dimensional reflectivity function, $\xi(x', y', z')$, is mapped to a two-dimensional reflectivity estimate, $\hat{\xi}_i(x_i, r_i)$. The (x', y', z') coordinates represent the along-track position, cross-track position, and altitude in the earth's reference frame, respectively, and the (x_i, r_i) coordinates represent the cross-range and range relative to the receiver array for the physical aperture image with index i. By analyzing the transformation of a grid of points in (x', y', z') into successive physical aperture image coordinates under realistic research vehicle operating conditions, a model is formulated that relates the coordinates in a sequence of physical aperture images. Figure 2 depicts overlapping physical aperture images and the different coordinate systems for each physical aperture image for a sonar system with minimum range of 20 meters and maximum range of 100 meters.

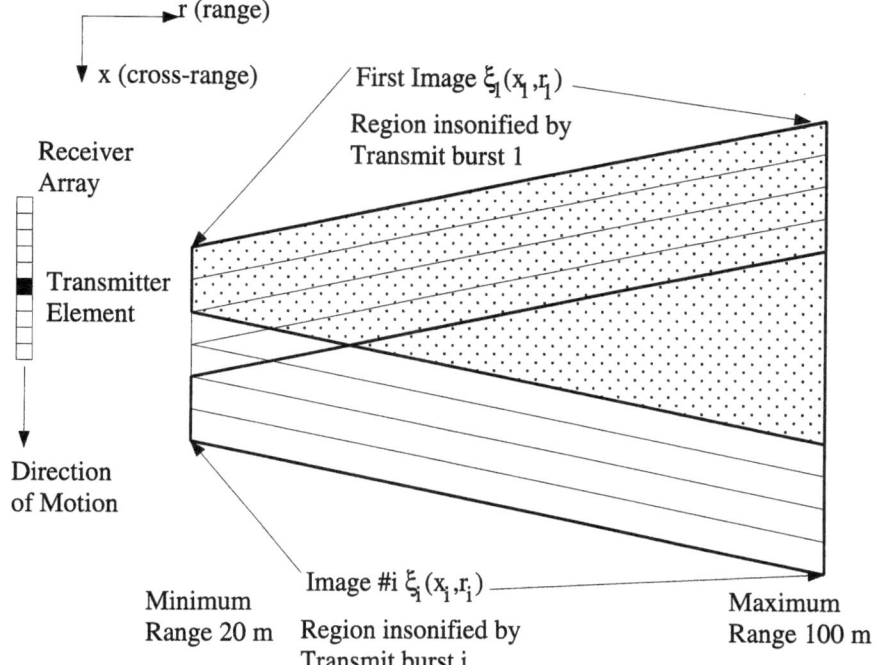

Figure 2 — Overlapping physical aperture images.

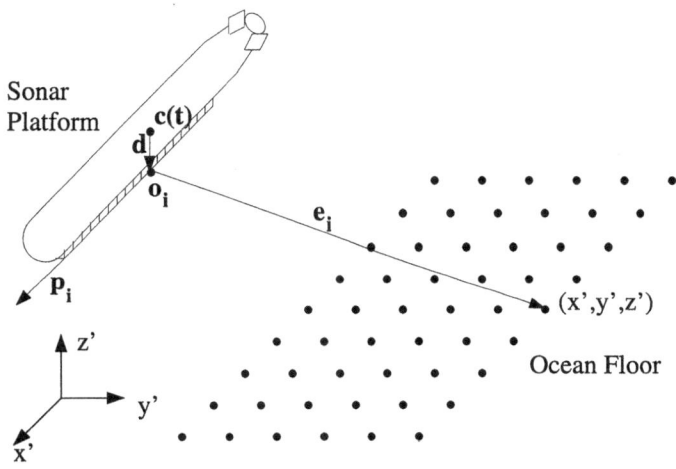

Figure 3 — Position and displacement vectors for images.

Figure 3 shows the position vectors and displacement vectors necessary in the calculation of the coordinates for each physical aperture image. The position of the center of mass of the research vehicle is a vector function of time, $c(t)$. The displacement from the center of mass to the center of the receiver array in the research vehicle's coordinate system is described by a vector, d. A rotation matrix, $M(t)$, which depends upon the attitude of the sonar platform, converts the displacement vector in the research vehicle's coordinate system to the displacement vector in the earth's coordinate system. The time-dependent rotation matrix is described by

$$M(t) = \begin{vmatrix} 1 & 0 & 0 \\ 0 & \cos\theta_r(t) & -\sin\theta_r(t) \\ 0 & \sin\theta_r(t) & \cos\theta_r(t) \end{vmatrix} \begin{vmatrix} \cos\theta_p(t) & 0 & -\sin\theta_p(t) \\ 0 & 1 & 0 \\ \sin\theta_p(t) & 0 & \cos\theta_p(t) \end{vmatrix} \begin{vmatrix} \cos\theta_y(t) & -\sin\theta_y(t) & 0 \\ \sin\theta_y(t) & \cos\theta_y(t) & 0 \\ 0 & 0 & 1 \end{vmatrix} \quad (1)$$

where $\theta_p(t)$ = pitch angle, $\theta_y(t)$= yaw angle, and $\theta_r(t)$= roll angle. [2]

The coordinates of the origin of the receiver array are related to the center of mass and the rotation matrix by

$$\mathbf{o_i} = \mathbf{c}(t_i) + \mathbf{M}(t_i)\mathbf{d}. \quad (2)$$

The unit vector in the direction of the x-axis, $\mathbf{p_i}$, is given by

$$\mathbf{p_i} = \mathbf{M}(t_i)\begin{bmatrix} 1 \\ 0 \\ 0 \end{bmatrix}. \quad (3)$$

The vector from the origin to (x', y', z') is given by:

$$\mathbf{e_i} = \begin{bmatrix} \mathbf{x}' \\ \mathbf{y}' \\ \mathbf{z}' \end{bmatrix} - \mathbf{o_i}. \quad (4)$$

Finally, the cross-range and range coordinates are given by:

$$\begin{aligned} x_i &= \mathbf{p_i} \cdot \mathbf{e_i} \\ r_i &= \|\mathbf{p_i} \times \mathbf{e_i}\|_2 \end{aligned} \quad (5)$$

Motion decomposition

Since the target field is illuminated several times, there is sufficient redundancy in the data to extract motion parameters from each physical aperture image. These motion parameters can then be estimated and compensated during the imaging process. The motion between successive physical aperture images can be modeled by three motion parameters: constant cross-range error, linear phase error, and constant phase error. Under ideal operating conditions with no pitch, yaw, or roll, and a constant squint angle (indicating that the platform is either operating in no current, or its velocity is parallel or antiparallel to the current), the coordinates of the target field are offset by a constant cross-range factor. Figure 4 shows a plot of physical aperture image coordinates of a set of grid points that were insonified on both the first and sixth transmit bursts under ideal conditions.

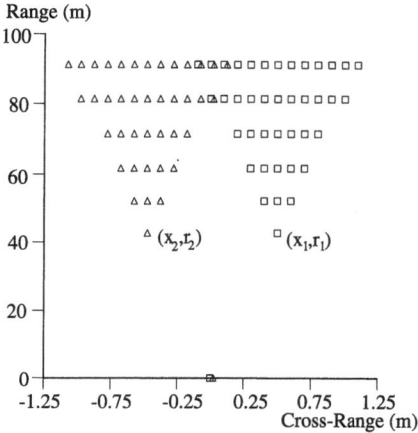

Figure 4 Physical aperture image coordinates under ideal conditions.

The squares represent the coordinates, (x_1, r_1) for the set of grid points for the first physical aperture image, and the triangles represent the coordinates, (x_6, r_6) for the same set of points for the sixth physical aperture image. Under ideal conditions, the coordinates for each point differ only in cross-range. Figure 5 shows the same physical aperture images under worst case conditions of pitch, yaw, and roll, and with a squint angle of 0.025. The physical aperture image coordinates are now offset by a range difference in addition to the cross-range difference. For all of the grid points common to the two images, the coordinates in the first image (x_1, r_1), and the coordinates in the ith image, (x_i, r_i) hold the following relationship:

$$x_i = x_1 + \delta x(r_1)$$
$$r_i = r_1 + \delta r_1(r_1) + \delta r_2(r_1)x_1 \tag{6}$$

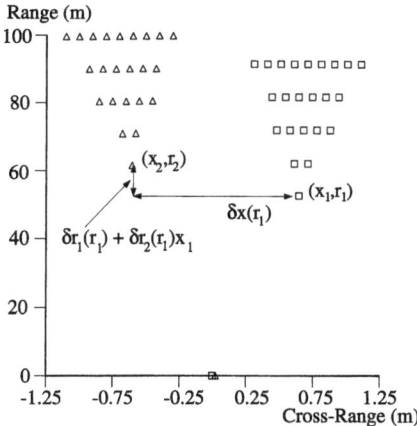

Figure 5 Physical aperture image coordinates under worst case conditions.

749

MOTION ESTIMATION AND COMPENSATION ALGORITHM

The motion estimation and compensation algorithm is based on the fact that each point on the ocean bottom is imaged multiple times. The algorithm relies on two assumptions. The first assumption is that the reflectivity of each point on the ocean bottom is similar for all transmit bursts that insonify it. This assumption is easily justified if the transmitter beamwidth is small so that the angle dependency of the reflectivity is minimal. The second assumption is that the majority of the reverberation occurs at the surface of the ocean floor. This assumption is justified when the insonification frequency is high and bottom penetration is limited.

Using the constant reflectivity assumption, successive physical aperture images represent the same reflectivity function, $\xi(x', y', z')$, mapped into different coordinates, $(x_1, r_1), (x_2, r_2), ... (x_i, r_i)$, in the sequence of physical aperture images, $\hat{\xi}_1(x_1, r_1), \hat{\xi}_2(x_2, r_2), ... \hat{\xi}_i(x_i, r_i)$. The first image in a sequence is chosen to be the reference image, and the other images are registered with the first to allow coherent superposition. From the motion decomposition, the relationship between physical aperture images can be expressed as

$$\hat{\xi}_1(x_1, r_1) \approx \hat{\xi}_i(x_i, r_i) \approx \hat{\xi}_i([x_1 + \delta x(r_1)], [r_1 + \delta r_1(r_1) + \delta r_2(r_1)x_1]). \tag{7}$$

In the physical aperture images, the range differences are small relative to the pixel spacing, but large relative to the acoustic wavelength. As a result, the range differences cause phase errors rather than registration errors. Therefore, the images in a sequence of physical aperture images are related by

$$\hat{\xi}_1(x_1, r_1) \approx \hat{\xi}_i(x_i, r_i) \approx \hat{\xi}_i([x_1 + \delta x(r_1)], r_1) \exp j[\phi_1(r_1) + \phi_2(r_1)x_1]. \tag{8}$$

The motion parameters $\phi_1(r_1), \phi_2(r_1) x_1$, and $\delta x(r_1)$, are then estimated between the two images. The physical aperture image is then compensated and coherently superimposed upon the composite image forming the composite image for the next physical aperture image. Figure 6 displays a flow chart for the motion compensation algorithm.

Estimating constant cross-range error

Assuming that the linear phase term is small, the cross-range error is estimated by the peak of the cross-correlation of the reference image and the physical aperture image in cross-range:

$$\hat{\delta x}(r_1) = \underset{\delta x_{min} < \delta x < \delta x_{max}}{argmax} \int_{x_1} \hat{\xi}_c(x_1, r_1)^* \hat{\xi}_i(x_1 + \delta x, r_1) \, dx_1 \tag{9}$$

The physical aperture image is then reregistered to align it in cross-range with the reference image. The new image is then represented as

$$\hat{\xi}'_i(x_1, r_1) = \hat{\xi}_i\left(x_i + \hat{\delta x}(r_1), r_i\right). \tag{10}$$

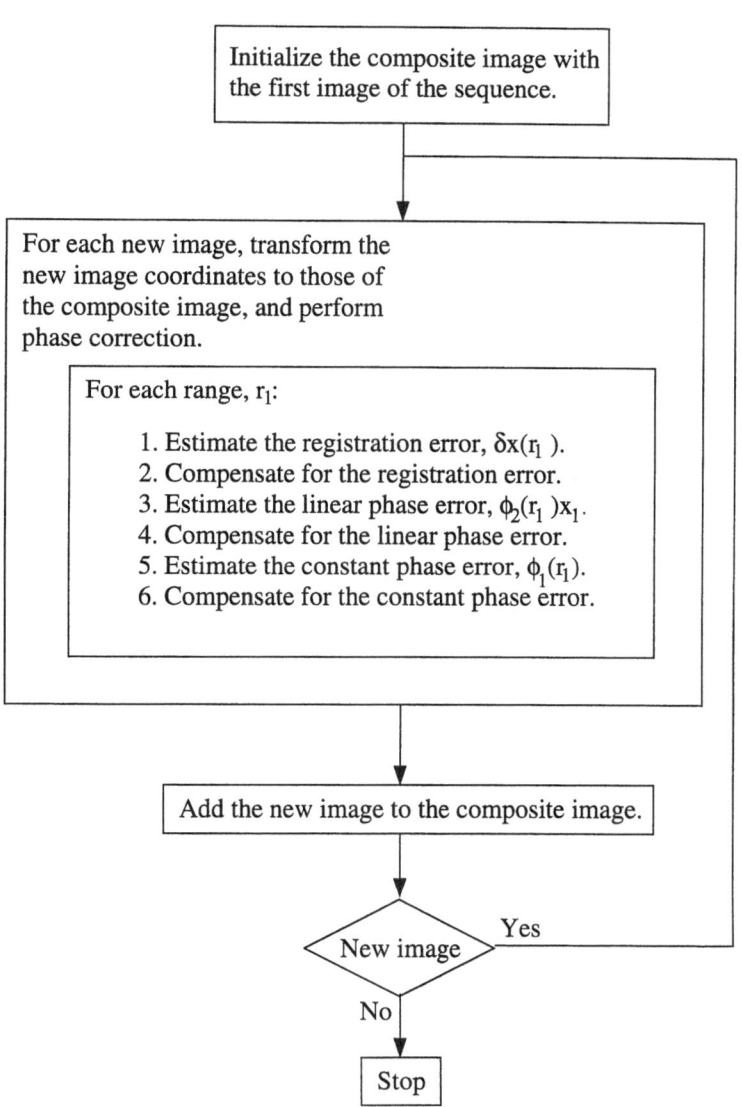

Figure 6 — Iterative synthetic aperture image formation procedure.

This algorithm step is highly dependent on the range parameter. Figure 7 shows the sonar platform and the associated ranges to the ocean floor. For ranges less than R_{min}, the sonar data is primarily composed of volume reverberation and noise, which would be uncorrelated between transmit bursts. The technique applied in this range which result in a cross-range displacement of zero. As the range approaches R_{max}, attenuation, spreading, and lower reflectivity due to a shallower grazing angle decrease the signal strength and degrade the performance of this cross-correlation technique for estimating the cross-range error. Therefore, the best estimate of the cross-range error should be determined only from regions where the received signal is from ocean bottom reflection.

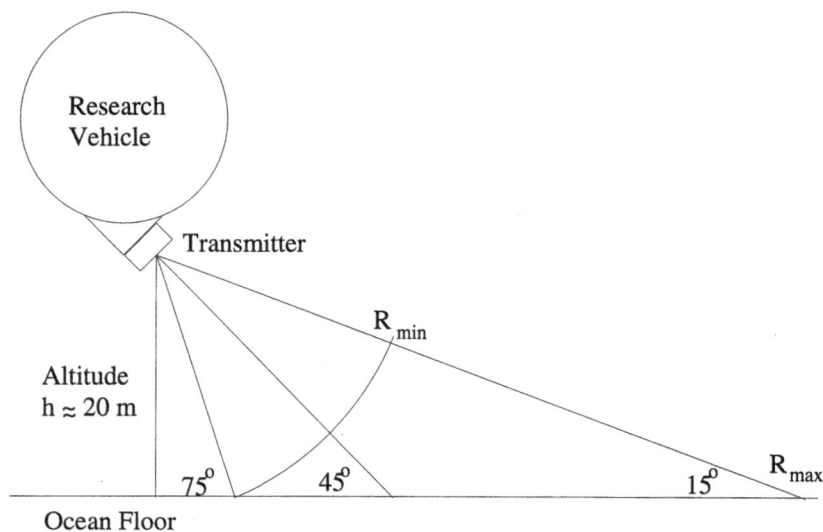

Figure 7 — Ranges associated with synthetic aperture sonar for a 60° vertical beam width.

Estimating linear phase error

The linear phase error, is estimated by determining the argument that maximizes the following integral,

$$\underset{\phi_{2_{min}} < \phi_2 < \phi_{2_{max}}}{argmax} \int_{x_1} \hat{\xi}_c(x_1, r_1)^* \hat{\xi}_i'(x_1, r_1) \exp(-j\phi_2 x_1)\, dx_1. \tag{11}$$

This is essentially a Fourier transform in cross-range on the product of the registered image and the conjugate of the composite image. The linear phase error correction term is found for each range and is applied to the previously corrected physical aperture image to form a new physical aperture image. This corrected image is represented as

$$\hat{\xi}_i''(x_1, r_1) = \hat{\xi}_i'(x_1, r_1) \exp\left(-j\hat{\phi}_2(r_1)x_1\right). \tag{12}$$

Estimating constant phase error

Assuming that the first two motion compensation operations are performed without error, the registered, partially phase-corrected image, $\hat{\xi}_i''(x_1, r_1)$ and the composite image

752

will have the following relationship:

$$\hat{\xi}_c(x_1, r_1) = \hat{\xi}_i''(x_1, r_1)\exp(j\phi_1(r_1)). \tag{13}$$

The constant-phase term is estimated by evaluating the phase of the inner product of the composite image and the partially phase-corrected image

$$\exp\left[-j\left(\hat{\phi}_1(r_1)\right)\right] = \frac{\int \hat{\xi}_c(x_1, r_1)^*\hat{\xi}_i''(x_1, r_1)\,dx_1}{\left|\int \hat{\xi}_c(x_1, r_1)^*\hat{\xi}_i''(x_1, r_1)\,dx_1\right|}. \tag{14}$$

The final correction is then applied to the partially corrected physical aperture image to form

$$\hat{\xi}_i'''(x_1, r_1) = \hat{\xi}_i''(x_1, r_1)\exp(j\phi_1(r_1)). \tag{15}$$

This image is then coherently superimposed with the composite image forming the new composite image

$$\hat{\xi}_c(x_1, r_1)_{new} = \hat{\xi}_c(x_1, r_1)_{old} + \hat{\xi}_i'''(x_1, r_1). \tag{16}$$

CONCLUSION

This paper presents a motion estimation and compensation algorithm for synthetic aperture sonar imaging utilizing autofocus techniques. This algorithm was tested during an open ocean test, where the worst case quantities of pitch, yaw, and roll were known. Synthetic aperture images were formed with the motion compensation algorithm and compared to single look and multilook images of the same target. The synthetic aperture images demonstrated superior cross-range resolution improvement over the other techniques. Figures 8 shows a sunken plane imaged via the synthetic aperture technique.

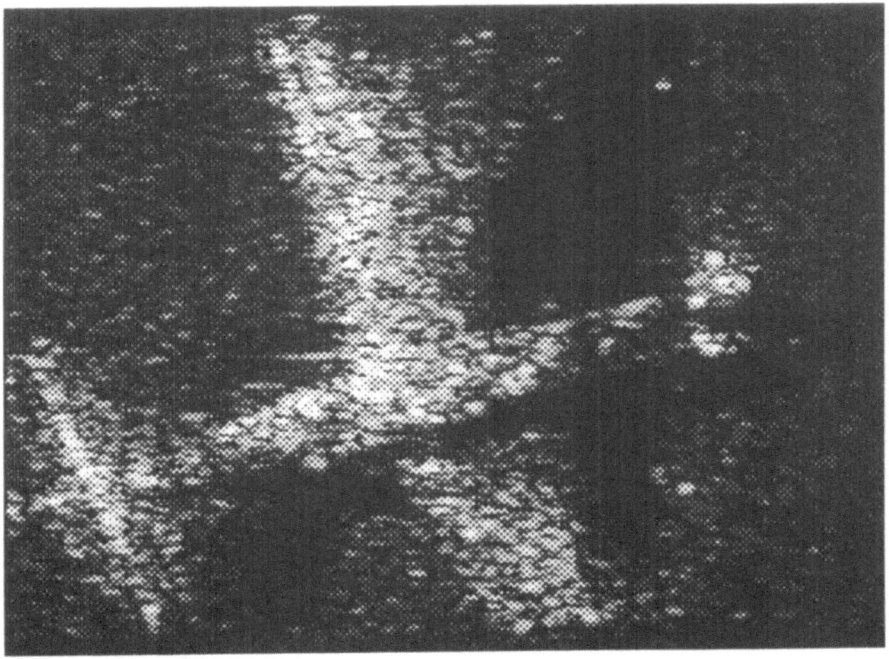

Figure 8 — Synthetic aperture image of a sunken plane.

ACKNOWLEDGMENTS

This research is supported by the UC MICRO program and Sonatech, Inc.

REFERENCES

1. R. E. Williams. Creating and acoustic synthetic aperture in the ocean. *Journal of the Acoustical Society of America*, 60(1):60–73, 1976.

2. Jr. John C Kirk. Motion compensation for synthetic aperture radar. *IEEE Transactions on Aerospace and Electronic Systems*, AES-11(3), 1975.

3. J. Moreira. A new method of aircraft motion error extraction from radar raw data for real time motion compensation. In *Proceedings of the International Geoscience and Remote Sensing Symposium, IGARSS '89*, pages 2217–2220, 1989.

4. J. Dall. A new frequency domain autofocus algorithm for SAR. In *Proceedings of the International Geoscience and Remote Sensing Symposium, IGARSS '91*, volume 2, pages 1069–1072, 1991.

5. R. E. Williams and H. F. Battesin. Time coherence of acoustic signals transmitted over resolved paths in the deep ocean. *Journal of the Acoustical Society of America*, 59(2), 1976.

6. L. J. Cutrona. Additional characteristics of synthetic-aperture sonar systems and a further comparison with nonsynthetic-aperture sonar systems. *Journal of the Acoustical Society of America*, 61(5):336–348, 1977.

7. L. J. Cutrona. Comparison of sonar system performance achievable using synthetic-aperture techniques with the performance achievable by more conventional means. *Journal of the Acoustical Society of America*, 58(2):336–348, 1975.

HOLOGRAPHIC 3-D IMAGING SYSTEM USING ENCODED WAVEFRONT : THEORETICAL ANALYSIS ON REPETITIVE TRANSMISSION

Yasutaka Tamura, Wataru Itoh, Takao Akatsuka,
Osamu Takano* and Nori Ishii*

Faculty of Engineering, Yamagata University,
Jyonan 4-3-16, Yonezawa 992, Japan
* Akishima Laboratories Inc.
 (Mitsui Engineering and Shipbuilding)
1-50 Tsutsujigaoka 1-chome, Akishima, Tokyo 196, Japan

ABSTRACT

This paper describes a theoretical analysis of a holographic imaging system which uses a Walsh-function-modulated encoded wavefront.

In the paper, the cross-correlation functions for the set of transmitting signals are derived theoretically. The range of the area where the zero-correlation holds is also examined for the number of repetitions.

Simulations and experiments using an airborne 3-D sonar are carried out, and the relative levels of artifacts are evaluated experimentally. Finally, the optimal radius ratio for coaxial circular arrays is derived theoretically, and is confirmed by a simulation.

INTRODUCTION

The authors have proposed a high-speed imaging system using Walsh-function-modulated transmission pulses[1]. The system has an array of acoustic transmitters and receivers. The transmitters are simultaneously driven by Walsh-function-modulated sinusoidal signals. The transmitted pulses generate an ultrasonic encoded wavefront while the echo waveforms reflected from distinct points are approximately uncorrelated. With a numerical decoding process, the transmitted beams are synthesized using the digitized echo waveforms.

While the system can reconstruct images with a single cycle of transmission and receiving, artifacts emerge in the images. To reduce the artifacts in the one-shot images, a repetitive transmitting and receiving method has been proposed[2].

In each cycle, the transmitters have a one-to-one correspondence with the row numbers of a Hadamard matrix. The row number is determined with modulo 2 addition for every corresponding bit of two binary-represented numbers, the number of the transmitting cycle and the number of the transmitter.

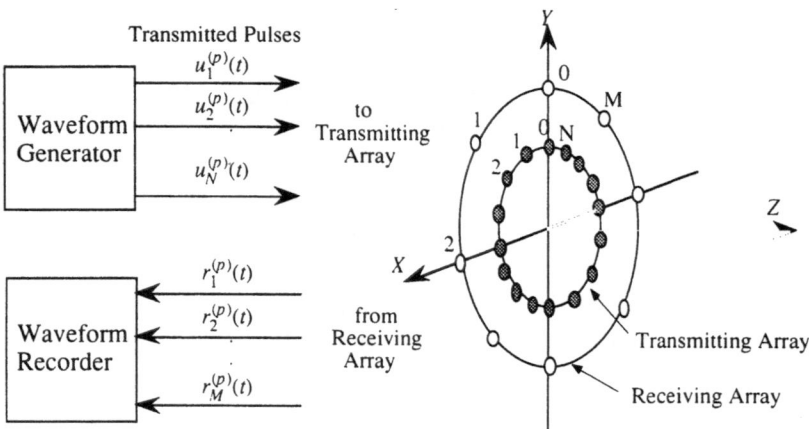

Fig.1. Schematic drawing of the data acquisition system

PRINCIPLE

Data acquisition system

Figure 1 shows a schematic drawing of the data acquisition system. Ultrasonic waves are simultaneously transmitted from N transmitters, where $N = 2^K$. We use sinusoidal waves of frequency f_0 modulated by a system of Walsh functions which is synchronized with a clock signal. The period of the clock signal, Δt, is equal to an integral multiple of the sine wave period, $1/f_0$.

The transmitting and receiving process is repeated P times. At the p-th transmitting and receiving cycle, the transmitting signal corresponding to the n-th ($n = 0, 1, 2, \ldots, N-1$) transmitter and the waveform detected by the m-th receiver are denoted by $u_n^{(p)}(t)$ and $r_m^{(p)}(t)$, respectively. The origins ($t = 0$) of these functions are fixed at the starting positions of each transmitting pulse.

We have proposed a set of transmitting waveform functions for a repetitive transmission method; the transmitting waveform function is given by

$$u_n^{(p)}(t) = \sum_{i=0}^{N-1} w_{(p \oplus n)i} f(t - i \Delta t), \quad n = 0, 1, \ldots, N-1, \tag{1}$$

where

$$f(t) = \begin{cases} e^{j\,2\pi f_0 t} & \text{for} \quad 0 \le t < \Delta t \\ 0 & \text{for} \quad t < 0,\, t \ge \Delta t \end{cases}, \tag{2}$$

is a single sinusoidal pulse of frequency f_0, w_{ij} denotes a i-j component of the $N \times N$ Hadamard matrix, $W^{(K)}$, which is generated by the repetition of Kronecker product,

$$W^{(n)} = \begin{pmatrix} W^{(n-1)} & W^{(n-1)} \\ W^{(n-1)} & -W^{(n-1)} \end{pmatrix}, \tag{3}$$

starting with

$$W^{(1)} = \begin{pmatrix} 1 & 1 \\ 1 & -1 \end{pmatrix}, \tag{4}$$

to simplify the equations, the column and row numbers are indexed from 0 to N -1, and $\oplus$ expresses modulo 2 addition for every corresponding bit of binary represented numbers. Thus, we have called the method "mod 2 repetition".

Reconstruction of image

The reconstructed object function $S(x)$ corresponding to the position x is calculated by the sum of the correlations between the received echo functions and expected ones for a single point target located at x. When x is in the para-axial region of the array, the expected received signal $h_m^{(p)}(x,t)$ is given by

$$h_m^{(p)}(x,t) = \sum_{n=0}^{N-1} u_n^{(p)}\left(t - \frac{|x - x_{Tn}| + |x - x_{Rm}|}{C}\right), \tag{5}$$

where x_{Tn} and x_{Rm} are position vectors of the n-th transmitter and the m-th receiver, respectively.

Thus, $S(x)$ is calculated by

$$s(x) = \left| \sum_{p=0}^{P-1} \sum_{m=0}^{M-1} \int_0^{T_p} r_m^{(p)}(t) \cdot h_m^{(p)}(x,t)^* dt \right|^2 \tag{6}$$

$$= \left| \sum_{p=0}^{P-1} \sum_{n=0}^{N-1} \sum_{m=0}^{M-1} \varphi_{mn}^{(p)}\left(\frac{|x - x_{Tn}| + |x - x_{Rm}|}{C}\right) \right|^2 \tag{7}$$

where

$$\varphi_{mn}^{(p)}(\tau) = \int_0^{T_p} r_m^{(p)}(\tau) u_n^{(p)}(t - \tau)^* dt \tag{8}$$

is a cross-correlation function between transmitted signal $u_n^{(p)}(t)$ at the n-th transmitter and received echo $r_m^{(p)}(t)$ at the m-th receiver. Suffix "$*$" denotes a complex conjugate operator.

ANALYSIS

Cross-correlation functions

In this section, the cross correlations of the transmitting pulses are calculated and the orthogonal property of the pulses will be demonstrated. The cross-correlation function for the i-th transmitting signal and the j-th one is expressed by

$$\varphi_{ij}(\tau) = \sum_{p=0}^{P-1}\int_0^{T_p} u_i^{(p)}(t) \cdot u_j^{(p)}(t-\tau)^* dt \, , \tag{9}$$

for $\tau < T_p - N\Delta t$, where T_p is the period of repetition.

Substituting equation (1), and using a property of the Hadamard matrix,

$$w_{(p \oplus i)k} = w_{p\,k} w_{i\,k} \, , \tag{10}$$

eq.(9) is transformed to

$$\varphi_{ij}(\tau) = \sum_{p=0}^{P-1}\int_0^{T_p} \sum_{k=0}^{N-1} w_{(p \oplus i)k} f(t-k\,\Delta t) \cdot \sum_{l=0}^{N-1} w_{(p \oplus j)l} f(t-l\,\Delta t-\tau)^* dt \tag{11}$$

$$= \sum_{p=0}^{P-1}\sum_{k=0}^{N-1}\sum_{l=0}^{N-1}\int_0^{T_p} w_{p\,k} w_{i\,k} w_{p\,l} w_{j\,l} f(t-k\,\Delta t) \cdot f(t-l\,\Delta t-\tau)^* dt \tag{12}$$

$$= \sum_{p=0}^{P-1}\sum_{k=0}^{N-1}\sum_{l=0}^{N-1} w_{p\,k} w_{i\,k} w_{p\,l} w_{j\,l} \, \varphi((l-k)\Delta t + \tau) \, , \tag{13}$$

where

$$\varphi(\tau) = \int_0^{T_p} f(t) \cdot f(t-\tau)^* dt = \int_{-\infty}^{+\infty} f(t) \cdot f(t-\tau)^* dt \tag{14}$$

is the auto-correlation function of the sinusoidal pulse, $f(t)$.

Using a property of the Hadamard matrix which is generated with the Kronecker product, the following orthogonal relation is derived for the rows of block matrices which constitute the $W^{(K)}$:

$$\sum_{p=0}^{P-1} w_{pk} \cdot w_{pl} = \begin{cases} P & \text{for } k=l \ mod \ P \\ 0 & \text{otherwise} \end{cases} = P \cdot \sum_{n=-(N/P-1)}^{N/P-1} \sum_{k=0}^{N-1} \delta_{(l-k)(n \cdot P)} \, , \tag{15}$$

where P is a power of 2. Then eq.(13) is transformed into

$$\varphi_{ij}(\tau) = \sum_{k=0}^{N-1}\sum_{l=0}^{N-1} w_{ik} \cdot w_{jl} \left(\sum_{p=0}^{P-1} w_{pk} \cdot w_{pl} \right) \cdot \varphi((l-k) \cdot \Delta t + \tau) \tag{16}$$

$$= N^2 \cdot \delta_{ij}\, \varphi(\tau) \qquad \text{for } P = N , \tag{17}$$

$$= P \cdot N \cdot \delta_{i\,j} \, \varphi\left(\tau\right) + P \sum_{n \neq 0,\, n = -(N/P - 1)}^{N/P - 1} \left\{ \sum_{l - k = n \cdot P} w_{ik} \cdot w_{jl} \, \varphi\left(n \cdot P \cdot \Delta t + \tau\right) \right\} \text{ for } \begin{matrix} P = 2^{\alpha} \\ \neq N \end{matrix}. \quad (18)$$

Equation (17) indicates that the transmitted pulses are orthogonal when the repetition number, P, is equal to the number of transmitters, N.

We also obtained eq.(18) when P is equal not to N but to power of 2. While the first term of eq.(18) becomes zero for $i \neq j$, the second is a nonzero one. Considering that $\varphi(\tau)$ is zero for $|\tau| > \Delta t$, however, the second term of eq.(18) becomes zero for $|\tau| \leq (P - 1) \cdot \Delta t$.

When P is not a power of 2 but an even number, P consists of several components that are equal to powers of 2. It is also indicated that the second term becomes zero for $|\tau| \leq (P_0 - 1) \cdot \Delta t$, where P_0 is the minimum component of P. For example, when $P = 12$ then $P = 2^3 + 2^2$ and $P_0 = 4$, and transmitted pulses are orthogonal for $|\tau| \leq 3\Delta t$.

The discussion in this section can be summarized as follows: the orthogonal property holds when $|\tau| \leq T_P - N\Delta t$ for $P = N$, and $|\tau| \leq (P_0 - 1) \cdot \Delta t$ for P is even. The "mod 2 repetition" is equivalent to a time-divided transmitting method in which an array sequentially transmits a single pulse from each transmitter, when the delay times involved meet the above conditions. The "mod 2 repetition", however, has the advantage of high signal energy to the "time-divided repetition".

Spatial range of zero cross-correlations

The spatial region where the zero cross-correlation holds is theoretically derived in this section. In the obtained region, the propagated waves from each transmitter are not correlated, and the total energy of the waveforms distributes uniformly.

The following condition is considered: D_T is the diameter of the transmitting array, the number of repetitions, P, is a power of 2, and the region is far from the array.

We derive the angle of view, where the propagated waveforms are uncorrelated. Consider the point at view angle of θ from the array's axis. The maximum value, τ_{max}, of the mutual delay times of reached waveforms is approximately given by

$$\tau_{max} = \frac{D_T \cdot \sin\theta}{C}.$$

Since τ_{max} is smaller than $(P_0 - 1) \cdot \Delta t$ in this region, the range of view angle, θ_{max}, is given by

$$\theta_{max} = \sin^{-1}\left[\frac{C \cdot (P_0 - 1)\Delta t}{D_T} \right] \quad (19)$$

$$\approx \frac{C \cdot (P_0 - 1)\Delta t}{D_T} \quad \text{for} \quad \frac{C \cdot (P_0 - 1)\Delta t}{D_T} \ll 1. \quad (20)$$

The field of view, θ_{max}, of the region involved is given by eqs.(19) and (20). Within this field of view, the "mod 2 repetition" is equivalent to the "time-divided repetition" in terms of the energy distribution of the transmitted wavefront.

SIMULATIONS AND EXPERIMENTS

Computer simulations and experiments using an air-borne 3-D sonar were carried out in order to confirm the results of the theoretical analysis.

The airborne sonar has a coaxial circular array transducer; the transmitting array consists of 16 transmitters, and the receiving array consists of 8 receivers. The radii of the transmitting and the receiving arrays are 60mm and 90mm, respectively. The frequency of the carrier wave, f_0, is 50kHz (wavelength =6.9mm), and the clock period, Δt , for the Walsh functions is $2 \times 1 / f_0 = 40 \mu sec$. The details of the airborne sonar have been reported by the authors[2].

Zero cross-correlation property

Figure 2 demonstrates the zero cross-correlation property of the transmitted pulses which uses the mod 2 repetition. The cross-correlation functions are calculated for a single shot and 16 repetitions, respectively. Undesired peaks observed in the single shot case are completely eliminated after 16 repetitions.

Figure 3 shows the spatial distributions of the energy of the transmitted wavefront for various repetition numbers, P. The energy profiles in $440 \times 440 mm^2$ square area 500mm from the array are calculated. Note that the region of uniform energy appears at $P = 4$, and disappears at $P = 5$.

$$\left| \varphi_{7j}(\tau) \right| = \left| \int u_7^{(0)}(t) u_j^{(0)}(t - \tau)^* dt \right| \qquad\qquad \left| \varphi_{7j}(\tau) \right| = \left| \sum_{p=0}^{15} \int u_7^{(p)}(t) u_j^{(p)}(t - \tau)^* dt \right|$$

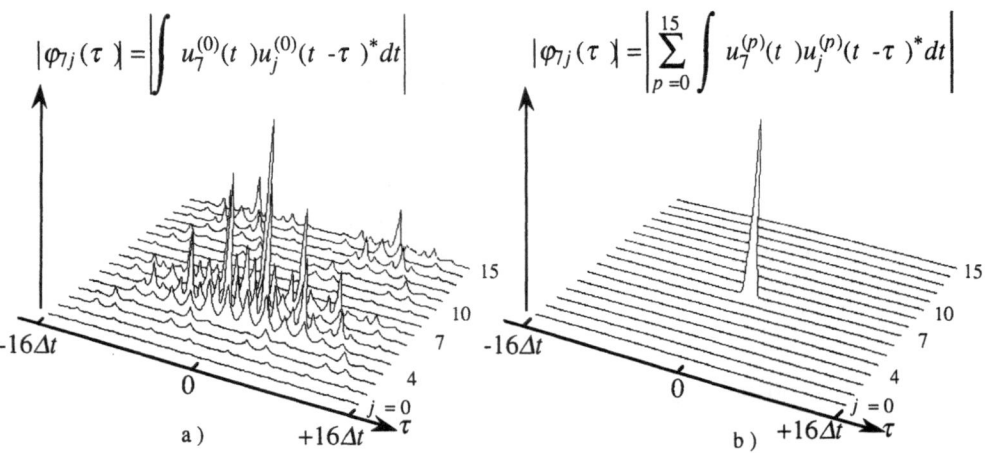

Fig.2. Autocorrelation and cross-correlation functions of transmitted pulses
a) with single shot, b) with "mod 2 repetition"

Artifacts in the reconstructed images

Photo 1 shows a view of the experimental setup; a 30mmϕ plastic sphere was placed at a distance of 500mm from the array. Figure 4 shows the reconstructed 3-D images of the sphere. The artifacts observed in the single shot image (Fig.4-a) are reduced after 16 repetitions (Fig.4-b).

The spatial divergences of the images of the sphere were 3.5deg. in azimuth and 10.7mm in depth. These values approximately agreed with those of a point target obtained

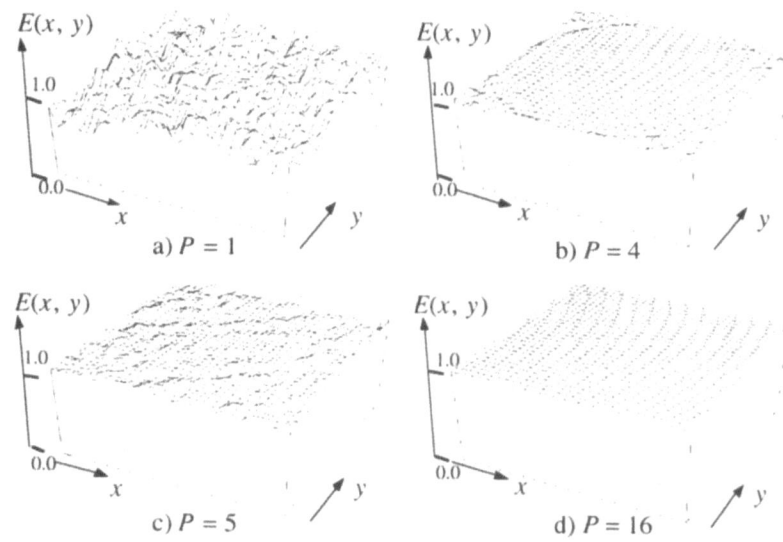

Fig.3. Spatial distributions of transmitted energy for various numbers of repetitions, P

Photo 1. View of the experimental setup

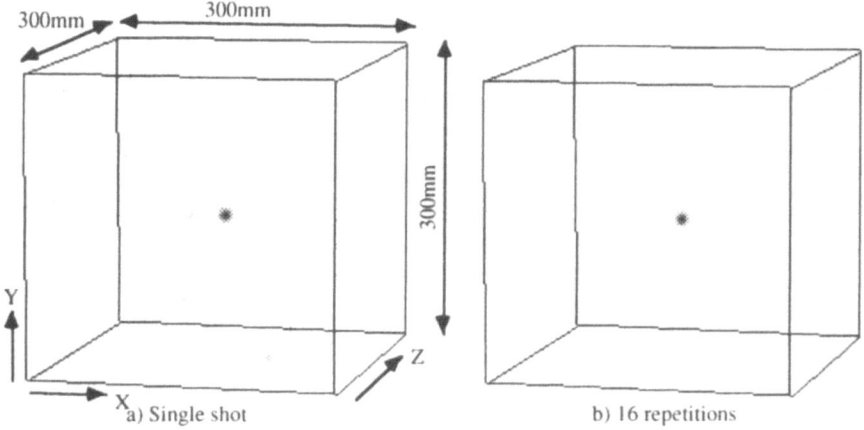

Fig.4. Reconstructed 3-D images of a 30mm ϕ plastic sphere 500mm from the array

by computer simulation. Hence, images of the sphere represented the point spread functions (PSFs) of the sonar.

The artifacts are caused by undesired spurious peaks of the PSF. The relative levels of maximum and averaged spurious peaks, $\max(S_{\mathrm{sub}})/S_{\mathrm{main}}$ and $\overline{S_{\mathrm{sub}}}/S_{\mathrm{main}}$, are shown in Fig.5. While the maximum level of the spurious peaks shows a dyadic behavior to the numbers of repetitions, P, the averaged level of spurious peaks increased linearly with P.

Figure 6 demonstrates a 3-D image of a wooden plate after 16 repetitions. The target was 0.5m from the array. The plate was formed into the Chinese character "山". This is the first character of "山形大学", which means "Yamagata University".

Optimal geometry of array

We have adopted a coaxial circular array, because high angular resolving power is obtained with a small numbers of transducers when this geometry is used. Artifacts, however, emerge when sparse arrays are used.

In this section, the optimal geometry of the array is derived in terms of the maximum level of the artifacts. We considered the array as shown in Fig.1, which has a circular transmitting array of radius r_{T} and a receiving array of radius r_{R}. The optimal ratio of the two radii is obtained by theoretical consideration and computer simulation.

When the target is far from the array, the directive pattern, $d(\Delta\theta)$, of the imaging system is approximately expressed by

$$d(\Delta\theta) \approx J_0 \left(\frac{2\pi f_0 r_{\mathrm{T}}}{C}\Delta\theta\right) \cdot J_0 \left(\frac{2\pi f_0 r_{\mathrm{R}}}{C}\Delta\theta\right), \tag{21}$$

where $\Delta\theta$ is the difference of azimuthal angle, and J_0 is the Bessel function of 0-th order.

The directivity pattern is a product of ones formed by transmitting and receiving arrays. From this consideration, one choice of the array geometry is to make the first zero of the directivity due to one array coincide with the first peak of the one due to another array. Since the first zero and first local peak of $J_0(x)$ are at $x=2.40$ and $x=3.83$, respectively, the optimal ratio of r_{T} to r_{R} is given by $r_{\mathrm{T}}/r_{\mathrm{R}} = 1.6$ (=3.83/2.40) or 0.63 (= 2.40/3.83). Using the previous discussion on the range of zero cross-correlations of transmitted pulses, the smaller radius of transmitting array is preferable. Thus $r_{\mathrm{T}}/r_{\mathrm{R}} = 0.63$ is used.

A computer simulation was performed to confirm this consideration. The simulation assumed an underwater 3-D imaging system. The center frequency of acoustical wave was 200kHz, and clock period for the Walsh functions was $2/(200\times10^3)$ sec. The system had 16 transmitters and 16 receivers. The radius of the receiving array, r_{R}, was 0.6m. Under these conditions, PSFs at the distance of 20m were calculated.

Figure 7 shows the maximum levels of the artifacts against the radius of the transmitting array, r_{T}, while the radius of the receiving array, r_{R}, is fixed at 0.6m. The graph shows a minimum value at $r_{\mathrm{T}} = 0.38 = 0.63\times r_{\mathrm{R}}$. An another local minimum is observed at $r_{\mathrm{T}} = 0.28$, which is considered to be caused by spurious peaks due to the sparse arrangement of the arrays.

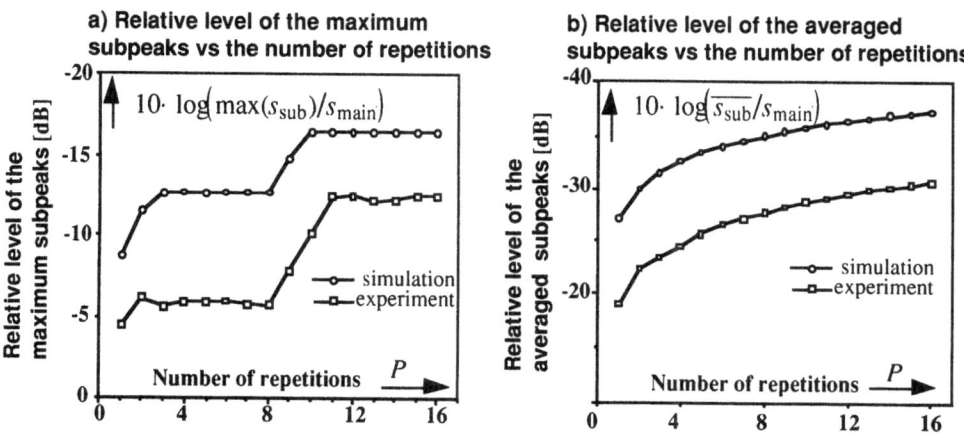

a) Relative level of the maximum subpeaks vs the number of repetitions

$$10 \cdot \log\left(\max(s_{sub})/s_{main}\right)$$

Relative level of the maximum subpeaks [dB]

Number of repetitions P

- simulation
- experiment

b) Relative level of the averaged subpeaks vs the number of repetitions

$$10 \cdot \log\left(\overline{s_{sub}}/s_{main}\right)$$

Relative level of the averaged subpeaks [dB]

Number of repetitions P

- simulation
- experiment

Fig.5. The relationship between relative level of spurious peaks and number of repetitions

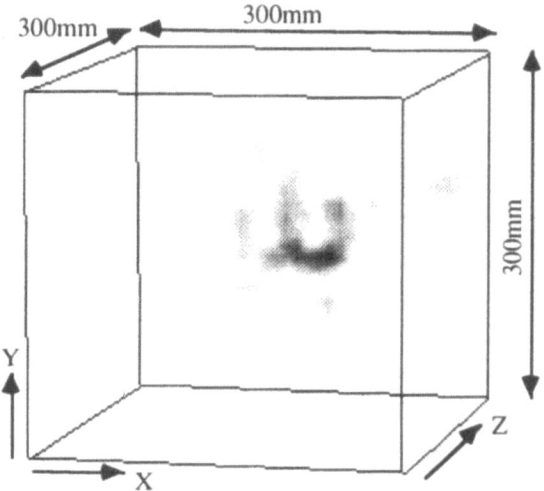

300mm

300mm

300mm

Y

X

Z

Fig. 6. Reconstructed 3-D image of a wooden plate forming the Chinese character "川"

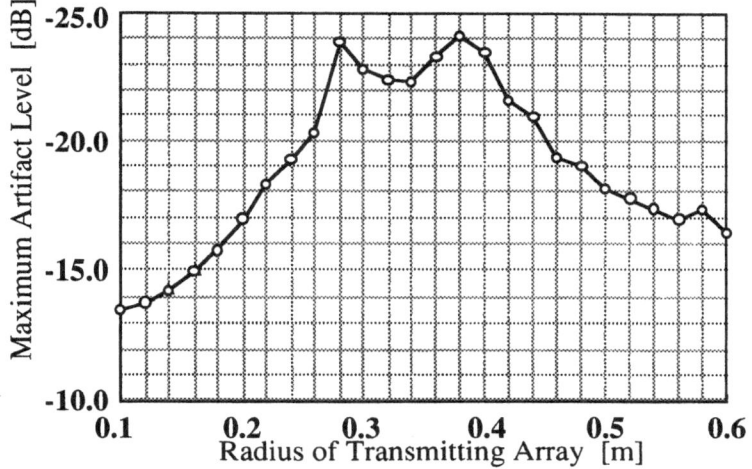

Maximum Artifact Level [dB]

Radius of Transmitting Array [m]

Fig.7. The maximum level of artifacts vs the radius of the transmitting array, while the radius of the receiving array is fixed at 0.6m

CONCLUSIONS

A theoretical analysis was carried out on a repetitive transmitting method for the holographic imaging system which uses Walsh functions. The cross-correlation functions of the set of transmitting signals was derived theoretically. The zero cross-correlations holds for a delay time shorter than $(P_0 - 1)\Delta t$, where P_0 corresponds to the minimum binary component of repetition number, P, and Δt is the clock period for the Walsh functions. The view angle where all the transmitted waveforms are uncorrelated is in proportion to the products of the delay time, $(P_0 - 1)\Delta t$, and the radius of the transmitting array.

The results of the analysis were confirmed by simulations and experiments. Finally, optimal radius ratio for coaxial circular arrays was also derived. The relative levels of undesired PSF peaks smaller than -24dB was obtained from the simulation of the underwater 3-D imaging system with 16 transmitters and 16 receivers.

REFERENCES

1. Y.Tamura and T.Akatsuka: Holographic sonar using orthogonal transmitting pulses, Acoustical Imaging, 17,753/760(1989)
2. Y.Tamura, Y.Aochi, S.Terasaki, O.Takano, C.Ishihara and N.Ishii: A multiple shots 3-D holographic sonar using a set of orthogonalized modulating signals, Acoustical Imaging, 20,737/743(1993)

HOLOGRAPHIC 3-D IMAGING SYSTEM USING ENCODED WAVEFRONT: AN UNDERWATER IMAGING SYSTEM

Chiaki Ishihara, Takashi Aoki, Osamu Takano, Norio Ishii,
Syuzo Hisamoto, Hajime Yuasa and Yasutaka Tamura*

Akishima Laboratories Inc.
(Mitsui Engineering and Shipbuilding)
1-50 Tsutsujigaoka 1-chome, Akishima, Tokyo 196, Japan
*Faculty of Engineering, Yamagata University
3-16 Jyonan 4-chome, Yonezawa, Yamagata 992, Japan

INTRODUCTION

Optical imaging, e.g., using a television camera, gives us detailed information on shape, size, and color clearly and continuously. However, even in clear water, optical imaging are not so useful in detecting an object placed up to 10m away. Moreover, in muddy water, optical imaging cannot be used at all.

In these cases, acoustical observations have been used to detect underwater objects. However, conventional systems, they use a narrow spearlike ultrasonic beam, have fundamental defects; the resolution of the image is limited by the sharpness of the beam and they cannot observe objects simultaneously. These disadvantages of the conventional systems, including intermittent observations, make it impossible to observe objects with some motion.

Holographic imaging by "encoded wavefront technique" has the advantages of high-speed data acquisition and simple hardware. This technique gives us a new three-dimensional observation method which makes it possible to visualize objects even with some relative motion. The authors developed an underwater imaging system based on this technique.

In this paper, we describe the principles of imaging first; then we report the outline of the developed system and the results of the performance test.

PRINCIPLES OF IMAGING AND MOD2 TRANSMISSION

Priciples of Imaging [1]

The geometry of the transducer array is shown in Fig. 1. The transmission array is composed of multiple transmitters arranged at even intervals on the circumference. Signals generated by the modulation of the carrier with the Walsh function $Wal_k(t)$, which handles the row or column elements of the Walsh-Hadamard matrix as wave functions, are transmitted simultaneously. The transmission signal used is the carrier of frequency f_0 with its phase modulated every n periods. Assuming an element at row i and column j of the Walsh-Hadamard matrix as W_{ij}, the signal transmitted from the k-th transmitter is expressed as

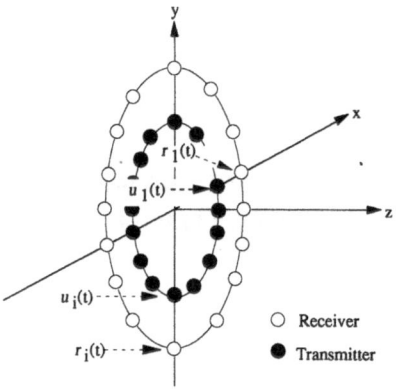

Figure 1. Geometry of array and co-ordinates

$$u_k(t)=Wal_k(t)\cdot e^{j2\pi f_0 t} = \sum_{j=1}^{N} W_{ij}\cdot f(t-(j-1)\Delta t) \qquad (1)$$

The signals reflected on an object are observed by receivers. These receivers are arranged at even distances on a concentric circle with that of transmission arrays. Assuming the signal observed by the i-th receiver as $r_i(t)$, the function $S(x)$ presenting an image is defined as follows:

$$s(x)=\left| \int_{-\infty}^{\infty} \sum_{i=1}^{M} r_i(t+\tau(v_i,x))\sum_{k=1}^{N} u_k(t-\tau(\mu_k,x))^* dt \right|^2 \qquad (2)$$

where v_i and μ_k are the vectors representing the positions of the i-th receiver and the k-th transmitters, respectively, and $\tau(m, x)$ is the time taken for sound to propagate from point m to point x.

Principles of Mod2 Transmission [2]

To reduce artifacts is to improve quality of images. One of the major reasons for the generation of an artifact is the decrease in the orthogonality between signals. Tamura et al., focusing on the features of the Walsh function, developed a method to compensate the orthogonality between signals by repetitions of transmissions and receptions. We call this the mod2 transmission method.

The signal $u^{(l)}_k(t)$ output from the k-th transmitter in the l-th transmission may be expressed as

$$u_k^{(l)}(t)=\sum_{j=1}^{N} W_{\zeta(k,l)j} \cdot f(t-(j-1)\cdot\Delta t) \qquad (3)$$

where $\zeta(i, j)$ implies the bit-by-bit modulo 2 addition when i and j are expressed as binary values. We can prove that the correlation $\phi_{kn}(\tau)$ between the signal $u^{(l)}_k(t)$ and that transmitted from the n-th transmitter after N transmissions and receptions may be calculated as in Equation (4).

$$\phi_{kn}(\tau) = \int_{-\infty}^{\infty} \sum_{l=1}^{N} u^{(l)}_k(t) \cdot \sum_{l=1}^{N} u^{(l)}_k(t-\tau)dt \quad \begin{aligned} &= N \int_{-\infty}^{\infty} f(t) \cdot f(t-\tau)dt \quad &(k=n) \\ &= 0 \quad &(k \neq n) \end{aligned} \tag{4}$$

CONFIGURATION OF THE DEVELOPED SYSTEM

Configuration of Imaging System

A front view of the system is shown in Fig.2. The measuring unit is shown on the left-hand side, and the transducer array is shown on the right-hand side. The measuring unit consists of transmission/reception system and control/display system. The transmission/reception system has a 16ch. arbitrary waveform generator, a 16ch. power amplifier and a 16ch. A/D converter which has transient wave memories and the gain-controllable preamplifiers. The control of the whole system and the display of the images are on the engineering work station. A processor, which is composed of digital signal processors, enables parallel processing. The A/D converter and the processor are connected with an optical data bus link for high-speed data transfer.

Sixteen transmitters are arranged at even pitch on a circumference of radius 0.45m, and 16 receivers are located at even pitch on a concentric circle of radius 0.6m.

Transduces

An exterior view of the transducers are shown in Fig.3. We selected piezo-rubber, a composite of rubber and ceramic, as a piezoelectric material of the transmitter. The transducer composed of piezo-rubber have a broad bandwidth, but the sensitivity is inferior than the ceramic used. Therefore multiple laminations of piezo-rubber films are used for higher sensitivity.

The maximum sensitivity is +152dB (dB re. 1 μ Pa/volt) at the frequency of 225kHz. The directivity was evaluated as 6 degrees with a half-value angle.

A polymer film called PVDF is used as a piezoelectric material of the receiver. Evidently, PVDF has high reception sensitivity, but its high electrical impedance is undesirable. Therefore a preamplifier is mounted close to the piezoelectric material to perform impedance matching and maintain a high signal-noise ratio.

The maximum sensitivity is -174dB (dB re. 1volt/ μ Pa) at 400kHz. A broad bandwidth is realized to set the resonance frequency lower than the carrier's frequency. The small aperture makes the spread of reception sensitivity wide. The directivity was evaluated as 20 degrees.

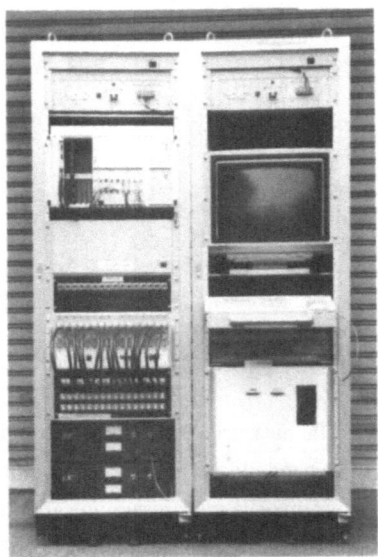 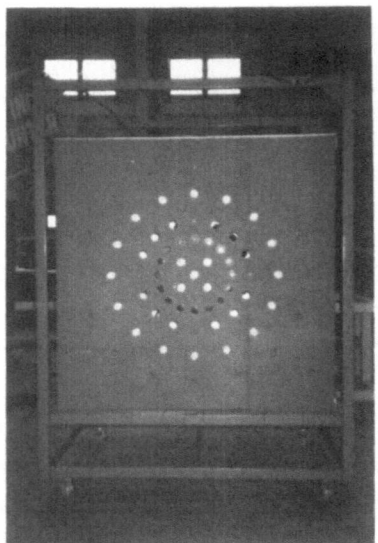

measuring unit   transducer array

Figure 2. Front view of developed system

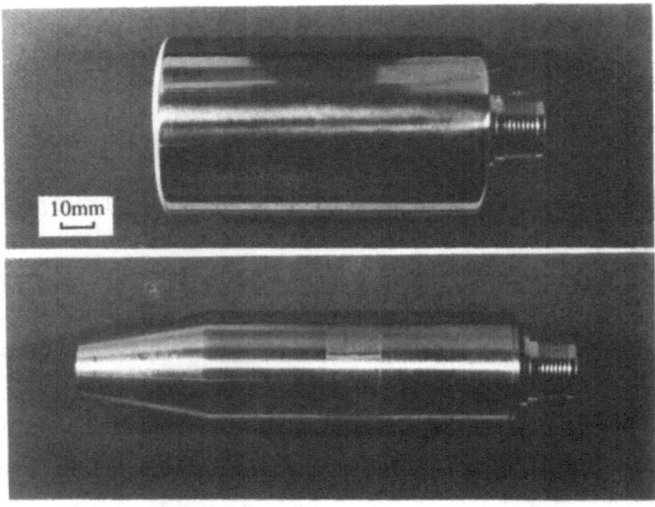

Figure 3. Transmitter (top) and receiver (bottom)

RESULTS OF PERFORMANCE TEST

The experiments for checking the performance of the system were carried out in a large tank at Akishima Laboratories of Mitsui Engineering and Shipbuilding Co. This tank has the dimensions of 220m length, 14m width, and 6m water depth. The measuring unit was set on a self-propelled carriage. The transducer array was lifted down from the carriage to be fixed at a vertical position 3m below the water surface. The target was suspended from the sub-carriage.

Check of Visibility

A long steel pipe with diameter 50mm was used as a target to check the visibility of the system. The targets at the range of one meter, 50m and 100m are reconstructed as shown in Fig.4. The images shown on the left-hand side are called B-mode images; the C-mode images are shown on the right-hand side. We can recognize that the increment of the range of the target makes the spread of the image in the traverse direction wide. The upper part and lower part of the target are not displayed. These phenomena will be explained in terms of the characteristics of reflection and interference of acoustical waves and the resolution of this imaging system.

A steel pipe having a rectangular cross section of side 50mm was detected at the range of 170m as shown in Fig.5. However, the image of target reconstructed in three-dimensional space is like the image of a point target shown on the right-hand side. The range of 170m is the maximum distance at which the target can be located in this tank.

As a result, we evaluate the visibility of the imaging system as 100m. The performance of high-speed imaging was checked at the same time. It was found that, including the sound propagation time, it takes less than one second to reconstruct and display the image.

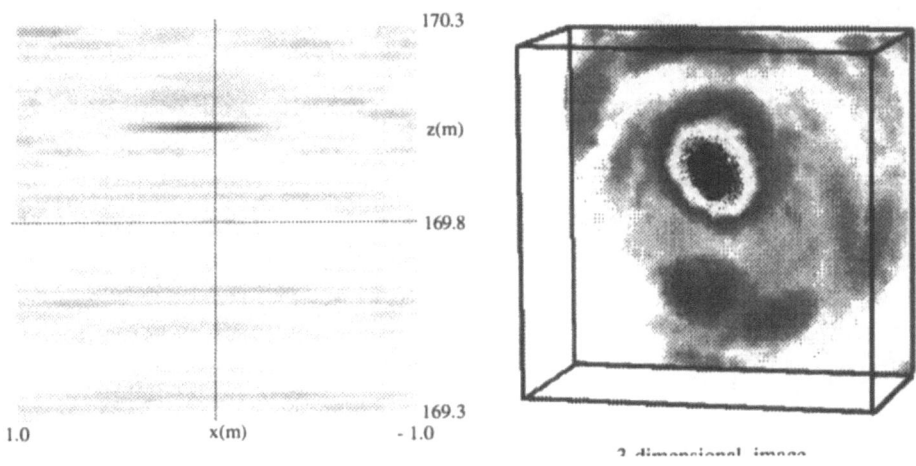

Figure 5. Detected rectangular cross-section pipe at 170m distance

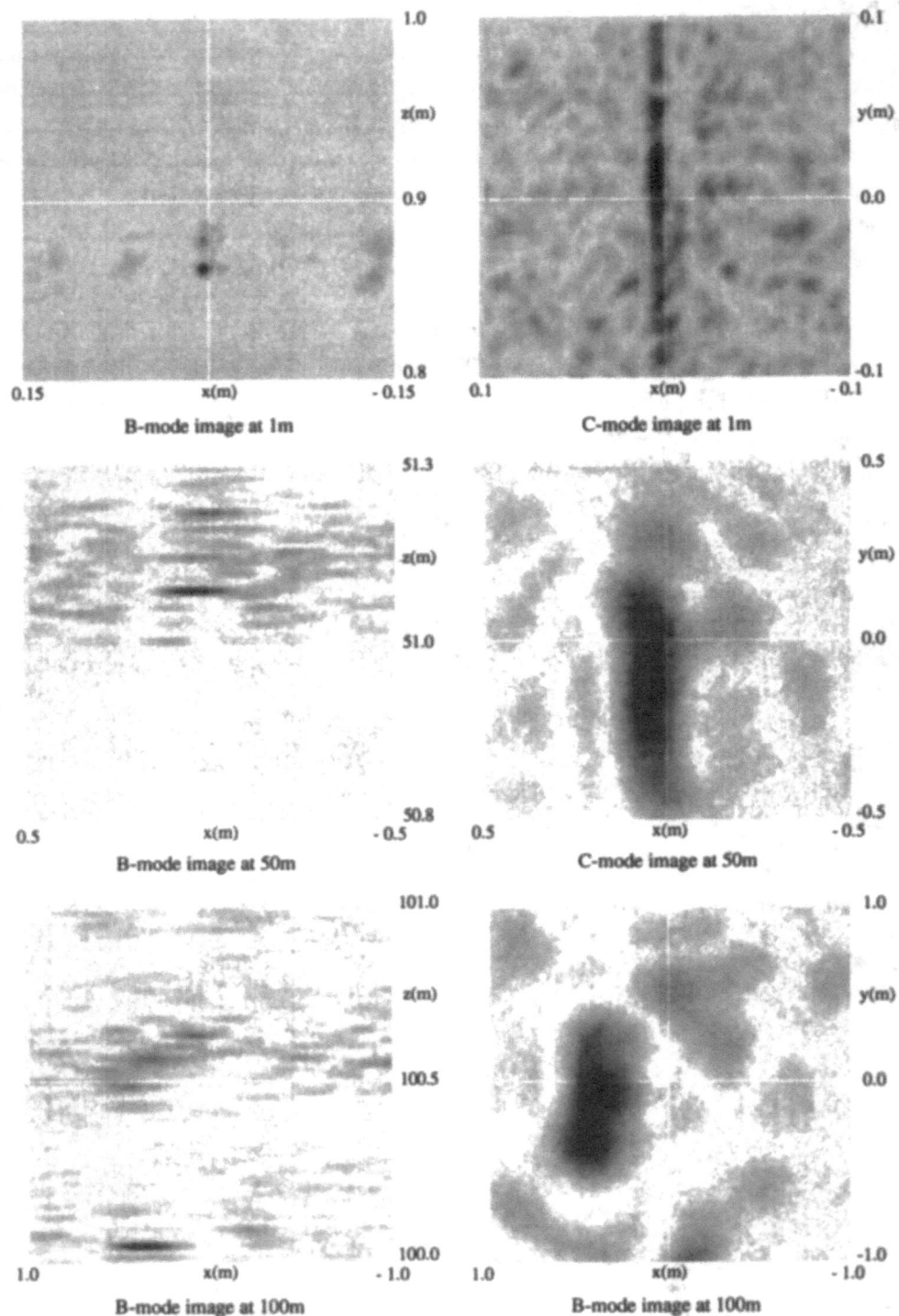

Figure 4. Check of visivility (target: 50mm diameter steel pipe)

Check of Resolution and Effectiveness of Mod2 Transmission

The range resolution was checked by using two steel pipes of 50mm diameter separate from each other by 30mm in the longitudinal direction and 220mm in the azimuthal direction at a location 20m forward from the array. The reconstructed B-mode image is shown on the left-hand side in Fig.6, and C-mode images are shown on the right-hand side. Certainly, we can recognize the existence of two scattering objects on B-mode image, but the quality of the C-mode image is not so fine with an unexpected object, which is considered to be another target at separate location.

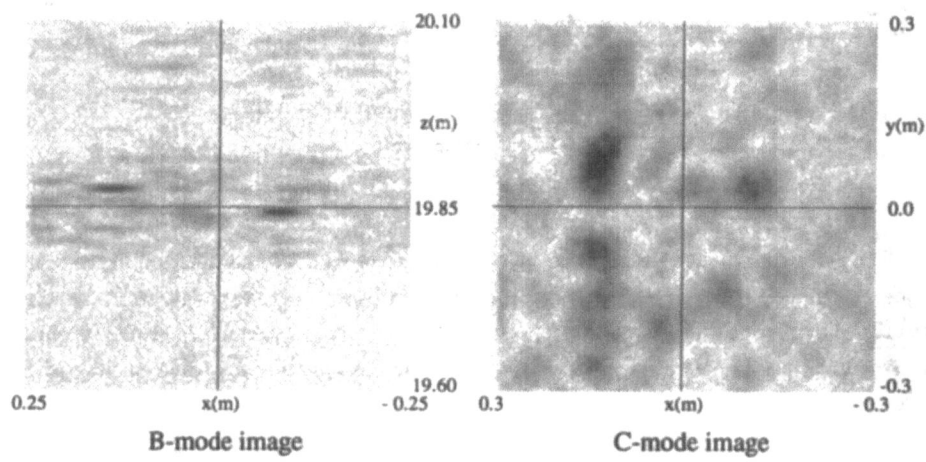

B-mode image   C-mode image

Figure 6. Check of resolution of imaging system

A rectangular frame of 220mm height and width configured by steel pipes of 50mm diameter is formed as a target. The height and width of the frame are equal to the 0.7 degree of azimuthal resolution at the range of 20m .

Three-dimensional imaging, which does not mean measurement of the thickness of a target, is a superior performance of this imaging system. The images shown in Fig.7 are reconstructed in three-dimensional space. These pseudo three-dimensional images were reconstructed as follows; a three-dimensional space data were constructed through the integration of planar images, e.g., the C-mode images, in the longitudinal direction and then the images were projected on a plane.

The artifacts are reduced and the quality of the image is improved by the mod2 method. We find that the azimuthal resolution is less than 0.7 degree and the mod2 transmission method is effective in improving the quality of the reconstructed image.

CONCLUSIONS

I now summarize the results of development.

We have developed an underwater imaging system using the acoustic holography technique with a transducer array and evaluated the performance through several experiments.

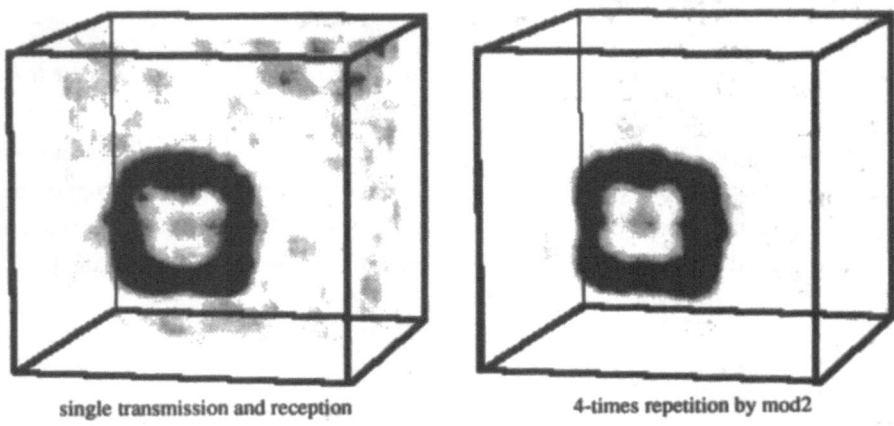

single transmission and reception   4-times repetition by mod2

Figure 7. check of azimuthal resolution and effectiveness mod2
(area : W=0.6m, H=0.6, D=0.6m)

The following matters were clarified: (1) One image will be reconstructed in the real time manner; it takes less than 1 second to reconstruct and display the image of target at the range of 20m; (2) Maximum visibility is 100m for a steel pipe of 50mm diameter; (3) Range resolution is less than 30mm; as a result of detailed analysis, the range resolution was evaluated as about 5mm; (4) Azimuthal resolution is less than 0.7 degree; detailed analysis showed the azimuthal resolution to be 0.3 degrees; (5) The "mod2" repetition transmission method considerably improves the quality of the image; (6) Technical possibility of a new type of underwater acoustic imaging without recourse to conventional means of electrical or mechanical scanning was elucidated.

ACKNOWLEDGMENT

This research and development was carried out under the trust of the Research Development Corporation of Japan. The authors would like to thank the staff of the Research Development Corporation of Japan and the staff of Kureha Chemical Industry Co.,Ltd. and NGK Spark Plug Co.,Ltd. for their help in developing the ultrasonic transducers. We would like to express our thanks to Dr. Hiro Yamasaki, professor of Tokyo University, and Dr. Kiyoto Koyama, professor of Yamagata University, for their help and useful advice in development.

REFERENCES

1. Y.Tamura et al., A multiple shots 3-D holographic sonar using a set of orthogonalized modulating signals, Acoustical Imaging, Vol.20, 737/743, Plenum, New York (1993)
2. Y.Tamura et al., Holographic 3-D imaging system using encoded wavefront : theoretical analysis on repetitive transmission, Acoustical Imaging, Vol.21, Plenum, New York

MULTI-VIEW IMAGING IN OCEANIC WAVEGUIDES BY DARK FIELD METHOD

E. L. Borodina, N. V. Gorskaya, S. M. Gorsky,
A. I. Khil'ko, V. N. Shirokov

Institute of Applied Physics,
46 Uljanov Str., 603600 Nizhny Novgorod, Russia

INTRODUCTION

A full understanding of the problem of visualization the acoustic sources ("an acoustic vision")[1,2], i.e. forming the spatial signal intensity distribution pattern, is fundamental to most developments in a large number of practical fields, such as underwater engineering, ecologic monitoring of extensive oceanic region state, navigation.

In this paper we transfer the optical methodology to geophysical waveguide conditions (it is assumed, that the distances of observation are large enough, such that the waveguiding conditions of the sound propagation are fully shown). It should be mentioned, firstly, that the oceanic medium in common is inhomogeneous that complicates the process of local inhomogeneity inverse reconstruction, because such medium does not "transmit" images[3].

Secondly, the oceanic medium is unsteady and randomly inhomogeneous. It results in illuminating source fluctuations, that restricts the use of sonovision methods, developed for homogeneous media. And, thirdly, usually observed objects have large wave dimensions (for example, an intrathermocline lens), so that the most part of scattered field energy is concentrated in a small angle around the illumination direction.

The previous investigations allow us to determine that the scheme in which the observed inhomogeneities are situated between the source and a receiving array is an optimal hydroacoustic vision scheme (this proposition has theoretical and experimental proof). This scheme is analogous to optical schemes of spatial filtration. But in the acoustic scheme the spatially-distributed receiving system (the array consisting of remote hydrophones or the antenna of synthesized aperture) plays a role of an image forming lens in coupling with the system of reconstruction. Moreover, the important part of the algorithm is the dark field method, that provides a filtration of strong illumination source field[4].

PRINCIPLES OF THE IMAGING CONCEPT

Consider now the image formation in the Fresnel zone using analogy of the optical lens and the antenna. It is assumed that the observed rigid inclusion of horizontal

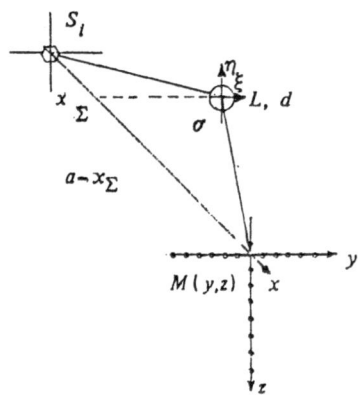

Figure 1. Basic scheme of the hydroacoustic vision system and the measurement geometry.

and vertical dimensions L_h and L_v, respectively, is situated in the waveguide between the source and the receiving system (i.e. horizontally and vertically stretched array of hydrophones), see Fig.1.

The common velocity potential $u(\vec{R})$ in the region of observation $\vec{R}(x, y, z)$ is determined from the Helmholtz-Kirchhof equation[5]:

$$u_0(\vec{R}) + \frac{1}{4\pi} \oint_S \left(\frac{\partial u(\vec{R}_s)}{\partial n} G(\vec{R}_s, \vec{R}) + u(\vec{R}_s) \frac{\partial G(\vec{R}_s, \vec{R})}{\partial n} \right) d\vec{R}_s = u(\vec{R}) \tag{1}$$

where $u_0(\vec{R})$ - is the velocity potential of the direct illumination field, n - is the outer normal of an inhomogeneity surface S, $G(\vec{R}_s, \vec{R})$ - is the Green function of the undisturbed medium. The problem consists in the reconstruction of an inhomogeneity location and shape (by spatial distribution of secondary sources $\frac{\partial u(\vec{R}_s)}{\partial n}$ and $u(\vec{R}_s)$ from fields $u(\vec{R}_s)$ measured on an array aperture $M(y, z)$.

In addition we assume that: 1) the horizontal wave dimension of the inhomogeneity is large $L_h \gg \lambda$ (λ - is the wavelength of the illuminating source); 2) the small angle approximation is realized; 3) distances between the source, the inhomogeneity and the receiver are large with respect to the waveguide thickness. In this case the equation (1) is as follows[6]:

$$u_0(\vec{R}) - \frac{1}{2\pi} \int_\sigma \int \sigma(\xi, \eta) \frac{\partial u_0(\xi, \eta)}{\partial n} G(\vec{R}_s, \vec{R}) \, d\xi \, d\eta = u(\vec{R}) \tag{2}$$

where $\sigma(\xi, \eta)$ - is a part of plane limited by a line dividing light and dark sides of the inhomogeneity situated at a distance x_Σ from the source and $a - x_\Sigma$ from the receiver. In the mode approach of the acoustic field in the oceanic waveguide[5] expressions for the incident field and Green function can be written as

$$u_0(x, y, z) = \sum_{n=1}^{N} A_0 \varphi_n(z) \varphi_n(z_i) \frac{\exp[i(a\, h_n + \frac{y^2}{2\,a} h_n - \frac{\pi}{4})]}{\sqrt{a h_n}} \tag{3}$$

$$G(\xi, \eta, 0; x, y, z) = \sum_{m=1}^{N} \varphi_m(y) \varphi_m(\eta) \frac{\exp(i((a - x_\Sigma) h_m + \frac{\eta^2 + y^2}{2(a - x_\Sigma)} h_m - \frac{y \eta h_m}{(a - x_\Sigma)} - \frac{\pi}{4}))}{\sqrt{h_m(a - x_\Sigma)}} \tag{4}$$

where N - is the number of propagated waveguide modes, h_n and φ_n - are eigenvalues and eigenfunctions of the waveguide, respectively, $A_0 = const$.

Combining (3), (4) and (2), we obtain the integral equation with respect to the position and shape of the inhomogeneity. For the arbitrary shape σ the vertical and horizontal coordinates (ξ, η) in (2) are interrelated. It is convenient to investigate these dependencies separately, because they contain the different physical information. In the vertical direction at large distances only waveguide modes take part in the scattering, in which a transformation of a spectrum occurs[6]. In the horizontal direction the diffraction of each mode on the inhomogeneity is analogous to the diffraction in the free space. To simplify the investigations the assumption is done $\sigma(\xi, \eta) = L(\eta)\, T(\xi)$.

The first component of the sum (2) represents the incident field in the observed region in mode presentation, the second component corresponds to the scattered field. The scattering in vertical direction is described by mode spectrum transformation and is defined by scattering matrix component

$$T_{nm} = \int\limits_{-\infty}^{\infty} T(\xi)\, \varphi_n(\xi + z_\Sigma)\, \varphi_m^*(\xi + z_\Sigma)\, d\xi \tag{5}$$

In this case the possible ways of reconstruction require a waveguide mode selection that demands the use of vertical arrays, time gating, etc. All of these methods are connected both with technical difficulties and a complication of processing algorithms[7,8].

We do not discuss this problem and consider in detail images providing an information on horizontal distribution of inhomogeneities. The multi-mode structure defines a complementary field modulation. In paper[6] the conditions are obtained for which the interference modulation spatial spectrum and the spatial spectrum corresponding to the inhomogeneity variations substantially differ:

$$\left(\frac{4\pi r_a}{\Delta h_{ij}}\right)^{\frac{1}{2}} \approx \frac{r_a \pi}{\langle h_{ij}\rangle L} \tag{6}$$

where $r_a = a - x_\Sigma$, Δh_{ij} - is the difference of horizontal projections of eigenvectors, $< h_{ij} >$ - is the average difference value. For $r_a \gg \tilde{r}_a$ ($\tilde{r}_a$ is a distance at which the condition (6) is realized) the frequency of an interference modulation is more, and for $r_a \ll \tilde{r}_a$ -it is less then the "useful" variations. That allows the easy filtration of them using a prior information both on waveguide parameters and on inhomogeneities. At middle distances (determined from (6)) the summary image interpretation is rather difficult.

Consider one mode image now (i.e. $n = m$). To this end we assume that the mode selection is carried out, or one mode is differentiated due to the dissipation loss in waveguiding propagation, or the mode interference component is filtered in the imaging process[9]. In one-mode approximation (2) can be written as:

$$u(a, y, z_a) \simeq S_n^0 e^{i\frac{y^2}{2a}h_n} - S_n e^{i\frac{y^2}{2(a-x_\Sigma)}} \int_{-\infty}^{\infty} L(\eta) \exp\left(-i\left(\frac{y_0}{s'} + \frac{y}{r'}\right)h_n\eta\right) d\eta \tag{7}$$

$$S_n^0 = A_0 \varphi_n(z_a) \varphi_n(z_i)(r_0 h_n)^{-\frac{1}{2}} \exp[i(ah_n - \frac{\pi}{4})]$$

$$S_n = \frac{A_0}{2\pi} \varphi_n(z_a) \varphi_n(z_i)(-ih_n)\, T_{nn}(h^2 r' s')^{-\frac{1}{2}} \exp[i(h_n x_\Sigma + h_n(a - x_\Sigma) - \frac{\pi}{2})]$$

$$r_0 \simeq a + \frac{y^2}{2a}, \quad r' \simeq a - x_\Sigma + \frac{y^2}{2(a - x_\Sigma)}, \quad s' = x_\Sigma + \frac{y^2}{2x_\Sigma}$$

Assuming $y_0 = 0$, we obtain the expression for the second component in equation (7): $S_n \int_{-\infty}^{\infty} L(\eta) exp[i(-h_n \frac{\eta}{r'} y + h_n \frac{y^2}{2(a-x_\Sigma)})]d\eta$. The Fresnel integral is obtained, where the exponential factors represent complete orthonormalized basis. Integration of (7) multiplied by $M(y_A) exp[i(h_n y_A \sin \alpha - h_n \frac{y_A^2}{R})]$ (where (α, R) - are polar coordinates of a point (x, y)) within the receiving aperture as $M(y_A)$ yields

$$\Phi_n(\alpha, R) \simeq S_n^0 F_n(\alpha, ((2a)^{-1} - R^{-1})^{-1}) - S_n \int_{-\infty}^{\infty} L(\eta) F_n(\alpha - \frac{\eta}{r'}, \varepsilon^{-1}) \, d\eta \qquad (8)$$

where $\varepsilon = \frac{1}{2(a-x_\Sigma)} - \frac{1}{R}$ - is a parameter of vision system focusing, $\sin \alpha \simeq \alpha$, $\Phi_n(\alpha, R)$ determines an algorithm of image reconstruction from measured data and F_n - is the pulse-transient characteristic of the reconstructing system:

$$F_n(\alpha, R) = \int_{-\infty}^{\infty} M(y_A) \exp(ih_n[(\frac{\eta}{r'} - \alpha)y_A + \varepsilon y_A^2]) \, dy_A \qquad (9)$$

For the rectangular aperture of length D at $\varepsilon = 0$ (focused image) $F_n(\alpha, a - x_\Sigma) = DSinc(h_n \alpha D)$, and a transverse dimension of resolution element (on y axis) is $\Delta x = \frac{\lambda(a-x_\Sigma)}{2D}$. The point source image is much wider in a longitudinal direction: $\Delta y \sim (5 \div 10) \Delta x$ (an exact estimation is determined by Fresnel integral).

Equation (8) defines the source and the scatterer images. Generally the strong direct signal fluctuations masks inhomogeneities images. To suppress the source signal the general dark field method was developed[4,9]. This method is based on a selection of scattered and direct signal spectra, corresponding the sources differently remote from an array. In the first case of the dark field method realization the signals of two neighboring receivers are subtracted and resulting spatial components of masked source signal are filtered[9]. This operation realizes the spatial mask in a focal plane matched according to prior information of the source location. Another way is based on spatial filtration of Fresnel images by two-dimensional filters adjusted to the source image. Both methods require prior information on the source location, however, the second way may be readily used in complex-structure media, when a filter is constructed using the empirical data in the absence of inhomogeneities.

Some sonovision schemes realized in practice by means of sparse transducer array and also by two remote arrays at an angle to each other ("binocular scheme") are discussed in the next section.

NUMERICAL MODELLING

Dark Field Method

As it was mentioned, the strong direct signal fluctuations mask inhomogeneity images and decrease dynamic range of registered signal. To overcome these difficulties the general dark field method for source field suppression was developed[4,9]. This method is based on the selection of scattered and direct signal spectra, corresponding the sources differently remote from an array.

Consider briefly one view reconstruction in one-mode approximation. The aperture function of uniformly spaced linear receiving array with omnidirectional sensors can be written as:

$$M(y_A) = \sum_{n=1}^{N} A_0 \delta(nd - y_A) \Pi_{y_A} \qquad (10)$$

776

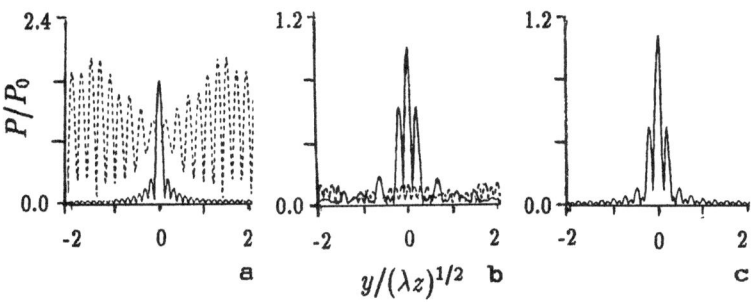

Figure 2. Reconstruction of point source image in the scatterer plane before (a) and after the filtration by meander mask (b) and sinusoidal mask (c) using 16-element array.

where

$$\Pi_{y_A} = \begin{cases} 1 & |y_A| \le D/2 \\ 0 & |y_A| > D/2 \end{cases}$$

$A_0 = const$, d is the spacing between neighboring sensors, N - is the number of elements.

Fig.2 shows one period of restored multiplied images of the point scatterer and the source in the plane of scatterer localization.

The received signal is processed to focus the source image leaving the scatterer unfocused. Combining (9) and (10) for $r = 2a$ we obtain the focused image of the illuminating source $S_n^0 \sin(Nkyd/(2a)) \sin^{-1}(kyd/(2a))$, where $k = 2\pi/\lambda$.

For direct signal suppression a filter enclosing main and two neighboring maxima of each source image is used, and the obtained signal is restored into vision area (Fig.2b). Besides that a more successive filtration is possible, when the matched filter multiplier is $\theta = \sin(kyd/(2a))$. Taking off the spectral components of number $\pm N$ we obtain pure scatterer signal (Fig.2c).

Spatial filtration can be carried out also using two-dimensional filters adjusted to the source position[4]. Both methods require prior information on the source location, however, the second way can be readily employed in complex media, when the filter is constructed from the empirical data in the absence of inhomogeneities. An example of diffracted signal differentiation from high background (when the direct signal is by 15 dB larger than the diffracted signal) is given in Fig.3. Fig.3b shows the scatterer image obtained using two-dimensional spatial spectrum filter $\theta(\xi_x, \xi_y) = |F^0(\xi_x, \xi_y)|^{-1}$, where $F^0(\xi_x, \xi_y)$ - is the source spectrum, ξ_x, ξ_y - are spatial frequencies. In the considered example the aperture was about 10 Fresnel zones (concerning the source).

It should be mentioned for an interpretation of images reconstructed by sparse-element arrays, that the fewness of an array limits the vision area and makes it possible to fix an inhomogeneity position within a multiplication period. In the considered case a number of individual elements of the image in a transverse direction (on y-axis) is of order D/d. The resolution in the longitudinal direction is limited both by dimensions of measurement region (as for the continuous aperture) and by a multiplication effect.

Computer Simulation of Multi-View Reconstruction Scheme

Consider now measurement schemes where the angle of illumination and the receiving array location angle shift at the same value in opposite directions. In an observation

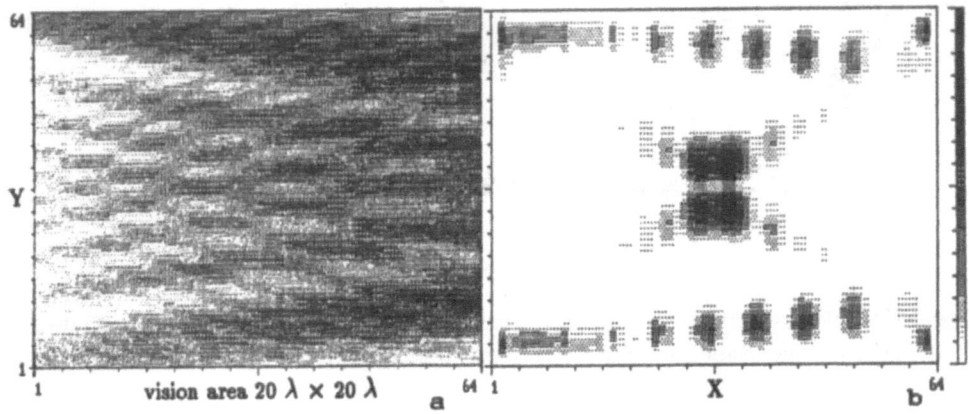

Y
vision area 20 λ × 20 λ   a   X   b

Figure 3. Unfiltered summary signal (a), in which only the illuminating source is seen, and filtered signal (b), where the scattered image appears.

of stationary distributions the sequence of partial images at each view may be fixed in succession, in the opposite case simultaneous measurements are required. As mentioned, the resulting image may be obtained both by coherent and incoherent summation of separated projections. The coherent summation is connected with interference effects, which can essentially distort the observed object image[10]. At the same time this way of processing admits the summation with correcting complex weight multipliers, that improves the image characteristics. In certain cases, when an estimation of an average inhomogeneity distribution is required, or partial images are incoherent due to random inhomogeneities, the incoherent summation is carried out.

As seen from (8),(9) the spatial resolution significantly depends on the position in vision area. In multi-view imaging the best resolution appears in the area center, with a displacement from center an individual element of resolution becomes more indistinct and decreases in its amplitude. Thus, the summary image becomes illegible. Numerical estimations show that for multi-view system of reconstruction in Fresnel zone, apparently, there is a circle of a diameter $0.75\,a$ and center in a point $(0.5\,a,\,0)$.

Fig.4 shows the image of a letter Π as a result of incoherent summation of images reconstructed from 32 projections equidistant within the interval $0° \div 180°$.

In the considered example the image composed of illuminating points was calculated, that allowed to investigate possibilities of Fresnel image reconstruction without reference to the problem of the source field suppression. It is seen, that more adequate information on the source spatial distribution is obtained by an incoherent summation. In a coherent summation one of lines forming the letter is weakly visible. This suppression is evidently caused by the interference, because the effect disappears with a shift of the letter. A comparison of these two images shows, that in the case of incoherent summation the noise appears in the form of some averaged "halo", which may be removed by a low frequency filter. In another case, when partial images are summed in a coherent way an interference speckle-noise appears whose wide spatial spectrum hinders the filtration.

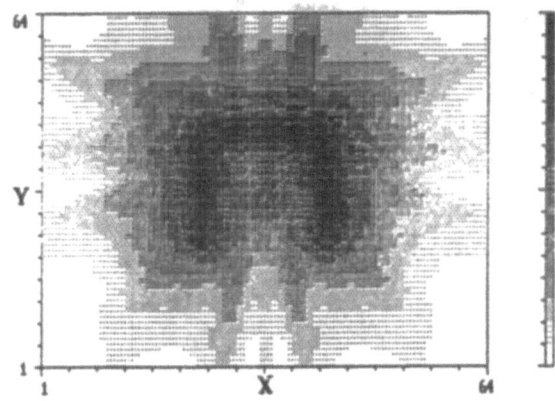

Figure 4. The result of incoherent multi-view reconstruction of the sign Π.

EXPERIMENTAL RECONSTRUCTION RESULTS

For the verification of image reconstruction algorithms an ultrasonic experimental set was designed (Fig.5), which allowed modeling of propagation and scattering of hydroacoustic signals in oceanic waveguides. The system of modeling measurements included homogeneous water layer of a thickness 3 cm and a sound velocity 1485 m/s on a rubber bottom. The directivity pattern of the piezoceramic source was specially arranged in a horizontal plane, that allowed to avoid the reflection at basin walls. We realized the conditions of scanty mode propagation for the pulse signals 300 μs in duration at frequencies 140 Hz and 512 Hz. Waveguide modes were weakened due to losses in rubber layer (a tangent of the loss angle in the rubber was 0.28). Signals were recorded by two channels.

Three scattering vertical steel cylinders of a diameter 1 cm, 2 cm and 3 cm were mounted on a mechanical rotatable frame, so that the distances between them were 9 cm, 5 cm, 6 cm. The length of the antenna synthesized by moving at the depth $z_A = 0.3$ cm receiver was 36.5 cm. The depth of the source 7 cm long was 1.7 cm, $a - x_\Sigma = 25$ cm, $a = 152$ cm.

The filter was formed inversely to experimental amplitude spectrum of the signal in the absence of the cylinder accounting the mode interference. The results of reconstruction of the inhomogeneity spatial distribution from 32 projections are given in Fig.6.

As it is seen from the comparison of images the result of incoherent summation is preferable. But in both cases the restored cylinders are contoured and almost overlapped by significant interference disfigurations (among them in the spatial frequency range comprising the useful signal). It prevents a suppression of the interference by simple methods.

CONCLUSION

In this work the problem of image reconstruction of oceanic large-scale inhomogeneities by scanty-view systems was discussed. The peculiarities of the hydroacous-

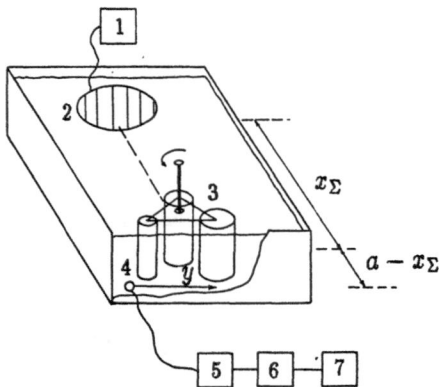

Figure 5. Experimental arrangement of multi-view reconstruction (1 - pulse generator, 2 - transducer, 3 - rotatable object, 4 - moving receiver, 5 - amplifier, 6 - filter, 7 - computer).

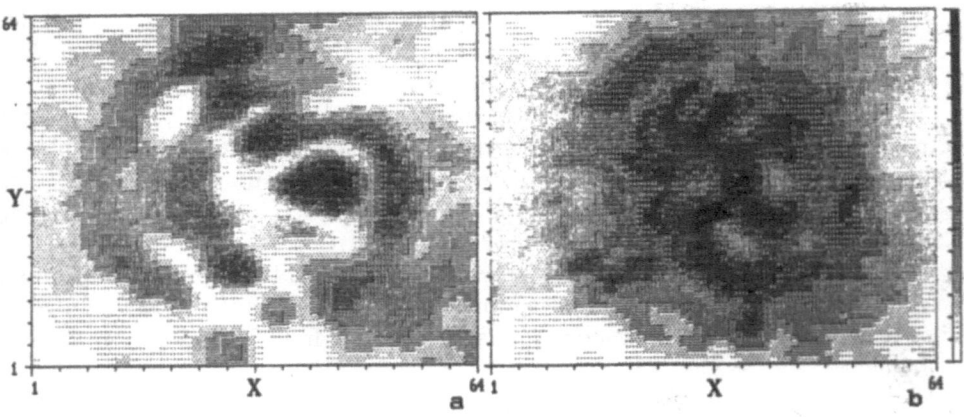

Figure 6. Image of three cylinders reconstructed from experimental data as a result of coherent (a) and incoherent (b) summation after illumination source suppression.

tic imaging in Fresnel zone in oceanic waveguides were investigated. The analysis of multi-view schemes was fulfilled both theoretically (analytically and numerically) and experimentally by physical modeling.

The observed particular problem is one of the more general field of inverse problems of scattering (including tomographic methods). In this paper we investigated only the problem of vision, i.e. a reconstruction of the secondary source spatial distribution on an inhomogeneity surface without reconstruction its physical structure. So that one can obtain the information on inhomogeneity localization in observation area and its shape.

REFERENCES

1. P. Greguss, "Ultrasonic Imaging", Focal Press Inc., New York (1980).
2. A. Makovski, Ultrasonic imaging using arrays, Proc. IEEE, 67:484 (1979).
3. Yu.A. Kravtsov, V.M. Kuz'kin, and V.G. Petnikov, Diffraction of waves on regular scatterers in multimode waveguides, *Sov. Phys. Acoust.* , 3:339 (1984).
4. E.L. Borodina, N.V. Gorskaya, S.M. Gorsky, et al., Possibilities of shadow methods in studying of diffracted sound fields in waveguides, in: "Forming of Acoustic Fields in Oceanic Waveguides", V.A. Zverev, ed., IAP RAS, Nizhny Novgorod, (1991).
5. L.M. Brechovskich and Yu.P. Lysanov, "Theoretical Principles of Oceanic Acoustics", Gidrometeoizdat, Moscow (1982).
6. N.V. Gorskaya, S.M. Gorsky, V.A. Zverev, et al., Peculiarities of short-waves diffraction of sound in multimode layered waveguides, in: "Acoustics in the Ocean", I. B. Andreeva, L. M. Brechovskich, ed., Nauka, Moscow (1992).
7. "Underwater Acoustics and Signal Processing", L. Bjorno, ed., Reidel Publ. Comp., Dordrecht (1981).
8. A.G. Nechaev, A.I. Khil'ko, Differential acoustic diagnostics of random oceanic inhomogeneities, *Sov. Phys. Acoust.* , 2:285 (1988).
9. E.L. Borodina, N.V. Gorskaya, S.M. Gorsky, et al., Spatial filtration of images in ultrasonic visualization of large inhomogeneities, *Sov. Phys. Acoust.* , 6:1004 (1992).
10. M.F. Adams and A.P. Anderson, Tomography from multiview ultrasonic diffraction data: comparison with image reconstruction from projections, *Acoustical Imaging* , 10:365 (1980).

THE APPLICATION OF FRESNEL INTERGRALS TO THE STUDY OF UNDERWATER NOISE SOURCES

Yang Desen

Department of Acoustical Engineering
Harbin Shipbuilding Engineering Institute (HSEI)
Harbin 150001, People's Republic of China

In this paper a sound imaging system is constucted to investigate the characteristics of underwater noise, especially that of moving objects. Space transformation of the sound field is performed by the Fresnel integral. It is shown how to calculate the Fresnel integral with the Fourier transformation, and the selection of parameters in the Fresnel integral is also presented. Three experiments are described in the paper: the sound image of a cross array with five elements, the sound image of a moving ship model and that of an underwater engine. The method of scanning in experiments is discussed. It is shown that the position of the main sources can be easily recognized from the sound images of measured objects.

1. INTRODUCTION

Image reconstuction has proven to be a useful tool for studying underwater noise sources. Characteristics such as the position and intensity distribution of the noise sources may be obtained by this method. In this paper the sound field transformation from the measured plane to the source plane is performed by the Fresnel integral, therefore the sound wave mutual intensity in the source plane is obtained from which the radiation sound images are made.

2. THEORY OF ACOUSTICAL IMAGING

Consider the plane noise sources distributed on the surface of source plane S, see Fig. 1. The measurement plane Σ is parallel with S. The distance between S and Σ is R, A(x,y,t) is the pressure connected with the source intensity on the source plane. For the convinience of both

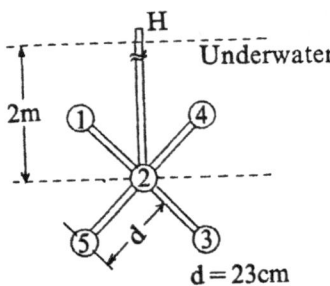

Figure 1. A typical model for source plane S and measurement plane Σ

mathematics and measurement we consider a reference point $M_o(\xi_o, \eta_o, R)$ on the measurement plane. The intensity of the sound wave between any point and $M_o(\xi_o, \eta_o, R)$ on the measurement plane is defined as

$$J(\xi, \eta, R) = <P(\xi, \eta, R, t) \bullet P^*(\xi_o, \eta_o, R, t)> \tag{1}$$

where $< \bullet >$ is the time average, " * " is the complex conjugate. $P(\xi, \eta, R, t)$ and $P(\xi_o, \eta_o, R, t)$ can be written as

$$P(\xi, \eta, r, t) = \iint_s \frac{1}{R_1} A\left(x, y, t - \frac{R_1}{C}\right) dxdy \tag{2}$$

$$P(\xi_o, \eta_o, R, t) = \iint_s \frac{1}{R_2} A\left(x, y, t - \frac{R_2}{C}\right) dxdy \tag{3}$$

$R^2_1 = (x-\xi)^2 + (y-\eta)^2 + R^2, R^2_2 = (x-\xi_o)^2 + (y-\eta_o)^2 + R^2$. As shown by Ghatak[1], the Fresnel nearfield limit is $R_F, R_F < (2d)^2/\lambda$, where λ is the wavelength, (2d) is the largest dimension of the source, the sound signal is narrow band, and the measurement plane is infinite. The relation of mutual intensities between the measurement plane and the source plane can be writen as

$$J(\xi,\eta,R) = -\frac{1}{i\lambda R}\exp(-iKR)\int_{-\infty}^{+\infty}\int I(x,y,o)$$

$$\bullet\exp\left\{-\frac{ik}{2R}\left[(\xi-x)^2+(\eta-y)^2\right]\right\}dxdy$$

$$= -\frac{1}{i\lambda R}\exp(-iKR)\iint_s I(x,y,o)\exp$$

$$\left\{-\frac{ik}{2R}\left[(\xi-x)^2+(\eta-y)^2\right]\right\}dxdy \tag{4}$$

$$I(x,y,o) = \frac{1}{i\lambda R}\exp(iKR)\iint_\Sigma J(\xi,\eta,R)$$

$$\exp\left\{\frac{ik}{2R}\left[(x-\xi)^2+(y-\eta)^2\right]\right\}d\xi d\eta \tag{5}$$

Where $K = 2\pi/\lambda$, $J(\xi,\eta,R)$ and $I(x,y,o)$ are mutual intensities in the measurement and source plane respectively. Integrals (4) and (5) are called pairs of Fresnel transformation. As shown in Ghatak [1] and Ueha [2], it is difficult to calculate equations (4) and (5), but they can be calculated by the Fourier transformation. Equation (4) can be rewritten in another way:

$$J(\xi,\eta,R) = -\frac{1}{i\lambda R}\exp(-iKR)\iint_s I(x,y,o)\bullet\exp\left[-\frac{iK}{2R}(x^2+y^2)\right] \bullet$$

$$\exp\left[-\frac{iK}{2R}(\xi^2+\eta^2)\right] \bullet\exp\left[\frac{iK}{R}(\xi x+\eta y)\right]dxdy \tag{6}$$

or

$$J(\xi,\eta.R)\exp\left[\frac{iK}{2R}(\xi^2+\eta^2)\right] = -\frac{1}{i\lambda R}\exp(-iKR)\iint_s I(x,y,o)$$

$$\exp\left[-\frac{iK}{2R}(x^2+y^2)\right] \bullet\exp\left[\frac{iK}{R}(\xi x+\eta y)\right]dxdy \tag{7}$$

Integral (5) can also be written as:

$$I(x,y,o)\exp\left[-\frac{iK}{2R}(x^2+y^2)\right]=\frac{1}{i\lambda R}\exp(iKR)\iint_{\Sigma} J(\xi,\eta,R)\bullet$$

$$\bullet\exp\left[\frac{iK}{2R}(\xi^2+\eta^2)\right]\exp\left[-\frac{iK}{R}(x\xi+y\eta)\right]d\xi y\eta \tag{8}$$

Where

$$W(\xi,\eta)=J(\xi,\eta,R)\exp\left[\frac{iK}{2R}(\xi^2+\eta^2)\right] \tag{9}$$

$$\tilde{W}\left(\frac{Kx}{R},\frac{Ky}{R}\right)=I(x,y,o)\exp\left[-\frac{iK}{2R}(x^2+y^2)\right] \tag{10}$$

Integrals (7) and (8) can be represented as

$$W(\xi,\eta)=-\frac{1}{i\lambda R}\exp(-iKR)\iint_{s}\tilde{W}\left(\frac{Kx}{R},\frac{Ky}{R}\right)\exp\left[\frac{iK}{R}(\xi x+\eta y)\right]dxdy \tag{11}$$

$$\tilde{W}\left(\frac{Kx}{R},\frac{Ky}{R}\right)=\frac{1}{i\lambda R}\exp(iKR)\iint_{\Sigma} W(\xi,\eta)\exp\left[-\frac{iK}{R}(x\xi+y\eta)\right]d\xi y\eta \tag{12}$$

Since these two integrals are pairs of the Fourier transformation, then the Fresnel integral can be calculated conveniently by the Fourier transformation.

The spatial resolution of image reconstruction is an important problem. In integrals (11) and (12) the measurement interval in the experiment plane for getting higher resolution images, both in X and Y directins, must be staisfied with:

$$\Delta\xi<\frac{\lambda R}{a} \tag{13}$$

$$\Delta\eta<\frac{\lambda R}{b} \tag{14}$$

where a and b are the length of the source plane in the X and Y directions respectively.

3. EXPERIMENTAL RESULTS

As mentioned before, equations (13) and (14) should be satisfied to get the image of underwater sound sources so that $R > (2d)$, $R < R_F$, $\Sigma \to \infty$. To test this theory, two experiments were performed in a tank. The three dimensions are 11m x 5m x4m at HSEI. For the actual application we selected two kinds of sources, one a fixed five element cross array, and the other a small moving ship. The third experiment, an application of noise control in an underwater engine, was accomplished in a reservoir. The images are distributed from 0-35 in an arbitrary scale. The imge was obtained with full scale intensities. If the level was higher than 30, the level was changed to 35 and if the level was lower than 10, the level was changed to 0. The images were drawn by a thermal sensitive plotter. To locate the sound sources, the shape and the position are expressed in the reconstruction imges.i.e the whiter position means the higher intensity.

3.1 IMAGE OF FIXED SOUND SOURCE

The sound source system is shown in Fig.2. It is a five point source,

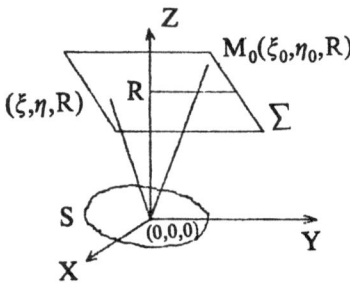

Figure 2. Five element cross sound source system
Numbers 1-4 are radiation transducers and number 5 is a diffraction transducer

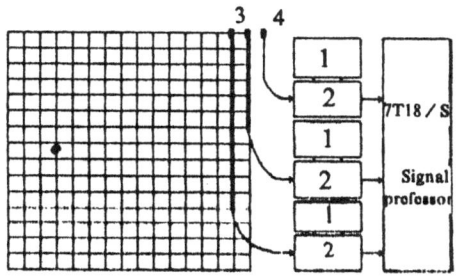

Figure 3. Measurement plane and block diagram of measurement setup
for $\Sigma = 60$cm x 60cm in fixed source image experiment, $\Delta\zeta = \sigma\eta = 4$cm

source 5 is not a real source but rather, a diffraction source. The signal is a narrow band white noise and its canter frequency f_0 is 50KHz. The array is 2m deep in the water. Other parameters are $R = 1m$, $\Delta\xi = 4cm$ and $\Sigma = 60cm \times 60cm$. The measurement plane and block diagram are shown in Fig.3. A fixed hydrophone was used to meeasure the reference signal and two hydrophones were moved to measure in the experimental plane. The reconstructed imge of underwater sound sources is shown in Fig.4. Compared with the sources in Fig.2 the image shows clearly the face of the sound sources. Although the fifth source is not a radiation sound, it shows the lower grey scale because of its diffraction.

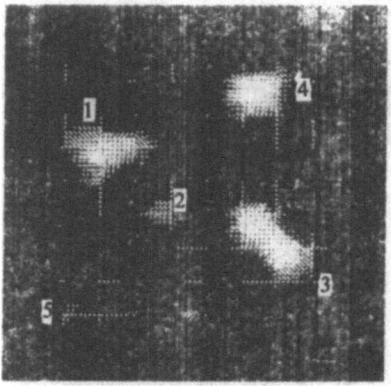

Figure 4. The sound image of the five element
cross array underwater.

3.2   Image of Moving Ship

It is important to reconstruct the image of a moving ship. In this experiment the shape of the ship is shown in Fig.5. The bottom of the ship, which is 40cm x 20cm, is driven by an exciter. The experimental plane is formed by a line measurement array and the direction in which the ship is moving. The line array is 31 cm deep in water and consisted of 13 hydrophones; the interval between them is 6cm. The reference hydrophone moves with the ship. Its speed is 16.4cm/sec. The signal is a single frequency at 6.1KHz. When the ship is moving the signals are being measued continually by the line array. The signals are divided into discrete point signals in the measurement plane. The area of the measurement plane is 72cm x 60cm. The block diagram is shown in Fig.6. Fig.7 is the reconstructed image of the sound source.

As the bottom of the small ship is driven by the exciter, all surfaces of the ship radiate noise in the water. In Fig.7, it is easy to see the shape of the ship's bottom.

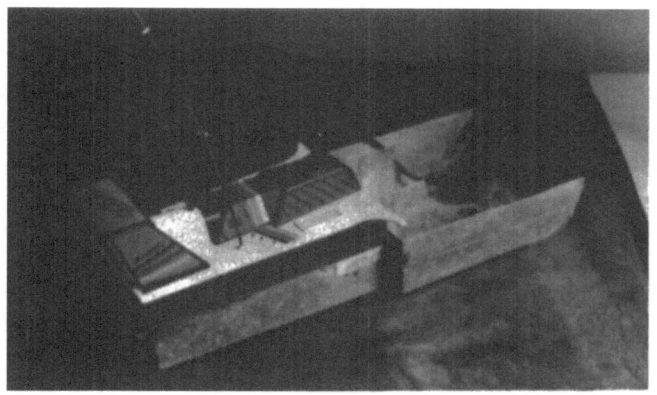

Figure 5. Samll ship with an exciter

3.3   IMAGE OF UNDERWATER STATIC ENGINE

This is an actual application in noise control engineering. The noise source is an underwater engine. For controlling its noise level, it must be known where the sound is radiated efficiently. Naturally, the reconstructed image is considered first. The dimensions of the engine are 6.0 meters long and 0.58 meters in diameter. The special difficulty is that the engine cannot work too long for the complex measurements. To solve this problem, the experiment was carried out by automaticly scanning using a vertical array of 11 hydrphones at an interval f 10cm. The scanning device consisted of a pulley, a pair of tracks and other parts. For reducing vibrations the pulley and tracks were all laid with rubber. The hydrophone array moves along the major axis of the engine when the measurement is made. The vertical array and the engine's major axis meet at

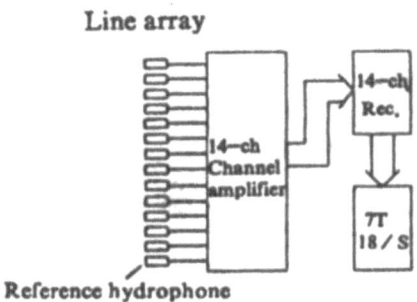

Figure 6. Block diagram for sound measurment of the moving ship

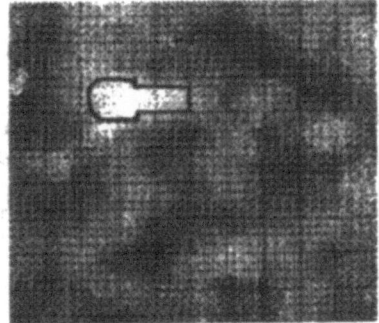

Figure 7. Sound image of the moving ship in water

right angles and the distance between them is 1 meter. All measurements were carried out 8.63 meters deep in the water. The experiment was made in a large resevoir in the Southern part of China. The devices of the experiment are shown in Fig.8. As the measurement array was continuosly moving and the engine was static, the measurement array should not move too long for the A/D to be completed by the computer. For this reason the scanning speed of the array cannot be too fast, (8.63cm/sec). In this experiment the scanning length is 11.8 meters so the area of the measurement is 11.8m x 1m. The details of the experiment are seen in Fig.8. By controlling the computer the signals are divided into discrete data blocks when the array is scanning. As shown in equations 11 to 14, the higher the frequency the more the sample points, and the more data blocks are needed. Thus, the time for calculating the Fresnel integral is very long.

The sound images of an underwater engine, at center frequency $f_0 = 315Hz \sim 500Hz$, 1/3 octave band are shown in Fig.9. The white area in the images represent strong radiations. The images show that the middle of the engine is white for $f_0 = 315Hz$ and 400Hz. That is to say stronger noise is radiated there. This is correct because many of the vibrating parts of the engine

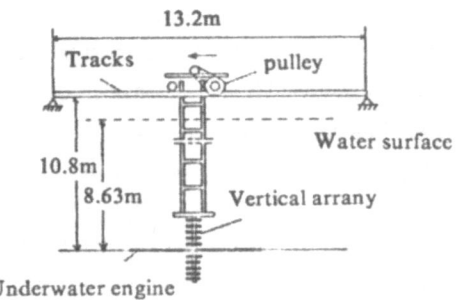

Figure.8. Diagram of scanning measurement devices for image of underwater engine.

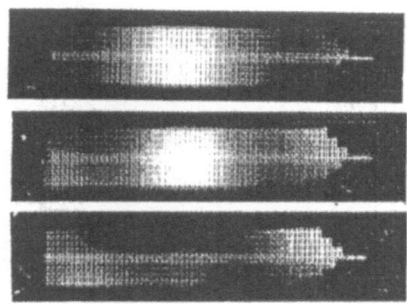

Figure.9. The sound images of an underwater engine

a : $f_0 = 315Hz$, 1/3Oct. band,   b : $f_0 = 400Hz$, 1/3Oct. band,   c : $f_0 = 500Hz$, 1/3Oct. band

are there. When the frequency is $f_0 = 500Hz$, the white area is changed from the middle of the front and back of the engine. For proving these results, some accelerometers on the surface of the engine and hydrophones all around the engine were used to measure the surface vibration and radiating noise in this experiment. The image results are proven to be right compared with the results from acceleration measurements and noise radiation measurements in the water, see [4]. The strong and weak sound levels are clearly seen in the images.

These results have been used to control the engine's noise.

4. CONCLUSIONS

A method of sound reconstructed imaging has been shown using the Fresnel integral. The mutual intensity in the measurement plane is transformed to the source plane. As it is important to know the main noise sources for underwater noise control, the practicalness of the method is considered especially for moving targets in water. In this paper, the relative motion between the measurement array and the target is considered, and the continuos signals are divided into discrete data blocks using a computer. The measurement plane is formed in Fig.3. This technique reduced the measurement instruments and quantity required. The Fourier transformation is convenient to calculate the Fresnel integral especially for the large measurement plane. In this paper three experiments are introduced for sound image reconstruction. The method shown in this paper may be applied for getting clear images of underwater sound sources, if the parameters satisfy the condition of the Fresnel integral.

5. ACKNOWLEDGEMENTS

The author wishes to thank professor He-zuoyong and Tang-WeiLin at the Acoustical Engineering Department and senior Engineer Wang-JinTang for helping with the theory and the model experimental devices.

REFERENCES

1. A.K. Ghatak and K. Thyagarajan 1978, Plenum Press, New York, Contemporary Optics.
2. S. Ueha 1976. Optica ACTA, Vol. 23, No. 2. pp107-114. Imaging Of Accoustic Radiation Sources With Accoustical Holography.
3. W.A. Veronesi and J.D. Maynard 1989 J. Acoust. Soc. Am. Vol.85, No. 2, February 588-598. Digital Holographic Reconstruction of Sources with Arbitrarily Shaped Surfaces.
4. Desen. Y 1992. The Study Report of HSEI, Harbin, China. The study of Vibration and Noise For Underwater Engine.

SUPPRESSION OF SURFACE AND BOTTOM REVERBERATION IN MULTI-BEAM SONAR IMAGES

N. Rajpal, R. Bahl and A. Arora

Centre for Applied Research in Electronics
Indian Institute of Technology Delhi
New Delhi-110016, India

INTRODUCTION

Underwater acoustic images obtained from multi-beam sonar are contaminated with ambient noise and reverberation. These contaminations are required to be removed before extraction of target parameters. Conventional techniques such as normalisation, doppler-OR-ing and thresholding do not completely eliminate the noise, especially when the target is near one of the reverberation boundaries.

In this paper we present a novel technique for suppressing surface and bottom reverberation from 4D image data from a multi-beam sonar that provides range, bearing, elevation and doppler information. The method is based on dual-frame processing algorithms operating on a pair of consecutive 4D images.

PHILOSOPHY OF DUAL FRAME PROCESSING

Typical multi-beam imaging sonars mounted on moving platforms operate in shallow channels or near the surface/bottom. In such situations the sonar is faced with the daunting task of automatic extraction of the target from amongst the surface and/or bottom reverberation. While the number of false alarms due to ambient noise can be kept as small as required by raising the threshold, it may not be possible to completely eliminate strong reverberation "edges" thus jeopardising target extraction. We propose to discriminate targets from reverberation on the basis of the fact that the reverberation appears at almost the same location and with the same doppler in the 4D image in consecutive frames, because the relative location of the sea boundaries do not change over the small distance travelled by the sonar between frames. On the other hand a target would appear to change its location between frames because of non-zero relative velocity. It is this underlying principle that forms the basis of extraction of a target from a reverberation background. At the starting point, we have binary intensity images obtained from thresholding the raw 4D data. A flow-chart showing the various procedures is shown in figure 1.

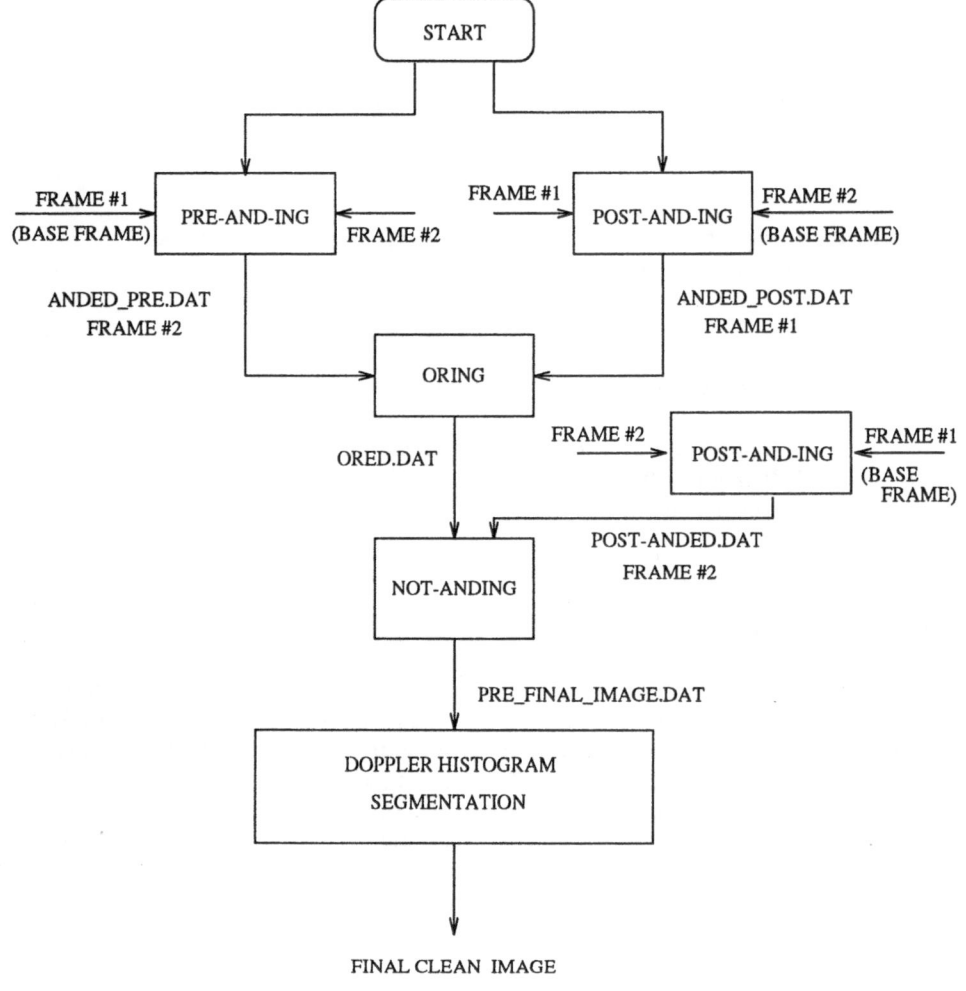

Figure 1. DUAL FRAME IMAGE CLEANING FLOW CHART

REJECTION OF SEA SURFACE AND BOTTOM REVERBERATION

In order to cleanse the images from all surface and/or bottom reverberation and ambient noise that have crossed the threshold, we adopt a number of processing steps as discussed below.

Step 1 (refer figure 2)

Two frames are required to get a clean image. (We shall call the most recent frame as the second frame and the previous frame as the first frame). To do so, we first use the first frame as base frame and do the pre-AND-ing operation. In this operation, we pick up a point in the second frame and look for a point in the first frame in 4 earlier range bins (depends on the sonar platform movement between the two frames). If there is a corresponding point at the same doppler (within ±1 doppler cells) then in the second frame that particular point is retained. The algorithm is shown below.

```
procedure-pre-AND-ing
    begin
     for φ = 1 to m
       for θ = 1 to n do
         for r = 1 to R  do
           begin
             for r1 = r-3 to r do
               begin
                 ANYOF(RETURN_base[r1][θ][φ][dORd+/-1]) !=0;
                 RETAIN(RETURN[r][θ][φ][d]);
               end;
           end;
    end.
```

The algorithm is based on the fact that the target speed is less than the sonar platform's speed (which is generally the case) and it is being approached by the sonar. Thus, it appears closer in the second frame. Another assumption, which has been made here, is that the reverberation is more or less stationary (it shows slight movement because of randomness). The above mentioned assumptions lead to a pre-AND-ed picture (this is the processed second frame and the resultant image is called as **anded-pre.dat**). Since we "look" for a return (echo) in those range bins where we do not expect a target echo, this image has the following features:

a) very low probability of having target points,
b) almost no ambient noise, and
c) most of the reverberation points.

Step 2 (refer figure 3)

In the next step we consider the second frame as base frame and process the first frame to perform post-AND-ing operation. In this operation, we pick up a point in the first frame and look for a point in the second frame in 4 forthcoming ranges. If there is a corresponding point at the same doppler (within ±1 doppler cells) then in the first frame that particular point is retained. The algorithm is shown below.

```
procedure-post-AND-ing
    begin
     for φ = 1 to m
       for θ = 1 to n do
         for r = 1 to R  do
           begin
             for r1 = r to r+3 do
               begin
                 ANYOF(RETURN_base[r1][θ][φ][d OR d(+/-)1]) !=0;
                 RETAIN(RETURN[r][θ][φ][d]);
               end;
           end;
    end.
```

This algorithm also makes the same assumptions as made in **STEP 1** and a post-AND-ed image is obtained. (This is actually the first frame processed relative to the second frame.) The new image is called **anded-post.dat**. Once again, since we are looking in those range bins where we do not expect the target to lie, this image has similar features as **anded-pre.dat**.

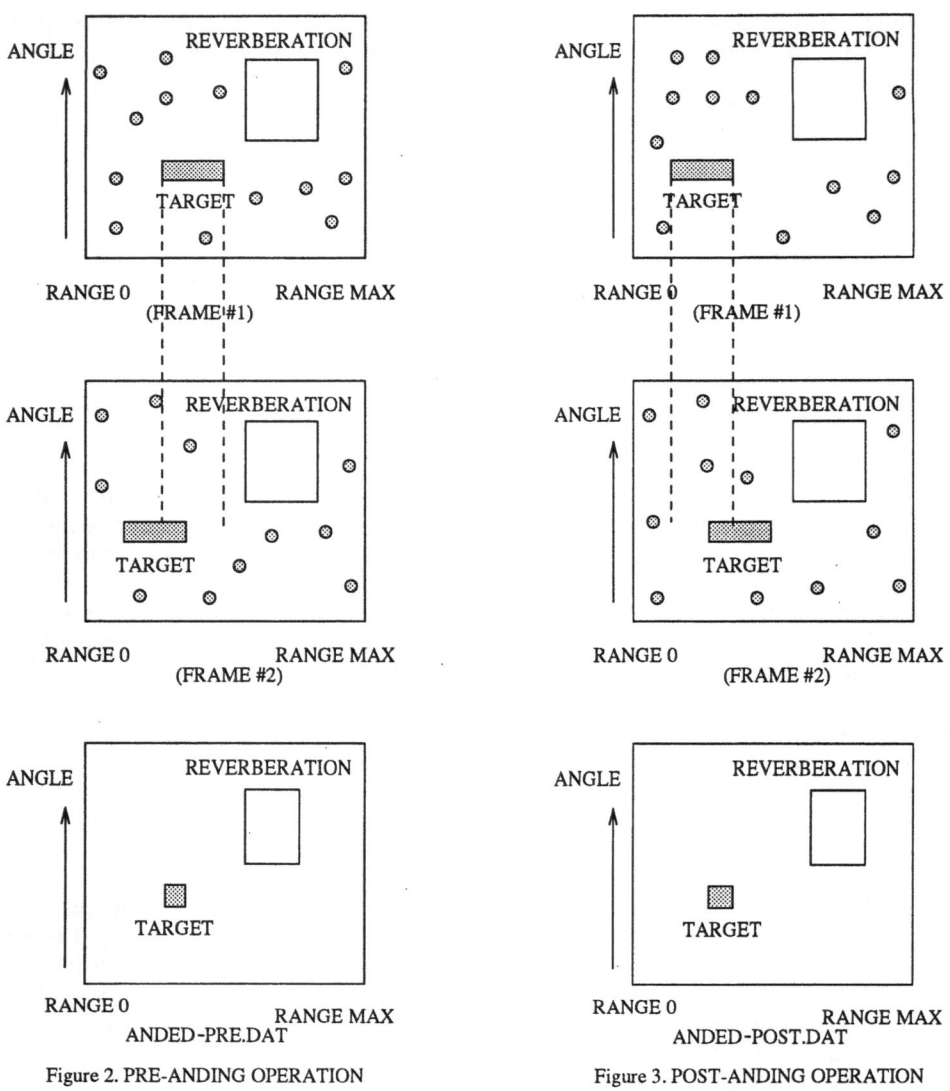

Figure 2. PRE-ANDING OPERATION
ON FRAME #2

Figure 3. POST-ANDING OPERATION
ON FRAME #1

Step 3 (refer figure 4)

We now have two images from the previous two steps, that have between them, most of the reverberation points. In order to generate a composite reverberation window, we 'OR' the previously obtained two images i.e **anded-pre.dat** and **anded-post.dat** as explained in the following algorithm.

```
procedure -OR-ing
  begin
    for φ = 1 to m
      for θ = 1 to n do
```

```
for r = 1 to R  do
    begin
    RETURN_pre[r1][θ][φ][d]+RETURN_post[r][θ][φ][d];
    RETAIN-DOPPLER-OF(RETURN_pre OR RETURN_post);
    IF(RETURN_pre NE 0 && RETURN_post NE 0)
    THEN
    RETAIN-DOPPLER-OF(CLOSE TO STATIONARY DOPPLER);
    end;
end.
```

With this method a new image is created which consists of a reverberation window only. This is used in the next section to mask the reverberation in order to recover the target.

TARGET RECOVERY

In order to recover the target, the reverberation window, obtained by the previous step 3, is used to mask the reverberation so that a noise-free, reverberation-free image of the target can be obtained. This is done in following steps:

Step 4 (refer figure 5)

In this step, the first frame is taken as the base frame and the second frame is processed with respect to it to perform post-AND-ing operation. As explained in step 2, in this operation, we pick up a point in the second frame and look for a point in the first frame in 4 forthcoming ranges. If there is a corresponding point at the same doppler (within ±1 doppler cells) then in the second frame (which is most likely the case for both target and reverberation) that particular point is retained. Thus, the post-AND-ed image of second frame (**post-anded.dat**) has the following features:
a) most of the target points are present,
b) reverberation patch is present, and
c) negligible ambient noise.

Step 5 (refer figure 6)

The OR-ed image obtained earlier in step 3 is now used to mask the reverberation patch in the post-AND-ed image obtained in step 4 by NOT-AND-ing the two images (OR-ed as base and post-AND-ed as the one to be processed) as per the following algorithm :
```
procedure-NOT-AND-ing
    begin
    for φ = 1 to m
      for θ = 1 to n do
        for r = 1 to R  do
        begin
        RETURN_base[r1][θ][φ][d] !=0;
        DONOTRETAIN(RETURN[r][θ][φ][d];
            end;
    end.
```
The above process filters out the reverberation patch from the post-AND-ed image and thus only the target points are allowed to pass through. The resultant final image is **pre-final-image.dat**

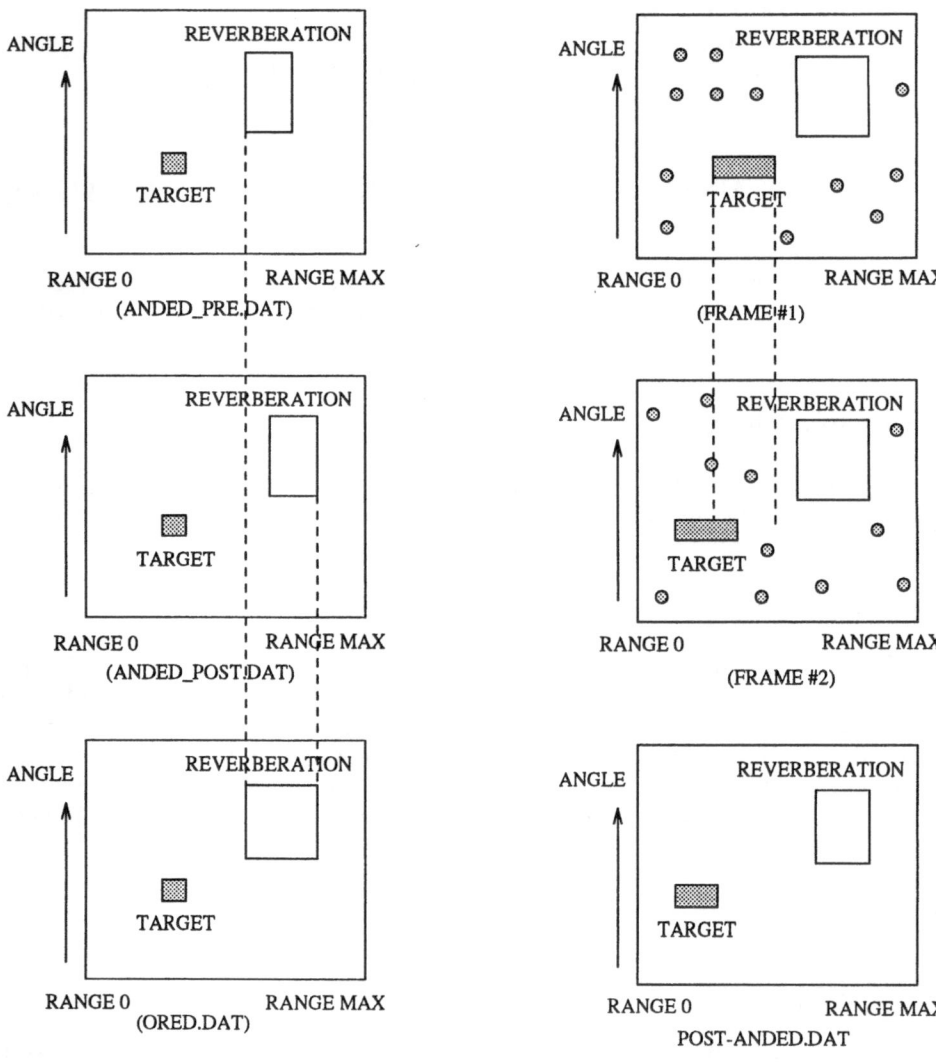

Figure 4. ORING OPERATION

Figure 5. POST-ANDING OPERATION
ON FRAME #2

DOPPLER HISTOGRAM

In order to do the final cleaning, i.e. if at all some of the reverberation points make entry in the cleaned image, the image is passed through a histogram routine which reads the doppler tags of all the points in the image and makes a doppler histogram from it.

The centre of those points that have doppler equal to the maximum occurring doppler cell is determined. Then all points within a certain range band around the centre and with doppler within a certain doppler band around the maximum occurring

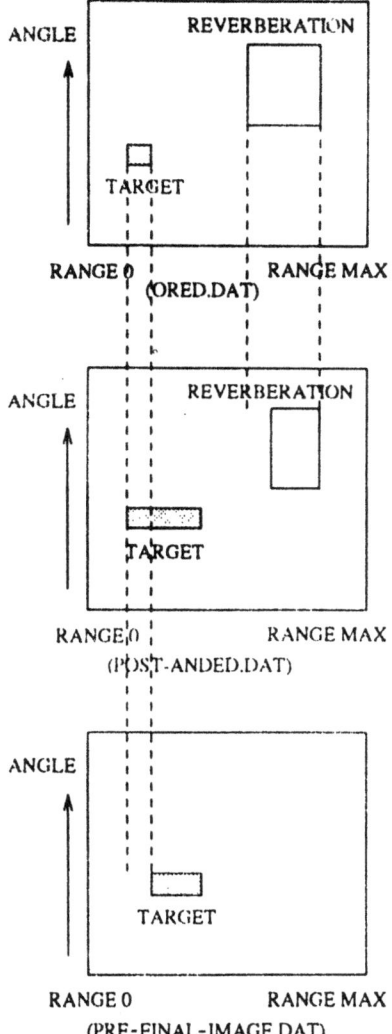

Figure 6. NOT-ANDING OPERATION

Figure 9. ANDED-PRE.DAT.

Figure 10. ANDED-POST.DAT.

Figure 11. ORED.DAT.

Figure 12. POST-ANDED.DAT.

Figure 13. PRE-FINAL-IMAGE.DAT.

Figure 7. FRAME #1.

Figure 8. FRAME #2.

doppler are retained. These bands allow for expected target range and doppler spread. The resultant image is called **post-hist.dat**.

RESULTS AND CONCLUSTION

The algorithm has been tested on ADSP-2100 DSP processor using simulated data generated by THIRISM (3-D high resolution sonar data simulator) designed by Signal Processing Group CARE, IIT Delhi[1] and is shown to provide real-time operation. Figures 7 and 8 shows frame #1 and frame #2 of input test data obtained after conventional normalization, doppler-OR-ing and thresholding techniques. Figure 9 shows result of pre-AND-ing operation operation on frame #2 and figure 10 shows result of post-AND-ing operation on frame #1. Figure 11 shows result of ORing operation on the resultant images shown in figures 9 and 10. Figure 12 shows the result of post-AND-ing operation on frame #2. Figure 13 shows result of NOT-AND-ing operation on the resultant images shown in figures 11 and 12. The algorithm has been tested on variety of other scenarios also with target moving at different velocities against bottom as well as surface reverberation and has provided satisfactory results.

REFERENCES

1. O.George, N.Rajpal, R.Bahl, T.B.Rao and V.Natarajan, "3-D High Resolution Sonar Imaging", proceedings of 19th International Symposium on Acoustical Imaging, held April 1991, at Bochum Germany, pp. 903-908.

SEGMENTATION OF MULTISPECTRAL SONAR IMAGES

N. Rajpal, R. Bahl and E.P. Selvan

Centre for Applied Research in Electronics
Indian Institute of Technology Delhi
New Delhi-110016, India

INTRODUCTION

Acoustic images obtained from sonar may consist of multiple objects like undersea vehicles, ship wrecks, rocks, mines etc. Sometimes, strong echoes due to surface or bottom reverberation may also cause confusion in interpretation of such images. Such complex scenarios with multiple objects would first require segmentation of the acoustic image to separate out the different objects, to be followed by techniques to find their relative positions, motion and extent parameters. These parameters are useful for manoeuvering autonomous underwater vehicles to avoid obstacles in their path. Algorithms for extraction of motion and extent parameters from a sonar image consisting of a single object have already been developed by us[1].

In this paper, we discuss a technique for segmentation of multispectral sonar images. The technique is based on unsupervised classification of patterns where, in addition to geometrical coordinates, doppler (spectrum) information available in sonar data is also used. Dividing an image into meaningful regions is known as segmentation. There are many techniques used for segmentation viz. Clustering, Region Growing, Relaxation etc. This paper deals with segmentation of sonar images using clustering. The algorithms have been tested for various scenarios generated by 3-d high resolution sonar simulator (THIRISM) designed by Signal Processing Group CARE, IIT Delhi[2].

CLUSTERING

Assigning patterns to predetermined classes is called Supervised Learning. In some of the problems even the number and the nature of the classes, if any, are unknown. A large amount of data may often be analyzed by a digital computer in such a way as to give an understanding of the structure of the data and useful numerical measurements of certain characteristics of the data. These problems form the object of unsupervised learning, wherein, without having any classification information, we learn something of the nature of the process for which the data is descriptive. The subject of unsupervised learning encompasses an attempt to apply recognition techniques to unclassified data. We can define clustering to be the unsupervised classification of objects which amounts to the process of generating classes without any knowledge of prototype classification. The essential characteristic is the sorting of the data into subsets such that each subset contains data points that are as much alike as possible.

DISTANCE AND SIMILARITY MEASURES

In order to do any clustering it is necessary to define a measure of similarity between two samples, usually with properties of a distance function. The simplest distance measure between two points, X_i and X_j, is probably the Euclidean or Cartesian distance $d^2(X_i,X_j) = (X_i-X_j)^t(X_i-X_j)$, where X_i and X_j are multidimensional points in the observation space. The Euclidean distance measure is particularly susceptible to the problem of multiple dimensions of a variety of units. Certain other distance or similarity measures are listed below:

Weighted Euclidean Distance:
$$d^2(X_i,X_j) = \Sigma w_n(X_{ni}-X_{nj})^2$$
Where, w_n are the weights for each dimension.

Correlation:
$$d(X_i,X_j) = X_i^t X_j$$

Similarity Ratio:
$$d(X_i,X_j) = X_i^t X_j/(X_i^t X_i + X_j^t X_j - X_i^t X_j)$$

Normalized Correlation:
$$d(X_i,X_j) = X_i^t X_j/((X_i^t X_i)(X_j^t X_j))^{1/2}$$

In our work, we are using 4 dimensional sonar data for the clustering. The 4 dimensions are range, bearing, elevation and doppler of each data point. The resolution of these dimensions depends upon the sonar parameters. The Weighted Euclidean Distance is better suited in comparison to other similarity measures for this application as the significance of each dimension depends upon its degree of precision. Weights for a particular dimension can also be changed on line by analyzing the spread of data points in that dimension.

CLUSTERING CRITERION AND METHOD

Once an appropriate similarity has been selected it becomes necessary to develop a criterion for cluster formation and clustering technique. Various criterion functions have been discussed in the literature[3] like Sum-of-Square-Error Criterion, Related Minimum Variance Criterion, Scattering Criterion, Mutual Neighborhood Criterion, Nearest Neighbor Criterion, Furthest Neighbor Criterion, etc. Based on these criteria there are various techniques for clustering like Iterative Optimization, Hierarchical Clustering, Graph Theoretic Methods etc. Iterative Optimization technique[3] using Sum-of-Squared-Error Criteria or Related Minimum Variance Criteria or any of Scattering criteria is computationally intensive and hence not suitable for real-time applications. Graph Theoretic Method using Mutual Neighborhood Criteria[4,5] is also computationally intensive for real-time applications. Hierarchical Clustering or Graph Theoretic method using Furthest Neighbor Criteria[3] discourage the growth of elongated clusters, whereas most of the underwater objects are of elongated type. Due to the above mentioned problems Graph Theoretic method based on Nearest Neighbor Criteria is best suited for our problem.

CLUSTERING ALGORITHM

We have our pattern space to be four dimensional; the dimensions being range, bearing, elevation and doppler. In the clustering technique we do not use the intensity information. The range, bearing, elevation and doppler of thresholded data is given as input to the clustering module. The similarity measure used in the algorithm is the Weighted Euclidean Distance. A threshold distance d_0 is selected depending on the weights of the dimensions which further depends on the sonar parameters. An undirected graph is defined in which nodes correspond to points and an edge joins the node i and j if and only if the weighted euclidean distance between these points is less than threshold d_0. Two points X_i and X_j are in the same cluster if there is a path between the corresponding nodes in the graph. Thus, clustering corresponds to finding connected components in the graph. The algorithm for finding connected components in the graph has been implemented using Set Union and Find procedures given in literature[6]. The sets will be represented by trees. Nodes are linked on the parent relationship, i.e each node in a set other than root is linked to its parent. The operation FIND(i) determines the root of the tree containing the element i and UNION(i,j) requires two trees with root i and j to be joined. Nodes in the trees are numbered 1 to N so that node index corresponds to the point index. Consequently, each node needs only PARENT field to link to its parent. A CLUSTER field is also assigned to each node which represents cluster number to which that node belongs. Points belonging to a cluster having less than 10% of total points are assigned cluster number 0 so that these can easily be removed as noisy points. As root of a tree does not require parent field it is used to keep the number of nodes in the tree. To distinguish it from other parent field, the number of nodes in a tree are maintained as a negative number in the parent field of root. The algorithm for clustering is given as follows. Code used is similar to that of PASCAL language.

Algorithm CLUSTERING
Notation: Array PARENT[] contains parent field of nodes. Array, TH[], FI[], DP[] represents range, bearing, elevation and doppler respectively of data points. DIST represents Weighted Euclidean distance between points. THRES represents threshold to decide whether edge is present between points or not. WR, WT, WF, WD represents weights for range, bearing, elevation and doppler respectively. N represents number of data points. Array CLUSTER[] contains cluster number of the node to which it belongs.

```
PROCEDURE UNION(I,J : INTEGER);
BEGIN
    UI := PARENT[I] + PARENT[J];
    IF PARENT[I] > PARENT[J]
    THEN
    BEGIN
      PARENT[J] := UI;
      PARENT[I] := J;
    END
    ELSE
    BEGIN
      PARENT[I] := UI;
      PARENT[J] := I;
    END;
END;
```

```
FUNCTION  FIND(I : INTEGER) : INTEGER;
BEGIN
   FJ := I;
   WHILE (PARENT[FJ] > 0) DO
      FJ := PARENT[FJ];

   FK := I;
   WHILE (FK < > FJ) DO
   BEGIN
      FT := PARENT[FK];
      PARENT[FK] := FJ;
      FK := FT;
   END;
   FIND := FJ;
END;


BEGIN
   FOR I := 1 TO N DO
    PARENT[I] := -1 ;

   FOR I := 1 TO N DO
   BEGIN
    FOR J := I+1 TO N DO
     BEGIN
        DIST :=  WR*(R[I]-R[J])*(R[I]-R[J]) +
                        WT*(TH[I]-TH[J])*(TH[I]-TH[J]) +
                        WF*(FI[I]-FI[J])*(FI[I]-FI[J]) +
                        WD*(DP[I]-DP[J])*(DP[I]-DP[J]);
        IF( DIST < THRESH)
        THEN
          BEGIN
           L := FIND(I);
           M := FIND(J);
           IF (L < > M)
           THEN
              UNION(L,M);
          END;
     END;
   END;
   END;

   K := 0;
   M := -1*(N DIV 10);
   FOR I := 1 TO N DO
   BEGIN
    IF(PARENT[I] < M)
     THEN
      BEGIN
       K := K+1;
       CLUSTER[I] := K ;
        END
     ELSE
       CLUSTER[I] := 0;
   END;
```

```
      FOR I := 1 TO N DO
      BEGIN
       IF(PARENT[I] > 0)
       THEN
        BEGIN
        J := FIND(I);
          CLUSTER[I] := CLUSTER[J];
       END;
        END;
END.
```

RESULTS AND CONCLUSION

Figure 1 shows an input scenario simulated using THIRISM after preprocessing. It consists of an elongated moving object, surface reverberation patch and some noise points which could not be removed by preprocessing steps like normalization, doppler OR-ring and thresholding. Figure 2 shows output of the clustering technique applied on this scenario. It removes noise points and separates reverberation cluster (shown by small rectangles) and object cluster (shown by small triangles). For proper representation of 3D scenario on 2D space we have represented the image as four B-

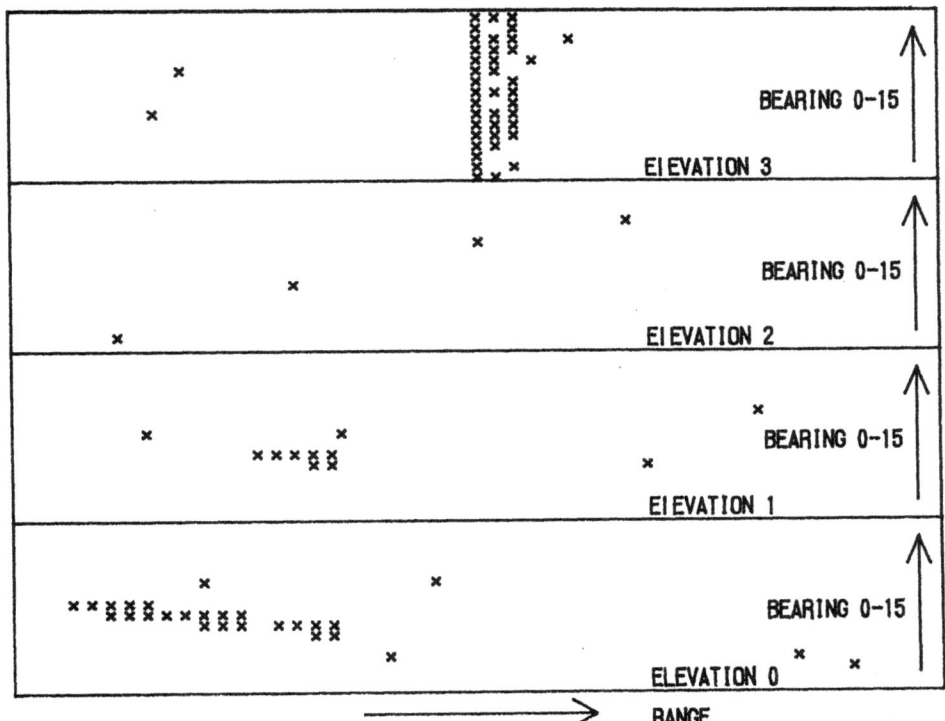

Figure 1. INPUT SCENARIO 1.

Scan (bearing vs range) images each corresponding to an elevation cell as shown in figures 1 and 2. Figure 3 shows another input scenario consisting of two cylindrical objects close to each other but moving with different velocities. Figure 4 shows output of clustering technique applied on this image.

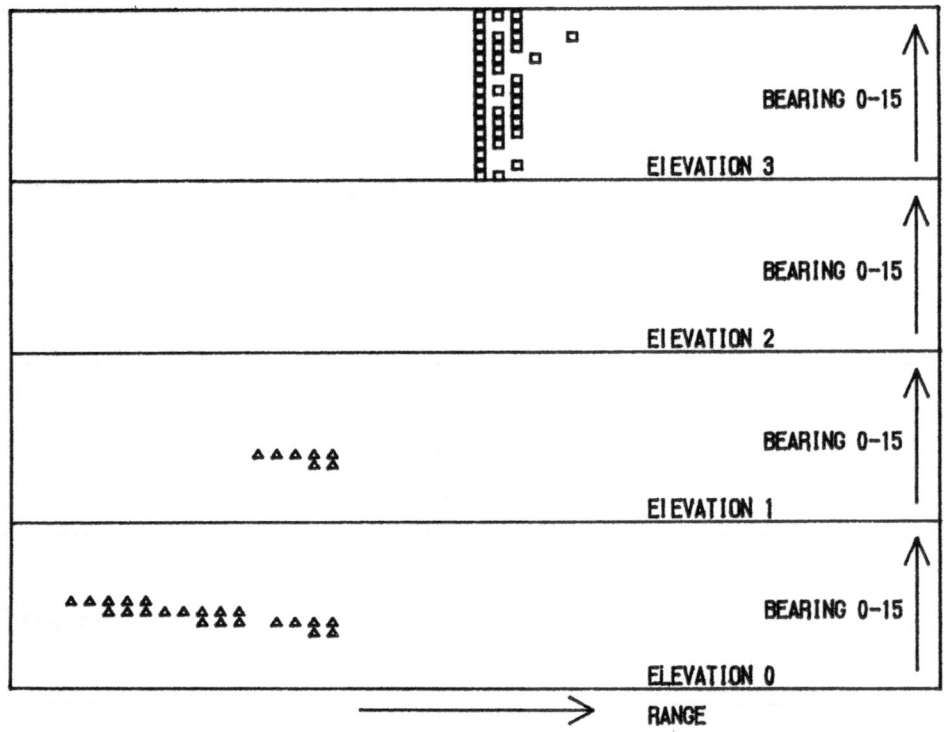

Figure 2. OUTPUT OF SCENARIO 1.

The technique can use intensity information also as the fifth dimension, while calculating the Weighted Euclidean Distance. This segmentation technique combined with the algorithm for extraction of motion and extent is potentially very useful for real-time systems like autonomous underwater vehicles to intelligently avoid obstacles in their path.

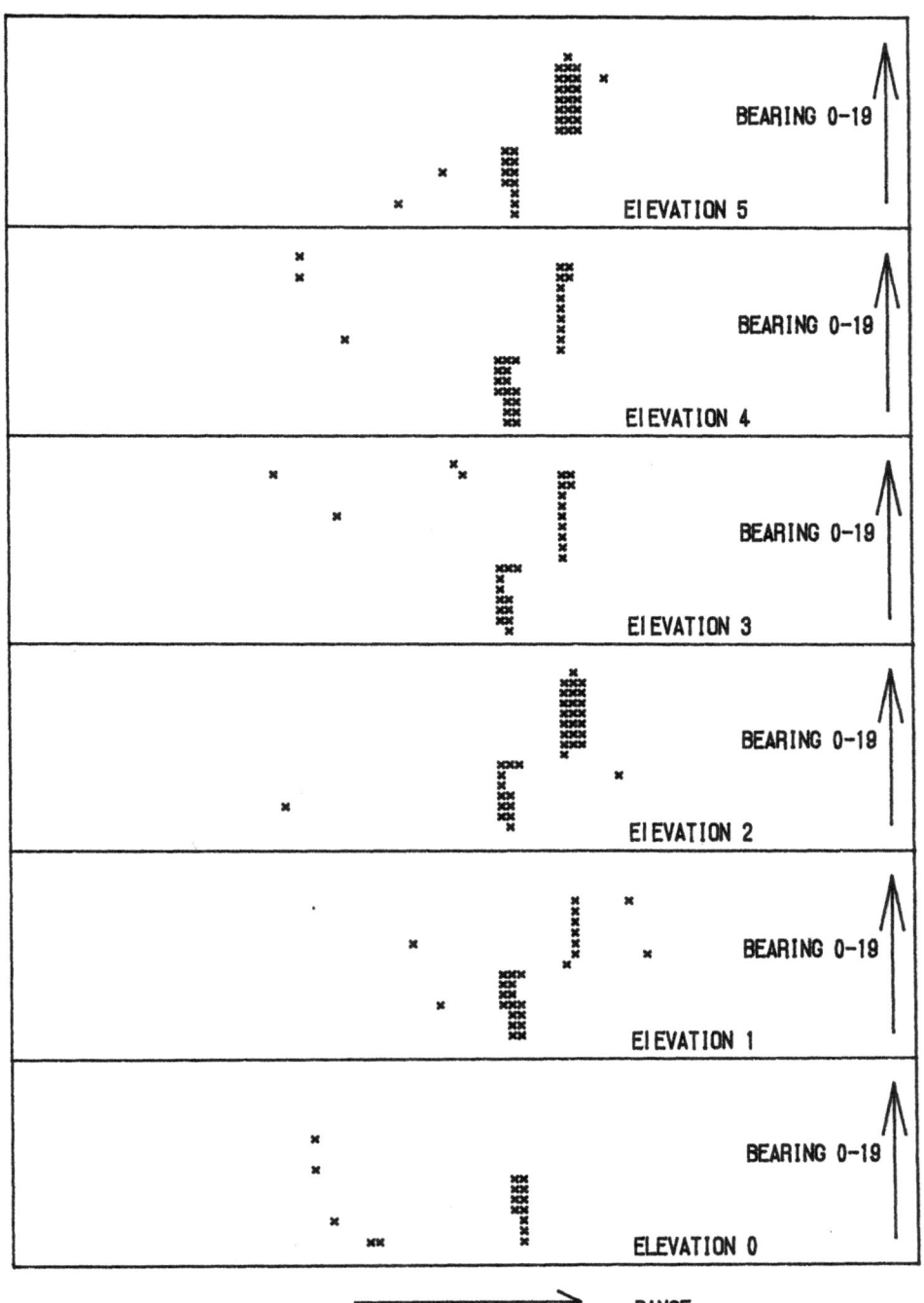

Figure 3. INPUT SCENARIO 2.

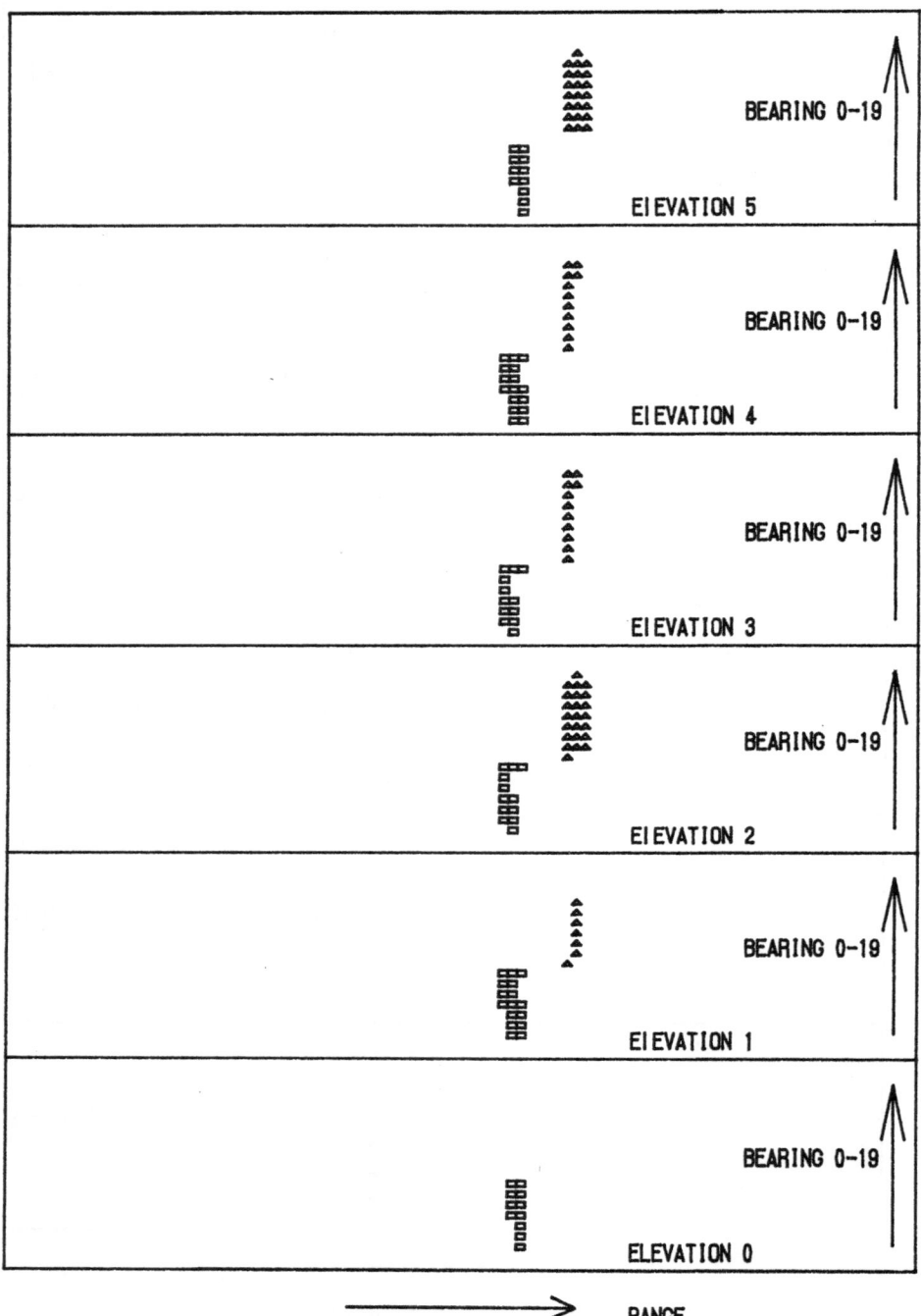

Figure 4. OUTPUT OF SCENARIO 2.

REFERENCES

1. R.Venugopal, N.Rajpal and R.Bahl, "Algorithms for Extraction of Motion and Extent of Obstacles from 3-D High Resolution Sonar Data", Proceedings of Workshop on Artificial Intelligence Control and Advanced Technology in Marine Automation (CAMS 92) at Genova, Italy, April 1992.
2. O.George, N.Rajpal, R.Bahl, T.B.Rao and V.Natrajan, "3-D High Resolution Sonar Imaging", Proceedings of 19th International Symposium on Acoustical Imaging at Bochum Germany, April, 1991.
3. R.O.Duda and P.E.Hart, "Pattern Classification and Scene Analysis", Wiley, New York, 1973.
4. K.C.Gowda and G.Krishna, "Disaggregative Clustering Using Concept of Mutual Nearest Neighborhood", IEEE Trans. on Syst. Man Cybern, Vol. SMC-8, no. 12, pp. 888-895, Dec. 1978.
5. S.P.Smith, "Threshold Validity of Mutual Neighborhood Clustering", IEEE Trans. on Patt. Anal. and Machine Intell., Vol. PAMMI-15, no. 1, Jan. 1993.
6. E.Horowitz and S.Sahni, "Fundamentals of Computer Algorithms", Computer Science Press, Potomac, Maryland, 1984.

CONTRIBUTORS

Prof. Pierre Alais
Laboratoire Mecanique Physique
Universite Paris 6
2 Place de la Gare de Ceintur
78210 Saint-Cyr France

Forrest Anderson
Impulse Imaging Inc.
P.O. Box 1400
Bernalillo, NM 87004 USA

Dr. Michael P. Andre
Department of Radiology (9114)
University of California at San Diego
La Jolla, CA 92093 USA

Professor Max Anliker
University of Zurich and the Swiss
Federal Institute of Technology
Zurich, Switzerland

Prof. Yoshinao Aoki
Dept. of Information Engineering
Hokkaido University, N-13, W-8
Sapporo 060 Japan

Prof. Walter Arnold
Fraunhofer-Institut, IZfP
Universitat Geb. 37
D-66123 Saarbrucken Germany

Dr. Michel Bertrand
Institute of Biomedical Engineering
University of Montreal
Ecole Polytechnique
P.O. Box 6079, Station Centre-Ville
Montreal, Quebec, H3C 3A7 Canada

Mark Betts
Dept. of Medical Engineering and
    Physics
King's College School of Medicine
    and Dentistry
Dulwich Hospital
London SE22-9PT UK

Dr. Elena Biagi
Department of Electrical Engineering
University of Florence
Via Santa Marta 3
50139 Firenze Italy

Laurence Bohs
Dept. of Biomedical Engineering
Duke University
136 Engineering Bldg.
Durham, NC 27706 USA

Mrs. Odile Bonnefous
Laboratoires D'Electronique Philips
22 Ave. Descartes - B.P. 15
94453 Limeil Brevannes France

Elena Borodina
Institute of Applied Physics,
Russian Academy of Sciences
46 Uljanova St.
603600 Nizhny Novgorod, Russia

Prof. Valentin Burov
Department of Physics
Moscow State University
Moscow 119899 Russia

Dr. Lorenzo Capineri
Dept. of Electrical Engineering
University of Florence
Via Santa Marta 3
50139 Firenze Italy

Didier Cassereau
Laboratoire Ondes et Acoustique
ESPCI
10 rue Vauquelin
75005 Paris France

Dr. P.A.N. Chandraratna, M.D.
Section of Cardiology
LA County/USC Medical Center
2025 Zonal Ave.
Los Angeles, CA 9003 USA

Dr. Konstantin Chishko
Institute of Low Temperature Physics
    and Engineering
47 Lenin Ave.
310164 Kharkov, Ukraine

Bernard Cretin
LPMO/CNRS
32 Ave. de l'Observatoire
25000 Besancon France

Dr. Hamid Dabirikhah
Dept. of Electronic and
    Electrical Engineering
King's College
The Strand
London WC2R-2LS UK

Dr. Richard J. Dewhurst
Dept. of Instrumentation and
    Analytic Science
University of Manchester
Institute of Science & Technology
P.O. Box 88
Manchester M60 1QD UK

Dr. Stanislav Emelianov
Bioengineering Program
University of Michigan
3304 G.G. Brown, 2350 Hayward
Ann Arbor, MI 48109-2125 USA

Detlef Englich
Department of Physics
University of Oldenburg
Carl von Ossietzky Str.
26111 Oldenburg, Germany

Prof. Helmut Ermert
Institut fur Hochfrequenztechnik
Ruhr-Universitat Bochum
Bldg. IC 6/132
D-44780 Bochum, Germany

Dr. Kathy Ferrara
Riverside Research Institute
330 West 42nd St.
New York, NY 10036 USA

Prof. Leonard Ferrari
Dept. of Electrical & Computer
    Engineering
University of California, Irvine
Irvine, CA 92717 USA

Prof. Mathias Fink
Laboratoire Ondes et Acoustique
ESPCI
10 rue Vauquelin
75005 Paris France

Dr. Flemming Forsberg
Dept. of Radiology
Thomas Jefferson University
Philadelphia, PA 19107 USA

Dr. Valerio Gabbani
Dept. of Electrical Engineering
University of Florence
Via Santa Marta 3
50139 Firenze Italy

Dr. Jacqueline Gallet
Dept. of Radiological Sciences
University of California, Irvine
Irvine, CA 92717 USA

Woon Siong Gan
Acoustical Services Pte. Ltd.
29 Telok Ayer St.
Singapore 0104, Republic of Singapore

Cindy Gillespi
Dept. of Geological Sciences
University of Texas El Paso
El Paso, TX 79968-0555 USA

George Goebel
475 Ranchito Vista Road
Santa Barbara, CA 93108 USA

Prof. Xiu-Fen Gong
Institute of Acoustics
Nanjing University
Nanjing, 210008 P.R. China

Vladimir Gorentsveig
Institute for Low Temperature Physics
    and Engineering
47 Lenin Ave.
Kharkov, 310164 Ukraine

Dr. James E Greenleaf
Biodynamics Research Unit
Mayo Clinic
Rochester, MN 55905 USA

Dr. Dan Grolemund
Dept. of Radiological Sciences
University of California, Irvine
Irvine, CA 92717 USA

Christian U. Grosse
FMPA-Innovationsforschung
Pfaffenwaldring 4
D70569 Stuttgart Germany

Andy Healy
Dept. of Medical Eng. and
    Physics
Dulwich Hospital
East Dulwich Grove
London SE22-8PT UK

Steve Isakson
Computer Science
University of California,
    Santa Barbara
Santa Barbara, CA 93106 USA

Chiaki Ishihara
Akishima Laboratories Inc.
Mitsui Engineering &
    Shipbuilding
Akishima, Tokyo 196 Japan

Prof. Joie Jones
Dept. of Radilogical Science
University of California, Irvine
Irvine, CA 92717 USA

Faouzi Kallel
Institute of Biomedical Enginerring
University of Montreal
Ecole Polytechnique
P.O. Box 6079, Station Centre-Ville
Montreal, Quebec H3C 3A7 Canada

Dr. Keisuke Kameyama
Interdisciplinary Graduate School of
Science and Engineering
Tokyo Institute of Technology
4259 Nagatsuta, Midori-ku
Yokohama 227 Japan

Matsaru Kato
Interdisciplinary Graduate School of
    Science and Engineering
Tokyo Institute of Technology
4259 Nagatsuta, Midori-ku
Yokohama 227 Japan

Dr. Alexander Khil'ko
Institute of Applied Physics,
Russian Academy of Sciences
46 Uljanova St.
Nizhny, Novgorod 603600 Russia

Dr. Bernd Kohler
Fraunhofer Institute, IZfP
Dept. EADQ
Krugerstr. 22
D-012326 Dresden Germany

Dr. Elfgard Kuehnicke
Dresden University of Technology
Institute of Technical Acoustics
D-01062 Dresden Germany

Andrzej Kulik
EPFL
Institut de Genie Atomique
Ecublens
Ch-1015 Lausanne, Switzerland

Dr. T. Kundu
Dept. of Civil Engineering and
        Engineering Mechanics
University of Arizona
Tucson, AZ 85721 USA

N. Lamberti
D.I.I.G
University of Salerno
Via Ponte Don Melillo
84084 Fisciano (SA) Italy

Prof. Hua Lee
Dept. of Electrical & Computer
        Engineering
University of California,
Santa Barbara
Santa Barbara, CA 93106-9560 USA

Dr. Sidney Leeman
Dept. of Medical Engineering and
        Physics
Dulwich Hospital, King's College
East Dulwich Grove
London SE22-8PT UK

Dr. Sidney Lees
Bioengineering Department
Forsyth Dental Center
140 Fenway
Boston, MA 02115 USA

Pai-Chi Li
3304 G.G. Brown, 2350 Hayward
EECS Department
University of Michigan
Ann Arbor, MI 48109 USA

Mrs. Evelin Lieback
Deutsches Herzzentrum Berlin
Augustenburger Platz 1
D133353 Berlin Germany

Oleg Lobkis
Dept. of Civil Engineering and
        Engineering Mechanics
University of Arizona
Tucson, AZ 85721 USA

Jian-yu Lu
Biodynamics Research Unit
Mayo Clinic
Rochester, MN 55905 USA

Prof. Volker Mellert
Department of Physics
University of Oldenburg
Carl von Ossietzky Str.
26111 Oldenburg, Germany

Urs Moser
Institute of Biomedical Engineering
        and Medical Informatics
Moussonstr 18
CH 8044 Zurich Switzerland

Keinosuke Nagai
Institute of Applied Physics
University of Tsukuba
Tsukuba, Ibaraki, 305, Japan

Dr. Grey Otto
Thermo Trex Corporation
9550 Distribution Ave.
San Diego, CA 92121-2306 USA

Dr. David Palmer
AOML/Office of the Director
4301 Rickenbacker Cswy
Miami, FL 33149 USA

Joe Piel Jr.
GE Corp. R and D
P.O. Box 8
Schenectady, NY 12301 USA

Nick Pingitore
Dept. of Geological Sciences
University of Texas at El Paso
El Paso, TX 79968-0555 USA

Timothy J. Pitt
Medical Physics Department
Hammersmith Hospital
Du Cane Road
London W12-0HS UK

Dr. John Powers
Naval Postgraduate School
Code EC/Po, ECE Dept.
833 Dyer Road, Room 437
Monterey, CA 93943-5121 USA

David Purcell
Dept. of Mechanical Engineering
University of Waterloo
Waterloo, Ontario N2L 3G1 Canada

Navin Rajpal
Center for Applied Research in
        Electronics,
IIT Delhi
Hauz Khas
New Delhi 110016 India

Dr. Navalgund Rao
Chester F. Carlson Center for Imaging
        Science
54 Lomb Memorial Drive
Rochester Institute of Technology
Rochester, NY 14623 USA

John M. Richardson
5911 Busch Drive
Malibu, CA 90265 USA

Dr. Yoshifumi Saijo
Dept. of Medical Engineering and
        Cardiology
Institute of Development, Aging
        and Cancer
Tohoku University
4-1 Seiryomachi Aoba-ku
Sendai 980, Japan

Prof. Takuso Sato
Interdisciplinary Graduate School of
        Science and Engineering
Tokyo Institute of Technology
4259 Nagatsuta, Midori-ku
Yokohama 227 Japan

Peter Schumacher
Institute of Biomedical Enginering and
        Medical Informatics
Moussonstr. 18
CH 8044 Zurich Switzerland

Jianzhong Shen
Institute of Acoustics, Academia
Sinica
P.O. Box 2712
Beijing 100080
People's Republic of China

Dr. Wallace Arden Smith
Office of Naval Research, Materials
        Div.
ONR 331
800 N. Quincy Street
Arlington, VA 22217-5660 USA

Dr. Tai K. Song
Dept. of Electrical Engineering
KAIST
P.O. Box 150, Cheongryang 130-650
Seoul, Korea

Dr. E.S Syrkin
Institute for Low Temperature Physics
    and Engineering
Lenin Ave. 47
310164 Kharkov, Ukraine

Yasutaka Tamura
Yamagata University
Yonezawa 992 Japan

Xiu-ping Tao
Institute of Acoustics
Nanjing University
Nanjing 210008
People's Republic of China

Dr. Johan M. Thijssen
Dept. of Ophthamology, Biophysics
    Lab
University Hospital
P.O. Box 9101
6500HB Nijmegen The Netherlands

Prof. Piero Tortoli
Dept. Of Electrical Engineering
University of Florence
Via S. Marta 3
50139 Firenze Italy

Dr. M.V. Voinova
Dept. Theoretical Physics
Kharkov State University
Svobody Sq. 4
Kharkov, Ukraine

Xiao-Liang Xu
Biodynamics Research Unit
Mayo Clinic
Rochester, MN 55905 USA

Tsuyoshi Yamamoto
Hokkaido University
North 11, West 5
Sapporo 060 Japan

Desen Yang
Harbin Shipbuilding Eng.
    Institute
Harbin, Heilongjiang 150001
People's Republic of China

Dr. Feng Yin
Dept. of Geophysics
Stanford University
Stanford, CA 94305-2215 USA

INDEX